INVESTIGATING
OCEANOGRAPHY

INVESTIGATING
OCEANOGRAPHY

Keith A. Sverdrup
UNIVERSITY OF WISCONSIN–MILWAUKEE

Raphael M. Kudela
UNIVERSITY OF CALIFORNIA, SANTA CRUZ

Mc
Graw
Hill

INVESTIGATING OCEANOGRAPHY

2 3 4 5 6 7 8 9 0 QVS/QVS 1 0 9 8 7 6 5 4 3

ISBN 978–0–07–802291–3
MHID 0–07–802291–6

Senior Vice President, Products & Markets: Kurt L. Strand
Vice President, General Manager, Products & Markets: *Marty Lange*
Vice President, Content Production & Technology Services: *Kimberly Meriwether David*
Managing Director: *Thomas Timp*
Brand Manager: *Michelle Vogler*
Development Editor: *Jodi Rhomberg*
Director of Digital Content: *Andrea M. Pellerito, Ph.D.*
Associate Marketing Manager: *Matthew Garcia*
Content Project Manager: *Katie L. Fuller*
Buyer: *Susan K. Culbertson*
Designer: *Tara McDermott*
Cover/Interior Designer: *Elise Lansdon*
Cover Image: *© Ingram Publishing*
Content Licensing Specialist: *Shawntel Schmitt*
Photo Research: *Jerry Marshall*
Compositor: *ArtPlus Ltd.*
Typeface: *10/12 Times LT Std Roman*
Printer: *Quad/Graphics*

All credits appearing on page or at the end of the book are considered to be an extension of the copyright page.

Library of Congress Cataloging-in-Publication Data

Sverdrup, Keith A.
    Investigating oceanography / Keith A. Sverdrup, Raphael Kudela. – 1st ed.
        p. cm.
    Includes index.
    ISBN 978–0–07–802291–3 — ISBN 0–07–802291–6 (hard copy : alk. paper) 1. Oceanography.
I. Kudela, Raphael. II. Title.
    GC11.2.S94 2013
    551.46–dc23
                                                                        2012032920

www.mhhe.com

Dedicated to

Barbara Sverdrup Stone and
Stephanie Sverdrup Stone

and

Robert, Eleanor, and Sarah Kudela

About the Authors

Keith A. Sverdrup is a Professor of Geophysics at the University of Wisconsin-Milwaukee (UWM), where he has taught oceanography for thirty years and conducts research in tectonics and seismology. He is a recipient of UWM's Undergraduate Teaching Award and is a Fellow of the Geological Society of America.

Keith received his B.S. in Geophysics from the University of Minnesota and his Ph.D. in Earth Science, with a dissertation on seismotectonics in the Pacific Ocean basin from the Scripps Institution of Oceanography at the University of California-San Diego. Keith has participated in a number of oceanographic research cruises throughout the Pacific Ocean including the far Western Pacific, from Guam to the Philippines and Taiwan; the South Central Pacific in regions of French Polynesia including the Society Islands, the Line Islands and the Marquesas; and in the Eastern Pacific off the coast of Mexico.

Keith has been active in oceanography education throughout his career, serving on committees of the American Geophysical Union (AGU), the American Institute of Physics (AIP), and the Geological Society of America (GSA). He was a member of AGU's Education and Human Resources Committee for twelve years (chairing it for four years), and also chaired AGU's Excellence in Geophysical Education Award Committee, the Editorial Advisory Committee for the journal *Earth in Space*, and the Sullivan Award Committee for excellence in science journalism. Keith served as a member of AIP's Physics Education Committee for six years.

Keith was the Geosciences Program Officer for the Division of Undergraduate Education at the National Science Foundation from 2005–2007.

Dr. Raphael M. Kudela is a Professor of Ocean Sciences at the University of California, Santa Cruz (UCSC), where he teaches and conducts research on biological oceanography. He received his B.S. in Biology with a Marine Science emphasis at Drake University and his Ph.D. in Biology from the University of Southern California.

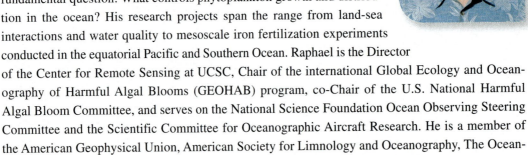

Raphael is a phytoplankton ecologist who wishes to understand the fundamental question: What controls phytoplankton growth and distribution in the ocean? His research projects span the range from land-sea interactions and water quality to mesoscale iron fertilization experiments conducted in the equatorial Pacific and Southern Ocean. Raphael is the Director of the Center for Remote Sensing at UCSC, Chair of the international Global Ecology and Oceanography of Harmful Algal Blooms (GEOHAB) program, co-Chair of the U.S. National Harmful Algal Bloom Committee, and serves on the National Science Foundation Ocean Observing Steering Committee and the Scientific Committee for Oceanographic Aircraft Research. He is a member of the American Geophysical Union, American Society for Limnology and Oceanography, The Oceanography Society, and the International Society for the Study of Harmful Algae.

Raphael teaches at both the undergraduate and graduate levels, including participation in the NASA Student Airborne Research Program.

Brief Contents

Contents

CHAPTER **1**

The Water Planet 24

CHAPTER **2**

Earth Structure and Plate Tectonics 48

CHAPTER **3**

The Sea Floor and Its Sediments   82

CHAPTER **4**

The Physical Properties of Water   110

CHAPTER **5**

The Chemistry of Seawater   130

CHAPTER **6**

The Atmosphere and the Oceans   150

Oceanography from SPACE   OS-1

CHAPTER **14**

The Benthos: Living on the Sea Floor 366

CHAPTER **15**

Environmental Issues 394

CHAPTER **16**

The Oceans and Climate Disruption 420

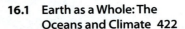

Preface

A Basic Knowledge of Ocean Science Is Crucial for Understanding and Protecting the Environment

Human beings have been curious about the oceans since they first walked along their shores. As we have learned more about the oceans, the tremendous influence these bodies of salt water have on our lives has become increasingly clear.

- The oceans cover over 70% of Earth's surface, creating a habitat for thousands of known species and countless others still to be discovered.
- We increasingly rely on the oceans as a source of food as well as a treasure trove of natural products used in pharmaceuticals and other products.
- The oceans contain vast quantities of diverse natural resources in the water and on the sea floor; some are actively exploited today, and many more may be recovered in the future with improved technology and greater demand.
- Global climate and weather are strongly influenced by the oceans as they interact with the atmosphere through the transfer of moisture and heat energy.
- The ocean basins are the location of great geologic processes and features such as earthquakes, volcanoes, massive mountain ranges, and deep trenches, all of which are related to the creation and destruction of sea floor in the process of plate tectonics.
- Coastal populations in some areas are at risk from the danger of tsunamis generated by offshore earthquakes and severe storms accompanied by coastal flooding. Rising sea level is a real and serious threat for these same coastal populations.

Much of what happens in the oceans and on the sea floor is hidden from direct observation. Although the *Hubble Space Telescope* can capture images from light that has traveled over 10 billion trillion kilometers, we cannot see more than a few tens of meters below the ocean's surface, even under the most favorable conditions, because of the efficient scattering and absorption of light by seawater. We have more detailed images of the surface of the Moon than we have of the floor of the oceans. Consequently, most of what we know about the oceans comes from indirect, or remote, methods of observation. With constantly improving technology and innovative applications of that technology, we continue to learn more about the geological, physical, chemical, and biological characteristics of the ocean environment.

Although careful scientific study of the oceans is often difficult and challenging, it is both necessary and rewarding. Our lives are so intimately tied to the oceans that we benefit from each new fact that we discover. Although it is critical that we continue to train marine scientists to study the oceans, it is no less important for people in all walks of life to develop a basic understanding of how the oceans influence our lives and how our actions influence the oceans. In studying oceanography, you are preparing yourself to be an informed global citizen. It is likely that at some point in the future you will have the opportunity to voice your concern about the health of the oceans, either directly or through the governmental process. **Your interest in, and study of, oceanography will help you participate in future discussions and decision-making processes in an informed manner.**

Solid Science and Sound Pedagogy

Investigating Oceanography is the product of decades of experience in oceanographic research and teaching in the classroom. Concepts and processes are presented in clear language without any loss of scientific accuracy or rigor.

The information in this book is enhanced by beautiful art and an exciting design that capture both the eye and the imagination.

In each chapter, the discussion of new ocean science issues scaffolds on knowledge gained in previous chapters. Information in this book is both relevant to students' lives and up-to-date with the latest discoveries in oceanography.

Our goal is to help students understand the intimate relationship between the oceans and their own well-being, and that they know the importance of having a combination of dedicated scientists who continue to explore the oceans and a knowledgeable public committed to protecting the ocean environment. We also hope that this book can help each student discover ways to become engaged in their own lifelong interest and involvement in the oceans—whether their strengths are in science, politics, journalism, art, or any other field.

Chapter Elements

Investigating Oceanography offers numerous aids to engage students and help them identify and understand key concepts in ocean science.

Learning Outcomes

After studying the information in this chapter students should be able to:

1. *outline* and *discuss* Earth's heat budget,
2. *distinguish* between specific heat and heat capacity,
3. *list* the layers of the atmosphere in order of ascending height and *sketch* a plot of temperature vs. elevation in each layer,
4. *argue* that global warming is enhanced by the accumulation of greenhouse gases, such as carbon dioxide, in the atmosphere,
5. *explain* the characteristics of the Antarctic ozone hole and *relate* its size to cloud formation and temperature,

QUICK REVIEW

1. How is sea ice formed?
2. What limits the thickness of seasonal sea ice?
3. How are icebergs formed, and where are they most commonly found?
4. What is the difference between fast ice and drift ice?
5. What is pancake sea ice?

▲ Quick Review Questions

Each major subsection of a chapter ends with a list of quick review questions. These help students check their identification and understanding of the major concepts and processes discussed in the subsection. These questions also challenge the students to think in greater depth about what they have learned.

▲ Learning Outcomes

Each chapter opens with learning outcomes students can use to guide their study and focus on critical concepts presented in the chapter. These outcomes specify what students are expected to know, understand, and be able to do after studying the chapter.

◀ Summary

An end-of-chapter summary briefly outlines the major concepts and processes discussed in the chapter. Students can get a quick overview of the chapter by reading the summary and seeing how different ideas are linked to one another and build on each other.

◀ Key Terms

In order to understand the constant barrage of information concerning our planet and marine issues, students must have a basic command of the language of marine science in addition to understanding processes and principles. For this reason we maintain an emphasis on critical vocabulary. All terms are defined in the text; key terms are printed in boldface. A list of key terms appears at the end of each chapter, and a glossary is included at the end of the book.

▲ Study Problems

Students need to become comfortable with graphs, data, and comparing numbers. At the end of each chapter we provide study questions that allow students to practice and improve their quantitative skills. This gives them a deeper understanding of concepts and builds confidence in their ability to quantitatively analyze and interpret data.

In-Depth Essays Written by Scientists ▼ and Educators

We have invited outstanding scientists and educators to write guest essays in their fields of specialization. We call these ***Diving In*** essays and they are found throughout the book. *Diving In* essays engage students in a deeper study of intriguing topics and also encourage them to explore in greater detail other topics of particular interest to them.

Oceanography From Space Section

Because so much of modern oceanographic research relies on remote sensing satellite data, we have also prepared a special center spread on the use of satellites in oceanography. Today's students rely daily on satellite data for weather forecasts, the global positioning satellite system, and communications satellites to let us talk with friends and colleagues around the globe. Satellites provide us with an unprecedented ability to monitor the surface ocean and are routinely used in all aspects of oceanography. This section can be referenced as part of the other chapters or used as a stand-alone section. In this section we provide an overview of the various sensors—past, current, and future—and a brief tutorial on how they operate. Specific examples are given for how satellites are used in geological, physical, chemical, and biological oceanography, while additional examples are included throughout the text. ▶

Striking Photos and Illustrations

This first edition contains over 600 photos and illustrations, all carefully designed and selected to compliment and reinforce the text while engaging your students. This extensive art program is available for download in addition to an interactive eBook where instructors can quickly access images for presentations.

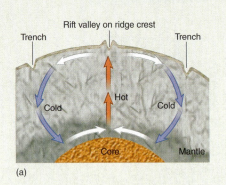

(a)

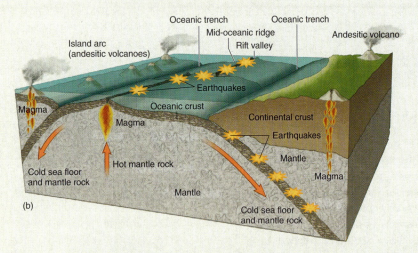

(b)

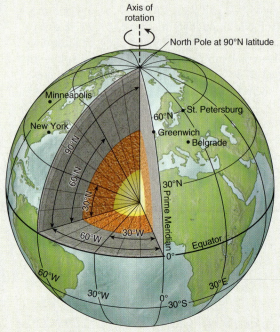

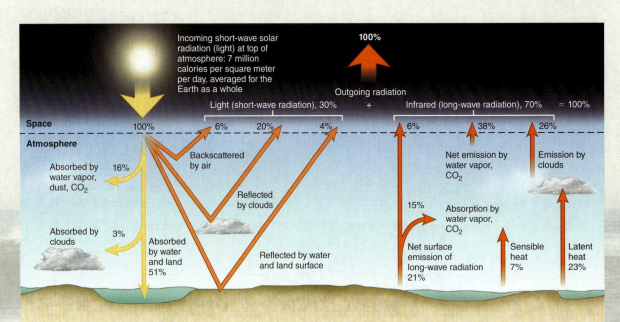

Digital Resources

McGraw-Hill offers various tools and technology products to support *Investigating Oceanography*, 1st edition.

McGraw-Hill's ConnectPlus™

McGraw-Hill's Connect Plus (www.mcgrawhillconnect.com/oceanography) is a web-based assignment and assessment platform that gives students the means to better connect with their coursework, with their instructors, and with the important concepts that they will need to know for success now and in the future. The following resources are available in Connect:

- Auto-graded assessments
- LearnSmart, an adaptive diagnostic tool
- Powerful reporting against learning outcomes and level of difficulty
- McGraw-Hill Tegrity Campus, which digitally records and distributes your lectures with a click of a button
- The full textbook as an integrated, dynamic eBook that you can also assign.
- Instructor Resources such as an Instructor's Manual, PowerPoints, and Test Banks.
- Image Bank that includes all images available for presentation tools.

With ConnectPlus, instructors can deliver assignments, quizzes, and tests online. Instructors can edit existing questions and author entirely new problems; track individual student performance—by question, assignment; or in relation to the class overall—with detailed grade reports; integrate grade reports easily with Learning Management Systems (LMS), such as WebCT and Blackboard; and much more.

By choosing Connect, instructors are providing their students with a powerful tool for improving academic performance and truly mastering course material. Connect allows students to practice important skills at their own pace and on their own schedule. Importantly, students' assessment results and instructors' feedback are all saved online, so students can continually review their progress and plot their course to success.

LearnSmart™

Built around metacognition learning theory, LearnSmart provides your students with a GPS (Guided Path to Success) for your oceanography course. Using artificial intelligence, LearnSmart intelligently assesses a student's knowledge of course content through a series of adaptive questions. It pinpoints concepts the student does not understand and maps out a *personalized study plan* for success. Available as an integrated feature of McGraw-Hill's Connect, you can incorporate LearnSmart into your course in a number of ways to

- Gauge student knowledge before a lecture
- Reinforce learning after lecture
- Prepare students for assignments and exams

Visit www.mhlearnsmart.com to discover for yourself how the LearnSmart diagnostic ensures students will connect with the content, learn more effectively, and succeed in your course.

TEGRITY™

Tegrity Campus is a service that makes class time available all the time by automatically capturing every lecture in a searchable format for students to review when they study and complete assignments. With a simple one-click start and stop process, you capture all computer screens and corresponding audio. Students replay any part of any class with easy-to-use, browser-based viewing on a PC or Mac.

Educators know that the more students can see, hear, and experience class resources, the better they learn. With Tegrity Campus, students quickly recall key moments by using Tegrity Campus's unique search feature. This search helps students efficiently find what they need, when they need it, across an entire semester of class recordings. Help turn your students' study time into learning moments immediately supported by your lecture.

To learn more about Tegrity, watch a 2-minute Flash demo at http://tegritycampus.mhhe.com.

Customizable Textbooks: Create™

Create what you've only imagined. Introducing McGraw-Hill Create—a new, self-service website that allows you to create custom course materials—print and eBooks—by drawing upon McGraw-Hill's comprehensive, cross-disciplinary content. Add your own content quickly and easily. Tap into other rights-secured third party sources as well. Then, arrange the content in a way that makes the most sense for your course. Even personalize your book with your course name and information. Choose the best format for your course: color print, black and white print, or eBook. The eBook is now viewable on an iPad! And when you are finished customizing, you will receive a free PDF review copy in just minutes! Visit McGraw-Hill Create at www.mcgrawhillcreate.com today and begin building your perfect book.

CourseSmart eBook

CourseSmart is a new way for faculty to find and review eBooks. It's also a great option for students who are interested in accessing their course materials digitally and saving money. *CourseSmart* offers thousands of the most commonly adopted textbooks across hundreds of courses. It is the only place for faculty to review and compare the full text of a textbook online, providing immediate access without the environmental impact of requesting a print exam copy. At *CourseSmart*, students can save up to 50% off the cost of a print book, reduce their impact on the environment, and gain access to powerful Web tools for learning including full text search, notes and highlighting, and email tools for sharing notes between classmates.

To review comp copies or to purchase an eBook, go to www.coursesmart.com.

Acknowledgments

As a book is the product of many experiences, it is also the product of many people working hard to produce an excellent text. We owe very special thanks to the McGraw-Hill team of professionals who created the book you have in your hands:

Managing Director: Thomas Timp
Brand Manager: Michelle Vogler
Developmental Editor: Jodi Rhomberg
Director of Development: Rose Koos
Marketing Manager: Matthew Garcia
Content Project Manager: Kelly Heinrichs, Katie Fuller
Buyer: Susan Culbertson
Designer: Tara McDermott
Photo Researcher: Shawntel Schmitt

Reviewers

In addition we would like to particularly thank the following people who authored *Diving In* special interest boxes.

American Public University System, Robert McDowell
Metropolitan State University of Denver, Beth Simmons
Palm Beach State College, Waweise Schmidt
University of Findlay, Gwynne Stoner Rife
University of Texas, Austin, G. Christopher Shank
University of Washington, Eddie Bernard

We would also like to thank the following individuals who wrote and/or reviewed learning goal-oriented content for **LearnSmart**.

American Public University System, Robert McDowell
Broward College, Nilo Marin
Broward College, David Serrano
Northern Arizona University, Sylvester Allred
Jessica Miles
Roane State Community College, Arthur C. Lee
University of North Carolina at Chapel Hill, Trent McDowell
University of Wisconsin, Milwaukee, Gina S. Szablewski

Input from professors teaching this course is invaluable to the development of a new text. Our thanks and gratitude go out to the following individuals who either completed detailed chapter reviews or provided market feedback vital for this text.

American Public University System, Robert McDowell
Appalachian State University, Ellen A. Cowan
Bellevue College & North Seattle Community College, Gwyneth Jones
Blinn College, Michael Dalman
Blinn College, Amanda Palmer Julson
Boise State University, Jean Parker
Bowie State University, William Lawrence

Brevard Community College, Eric Harms
Brevard Community College, Ashley Spring
Broward College, Jay P. Muza
California Polytechnic State University, Randall Knight
California State University Fullerton, Wayne Henderson
California State University Northridge, Karen L. Savage
Citrus College, George M. Hathaway
City University of Seattle, Gregory W. Judge
Coastal Carolina University, Erin J. Burge
College of San Mateo, John Paul Galloway
College of the Atlantic, Sean Todd
College of the Desert, Robert E. Pellenbarg
Columbia College, Glen White
Columbia River Estuary Study Taskforce, Micah Russell
Columbus State University, William James Frazier
Concordia University, Sarah B. Lovern
Dickinson College, Jeff Niemitz
Dixie State College of Utah, Jennifer Ciaccio
East Los Angeles College, Steven R. Tarnoff
East Stroudsburg University, James C. Hunt
Eastern Illinois University, Diane M. Burns
Eastern Illinois University, James F. Stratton
Eastern Kentucky University, Walter S. Borowski
Eckerd College, Joel Thompson
Edgewood College, Dan E. Olson
Edison State College, Rozalind Jester
Fairleigh Dickinson University, Shane V. Smith
Farmingdale State College, Michael C. Smiles
Fitchburg State University, Elizabeth S. Gordon
Florida Gulf Coast, Toshi Urakawa
Florida State College, Rob Martin
George Mason University, E. C. M. Parsons
Hartnell College, Phaedra Green Jessen
Howard Community College, Jill Nissen
Indiana State University, Jennifer C. Latimer
Johnson State College, Tania Bacchus
Kent State University, Elizabeth M. Griffith
Kent State University, Carrie E. Schweitzer
Lincoln Land Community College, Samantha Reif
Lock Haven University, Khalequzzaman
Louisana State University, Lawrence J. Rouse
Lynchburg College, David Perault
Madison College, Steven Ralser
Manhattanville College, Wendy J. McFarlane
Marian University, Ronald A. Weiss
Massachusetts Institute of Technology, Raffaele Ferrari
Metropolitan State University of Denver, Thomas C. Davinroy
Metropolitan State University of Denver, Beth Simmons

MiraCosta College, Oceanside, Patty Anderson

Monterey Peninsula College, Alfred Hochstaedter

New College of Florida, Sandra L. Gilchrist

Oberlin College, Steven Wojtal

Palm Beach Atlantic University, Donald W. Lovejoy

Palm Beach State College, Waweise Schmidt

Pasadena City College/California Institute of Technology, Martha House

Philadelphia University, Jeffrey Ashley

Pima Community College, Nancy Schmidt

Princeton University, Danielle M. Schmitt

Purdue University, Jim Ogg

Rollins College, Kathryn Patterson Sutherland

Saint Paul College, Maggie Zimmerman

Southwestern College, Ken Yanow

Spring Hill College, Charles M. Chester

Suffolk County Community College, Jean R. Anastasia

SUNY Maritime College, Marie de Angelis

Tarrant County College – Southeast Campus, Christina L. Baack

Texas Southern University, Astatkie Zikarge

Texas State University, San Marcos, Richard W. Dixon

University of California, Davis, Tessa M. Hill

University of Findlay, Gwynne Stoner Rife

University of Hawaii, Evelyn F. Cox

University of Kansas, G. L. Macpherson

University of Maryland University College; Northern Virginia Community College; The Johns Hopkins University, T. John Rowland

University of Michigan, Michela Arnaboldi

University of Michigan, Robert M. Owen

University of Nebraska, Lincoln, John R. Griffin

University of New England, Stephan I. Zeeman

University of Northern Colorado, William H.Hoyt

University of Puget Sound, Michael Valentine

University of Richmond, Roni J. Kingsley

University of Seattle, Gregory W. Judge

University of South Florida, Chantale Bégin

University of Texas, Austin, G. Christopher Shank

University of Texas, Permian Basin, Lori L. Manship

University of Wisconsin-Madison, Harold J. Tobin

Victor Valley College, Walter J. Grossman

Waubonsee Community College, George Bennett

Wenatchee Valley College, Rob Fitche

Wester Connecticut State University, J. P. Boyle

Western Michigan University, Michelle Kominz

Western New England University, Alexander Wurm

Whatcom Community College, Doug McKeever

Wheaton College, Stephen O. Moshier

Windward Community College, Michelle Smith

Wright State University, Paul J. Wolfe

The History of Oceanography

Learning Outcomes

After studying the information in this chapter students should be able to:

1. *discuss* the interaction of early civilizations with the oceans,

2. *sketch* the major seafaring routes of the great voyages of discovery in the fifteenth and sixteenth centuries, James Cook's voyages of discovery, and the scientific voyages of Charles Darwin and the *Challenger* expedition,

3. *list* the major discoveries of the *Challenger* expedition,

4. *compare* and *contrast* the methods of making scientific measurements in the nineteenth and twentieth centuries,

5. *describe* the difference in both the quantity of oceanographic data and the density of that data available to oceanographers now compared to the nineteenth century.

A sextant and marine charts. The sextant is an early navigational aid first constructed by John Bird in 1759.

Oceanography is a broad field in which many sciences are focused on the common goal of understanding the oceans. Geology, geography, geophysics, physics, chemistry, geochemistry, mathematics, meteorology, botany, and zoology have all played roles in expanding our knowledge of the oceans. Oceanography today is usually broken down into a number of subdisciplines because the field is extremely interdisciplinary.

Geological oceanography includes the study of Earth at the sea's edge and below its surface, and the history of the processes that form the ocean basins. Physical oceanography investigates the causes and characteristics of water movements such as waves, currents, and tides and how they affect the marine environment. It also includes studies of the transmission of energy such as sound, light, and heat in seawater. Marine meteorology (the study of heat transfer, water cycles, and air-sea interactions) is often included in the discipline of physical oceanography. Chemical oceanography studies the composition and history of the water, its processes, and its interactions. Biological oceanography concerns marine organisms and the relationship between these organisms and the environment in the oceans. Ocean engineering is the discipline that designs and plans equipment and installations for use at sea.

The study of the oceans was promoted by intellectual and social forces as well as by our needs for marine resources, trade and commerce, and national security. Oceanography started slowly and informally; it began to develop as a modern science in the mid-1800s and has grown dramatically, even explosively, in the last few decades. Our progress toward the goal of understanding the oceans has been uneven and progress has frequently changed direction. The interests and needs of nations as well as the scholarly curiosity of scientists have controlled the ways we study the oceans, the methods we use to study them, and the priority we give to certain areas of study. To gain perspective on the current state of knowledge about the oceans, we need to know something about the events and incentives that guided people's previous investigations of the oceans.

P.1 The Early Times

People have been gathering information about the oceans for millennia, accumulating bits and pieces of knowledge and passing it on by word of mouth. Curious individuals must have acquired their first ideas of the oceans from wandering the seashore, wading in the shallows, and gathering food from the ocean's edges.

During the Paleolithic period, humans developed the barbed spear, or harpoon, and the gorge. The gorge was a double-pointed stick inserted into a bait and attached to a string. At the beginning of the Neolithic period, the bone fishhook was developed and later the net. By 5000 B.C., copper fishhooks were in use.

As early humans moved slowly away from their inland centers of development, they were prepared to take advantage of the sea's food sources when they first explored and later settled along the ocean shore. The remains of shells and other refuse, in piles known as kitchen middens, have been found at the sites of ancient shore settlements. These remains show that our early ancestors gathered shellfish, and fish bones found in some middens suggest that they also used rafts or some type of boat for offshore fishing. Some scientists think that many more artifacts have been lost or scattered as a result of rising sea level. The artifacts that have been found probably give us only an idea of the minimum extent of ancient shore settlements. Drawings on ancient temple walls show fishnets; on the tomb of the Egyptian Pharaoh Ti, Fifth Dynasty (5000 years ago), is a drawing of the poisonous pufferfish with a hieroglyphic description and warning. As long ago as 1200 B.C. or earlier, dried fish were traded in the Persian Gulf; in the Mediterranean, the ancient Greeks caught, preserved, and traded fish, while the Phoenicians founded fishing settlements, such as "the fisher's town" Sidon, that grew into important trading ports.

Early information about the oceans was mainly collected by explorers and traders. These voyages left little in the way of recorded information. Using descriptions passed down from one voyager to another, early sailors piloted their way from one landmark to another, sailing close to shore and often bringing their boats up onto the beach each night.

Some historians believe that seagoing ships of all kinds are derived from early Egyptian vessels. The first recorded voyage by sea was led by Pharaoh Snefru about 3200 B.C. In 2750 B.C., Hannu led the earliest documented exploring expedition from Egypt to the southern edge of the Arabian Peninsula and the Red Sea.

The Phoenicians, who lived in present-day Lebanon from about 1200 to 146 B.C., were well-known as excellent sailors and navigators. While their land was fertile it was also densely populated, so they were compelled to engage in trade to acquire many of the goods they needed. They accomplished this by establishing land routes to the east and marine routes to the west. The Phoenicians were the only nation in the region at that time that had a navy. They traded throughout the Mediterranean Sea with the inhabitants of North Africa, Italy, Greece, France, and Spain. They also ventured out of the Mediterranean Sea to travel north along the coast of Europe to the British Isles and south to circumnavigate Africa in about 590 B.C.

Evidence from anthropology and archaeology suggests that the people who would first explore and populate the Pacific Ocean Basin migrated from Asia to the island of Taiwan, approximately 10,000–6000 B.C. These people are known as the Austronesians. Sometime between 5000–2500 B.C., a growth in population led the Austronesians to travel to the main island of Luzon in the north Philippines. Over the next 1000 years they moved progressively southeast through the rest of the Philippines and the nearby islands of the Celebes Sea, Borneo, and Indonesia. This was relatively easy because of the comparatively short distance between islands in the far southwestern Pacific region. The Austronesians traveled to the islands of Melanesia and Micronesia between 1200 B.C. and A.D. 500, respectively (fig. P.1). The Austronesians discovered and began populating Polynesia by 1000 B.C. and the distinctive Polynesian culture began to develop. Polynesians embarked on more extensive voyages eastward, where the distance between islands grew from tens of kilometers in the western Pacific to thousands of kilometers in the case of voyages to the Hawaiian Islands and Easter Island. There is some uncertainty about when they settled specific regions. It is thought they reached and colonized the Hawaiian Islands by about A.D. 400, arrived in New Zealand around A.D. 800, and settled Easter Island around A.D. 1200. By the early thirteenth century, Polynesians had colonized every habitable island in a triangular region roughly twice the size of the United States bound by Hawaii on the north, New Zealand in the southwest, and Easter Island to the east (fig. P.1).

A basic component of navigation throughout the Pacific was the careful observation and recording of where prominent stars rise and set on the horizon. Observed near the equator,

the stars appear to rotate from east to west on a north-south axis. Some rise and set farther to the north and some farther to the south, and they do so at different times. Navigators created a "star structure" by dividing the horizon into thirty-two segments where their known stars rose and set. These directions form a compass and provide a reference for recording information about the direction of winds, currents, waves, and the relative positions of islands, shoals, and reefs (fig. P.2). The Polynesians also navigated by making close observations of waves and cloud formations. Observations of birds and distinctive smells of land such as flowers and wood smoke alerted them to possible landfalls. Once islands were discovered, their locations relative to one another and to the regular patterns of sea swell and waves bent around islands could be recorded with stick charts constructed of bamboo and shells (fig. P.3).

As early as 1500 B.C., Middle Eastern peoples of many different ethnic groups and regions were exploring the Indian Ocean. In the seventh century A.D., they were unified under Islam and controlled the trade routes to India and China and consequently the commerce in silk, spices, and other valuable goods. (This monopoly wasn't broken until Vasco da Gama defeated the Arab fleet in 1502 in the Arabian Sea.)

The Greeks called the Mediterranean "Thalassa" and believed that it was encompassed by land, which in turn was surrounded by the endlessly circling river Oceanus. In 325 B.C., Alexander the Great reached the deserts of the Mekran Coast, now a part of Pakistan. He sent his fleet down the coast in an apparent effort to probe the mystery of Oceanus. He and his troops had expected to find a dark, fearsome sea of whirlpools

Figure P.1 The migration of people across the Pacific Ocean basin. Austronesians, the ancestors of the Polynesians, migrated from Asia into the far western Pacific by 2500 B.C. Polynesian culture began about 1000 B.C. Migration throughout Polynesia first was to the northeast to the Hawaiian Islands, then southwest to New Zealand, and eventually east to Easter Island.

Figure P.2 On Satawal Island, master navigator Mau Piailug teaches navigation to his son and grandson with the help of a star compass. The compass consists of an outer ring of stones, each representing a star or a constellation when it rises or sets on the horizon, and an inner ring of pieces of palm leaf representing the swells, which travel from set directions, and together with the stars, help the navigator find his way over the sea. In the center of the ring, the palm leaves serve as a model outrigger canoe.

Figure P.3 A navigational chart (*rebillib*) of the Marshall Islands. Sticks represent a series of regular wave patterns (swells). Curved sticks show waves bent by the shorelines of individual islands. Islands are represented by shells.

and water spouts inhabited by monsters and demons; they did find tides that were unknown to them in the Mediterranean Sea. Pytheas (350–300 B.C.), a navigator, geographer, astronomer,

and contemporary of Alexander, made one of the earliest recorded voyages from the Mediterranean to England. From there, he sailed north to Scotland, Norway, and Germany. He recognized a relationship between the tides and the Moon, and made early attempts at determining latitude and longitude. These early sailors did not investigate the oceans; for them, the oceans were only a dangerous road, a pathway from here to there, a situation that continued for hundreds of years. However, the information they accumulated slowly built into a body of lore to which sailors and voyagers added each year.

While the Greeks traded and warred throughout the Mediterranean, they observed the sea and asked questions. Aristotle (384–322 B.C.) believed that the oceans occupied the deepest parts of Earth's surface; he knew that the Sun evaporated water from the sea surface, which condensed and returned as rain. He also began to catalog marine organisms. The brilliant Eratosthenes (c. 276–195 B.C.) of Alexandria, Egypt, invented the study of geography as well as a system of latitude and longitude. He was the first person to calculate the tilt of Earth's axis. One of his greatest achievements was his calculation of Earth's circumference; he accomplished this without ever leaving Egypt (fig. P.4). Eratosthenes knew that at local noon on the summer solstice, the Sun would be directly overhead in the city of Syene, located on the Tropic of Cancer at 23½°N. On that day and at that time, the Sun's rays would shine down into a well in the city and illuminate the bottom of the well. At the same time and day, the Sun's rays would cast a shadow behind a pole in his home city of Alexandria due north of Syene. By measuring the height of the pole and the length of the shadow, Eratosthenes could calculate the angle away from perpendicular of the Sun's elevation in Alexandria, which he determined to be about 7.2°, or roughly 1/50 of a circle. Assuming that the distance between the Sun and Earth is so large that all of the Sun's rays of light are parallel to each other when they reach Earth, Eratosthenes could say that the angle between Syene

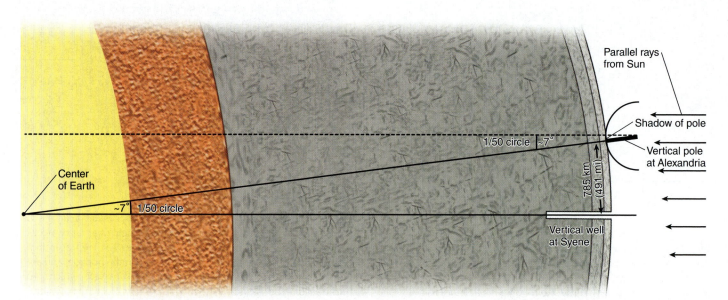

Figure P.4 Eratosthenes used geometry to calculate Earth's circumference. By careful measurement, he was able to estimate Earth's circumference to within about 1% of today's value. (The diagram is not drawn to scale.)

and Alexandria was also about 7.2°, or roughly 1/50 of a circle. Repeated surveys of the distance between Syene and Alexandria yielded a distance of 5000 stadia, so Eratosthenes concluded that Earth's circumference was 252,000 stadia. The length of an Egyptian stadion was 157.5 m, so Eratosthene's estimate of Earth's circumference was 39,690 km (24,662 mi), an error of only about 1% compared to today's accepted average value of about 40,030 km (24,873 mi). Posidonius (c. 135–50 B.C.) reportedly measured an ocean depth of about 1800 m (6000 ft) near the island of Sardinia, according to the Greek geographer Strabo (c. 63 B.C.–A.D. 21). Pliny the Elder (c. A.D. 23–79) related the phases of the Moon to the tides and reported on the currents moving through the Strait of Gibraltar. Claudius Ptolemy (c. A.D. ~85–161) produced the first world atlas and established world boundaries: to the north, the British Isles, Northern Europe, and the unknown lands of Asia; to the south, an unknown land, "Terra Australis Incognita," including Ethiopia, Libya, and the Indian Sea; to the east, China; and to the west, the great Western Ocean reaching around Earth to China. His atlas listed more than 8000 places by latitude and longitude, but his work contained a major flaw. He had accepted a value of 29,000 km (18,000 mi) for Earth's circumference. This value was much too small and led Columbus, more than 1000 years later, to believe that he had reached the eastern shore of Asia when he landed in the Americas.

QUICK REVIEW

1. Name the subfields of oceanography.
2. What did early sailors use for guidance during long ocean voyages?
3. What kind of "compass" did the Polynesians use for navigation?
4. How long ago was Earth's circumference first calculated and how was it done?
5. How did Ptolemy's atlas contribute to a greater understanding of world geography, and how did it produce confusion?

P.2 The Middle Ages

After Ptolemy, intellectual activity and scientific thought declined in Europe for about 1000 years. However, shipbuilding improved during this period; vessels became more seaworthy and easier to sail, so sailors could make longer voyages. The Vikings (Norse for *piracy*) were highly accomplished seamen who engaged in extensive exploration, trade, and colonization for nearly three centuries from about 793 to 1066 (fig. P.5). During this time, they journeyed inland on rivers through Europe and western Asia, traveling as far as the Black and Caspian Seas. The Vikings are probably best known for their voyages across the North Atlantic Ocean. They sailed to Iceland in 871 where as many as 12,000 immigrants eventually settled. Erik Thorvaldsson (known as Erik the Red) sailed west from Iceland in 982 and discovered Greenland. He lived there for three years before returning to Iceland to recruit more settlers. Icelander Bjarni Herjolfsson, on his way to Greenland to join the colonists in 985–86, was blown off course, sailed south of Greenland, and is believed to have come within sight of Newfoundland before turning back and reaching Greenland. Leif Eriksson, son of Erik the Red, sailed west from Greenland in 1002 and reached North America roughly 500 years before Columbus.

To the south, in the region of the Mediterranean after the fall of the Roman Empire, Arab scholars preserved Greek and Roman knowledge and continued to build on it. The Arabic writer El-Mas'údé (d. 956) gives the first description of the reversal of the currents due to the seasonal monsoon winds. Using this knowledge of winds and currents, Arab sailors established regular trade routes across the Indian Ocean. In the 1100s, large Chinese junks with crews of 200 to 300 sailed the same routes (between China and the Persian Gulf) as the Arab dhows.

During the Middle Ages, while scholarship about the sea remained primitive, the knowledge of navigation increased. Harbor-finding charts, or *portolanos,* appeared. These charts carried a distance scale and noted hazards to navigation, but they did not have latitude or longitude. With the introduction of the magnetic compass to Europe from Asia in the thirteenth century, compass directions were added.

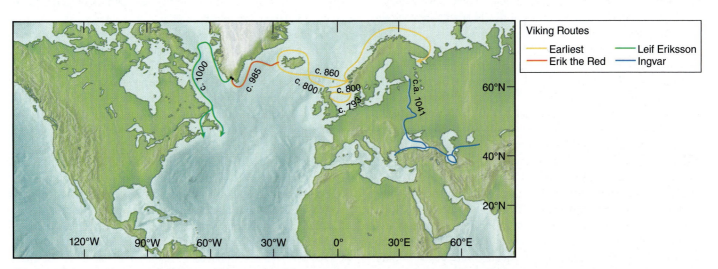

Figure P.5 Major routes of the Vikings to the British Isles, to Asia, and across the Atlantic to Iceland, Greenland, and North America.

Although tides were not understood, the Venerable Bede (673–735) illustrated his account of the tides with data from the British coast. His calculations were followed in the tidal observations collected by the British Abbot Wallingford of Saint Alban's Monastery in about 1200. His tide table, titled "Flod at London Brigge," documented the times of high water. Sailors made use of Bede's calculations until the seventeenth century.

As scholarship was reestablished in Europe, Arabic translations of early Greek studies were translated into Latin and thus became available to European scholars. The study of tides continued to absorb the medieval scientists, who were also interested in the saltiness of the sea. By the 1300s, Europeans had established successful trade routes, including some partial ocean crossings. An appreciation of the importance of navigational techniques grew as trade routes were extended.

QUICK REVIEW

1. What advances occurred during the Middle Ages that allowed longer ocean voyages?

2. During the tenth century, which oceans were explored and by which cultures?

3. Where did the Vikings establish a large colony in the North Atlantic?

P.3 Voyages of Discovery

From 1405 to 1433, the great Chinese admiral Zheng He conducted seven epic voyages in the western Pacific Ocean and across the Indian Ocean as far as Africa. Zheng He's fleet consisted of over 300 ships. The fleet is believed to have included as many as sixty-two "treasure ships" thought to have been as much as 130 m (426 ft) long and 52 m (170 ft) wide; this was ten times the size of the ships used for the European voyages of discovery during this period of time (fig. P.6). The purpose of these voyages remains a matter of debate among scholars. Suggested reasons include the establishment of trade routes, diplomacy with other governments, and military defense. The voyages ended in 1433, when their explorations led the Chinese to believe that other societies had little to offer, and the government of China withdrew within its borders, beginning 400 years of isolation.

In Europe, the desire for riches from new lands persuaded wealthy individuals, often representing their countries, to underwrite the costs of long voyages to all the oceans of the world. The individual most responsible for the great age of European discovery was Prince Henry the Navigator (1394–1460) of Portugal. In 1419, his father, King John, made him governor of Portugal's southernmost coasts. Prince Henry was keenly interested in sailing and commerce, and studied navigation and mapmaking. He established a naval observatory for the teaching of navigation, astronomy, and cartography about 1450. From 1419 until his death in 1460, Prince Henry sent expedition after expedition south along the west coast of Africa to secure trade routes and establish colonies. These expeditions moved slowly due to the mariners' belief that waters at the equator were at the boiling point and that sea monsters would engulf ships. It wasn't until

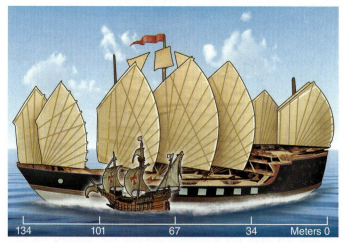

(a)

(b)

Figure P.6 (a) Admiral Zheng He's "treasure ships" were over 130 m long. In comparison, Christopher Columbus's flagship, the *Santa Maria*, is estimated to have been about 18 m long. (b) Zheng He's probable route from China along the coast of the Indian Ocean to Africa.

twenty-seven years after Prince Henry's death that Bartholomeu Dias (1450?–1500) braved these "dangers" and rounded the Cape of Good Hope in 1487 in the first of the great voyages of discovery (fig. P.7). Dias had sailed in search of new and faster routes to the spices and silks of the East.

Portugal's slow progress along the west coast of Africa in search for a route to the east finally came to fruition with Vasco da Gama (1469–1524) (fig. P.7). In 1497, he followed Bartholomeu Dias's route to the Cape of Good Hope and then continued beyond along the eastern coast of the African continent. He successfully mapped a route to India but was challenged along the way by Arab ships. In 1502, da Gama returned with a flotilla of fourteen heavily armed ships and defeated the Arab fleet. By 1511, the Portuguese controlled the spice routes and had access to the Spice Islands. In 1513, Portuguese trade extended to China and Japan.

Christopher Columbus (1451–1506) made four voyages across the Atlantic Ocean in an effort to find a new route to the East Indies by traveling west rather than east. By relying on inaccurate estimates of Earth's size, he badly underestimated the distances involved and believed he had found islands off the coast of Asia when, in fact, he had reached the New World (fig. P.7).

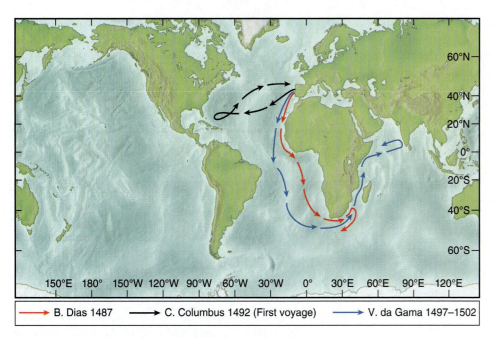

→ B. Dias 1487	→ C. Columbus 1492 (First voyage)	→ V. da Gama 1497–1502

Figure P.7 The routes of Bartholomeu Dias and Vasco da Gama around the Cape of Good Hope and Christopher Columbus's first voyage.

Italian navigator Amerigo Vespucci (1454–1512) made several voyages to the New World (1499–1504) for Spain and Portugal, exploring nearly 10,000 km (6000 mi) of South American coastline. He accepted South America as a new continent not part of Asia, and in 1507, German cartographer Martin Waldseemüller applied the name "America" to the continent in Vespucci's honor. Vasco Núñez de Balboa (1475–1519) crossed the Isthmus of Panama and found the Pacific Ocean in 1513, and in the same year, Juan Ponce de León (1460?–1521) discovered Florida and the Florida Current. All claimed the new lands they found for their home countries. Although these men had sailed for fame and riches, not knowledge, they more accurately documented the extent and properties of the oceans, and the news of their travels stimulated others to follow.

Ferdinand Magellan (1480–1521) left Spain in September 1519 with 270 men and five vessels in search of a westward passage to the Spice Islands. The expedition lost two ships before finally discovering and passing through the Strait of Magellan and rounding the tip of South America in November 1520. Magellan crossed the Pacific Ocean and arrived in the Philippines in March 1521, where he was killed in a battle with the natives on April 27, 1521. Two of his ships sailed on and reached the Spice Islands in November 1521, where they loaded valuable spices for a return home. In an attempt to guarantee that at least one ship made it back to Spain, the two ships parted ways. The *Victoria* continued sailing

west and successfully crossed the Indian Ocean, rounded Africa's Cape of Good Hope, and arrived back in Spain on September 6, 1522, with eighteen of the original crew. This was the first circumnavigation of Earth (fig. P.8). Magellan's skill as a navigator makes his voyage probably the most outstanding single contribution to the early charting of the oceans. In addition, during the voyage, he established the length of a degree of latitude and measured the circumference of Earth. It is said that Magellan tried to test the mid-ocean depth of the Pacific with a hand line, but this idea seems to come from a nineteenth-century German oceanographer; writings from Magellan's time do not support this story.

By the latter half of the sixteenth century, adventure, curiosity, and hopes of finding a trading shortcut to China spurred efforts to find a sea passage around North America. Sir Martin Frobisher (1535?–94) made three voyages in 1576, 1577, and 1578, and Henry Hudson (d. 1611) made four voyages (1607, 1608, 1609, and 1610), dying with his son when set adrift in Hudson Bay by his mutinous crew. The Northwest Passage continued to beckon, and in 1615 and 1616, William Baffin (1584–1622) made two unsuccessful attempts.

While European countries were setting up colonies and claiming new lands, Francis Drake (1540–96) set out in 1577 with 165 crewmen and five ships to show the English flag around the world (fig. P.8). He was forced to abandon two of his ships off the coast of South America. He was separated from the other

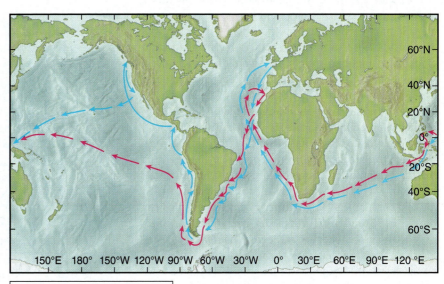

→ F. Magellan 1519–22
→ F. Drake 1577–80

Figure P.8 The sixteenth-century circumnavigation voyages by Magellan and Drake.

two ships while passing through the Strait of Magellan. During the voyage Drake plundered Spanish shipping in the Caribbean and in Central America and loaded his ship with treasure. In June 1579, Drake landed off the coast of present-day California and sailed north along the coast to the present United States–Canadian border. He then turned southwest and crossed the Pacific Ocean in two months' time. In 1580, he completed his circumnavigation and returned home in the *Golden Hind* with a cargo of Spanish gold, to be knighted and treated as a national hero. Queen Elizabeth I encouraged her sea captains' exploits as explorers and raiders because, when needed, their ships and knowledge of the sea brought military victories as well as economic gains.

QUICK REVIEW

1. Why do you think Zheng He's epic voyage followed a route along the coast?
2. What stimulated the long voyages of the fifteenth and sixteenth centuries?
3. Who was Amerigo Vespucci, and how was he honored?
4. Why was Magellan's voyage of such great importance?
5. What is the Northwest Passage and why was there so much interest in finding it?

P.4 The Importance of Charts and Navigational Information

As colonies were established far from their home countries and as trade, travel, and exploration expanded, interest was renewed in developing more accurate charts and navigational techniques. The first hydrographic office dedicated to mapping the oceans was established in France in 1720 and was followed in 1795 by the British Admiralty's appointment of a hydrographer.

As early as 1530, the relationship between time and longitude had been proposed by Flemish astronomer Gemma Frisius, and in 1598, King Philip III of Spain had offered a reward of 100,000 crowns to any clockmaker building a clock that would keep accurate time onboard ship (see the discussion of longitude and time in chapter 1, section 1.4). In 1714, Queen Anne of England authorized a public reward for a practical method of keeping time at sea, and the British Parliament offered 20,000 pounds sterling for a seagoing clock that could keep time with an error not greater than two minutes on a voyage to the West Indies from England. A Yorkshire clockmaker, John Harrison, built his first chronometer (high-accuracy clock) in 1735, but not until 1761 did his fourth model meet the test, losing only fifty-one seconds on the eighty-one-day voyage. Harrison was awarded only a portion of the prize after his success in 1761, and it was not until 1775, at the age of eighty-three, that he received the remainder from the reluctant British government. In 1772, Captain James Cook took a copy of the fourth version of Harrison's

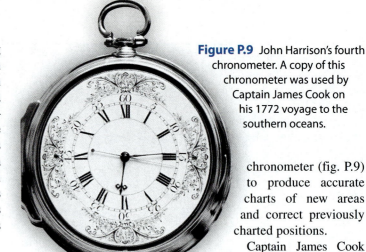

Figure P.9 John Harrison's fourth chronometer. A copy of this chronometer was used by Captain James Cook on his 1772 voyage to the southern oceans.

chronometer (fig. P.9) to produce accurate charts of new areas and correct previously charted positions.

Captain James Cook (1728–79) made his three great voyages to chart the Pacific Ocean between 1768 and 1779 (fig. P.10). In 1768, he left England in command of the *Endeavour* on an expedition to chart the transit of Venus; he returned in 1771, having circumnavigated the globe, and explored and charted the coasts of New Zealand and eastern Australia. Between 1772 and 1775, he commanded an expedition of two ships, the *Resolution* and the *Adventure,* to the South Pacific. On this journey, he charted many islands, explored the Antarctic Ocean, and, by controlling his sailors' diet, prevented vitamin C deficiency and scurvy, the disease that had decimated crews that spent long periods of time at sea. Cook sailed on his third and last voyage in 1776 in the *Resolution* and *Discovery.* He spent a year in the South Pacific and then sailed north, discovering the Hawaiian Islands in 1778. He continued on to the northwest coast of North America and into the Bering Strait, searching for a passage to the Atlantic. He returned to Hawaii for the winter and was killed by natives at Kealakekua Bay on the island of Hawaii in 1779. Cook takes his place not only as one of history's greatest navigators and seamen but also as a fine scientist. He made soundings to depths of

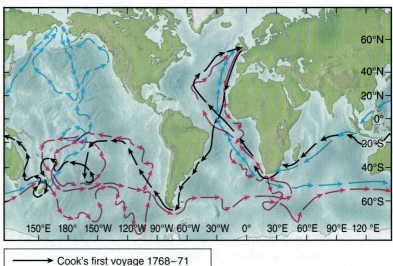

Cook's first voyage 1768–71
Cook's second voyage 1772–75
Cook's third voyage 1776–79

Figure P.10 The three voyages of Captain James Cook.

400 m (1200 ft) and accurate observations of winds, currents, and water temperatures. Cook's careful and accurate observations produced much valuable information and made him one of the founders of oceanography.

In the United States, Benjamin Franklin (1706–90) became concerned about the time required for news and cargo to travel between England and America. With Captain Timothy Folger (Franklin's cousin and a whaling captain from Nantucket) he constructed the 1769 Franklin-Folger chart of the Gulf Stream current (fig. P.11). When published, the chart encouraged captains to sail within the Gulf Stream en route to Europe and return via the trade winds belt and follow the Gulf Stream north again to Philadelphia, New York City, and other ports. Since the Gulf Stream carries warm water from low latitudes to high latitudes, it is possible to map its location with satellites that measure sea surface temperature. Compare the Franklin-Folger chart in figure P.11 to a map of the Gulf Stream shown in figure P.12 based on the average sea surface temperature during 1996. In 1802, Nathaniel Bowditch (1773–1838), another American, published the *New American Practical Navigator*. In this book, Bowditch made the techniques of celestial navigation available for the first time to every competent sailor and set the stage for U.S. supremacy of the seas during the years of the Yankee clippers.

In 1807, the U.S. Congress, at the direction of President Thomas Jefferson, formed the Survey of the Coast under the Treasury Department, later named the Coast and Geodetic Survey and now known as the National Ocean Survey. The U.S. Naval Hydrographic Office, now the U.S. Naval Oceanographic Office, was set up in 1830. Both were dedicated to exploring the oceans and producing better coast and ocean charts. In 1842, Lieutenant Matthew F. Maury (1806–73), who had worked with the Coast and Geodetic Survey, was assigned to the Hydrographic Office and founded the Naval Depot of Charts. He began a systematic collection of wind and current data from ships' logs. He produced his first wind and current charts

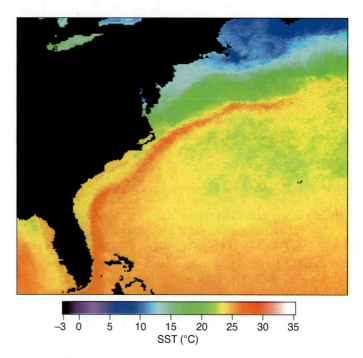

Figure P.12 Annual average sea surface temperature for 1996. The *red-orange streak* of 25° to 30°C water shows the Gulf Stream. Compare with figure 7.26.

of the North Atlantic in 1847. At the 1853 Brussels Maritime Conference, Maury issued a plea for international cooperation in data collection, and from the ships' logs he received, he produced the first published atlases of sea conditions and sailing directions. His work was enormously useful, allowing ships to sail more safely and take days off their sailing times between major ports around the world. The British estimated that Maury's sailing directions took thirty days off the passage from the British Isles to California, twenty days off the voyage to Australia, and ten days off the sailing time to Rio de Janeiro. In 1855, he published *The Physical Geography of the Sea*. This work includes chapters on the Gulf Stream, the atmosphere, currents, depths, winds, climates, and storms, as well as the first bathymetric chart of the North Atlantic with contours at 6000, 12,000, 18,000, and 24,000 ft. Many marine scientists consider Maury's book the first textbook of what we now call oceanography and consider Maury the first true oceanographer. Again, national and commercial interests were the driving forces behind the study of the oceans.

QUICK REVIEW

1. Why was John Harrison's clock important to open ocean navigation?
2. What did Captain James Cook's voyages contribute to the science of oceanography?
3. Why was Benjamin Franklin interested in the Gulf Stream?
4. Who was Matthew F. Maury, and what was his contribution to ocean science?

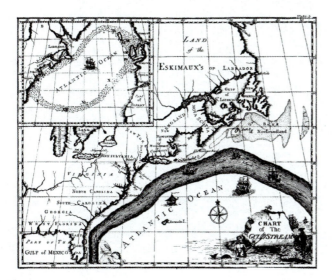

Figure P.11 The Franklin-Folger map of the Gulf Stream, 1769. Compare this map with figure P.12.

P.5 Ocean Science Begins

As charts became more accurate and as information about the oceans increased, the oceans captured the interest of naturalists and biologists. Baron Alexander von Humboldt (1769–1859) made observations on a five-year (1799–1804) cruise to South America; he was particularly fascinated with the vast numbers of animals inhabiting the current flowing northward along the western coast of South America, the current that now bears his name. Charles Darwin (1809–82) joined the survey ship *Beagle* and served as the ship's naturalist from 1831–36 (fig. P.13). He described, collected, and classified organisms from the land and sea. His theory of atoll formation is still the accepted explanation.

At approximately the same time, another English naturalist, Edward Forbes (1815–54), began a systematic survey of marine life around the British Isles and in the Mediterranean and Aegean Seas. He collected organisms in deep water and, on the basis of his observations, proposed a system of ocean depth zones, each characterized by specific animal populations. However, he also mistakenly theorized that there was an azoic, or lifeless, environment below 550 m (1800 ft). His announcement is curious, because twenty years earlier, the Arctic explorer Sir John Ross (1777–1856), looking again for the Northwest Passage, had taken bottom samples at over 1800 m (6000 ft) depth in Baffin Bay with a "deep-sea clamm," or bottom grab, and had found worms and other animals living in the mud. Ross's nephew, Sir James Clark Ross (1800–62), took even deeper samples from Antarctic waters and noted their similarity to the Arctic species recovered by his uncle. Still, Forbes's

systematic attempt to make orderly predictions about the oceans, his enthusiasm, and his influence make him another candidate as a founder of oceanography.

Christian Ehrenberg (1795–1876), a German naturalist, found the skeletons of minute organisms in seafloor sediments and recognized that the same organisms were alive at the sea surface; he concluded that the sea was filled with microscopic life and that the skeletal remains of these tiny organisms were still being added to the sea floor. Investigation of the minute drifting plants and animals of the ocean was not seriously undertaken until German scientist Johannes Müller (1801–58) began his work in 1846. He used an improved fine-mesh tow net similar to that used by Charles Darwin to collect these organisms, which he examined microscopically. This work was continued by Victor Hensen (1835–1924), who improved the Müller net, introduced the quantitative study of these minute drifting sea organisms, and gave them the name *plankton* in 1887.

Although science blossomed in the seventeenth and eighteenth centuries, there was little scientific interest in the sea except as we have seen for the practical reasons of navigation, tide prediction, and safety. In the early nineteenth century, ocean scientists were still few and usually only temporarily attracted to the sea. Some historians believe that the subject and study of the oceans were so vast, requiring so many people and such large amounts of money, that government interest and support were required before oceanography could grow as a science. This did not happen until the nineteenth century in Great Britain.

In the last part of the nineteenth century, laying transatlantic telegraph cables made a better knowledge of the deep sea a

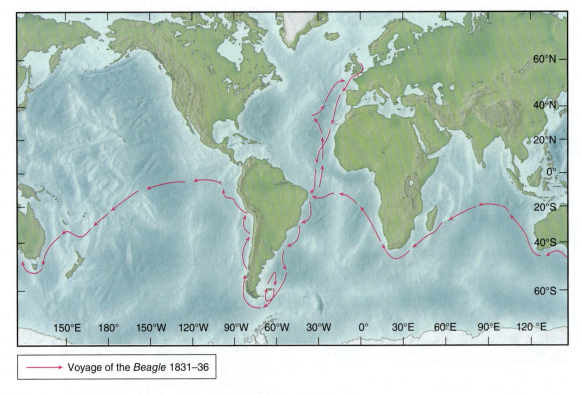

Voyage of the *Beagle* 1831–36

Figure P.13   The voyage of Charles Darwin and the survey ship HMS *Beagle,* 1831–36.

necessity. Engineers needed to know about seafloor conditions, including bottom topography, currents, and organisms that might dislodge or destroy the cables. The British began a series of deep-sea studies stimulated by the retrieval of a damaged cable from more than 1500 m (5000 ft) deep, well below Forbes's azoic zone. When the cable was brought to the surface, it was found to be covered with organisms, many of which had never been seen before. In 1868, the *Lightning* dredged between Scotland and the Faroe Islands at depths of 915 m (3000 ft) and found many animal forms. The British Admiralty continued these studies with the *Porcupine* during the summers of 1869 and 1870, dredging up animals from depths of more than 4300 m (14,000 ft). Charles Wyville Thomson (1830–82), like Forbes, a professor of natural history at Edinburgh University, was one of the scientific leaders of these two expeditions. On the basis of these results, he wrote *The Depths of the Sea,* published in 1873, which became very popular and is regarded by some as the first book on oceanography.

English biologist Thomas Henry Huxley (1825–95), a close friend and supporter of Charles Darwin, was particularly interested in studying the organisms that inhabit the deep sea. Huxley was a strong supporter of Darwin's theory of evolution and believed that the organisms of the deep sea could supply evidence of its validity. In 1868, while examining samples of mud recovered from the deep-sea floor of the Atlantic eleven years before and preserved in alcohol, Huxley noticed that the surface of the samples was covered by a thick, mucus-like material with small embedded particles. Under the microscope it appeared as if these particles moved, leading him to conclude that the mucus was a form of living protoplasm. Huxley named this "organism" *Bathybius haeckelii* after the noted German naturalist Ernst Haeckel (fig. P.14). Haeckel, also a strong supporter of the theory of evolution, viewed this protoplasm as the primordial ooze from which all other life evolved. He believed it blanketed the deep-sea floor and provided an inexhaustible supply of food for higher order organisms of the deep ocean. One of the primary scientific objectives of the *Challenger* expedition, described in the next section, was to study the distribution of *Bathybius haeckelii.*

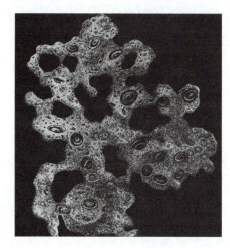

Figure P.14 *Bathybius haeckelii.* English biologist Thomas Huxley believed this to be a primoridial protoplasm covering the ocean floor.

QUICK REVIEW

1. How did Edward Forbes advance understanding of the oceans, and in what way was he mistaken?
2. How were the discoveries of John Ross and James Clark Ross similar? How did they differ?
3 How did the desire to improve communication between the United States and Europe help to advance oceanography?

P.6 Early Expeditions of the Nineteenth and Twentieth Centuries

The *Challenger* Expedition

With public interest running high, the Circumnavigation Committee of the British Royal Society was able to persuade the British Admiralty to organize the most comprehensive oceanographic expedition yet undertaken, using the *Challenger.* The *Challenger* sailed from Portsmouth, England for a voyage that was to last nearly three-and-a-half years (fig. P.15). The first leg of the voyage took the vessel to Bermuda, then to the South Atlantic island of Tristan da Cunha, around the Cape of Good Hope, and east across the southernmost part of the Indian Ocean. It continued on to Australia, New Zealand, the Philippines, Japan, and China. Turning south to the Marianas Islands, the vessel took its deepest sounding at 8180 m (26,850 ft) in what is now known as the Challenger Deep in the Marianas Trench. Sailing across the Pacific to Hawaii, Tahiti, and through the Strait of Magellan, it returned to England on May 24, 1876.

The numerous dredges of deep-sea sediments obtained during the expedition failed to find any evidence of *Bathybius haeckelii* in fresh samples. One of the naturalists noticed, however, that when alcohol was added to a sample, something similar to *Bathybius haeckelii* was produced. Rather than being the primitive life form from which all other organisms evolved, *Bathybius haeckelii* was shown to be a chemical precipitate produced by the reaction of the sediment with alcohol.

The work of organizing and compiling information continued for twenty years. William Dittmar (1833–92) prepared the information on seawater chemistry. He identified the major elements present in the water and confirmed the findings of earlier chemists that in a seawater sample, the relative proportions of the major dissolved elements are constant. Oceanography as a modern science is usually dated from the *Challenger* expedition. More information about this historic expedition is given in "Diving In: The Voyage of the *Challenger* 1872–76."

The Voyage of the *Fram*

During the late nineteenth and early twentieth centuries, oceanography was changing from a descriptive science to a quantitative one. Oceanographic cruises now had the goal of testing hypotheses by gathering data. Theoretical models of ocean circulation

Diving in

The Voyage of the *Challenger*, 1872–1876

BY BETH SIMMONS

Dr. Beth Simmons, professor at Metropolitan State College of Denver, teaches basic oceanography to landlubbers in Colorado. Enrolled in over a dozen sections per semester, both online and in the classroom, Metro students—some from as far away as Hawaii, Florida, and California—take their first step toward a lifelong dream of a career in some phase of marine science.

During the Victorian Era, the British ruled the seas. After Charles Darwin made famous the four-year voyage of Her Majesty's Ship [HMS] *Beagle* (1832–36), British scientists clamored the government for more voyages in the name of oceanic science (box fig. 1*a*). So, much like the American space program of the twentieth century, the British government initiated an oceanographic research program beginning with two test cruises. Scientific probes made during the voyages of the HMS *Lightning*, an old naval paddle-steamer, in 1868, and on four voyages of the HMS *Porcupine* in 1869 and 1870 throughout the North Atlantic and Mediterranean Seas, tested equipment and sampling techniques in preparation for a global expedition.

On earlier experimental voyages in the 1840s, pioneering oceanographers had noticed the existence of life zones (layers of life) along the shorelines and into the depths of the seas. One scientist, Edwin Forbes, predicted that because of lack of sunlight and intense pressures, life could not exist below 300 fathoms (1800 feet). Dredge hauls from the *Porcupine* and *Lightning* proved that life, albeit in strange forms, existed at all depths in the sea. The scientists wanted to know more!

To satisfy their interest and further map the ocean, the Royal Navy replaced sixteen of the guns on HMS *Challenger*, a Pearl-class corvette (a small, maneuverable, lightly armed warship), with miles of sampling rope, wire, thermometers, water bottles, and bottom samplers. The ship sported a steam engine capable of holding the ship in position during sampling and when there was a lack of wind in her sails. Below deck were state-of-the-art natural history, chemical, and physical testing laboratories (box fig. 1*b*). Charles Wyville Thomson, one of Forbes's

Box Figure 1a "H.M.S. *Challenger*—Shortening Sail to Sound," decreasing speed to take a deep-sea depth measurement.

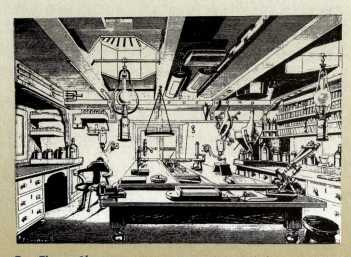

Box Figure 1b Zoological laboratory on the main deck.

and water movement were developed. The Scandinavian oceanographers were particularly active in the study of water movement. One of them, Fridtjof Nansen (1861–1930) (fig. P.16*a*), a well-known athlete, explorer, and zoologist, was interested in the currents of the polar seas. This extraordinary man decided to test his ideas about the direction of ice drift in the Arctic by freezing a vessel into the polar ice pack and drifting with it to reach the North Pole. To do so, he had to design a special vessel that would be able to survive the great pressure from the ice;

the 39 m (128 ft) wooden *Fram* ("to push forward"), shown in figure P.16, was built with a smoothly rounded hull and planking over 60 cm (2 ft) thick.

Nansen departed with thirteen men from Oslo in June 1893. The ship was frozen into the ice nearly 1100 km (700 mi) from the North Pole and remained in the ice for thirty-five months. During this period, measurements were made through holes in the ice that showed that the Arctic Ocean was a deep-ocean basin and not the shallow sea that had been expected. Water and

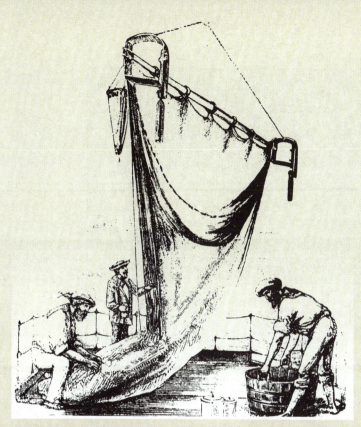

Box Figure 1c A biological dredge used for sampling bottom organisms. Note the frame and skids that keep the mouth of the net open and allow it to slide over the sea floor.

ocean water constant? What was the green mud that Ross had brought up from the depths of the Antarctic Ocean? What was its extent on the ocean floor? And, of course, the Royal Navy wanted to know—where did the oceanic currents flow? Where should ships navigate to avoid hitting reefs or underwater volcanoes?

To keep the British public informed about the progress of the three-year-long adventure into this new scientific realm, ships returning to England from ports along the way toted samples, specimens, and reports back to the motherland. As soon as the communiqués reached London, well-established scientist and author Thomas Huxley spread the facts about the new finds and their possible interpretations like a television news reporter, through newspaper essays and lectures. Recognizing just one of the long-lasting impacts of the voyage's data, in 1875, a year before the ship returned to port, Huxley analyzed the seafloor sediments. He wrote:

> *The discoveries made by the Challenger expedition, like all recent advances in our knowledge of the phenomena of biology, or of the changes now being effected in the structure of the surface of the earth, are in accordance with, and lend strong support to, that doctrine of Uniformitarianism, which fifty years ago was held only by a small minority of English geologists . . . but now . . . has gradually passed from the position of a heresy to that of catholic doctrine.*

The *Challenger* docked in Spithead, England, on May 24, 1876, having traversed almost 70,000 nautical miles. Scientists around the world stepped forward to identify and describe 4717 new species taken during 133 dredge samples and 151 trawl hauls (box fig. 1c). Physical data from almost 500 deep-sea soundings and over 250 water samplings from depths down to and below 6000 feet showed the uniformity of ocean waters. Unfortunately, C. W. Thomson died shortly after the return to England, but his copious notes, letters, and specimens kept scientists occupied for twenty years. Sketches of graceful, delicate glass sponges; of clams; and of microscopic radiolarians, along with descriptions of "phosphatic" glowing fish, filled fifty volumes of reports that Thomson's assistant, John Murray, compiled from the voyage of HMS *Challenger*.

The voyage of HMS *Challenger* established protocols followed by future explorers of ocean and space—planning, development of new methods and tools and techniques, practice using new equipment, publicity, and post-mission publications. Many countries—Norway, Germany, France, Austria, the United States, Italy, and Russia—followed suit with their own research missions. A new era of Earth and space exploration had begun. In honor of this famous oceanographic voyage, NASA named the second space shuttle *Challenger*.

former students—who was a professor of natural history at University of Scotland at Edinburgh and the lead scientist for the previous oceanic research voyages—was the logical scientist to head the extensive *Challenger* expedition. The unique scientific equipment developed for the voyage was as technically advanced for its time as one of NASA's space probes 100 years later.

Sailing into familiar waters, HMS *Challenger* cast off from the dock at Sheerness, England, on December 21, 1872, with six scientists; Charles W. Thomson, Henry Moseley, Rudolf von Willemoes-Suhm, John Buchanan, John Murray, and official artist, J. J. Wild, aboard the world's most modern floating laboratory. The ship's commanding officer, Captain George Nares; twenty naval officers, including surgeons and engineers; and 200 crew members supported the operation.

The *Challenger*'s mission was to answer fundamental questions about the ocean. How deep is the sea? Is there really a mountain range under the ocean? What is the salinity of the oceans? Is the chemistry of

air temperatures were recorded, water chemistry was analyzed, and the great plankton blooms of the area were observed. Nansen became impatient with the slow rate of drift and, with F. H. Johansen, left the *Fram* locked in the ice some 500 km (300 mi) from the pole. They set off with dogsleds toward the pole, but after four and a half weeks, they were still more than 300 km (200 mi) from the pole, with provisions running low and the condition of their dogs deteriorating. The two men turned away from the pole and spent the winter of 1895–96 on the ice, living

on seals and walrus. They were found by a British polar expedition in June 1896 and returned to Norway in August of that year. The crew of the *Fram* continued to drift with the ship until they freed the vessel from the ice in 1896 and returned home. Nansen's expedition had laid the basis for future Arctic work.

After the expedition's findings were published, Nansen continued to be active in oceanography, and his name is familiar today from the Nansen bottle, which he designed to collect water samples from deep water. In 1905, he turned to a career

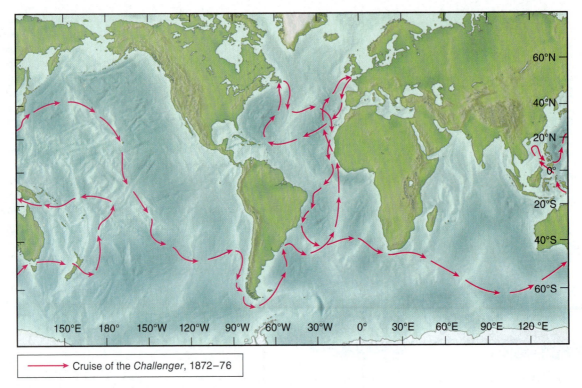

→ Cruise of the *Challenger*, 1872–76

Figure P.15 The cruise of the *Challenger,* the first major oceanography research effort.

Figure P.16 (a) Fridtjof Nansen, Norwegian scientist, explorer, and statesman (1861–1930), using a sextant to determine his ship's position. (b) The *Fram* frozen in ice. As the ice pressure increased, it lifted its specially designed and strengthened hull so that the ship would not be crushed.

(a)

(b)

as a statesman, working for the peaceful separation of Norway from Sweden. After World War I, he worked with the League of Nations to resettle refugees, for which he received the 1922 Nobel Peace Prize.

The *Meteor* Expedition

The reliable and accurate measurement of ocean depths had to wait until the development of the echo sounder, which was given its first scientific use on the 1925–27 German cruise of the *Meteor*. The *Meteor* expedition zigzagged between Africa and South America, exploring the region from 20°N to 60°S. Measurements of water temperature, atmospheric observations, water sampling, and studies of marine life were all conducted. However, the expedition is best known for its first use of the echo sounder. Over 67,000 depth soundings were made, demonstrating that the South Atlantic sea floor was not featureless, as previously thought. Instead, the depth soundings showed that the Mid-Atlantic Ridge was a continuous feature in the South Atlantic and continued into the Indian Ocean.

QUICK REVIEW

1. Discuss the significance of the *Challenger* expedition and the *Challenger* reports.
2. What did Nansen discover about the depth of the Arctic Ocean?
3. How was the *Meteor* expedition different from the *Challenger* expedition?
4. What is the *Meteor* expedition best known for?

P.7 Ocean Science in Modern Times

Establishing Oceanographic Institutions

In the United States, government agencies related to the oceans proliferated during the nineteenth century. These agencies were concerned with gathering information to further commerce, fisheries, and the navy. After the Civil War, the replacement of sail by steam lessened government interest in studying winds and currents and in surveying the ocean floor. Private institutions and wealthy individuals took over the support of oceanography in the United States. Alexander Agassiz (1835–1910), mining engineer, marine scientist, and Harvard University professor, financed a series of expeditions that greatly expanded knowledge of deep-sea biology. Agassiz served as the scientific director on the first ship built especially for scientific ocean exploration, the U.S. Fish Commission's *Albatross,* commissioned in 1882. He designed and financed much of the deep-sea sampling equipment that enabled the *Albatross* to recover more specimens of deep-sea fishes in one haul than the *Challenger* had collected during its entire three-and-a-half years at sea.

One of Agassiz's students, William E. Ritter, became a professor of zoology at the University of California–Berkeley. From 1892–1903, Ritter conducted summer field studies with his students at various locations along the California coast. In 1903, a group of business and professional people in San Diego established the Marine Biological Association and invited Ritter to locate his field station in San Diego permanently. With financial support from members of the Scripps family, who had made a fortune in newspaper publishing, Ritter was able to do this. This was the beginning of the University of California's Scripps Institution of Oceanography (fig. P.17a). The property and holdings of the Marine Biological Association were formally transferred to the University of California in 1912.

Also, in the first twenty years of the century, the Carnegie Institute funded a series of exploratory cruises, including investigations of Earth's magnetic field, and maintained a biological laboratory. In 1927, a National Academy of Sciences committee recommended that ocean science research be expanded by creating a permanent marine science laboratory on the East Coast. This led to the establishment of the Woods Hole Oceanographic Institution in 1930 (fig. P.17b). It was funded largely by a grant from the Rockefeller Foundation. The Rockefeller Foundation allocated funds to stimulate marine research and to construct additional laboratories, and oceanography began to move onto university campuses. Teaching oceanography required that the subject material be consolidated, and in 1942, *The Oceans,* by Harald U. Sverdrup, Martin W. Johnson, and Richard H. Fleming, was published. It captured nearly all the world's knowledge of oceanographic processes and was used to train a generation of ocean scientists.

Oceanography mushroomed during World War II, when practical problems of military significance had to be solved quickly. The United States and its allies needed to move personnel and materials by sea to remote locations, to predict ocean and shore conditions for amphibious landings, to know how explosives behaved in seawater, to chart beaches and harbors from aerial reconnaissance, and to use underwater sound to find submarines. Academic studies ceased as oceanographers pooled their knowledge in the national war effort.

Large-Scale, Direct Exploration of the Oceans

After World War II, oceanographers returned to their classrooms and laboratories with an array of new, sophisticated instruments, including radar, improved sonar, automated wave detectors, and

(a)

(b)

Figure P.17 (a) The Scripps Institution of Oceanography in La Jolla, California. Established in 1903 by William Ritter, a zoologist at the University of California–Berkeley, with financial support from E. W. Scripps and his daughter Ellen Browning Scripps. The first permanent building was erected in 1910. (b) The Woods Hole Oceanographic Institution in Woods Hole, Massachusetts. In a rare moment, all three WHOI research vessels are in port. *Knorr* is in the foreground, with *Oceanus* bow forward and *Atlantis* stern forward, on the opposite side of the pier.

temperature-depth recorders. International cooperation brought about the 1957–58 **International Geophysical Year (IGY)** program, in which sixty-seven nations cooperated to explore the sea floor and made discoveries that changed the way geologists thought about continents and ocean basins. As a direct result of the IGY program, special research vessels and submersibles were built to be used by federal agencies and university research programs. In the 1960s, electronics developed for the space program were applied to ocean research. Computers went aboard research vessels, and for the first time, data could be sorted, analyzed, and interpreted at sea. This made it possible for scientists to adjust experiments while they were in progress. Government funding allowed large-scale ocean experiments. Fleets of oceanographic vessels from many institutions and nations carried scientists who studied all aspects of the oceans.

In 1968, the **Deep Sea Drilling Program (DSDP)**, a cooperative venture among research institutions, began to sample Earth's crust beneath the sea using the specially built drill ship *Glomar Challenger* (fig. P.18*a*) (see chapter 2). It was finally retired in 1983 after fifteen years of extraordinary service. The *Glomar Challenger* was named after the ship used during the *Challenger* expedition. In 1983, DSDP became the **Ocean Drilling Program (ODP)**. The ODP was managed by an international partnership of fourteen U.S. science organizations and twenty-one international organizations called the Joint Oceanographic Institutions for Deep Earth Sampling (JOIDES). The ODP replaced the retired drilling ship *Glomar Challenger* with a larger vessel, the *JOIDES Resolution* (fig. P.18*b*), named after the HMS *Resolution*, used by Captain James Cook to explore the Pacific Ocean basin over 200 years ago. The JOIDES Resolution started drilling operations in 1985 and was in continuous service until ODP ended in 2003. The successor program to ODP is the **Integrated Ocean Drilling Program (IODP)**, an international marine drilling program involving sixteen countries and hundreds of scientists. The U.S. involvement is directed by a group of institutions called the Joint Oceanographic Institutions (JOI). IODP utilizes two major drilling vessels: the *JOIDES Resolution* (which was extensively renovated and returned to service in February 2009), operated by the United States; and a new vessel, the *Chikyu* (fig. P.18*c*), built and operated by Japan. The *Chikyu* hull was launched in January 2002, and initial sea trials began in December 2004. Scientific drilling cruises began in 2007. The deepest hole drilled by the *JOIDES Resolution* to date is 2111 m (6926 ft, about 1.3 mi). The *Chikyu* is designed to drill up to 7500 m (24,600 ft) beneath the sea floor. The *Chikyu*'s derrick height is 110 m (361 ft) above water level.

In 1970, the U.S. government reorganized its earth science agencies. The National Oceanic and Atmospheric Administration (NOAA) was formed under the Department of Commerce. NOAA combined several formerly independent agencies, including the National Ocean Survey, National Weather Service, National Marine Fisheries Service, Environmental Data Service, National Environmental Satellite Service, and Environmental Research Laboratories. NOAA also administers the National Sea Grant College Program. This program consists of a network of twenty-nine individual programs located in each of the coastal and Great

Figure P.18 (a) The *Glomar Challenger,* the Deep Sea Drilling Program drill ship used from 1968–83. (b) The *JOIDES Resolution,* the Ocean Drilling Program drill ship in use since 1985. (c) The drill ship *Chikyu,* the newest ocean drilling vessel.

Lakes states. The Sea Grant College Program encourages cooperation in marine science and education among government, academia, and industry.

The International Decade of Ocean Exploration (IDOE) in the 1970s was a multinational effort to survey seabed mineral resources, improve environmental forecasting, investigate coastal ecosystems, and modernize and standardize the collection, analysis, and use of marine data.

At the end of the 1970s, oceanography faced a reduction in funding for ships and basic research, but the discovery of deep-sea hot-water vents and their associated animal life and mineral deposits renewed excitement over deep-sea biology, chemistry, geology, and ocean exploration in general. Instrumentation continued to become more sophisticated and expensive as deep-sea mooring, deep-diving submersibles, and the remote sensing of the ocean by satellite became possible. Increased cooperation among institutions led to the integration of research at sea among subdisciplines and resulted in large-scale, multifaceted research programs. This model for coordinated, interdisciplinary study of the oceans continues to dominate oceanographic research today.

There are many additional examples of large-scale research efforts. The **Tropical Atmosphere Ocean (TAO)** project collects real-time data from a massive array of nearly seventy moored buoys stretching across the tropical Pacific Ocean basin. These buoys support instruments that measure near-surface atmospheric conditions and subsurface water temperature at ten depths in the upper 500 m (1640 ft). The first TAO buoys were deployed in 1984 and the array was finished in 1994. The mission of the TAO project is to monitor tropical conditions in the Pacific in order to better predict the occurrence of El Niño (see chapter 6). The **Acoustic Thermometry of Ocean Climate (ATOC)** project measured the travel time of sound waves over distances of the order of 5000 km (3100 mi). Because sound speed is a function of water temperature, the travel time is a measure of the temperature of the water along the transmission path. Changes in heat content of the northeast Pacific (an important climate parameter) were monitored by acoustic transmission from sources off California and Hawaii to eleven receivers in the eastern Pacific over a ten-year period from 1996 to 2006. This project proved to be very successful in measuring sound travel times with great accuracy, resulting in detailed models of temperature. The project was also plagued by concerns from the public about the possible adverse effects of the sound transmissions on marine mammals. Detailed studies of the potential effects ultimately concluded that the ATOC transmissions had "no significant biological impact." The U.S. **Joint Global Ocean Flux Study (JGOFS)** is one of the largest and most complex ocean biogeochemical research programs ever organized. Begun in 1988, the JGOFS program resulted in over 3000 ship days (more than eight years) of research and 343,000 nautical miles of ship travel (almost sixteen times around the globe) before ending in 2003. The goal of JGOFS research efforts was to measure and understand on a global scale the processes controlling the cycling of carbon and other biologically active elements among the ocean, atmosphere, and land. The **World Ocean Circulation Experiment (WOCE)** ran from 1990–2002.

The WOCE involved using chemical tracers to model circulation in the ocean and obtain data to help predict the response of seawater circulation to long-term changes in atmospheric circulation. **Ridge Interdisciplinary Global Experiments** and its successor **Ridge 2000 (RIDGE/R2K)** was a roughly twenty-year program that started in 1989. It supported research into the physical, chemical, and biological processes that occur at the global mid-ocean ridge system. This research was driven by the discovery of hydrothermal vents on mid-ocean ridges in 1977 and the discovery of black-smoker vents in 1979.

In 1991, the Intergovernmental Oceanographic Commission (IOC) recommended the development of a Global Ocean Observing System (GOOS) to include satellites, buoy networks, and research vessels. The goal of this program is to enhance our understanding of ocean phenomena so that events such as El Niño (see chapter 6) and its impact on climate can be predicted more accurately and with greater lead time. The successful prediction of the 1997–98 El Niño six months in advance made planning possible prior to its arrival.

An integral part of the GOOS is an international project called Argo, named after the mythical vessel used by the ancient Greek seagoing hero Jason. Argo consists of an array of 3000 independent instruments, or floats, throughout the oceans (fig. P.19a,b). Each float is programmed to descend to a depth of up to 2000 m (6560 ft, or about 1.25 mi), where it remains for ten to fourteen days (fig. P.19c). It then ascends to the surface, measuring temperature and salinity as it rises. When it reaches the surface, it relays the data to shore via satellite and then descends once again and waits for its next cycle. The floats drift with the currents as they collect and transmit data. In this fashion, the entire array will provide detailed temperature and salinity data of the upper 2000 m of the oceans every ten to fourteen days. Argo floats have a life expectancy of four years. Roughly one-quarter of the floats will have to be replaced each year.

The United Nations designated 1998 as the Year of the Ocean. Goals included: (1) a comprehensive review of national ocean policies and programs to ensure coordinated advancements leading to beneficial results and (2) raising public awareness of the significance of the oceans in human life and the impact that human life has on the oceans. Also in 1998, the National Research Council's Ocean Studies Board released a report highlighting three areas that are likely to be the focus of future research: (1) improving the health and productivity of the coastal oceans, (2) sustaining ocean ecosystems for the future, and (3) predicting ocean-related climate variations.

One of the largest and most ambitious research and education programs under development is the National Science Foundation's **Ocean Observatories Initiative (OOI)**. The OOI is designed to provide up to thirty years of continuous ocean measurements to study such fundamental scientific problems as climate variability, ocean circulation, ecosystem dynamics, air-sea interaction, seafloor processes, and plate tectonics. The OOI will consist of a network of instruments, undersea cables, and instrumented moorings located in multiple regions of the Western Hemisphere. The OOI will be one, fully integrated

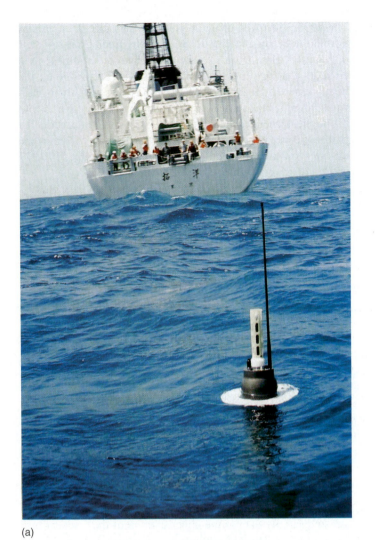

(a)

system that will measure physical, chemical, geological, and biological phenomena in selected global, regional, and coastal areas. The global component will consist of a network of buoys with sensors for measurement of air-sea fluxes of heat, moisture, and momentum; physical, biological, and chemical properties throughout the water column; and geophysical observations made on the seafloor. The regional observatories are designed to monitor processes along a single tectonic plate, using cables to transmit both power and data between the instruments and shore. The coastal OOI is comprised of two arrays: The Endurance Array will be located in the Northeast Pacific off the coast of Washington and Oregon; the Pioneer Array will be located in the Northwest Atlantic off the coast of New England. These arrays will be designed to monitor processes that occur between land and sea, such as beach erosion, modifications of coastlines, nutrient and carbon fluxes across the continental shelves, and coastal circulation.

Although large-scale, federally funded studies are presently in the forefront of ocean studies, it is important to remember that studies driven by the specific research interests of individual scientists are essential to point out new directions for oceanography and other Earth sciences.

Figure P.19 (a) The Japanese coast guard cutter *Takuyo* prepares to retrieve an Argo float (photo courtesy of the International Argo Steering Team). (b) Schematic of an Argo float. The float's buoyancy is controlled by a hydraulic pump that moves hydraulic fluid between an internal reservoir and an external flexible bladder that expands and contracts. (c) A single measurement cycle involves the slow descent of the float to a depth of up to 2000 m (6560 ft) where it remains for about ten days before hydraulic fluid is pumped into the external bladder and the float rises again to the surface to transmit data before descending again.

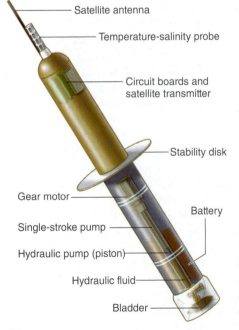

- Satellite antenna
- Temperature-salinity probe
- Circuit boards and satellite transmitter
- Stability disk
- Gear motor
- Single-stroke pump
- Hydraulic pump (piston)
- Hydraulic fluid
- Bladder
- Battery

(b)

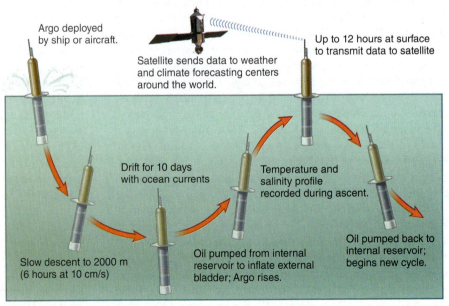

Argo deployed by ship or aircraft.

Satellite sends data to weather and climate forecasting centers around the world.

Up to 12 hours at surface to transmit data to satellite

Drift for 10 days with ocean currents

Temperature and salinity profile recorded during ascent.

Slow descent to 2000 m (6 hours at 10 cm/s)

Oil pumped from internal reservoir to inflate external bladder; Argo rises.

Oil pumped back to internal reservoir; begins new cycle.

(c)

Diving in

"FLIP," the Floating Instrument Platform

BY GWYNNE RIFE

Dr. Gwynne Rife is a science educator and aquatic scientist at The University of Findlay. Her interests include invertebrate behavior and bioacoustics of the marine environment.

As an undergraduate I was dedicated to pursuing studies in bioacoustics. I was sure that there was a way to decode the amazing vocalizations made by whales. As I read and studied, I learned of a research vessel called the Floating Instrument Platform, or FLIP, that was specially designed to aid in the study of acoustical signals in the oceans (box fig. 1).

FLIP is owned by the U.S. Navy and was conceived and developed at the Marine Physical Laboratory (MPL) of the Scripps Institution of Oceanography. FLIP was designed out of a need to provide a stable platform for research activities by two Scripps scientists, Dr. Fred Fisher and Dr. Fred Spiess, who wanted a more stable open-ocean marine laboratory than a conventional research ship could provide. FLIP was officially launched in June 1962, and underwent an extensive refurbishing in 1995.

I had read about FLIP, but was not convinced that a vessel like this could really flip. I was overjoyed to get to tour this amazing vessel on a trip to visit my mentor while she was on sabbatical in San Diego, California. When I pulled up to the slip where FLIP was moored, it was not a huge, impressive ship at all. I was not sure I was going to enjoy the tour until I stepped on deck and through a hatch to see two sinks at a 90° angle, one sideways on the wall. Soon I wondered if I was now standing on the floor or on the wall.

I must admit, my first question was not at all inspired; it was "How does it flip?" The crew member just smiled and started to describe what I am sure he had many times before; the amazing process of how the 335-ft-long vessel takes on ballast water in 20 to 28 minutes to turn

the vessel into a buoy that can remain stable and float freely at sea for up to 35 days. It offers a unique resource for obtaining data and provides a stable platform for researchers to work from. To switch to the vertical position, FLIP is towed to a research site where ballast tanks are flooded with 1500 tons of seawater to "flip" it while scientists and crew literally "walk up the walls" to stay upright (box fig. 2). With only 55 ft (17 m)

Box Figure 2a   Seawater is pumped into FLIP's ballast tanks to start the 90° flip from horizontal to vertical.

Box Figure 2b   FLIP in the stable vertical position, ready for a variety of scientific missions.

Box Figure 1   Scripps Institution of Oceanography's Floating Instrument Platform, or FLIP, conducts sea trials off San Diego in May 2009.

Continued next page—

of the vessel above the surface and the remaining 300 ft (91 m) extending beneath the surface, scientists can conduct their research with minimal surface motion disruption.

When ballast water is pumped into the tanks, and as it flips, the crew of five oversees the inner workings and checks all the equipment while it is moving. I was interested in the bunk beds, as they hung from a swinging platform, "swivels and gimbals" I was told, so they will turn along with FLIP. Things that would not rotate as easily, like sinks, are built both horizontally and vertically in each room (box fig. 3).

During the flip, most things inside and out slowly swing along with the change. I was intrigued at the refrigerator and stove combo in the galley that was also on a platform that would swing, but I was incredulous when I was told that the heads (what you call toilets on a ship) also swiveled. With so much on the move, I asked if this transition always was smooth with all the planning and preparation for the big flip. The answer was that it almost always went smoothly unless someone forgot to loosen any of the gimbals before the flip, or, if someone forgot to re-tighten them afterwards. Apparently, the sound of a galley full of pots and pans spilling out of their cabinets was a tell-tale sign of a distracted crew member. And after all, who wouldn't occasionally get distracted by the wall quickly becoming the floor?

Each new area of FLIP showed its compact and efficient design. For example, the scientific instruments are built into the walls. As FLIP flips, so do the instruments, so that they are in a normal position when FLIP becomes vertical. Most rooms on FLIP have two doors—one to use when horizontal and one to use when flipped vertical.

Box Figure 3 Two sinks at a 90° angle aboard FLIP for both vertical and horizontal use.

This stable platform allows an incredible array of precise data to be taken in an otherwise impossible setting. Along with the crew, it can support a research team of up to eleven scientists. Two of the most valuable characteristics of FLIP to these research teams are: (1) FLIP allows for at-sea stability that allows for the deployment of extremely large sensor arrays (one reported up to 2 miles in length), and (2) FLIP provides a high-profile (25 m) observation post with 360° coverage for simultaneous visual and acoustical observations of marine mammals. Other research projects supported by FLIP include the strange nighttime chorusing behavior of fish, the behavior of earthquake T-phases (P-waves that travel through water) to study their propagation through the ocean, the mirage effect caused by bathymetry changes in shallow water, and the diving behavior of blue whales determined by recordings of their vocalizations. Research studies with titles that range from "Classification of behavior using vocalizations of Pacific white-sided dolphins (*Lagenorhynchus obliquidens*)" to "Dynamical coupling of wind and ocean waves through wave-induced air flow" could only have been accomplished with the quiet and steady FLIP and its crew to provide the conditions needed for arrays of hydrophones and deployment of sensors that offer a precision no other research vessel can.

Flip has reached 50-plus years of amazing service to the Navy and scientific community. With over 300 operations accomplished in many different areas of the World Ocean, FLIP shows no sign of becoming obsolete and remains the most unusual ocean research resource of its kind.

Satellite Oceanography, Remote Sensing of the Oceans

The ability to collect data by remote sensing from satellites has increasingly allowed researchers to observe the sea surface and ocean processes on a global scale. Currents, eddies, biological productivity, sea-level changes, waves, sea surface temperature, and air-sea interactions are all monitored via satellite, allowing scientists to develop computerized prediction models and to test them against natural phenomena.

Over the past decades, satellite sensors used for oceanography have continued to evolve. NASA's *NIMBUS-7* satellite, launched in 1978, carried a sensor package called the Coastal Zone Color Scanner (CZCS), which gave us the first truly global view of the ocean's chlorophyll distributions and productivity (fig. P.20). Lasting until 1986, CZCS was eventually replaced by the Sea-viewing Wide Field Sensor (SeaWiFS; 1997–2010) and the Moderate Resolution Imaging Spectrometer (MODIS; launched in 1999 and 2002). These sensors provide not only ocean color information, but also details about Earth's cloud cover, radiation budget, and dynamic changes in the oceans.

Other satellites provide valuable information about the topography and currents of the ocean's surface. TOPEX/Poseidon (1992–2005) was launched as a joint U.S.-French mission, and was able to measure the surface height of 95% of the ice-free ocean to an accuracy of 3.3 cm (1.3 in). A detailed understanding of variations in sea surface elevation provides invaluable information about ocean circulation. A series of follow-on missions began with Jason-1 in 2001, Jason-2 in 2008, and the planned launch of Jason-3 in 2014.

Satellite oceanography began in 1978 and will continue long into the future. Today, Earth is orbited by a constellation of satellites providing valuable information for all aspects of oceanography. As the scientific community develops new and improved sensors, methods, and questions, we continue to add to that constellation, making it easier than ever before to track the pulse of the planet. More details can be found in the "Oceanography from Space" section in the center section of the book.

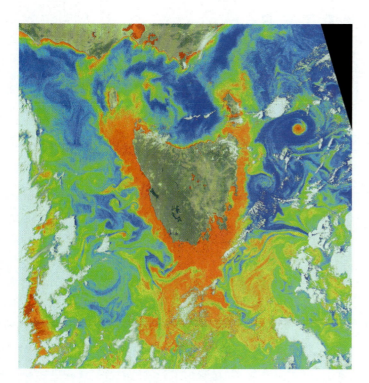

Figure P.20 False color image centered on the island of Tasmania. Tasmania is located south of the eastern coast of Australia. *Yellow* and *reds* indicate high concentrations of phytoplankton, *greens* and *blues* low concentrations, and *dark blue* and *purple* very low concentrations. The complex current interactions, indicated by swirling color patterns, around the island have significant influence on the distribution of phytoplankton.

QUICK REVIEW

1. Compare the U.S. government's support of oceanography before and after World War II.
2. Cite some examples of large-scale, direct measurement research programs and describe their mission.
3. Why are oceanographers interested in a global approach to ocean science?
4. Why are oceanographers interested in data collected by satellites as well as data collected from research vessels?
5. How do we collect deep samples of marine sediment?

Summary

Oceanography is a multidisciplinary field in which geology, geophysics, chemistry, physics, meteorology, and biology are all used to understand the oceans. Early information about the oceans was collected by explorers and traders such as the Phoenicians, the Polynesians, the Arabs, and the Greeks. Eratosthenes calculated the first accurate circumference of Earth, and Ptolemy produced the first world atlas.

During the Middle Ages, the Vikings crossed the North Atlantic, and shipbuilding and chartmaking improved. In the fifteenth and sixteenth centuries, Dias, Columbus, da Gama, Vespucci, and Balboa, as well as several Chinese explorers, made voyages of discovery. Magellan's expedition became the first to circumnavigate Earth. In the sixteenth and seventeenth centuries, some explorers searched for the Northwest Passage, while others set up trading routes to serve developing colonies.

By the eighteenth century, national and commercial interests required better charts and more accurate navigation techniques. Cook's voyages of discovery to the Pacific produced much valuable information, and Franklin compiled a chart of the Atlantic's Gulf Stream. A hundred years later, the U.S. Navy's Maury collected wind and current data to produce current charts and sailing directions and then wrote the first book on oceanography.

Ocean science began with the nineteenth-century expeditions and research of Darwin, Forbes, Müller, and others. The three-and-a-half-year *Challenger* expedition laid the foundation for modern oceanography with its voyage, which gathered large quantities of data on all aspects of oceanography. Exploration of the oceans in Arctic and Antarctic regions was pursued by Nansen and Amundsen into the beginning of the twentieth century.

In the twentieth century, private institutions played an important role in developing U.S. oceanographic research, but the largest single push came from the needs of the military during World War II. After the war, large-scale government funding and international cooperation allowed oceanographic projects that made revolutionary discoveries about the ocean basins. Development of electronic equipment, deep-sea drilling programs, research submersibles, and use of satellites continued to produce new and more detailed information of all kinds. At present, oceanographers are focusing their research on global studies and the management of resources as well as continuing to explore the interrelationships of the chemistry, physics, geology, and biology of the sea.

CHAPTER 1

The Water Planet

Learning Outcomes

After studying the information in this chapter students should be able to:

1. *explain* the "Big Bang" theory of the origin of the universe, and *describe* its structure,

2. *describe* the origin of the solar system,

3. *list* two possible sources of the water in the oceans,

4. *review* how we have come to estimate Earth's age as 4.5 to 4.6 billion years,

5. *rank* the eons and eras of geologic time in chronological order,

6. *list* and *date* the three major mass extinctions,

7. *define* and *sketch* lines of latitude and longitude,

8. *calculate* the difference in time between two locations of known longitude,

9. *diagram* the hydrologic cycle, and

10. *calculate* the mean depth of the oceans using data in table 1.4.

Pier at the Scripps Institution of Oceanography, University of California–San Diego.

cientists make discoveries about the natural world by gathering data through observation and experimentation. Scientific data must be reproducible and must include an estimate of error. After obtaining scientific data, scientists can propose an initial explanation of the data. This is called a **hypothesis**. If the hypothesis is supported consistently by different observations or experiments it may be advanced to the level of a **theory**. The great value of a theory is its ability to predict the existence of phenomena or relationships that had not previously been recognized.

In this chapter, you will learn about the most widely accepted theories for the birth of the universe and the formation of our planet, the nature and structure of Earth's interior, and the methods used to measure geologic time and calculate Earth's age.

Earth is unique in our solar system by having a surface that is now covered mostly by liquid water. In this chapter, you will also investigate our water planet (fig. 1.1) by learning about the distribution of water, the movement of water from one location to another, and the oceans where most of it is found. You will also discover how people find their way about the oceans.

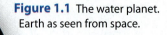

Figure 1.1 The water planet. Earth as seen from space.

universe as having initially been concentrated in an extremely hot, dense singularity much smaller than an atom. Roughly 13.7 billion years ago, this singularity experienced a cataclysmic explosion that caused the universe to rapidly expand and cool as it grew larger. One second after the Big Bang, the temperature of the universe was about 10 billion K (roughly 1000 times the temperature of the Sun's interior). The Kelvin (K) temperature scale is an absolute temperature scale, so 0 K is absolute zero, the coldest possible temperature. At this temperature, all atoms and molecules would stop moving. Room temperature is about 300 K (see appendix B for conversions from K to °C and °F). At this time, the universe consisted mostly of elementary particles, light, and other forms of radiation. The elementary particles, such as protons and electrons, were too energetic to combine into atoms. One hundred seconds after the Big Bang, the temperature had cooled to about 1 billion K (roughly the current temperature in the centers of the hottest stars).

1.1 Cosmic Beginnings

Origin of the Universe

For centuries, our concept of the nature of the universe was governed by visual observations from Earth's surface. Our current understanding of its history and structure has been greatly enhanced by observations made with instruments that are sensitive to energy, from long wavelength radio waves to short wavelength gamma rays. An example of one of these instruments is the *Hubble Space Telescope (HST)* (fig. 1.2).

The *HST* was deployed in April 1990, in low-Earth orbit at an altitude of 595 km (370 mi), and it circles Earth every ninety-seven minutes. Because of its location above the atmosphere, the 2.4 m (94.5 in) reflecting telescope (the size of a reflecting telescope refers to the diameter of its mirror) has an optical resolution, or image clarity, that is about ten times better than the best ground-based telescopes can achieve. It can detect objects one-billionth as bright as the human eye is capable of detecting.

In recent years, observational data have provided increasing evidence that the universe originated in an event known as the **Big Bang**. The Big Bang model envisions all energy and matter in the

Figure 1.2 The *Hubble Space Telescope (HST)* is a joint venture between the European Space Agency and the National Aeronautics and Space Administration. It was proposed in the 1940s, designed and constructed in the 1970s and 1980s, and began its operational life with its launch in 1990.

The universe was now cool enough for protons and neutrons, the nuclei of hydrogen atoms, as well as nuclei of deuterium, helium, and lithium, to begin to form. While the temperature was still very high, the universe was dominated by the radiation. Later, as the universe cooled, matter took over. Eventually, when the temperature had dropped to a few thousand degrees K, electrons and nuclei would have started to combine to form atoms, and strong interactions between matter and radiation ceased. It was then possible for small concentrations of matter to begin to grow gravitationally. Denser, cooler regions pulled in additional matter gravitationally, increasing their density even further. An extraordinary composite image of the early universe is shown in figure 1.3. This image was compiled from data obtained by the *Wilkinson Microwave Anisotropy Probe (WMAP)* satellite over a period of more than one year. It is essentially a temperature map of the universe 380,000 years after the Big Bang, over 13 billion years ago. The spatial variation in temperature reflects the clumping of mass in the universe at that time. The *WMAP* is so sensitive that it can resolve differences in temperature of only a millionth of a degree. Roughly 200 million years after the Big Bang, gravity began to pull matter into the structures we see in the universe today, and the first stars were formed.

The universe has a distinct structure. On a small scale, there are individual stars. Stars are responsible for the formation of elements heavier than lithium. Stars fuse hydrogen and helium in their interiors to form heavier elements such as carbon, nitrogen, and oxygen. The higher temperatures of more massive stars continue the nuclear fusion process to create elements as heavy as iron. These elements, so important in oceanographic processes, are created in stars and were not made by the Big Bang at the formation of the universe. Some of these stars are at the center of solar systems like our own, with planets that orbit them.

Galaxies are composed of clumps of stars. Our galaxy, the Milky Way, is composed of about 200 billion stars. It is shaped like a flattened disk, with a thickness of about 1000 **light-years** and a diameter of about 100,000 light-years. A light-year is equal to the distance light travels in one year, which is 9.46×10^{12} km (5.87×10^{12} mi). The observable universe contains from 10 billion to 100 billion galaxies. Galaxies are preferentially found in groups called **clusters**. A single cluster may contain thousands of galaxies. Clusters typically have dimensions of 1 million to 30 million light-years. Individual clusters tend to group in long, string-like or wall-like structures called superclusters. Superclusters may contain tens of thousands of galaxies. The largest supercluster known is about 500 million light-years across. At very large scales, the universe looks something like a sponge, with galaxies arranged in interconnected lines and sheets interspersed with huge regions in which very few galaxies are seen.

Throughout the universe, some stars are burning out or exploding; others are still forming, incorporating original matter from the Big Bang and recycling matter from older generations of stars. A widely accepted model of the universe is one in which the vast majority of the energy is in a form that cannot be detected. Roughly 30% of all energy is thought to be dark matter that does not interact with light. Sixty-five percent is thought to consist of an even more mysterious dark energy that is causing the expansion of the universe to accelerate. The familiar matter, the kind found in planets and stars, makes up only 5% of the energy in the universe.

Origin of Our Solar System

Present theories attribute the beginning of our solar system to the collapse of a single, rotating interstellar cloud of gas and dust that included material that was produced within older stars and dispersed into space when the old stars exploded. This rotating cloud, or **nebula,** appeared about 5 billion years ago. The shock wave from a nearby exploding star, or supernova, is thought to have imparted spin to the cloud, pushing it together and causing it to compress from its own gravitational pull. As the nebula collapsed, its speed of rotation increased, and, heated by compression, its temperature rose. The gas and dust, spinning faster and faster, flattened perpendicular to the axis of spin, forming a disk. At the center of the disk a star, our Sun, was formed. Self-sustaining nuclear reactions kept the Sun hot, but the outer regions began to cool. In this cooler outer portion of the rotating disk, molecules of gas and dust began to collide, accrete (or stick together), and chemically interact. The collisions and interactions produced particles that grew from further accretion and became large enough to have sufficient gravity to attract still other particles. The planets of our solar system had begun to form. After a few million years, the Sun was orbited by eight planets (in order from the Sun): Mercury, Venus, Earth, Mars, Jupiter, Saturn, Uranus, and Neptune (Fig. 1.4).

If Mercury, Venus, Earth, and Mars are compared to Jupiter, Saturn, Uranus, and Neptune, the four planets closer to the Sun are seen

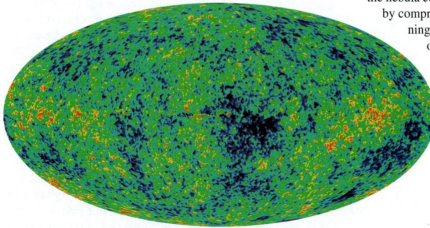

Figure 1.3 The structure of the early universe about 380,000 years after the Big Bang, as seen by the *Wilkinson Microwave Anisotropy Probe* satellite. Colors indicate "warmer" (red) and "cooler" (blue) spots. These patterns are extremely small temperature differences within an extraordinarily evenly dispersed microwave light bathing the universe, which now averages only 2.73 K. The difference in temperature between the warmest and coolest regions in this image is only 400 μK (400×10^{-6} K).

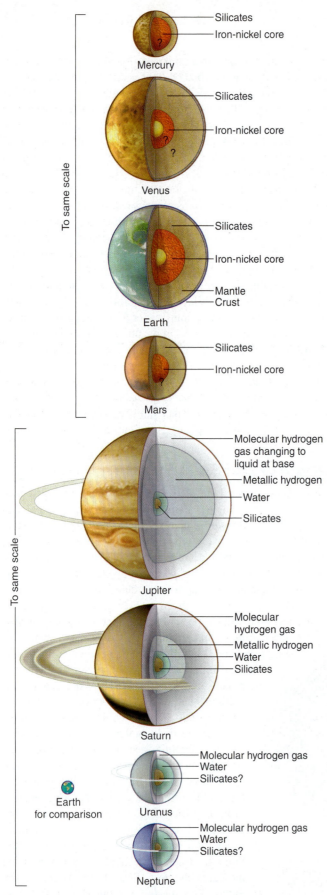

Mercury
— Silicates
— Iron-nickel core

Venus
— Silicates
— Iron-nickel core

Earth
— Silicates
— Iron-nickel core
— Mantle
— Crust

Mars
— Silicates
— Iron-nickel core

Jupiter
— Molecular hydrogen gas changing to liquid at base
— Metallic hydrogen
— Water
— Silicates

Saturn
— Molecular hydrogen gas
— Metallic hydrogen
— Water
— Silicates

Earth for comparison

Uranus
— Molecular hydrogen gas
— Water
— Silicates?

Neptune
— Molecular hydrogen gas
— Water
— Silicates?

To same scale

Figure 1.4 The composition of the planets. The inner rocky planets have an iron core surrounded by a silicate shell. The outer gas planets feature a rocky core of unknown composition surrounded by a large shell of hydrogen gas.

to be much smaller in diameter and mass. (See table 1.1. Note use of metric units; see appendix B for further information.) These four inner planets are rich in metals and rocky materials. The four outer planets are cold giants, dominated by ices of water, ammonia, and methane. Their atmospheres are made up of helium and hydrogen; the planets located nearer the Sun lost these lighter gases because the higher temperature and intensity of solar radiation tend to push these gases out and away from the center of the solar system. In addition, these inner planets are not massive enough for their gravitational fields to prevent these lighter gases from escaping. If the mass of each planet in table 1.1 is divided by its volume, the results will show that the outer planets are composed of lighter, or less dense, materials than the inner planets.

Extraterrestrial Oceans

Data obtained by NASA's *Voyager* and *Galileo* spacecraft indicate that two of Jupiter's moons, Europa and Callisto, may have oceans beneath their ice-covered surfaces (fig. 1.5a). Liquid oceans are believed to be possible despite extremely cold surface temperatures, $-162°C$ ($-260°F$) on Europa, because of heat generated by friction due to the continual tidal deformation by Jupiter's strong gravitational force.

Some of the most compelling evidence for the presence of these oceans has come from magnetic measurements made by the *Galileo* spacecraft. Neither moon has a strong internal magnetic field of its own, but *Galileo* detected induced magnetic fields around both moons, indicating that they both consist partly of strongly conducting material.

It is unlikely that the ice covering the moons can account for the induced magnetic fields because ice is a poor electrical conductor. Fresh water is also a relatively poor conductor, but water with a high concentration of dissolved ions, such as seawater, is a very good conductor. The most plausible explanation for the observed magnetic effect is that Europa and Callisto have liquid oceans containing salts beneath their surfaces. It is believed that magnesium sulfate might be a major component of Europa's water rather than sodium chloride, as is the case in Earth's oceans.

One proposed model for Europa includes a surface ice layer 15 km (10 mi) thick, covering a 100 km (62 mi) deep ocean. If this is the case, the Europan ocean would contain twice as much water as Earth's oceans, and it would be roughly ten times deeper than the greatest depths below sea level on Earth. In contrast, one model proposed for Callisto is a surface ice layer 100 km (62 mi) thick, covering a 10 km (6.2 mi) deep ocean. If such oceans exist, they may provide a possible environment for life.

In January 2004, two robotic rovers, *Spirit* and *Opportunity*, landed on opposite sides of the planet Mars. Their mission has been to probe the rocks of Mars for signs of past or current deposits of water, and they have succeeded in collecting both chemical and physical data that strongly support the hypothesis that water was once present.

Chemical analyses have discovered various kinds of salts in some Martian rocks. On Earth, rocks that contain equally large amounts of these salts either formed in water or, after formation, were altered by long exposures to water. Photographs of some

Origin of the Oceans

The oldest sedimentary rocks found on Earth, rocks that formed by processes requiring liquid water at the surface, are about 3.9 billion years old. This indicates that there have been oceans on Earth for approximately 4 billion years. Where did the water in the oceans come from? There are two possible sources for this water: the interior of Earth and outer space.

Traditionally, scientists have suggested that the water in the oceans and atmosphere originated in the interior of Earth in a region called the mantle and was brought to the surface by volcanism, a process that continues to this day (box fig. 1). The rock that makes up the mantle is thought to be similar in composition to meteorites, which contain from 0.1%–0.5% water by weight. The total mass of rock in the mantle is roughly 4.5×10^{27}g; thus, the original mass of water in the mantle would have been approximately 4.5×10^{24} to 2.25×10^{25}g. This is from three to sixteen times the amount of water currently in the oceans, so it is clear that the mantle is an adequate source for the water in the oceans; but is enough water brought to the surface through volcanism to actually fill the oceans? Magmas erupted by volcanoes contain dissolved gases that are held in the molten rock by pressure. Most magmas consist of 1%–5% dissolved gas by weight, most of which is water vapor. The gas that escapes from Hawaiian magmas is about 70% water vapor, 15% carbon dioxide, 5% nitrogen, and 5% sulfur dioxide, with the remainder consisting mostly of chlorine, hydrogen, and argon. It is estimated that thousands of tons of gas are ejected in volcanic eruptions each day. Undoubtedly, the rate of volcanic eruptions on Earth has varied with time, probably being much greater earlier in Earth's history when the planet was still very hot. However, if we conservatively assume that the present rate of ejection of water vapor by volcanism has been roughly constant over the last 4 billion years, then the volume of water expelled by volcanoes would have produced roughly 100 times the volume of water in the oceans.

The traditional view of the interior of Earth serving as the source of ocean water has recently been challenged by a bold new suggestion that large volumes of water are continually being added from outer space. Evidence for this idea comes from data collected by a polar-orbiting satellite called the *Dynamics Explorer 1 (DE-1)*. The *DE-1* carried an ultraviolet photometer capable of taking pictures of Earth's **dayglow**. Dayglow is ultraviolet light, invisible to the naked eye, emitted by atomic oxygen in the upper atmosphere when it absorbs and reradiates electromagnetic energy from the Sun. In many of the dayglow images of Earth obtained by the satellite, there are distinct dark spots, roughly 48 km (30 mi) in diameter, that appear to move across the face of Earth, suggesting that they were caused by moving objects (box fig. 2). The direction of motion of the dark spots matches the direction of motion of meteoritic material as it approaches Earth. Atmospheric physicist Louis Frank has suggested that these dark spots are created when small, icy comets vaporize in the outer atmosphere, creating clouds of water vapor that absorb the ultraviolet radiation of Earth's dayglow over a small area, thus creating a dark spot in the bright ultraviolet background. The size of the spots implies that the average mass of the comets is about 10 kg (22 lb). He estimates that an average of twenty of these comets enter the atmosphere each minute, or a staggering 10 million each year. If all of the water in these comets condensed to form a layer on the surface of Earth, it would be roughly 0.0025 mm (0.0001 in) deep. While this doesn't seem like a significant amount of water, over 4 billion years this rate of accumulation would fill the oceans two to three times.

Box Figure 1 Volcanic eruptions release gases including water vapor, carbon dioxide, and sulfur dioxide to the atmosphere and oceans. Even if it is not erupting, an active volcano can release thousands of tons of sulfur per day.

Some debate continues about the role of comet impacts in the formation of the oceans. Additional study may give us further insight to the relative importance of volcanism and the impact of extraterrestrial objects in creating the oceans. It is very likely that both processes have contributed to their formation.

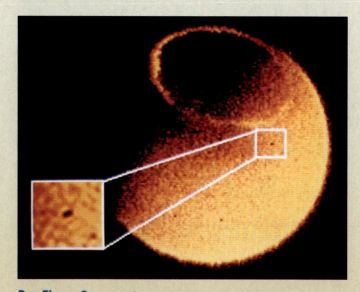

Box Figure 2 Image of Earth's dayglow at ultraviolet wavelengths taken from an altitude of 18,500 km (11,500 mi). The dayglow is due to the excitation of atomic oxygen by solar radiation. Inset shows a magnified view of a dark spot, or "hole," in the dayglow thought to be caused by the vaporization of small cometlike balls of ice.

Table 1.1   Features of the Planets in the Solar System

Planet	Mean Distance from Sun (10^6 km)	Diameter (km)	Mass Relative to Earth's Mass	Rotation Period[1] (hours, days)	Orbit Period (years)	Mean Temperature of Surface (°C)	Principal Atmospheric Gases[2]
Mercury	57.9	4,878	0.055	58.6 d	0.24	−170 night 430 day	Essentially a vacuum
Venus	108.2	12,104	0.815	243 d*	0.62	−23 clouds 480 surface	CO_2, N_2
Earth	149.6	12,756	1.000	23.94 h	1.00	16	N_2, O_2
Mars	227.9	6,787	0.107	24.62 h*	1.88	−50 (average)	CO_2, N_2, Ar
Jupiter	778.3	142,800	317.8	9.93 h	11.86	−150	H_2, He, CH_4, NH_3
Saturn	1429	120,000	95.2	10.5 h	29.48	−180	H_2, He, CH_4, NH_3
Uranus	2875	50,800	14.5	17.24 h*	84.01	−210	H_2, He, CH_4
Neptune	4504	48,600	17.2	16 h	164.8	−220	H_2, He, CH_4

1. * by rotation period indicates rotation in a direction opposite that of Earth.
2. CO_2 = carbon dioxide; N_2 = nitrogen; O_2 = oxygen; Ar = argon; H_2 = hydrogen; He = helium; CH_4 = methane; NH_3 = ammonia.

(a)   (b)

Figure 1.5 (a) View of a small region of the thin, disrupted ice crust of Jupiter's moon Europa. North is to the top of the picture. The image covers an area approximately 70 km (43 mi) east-west by 30 km (19 mi) north-south. The *white* and *blue colors* outline areas that have been blanketed by a fine dust of ice particles ejected at the time of formation of a large impact crater roughly 1000 km (620 mi) to the south. A few small craters of less than 500 m (1600 ft) in diameter can be seen. These were probably formed by large, intact blocks of ice thrown up in the impact explosion that formed the large crater to the south. (b) Image of small holes or cavities taken by the microscopic imager on the Mars Exploration Rover *Opportunity* in a region called "El Capitan" on a rock outcrop at Meridiani Planum, Mars. Several cavities have disk-like shapes with wide midpoints and tapered ends. This feature is consistent with salt minerals that crystallize within a rock matrix, either pushing the matrix grains aside or replacing them. These crystals are then either dissolved in water or eroded by wind activity to produce the cavities.

rocks show the presence of small holes or indentations similar to those found in some Earth rocks when crystals of salt minerals that were originally formed in the rocks were later dissolved by the flow of fresh water through the rocks (fig. 1.5*b*). The rover

Opportunity detected the presence of an iron-bearing mineral called gray hematite in some Martian rocks; on Earth, hematite containing crystalline grains of the size found in the Martian rocks typically forms in the presence of water.

Of particular interest was *Opportunity*'s discovery of sedimentary rocks with ripple marks, structures that are formed by waves or currents of water. Taken together, the data accumulated by the rovers strongly support the hypothesis that there was once a large body of salt water on the Martian surface and that *Opportunity*'s landing site was once the shoreline of a salty sea.

In August 2007, NASA launched the *Phoenix Probe* on a mission to sample ice beneath the surface in the north polar region of Mars and test it for the presence of water. The craft carried an oven-like instrument called the "Thermal and Evolved Gas Analyzer (TEGA)," designed to bake soil and ice samples and analyze the chemistry of released vapors. The *Phoenix Probe* landed on Mars in May 2008 and, in July 2008, confirmed the presence of water on Mars. This is the first time water has been directly identified on another planet.

Early Planet Earth

During the first billion years of Earth's existence, Earth is thought to have been a mixture of silicon compounds, iron and magnesium oxides, and small amounts of other elements. According to this model, Earth formed originally from cold matter, but events occurred that raised Earth's temperature and initiated processes that obliterated its earlier history and resulted in its present form. Early Earth was bombarded by particles of all sizes, and a portion of their energy was converted into heat on impact. Each new layer of accumulated material buried the material below it, trapping the heat and raising the temperature of Earth's interior. At the same time, the growing weight of the accumulating layers compressed the interior, and the energy of compression was converted to heat, raising Earth's internal temperature to approximately 1000°C. Atoms of radioactive elements, such as uranium and thorium, disintegrated by emitting subatomic particles that were absorbed by the surrounding matter, further raising the temperature.

Some time during the first few hundred million years after Earth formed, its interior reached the melting point of iron and nickel. When the iron and nickel melted, they migrated toward the center of the planet. Frictional heat was generated, and lighter substances were displaced. In this way, the temperature of Earth was raised to an average of 2000°C. The less dense material from the partially molten interior moved upward and spread over the surface, cooling and solidifying. The melting and solidifying probably happened repeatedly, separating the lighter, less dense compounds from the heavier, denser substances in the interior of the planet. In this way, Earth became completely reorganized and differentiated into a layered system.

Earth's oceans and atmosphere are probably both, at least in part, by-products of this heating and differentiation. As Earth warmed and partially melted, water, hydrogen, and oxygen locked in the minerals were released and brought to the surface along with other gases in volcanic eruptions. As Earth's surface cooled, water vapor condensed to form the oceans. Another possible source of water for the oceans is from space objects,

such as comet-like balls of ice or ice meteorites, that have collided with Earth throughout its history. The origin of the oceans is discussed further in the Diving In box titled "Origin of the Oceans," p. 29.

At first, Earth must have had too little gravity to have accumulated an atmosphere. It is generally believed that, during the process of differentiation, gases released from Earth's hot, chemically active interior formed the first atmosphere, which was primarily made up of water vapor, hydrogen gas, hydrogen chloride, carbon monoxide, carbon dioxide, and nitrogen. Any free oxygen present would have combined with the metals of the crust to form compounds such as iron oxide. Oxygen gas could not accumulate in the atmosphere until its production exceeded its loss by chemical reactions with Earth's crust. Oxygen production did not exceed loss until life evolved to a level of complexity in which photosynthetic organisms could convert carbon dioxide and water with the energy of sunlight into organic matter and free oxygen.

QUICK REVIEW

1. How and when did the universe form?
2. How and when did our solar system form?
3. What processes added heat to the early Earth, and how was Earth's interior changed by that heat?
4. What were the sources of the early Earth's oceans and atmosphere?

1.2 Earth's Age and Time

Earth's Age

Over the centuries, people have asked the question, How old is Earth? In the seventeenth century, Archbishop Ussher of Ireland attempted to answer the question by counting the generations listed in the Bible; he determined that the first day of creation was Sunday, October 23, 4004 B.C. In 1897, the English physicist Lord Kelvin calculated the time necessary for a molten Earth to cool to present temperatures and dated Earth as 20 million to 40 million years old. In 1899, John Joly calculated the age of Earth to be 90–100 million years based on the rate of addition of salt to the oceans from rivers. In 1896, Antoine Henri Becquerel discovered radioactivity. With this discovery and an understanding of radioactive decay, scientists were able to accurately date rocks and minerals. In 1905, Ernest Rutherford and Bertrum Boltwood used radioactive decay to date rock and mineral samples 500 million years old. Two years later, in 1907, Boltwood calculated an age of 1.64 billion years for a mineral sample rich in uranium.

The method pioneered by Rutherford and Boltwood, known as **radiometric dating,** uses radioactive **isotopes.** Each atom of a radioactive isotope has an unstable nucleus. This unstable nucleus changes, or decays, and emits one or more subatomic particles plus energy. For example, the radioactive isotope carbon-14 decays or changes to nitrogen-14; uranium-235 decays to lead-207; and potassium-40 decays to argon-40. The time at which any single nucleus will decay is unpredictable, but if large numbers of atoms of the same radioactive isotope are present, it is possible to predict that a certain fraction of the isotope will decay over a certain period of time. In this process of decay, the atom changes from one element (the parent element) to another (the daughter product). The time over which one-half of the atoms of a radioactive isotope decay is known as the isotope's **half-life** (fig. 1.6). The half-life of each radioactive isotope is characteristic and constant. For example, the half-life of carbon-14 is 5730 years; that of uranium-235 is 704 million years; and that of potassium-40 is 1.3 billion years. Therefore, if a substance is found that was originally made up only of atoms of uranium-235, after 704 million years, the substance is one-half uranium-235 and one-half lead-207. In another 704 million years, three-quarters of the substance will be lead-207 and only one-quarter uranium-235. Because each radioactive isotopic system behaves uniquely in nature, data must be carefully tested, compared, and evaluated. The best data are those in which different radioactive isotopic systems give the same date.

The most widely accepted age of Earth is between 4.5 billion and 4.6 billion years. This age is based on lead isotope studies of meteorites and samples of lead minerals from rocks. Heat generated by the formation of Earth probably created a molten surface. The original solid surface that formed as Earth cooled is believed to have been destroyed by a variety of geologic processes. The oldest minerals found are between 4.1 billion and 4.2 billion years old, providing a minimum age for Earth. The accepted age of Earth is based on dating objects in the solar system that are believed to have formed at the same time as the planets but are not geologically active (they have not changed significantly over time), such as meteorites. Several different dating methods on multiple meteorites have all resulted in estimated ages of about 4.5 billion years.

Geologic Time

To refer to events in the history and formation of Earth, scientists use geologic time (table 1.2). The principal divisions are the four eons: the Hadean (4.6 to 4.0 billion years ago), the Archean (4.0 to 2.5 billion years ago), the Proterozoic (2.5 billion to 570 million years ago), and the Phanerozoic (since 570 million years ago). The first three eons are known as the Precambrian; it represents nearly 88% of all geologic time. Fossils are known from other eons but are common only from the Phanerozoic. The Phanerozoic eon is divided into three eras: the Paleozoic era of ancient life; the Mesozoic era of middle life (popularly called the Age of Reptiles); and the Cenozoic era of recent life (the Age of Mammals). Each of these eras is subdivided into periods and epochs; today, for instance, we live in the Holocene epoch of the Quaternary period of the Cenozoic era. The appearance or disappearance of fossil types was used to set the boundaries of the time units before radiometric dating enabled scientists to set these time scale boundaries more accurately. The accurate calibration of radiometric dates and relative time determined from fossils is an ongoing process, and the exact dates defining the time units are constantly being adjusted with the acquisition of new data. Dating time units in the Precambrian eons is particularly difficult because of the lack of very old marine sediments and fossils.

Very long periods of time are incomprehensible to most of us. We often have difficulty coping with time spans of more than ten years. What were you doing exactly ten years ago today? We have nothing with which to compare the 4.6-billion-year age of Earth or the 500 million years since the first **vertebrates** (animals with a spinal column) appeared on this planet. To place geologic time in a framework we can understand, let us divide Earth's age by 100 million. If we do so, then we can think of Earth as being just forty-six years old. What has happened over that forty-six years?

No record remains of events during the first three years. The earliest history preserved can be found in some rocks of Canada, Africa, and Greenland that formed forty-three years ago. Sometime between thirty-five and thirty-eight years ago, the first primitive living cells of bacteria-type organisms appeared. Oxygen production by living cells began about twenty-three years ago, half the age of the planet. Most of this oxygen combined with iron in the early oceans and did not accumulate in the atmosphere. It took about eight years, or until roughly fifteen years ago, for enough oxygen to accumulate in the atmosphere to support significant numbers of complex oxygen-requiring cells. Oxygen reached its present concentration in the atmosphere

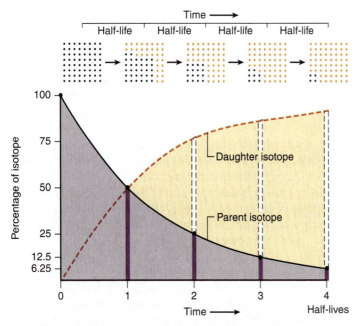

Figure 1.6 Parent isotopes decay into daughter isotopes at a rate that follows a curved line. Mathematically, this curved line is described by a decreasing exponential function—that is, e^{-x}. The exponential decay of radioisotopes allows geologists to calculate very precisely the age of rocks. The half-life of the element, indicated by dark purple bars, indicates the amount of the parent isotope remaining after one, two, three, and four half-lives. The red dashed line indicates the amount of daughter isotope formed over time. Measurements of the ratio of daughter isotopes to parent isotopes allow calculation of the rock's age when the half-life of the element is known.

Table 1.2 The Geologic Time Scale

Eon	Era	Period	Epoch	Time (millions of years ago)	Life Forms/Events
Phanerozoic	Cenozoic "Age of Mammals"	Quaternary	Holocene	0.0	Modern humans
				0.01	
			Pleistocene	1.6	Earliest humans
		Tertiary	Pliocene	5.3	
			Miocene	23.7	Earliest hominids
			Oligocene	36.6	Flowering plants
			Eocene		Earliest grasses
				57.8	Mammals, birds, and insects dominant
			Paleocene	65	
		Cretaceous-Tertiary boundary: The extinction of 50% of all species, including the dinosaurs, at the end of the Mesozoic era (65 million years ago).			
	Mesozoic "Age of Reptiles"	Cretaceous			Earliest flowering plants (115) Dinosaurs in ascendence
				144	First birds (155)
		Jurassic		208	Dinosaurs abundant
		Triassic-Jurassic boundary: The extinction of over 50% of all species, including the last of the mammal-like reptiles, leaving mainly dinosaurs on land (208 million years ago).			
		Triassic			First turtles (210) First mammals (221) First dinosaurs (228) First crocodiles (240)
				245	
		Permian-Triassic boundary: The greatest mass extinction of all time; 96% of all species perish at the end of the Paleozoic era (245 million years ago).			
	Paleozoic	Permian			Extinction of trilobites and many other marine animals
		Carboniferous	Pennsylvanian	286	Large coal swamps "Age of Amphibians" First reptiles (330)
			Mississippian	320	Amphibians abundant
		Devonian		360	First seed plants (365) First sharks (370) "Age of Fishes" First insect fossils (385) Fishes dominant
		Silurian		408	First vascular land plants (430)
		Ordovician		438	First land plants similar to lichen (470) First fishes (505) "Age of Marine Invertebrates" Earliest corals Marine algae
		Cambrian		505	Abundant shelled invertebrates Trilobites dominant
				570	
Proterozoic		Collectively, these are popularly known as the Precambrian.			First invertebrates (700) Earliest shelled organisms (~ 750) Oxygen begins to accumulate in the atmosphere (1500). First fossil evidence of single-celled life with a cell nucleus: eukaryotes (1500)
Archean				2500	First evidence of by-products of eukaryotes (2700) Earliest primitive life, bacteria and algae: prokaryotes (3500–3800) These will dominate the world for the next 3 billion years.
Hadean				4000	Oldest surface rocks (4030) Oldest single mineral (~ 4200) Oldest Moon rocks (4440) Oldest meteorites (4560)
				~ 4600	

approximately eleven years later. The first invertebrates (animals without backbones) developed seven years ago, and two years later, the first vertebrates appeared. Primitive fish first swam in the oceans, and corals appeared just five years ago. Three years, eight and one-half months ago, the first sharks could be found in the oceans, and roughly five months later, reptiles could be found on land. A massive extinction struck the planet just two and one-half years ago, killing 96% of all life. Following this catastrophic event, the dinosaurs appeared just two years, three months, and ten days ago. Three and one-half weeks later, the first mammals developed. A second major extinction occurred roughly two years ago. This event killed over half of the species on Earth, leaving mostly dinosaurs on land. Just eighteen and one-half months ago, the first birds flew in the air, and they would have seen the first flowering plants a little less than five months later. A third major extinction struck Earth 237 days ago, killing off the dinosaurs as well as many other species. Two hundred eleven days ago the mammals, birds, and insects became the dominant land animals. Our first human ancestors (the first identifiable member of the genus *Homo*) appeared just a little less than six days ago. About half an hour ago, modern humans began the long process we know as civilization, and only one minute ago, the Industrial Revolution began changing the Earth and our relationship with it.

Natural Time Periods

People first defined time by the natural motions of the Earth, Sun, and Moon. Later, people grouped natural time periods for their convenience, and still later, artificial time periods were created for people's special uses. Time is used to determine the starting point of an event, the event's duration, and the rate at which the event proceeds. An accurate measurement of time is required to determine location or position; this use of time is discussed in section 1.4.

The year is the time required by Earth to complete one orbit about the Sun. This time is 365¼ days, adapted for convenience to 365 days with an extra day added every four years, except for years ending in hundreds and not divisible by 400. As Earth orbits around the Sun, those who live in temperate zones and polar zones are very conscious of the seasons and of the differences in the lengths of the periods of daylight and darkness. The reason for these seasonal changes is seen in figure 1.7. Earth moves along its orbit with its axis tilted 23½° from the vertical. During the year, Earth's North Pole is sometimes tilting toward the Sun and sometimes tilting away from it. The Northern Hemisphere receives its maximum hours of sunlight when the North Pole is tilted toward the Sun; this is the

Northern Hemisphere's summer. During the same period, the South Pole is tilted away from the Sun, so the Southern Hemisphere receives the least sunlight; this period is winter in the Southern Hemisphere (note in fig. 1.7 that summer in the Northern Hemisphere and winter in the Southern Hemisphere occur when Earth is farthest from the Sun). When Earth is closest to the Sun, the North Pole is tilted away from the Sun, creating the Northern Hemisphere's winter, and the South Pole is inclined toward the Sun, creating the Southern Hemisphere's summer. Note also that during summer in the Northern Hemisphere, the periods of daylight are longer around the North Pole and shorter around the South Pole; the opposite is true during the Northern Hemisphere's winter.

As we live on the spinning Earth orbiting the Sun, we are not conscious of any movement. What we sense is that the Sun rises in the east and sets in the west daily and slowly moves up in the sky from south to north and back during one year. The periods of daylight in the Northern Hemisphere increase as the Sun moves north to stand above 23½°N, the **Tropic of Cancer.** It reaches this position at the **summer solstice,** on or about June 22, the day with the longest period of daylight and the beginning of summer in the Northern Hemisphere. On this day, the Sun does not sink below the horizon above the **Arctic Circle,** 66½°N latitude, nor does it rise above the **Antarctic Circle,** 66½°S. Following the summer solstice, the Sun appears to move southward until on or about September 23, the **autumnal equinox,** when it stands directly above the equator.

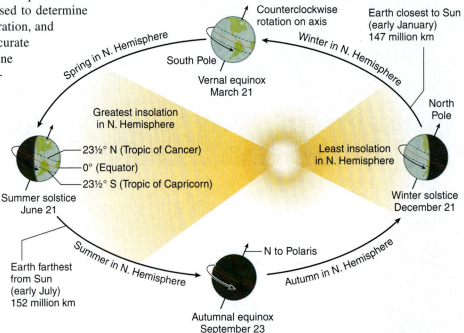

Figure 1.7 The annual orbit of the Earth around the Sun. The 23.5° tilt of the Earth on its axis relative to its plane of orbit around the Sun is responsible for the seasons. Changes in the overhead position of the Sun cause variations in the amount of solar radiation incident on a unit area of the Earth's surface. Thus, on a seasonal basis, the area of Earth's surface heated most directly changes (day to day, in fact).

On this day, the periods of daylight and darkness are equal all over the world. The Sun continues its southward movement until about December 21, when it stands over 23½°S, the **Tropic of Capricorn;** this position marks the **winter solstice** and the beginning of winter in the Northern Hemisphere. On this day, the daylight period is the shortest in the Northern Hemisphere; above the Arctic Circle, the Sun does not rise, and south of the Antarctic Circle, the Sun does not set. The Sun then begins to move northward, and on about March 21, the **vernal equinox,** it stands again above the equator; spring begins in the Northern Hemisphere, and the periods of daylight and darkness are once more equal around the world. Follow figure 1.7 around again, checking the position of the South Pole, and note how the seasons of the Southern Hemisphere are reversed from those of the Northern Hemisphere.

The greatest annual variation in the intensity of direct solar illumination occurs in the temperate zones. In the polar regions, the seasons are dominated by the long periods of light and dark, but the direct heating of Earth's surface is small because the Sun is always low on the horizon. Between the Tropics of Cancer and Capricorn, there is little seasonal change in solar radiation because the Sun is always nearly directly overhead at midday hours.

Weeks and months, as they presently exist, modify natural time periods. The Moon requires 27⅓ days to orbit Earth, but a period of 29½ days defines the **lunar month.** In the lunar month the Moon passes through four phases: new moon, first quarter, full moon, and last quarter. The four phases approximately match the four weeks of the month. Days are grouped into twelve months of unequal length in order to form one calendar year. The present arrangement is known as the Gregorian calendar after Pope Gregory XIII, who in the sixteenth century made the changes necessary to correct the old Julian calendar, adopted in 46 B.C. and named after Julius Caesar. The Gregorian calendar was adopted in the United States in 1752, by which time the Julian calendar was eleven days in error. In that year, by parliamentary decree, in both Great Britain and the United States, September 2 was followed by September 14. People rioted in protest, demanding their eleven days back. Also in 1752, the beginning of the calendar year was changed from the original date of the vernal equinox, March 21, to January 1. The year 1751 had no months of January and February.

The day is derived from Earth's rotation. The average time for Earth to make one rotation relative to the Sun is twenty-four hours; this is the average, or mean, **solar day**—our clock day. Another measure of a day is the time required for Earth to make a complete rotation with respect to a far-distant point in space. This is known as the **sidereal day** and is about four minutes shorter than the mean solar day; it gives the true rotational period of Earth. The sidereal day is useful in astronomy and navigation.

Living organisms respond to these natural cycles. In temperate zones, as the periods of daylight lengthen and the temperatures increase, flowers bloom and trees produce new leaves; then, as periods of darkness increase and temperatures fall, flowers die back, and trees lose their leaves and enter dormancy. In tropical areas, however, forests remain lush year-round. Some animals migrate and alternate periods of activity and hibernation or estivation with the seasons. Other animals set their internal clocks to the day-night pattern, hunting in the dark and sleeping in the light; still others do the reverse. Plants and animals of the sea also react to these rhythms, as do the physical processes that stir the atmosphere and circulate the water in the oceans. Understanding these cycles helps us understand processes that occur at the ocean surface and are discussed in later chapters: climate zones, winds, currents, vertical water motion, plant life, and animal migration.

QUICK REVIEW

1. How old is Earth?
2. How and why have estimates of the age of Earth changed over the past few hundred years? Do you think the present estimate of Earth's age will change in the future?
3. How are rocks dated?
4. How much of a radioactive isotope would be left after two half-lives had passed?
5. How were the divisions of geologic time established before age-dating of rocks was possible?
6. Why are the Arctic and Antarctic Circles located at 66½°N and 66½°S, respectively?

1.3 Earth's Shape

As Earth cooled and turned in space, gravity and the forces of rotation produced its nearly spherical shape. If Earth was a perfect sphere, it would have a radius of 6371 km (3959 mi). But Earth is not rigid and, as it spins, it tends to flatten at the poles and bulge along the equator (fig. 1.8). Consequently, it

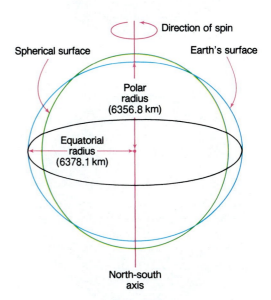

Figure 1.8 Rotation makes Earth bulge outward at the equator. Notice that the equatorial radius is larger than the polar radius.

has a shorter polar radius (6356.8 km; 3950 mi) and a longer equatorial radius (6378.1 km; 3963 mi). This is a difference of 21.3 km (about 13 mi).

Earth is also quite smooth. The top of Earth's highest mountain, Mount Everest in the Himalayas, is about 8840 m (29,000 ft) above sea level; the deepest ocean depth, the bottom of the Challenger Deep in the Mariana Trench of the Pacific Ocean, is about 11,000 m (36,000 ft). On a scale model of Earth the size of a basketball, the vertical distance from the height of Mount Everest to the depth of the Challenger Deep would be about the thickness of an average human fingernail. The topographic relief of Earth's surface—its high mountains and deep oceans—is minor compared to the size of the planet.

QUICK REVIEW

1. What is the shape of Earth?
2. Why isn't Earth a perfect sphere?
3. Relate Earth's highest elevations and greatest depths to its overall size.

1.4 Where on Earth Are You?

Latitude and Longitude

To find our way from one place to another, we need a reference (or location) system that allows us to uniquely identify where a particular place is. Most of us use such a system daily. Your home address is an example of a reference system that uses city name, street name or number, and building number. Armed with a city map, we can confidently navigate to places we have never been before. While this works well within a city, specifying unique locations on the surface of Earth requires a much different system that uses a grid of reference lines that cross at right angles. These grid lines are called lines of **latitude** and **longitude**.

Lines of latitude, also known as **parallels,** are referenced to the **equator** (Fig 1.9a). The equator is created by passing a plane through Earth halfway between the poles and at right angles to Earth's axis. This process is much like cutting an orange in two pieces halfway between the depressions marking the stem and the navel. The equator is marked as 0° latitude, and other latitude lines are drawn parallel to the equator, northward

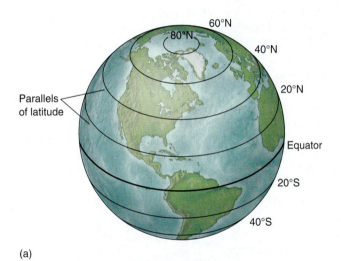

(a)

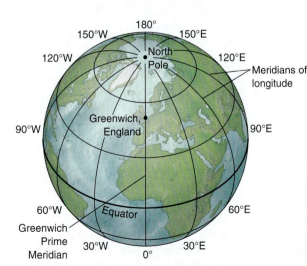

(b)

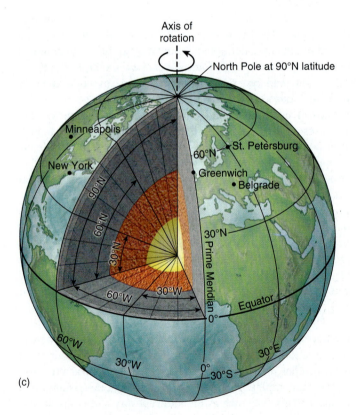

(c)

Figure 1.9 (a) The grid system: parallels of latitude. Note that the parallels become increasingly shorter closer to the poles. On the globe, the 60th parallel is only half as long as the equator. (b) The grid system: meridians of longitude. East-west measurements range from 0° to 180°—that is, from the prime meridian to the 180th meridian in each direction. Because the meridians converge at the poles, the distance between degrees of longitude becomes shorter as one moves away from the equator. (c) The earth grid, or graticule, consisting of parallels of latitude and meridians of longitude.

to 90°N, or the North Pole, and southward to 90°S, or the South Pole. Notice that the parallels of latitude describe increasingly smaller circles as the poles are approached. Notice also that latitudes are designated as either north or south of the equator. The latitude value is determined by the angle between the latitude line and the equatorial plane at Earth's center (fig. 1.9c). The Tropics of Cancer and Capricorn (see fig. 1.7) correspond to latitudes 23½°N and S, respectively. Latitudes 66½°N and S, respectively, correspond to the Arctic and Antarctic Circles.

Lines of longitude, or **meridians,** are formed at right angles to the latitude lines (fig. 1.9b). Longitude is referenced to an arbitrarily chosen point: the 0° longitude line on Earth's surface extends from the North Pole to the South Pole and passes directly through the Royal Naval Observatory in Greenwich, England, just outside London (fig. 1.10). On the other side of Earth, 180° longitude is directly opposite 0°. The 0° longitude line is known as the **prime meridian.** The 180° longitude line approximates the **international date line.** Longitude lines are

Figure 1.10 The Royal Naval Observatory at Greenwich, England. The fiber optic line running into the observatory's door marks the prime meridian, the division between east and west longitudes. The prime meridian is defined by the position of the telescope in the Observatory's Meridian Building. This was built by Sir George Biddell Airy, the seventh Astronomer Royal, in 1850. The crosshairs in the eyepiece of the telescope define longitude 0° for the world. In this night image of the observatory, its roof has been opened to allow use of the telescope.

identified by their angular displacement to the east and west of 0° longitude, as shown in figure 1.9c. Thus, longitude may be reported as either 0° to 180° east (also known as positive longitude) and 0° to 180° west (also known as negative longitude) or 0° to 360° east (always positive). In this manner −90°, 90°W, and 270° longitude all mark the same meridian. Note in figure 1.9b that meridians are the same size, much like the lines marking the segments of an orange. The meridians mark the intersection of Earth's surface with a plane passing through Earth's rotational axis at right angles to the parallels of latitude. Any circle at Earth's surface with its center at Earth's center is a **great circle.** All longitude lines form great circles; only the equator is a great circle of latitude. A great circle connecting any two points on Earth's surface defines the shortest distance between them. See appendix C for an example calculation of the great circle distance between two points.

To identify any location on Earth's surface, we use the crossing of the latitude and longitude lines; for example, 158°W, 21°N is the approximate location of the Hawaiian Islands, and 20°E, 33°S identifies the Cape of Good Hope at the southern tip of Africa. Because the distance expressed in whole degrees is large (1 degree of latitude equals 60 nautical miles), each degree (1°) of arc is divided into 60 minutes (60′), each minute into 60 seconds (60″), and each second into tenths of a second. The location of Honolulu Bay, Hawaii, is 21° 18′ 34″N, 157° 52′ 21″W. One **nautical mile** is equal to one minute of arc length of latitude or longitude at the equator, or 1852 m (1.15 land mi; see appendix B). Positions on Earth can be specified with great accuracy when this system is used.

Measuring Latitude

Early voyages often remained in sight of land because of the difficulty sailors had measuring latitude and longitude accurately. The problem of determining latitude is simpler than determining longitude, and methods to determine latitude have been known since ancient times. The Greeks knew that Earth's axis of rotation through the North and South Poles would point to a particular place in the sky directly above the North Pole. Latitude could be determined by measuring the elevation of this point in the sky above the horizon. Earth's axis of rotation through the North and South Poles rotates slowly, making one complete revolution in approximately 26,000 years. Because of this rotation, the point in the sky directly above the North Pole changes slowly over time. For the past 1000 years this point has coincided closely with a bright star, **Polaris,** often called the North Star. Consequently, since the Middle Ages sailors have been able to accurately estimate their latitude by measuring the elevation of Polaris above the horizon (fig. 1.11).

Longitude and Time

Determining longitude was a much more difficult task. Because the longitude lines rotate with Earth, 360° in twenty-four hours, it becomes necessary to know the time of day and the position of the Sun or the stars relative to one's longitude line. Although the

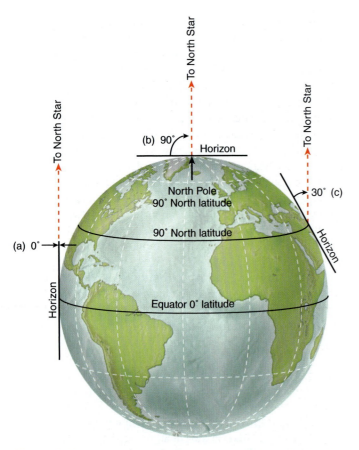

Figure 1.11 A person's latitude is equal to the angle of the North Star above the northward horizon. Note in particular that (a) on the equator (0° latitude), it is on the horizon (0° above the horizon), (b) at the North Pole (90° N latitude), it is directly overhead (90° above the horizon), and (c) at 30° N latitude, it is 30° above the horizon.

theory for using time to determine longitude had been proposed by Flemish astronomer Gemma Frisius in 1530, early clocks did not work satisfactorily on ships, and precise longitude measurements were not possible until the construction of an accurate ship's chronometer in the eighteenth century. See chapter 1 for the history of this achievement.

If a clock is set to exactly noon at some initial location when the Sun is at its **zenith,** or highest elevation above a reference longitude, and if that clock is then carried to a new location and the time on the clock is noted at the zenith time (noon) of the Sun at the new location, the difference in time between the time on the clock and noon at the new location can be used to determine the difference in longitude between the initial and new locations. When this technique is used, a position that is 15° of longitude west of the reference longitude is directly *under the Sun* one hour later, or at 1 P.M. by the clock, because Earth has turned eastward 15° during that hour. A position 15° east of the reference

longitude is directly under the Sun at 11 A.M., because an hour is required to turn the 15° to bring the reference longitude directly under the Sun at its zenith (fig. 1.12).

The reference longitude in use today is the prime meridian, or 0° longitude. The clock time is set to noon when the Sun is at its zenith above the prime meridian. This is **Greenwich Mean Time (GMT),** now **Universal Time** or **ZULU Time** (zero meridian time). Because Sun time changes by one hour for each 15° of longitude, Earth has been divided into time zones that are 15° of longitude wide. The time zones do not exactly follow lines of longitude; they follow political boundaries when necessary for the convenience of the people living in those zones (fig. 1.13).

QUICK REVIEW

1. Sketch parallels of latitude and meridians of longitude on an orange. Include the equator and the prime meridian. Place a dot at the approximate location of the city where you live.

2. Why is the equator a unique parallel whereas the prime meridian is an arbitrary choice?

3. Why are the Arctic and Antarctic Circles displaced from the pole by 23½°?

4. How is time important in the calculation of longitude?

5. Will we always be able to use the North Star to determine latitude? Why or why not?

6. Find the latitude and longitude for each of the following: a. Chicago, IL; b. Montreal, Canada; c. Buenos Aires, Argentina; d. London, England; e. Vienna, Austria; f. Peking, China; g. Tokyo, Japan; h. Cape Town, South Africa; i. Nairobi, Kenya; j. Canberra, Australia; k. Papeete, Tahiti

1.5 Modern Navigation

Modern navigation uses the U.S. Navstar **Global Positioning System (GPS).** GPS is a worldwide radio-navigation system consisting of twenty-four navigational satellites—twenty-one operational and three active spares—and five ground-based

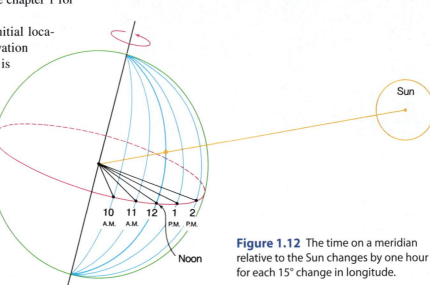

Figure 1.12 The time on a meridian relative to the Sun changes by one hour for each 15° change in longitude.

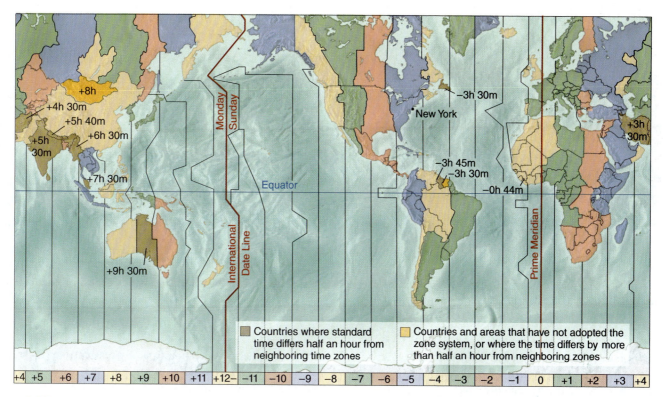

Figure 1.13 World time zones. Each time zone is about 15° wide, but variations occur to accommodate political boundaries. The figures at the bottom of the map represent the time difference in hours when it is 12 noon in the time zone centered on Greenwich, England. New York is in column −5, so the time there is 7 A.M. when it is noon at Greenwich. Modifications to the universal system of time zones are numerous. Thus, Iceland operates on the same time as Britain, although it is a time zone away. Spain, entirely within the boundaries of the GMT zone, sets its clocks at +1 hour, whereas Portugal conforms to GMT. China straddles five time zones, but the whole country operates on Beijing time (+8 hours). In South America, Chile (in the −5 hour zone) uses the −4 hour designation, whereas Argentina uses the −3 hour zone instead of the −4 hour zone to which it is better suited.

monitoring stations (fig. 1.14). The satellites orbit at an altitude of about 20,165 km (12,500 mi) and repeat the same track and relative configuration over any point about every twenty-four hours. At any given time, from five to eight satellites are visible from any point on Earth. The system uses this constellation of satellites as reference points for calculating positions on the surface with an accuracy of a few meters for commercial and private users and to better than a centimeter with advanced forms of GPS.

Each GPS satellite transmits a unique digital code that is sent as a radio signal to ground receivers. GPS receivers generate codes that are identical to those sent by the satellites at exactly the same time the satellites do. Because of the distance the satellite signals travel, there is a time lag between when the receiver generates a specific signal and when it receives the same signal generated at the same time by the satellite. This time delay is a function of the distance between the receiver and the satellite. The satellite signals travel at the speed of light, or approximately 300,000 km/s (186,000 mi/s), so measuring the arrival time of the signal accurately is critically important. The signal from a satellite directly overhead would reach a ground receiver in roughly 0.06 s. Precise measurements of distance between a receiver and four satellites will pinpoint the exact location of the receiver.

1.6 Earth Is a Water Planet

Water on Earth's Surface

As Earth and the other planets of the solar system cooled, the Sun's energy gradually replaced the heat of planet formation in maintaining their surface temperatures. Earth developed a nearly circular orbital path. Moving along this orbit, Earth is about 152×10^6 km (94×10^6 mi) from the Sun in June and about 147×10^6 km (91×10^6 mi) away in December, as shown in figure 1.7. At these distances from the Sun, Earth's orbit keeps the annual heating and cooling cycle within moderate limits. Earth's mean surface temperature is about 16°C (61°F), which allows water to exist as a gas, as a liquid, and as a solid.

Hydrologic Cycle

Earth's water occurs as a liquid in the oceans, rivers, lakes, and below the ground surface; it occurs as a solid in glaciers, snow packs, and sea ice; it occurs as water droplets and vapor in the atmosphere. The places in which water resides are called **reservoirs,** and each type of reservoir, when averaged over the entire Earth, contains a fixed amount of water at any one

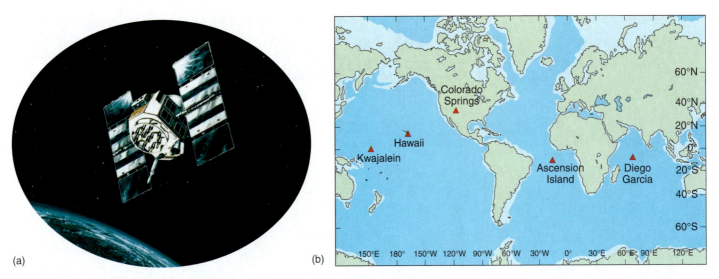

(a)

(b)

Figure 1.14 The Global Positioning System (GPS) consists of (a) twenty-four satellites and (b) five monitoring stations. The main control station is located at Falcon Air Force Base in Colorado Springs, Colorado.

instant. But water is constantly moving into and out of reservoirs. This movement of water through the reservoirs, diagrammed in figure 1.15, is called the **hydrologic cycle.**

Water is taken out of the oceans and moved into the atmosphere by evaporation. Most of this water returns directly to the sea by precipitation, but air currents carry some water vapor over the continents. Precipitation in the form of rain and snow transfers this water from the atmosphere to the land surface, where it percolates into the soil; is taken up by plants; runs off into rivers, streams, and lakes; or remains for longer periods as snow and ice in some areas. Some of this water returns to the atmosphere by evaporation, **transpiration** (the release of water by plants), and **sublimation** (the conversion of ice directly to water vapor). Melting snow and ice, rivers, groundwater, and land runoff move the water back to the oceans to complete the cycle and maintain the oceans' volume. For a comparison of the water stored in Earth's reservoirs, see table 1.3 and figure 1.16.

The properties of climate zones are principally determined by their surface temperatures and their evaporation-precipitation patterns: the moist, hot equatorial regions; the dry, hot subtropical deserts; the cool, moist temperate areas;

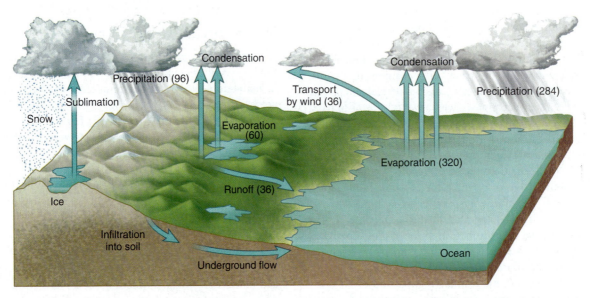

Figure 1.15 The hydrologic cycle and annual transfer rates for Earth as a whole. Precipitation transfer rate includes both snow and rain. Evaporation transfer rate from the continents includes evaporation of surface water, transpiration (the release of water to the atmosphere by plants), and sublimation (the direct change in state from ice to water vapor). Runoff from the continents includes both surface flow and underground flow. Annual transfer rates in thousands of cubic kilometers (10^3 km³).

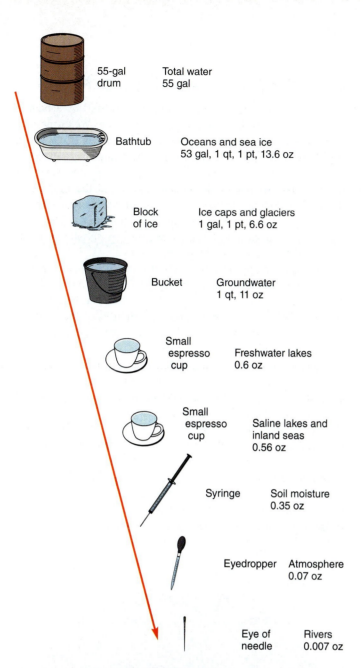

55-gal drum	Total water 55 gal
Bathtub	Oceans and sea ice 53 gal, 1 qt, 1 pt, 13.6 oz
Block of ice	Ice caps and glaciers 1 gal, 1 pt, 6.6 oz
Bucket	Groundwater 1 qt, 11 oz
Small espresso cup	Freshwater lakes 0.6 oz
Small espresso cup	Saline lakes and inland seas 0.56 oz
Syringe	Soil moisture 0.35 oz
Eyedropper	Atmosphere 0.07 oz
Eye of needle	Rivers 0.007 oz

Figure 1.16 Comparison of the amount of the world's water supply held in each of the major water reservoirs. For purpose of illustration, Earth's total water supply has been scaled down to the volume of a 55-gal drum.

and the cold, dry polar zones. Differences in these properties, coupled with the movement of air between the climate zones, move water through the hydrologic cycle from one reservoir to another at different rates. The transfer of water between the atmosphere and the oceans alters the salt content of the oceans' surface water, and, with the seasonal and latitudinal changes in surface temperature, determines many of the characteristics of the world's oceans.

Reservoirs and Residence Time

Because the total amount of water on Earth is essentially constant, the hydrologic cycle must maintain a balance between the addition and removal of water from Earth's water reservoirs. The rate of removal of water from a reservoir must equal the rate of addition to it. The average length of time that a water molecule spends in any one reservoir is called the **residence time** in that reservoir. Water's residence time can be calculated by dividing the volume of water in the reservoir by the rate at which the water is replaced. A large reservoir generally has a long residence time because of its large volume, whereas small reservoirs generally have a short residence time and the water in them can be replaced comparatively quickly. The size also determines how a reservoir reacts to changes in the rate at which the water is gained or lost. Large reservoirs show little effect from small changes, whereas small reservoirs may alter substantially when exposed to the same gain or loss. For example, if the ocean volume decreased by 6½% and that volume of water were added to the land ice, the result would be a 300% increase in the present volume of land ice and sea level would drop by 172 m (564 ft).

Table 1.3 Earth's Water Supply

Reservoir	APPROXIMATE WATER VOLUME		Approximate Percent of Total Water
	(km³)	(mi³)	
Oceans and sea ice	1,349,929,000	323,866,000	97.26
Ice caps and glaciers	29,289,000	7,000,000	2.11
Groundwater	8,368,000	2,000,000	0.60
Freshwater lakes	125,500	30,000	0.009
Saline lakes and inland seas	105,000	25,000	0.008
Soil moisture	67,000	16,000	0.005
Atmosphere	13,000	3,100	0.0009
Rivers	1,250	300	0.0001
Total water volume	1,387,897,750	332,940,400	100

This example reflects the changes that have occurred on Earth during the major ice ages, when more water was stored on land as ice and sea level was lower.

About 380,000 km³ (91,167 mi³) of water move through the atmosphere each year. Because the atmosphere holds the equivalent of 13,000 km³ (3119 mi³) of liquid water at any one time, a little arithmetic shows that the water in the atmosphere can be replaced twenty-nine times each year. Atmospheric water has a very short residence time. The residence time for water in the other, larger reservoirs is much longer. For example, it would take 4219 years to evaporate and pass all of the water in the oceans through the atmosphere, to the land as precipitation, and back to the oceans via rivers. Further study of water's movement shows us that annually 320,000 km³ (76,772 mi³) is evaporated from the oceans, and 60,000 km³ (14,395 mi³) is evaporated from land. When the water returns as precipitation, 284,000 km³ (68,135 mi³) is returned directly to the sea surface and 96,000 km³ (23,032 mi³) to the land. However, the excess gained by the land (36,000 km³ or 8637 mi³) flows back to the oceans in rivers, streams, and groundwater (see fig. 1.15).

Distribution of Land and Water

To understand the present distribution of land and water on Earth, consider Earth viewed from the north (fig. 1.17a) and from the south (fig. 1.17b). About 70% of Earth's landmasses are in the Northern Hemisphere, and most of this land lies in the middle latitudes. The Southern Hemisphere is the water hemisphere, with its land located mostly in the tropical latitudes and in the polar region.

One World Ocean Divided into Five

Looking at a globe, we can see Earth is covered by a blanket of seawater comprising a single world ocean. We commonly divide the world ocean into five separate oceans, or ocean basins (fig. 1.18). The boundaries of these ocean basins have developed over time for various historical, geographical, and scientific reasons. The shapes of the ocean basins and the mountains, trenches, and plains on the sea floor all influence ocean currents that transport heat, salt, nutrients, and pollutants. They also direct the propagation of energy released by submarine earthquakes that can create damaging tsunamis.

The three major oceans are the Pacific, the Atlantic, and the Indian. The smallest ocean geographically is the Arctic Ocean. The Southern Ocean is the "newest" ocean. The International Hydrographic Organization set the boundaries of the Southern Ocean in 2000. Each of these five oceans has its characteristic surface area, volume, and depth (table 1.4).

The Pacific Ocean (fig. 1.18a) has greater surface area, volume, and average depth than any of the other oceans. The deepest point in the world ocean is located in the Mariana Trench in the western Pacific, reaching a depth of 10,911 m (35,797 ft). The first European to sight the Pacific Ocean was the Spanish explorer Vasco Núñez de Balboa in 1513. Its name came from the Portuguese explorer Ferdinand Magellan who called it *Mar Pacifico*, meaning "peaceful sea," for the calm weather he and his crew enjoyed while crossing it in 1521. The Pacific covers a little over one-third of Earth's surface and nearly half of the world ocean's surface. At its maximum width near 5°N, the Pacific stretches 19,800 km (12,300 mi) from Indonesia to

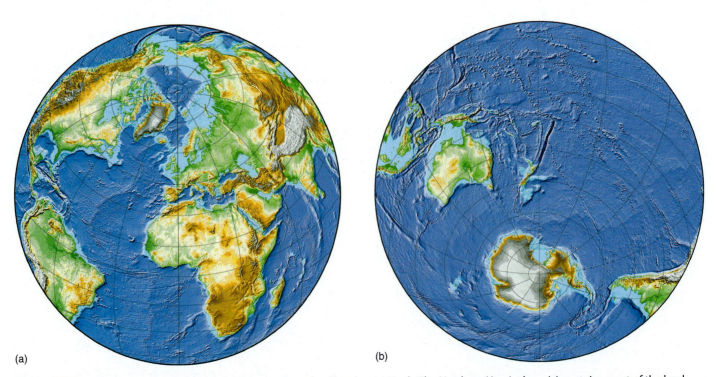

(a)   (b)

Figure 1.17 Continents and ocean basins are not distributed uniformly over Earth. The Northern Hemisphere (a) contains most of the land; the Southern Hemisphere (b) is mainly water.   Source of Data: National Oceanic and Atmospheric Administration (NOAA).

Table 1.4   Ocean Basin Areas, Volumes, and Depths

Ocean Basin	Area	Percent of Ocean Surface Area	Percent of Earth's Surface Area	Volume	Percent of Ocean Volume	Average Depth	Maximum Depth
Pacific	168,723,000 km² 65,143,950 mi²	46.6	33.1	669,880,000 km³ 160,704,212 mi³	50.1	3972 m 13,031 ft	10,911 m 35,797 ft
Atlantic	85,133,000 km² 32,869,851 mi²	23.5	16.7	310,410,900 km³ 74,467,575 mi³	23.3	3646 m 11,962 ft	8486 m 27,841 ft
Indian	70,560,000 km² 27,243,216 mi²	19.5	13.8	264,000,000 km³ 63,333,600 mi³	19.8	3741 m 12,274 ft	7906 m 25,938 ft
Southern	21,960,000 km² 8,478,756 mi²	6.1	4.3	71,800,000 km³ 17,224,820 mi³	5.4	3270 m 10,728 ft	7075 m 23,212 ft
Arctic	15,558,000 km² 6,006,944 mi²	4.3	3.1	18,750,000 km³ 4,498,125 mi³	1.4	1205 m 3953 ft	5567 m 18,264 ft
All Ocean Basins	361,934,000 km² 139,742,717 mi²	100	71	1,334,840,900 km³ 320,228,332 mi³	100	3688 m 12,100 ft	

Colombia. There are roughly 25,000 islands in the Pacific, the majority of which are south of the equator. This is more than the number of islands in the rest of the oceans combined. There are a number of marginal seas along the edges of the Pacific, including the Celebes Sea, Coral Sea, East China Sea, Sea of Japan, Sulu Sea, and Yellow Sea.

The Atlantic Ocean (figure 1.18b) is the second largest, roughly half the size of the Pacific. Its name is derived from Greek mythology and means "Sea of Atlas" (Atlas was a Titan who supported the heavens by means of a pillar on his shoulders). The first person known to have used the name "Atlantic" is the Greek geographer Herodotus around 450 B.C. The land area that drains fresh water into the Atlantic is four times larger than that of either of the two other major oceans, the Pacific and the Indian. This is because of the high mountain ranges that dominate the western coasts of North and South America as well as South Asia. There are relatively few islands in the Atlantic considering the size of the basin. The irregular coastline of the Atlantic includes a number of bays, gulfs, and seas. Some of the larger ones include the Caribbean Sea, Gulf of Mexico, Mediterranean Sea, North Sea, and Baltic Sea.

The Indian Ocean (figure 1.18c) is primarily a Southern Hemisphere ocean; it is a little smaller than the Atlantic but is quite deep. It is named after the nation of India. The northernmost extent of the Indian Ocean is in the Persian Gulf at about 30°N. The Indian Ocean is separated from the Atlantic Ocean to the west by the 20°E meridian and from the Pacific Ocean to the east by the 147°E meridian. It is nearly 10,000 km (6200 mi) wide between the southern tips of Africa and Australia and includes the world's fourth largest island, Madagascar. For centuries it has had tremendous strategic importance as a trade route between Africa and Asia.

The Southern Ocean (figure 1.18d) consists of the southernmost waters of the world ocean. There is still some debate about where the northern boundary of the Southern Ocean is. Some oceanographers define it as where northward flowing Antarctic waters meet the relatively warmer waters of the subantarctic, a region called the Antarctic Convergence. The location of the Antarctic Convergence varies between about 45°S and 60°S depending on longitude. The International Hydrographic Organization set the northern boundary at 60°S. It is the fourth largest ocean and the second shallowest. It is unique in that it completely encircles the globe, extending south to the shores of Antarctica. Ocean surface currents in the Southern Ocean can flow around the globe unimpeded by continents.

The Arctic Ocean (figure 1.18e) is the smallest and shallowest of the five, occupying a roughly circular basin over the North Pole region. It is connected to the Pacific Ocean through the Bering Strait and to the Atlantic Ocean through the Greenland Sea. Its floor is divided into two deep basins by an underwater mountain range. The major flow of water into and out of the Arctic Ocean is through the North Atlantic Ocean. The Arctic Ocean is partly covered by sea ice throughout the year.

Hypsographic Curve

Another method used by oceanographers to depict land-water relationships is shown in figure 1.19. This graph of depth or elevation versus Earth's area is called a **hypsographic curve.** Find the line indicating sea level and note that the elevation of land above sea level is given in meters along the left margin; the depth below sea level is given in meters along the right margin. The scale across the top of the figure indicates total Earth area in 100 million km² (10^8 km²). The scales along the bottom of

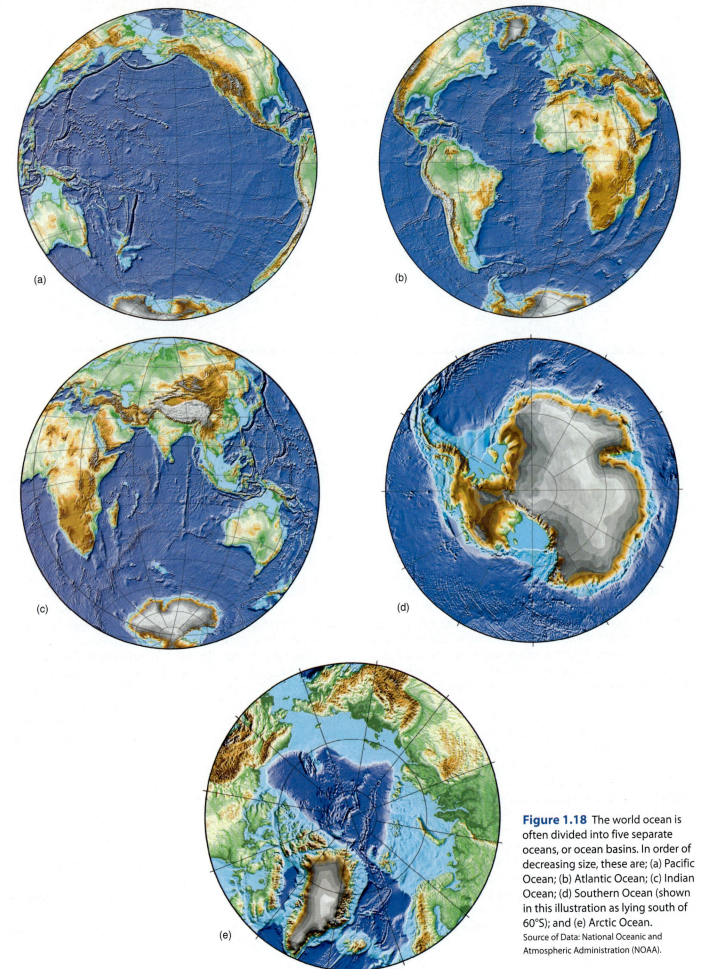

Figure 1.18 The world ocean is often divided into five separate oceans, or ocean basins. In order of decreasing size, these are; (a) Pacific Ocean; (b) Atlantic Ocean; (c) Indian Ocean; (d) Southern Ocean (shown in this illustration as lying south of 60°S); and (e) Arctic Ocean.
Source of Data: National Oceanic and Atmospheric Administration (NOAA).

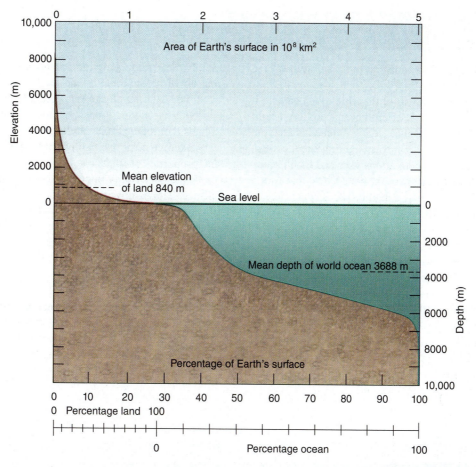

Figure 1.19 The hypsographic curve displays the area of Earth's surface at elevations above, and depths below sea level.

the figure indicate percentages of Earth's surface area; note that the curve crosses sea level at the 29% mark, showing that 29% of Earth's surface is above sea level and 71% is below sea level. The lower of the two scales gives land and ocean areas as separate percentages. Referring to the land area percentage scale, note that only 20% of all land areas are at elevations above 2 km. The ocean area scale shows that approximately 85% of the ocean floor is deeper than 2 km. The hypsographic curve helps us to see not only that Earth is 71% covered with water but also that the areas well below the sea surface are much greater than the areas well above it; there are basins beneath the sea that are about four times greater in area than the area of land in mountains. Mount Everest, the highest land peak, reaches 8.84 km (5.49 mi) above sea level, whereas the ocean's deepest trench descends 10.91 km (6.78 mi) below sea level.

Because the hypsographic curve is constructed as a plot of area versus height, an area of the diagram shows volume, which is the product of area times height. The mean elevation of the land is 840 m (2750 ft), and the entire land volume above sea level fits within a box 840 m high, covering 29% of Earth's surface. The mean depth of the world ocean is 3688 m (12,100 ft), and the entire ocean volume below sea level fits into a box 3688 m deep covering 71% of Earth's surface.

QUICK REVIEW

1. Why does the whole Earth hydrologic cycle shown in figure 1.15 not apply to a specific region of Earth?

2. What is the relationship between the hydrologic cycle and climate zones?

3. Trace several possible routes for a water molecule moving between a mountain lake and an ocean. In which reservoirs would the molecule spend the greatest amount of time and in which the least?

4. What percentages of Earth's surface are covered by water and by land?

5. How does the distribution of seawater and land differ in the Northern and Southern Hemispheres?

6. What does the hypsographic curve reveal about the differences in elevation and depth of continents and the ocean floor?

7. Name the five ocean basins in order of their area from largest to smallest.

Summary

The beginning expansion of the universe was followed by the first stars, the reactions that produced the elements, and billions of galaxies. Our solar system is part of the Milky Way galaxy; it began as a rotating cloud of gas. A series of events produced eight planets orbiting the Sun, each planet having unique characteristics. Over approximately 1.5 billion years, Earth heated, cooled, changed, and accumulated a gaseous atmosphere and liquid water.

Reliable age dates for Earth rocks, meteorites, and Moon samples are obtained by radiometric dating. The accepted age of Earth is 4.6 billion years. Geologic time is used to express the time scale of Earth's history.

The distance between Earth and the Sun, Earth's orbit, its period of rotation, and its atmosphere protect Earth from extreme temperature change and water loss. Because Earth rotates, its shape is not perfectly symmetrical. Its exterior is relatively smooth. Natural time periods (the year, day, and month) are based on the motions of the Sun, Earth, and Moon. Because of the tilt of Earth's axis as it orbits the Sun, the Sun appears to move annually between 23½°N and 23½°S, producing the seasons.

Latitude and longitude are used to form a grid system for the location of positions on Earth's surface.

To determine longitudinal position, one must be able to measure time accurately. This need required the development of accurate seagoing clocks for celestial navigation.

Modern navigational techniques make use of radar, radio signals, computers, and satellites. A satellite network provides very accurate position readings and maps storms, tides, sea level, and properties of surface waters.

Water is a vitally important compound on Earth. Of Earth's surface, 71% is covered by its oceans. There is a fixed amount of water on Earth. Evaporation and precipitation move the water through the reservoirs of the hydrologic cycle. Water's residence time varies in each reservoir and depends on the volume of the reservoir and the replenishment rate.

The Northern Hemisphere is the land hemisphere; the Southern Hemisphere is the water hemisphere. Earth has three large oceans extending north from Antarctica. Each has a characteristic surface area, volume, and mean depth. The hypsographic curve is used to show land-water relationships of depth, elevation, area, and volume. It is also used to determine mean land elevation and mean ocean depth.

Key Terms

hypothesis, 26
theory, 26
Big Bang, 26
galaxy, 27
light-year, 27
cluster, 27
nebula, 27
dayglow, 29
radiometric dating, 31
isotope, 31
half-life, 32
vertebrate, 32

Tropic of Cancer, 34
summer solstice, 34
Arctic Circle, 34
Antarctic Circle, 34
autumnal equinox, 34
Tropic of Capricorn, 35
winter solstice, 35
vernal equinox, 35
lunar month, 35
solar day, 35
sidereal day, 35
latitude, 36

longitude, 36
parallel, 36
equator, 36
meridian, 37
prime meridian, 37
international date line, 37
great circle, 37
nautical mile, 37
Polaris, 37
zenith, 38
Greenwich Mean Time
    (GMT), 38

Universal Time, 38
ZULU Time, 38
Global Positioning System
    (GPS), 38
reservoir, 39
hydrologic cycle, 40
transpiration, 40
sublimation, 40
residence time, 41
hypsographic curve, 43

Study Problems

1. Earth's mean radius can be defined as the radius of a sphere having the same volume as Earth. This is called the volumetric radius (R_v) and it is easily calculated from the equatorial radius (R_e) and the polar radius (R_p) using:

$$R_v = \text{cube root } (R_e^2 \times R_p)$$

   Using the values given in figure 1.8 for Earth's equatorial radius and polar radius, calculate Earth's volumetric radius, or mean radius, in kilometers and miles.

2. If it is 2:30 P.M. at your location when it is noon along the prime meridian, what is your longitude?

3. If it is 8:40 A.M. at your location when it is noon along the prime meridian, what is your longitude?

4. Use table 1.3 to determine the volume of water held in the atmosphere. Use figure 1.15 to determine how much water is removed from the atmosphere by precipitation over the ocean and the land each year. Using these two estimates, calculate how many times the water in the atmosphere is replaced in a year.

5. The average depth of the world ocean can be calculated from the average depths of the five ocean basins and the percent of the total world ocean area they cover. Use estimates of these values given in table 1.4 to verify the average depth of the world ocean in meters and feet.

6. Determine the distance between two locations: 110°W, 38½°N and 110°W, 45°N. Express this distance in nautical miles and kilometers.

7. Use the volume of the oceans and Earth's surface area to calculate the depth of a hypothetical ocean covering the entire globe. Express this depth in meters and feet.

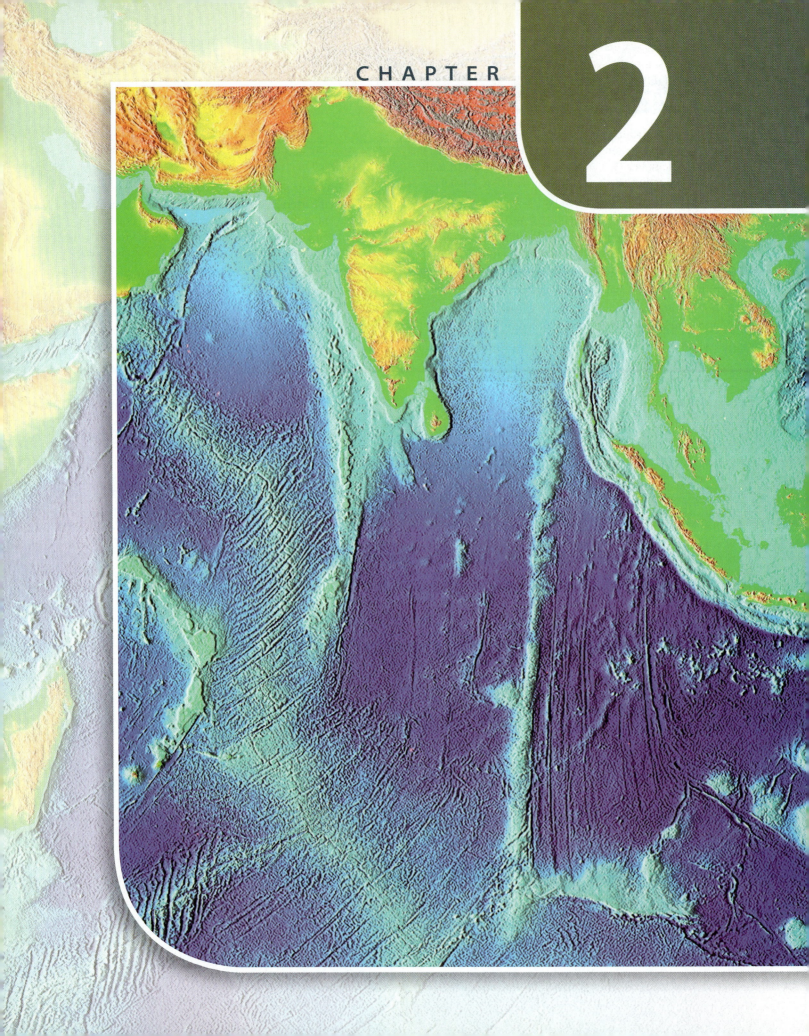

Earth Structure and Plate Tectonics

Learning Outcomes

After studying the information in this chapter students should be able to:

1. *sketch* Earth's internal structure and label the thickness of each region,

2. *explain* how seismology has provided critical data for modeling Earth's interior,

3. *differentiate* among the lithosphere, asthenosphere, and mesosphere,

4. *diagram* the three types of plate boundaries,

5. *review* the evidence in a written summary Alfred Wegener used to support his hypothesis of continental drift and the evidence supporting plate tectonics,

6. *distinguish* between continental drift and plate tectonics,

7. *describe* the formation of hydrothermal vents, and

8. *illustrate* the formation of magnetic stripes on the sea floor.

The aseismic Ninety East Ridge bisects the Indian Ocean Basin. The ridge consists of a series of seamounts increasing in age from about 43 Ma in the south to 82 Ma in the north.

Many features of our planet have presented Earth scientists with contradictions and puzzles. For example, the remains of warm water coral reefs are found off the coast of the British Isles, marine fossils occur high in the Alps and the Himalayas, and coal deposits that were formed in warm, tropical climates are found in northern Europe, Siberia, and northeastern North America. Great mountain ranges divide the oceans, the volcanoes known as the ring of fire border both the east and west coasts of the Pacific Ocean, and deep-ocean trenches are found adjacent to long island arcs. No single, coherent theory explained all these features, until the technology and scientific discoveries in the 1950s and early 1960s combined to trigger a complete reexamination of Earth's history.

In this chapter, we investigate Earth's interior, and we explore the history of the theory of plate tectonics, as well as the evidence that allowed this theory to become accepted. We review the research that continues to provide us with new insights into Earth's past, to comprehend and appreciate its present, and even to look forward into its future.

2.1 Earth's Interior

Earthquake Waves Reveal Earth's Layers

When earthquakes occur, energy is released that travels through Earth in the form of vibrations, or **seismic waves.** Most of what we know about Earth's interior has come from the recording and study of seismic waves. Two basic kinds of waves occur: surface waves and body waves.

Surface waves move over Earth's surface a little like ocean waves, although they travel much faster than ocean waves. These are the waves that generally do the most property damage in an earthquake, but they don't reveal much about what Earth is like deep beneath the surface.

Body waves typically cause less damage but they are far more useful in studying Earth's interior because they travel beneath the surface and through the "body" of the planet before returning to the surface some distance away. The speed and direction of body waves depend on the characteristics of the material they travel through, including changes in mineral or chemical composition, density, and changes in strength or physical state (solid, partially molten, or molten). As body waves travel from one type of material into another, their speed can change and they will change direction, or **refract**. Changes in speed and direction affect when and where a body wave will return to the surface. By carefully observing when and where body waves created by earthquakes all over the world are recorded, it is possible to create models of Earth's interior.

There are two kinds of body waves: primary-waves and secondary-waves. These two waves travel at different speeds and create different kinds of motion. Primary-waves are commonly known as **P-waves**. P-waves travel faster than any other kind of seismic wave and are the first waves to arrive at a location (fig. 2.1). P-waves are also known as compressional waves because they cause the material they pass through to alternately compress and stretch, parallel to the direction the waves travel. This is like the motion you can create by quickly pushing on the end of a stretched spring (fig. 2.2a). P-waves travel through all three states of matter: solid, liquid, and gas (sound travels through the air and the oceans as a compressional wave).

Secondary-waves are commonly known as **S-waves**. S-waves travel slower than P-waves but faster than surface waves. They are the second waves to arrive at a particular location (fig. 2.1). S-waves are also known as shear waves because they cause the material they pass through to shake from side-to-side, perpendicular to the direction they travel. This is the kind of motion that occurs if you shake a rope (fig. 2.2b). S-waves can only travel through solids; they cannot travel through liquids or gases.

Body waves travel downward from the location of an earthquake, reach some maximum depth, and then return to the surface. In general, the deeper they travel, the farther they will be from the earthquake when they return to the surface. If Earth was homogeneous, with constant density and no internal structure, body waves would travel straight paths and it would

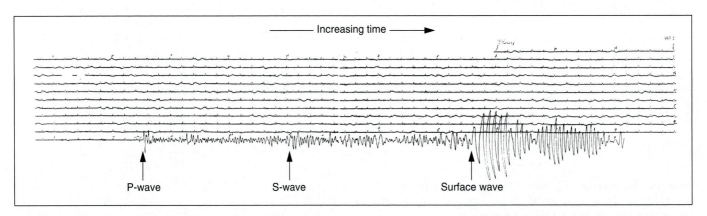

Figure 2.1 A seismogram of an earthquake that occurred in Taiwan recorded in Berkeley, California, 10,145 km (6300 mi) away. The faster body waves (P- and S-waves) arrive before the slower surface waves. Time increases from top to bottom and from left to right on the seismogram.

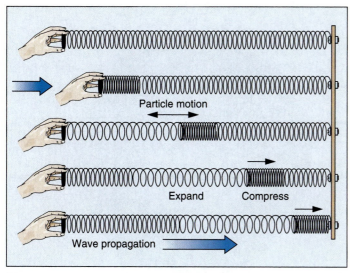

(a)

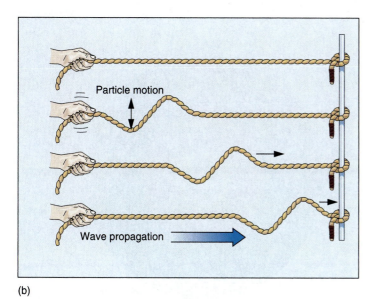

(b)

Figure 2.2 Particle motion in seismic waves. (a) P-wave motion can be illustrated with a sudden push on the end of a stretched spring. Vibration is parallel to the direction of propagation. (b) S-wave motion can be illustrated by shaking a rope to transmit a deflection along its length. Vibration is perpendicular to the direction of propagation.

be possible to record their arrival everywhere on the surface (fig. 2.3*a*). If Earth's density and rigidity increased smoothly with depth, body waves would travel curved paths but we would still be able to record their arrival everywhere on the surface (fig. 2.3*b*). However, we know from observation that body waves generated by an earthquake cannot be recorded everywhere at the surface. This is because density and rigidity are not constant and body waves encounter internal layers that have different physical and chemical properties. The areas on the surface where body waves do not travel directly are called shadow zones. The P-wave shadow zone is the area of Earth's surface from angular distances of roughly 104 to 140° from the location of the

earthquake (fig. 2.3*c*). It is caused by P-waves being refracted by the liquid outer core. The S-wave shadow zone is the area of Earth's surface at an angular distance greater than roughly 104° from the location of the earthquake (fig. 2.3*d*). The S-wave shadow zone results from S-waves being stopped entirely by the liquid outer core.

Scientists have used observations of where the P-wave and S-wave shadow zones appear, and how long it takes P-waves and S-waves to travel different distances from an earthquake, to create models of Earth's interior. These models demonstrate that Earth has a layered structure.

Earth's layered structure can be modeled two different ways. One way divides Earth into layers having different mineral and chemical compositions. The other way divides Earth into layers having different strengths and physical properties. These two models are both consistent with one another and both accurately describe different aspects of Earth's interior.

Model 1: Layers with Different Mineral and Chemical Compositions

Earth's interior can be described as consisting of three layers having different mineral and chemical compositions. From the surface downward, these are: the crust, the mantle, and the core. This model is illustrated on the right side of figure 2.4 and characteristics of the three layers are summarized in table 2.1.

Earth's surface consists of a thin, rocky layer called the **crust**. The crust comprises about 0.4% of Earth's mass and less than 1% of its volume. The crust is the product of partial melting at shallow depths in the mantle and volcanic eruptions throughout Earth's history. There are two kinds of crust: continental and oceanic. **Continental crust** is relatively low in density, about 2.7 g/cm³, and is generally about 30 to 50 km (20–30 mi) thick. Its composition and structure are highly variable. Continental crust consists primarily of the familiar rock **granite**, a light-colored rock rich in oxygen, silicon, sodium, potassium, and aluminum. **Oceanic crust** is denser than continental crust, about 2.9 g/cm³, and thinner, ranging from 5 to 10 km (3–6 mi) in thickness. Its composition and structure are relatively simple. Oceanic crust consists primarily of the dark-colored rock **basalt**. Compared to granite, basalt is poorer in oxygen and silicon, and rich in calcium, magnesium, and iron.

Beneath the crust is the **mantle**. Andrija Mohorovičić discovered the upper surface of the mantle in 1909. This surface is known as the **Mohorovičić discontinuity**, or simply the **Moho**. Like the crust, the mantle consists of rock. Mantle rock is denser than crustal rocks, ranging from about 3.2 to 5.6 g/cm³. The range in density of mantle rock is due primarily to the increase in pressure with depth rather than changes in rock type. Like basalt, mantle rock is relatively poor in oxygen and silicon but it is rich in magnesium and iron. The mantle is about 2900 km thick (1800 mi). It comprises about 84% of Earth's volume and roughly 68% of its mass.

At Earth's center is the **core**, discovered in 1906 by R. D. Oldham from his study of earthquake waves. The core has a radius of about 3480 km (2162 mi), making it larger than the

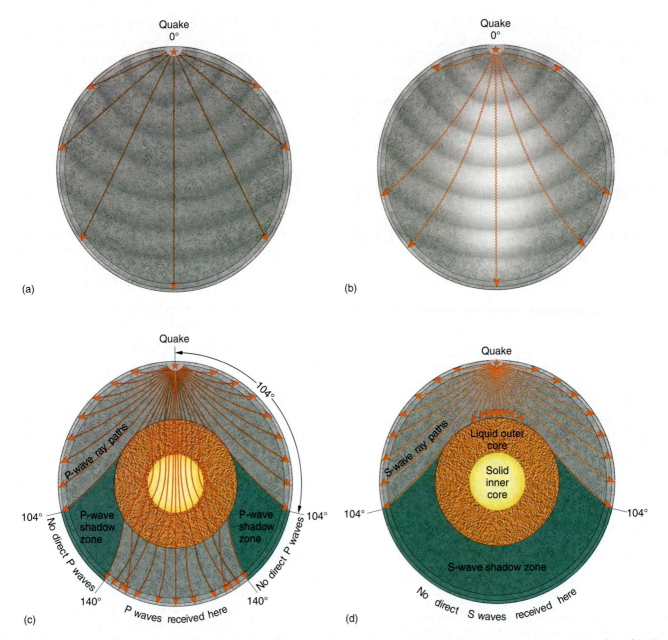

Figure 2.3 Movement of seismic waves through Earth. (a) Waves would move in straight lines and travel to all points on the surface if Earth was homogenous. (b) Waves would bend continuously, or refract, if Earth's interior steadily became more dense or rigid with depth. The waves would still travel to all points on Earth's surface. (c) Refraction of P-waves and P-wave shadow zone produced by Earth's internal layering. (d) Refraction of S-waves and S-wave shadow zone produced by the inability of S-waves to travel into the liquid outer core.

Table 2.1   Layers With Different Mineral and Chemical Compositions

Layer	Depth Range (km)	Layer Thickness (km)	Composition	Density (g/cm³)	Temperature (°C)
Crust					
Continental	0–50	40 (average)	Silicates rich in sodium, potassium, and aluminum	2.7	~0–1000
Oceanic	0–10	7 (average)	Silicates rich in calcium, magnesium, and iron	2.9	~0–1100
Mantle	Base of crust–2890	2866	Magnesium-iron silicates	3.2–5.6	1100–3200
Core	2890–6371	3481	Iron, nickel	9.9–13.1	3200–5500

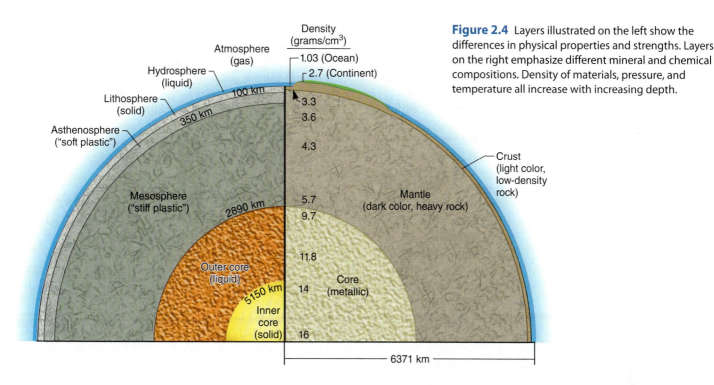

Figure 2.4 Layers illustrated on the left show the differences in physical properties and strengths. Layers on the right emphasize different mineral and chemical compositions. Density of materials, pressure, and temperature all increase with increasing depth.

Table 2.2 Layers With Different Strengths and Physical Properties

Layer	Depth Range (km)	Layer Thickness (km)	Characteristics
Lithosphere	0 to maximum ~100 in oceanic regions 0 to maximum ~150 in continental regions		Solid, rigid behavior
Asthenosphere	Base of lithosphere to ~350	Minimum of ~200	Solid, ductile behavior
Mesosphere	~350–2890	~2540	Solid, mobile
Outer core	2890–5150	2260	Liquid
Inner core	5150–6371	1221	Solid

planet Mars. The core is fundamentally different than the crust and mantle because it is metallic rather than rocky. The core accounts for about 15% of Earth's volume but over 30% of its mass because it consists of an iron alloy (about 90% iron and 10% nickel, sulfur, or oxygen, or some combination of these three elements) that is about five times as dense as the rock we walk on at the surface.

Model 2: Layers with Different Strengths and Physical Properties

Earth can also be modeled with five layers having different strengths and physical properties. From the surface downward, these are: the lithosphere, asthenosphere, mesosphere, outer core, and inner core. This model is illustrated on the left side of figure 2.4 and characteristics of the five layers are summarized in table 2.2. When rocks are subjected to a force that is less

than their breaking strength, they can behave in a rigid manner or a ductile manner, depending on their temperature and the pressure they are under. Rocks that behave rigidly do not permanently deform whereas rocks that behave in a ductile manner will deform, or flow, in response to the force.

The **lithosphere** (from the Greek word *lithos*, meaning "stone") is a rigid surface layer consisting of crust and shallow mantle rock fused together. It is the lithosphere, broken into roughly a dozen large pieces, that comprises the plates of plate tectonics, discussed later in this chapter.

In oceanic regions, the lithosphere thickens with increasing age of the sea floor. It reaches a maximum thickness of about 100 km (62 mi) at an age of 80 million years. In continental regions, the thickness of the lithosphere is slightly greater than in the ocean basins, varying from about 100 km (62 mi) beneath the geologically young margins of continents to about 150 km (93 mi) beneath old continental crust. The base of the lithosphere corresponds roughly to the region in the mantle where temperatures reach 650°C ± 100°C. At the higher temperatures found at these depths, mantle rock begins to lose some of its strength.

The lithosphere is underlain by a weak, deformable region in the mantle called the **asthenosphere** (from the Greek word *asthenes*, meaning "weak") where the temperature and pressure conditions lead to partial melting of the rock and loss of strength. The asthenosphere is often equated with a region of low seismic velocity called the low velocity zone (LVZ). Seismic waves travel more slowly through the asthenosphere, indicating that it may be as much as 1% melt. The asthenosphere behaves in a ductile manner, deforming and flowing slowly when stressed. It behaves roughly the way hot asphalt does.

The increase in pressure with increasing depth results in greater strength in the lower mantle, sometimes called the **mesosphere**, where the material is all solid but will convect slowly, moving upward in some regions and downward in others, because of temperature gradients and density differences. The depth to the base of the asthenosphere remains a matter of scientific debate. If it does correspond roughly to the LVZ, it would extend from the base of the lithosphere to about 350 km (217 mi). Some scientists believe it may be as shallow as 200 km (122 mi), while others think it may extend to depths of 700 km (435 mi). The lithosphere and underlying asthenosphere are shown in figure 2.5.

In this model, shown in the left side of figure 2.4, the core is divided into two regions: the **outer core** and the **inner core**. As early as 1926, studies of the tidal deformation of Earth made it clear that at least a portion of the core must behave like a fluid. In 1936, seismologist Inge Lehmann used earthquake data to establish the fact that the solid inner core is surrounded by a "fluid" outer core that S-waves cannot pass through, thus creating the S-wave shadow zone illustrated in figure 2.3*d*. Although it behaves like a fluid, the outer core may not be completely molten. It would behave like a fluid even if as much as 30% were composed of suspended metallic crystals. The inner core rotates about 1° per year faster than the mantle. It is thought that this increase in eastward rotation is caused by motion in the fluid outer core. The fluid motion in the outer core moves at a speed that is probably on the order of kilometers per year, generating Earth's magnetic field.

Detailed study of P-waves traveling in the vicinity of the mesosphere–outer core boundary has revealed that the upper surface of the outer core is not smooth but has peaks and valleys. These features extend as much as 11 km (7 mi) above and below the mean surface of the outer core. It is thought that the regions above a peak are areas where the mantle has excess heat and mantle rock rises, drawing the outer core upward. In other regions, cooler and denser mantle rock sinks and causes depressions in the outer core's surface. It is likely that these peaks and valleys last only as long as it takes a rising plume to lose its excess heat and sink back toward the outer core, perhaps a hundred million years.

Isostasy

The distribution of elevated continents and depressed ocean basins requires that a balance be kept between the internal pressures under the land blocks and those under the ocean basins. This is the principle of **isostasy**. The balance is possible because the greater thickness of low-density granitic crust in the continental regions is compensated for by the elevated higher-density mantle material under the thinner crust of the oceans (fig. 2.6). The situation is often compared to the floating of an iceberg. The top of the iceberg is above the sea surface, supported by the buoyancy of the displaced water below the surface. The deeper the ice extends below the surface, the higher the iceberg reaches above the water. The less dense continental land blocks float on the denser mantle in the same way, with most of the continental volume below sea level.

In other words, the asthenosphere offers buoyant support to a section of lithosphere sagging under the weight of a mountain range. Because the lithosphere is cooler, it is more rigid and stronger than the asthenosphere. If a thick section

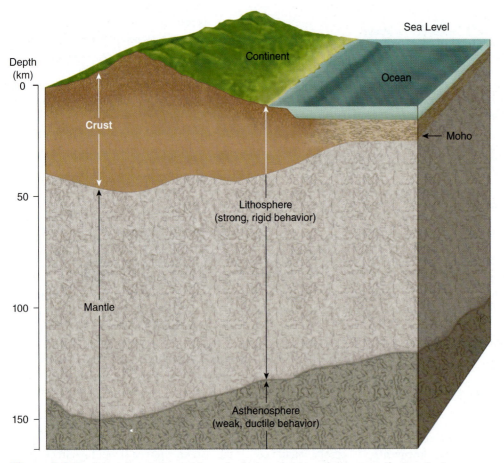

Figure 2.5 The lithosphere is formed from the fusion of crust and upper mantle. It varies in thickness; it is thinner in ocean basins and thicker in continental regions. The lithosphere rides on the weak, partially molten asthenosphere. Notice that the Moho is relatively close to Earth's surface under the ocean's basaltic crust but is depressed under the granitic continents.

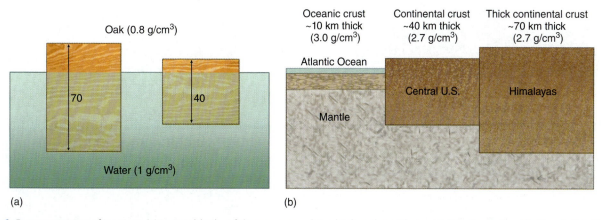

Figure 2.6 Demonstration of isostasy. (a) Larger blocks of the same wood rise higher above the water but also extend farther below the surface. (b) In the same way, thicker sections of crust below mountains have a deep crustal root that extends down farther into the mantle than do "normal" sections of continental crust.

of lithosphere has a large area and is mechanically strong, it depresses the asthenosphere slightly and is able to support a mountain range such as the Himalayas or the Alps. In other places where the crust is fractured and mechanically weak, a mountainous region must penetrate deeper into the mantle to provide buoyancy. The Andes are a mountain range with roots deep in the mantle.

If material is removed from or added to the continents, isostatic adjustment occurs. For example, parts of North America and Scandinavia continue to rise as the continents readjust to the lost weight of the ice sheets that receded at the close of the last ice age 10,000 years ago. Newly formed volcanoes protruding above the sea surface as islands often subside or sink back under the sea as their weight depresses the oceanic crust. The upper mantle gradually changes its shape in response to weight changes in the overlying, more rigid crust.

QUICK REVIEW

1. Through what types of materials will P-waves and S-waves travel?
2. How does continental crust differ from oceanic crust?
3. What and where is the Moho?
4. Describe the relationship between these two sets of layers: (a) crust and mantle, and (b) lithosphere, asthenosphere, and mesosphere.
5. Why is the crust thicker under mountain ranges?
6. How are the outer core and inner core similar? How are they different?

2.2 History of a Theory: Continental Drift

As world maps became more accurate, people became interested in the similarity in the shapes of continents on either side of the Atlantic Ocean. The possible "fit" of the east coast of South America and the west coast of Africa was noted by English scholar and philosopher Francis Bacon (1561–1626), French naturalist George Buffon (1707–88), German scientist and explorer Alexander von Humboldt (1769–1859), and others. In the 1850s, the idea was expressed that the Atlantic Ocean had been created by a separation of two landmasses during some unexplained cataclysmic event early in Earth's history.

As geologists continued to study Earth's crust, they saw greater similarities in patterns of rock formation, the distribution of plant and animal fossils, and the location of mountain ranges in the lands now separated by the Atlantic Ocean. In a series of volumes published between 1885 and 1909, Austrian geologist Edward Suess proposed that the southern continents had once been joined into a single continent he called **Gondwanaland**. He assumed that isostatic changes had caused portions of this continent to sink and create the oceans that now separate the southern continents. At the beginning of the twentieth century, Alfred Wegener and Frank Taylor independently proposed that the continents were slowly drifting about Earth's surface. Taylor soon lost interest, but Wegener, a German meteorologist, astronomer, and Arctic explorer, continued to promote this idea.

Wegener's theory, called **continental drift**, proposed the existence, at one point in Earth's history, of a single supercontinent called **Pangaea** (Greek, *pan*, "all"; *gaea*, "Earth, land") (fig. 2.7). This would result in a single world ocean he called **Panthalassa** (Greek, *pan*, "all"; *thalassa*, "ocean"). Wegener proposed that Pangaea broke apart, beginning about 200 million years ago, to form the continents as they are today. First, the northern portion composed of North America and Eurasia, which he called **Laurasia**, separated from the southern portion formed of Africa, South America, India, Australia, and Antarctica, for which he retained the earlier name Gondwanaland. The continents as we know them today then gradually separated from Laurasia and Gondwanaland and moved to their present positions. Wegener did not assume that the continents had always been present as a single landmass since the formation of Earth. He simply proposed a history of the movement of the continents since they had been joined together as Pangaea.

Pangaea, 250 MYA

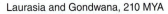

Laurasia and Gondwana, 210 MYA

Most modern continents had formed by 65 MYA

Figure 2.7 Continental drift. Wegener hypothesized that the supercontinent Pangaea, made up of a northern group of continents known as Laurasia and a southern group of continents known as Gondwana, broke up about 200–250 million years ago (MYA) and that the individual continents eventually drifted to their present positions.

Wegener based his theory of continental drift on a number of different observations including:

- **The geographic fit of the continents** is quite striking when the Atlantic Ocean is closed. Later work by Sir Edward Bullard in 1965 demonstrated that this fit is particularly good when the edges of the continents are defined as a water depth roughly midway between sea level and the deep ocean floor rather than the coastline (fig. 2.8).
- **Fossils of plants and animals** on Southern Hemisphere continents occur in linked bands (fig. 2.9a). The animals

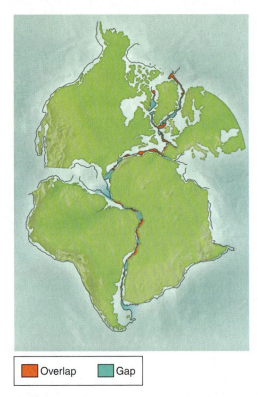

| Overlap | Gap |

Figure 2.8 The fit of the continents around the Atlantic Ocean using a water depth of 500 fathoms (~900 m, or ~3000 ft) as the edge of the continents, determined by Sir Edward Bullard in 1965.

were amphibians and other freshwater species that could not have crossed an open ocean from one continent to another. In particular, Wegener noted that fossils more than about 150 million years old collected on different continents were remarkably similar, implying land organisms were able to move freely from one location to another. Fossils less than about 150 million years old showed very different forms in different places, suggesting that the continents and their evolving organisms had separated from one another.

- **Mountain ranges of similar age, structure, and rock composition** in Northern Hemisphere continents align when the Atlantic is closed (fig. 2.9b).
- **Unusual sequences of rocks and rock units of similar age and chemistry** match along the coastlines of South America and Africa (fig. 2.9c).
- **Patterns of glaciation** found on Southern Hemisphere continents provide two kinds of evidence supporting the idea that they were once joined (fig. 2.9d). Wegener found identical glacial deposits on different continents that were all formed about 300 million years ago. These deposits indicate that some areas of these continents were once covered with thick ice sheets, as is the case on Antarctica today. This might have occurred if there had been a global ice age at that time, but the presence of coal deposits that formed in tropical swamps in the eastern United States and parts of Eurasia prove that wasn't the case. Wegener realized that the best explanation was that the southern continents were joined together and located near the South Pole. In addition, patterns of scratches on the rocks show the direction of movement of these ice sheets. With the present configuration of the continents, this

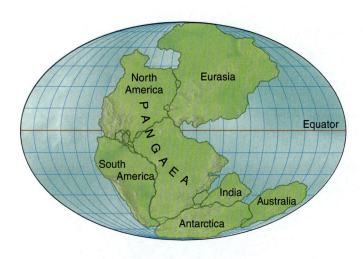

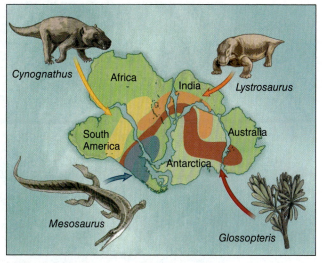

(a) Fossil distribution

(b) Match of mountain belts among North America, Europe, and Greenland

(c) Fit of continents, matching rock units

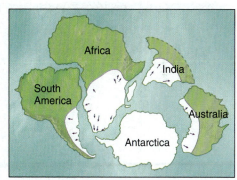

(d) Glacial deposits. Arrows illustrate direction of ice movement.

Figure 2.9 Wegener's evidence for Pangaea. Wegener used the similar shapes of some continental margins, matching rocks, mountains, and fossils, and evidence of glaciation on five continents, to infer that the continents must have been joined together as a single supercontinent between 300 and 200 million years ago. It broke apart about 200 million years ago. (a) Fossil distribution. (b) Match of mountain belts among North America, Europe, and Greenland. (c) Fit of continents, matching rock units. (d) Glacial deposits. Arrows illustrate direction of ice movement.

movement seems in many cases to be from the coast into the continental interior. This would suggest that the ice moved from sea level toward higher elevations, which would not be possible. The only reasonable explanation is that the continents were joined and the ice moved outward in different directions from the center of the ice sheet.

Wegener's theory provoked considerable debate in the 1920s. The most serious weakness in the theory was Wegener's inability to identify a mechanism that could cause the continents to break apart and drift through the ocean basins. When challenged on this point, he suggested that there were two possible driving mechanisms, centrifugal force from the rotation of Earth and the same tidal forces that raise and lower sea level. Both of these were quickly demonstrated to be too small to be capable of moving continental rock masses through the rigid basaltic crust of the ocean basins. With the lack of a viable driving mechanism, much of the scientific community remained skeptical of the continental drift theory, and little attention was given to it after Wegener's death.

QUICK REVIEW

1. Explain the evidence Wegener used to propose the theory of drifting continents.
2. What modern landmasses made up Pangaea?
3. What modern landmasses made up Laurasia and Gondwanaland?
4. What was the primary objection some scientists had to Wegener's theory?

2.3 Evidence for a New Theory: Seafloor Spreading

Armed with new sophisticated instruments and technologies developed during World War II, earth scientists were able to conduct detailed studies of the sea floor. In the 1950s, a worldwide effort was made to survey the sea floor, and for the first time,

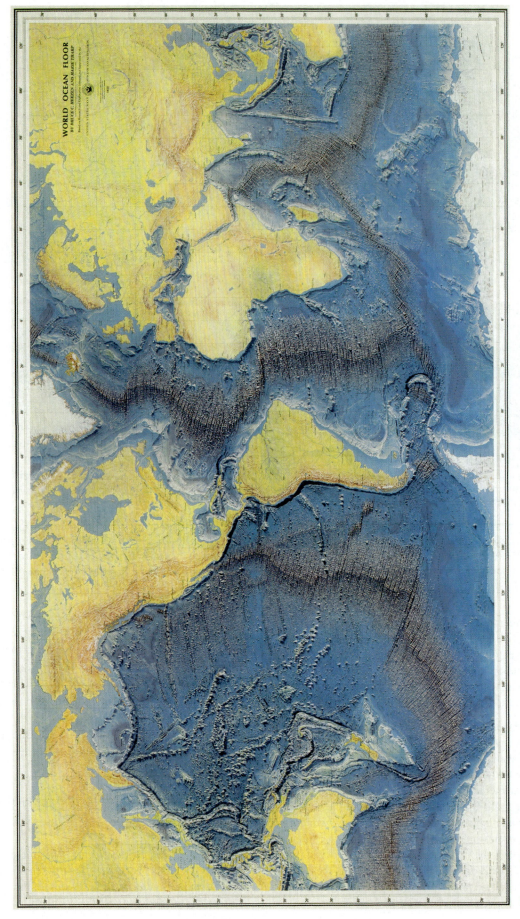

Figure 2.10 The first physiographic chart of the world's oceans, published in 1974. This map played a significant role in the acceptance of the theory of plate tectonics and greatly expanded public awareness and understanding of the major geologic features of the sea floor.

scientists were able to examine, in detail, the deep-ocean floor. These surveys clearly demonstrated that the deep-ocean floor is not a flat, featureless environment but includes steep-sided ocean trenches and a series of mountain ranges, periodically offset by long faults. The data from these surveys were used to construct the first physiographic map of the world's oceans, shown in figure 2.10. The characteristics of these major seafloor features are discussed in greater detail in chapter 3.

Wegener's theory of continental drift had not predicted the existence of such extensive mid-ocean features and could not explain them. However, a new theory that could explain the existence of deep trenches and long mountain ranges on the sea floor was advanced in the early 1960s by Harry H. Hess (1906–69) of Princeton University. Hess promoted the concept that deep within Earth's mantle there are large circulation cells of low-density molten material heated by Earth's natural radio-activity and heat from the core (fig. 2.11).

When the upward-moving mantle rock reaches the base of the lithosphere, it moves horizontally beneath the lithosphere, cooling and increasing in density. Hess suggested that the lithosphere rides atop the convecting mantle rock. At some point, the mantle rock is cool and dense enough to sink once again. These patterns of moving mantle material are called **convection cells** (fig. 2.11). There are two proposed models of mantle convection. Some scientists believe that two regions of convection exist, one cell confined to the upper mantle above a depth of 700 km (435 mi) and the other in the lower mantle. Other scientists believe that convection occurs throughout the entire mantle to the core-mantle boundary; this model is called whole-mantle convection.

In Hess's model, the upward-moving mantle rock would carry heat with it toward the surface. This heat would cause the overlying oceanic crust and shallow mantle rock to expand, thereby creating the mid-ocean ridge. Active volcanism along the axis of the ridge results in the extrusion of basaltic magma onto the sea floor, where it cools, hardens, and creates new sea floor and oceanic crust. If new sea floor is being produced in this manner, there must be some mechanism for removing old sea floor because no measurable change occurs in the total surface area of Earth. Hess proposed that the great, deep ocean trenches are areas where relatively old, cold, and dense oceanic lithosphere descends back into Earth's interior. Figure 2.11 shows this process of producing new lithosphere at the ridges and losing old lithosphere at the trenches in a system driven by the motion of convection cells.

Although some of the ascending molten material breaks through the crust and solidifies, most of the rising material is turned aside under the rigid lithosphere and moves away toward the descending sides of the convection cells, dragging pieces of the lithosphere with it. This lateral movement of the oceanic lithosphere produces **seafloor spreading** (fig. 2.11). Areas in which new sea floor and oceanic lithosphere are formed above rising magma are **spreading centers:** areas of descending older oceanic lithosphere are **subduction zones**. The seafloor spreading mechanism provides the forces causing movement of the lithosphere. The continents are not moving through the basalt of the sea floor; instead, they are being carried as passengers on the lithosphere, similar to boxes on a conveyor belt.

Hess's general model of seafloor spreading in which the sea floor is created along ocean ridges, moves away from the ridges, and eventually is subducted into the mantle at ocean trenches was correct. His proposal that there is convection in the mantle was also correct. However, his suggestion that mantle convection is the driving force for seafloor spreading is now believed to be wrong. The actual forces driving the motion of the sea floor are discussed in section 2.5.

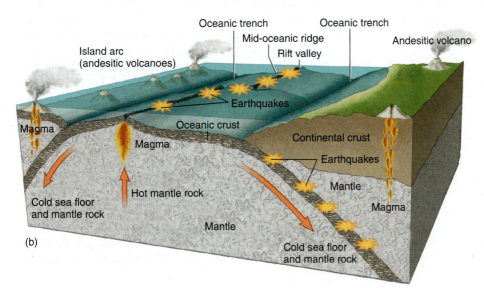

(a)

(b)

Figure 2.11 The model of mantle convection and seafloor spreading proposed by Harry Hess. (a) Hess proposed that convection extended throughout the mantle. (b) Rising hot mantle caused oceanic crust to elevate, forming a mid-ocean ridge. Ridges are characterized by high heat flow, shallow earthquakes, and active basaltic volcanism that creates new sea floor. Sea floor diverges at the ridge. As it moves away from the ridge it cools, thickens, and becomes denser. Cold, dense sea floor sinks back into the mantle at ocean trenches, producing deep earthquakes and andesitic volcanism.

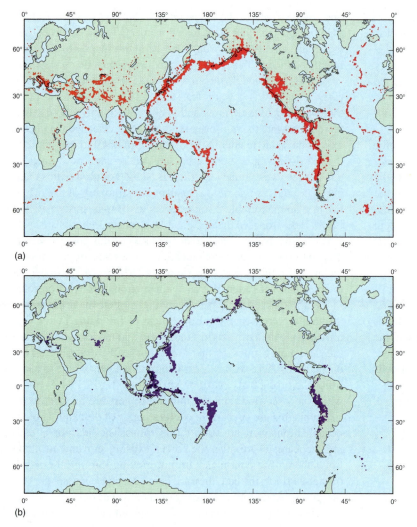

(a)

(b)

Figure 2.12 Earthquake epicenters, 1961–67. (a) Epicenters of earthquakes with depths less than 100 km (60 mi) outline regions of crustal movement. (b) Epicenters of earthquakes with depths greater than 100 km are related to subduction.

Figure 2.13 Recovering a heat flow probe from the deep-sea floor. The probe consists of a cylinder holding the measuring and recording instruments above a thin probe that penetrates the sediment. Thermisters (devices that measure temperature electronically) at the top and bottom of the probe measure the temperature difference over a fixed distance equal to the length of the probe. This temperature change with depth is used to compute the heat flow through the sediment.

Evidence for Crustal Motion

Additional evidence was needed to support the idea of seafloor spreading. As oceanographers, geologists, and geophysicists explored Earth's crust on land and under the oceans, evidence began to accumulate.

Epicenters of earthquakes were known to be distributed around Earth in narrow and distinct zones. Epicenters are the points on Earth's surface directly above the actual earthquake location, which is called the earthquake **focus** or **hypocenter**. These zones were found to correspond to the areas along the ridges, or spreading centers, and the trenches, or subduction zones. Earthquakes that are shallower than 100 km (60 mi) are prevalent in the ocean basins along ridges and rises and at trenches where the lithosphere bends and fractures as it is subducted into the mantle. These relatively shallow events are also found in continental regions that are undergoing significant deformation. Earthquakes deeper than 100 km are generally associated with the subduction of oceanic lithosphere.

The deeper the earthquakes are, the farther they are displaced inland away from the trench beneath continents or island arcs (fig. 2.12).

Researchers sank probes into the sea floor to measure the heat flowing from Earth's interior through the oceanic crust (fig. 2.13). The measured heat flow clearly shows a regular pattern related to distance from the crest of the ocean ridges (fig. 2.14). It is highest in the vicinity of the mid-ocean ridges over the ascending portion of a mantle convection cell, where the crust is more recent and thinner, and it decreases as the distance from the ridge crest and the crustal age increases (fig. 2.15).

Radiometric age dating of rocks from the land and sea floor shows that the oldest rocks from the oceanic crust are only about 200 million years old, whereas rocks from the continents are much older. The sea floor formed by convection-cell processes at the spreading centers is young and short-lived, for it is lost at subduction zones, where it plunges back down into the mantle.

Vertical, cylindric samples, or **cores**, were obtained by drilling through the sediments that cover the ocean bottom and into the rock of the ocean floor. Drilling cores through, on average, 500–600 m (1600–2000 ft) of sediments and into the ocean floor rock required the development of a new technology and a new kind of ship. In the late summer of 1968, the specially constructed drilling ship *Glomar Challenger* (see fig. P.18a) was used for a series of studies. Its specialized bow and stern thrusters and its propulsion system responded automatically to computer-controlled navigation, using acoustic beacons on the sea floor. This setup enabled the ship to remain for long periods of time in

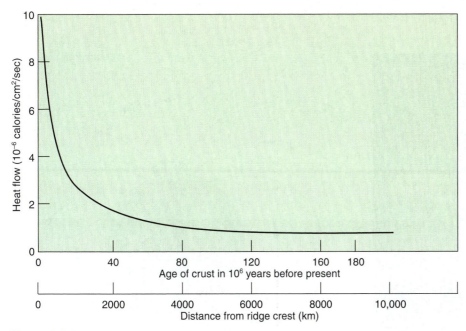

Figure 2.14 Heat flow through the Pacific Ocean floor. Values are shown against age of crust and distance from the ridge crest.

recovering a total of 96 km (60 mi) of deep-sea cores for study. A new deep-sea drilling program with a new vessel, the *JOIDES Resolution* (fig. 2.17), began in 1985. The drill pipe comes in sections 9.5 m (31 ft) long that can be screwed together to make a single pipe up to 8200 m (27,000 ft) long.

The cores taken by the *Glomar Challenger* provided much of the data needed to establish the existence of seafloor spreading. No ocean crust older than 180 million years was found, and sediment age and thickness were shown to increase with distance from the ocean ridge system (fig. 2.18). Note that the sediments closest to the ridge system are thin over the new crust, which has not had as long to accumulate its sediment load. The crust farther away from the ridge system is older and is more heavily loaded with sediments.

Although many pieces of evidence fit the theory that Earth's crust produced at the ridge system is new and young, the most elegant proof for seafloor spreading came from a study of the magnetic evidence locked into the ocean's floors.

The familiar geographic North (90°N) and South (90°S) Poles mark the axis about which Earth rotates. Earth also has a magnetic field similar to what would exist if there was a giant bar magnet, tilted roughly 11.5° away from the axis of rotation, embedded in its interior (fig. 2.19).

A magnetic field like Earth's, with two opposite poles, is called a **dipole**. Earth's magnetic field consists of invisible lines of magnetic force that are parallel to Earth's surface at the magnetic

Figure 2.15 Distribution of global heat flow in milliwatts per square meter. Heat flow in the ocean basins is highest in the vicinity of oceanic ridges and decreases with distance from the ridges.

a nearly fixed position over a drill site in water too deep to anchor. An acoustic guidance system made it possible to replace drill bits and place the drill pipe back in the same bore holes in water about 6000 m (20,000 ft) deep. This method is illustrated in figure 2.16.

In 1983, the *Glomar Challenger* was retired, after logging 600,000 km (375,000 mi), drilling 1092 holes at 624 drill sites, and

Figure 2.16 Deep-ocean drilling technique. Acoustical guidance systems are used to maneuver the drilling ship over the bore hole and to guide the drill string back into the bore hole.

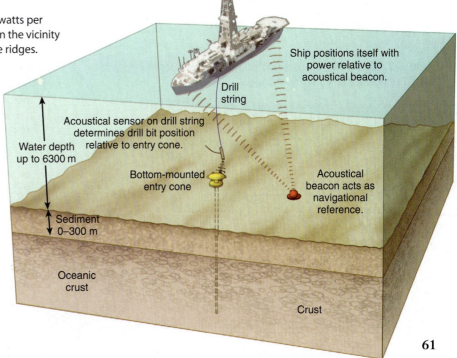

Figure 2.17 The deep-sea drilling vessel *JOIDES Resolution.*

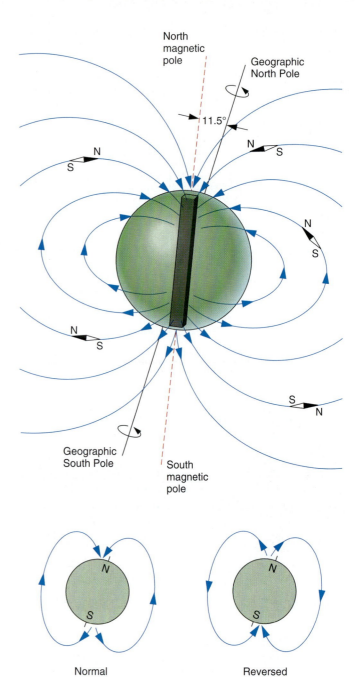

Figure 2.19 Lines of magnetic force surround Earth and converge at the magnetic poles. A freely suspended magnet aligns itself with these lines of magnetic force, with the north-seeking end of the magnet pointing to the north magnetic pole and the south-seeking end pointing to the south magnetic pole. The magnet hangs parallel to Earth's surface at the magnetic equator and dips at an increasing angle as it approaches the magnetic poles. N and S in the two small figures indicate the *geographic* poles.

equator and converge and dip toward Earth's surface at the magnetic poles. A small, freely suspended magnet (such as a compass needle) will align itself with these lines of magnetic force, with the north-seeking end pointing to the north magnetic pole and the south-seeking end pointing to the south magnetic pole. In addition, a suspended magnet dips toward Earth's surface by differing amounts depending on its distance from the magnetic equator. At the magnetic equator a suspended magnet would be horizontal, or parallel to Earth's surface. In the north magnetic hemisphere, the north-seeking end of the magnet would point downward at an increasing angle until it pointed vertically downward at the north magnetic pole. In the south magnetic hemisphere, the north-seeking end of the magnet would point upward

Figure 2.18 Variation in age and thickness of seafloor sediments with distance from the mid-ocean ridge.

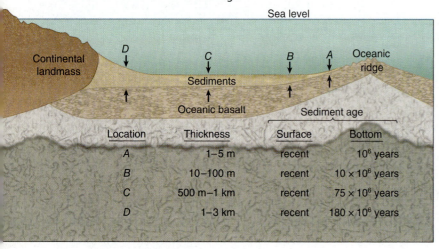

Location	Thickness	Surface	Bottom
A	1–5 m	recent	10^6 years
B	10–100 m	recent	10×10^6 years
C	500 m–1 km	recent	75×10^6 years
D	1–3 km	recent	180×10^6 years

at an increasing angle until it pointed vertically upward at the south magnetic pole. The task of locating the magnetic poles is difficult for many reasons: the area over which the dip is nearly 90° is often relatively large, the pole can move because of daily variations and magnetic storms, and it is difficult for survey crews to work in polar regions.

The location of the magnetic north pole was first determined in 1831 by Sir James Clark Ross. Its position didn't change significantly until 1904 when it began to move northeast at about 14 km (9 mi) per year. The speed began to increase and the direction of motion changed to northwest in about 1998. Currently, the magnetic north pole is moving to the northwest at about 60 km (37 mi) per year. The location for the north magnetic pole in 2010 was approximately 84.7°N, 129.1°W, just north of the Sverdrup Islands in the Canadian Arctic. The location of the south magnetic pole in 2010 was approximately 64.0°S, 137.2°E, near the coast of Wilkes Land in Antarctica.

Most igneous rocks contain particles of a naturally magnetic iron mineral called magnetite. Particles of magnetite act like small magnets. These particles are particularly abundant in basalt, the type of rock that makes up the oceanic crust. When basaltic magma erupts on the sea floor along ocean ridges, it cools and solidifies to form basaltic rock. During this time, the magnetite grains in the rock become magnetized in a direction parallel to the existing magnetic field at that time and place. When the temperature of the rock drops below a critical level called the **Curie temperature**, or Curie point (named in honor of the twice Nobel Prize–winning physicist Marie Curie), roughly 580°C, the magnetic signature (magnetic strength and orientation) of Earth's field is "frozen" into these magnetite grains, creating a "fossil" magnetism that remains unchanged unless the rock is heated again to a temperature above the Curie temperature. In this manner, rocks can preserve a record of the strength and orientation of Earth's magnetic field at the time of their formation. The investigation of fossil magnetism in rocks is the science of **paleomagnetism**.

Research on age-dated layers of volcanic rock found on land shows that the polarity, or north-south orientation, of Earth's magnetic field reverses for varying periods of geologic time. Thus, at different times in Earth's history, the present north and south magnetic poles have changed places. Each time a layer of volcanic material cooled and solidified, it recorded the magnetic orientation and polarity of Earth's field at that time. The dating and testing of samples taken through a series of volcanic layers have enabled scientists to build a calendar of these events (fig. 2.20). During these **polar reversals,** Earth's magnetic field gradually decreases in strength by a factor of about ten over a period of a few thousand years. Some evidence exists that during a reversal, Earth's field may not remain a simple dipole. The collapse of the magnetic field during reversals enables more intense cosmic radiation to penetrate the planet's surface, and recent work has suggested a correlation between such periods and a decrease in populations of delicate, single-celled organisms that live in the surface layers of the ocean. Nearly 170 reversals have been identified during the last 76 million years, and the present magnetic orientation has existed for 780,000 years. The interval between reversals has varied, with some notable periods of constant orientation for long intervals. During the Cretaceous period (from roughly 120 million to 80 million years ago), the field was stable and normally polarized. The field was stable and reversely polarized for about 50 million years in the Permian period, roughly 300 million years before present. The cause of reversals is unknown but is thought to be associated with changes in the

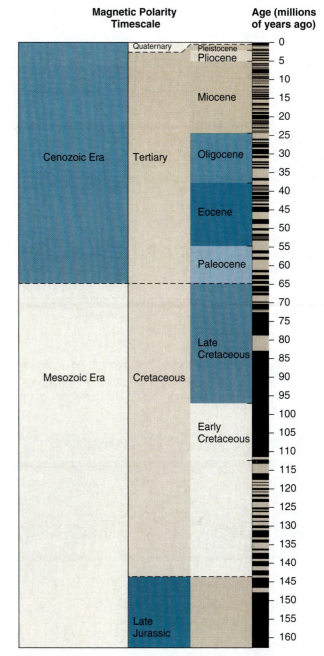

Figure 2.20 Worldwide magnetic polarity time scale for the Cenozoic and Mesozoic eras. *Black* indicates positive anomalies (and therefore normal polarity). *Tan* indicates negative anomalies (reverse polarity).

motion of the material in Earth's liquid outer core. How long our current period of normal polarity will last we do not know.

During the 1950s, oceanographers began to measure the strength of the magnetism of the oceanic crust during research cruises by towing marine magnetometers behind the ship, away from the magnetic noise of the vessel. From 1958–61, the results of several marine magnetic surveys were published as maps of high and low magnetic intensity that resembled the ridges and valleys of a topographic map. These alternating areas of high and

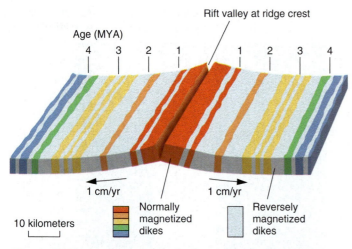

Figure 2.21 Reversals in the polarity of Earth's magnetic field cause a symmetrically striped pattern of magnetic anomalies on the sea floor. Correlation of these anomalies with the ages of known magnetic reversals allows the anomalies to be dated, revealing the age of the sea floor and the rate of seafloor spreading.

low intensity came to be known as "magnetic stripes." The stripes varied in width from a few kilometers to many tens of kilometers and were as much as several thousand kilometers long. When the stripes were compared to marine bathymetry, it was noted that the maps revealed a mirror image pattern of roughly parallel magnetic stripes on each side of the mid-ocean ridge (fig. 2.21). The significance of this pattern was not understood until 1963, when F. J. Vine and D. H. Matthews of Cambridge University proposed that these stripes represented a recording of the polar reversals of the vertical component of Earth's magnetic field, frozen into the rocks of the sea floor. As the molten basalt rose along the crack of the ridge system and solidified, it locked in the direction of the prevailing magnetic field. As seafloor spreading moved this material to both flanks of the ridge, it was replaced at the ridge crest by more molten materials. Each time Earth's magnetic field reversed, the direction of the magnetic field was recorded in the new crust. Vine and Matthews proposed that if such were the case, there should be a symmetric pattern of magnetic stripes centered at the ridges and becoming older away from the ridges. Their ideas were confirmed; the polarity and age of these stripes corresponded to the same magnetic field changes found in the dated layers on land. Their discovery dramatically demonstrated seafloor spreading.

The magnetic stripe data in the world's oceans has been used to produce a map of the age of the sea floor (fig. 2.22). Seafloor age increases away from the oceanic spreading centers, providing another verification of seafloor spreading. The present ocean basins are young features, created during the past 200 million years, or the last 5% of Earth's history.

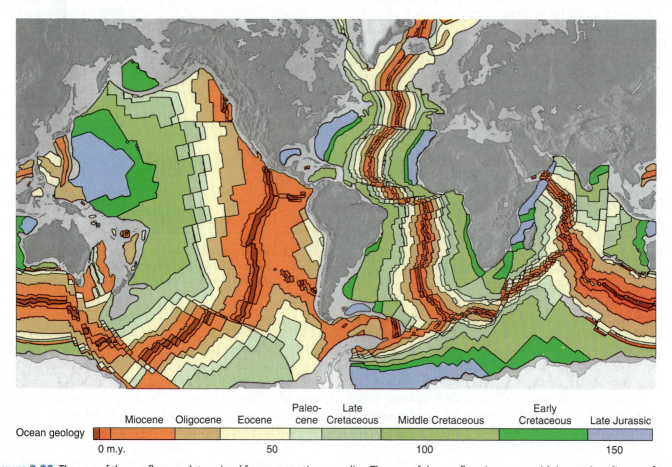

Figure 2.22 The age of the sea floor as determined from magnetic anomalies. The age of the sea floor increases with increasing distance from the mid-ocean ridge as a result of seafloor spreading. Oceanic fracture zones offset the ridge and age patterns.

QUICK REVIEW

1. What causes convection cells in Earth's mantle?
2. In the Hess model, how do convection cells cause seafloor spreading?
3. How do earthquake patterns, seafloor heat flow, age of Earth's crust, and thickness of seafloor sediments help to support the idea of seafloor spreading?
4. What is a magnetic reversal, and how is it recorded in seafloor rock?
5. How do the patterns of magnetic reversals in seafloor rocks verify seafloor spreading?

2.4 Plate Tectonics

The theory of **plate tectonics** incorporates the ideas of continental drift and seafloor spreading in a unified model. Drifting continents and the motion of the sea floor both result from the fragmentation and movement of the outer rigid shell of Earth called the lithosphere. The lithosphere is fragmented into seven major plates along with a number of smaller ones (fig. 2.23 and table 2.3). The plates are outlined by the major

Table 2.3   Approximate Plate Area (10^6 km^2)

Major Plates	
Pacific	105
African	80
Eurasian	70
North American	60
Antarctic	60
South American	45
Australian	45
Smaller Plates	
Nazca	15
Indian	10
Arabian	8
Philippine	6
Caribbean	5
Cocos	5
Scotia	5
Juan de Fuca	2

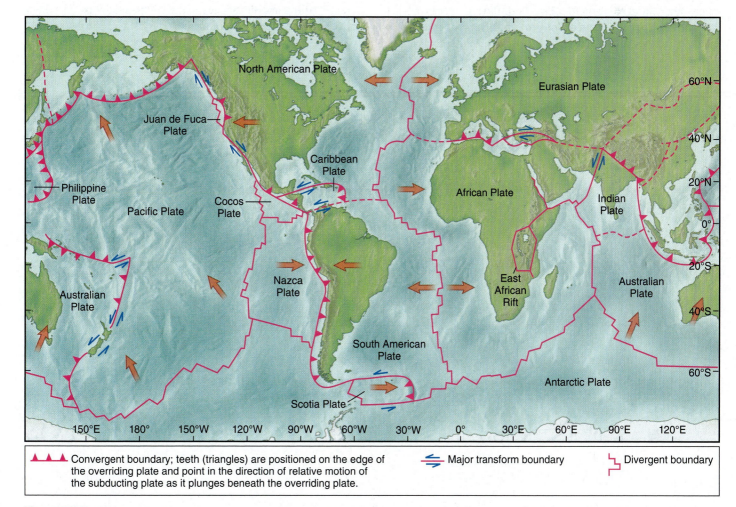

▲▲▲ Convergent boundary; teeth (triangles) are positioned on the edge of the overriding plate and point in the direction of relative motion of the subducting plate as it plunges beneath the overriding plate.

⇌ Major transform boundary   ⌐ Divergent boundary

Figure 2.23 Earth's lithosphere is broken into a number of individual plates. The edges of the plates are defined by three different types of plate boundaries. *Arrows* indicate direction of plate motion.

earthquake belts of the world (see fig. 2.12); this relation was first pointed out in 1965 by J. T. Wilson, a geophysicist at the University of Toronto.

Plates and Their Boundaries

Each lithospheric plate consists of the upper roughly 80–100 km (50–62 mi) of rigid mantle rock capped by either oceanic or continental crust. Lithosphere capped by oceanic crust is often simply called oceanic lithosphere, and, in a similar fashion, lithosphere capped by continental crust is referred to as continental lithosphere. Some plates, such as the Pacific Plate, consist entirely of oceanic lithosphere, but most plates, such as the South American Plate, consist of variable amounts of both oceanic and continental lithosphere with a transition from one to the other along the margins of continents. The plates move with respect to one another on the ductile asthenosphere below. As the plates move, they interact with one another along their boundaries, producing the majority of Earth's earthquake and volcanic activity.

The three basic kinds of plate boundaries are defined by the type of relative motion between the plates. Each of these boundaries is associated with specific kinds of geologic features and processes (table 2.4 and fig. 2.24). Plates move away from each other along **divergent plate boundaries**. Ocean basins and the sea floor are created along marine divergent boundaries marked by the mid-ocean ridges and rises, while continental divergent boundaries are often marked by deep rift zones. Plates move toward one another along **convergent plate boundaries**. Marine convergent boundaries are associated with deep-ocean trenches, the destruction of sea floor, and the closing of ocean basins, while continental convergent boundaries are associated with the creation of massive mountain ranges. The third type of plate boundary occurs where two plates are neither converging nor diverging but are simply sliding past one another. These are called **transform boundaries** and are marked by large faults called **transform faults**.

Divergent Boundaries

Ocean basins are created along divergent boundaries that break apart continental lithosphere (fig. 2.25). A series of successive divergent boundaries were responsible for the breakup of Pangaea, producing the individual continents and ocean basins we see today. Upwelling of hot mantle rock is thought to be responsible for the creation of divergent boundaries in continental lithosphere. Mantle upwelling heats the base of the lithosphere, thinning it and causing it to dome upward (fig. 2.25a). This weakens the lithosphere and produces extensional forces and stretching that lead to faulting and volcanic activity. Faulting along the boundary creates rifting and subsidence in the continental crust, forming a **rift zone** along the length of the boundary (fig. 2.25b). A present-day example of this early rifting is the continuing development

Table 2.4   Types and Characteristics of Plate Boundaries

Plate Boundary	Types of Lithosphere	Geologic Process	Geologic Feature	Earthquakes	Volcanism	Examples
Divergent (move apart)	Ocean-Ocean	New sea floor created, ocean basin opens	Mid-ocean ridge	Yes, shallow	Yes	Mid-Atlantic Ridge, East Pacific Rise
	Continent-Continent	Continent breaks apart, new ocean basin forms	Continental rift, shallow sea	Yes, shallow	Yes	East African Rift, Red Sea, Gulf of Aden, Gulf of California
Convergent (move together)	Ocean-Ocean	Old sea floor destroyed by subduction	Ocean trench	Yes, shallow to deep	Yes	Aleutian, Mariana, and Tonga Trenches (Pacific Ocean)
	Ocean-Continent	Old sea floor destroyed by subduction	Ocean trench	Yes, shallow to deep	Yes	Peru-Chile and Middle America Trenches (Eastern Pacific Ocean)
	Continent-Continent	Mountain building	Mountain range	Yes, shallow to intermediate	No	Himalaya Mountains, Alps
Transform (slide past each other)	Ocean	Sea floor conserved (neither created nor destroyed)	Transform fault (offsets segments of ridge crest)	Yes, shallow	No	Mendocino and Clipperton (Eastern Pacific Ocean)
	Continent	Sea floor conserved (neither created nor destroyed)	Transform fault (offsets segments of ridge crest)	Yes, shallow	No	San Andreas Fault, Alpine Fault (New Zealand), North and East Anatolian Faults (Turkey)

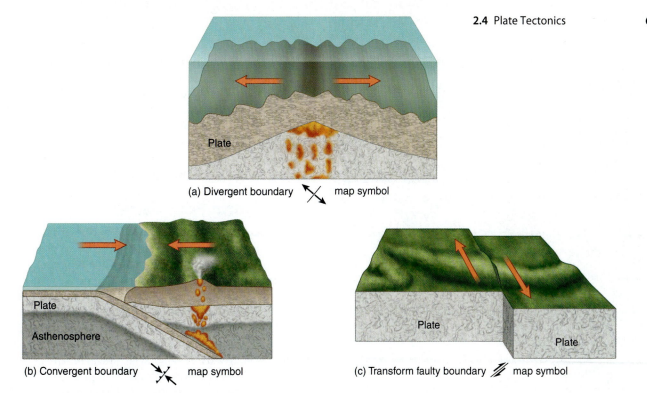

(a) Divergent boundary   map symbol

(b) Convergent boundary   map symbol

Plate

Asthenosphere

(c) Transform faulty boundary   map symbol

Plate

Plate

Figure 2.24 The three basic types of plate boundaries include (a) divergent boundaries where plates move apart, (b) convergent boundaries where plates collide, and (c) transform boundaries where plates slide past one another.

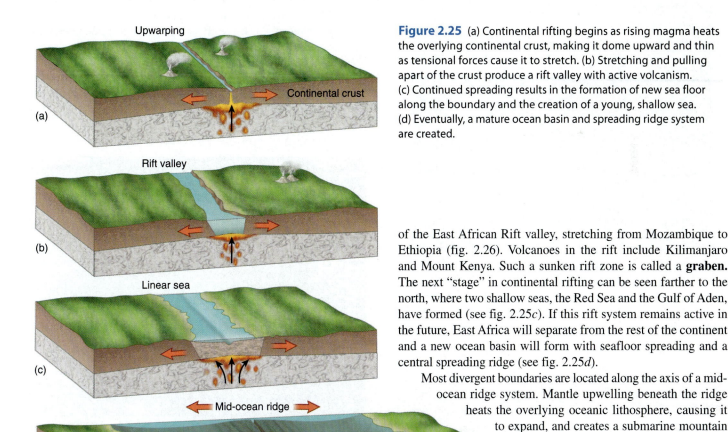

Upwarping

Continental crust

(a)

Rift valley

(b)

Linear sea

(c)

Mid-ocean ridge

Rift

Continental crust   Oceanic crust

(d)

Figure 2.25 (a) Continental rifting begins as rising magma heats the overlying continental crust, making it dome upward and thin as tensional forces cause it to stretch. (b) Stretching and pulling apart of the crust produce a rift valley with active volcanism. (c) Continued spreading results in the formation of new sea floor along the boundary and the creation of a young, shallow sea. (d) Eventually, a mature ocean basin and spreading ridge system are created.

of the East African Rift valley, stretching from Mozambique to Ethiopia (fig. 2.26). Volcanoes in the rift include Kilimanjaro and Mount Kenya. Such a sunken rift zone is called a **graben.** The next "stage" in continental rifting can be seen farther to the north, where two shallow seas, the Red Sea and the Gulf of Aden, have formed (see fig. 2.25c). If this rift system remains active in the future, East Africa will separate from the rest of the continent and a new ocean basin will form with seafloor spreading and a central spreading ridge (see fig. 2.25d).

Most divergent boundaries are located along the axis of a mid-ocean ridge system. Mantle upwelling beneath the ridge heats the overlying oceanic lithosphere, causing it to expand, and creates a submarine mountain range. As the plates on either side of the boundary move apart, cracks form along the crest of the ridge, allowing molten rock to seep up from the mantle and flow onto the ocean floor. In the early stages of

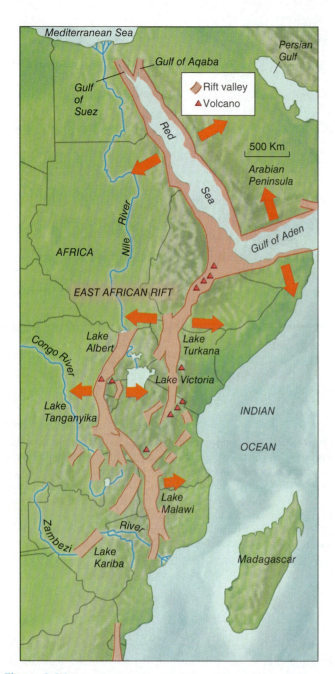

Figure 2.26 An active continent-continent divergent boundary, still in its early stage of development, has created the East African Rift system. This same boundary is more fully developed to the north, where it has created two shallow seas: the Red Sea and the Gulf of Aden.

form new ocean crust along the edges of each diverging plate. The thickness of the oceanic crust remains relatively constant from the time of its formation at the ridge. However, as the two plates diverge, the mantle rock immediately beneath the crust cools, fuses to the base of the crust, and behaves rigidly, causing the thickness of the oceanic lithosphere to increase with age and distance from the ridge. Plates move apart at different rates along the length of the ridge system.

Transform Boundaries

The ocean ridge system is divided into segments that are separated by transform faults (fig. 2.28). As two plates move away from the ridge crest, they simply slide past one another along the transform fault. In this manner, the ridge crest segments and transform faults form a continuous plate boundary that alternates between being a divergent boundary and a transform boundary. Differences in the age and temperature of the plates across a transform boundary can create significant and rapid changes in the elevation of the sea floor called **escarpments.** These changes in elevation are propagated into each plate by seafloor spreading, where they are preserved as "fossil" transform faults. These fossil faults, along with the active transform fault between the two segments of ridge crest, form a continuous linear feature called a **fracture zone.** The longest fracture zones can be up to 10,000 km (6200 mi) long, extending deep into each plate on either side of the ridge. It is important to remember that no movement or earthquake activity occurs along the fracture zone where it extends into either plate. However, there is relative motion and seismicity along the transform fault between adjacent segments of ridge crest. The direction of motion of the plates on either side of a transform fault is determined by the

Figure 2.27 Pillow lavas are mounds of elongated lava "pillows" formed by repeated oozing and quenching of extruding basalt magma on the sea floor. First, a flexible, glassy crust forms around the newly extruded lava, forming an expanded pillow. Next, pressure builds within the pillow until the crust breaks and new basalt extrudes like toothpaste, forming another pillow.

a volcanic eruption, basaltic magma can flow rapidly onto the sea floor in relatively flat flows called sheet flows that may extend several kilometers away from their source. Basaltic magma has a low viscosity, allowing it to flow freely, because the magma is very hot and has a relatively low content of silicon dioxide. As the eruption rate decreases, the magma is often extruded more slowly onto the sea floor, creating rounded flows called pillow lavas, or **pillow basalts** (fig. 2.27). The magma solidifies to

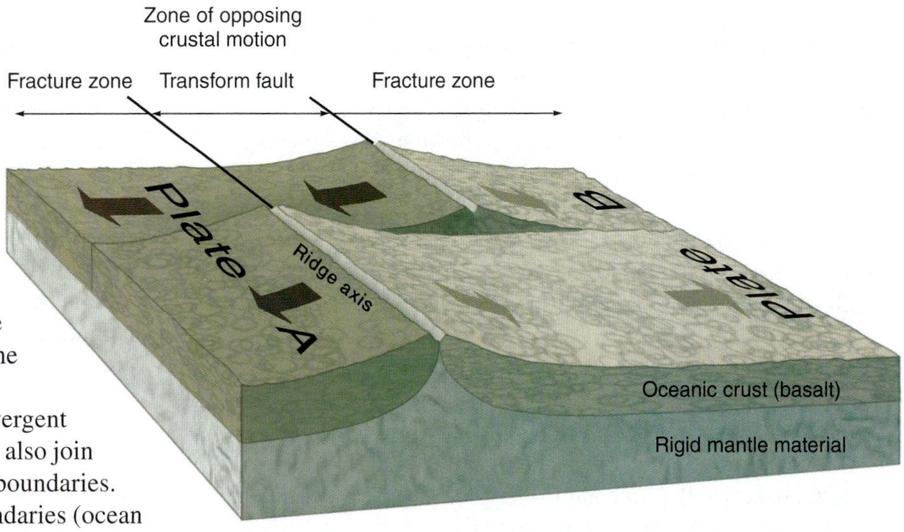

Figure 2.28 Transform boundaries are marked by transform faults. Transform faults usually offset segments of ocean ridges (divergent boundaries). Plates slide past one another along transform boundaries. The direction of motion is opposite the sense of displacement of the ridge segments.

direction of seafloor spreading in each plate and is opposite the sense of displacement of the ridge segments.

While most transform faults join two divergent boundaries (segments of ridge crest), they may also join different combinations of other types of plate boundaries. Transform faults can join two convergent boundaries (ocean trenches), as they do between the Caribbean and American Plates, or a convergent boundary and a divergent boundary, as they do on either side of the Scotia Plate (see fig. 2.23). The diverse settings in which transform faults play a role in marking the boundaries of plates can be seen along the west coast of the United States (fig. 2.29). Much of the boundary between the

North American and Pacific Plates is marked by a long transform fault; its southern portion is known as the San Andreas Fault, cutting through continental crust from the Gulf of California to Cape Mendocino, California. Offshore of Cape Mendocino, a transform fault known as the Mendocino Fault links a spreading center and a trench, forming a boundary between the Pacific and Juan de Fuca Plates. Farther to the north, additional transform faults offset segments of the Juan de Fuca Ridge.

The San Andreas Fault allows crustal sections on either side to move horizontally relative to each other. The land on the Pacific side, including the coastal area from San Francisco to the tip of Baja California, is actively moving northward with respect to the land on the east side of the fault system. The motion is not uniform; when the accumulated stress exceeds the strength of the rock, a sudden movement and displacement occur along the fault, causing an earthquake. Movement along this fault system caused the famous 1906 and 1989 (Loma Prieta) earthquakes in the San Francisco Bay area.

The reason for the many transform faults that are found associated with the ridge system is related to changes in the speed and direction of the plates as they move apart on a spherical surface. Variations in the strength or location of convection cells and collisions between sections of lithosphere may also result in transform faults.

Convergent Boundaries

Plates move toward each other along convergent plate boundaries. The specific kinds of geologic features and processes that occur along convergent boundaries depend on whether the colliding edges of the plates consist of oceanic or continental lithosphere. This is because of the difference in density between oceanic and continental lithosphere.

Roughly 20%–40% of the thickness of continental lithosphere consists of continental crust. Because the density of continental crust is relatively low compared to the density of the mantle (see table 2.1), continental lithosphere is buoyant and will not subduct into the mantle. The density of oceanic lithosphere

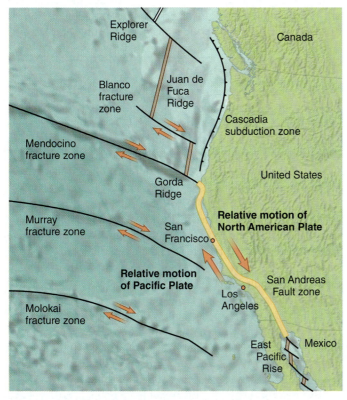

Figure 2.29 Transform boundaries can occur in continental or oceanic crust. In addition, they can join together either divergent or convergent boundaries. The San Andreas Fault cuts through continental crust, joining the northern end of the East Pacific Rise in the Gulf of California with the Gorda Ridge off the coast of northern California. Fracture zones, extensive inactive transform faults, often occur in association with active tranform boundaries.

is greater than the density of continental lithosphere and it increases with age as seafloor spreading carries the oceanic lithosphere away from the ridge crest, and the plate cools and thickens. The thickness of oceanic lithosphere increases at a rate proportional to the square root of age, as shown by the following equation and illustrated in figure 2.31.

$$\text{Thickness (km)} = 10 \times \sqrt{\text{(age in millions of years)}}$$

There are three possible kinds of convergent plate boundaries:

- **Ocean-ocean convergence** occurs when the edges of both colliding plates are composed of oceanic lithosphere (fig. 2.30*a*). Along this type of convergent boundary, one plate (often referred to as the downgoing plate) will subduct beneath the other plate (called the overriding plate). An ocean trench will form along the boundary. Shallow earthquakes will occur along the plate boundary; deeper earthquakes will occur in the downgoing plate, creating a dipping zone of seismicity called a **Wadati-Benioff zone** in honor of the geophysicists who first studied them. The deepest earthquakes in the world occur in Wadati-Benioff zones extending to depths of as much as 650 km (400 mi) in the western Pacific Ocean. As the downgoing plate descends deeper into the mantle, it will heat up and begin to partially melt, creating magma that will rise and produce a line of active volcanoes on the sea floor along the edge of the over-riding plate. These volcanoes may eventually grow large enough to reach the sea surface and form a volcanic **island arc**. Island arc volcanism can be explosive. Island arcs are most commonly found in the Pacific Ocean. Examples of island arcs include the Mariana Islands, Tonga, the Aleutian Islands, and the Philippines. There are only two island arcs in the Atlantic Ocean: the Sandwich Islands in the South Atlantic and the Lesser Antilles on the eastern side of the Caribbean Sea. In some cases, island arcs can form on pieces of continental crust that have broken away from the mainland. Examples of this include New Zealand and Japan.
- **Ocean-continent convergence** occurs when the edge of one of the colliding plates is composed of oceanic lithosphere and the edge of the other plate is composed of continental lithosphere (fig. 2.30*b*). Along this type of convergent boundary, the downgoing plate will be the oceanic lithosphere and the overriding plate will be the continental lithosphere. Once again, an ocean trench will form along the boundary and there will be both shallow earthquakes along the boundary and a Wadati-Benioff zone of deeper earthquakes tracing the position of the downgoing plate. Magma produced by partial melting of the downgoing plate will rise through the continental crust of the overriding plate, partially melting it. The two sources of magma will mix, forming a magma with a composition between basalt and granite. This magma produces a rock called **andesite**. Eventually this rising andesitic magma will produce active volcanoes along the edge of the overriding continental plate. These andesitic volcanoes often erupt explosively, ejecting large amounts of rock, ash, and gas that can rise high into the atmosphere. Examples of active continental volcanic chains include the Andes along the west coast of South America and the Cascade Range of the Pacific Northwest, including Mount St. Helens (fig. 2.32).
- **Continent-continent convergence** occurs when the edges of both colliding plates are composed of continental lithosphere (fig. 2.30*c*). These boundaries are the end result of the closing of an ocean basin by ocean-continent convergence. As the ocean basin closes, the continental lithosphere of the downgoing plate moves closer to the trench. When the sinking oceanic lithosphere of the downgoing plate is completely subducted and the ocean basin has closed, the continental lithosphere of the downgoing plate will collide with the continental lithosphere of the overriding plate, forming a **suture zone** that may include sediments scraped off the subducted portion of the downgoing plate. Along the edges of the two converging continents, the continental crust buckles, fractures, and thickens. What was once a well-defined convergent boundary marked by a narrow ocean trench becomes a broad zone of intense deformation and mountain building with no clear, discrete boundary. These boundaries are characterized by relatively shallow earthquakes and no active volcanism.

Continent-continent convergence created the Himalayas as India collided with Asia. India continues to push farther into Asia at a rate of about 5 cm (2 in) per year, and the Himalayas are still rising in elevation. Other examples of mountain ranges created in this manner are the Appalachians, Alps, and Urals.

Continental Margins

When a continent rifts and moves away from a spreading center, the resultant continental margin is known as a **passive,** or **trailing, margin.** Such margins are also referred to as Atlantic-style margins because they are found on both sides of the Atlantic Ocean as well as around Antarctica, the Arctic Ocean, and the Indian Ocean. Continental and oceanic lithosphere are joined along passive margins so there is no plate boundary at the margin. As passive margins move away from the ridge, the oceanic lithosphere cools, increases its density, thickens, and subsides. This causes the edge of the continent to slowly subside as well. Passive margins are modified by waves and currents; they may also be modified by coral reefs. These margins accumulate sediment eroded from the continents to a depth of about 3 km (1.8 mi). Currents move the sediments downslope to the sea floor, where thick deposits of sediments build on top of the oceanic crust. Old passive margins are not greatly modified by tectonic processes because of their distance from the ridge. These margins are broad, the water is shallow, and the sedimentary deposits are thick, as along the eastern coast of the United States. While passive margins begin at a divergent plate boundary, they move to a mid-plate position as a result of seafloor spreading and the opening of the ocean basin.

When a plate boundary is located along a continental margin, the margin is called an **active,** or **leading, margin.** Active continental margins are often marked by ocean trenches where oceanic lithosphere is subducted beneath the edge of the continent.

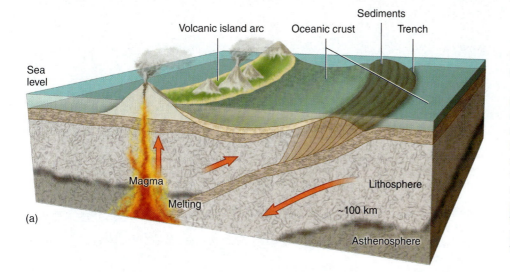

Sea level

Volcanic island arc   Oceanic crust   Sediments   Trench

Magma

Melting

Lithosphere

~100 km

Asthenosphere

(a)

Figure 2.30 The three different types of convergent plate boundaries are:

(a) **Ocean-ocean convergence:** The convergent edges of both plates consist of oceanic lithosphere. One plate is subducted beneath the other, forming a trench. Partial melting of the downgoing plate results in volcanism and the formation of an island arc along the edge of the overriding plate. Seafloor sediments scraped from the downgoing plate accumulate at the plate boundary. One example of this is the Aleutian Islands of Alaska.

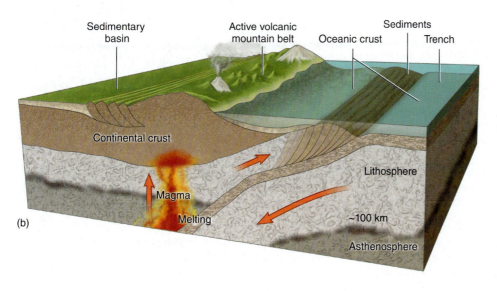

Sedimentary basin

Active volcanic mountain belt   Oceanic crust   Sediments   Trench

Continental crust

Magma

Melting

Lithosphere

~100 km

Asthenosphere

(b)

(b) **Ocean-continent convergence:** The edge of one convergent plate is relatively dense oceanic lithosphere while the edge of the other plate is relatively light continental lithosphere, forming a trench. Partial melting of the downgoing oceanic plate results in volcanism and the formation of an active volcanic mountain range along the edge of the overriding continental plate. Seafloor sediments scraped from the downgoing plate accumulate at the plate boundary. One example of this is the western coast of South America.

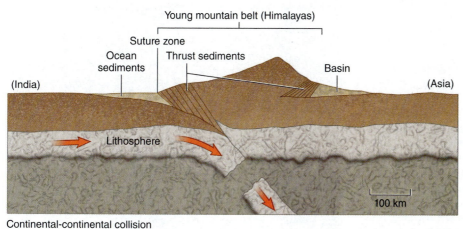

Young mountain belt (Himalayas)

Suture zone

Ocean sediments   Thrust sediments   Basin

(India)   (Asia)

Lithosphere

100 km

Continental-continental collision

(c)

(c) **Continent-continent convergence:** The convergent edges of both plates consist of continental lithosphere. Neither plate will subduct. The two plate edges will crumple, thicken, and form a mountain belt in a region known as a suture zone. The tallest mountains on Earth are found in such regions. One example is the Himalayas.

These margins are typically narrow and steep with volcanic mountain ranges, as along the west coast of South America as well as Oregon and Washington. Because sediments moving from the land into the coastal ocean move directly downslope into deeper water, into trenches, or into adjacent ocean basins, thick sediment deposits do not routinely accumulate to form broad shallow shelves at active margins. Active margins are found primarily in the Pacific Ocean.

Recovery of Black Smokers

In 1998, an ambitious expedition was undertaken by scientists at the University of Washington and the American Museum of Natural History in cooperation with the Canadian Coast Guard to recover black smokers from an active hydrothermal vent system on the Endeavour segment of the Juan de Fuca Ridge. The Endeavour, located about 300 km (180 mi) off Washington and Vancouver Island, is one of the most active hydrothermal vent regions ever discovered. There are at least four major hydrothermal vent fields along the Endeavour segment with hundreds of black smokers, some of which emit fluids at temperatures of 400°C (750°F). One giant smoker, Godzilla, stood roughly 45 m (150 ft) tall before it collapsed in 1996. The southernmost field in this area, the Mothra Hydrothermal field, was chosen as the recovery site because it contains abundant steep-sided smokers that host diverse macrofaunal communities. Venting sulfide structures in this area are up to 24 m (79 ft) high and extend linearly for over 400 m (1300 ft). The tops of black smokers targeted for recovery were initially identified during a preliminary survey conducted in 1997 using the ROV *Jason* and the staffed submersible *Alvin*. The smokers identified as possible targets for retrieval were relatively small; the largest was about 3 m (10 ft) tall and weighed roughly 6800 kg (15,000 lb).

The expedition used the Canadian Remotely Operated Platform for Ocean Science (ROPOS) to recover four sulfide chimneys from a depth of about 2250 m (7400 ft). ROPOS works out of a cage that acts as a garage for the vehicle while it is being lowered to the bottom or brought back to the surface (box fig. 1). The fiber-optic tether that connects the cage to the ship is also used to transmit images from cameras on ROPOS and other data, including piloting commands, to the vehicle. After the cage is lowered to the target area, ROPOS "swims" out on a separate tether connected to the cage. Because ROPOS is on its own tether, it is unaffected by any motion of the cage caused by ship motion at the surface.

Once the expedition had arrived at the site, ROPOS was lowered from the University of Washington's research vessel *Thomas G. Thompson* to the bottom, where it moved to the first target, an inactive black smoker called Phang. Its initial task was to take pictures of this structure to document its characteristics for before-and-after studies. It then cinched cables around the smoker using a recovery cage that had been specially designed to fit over it. The next step was for ROPOS to approach the smoker and cut into its base using a chain saw with carbide and diamond-embedded blocks in the chain (box fig. 2). After a series of cuts had been made, ROPOS was backed away from the chimney. The recovery cage and cables were attached to a line brought over by ROPOS from a previously deployed recovery basket on the bottom holding 2400 m (8000 ft) of line. ROPOS was then brought on deck, and the recovery line was floated to the surface, where it was brought onboard the Canadian Coast Guard's vessel *John P. Tully* and attached to a winch. The *Tully* then took up slack on the line, broke the smoker off its base, and brought it to the surface (box fig. 3), where the structure was cut in half in preparation for geologic and microbiological studies.

This process was repeated three times during the course of the expedition to recover a total of four sulfide chimneys (box fig. 4). The sulfide edifices were chosen in part for their diversity. Phang, the dead

Box Figure 1 The Canadian ROPOS (Remotely Operated Platform for Ocean Science) in its deployment cage.

Box Figure 2 With the recovery cage in place, ROPOS approaches the base of the chimney in preparation for cutting with a carbide and diamond-embedded chain saw.

Diving in

Box Figure 3 A sulfide chimney breaking the surface during recovery, with the recovery cage surrounding it.

chimney, did not have any water flowing through it, but another chimney, Finn, was very active, with 304°C (579°F) hydrothermal fluid flowing out. Neither chimney had many large organisms attached. The remaining two chimneys recovered, Roane and Gwenen, were venting fluids at temperatures of 20°–200°C (68°–392°F) and had diverse communities of organisms living on them, including tube worms, limpets, and snails (box fig. 5). These two structures remained hot when brought up from a water depth of over 2000 m (nearly 1.5 mi); Roane had temperatures of 90°C (194°F), and Gwenen had temperatures of 60°C (140°F). Their temperatures were measured from within the structures when they were on deck. The largest chimney recovered is 1.5 m (5 ft) tall and weighs 1800 kg (4000 lb); it is now on display at the American Museum of Natural History in New York City.

The smokers will be studied in detail by geologists, biologists, and chemists to learn more about the extreme chemical and thermal gradients that characterize these environments, the conditions under which sulfide structures grow and evolve, and how nutrients may be delivered to the organisms that live on and within the structures. Such studies are likely to find new species of microorganisms that thrive in these high-temperature, sunlight-free environments. Preliminary studies on these samples indicate that microorganisms living within the structures at temperatures of 90°C (194°F) derive their energy from carbon-bearing species in the hydrothermal fluids and from mineral-fluid reactions within the rocks. These and similar findings are exciting in that they show that microorganisms are capable of living in the absence of sunlight in volcanically active water-saturated regions of our planet. Other hydrothermally active planets may harbor similar life forms.

(a)

(b)

Box Figure 4 (a) Several segments of recovered chimneys secured on deck. (b) Section of a chimney that was cut in half for study of its interior.

Box Figure 5 Tube worm colony attached to one of the recovered chimneys.

(a) Age = 0

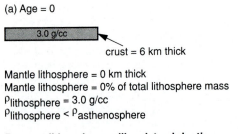

crust = 6 km thick

Mantle lithosphere = 0 km thick
Mantle lithosphere = 0% of total lithosphere mass
$\rho_{lithosphere}$ = 3.0 g/cc
$\rho_{lithosphere} < \rho_{asthenosphere}$

Buoyant lithosphere; will resist subduction

(b) Age = 10 Ma

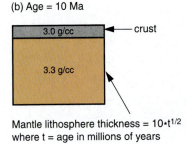 ← crust

Mantle lithosphere thickness = $10 \ast t^{1/2}$
where t = age in millions of years

Mantle lithosphere = 31.6 km thick
Lithosphere (crust + mantle) = 37.6 km thick
Mantle lithosphere = 84% of total
                         lithosphere thickness
$\rho_{lithosphere}$ = 3.25 g/cc
$\rho_{lithosphere} = \rho_{asthenosphere}$

Neutrally buoyant

(c) Age = 100 Ma

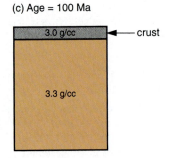 ← crust

Mantle lithosphere = 100 km thick
Lithosphere (crust + mantle) = 106 km thick
Mantle lithosphere = 94% of total
                         lithosphere thickness
$\rho_{lithosphere}$ = 3.28 g/cc
$\rho_{lithosphere} > \rho_{asthenosphere}$

**Negatively buoyant lithosphere;
will be easily subducted**

Constants and Considerations	
Crust = 6 km thick	ρ_{crust} = 3.0 g/cc
$\rho_{asthenophere}$ = 3.25 g/cc	$\rho_{mantle\ lithosphere}$ = 3.3 g/cc
$\rho_{lithosphere}$ = weighted mean of ρ_{crust} + $\rho_{mantle\ lithosphere}$	

Figure 2.31 Diagrammatic representation of how the thickness and density (ρ) of oceanic lithosphere increase with age.

Figure 2.32 Mount St. Helens erupted violently on May 18, 1980. The mountain lost nearly 4.1 km³ (1 mi³) from its once symmetrical summit, reducing its elevation from 2950 to 2550 m (a change of 1312 ft). The force of the lateral blast blew down forests over a 594 km² (229 mi²) area. Huge mud flows of glacial meltwater and ash moved down the mountain.

QUICK REVIEW

1. Locate the boundaries of the seven major plates listed in table 2.3.

2. What are the three main types of plate boundaries and what is the relative motion between plates on each one of them?

3. Where are most divergent plate boundaries found?

4. What is the difference between a fracture zone and a transform fault?

5. Most transform faults join what type of plate boundaries together?

6. What is a Wadati-Benioff zone?

7. Why can oceanic lithosphere be subducted but continental lithosphere cannot?

8. What is the difference between passive and active continental margins?

2.5 Motion of the Plates
Mechanisms of Motion

The mechanism that drives the plates is still not fully understood. Wegener's inability to identify a plausible mechanism for moving continents was a major reason why his ideas about continental drift were not widely accepted in his time. The basic question that must be answered in proposing any plate-driving mechanism is, why do plates diverge at mid-ocean ridges and sink at trenches? It is likely that the actual plate-driving mechanism is complex and involves multiple processes but there is little doubt that heat and gravity are involved.

One proposed mechanism is the **convection model** (fig. 2.33). In this model, heat loss from the core heats the base of the mantle, decreasing the density of deep mantle rock and causing it to rise. As the mantle rock rises, it cools and spreads out

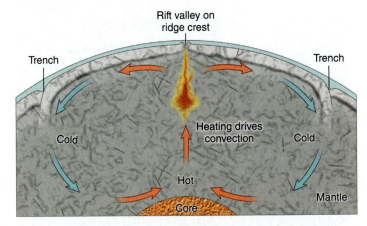

Figure 2.33 The convection model of plate tectonics proposes that plates are driven by convection cells in the mantle. Hot, upward-moving mantle material occurs beneath mid-ocean ridges and cold, downward-moving mantle material occurs beneath ocean trenches.

Figure 2.34 The ridge-push, slab-pull model of plate tectonics. Rising, hot mantle material beneath the ridge elevates the sea floor. The sea floor cools and subsides as it moves away from the ridge, eventually sinking into the mantle. Plate motion is caused by gravity. The plate slides down the incline caused by heat under the ridge. In addition, the cold, dense plate subducting into the mantle pulls on the plate.

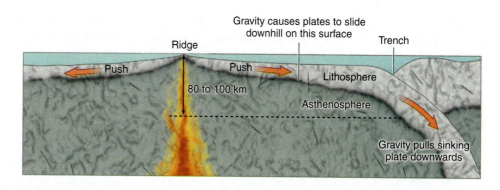

beneath the overlying plates where it moves horizontally for some distance. Eventually it cools enough that its density is greater than the material below it and it sinks, thus completing the convection cell. Frictional drag between the tops of convection cells and the bottoms of plates could drive the plates. The plates themselves would be relatively warm and buoyant above where the convection cells transport hot mantle rock upward. As the plates ride on top of the shallow, horizontally moving convection they would cool, thicken, and become denser. If they were sufficiently dense, they would sink into the mantle.

A second proposed mechanism is the **ridge-push**, **slab pull** model (fig. 2.34). This model also assumes some sort of convection in the mantle. In this model, hot, rising mantle material heats the overlying sea floor. This heating decreases the sea floor's density, causing it to expand, or elevate, creating a mid-ocean ridge. With increasing distance from the hot rising mantle material, the overlying sea floor cools, becomes denser, and sinks lower in elevation. In this way, the mantle convection results in the sea floor sloping downward away from the ridge. A ridge-push force is created by gravity, causing the sea floor to slide down this sloped surface. At the same time, the sea floor cools, its density increases, and it thickens as it moves away from the ridge. Eventually it becomes dense enough that it sinks into the mantle. The heavy, sinking end of the plate creates an additional slab-pull force that drags the plate toward the trench.

Rates of Motion

The sea floor, acting like a conveyor belt, moves away from the ridge system where it is created by volcanic activity. The rate at which each plate moves away from the axis of the ridge is known as the half spreading rate, while the rate at which the two plates move away from each other is called the full spreading rate, or simply the **spreading rate** (fig. 2.35).Spreading rates vary between about 1 and 20 cm (0.4 and 8 in) per year but are generally between about 2 and 10 cm (0.8 and 4 in) per year. The average spreading rate is about 5 cm (2 in) per year, roughly the rate at which

fingernails grow. In the course of a human lifetime of seventy-five years, a plate moving at an average spreading rate would travel 3.75 m (12.3 ft), or roughly the length of an automobile. Although these rates are slow by everyday standards, they produce large changes over geologic time. For example, if a plate moved at the rate of just 1.6 cm (0.6 in) per year, it would take 100,000 years for the plate to travel 1.6 km (1 mi); therefore, in the 200 million years since the breakup of Pangaea, it could move more than 3200 km (2000 mi), which is more than half the distance between Africa and South America. The spreading rate affects the physical structure of divergent plate boundaries. Slow spreading rates are found along ridges that have steep profiles and deep central valleys, such as the Mid-Atlantic Ridge. Fast spreading rates produce ridges with gentler slopes and shallower, or nonexistent, central valleys, such as the East Pacific Rise. Spreading rates are estimated at about 2.5–3 cm (1–1.2 in) per year for the Mid-Atlantic Ridge and about 8–13 cm (3–5 in) per year for the East Pacific Rise. Keep in mind that the process of spreading does not occur smoothly and continuously but goes on in increments, with varying time periods between

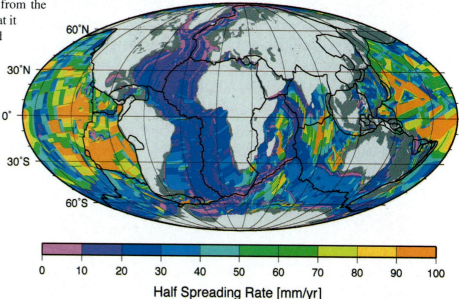

Figure 2.35 Half spreading rate of the plates in mm/yr. Fastest spreading rates are generally found in the Pacific Ocean; slower spreading occurs in the Atlantic Ocean.   Source of Data: National Geophysical Data Center.

occurrences. Compare the width of the age band centered on the East Pacific Rise with the width of the band centered on the Mid-Atlantic Ridge in figure 2.22; the wider the age band, the faster the spreading rate. Spreading rate is also believed to be related to the frequency of volcanic eruptions along the ridge. While very little is actually known about eruption frequency, it has been estimated that eruptions will occur approximately every 50–100 years at any given location along a fast-spreading ridge, while the rate may be only once every 5000–10,000 years on slow-spreading ridges.

Although seafloor spreading can be observed directly with great expense and difficulty at sea, there is one place where many of the processes can be seen on land—in Iceland, the only large island lying across a mid-ocean ridge and rift zone. Spreading in Iceland occurs at rates similar to those found at the crest of the Mid-Atlantic Ridge. Northeastern Iceland had been quiet for 100 years, until a volcanic rift opened in 1975; in six years, this rift widened by 5 m (17 ft) along an 80 km (50 mi) long stretch of the ridge's crest. Over 100 years, this spreading rate is 5 cm (2 in) per year, which is within the typical range. The spreading motion between the tectonic plates can now be monitored by satellite. The Global Positioning System of satellites can be used to determine the relative motion of tectonic plates directly by measuring the distance between two widely separated points to an accuracy of 1 cm (0.4 in) (fig. 2.36).

Investigations of ancient granitic rocks, formed during the Archean eon and exposed in West Greenland, eastern Labrador, Wyoming, western Australia, and southern Africa, indicate that, 3.5 billion years ago, crustal plates existed and moved granitic continental blocks at an average rate of about 1.7 cm (0.67 in) per year, again within today's average rates of plate motion.

Hotspots

Scattered around Earth are approximately forty areas of isolated volcanic activity known as **hotspots** (fig. 2.37). They are found under continents and oceans, in the center of plates, and at the mid-ocean ridges. These hotspots periodically channel hot material to the surface from deep within the mantle, possibly from the core-mantle boundary. Above these sites, a plume of mantle material may force its way through the lithosphere and form a volcanic peak, or seamount. If the hotspot does not break through, it may produce a broad swell on the ocean floor or the continent that subsides as the plate moves over and away from the magma source. Hotspots may also resupply magma to the asthenosphere, which cools and becomes attached to the base of the lithosphere, thickening the plates. Some people believe that the breakup of Pangaea began when a chain of hotspots developed under the supercontinent.

Hotspot plumes of magma are not uniform; they differ in chemistry, suggesting that they come from different mantle depths, and it has been suggested that their discharge rates may also vary. Hotspots may fade away and new ones may form. The life span of a typical hotspot appears to be about 200 million years. Although their positions may change slightly, hotspots tend to remain relatively stationary in comparison to moving plates, and they can be useful in tracing plate motions.

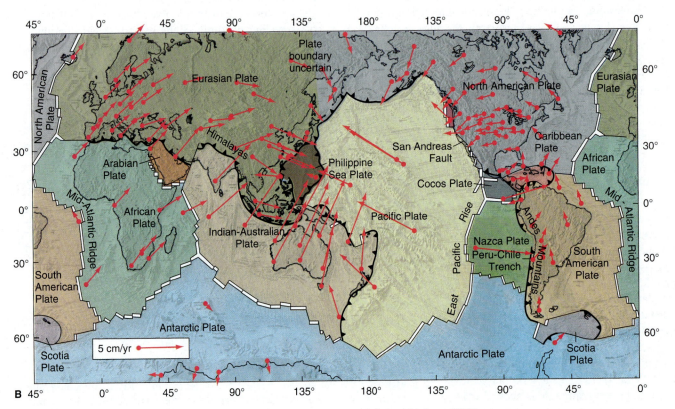

Figure 2.36 Annual plate motion vectors measured by GPS at stations around the world. From NASA.

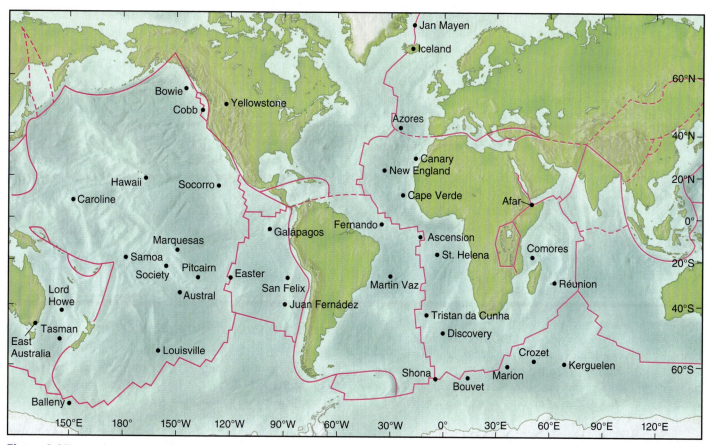

Figure 2.37 Location of major hotspots.

As the oceanic lithosphere moves over a hotspot, successive eruptions can produce a linear series of peaks, or seamounts, on the sea floor. The youngest peak is above the hot plume, and the seamounts increase in age as their distance from the hotspot increases (fig. 2.38). For example, in the islands and seamounts of the Hawaiian Islands system, the island of Hawaii, with its active volcanoes, is presently located over the hotspot. The newest volcanic seamount in the series is Loihi, found in 1981, 45 km (28 mi) east of Hawaii's southernmost tip and rising 2450 m (8000 ft) above the sea floor but still under water. At its current rate of growth, it should become an island in 50,000–100,000 years.

The land features of the island of Hawaii show little erosion, for it is comparatively young. To the west are the islands of Maui, Oahu, and Kauai, which have been displaced away from the magma source. Kauai's canyons and cliffs are the result of erosion over the longer period of time that it has been exposed to the winds and rains. Although these four islands are the most familiar, other islands and atolls attributed to the

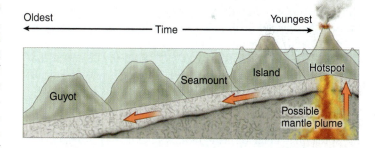

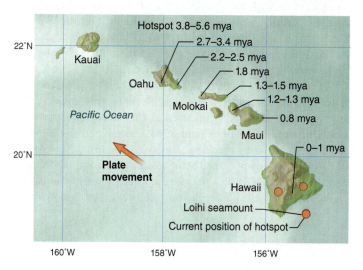

Figure 2.38 Hotspots beneath oceanic lithosphere can elevate the sea floor by heating it, and create a volcanically active island. As the plate moves away from the hotspot, the island ceases to be volcanically active, and sinks as the sea floor cools and subsides. This results in a chain of islands and seamounts that get progressively older with distance from the hotspot. One example is the Hawaiian Islands.

same hotspot stretch farther west across the Pacific. They are peaks of eroded and subsided seamounts formed above the same hotspot. Subsidence occurs when the heated and expanded plate moves slightly downhill and away from the mantle bulge over the hotspot. As the plate moves away from the hotspot, it cools and contracts, and combined with the weight of the seamount it is carrying, the result is a sinking of the sea floor that gradually carries the seamount below the ocean surface. When seamounts in tropical areas sank slowly, coral grew upward and coral atolls were the result. The formation of atolls is discussed further in chapter 3.

West of Midway Island, the chain of peaks changes direction and stretches to the north, indicating that the Pacific Plate moved in a different direction some 40 million years ago. This line of peaks is the Emperor Seamount Chain, volcanic peaks that once were above the sea surface as islands but have since eroded and subsided over time, resulting in many flat-topped seamounts known as **guyots** (or tablemounts) 1000 m (0.6 mi) below the surface (fig. 2.38). The northern end of the Emperor Seamount Chain is estimated to be 75 million years old, and Midway Island itself may be 28 million years old. Refer to figure 2.10 to follow this seamount chain. Notice in figure 2.38 that the peaks formed by the hotspot get older in the direction in which the plate is moving. The plate is presently moving westward; Midway Island is northwest of the main Hawaiian Islands, and it is also older.

It is possible to check the rate at which the plate is moving by using the distance between seamounts in conjunction with radiometric dating. For example, the distance between the islands of Midway and Hawaii is 2430 km (1509 mi). Midway was an active volcano 28 million years ago, when it was located above the hotspot currently occupied by Hawaii. In other words, Midway has moved 2430 km (1509 mi) in 28 million years, or 8.7 cm (3.4 in) per year.

There is another hotspot at 37° 27′S, in the center of the South Atlantic, marked by the active volcanic island of Tristan da Cunha (see fig. 2.37). Volcanic activity at this more slowly diverging plate boundary produces seamounts that can be carried either to the east or to the west, depending on the side of the spreading center on which the seamount was formed. The hotspot produces a continuous series of seamounts very close together, forming a **transverse,** or **aseismic, ridge:** the Walvis Ridge to the east, between the Mid-Atlantic Ridge and Africa, and the Rio Grande Rise to the west, between the Mid-Atlantic Ridge and South America. Refer to figures 2.10 and 3.11.

If a hotspot is located at a spreading center, the flow of material to the surface is intensified. The crust may thicken and form a platform. Iceland is an extreme example of this process in which the crust has become so thick that it stands above sea level.

Seamount chains, plateaus, and swellings of the sea floor, all products of hotspots, are being used to trace the motion of Earth's plates over known hotspot locations. Recently, hotspot tracks have been used to reconstruct the opening of the Atlantic

and Indian Oceans. Although the exact mechanism is not known, hotspots also appear to have formed huge basaltic plateaus: the Kerguelen Plateau in the southern Indian Ocean about 110 million years ago, and the Ontong Java Plateau in the western Pacific about 122 million years ago.

QUICK REVIEW

1. What is the average rate of seafloor spreading?
2. Describe the convection model of plate-driving force.
3. Describe the ridge-push, slab-pull model of plate-driving force.
4. What is a hotspot?
5. How can hotspots be used to measure the speed and direction of plates?

2.6 History of the Continents

The Breakup of Pangaea

Figure 2.39 traces the recent plate movements that led to the configuration of the continents and oceans as we know them today. In the early Triassic, when the first mammals and dinosaurs appeared, the continents were all joined in a single landmass called Pangaea. Most of the rest of the globe at this time was covered by the massive Panthalassic Ocean, also known as Panthalassa. A second, much smaller ocean called the **Tethys** occupied an indentation in Pangaea between what would eventually become present-day Australia and Asia. About 200 million years ago, Pangaea began to break apart into Laurasia and Gondwana. By about 150 million years ago, when the dinosaurs were flourishing, Laurasia and Gondwana were separated by a narrow sea that would grow to be the central Atlantic Ocean. At the same time, India and Antarctica were beginning to move away from South America and Africa. Water flooded the spreading rift between South America and Africa about 135 million years ago, and the South Atlantic Ocean began to form. As seafloor spreading opened the Atlantic Ocean, Panthalassa (what we now call the Pacific Ocean) was growing progressively smaller. At the same time, India was moving north and the southern Indian Ocean was forming as the Tethys Ocean was closing.

By the late Cretaceous, 97 million years ago, the North Atlantic Ocean was opening and the Caribbean Sea was beginning to form. India was continuing to move toward Asia, and the Tethys Ocean was being consumed from the north as the Indian Ocean was expanding from the south. By the end of the Cretaceous, shortly before the meteor impact that led to the extinction of the dinosaurs and many other species, the Atlantic Ocean was well developed and Madagascar had separated from India. During the middle Eocene, 50 million years ago, Australia had separated from Antarctica, India was on

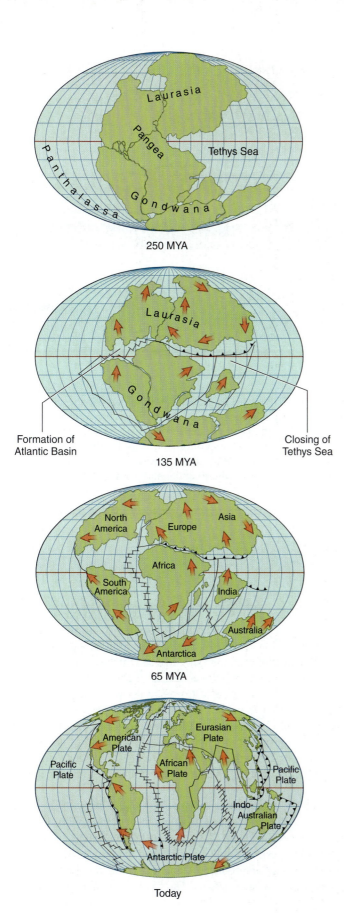

the verge of colliding with Asia, and the Mediterranean Sea had formed. About 40 million years ago, India collided with Asia and initiated a continent-continent convergent boundary, destroying the last of the Tethys Ocean and beginning the formation of the Himalayas. Twenty million years ago, Arabia moved away from Africa to form the Gulf of Aden and the new, still opening Red Sea.

Before Pangaea

There is ample evidence that Pangaea and its breakup into today's continents was not the only time when the sea floor spread and the continents changed position. We know that there have been landmasses, and consequently oceans, for much of Earth's history. Ancient rocks exceeding 3.5 billion years in age are found on all of Earth's continents. The oldest rocks on Earth found so far are located in northwestern Canada; these rocks are slightly over 4 billion years old. Ancient mountain ranges located in the interiors of today's continents are evidence of collisions between continents that existed in the past. Evidence from rock types, paleomagnetism, and fossils has allowed scientists to reconstruct an older supercontinent before Pangaea that is called **Rodinia**. Rodinia first formed about 1.1 billion years ago and began to break up about 750 million years ago. Because tectonic processes are continuous, there has probably never been a stable geography of Earth's surface. Instead, continents break apart and collide, and ocean basins open and close. These processes occur over and over again in a series of stages known as the **Wilson cycle** (fig. 2.40). The Atlantic Ocean is a good example of a mature ocean basin that continues to grow larger with seafloor spreading and essentially no subduction along its margins. The Pacific Ocean is an older basin that has reached its maximum size and will slowly close.

QUICK REVIEW

1. Compare the Panthalassa Ocean and the Tethys Sea.
2. As Panthalassa became smaller, what ocean did it eventually form?
3. Describe the stages of the Wilson cycle.

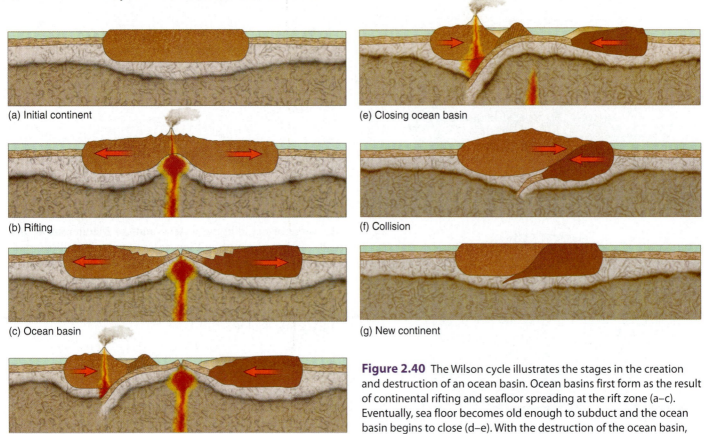

(a) Initial continent

(b) Rifting

(c) Ocean basin

(d) Subduction zone

(e) Closing ocean basin

(f) Collision

(g) New continent

Figure 2.40 The Wilson cycle illustrates the stages in the creation and destruction of an ocean basin. Ocean basins first form as the result of continental rifting and seafloor spreading at the rift zone (a–c). Eventually, sea floor becomes old enough to subduct and the ocean basin begins to close (d–e). With the destruction of the ocean basin, a suture zone is formed where continents collide and join (f–g).

Summary

Earth is made up of a series of concentric layers: the crust, the mantle, the liquid outer core, and the solid inner core. The evidence for this internal structure comes indirectly from studies of Earth's dimensions, density, rotation, gravity, and magnetic field and of meteorites. It also comes from the ways in which seismic waves change speed and direction as they move through Earth. Seismic tomography is being used to describe Earth's interior layers.

Continental crust is formed from granitic rocks, which are less dense than oceanic crust (formed of basalt). The top of the mantle is fused to the crust to form the rigid lithosphere. The lithosphere floats on the deformable upper mantle, or asthenosphere. The pressures beneath the elevated continents and depressed ocean basins are kept in balance by vertical adjustments of the crust and mantle, a process known as isostasy.

Alfred Wegener's theory of drifting continents was based on the geographic fit of the continents and the similarity of fossils collected on different continents. His ideas were ignored until the discovery of the mid-ocean ridge system and until the proposal of convection cells in the asthenosphere led to the concept of seafloor spreading. New lithosphere is formed at the ridges, or spreading centers. Old lithospheric material descends into trenches at subduction zones. Seafloor spreading is the mechanism of continental drift. Evidence for lithospheric motion includes the match of earthquake zones to spreading centers and subduction zones, the greater heat flow along ridges, age measurement of seafloor rocks, age and thickness measurements of sediment from deep-sea cores, and the magnetic stripes in the sea floor on either side of the ridge system. Rates of seafloor spreading are generally 1–20 cm (0.4–8 in) per year, averaging about 5 cm (2 in) per year. Plate tectonics is the unifying concept of lithospheric motion. Plates are made up of continental and oceanic lithosphere bounded by ridges, trenches, and faults. Plates move apart at divergent plate boundaries and together at convergent plate boundaries.

Rift zones separate ocean basins, and, in the past, they have separated landmasses and produced new ocean basins. Oceanic lithosphere is made up of layers of sediment, fine-grained basalt, vertical basalt dikes, igneous rock, and, under the Moho, mantle rock. Subduction produces island arcs, mountain ranges, earthquakes, and volcanic activity. The passive, or trailing, continental margin is closest to the ridge system; the active, or leading, margin borders a subduction, or collision, zone. Hotspots can be used to trace plate motions.

The mechanism that drives the plates is not fully known. Two mechanisms that have been proposed involve mantle circulation carrying the plates on top of large convection cells, and plates sliding down the elevated flanks of ridge crests and being pulled into the mantle by their own dense, subducting edges by gravity.

Throughout Earth's history, the mechanisms of plate tectonics have changed the locations of landmasses and ocean basins. Landmasses repeatedly collide and join, and ocean basins repeatedly open and close in a process known as the Wilson cycle. The most recent supercontinent, Pangaea, existed about 250 million years ago. At that time there was one major ocean, Panthalassa, and a smaller Tethys Sea. Pangaea has since broken apart to the form the geography of continents and oceans we see today.

Key Terms

seismic wave, 50
surface wave, 50
body wave, 50
P-wave, 50
S-wave, 50
refract, 50
mantle, 51
crust, 51
continental crust, 51
Mohorovičić discontinuity, 51
Moho, 51
granite, 51
oceanic crust, 51
basalt, 51
lithosphere, 53
asthenosphere, 53
mesosphere, 54

inner core, 54
outer core, 54
isostasy, 54
Gondwanaland, 55
continental drift, 55
Pangaea, 55
Panthalassa, 55
Laurasia, 55
convection cell, 59
seafloor spreading, 59
spreading center, 59
subduction zone, 59
epicenter, 60
focus, 60
hypocenter, 60
core, 51, 60
dipole, 61

Curie temperature, 63
paleomagnetism, 63
polar reversal, 63
plate tectonics, 65
divergent plate boundary, 66
convergent plate boundary, 66
transform boundary, 66
transform fault, 66
rift zone, 66
pillow basalt, 66
graben, 67
escarpment, 68
fracture zone, 68
ocean-ocean convergence, 70
ocean-continent convergence, 70

continent-continent convergence, 70
Wadati-Benioff zone, 70
island arc, 70
andesite, 70
suture zone, 70
passive/trailing margin, 70
active/leading margin, 70
convection model, 74
ridge-push, slab-pull, 75
spreading rate, 75
hotspot, 76
guyot, 78
transverse/aseismic ridge, 78
Tethys, 78
Rodinia, 79
Wilson cycle, 79

Study Problems

1. Estimate the thickness of oceanic lithosphere that is (a) 5 million, (b) 10 million, (c) 20 million, and (d) 50 million years old.

2. Given the results you obtained in the previous problem, how would you describe a plot of oceanic lithosphere thickness as a function of age of the lithosphere?

3. Near the Hawaiian Islands, the Pacific Plate is moving to the northwest at a speed of about 7 cm (~2.76 in) per year. How far will the Island of Oahu move in 30 million years? Give your answer in kilometers and miles. Compare this with the distance between Los Angeles, California and New York, New York.

The Sea Floor and Its Sediments

Learning Outcomes

After studying the information in this chapter students should be able to:

1. *review* the evolution of methods to measure ocean depth from the time of the ancient Greeks to the present,

2. *construct* a simple cross section of an ocean basin, including both passive and active continental margins,

3. *discuss* the formation of atolls,

4. *sketch* the location of ocean ridges and trenches,

5. *explain* three different ways to classify sediments,

6. *list* the organisms that produce the majority of calcareous and siliceous sedimentary particles,

7. *identify* where biogenous and lithogenous sediments are dominant on the sea floor,

8. *define* isotopes and describe how they can be used with marine sediments as historical records, and

9. *list* multiple seabed resources and *appraise* the extent to which they are currently being recovered.

A large eroded sandstone boulder on the coast.

E arly mariners and scholars believed that the oceans were large basins or depressions in Earth's crust, but they did not conceive that these basins held features that were as magnificent as the mountain chains, deep valleys, and great canyons of the land. As maps became more detailed and as ocean travel and commerce increased, measurement of water depths and recording of seafloor features in shallower regions became necessary to maintain safe travel and ocean commerce. The secrets of the deeper oceanic areas had to wait for hundreds of years until the technology of the late twentieth century made it relatively easy to map and sample the sea floor. It was only then that large numbers of survey vessels accumulated sufficient data to provide the details of this hidden terrain.

What we know about the sea floor and its covering of sediments comes almost entirely from observations by surface ships; more recently, submersibles, robotic devices, and satellites have added to our knowledge. Some areas of the sea floor have been measured in great detail; charts of other areas have been made from scanty data. The demand for more measurements to describe and explain the features of the sea floor continues.

In this chapter, we survey the world's ocean floors and discuss their topography and geology. We examine the sources, types, and sampling of sediments and discuss seabed mineral resources.

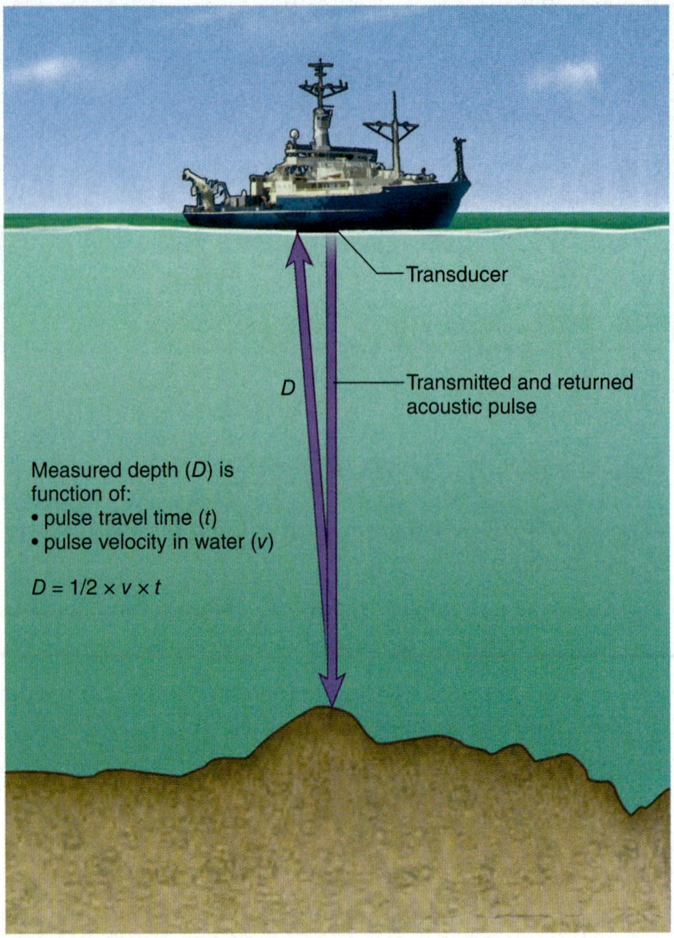

Measured depth (D) is function of:
• pulse travel time (t)
• pulse velocity in water (v)

$D = 1/2 \times v \times t$

Figure 3.1 Measuring water depth (D) with a precision depth recorder (PDR), or echo sounder. Sound travels a total distance of twice the water depth. Depth can be calculated knowing the speed of sound in water (v) and the time (t) it takes to hear the echo, or reflection, of the sound off the bottom.

3.1 Measuring the Depths

In about 85 B.C., a Greek geographer named Posidonius set sail, curious about the depth of the ocean. He directed his crew to sail to the middle of the Mediterranean Sea, where they eased a large rock attached to a long rope over the side. They lowered it nearly 2 km (1.2 mi) before it hit bottom and answered Posidonius's question. Crude as this method was, it continued with minor modifications as the means of obtaining **soundings**, or depth measurements, for the next 2000 years.

An early modification made by nineteenth-century surveyors was the use of hemp line or rope with a greased lead weight at its end. This line was marked in equal distances (usually **fathoms**; a fathom is the length between a person's fully outstretched hands, standardized at 6 ft). The change in line tension when the weight touched bottom indicated depth, and the particles from the bottom adhering to the grease confirmed the contact and brought a bottom sample to the surface. This method was quite satisfactory in shallow water, and the experienced captain used the properties of the bottom sample to aid in navigation, particularly at night or in heavy fog.

Later, piano wire with a cannonball attached was used in deep water. These nineteenth-century surveyors used a mechanical

sounding machine that allowed the rope or wire to free-fall. Using a clock, they carefully timed the rate at which premeasured lengths of the sounding line paid out over the side of the ship. This rate was a constant as long as the weight was free-falling toward the bottom but when the rate of payout decreased abruptly, they knew the weight had hit the bottom. In this way, they were able to obtain reasonably precise soundings in deep water even when the ship was moving at a slow speed. The time required (eight to ten hours) to let the weight free-fall to the bottom and then winch it back up in deep water was so great that by 1895, only about 7000 measurements had been made in water deeper than 2000 m (6600 ft) and only 550 measurements had been made of depths greater than 9000 m (29,500 ft).

It was not until the 1920s, when acoustic sounding equipment was invented, that deep-sea depth measurements became routine. The **echo sounder**, or **depth recorder**, which measures the time required for a sound pulse to leave the surface vessel, reflect off the bottom, and return, allows continuous measurements to be made easily and quickly when a ship is underway (fig. 3.1). The behavior of sound in seawater and its uses as an oceanographic tool are discussed in chapter 4. A trace from a

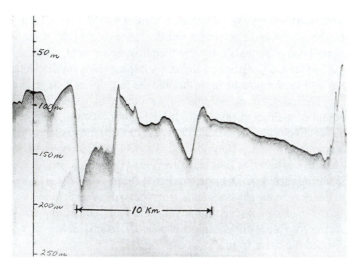

Figure 3.2 Depth recorder trace. A sound pulse reflected from the ocean floor traces a depth profile as the ship sails a steady course. The horizontal scale depends on ship speed.

depth recorder is shown in figure 3.2. (See figs. 4.18 and 4.19 for an illustration of how sound can be used to measure water depth and a picture of a precision depth recorder.)

In 1925, the German vessel *Meteor* made the first large-scale use of an echo sounder on a deep-sea oceanographic research cruise and detected the Mid-Atlantic Ridge for the first time. After this expedition, depth measurements gradually accumulated at an ever-increasing rate. As the acoustic equipment improved and was used more frequently, knowledge of the ocean floor's bathymetry expanded and improved, culminating in the 1950s with the first detailed mapping of the mid-ocean ridge and trench systems.

Today, a wide variety of methods are used to obtain even more detailed seafloor bathymetry at scales that range from centimeters (inches) to thousands of kilometers (thousands of miles) (fig. 3.3). The specific technique used depends on the amount of time that can be spent, the scale of the feature that is being examined, and the amount of detail that is required. When necessary, direct observation of small-scale structures is possible with the use of staffed submersibles and remotely operated vehicles (ROVs) carrying video cameras. These images can be transmitted to surface ships and relayed by satellite anywhere in the world in real time. Investigations by staffed submersibles or ROVs provide great detail, but they typically cover very small areas and are both time-consuming and expensive for the amount of sea floor surveyed. On large scales of tens or hundreds of square kilometers, sophisticated multibeam sonar systems can rapidly map extensive regions at relatively low cost with great accuracy.

Very-large-scale seafloor surveys use satellite measurements of changes in sea surface elevation caused by changes in Earth's gravity field due to seafloor bathymetry. These changes in sea surface elevation can be detected by radar altimeters that measure the distance between the satellite and the sea surface (fig. 3.4). The sea surface is not flat even when it is perfectly calm. Changes in gravity caused by seafloor topography create gently sloping hills and valleys in the sea surface. The excess mass of features such as seamounts and ridges creates a gravitational attraction that draws water toward them, resulting in a higher elevation of the sea surface. Conversely, the deficit of mass along deep-ocean trenches, and subsequent weaker gravitational attraction, results in a depression of the sea surface as water is drawn away toward surrounding areas with greater gravitational attraction. Sea level over large seamounts is elevated by as much as 5 m (16 ft) and over ocean ridges by about 10 m (33 ft); it is depressed over trenches by about 25–30 m (80–100 ft). These changes in elevation occur over tens to hundreds of kilometers, so the slopes are very gentle. The sea surface is always perpendicular to the local direction of gravity, so precise measurements of the slope of the surface can be used to determine the direction and magnitude of the gravitational field at any point. Because these changes in gravity are related to seafloor topography, it is possible to use them to reconstruct a bathymetry that produces the observed variations in sea surface topography (fig. 3.5). Tides, currents, and changes in atmospheric pressure can cause undulations of more than a meter (3 feet) in the ocean surface. These effects are filtered out

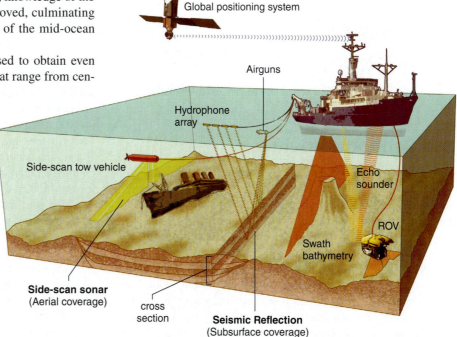

Figure 3.3 Mapping the sea floor using sound waves can be done with many different instruments depending on the level of detail required and whether the goal is to simply produce a map of the sea floor or to also map layers of sediment and rock beneath the sea floor. The various methods used differ in the intensity, frequency, and swath of the emitted sound signal, and in the method used to detect and record the echo.

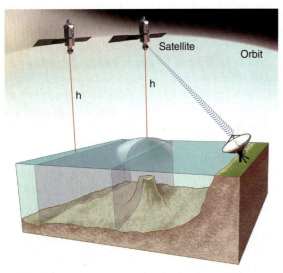

Figure 3.4 Satellite altimeters determine the elevation of the ocean surface by measuring the precise travel time of radar signals. The orbit of the satellite is known with a high degree of accuracy. The gravitational attraction of a seamount causes water to be drawn toward it, increasing the elevation of the sea surface above it.

to produce the bathymetric details. Bathymetric features with "footprints," or horizontal dimensions as small as about 10 km (6.2 mi), can be resolved with satellite altimetry data. Satellite maps are particularly valuable in the Southern Ocean, where the weather and sea conditions are frequently bad and it is difficult to conduct general bathymetric surveys to locate areas of scientific interest.

QUICK REVIEW

1. Why do oceanographers use sound rather than light to measure ocean depths?
2. Does satellite altimetry measure ocean depth directly or indirectly? Explain your answer.

3.2 Seafloor Provinces

The sea floor is as rugged as any land surface. The Grand Canyon, the Rocky Mountains, the desert mesas in the Southwest, and the Great Plains all have their undersea counterparts. In fact, the undersea mountain ranges are longer, the valley floors are wider and flatter, and the canyons are often deeper than those found on land. Features of land topography, such as mountains and canyons, are continually and aggressively eroded by wind, water, ice, changes in temperature, and the chemical alteration of minerals in rocks. The erosion of seafloor bathymetry is generally slow. Physical weathering occurs primarily by waves and currents, and chemical erosion occurs by the dissolution of minerals. More rapid erosion generally occurs closer to coastlines.

The most important agents of physical change on the deep-sea floor are the gradual burial of features by a constant rain of sediments falling from above, and volcanism associated with the mid-ocean ridge system, hotspots, island arcs, and active

seamounts and abyssal hills. Movements of Earth's crust can displace features and fracture the sea floor, and the weight of some islands and seamounts can cause them to subside, but the appearance of the bathymetric features of the ocean basins and sea floor has remained much the same through the last 100 million years.

The geology and structure of continents is generally very complex, making efforts to describe the cross section of a typical continent almost meaningless. In contrast, the geology and structure of the sea floor are relatively simple, allowing us to identify and describe four basic provinces, or depth zones, found in a generalized ocean basin (fig. 3.6). These provinces are continental margins, abyssal plains, mid-ocean ridges, and trenches. Whether or not you would find all of these provinces in a cross section of a specific ocean basin would depend on where you crossed the basin from shore to shore.

Continental Margins and Submarine Canyons

The edges of the landmasses below the ocean surface and the steep slopes that descend to the sea floor are known as the **continental margin**. There are two basic types of continental margins: passive margins and active margins (fig. 3.6). **Passive margins** are found where the continent-ocean transition is not a plate boundary. The transition from continental crust to oceanic crust occurs within a single plate. Passive margins have little seismic or volcanic activity and they tend to be relatively wide. They form after continents are rifted apart, creating a new and growing ocean basin between them. The continental margins in the Atlantic Ocean are of this type. **Active margins** are found where the continent-ocean transition is a plate boundary. In moving from continental crust to oceanic crust, you move from one tectonic plate to another. Active margins are often associated with earthquakes and volcanism and are often relatively narrow. Most active margins are associated with plate convergence and subduction of oceanic lithosphere beneath a continent, creating an ocean trench (review fig. 2.30*b*). Some active margins are created when two plates slide past each other along a transform fault (review fig. 2.29). The continental margins in the Pacific Ocean are generally active margins. The general model of a continental margin consists of four parts: continental shelf, shelf break, slope, and rise (fig. 3.7).

The **continental shelf** lies at the edge of the continent; continental shelves are the nearly flat borders of varying widths that slope very gently away from the shoreline. The continental shelves are geologically part of the continental crust; they are the submerged seaward edges of the continents. Shelf widths average about 65 km (40 mi) but are typically much narrower along active margins than along passive margins. In figure 3.5 you can see the striking difference between the narrow shelf of the active continental margin along the Pacific coast of South America and the broad shelf of the passive continental margin along South America's Atlantic coast. The width of the continental shelf can be as much as 1500 km (930 mi). Water depth at the outer edge of the continental shelf varies, but on average it is about 130 m (430 ft).

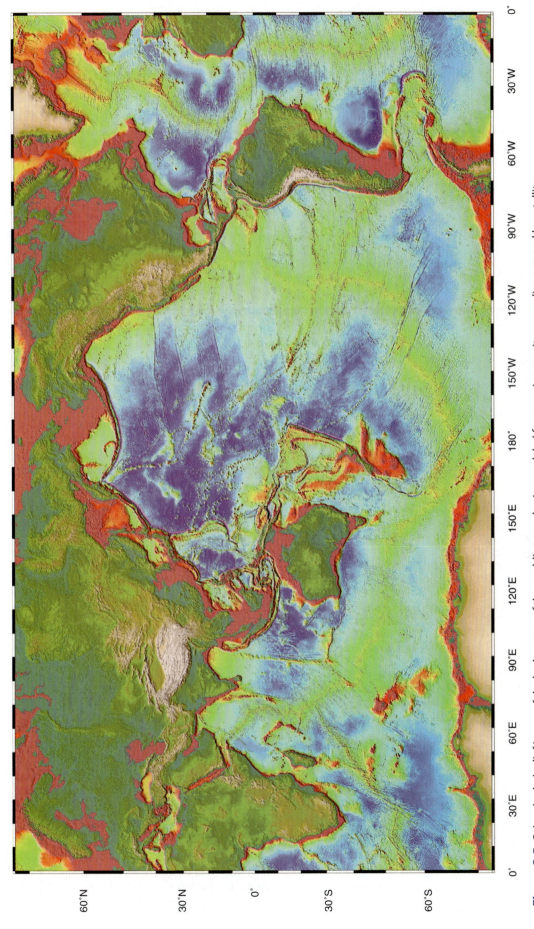

Figure 3.5 Color-shaded relief image of the bathymetry of the world's ocean basins modeled from marine gravity anomalies mapped by satellite altimetry and checked against ship depth soundings.

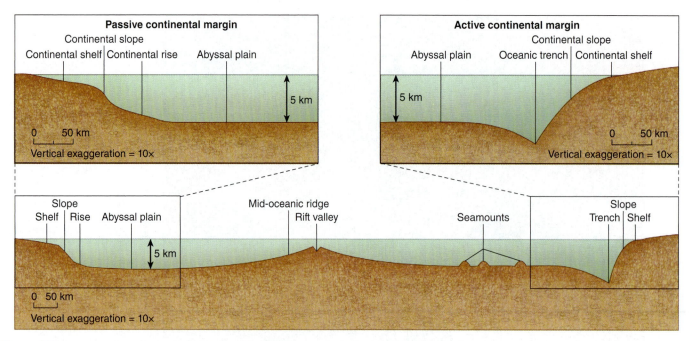

Figure 3.6 A cross section of a generalized ocean basin consists of four basic provinces: the continental margin (of some sort), the abyssal plain, a ridge, and a trench. Specific ocean basins can have different types of continental margins and may or may not have a trench.

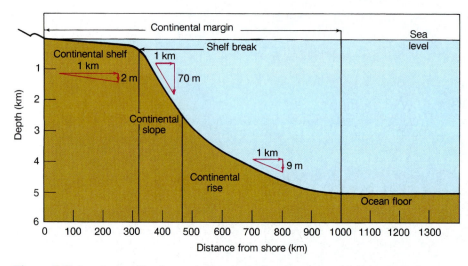

Figure 3.7 A typical profile of a passive continental margin. Notice both the vertical and horizontal extent of each subdivision. The average slope is indicated for the continental shelf, slope, and rise. The vertical scale is 100 times greater than the horizontal scale (V.E. = 100 times).

During past ages, the shelves have been covered and uncovered by fluctuations in sea level. During the glacial ages of the Pleistocene epoch, a number of short-term changes occurred in sea level, some of which were greater than 120 m (400 ft). When sea level was low, erosion deepened valleys, waves eroded previously submerged land, and rivers left sediments far out on the shelf. When the glacial ice melted, these areas were flooded, and sediments built up in areas closer to the new shore. At present, although submerged, these areas still show the scars of old riverbeds and glaciers acquired when the land was above water. Today, some continental shelves are covered with thick deposits of silt, sand, and mud sediments derived from the land: examples are offshore from the mouths of the Mississippi and Amazon Rivers, where large amounts of such sediments are deposited annually. Other shelves are bare of sediments, such as where the fast-moving Florida Current sweeps the tip of Florida, carrying the sediments northward to the deeper water of the Atlantic Ocean.

The **continental shelf break** is an abrupt change in the slope of the sea floor that occurs at the outer edge of the continental shelf. This marks the point at which there is a rapid increase in depth with distance from the coast.

The **continental slope** dips relatively steeply down to the ocean basin floor. The angle and extent of the slope vary from location to location. The slope can be short and steep, dropping to depths of around 3000 m (10,000 ft) along passive margins (as in fig. 3.7), or it might drop as far as 8000 m (26,000 ft) into a deep-ocean trench along an active margin (for example, off the western coast of South America, where the narrow continental shelf is bordered by the Peru-Chile Trench). The continental slope may show rocky outcroppings and be relatively bare of sediments because of its steepness, tectonic activity, or a low supply of sediment from land.

The most outstanding features of the continental slopes are **submarine canyons**. These canyons sometimes extend up, into, and across the continental shelf. A submarine canyon is

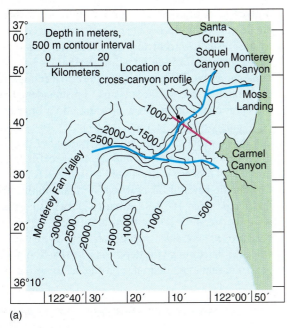

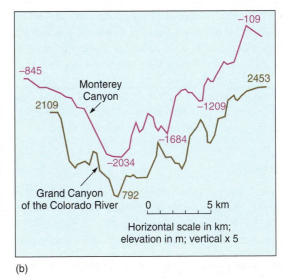

(a)

(b)

Figure 3.8 (a) Depth contours depict three submarine canyons off the California coast as they cut across the continental slope and continental shelf. The axes of the canyons, which merge seaward, are indicated by the *blue line*. (b) Cross-canyon profile, along the *red line* in (a), of the Monterey Canyon. Compare this profile to that of the Grand Canyon drawn to the same scale.

steep-sided and has a V-shaped cross section, with tributaries similar to those of river-cut canyons on land. Figure 3.8*a* shows the Monterey and Carmel Canyons off the coast of California. Figure 3.8*a* is a bathymetric chart; figure 3.8*b* compares the profile of the Monterey Canyon with the profile of the Grand Canyon of the Colorado River.

Many of these submarine canyons are associated with existing river systems on land and were apparently cut into the shelf during periods of low sea level, when the glaciers advanced and the rivers flowed across the continental shelves. Ripple marks on the floor of the submerged canyons and sediments fanning out at the ends of the canyons suggest that they were formed by moving flows of sediment and water called **turbidity currents**. Caused by earthquakes or the overloading of sediments on steep slopes, turbidity currents are fast-moving avalanches of mud, sand, and water that flow down the slope, eroding and picking up sediment as they gain speed. In this way, the currents erode the slope and excavate the submarine canyon. As the flow reaches the bottom, it slows and spreads, and the sediments settle. Because of their speed and turbulence, such currents can transport large quantities of materials of mixed sizes. The settling process produces graded beds of coarse material overlain (upward) by smaller particles. These graded deposits are called **turbidites**. Figure 3.9 shows a turbidite preserved in compacted seafloor sediments that have been uplifted and exposed by wave erosion. These large and occasional currents have never been directly observed, although similar but smaller and more continuous flows, such as sand falls, have been observed and photographed.

Research on turbidity currents began with laboratory experiments in the 1930s. Later analysis of a 1929 earthquake that broke transatlantic telephone and telegraph cables on the continental slope and rise off the Grand Banks of Newfoundland showed a pattern of rapid and successive cable breaks high on the continental slope, followed by a sequence of downslope breaks. These breaks were calculated to have been caused by a turbidity current that ran for 800 km (500 mi) at speeds of 40–55 km (25–35 mi) per hour. Later samples taken from the area showed a series of graded sediments at the end of the current's path. Searches of cable company records showed similar patterns of cable breaks in other parts of the world.

Figure 3.9 This beach cliff at Point Lobos, California shows ancient turbidite deposits that have been uplifted and then exposed by wave erosion. Turbidites are graded deposits, with the largest particles in the deposit at the bottom of the turbidite and the smallest at the top.

At the base of the steep continental slope may be a gentle slope formed by the accumulation of sediment. This portion of the sea floor is the **continental rise**, made up of sediment deposited by turbidity currents, underwater landslides, and any other processes that carry sands, muds, and silt down the continental slope. The continental rise is a conspicuous feature at passive margins in the Atlantic and Indian Oceans and around the Antarctic continent. Few continental rises occur in the Pacific Ocean, where active margins border the great seafloor trenches located at the base of the continental slope. Refer to figure 3.7 to see the relationship of the continental rise to the continental slope.

Abyssal Plains

The true oceanic features of the sea floor occur seaward of the continental margin. The deep-sea floor, between 4000 and 6000 m (13,000 and 20,000 ft), covers more of Earth's surface (30%) than do the continents (29%). In many places, the ocean basin floor is a vast plain extending seaward from the base of the continental slope. It is flatter than any plain on land and is known as the **abyssal plain.** The abyssal plain is formed by sediments that fall from the surface or are deposited by turbidity currents to cover the irregular topography of the oceanic crust. An area of the abyssal plain that is isolated from other areas by continental margins, ridges, and rises is known as a basin, and some basins may be divided into subbasins by ridge and rise subsections. The distribution of these basins and subbasins is shown in figure 3.5. Low ridges allow some exchange of deeper water between adjacent basins, but if the ridge is high, both the deep water and the deep-dwelling marine organisms within the basin are effectively cut off from other basins.

Abyssal hills and **seamounts** are scattered across the sea floor in all the oceans. Abyssal hills are less than 1000 m (3300 ft) high, and seamounts are steep-sided volcanoes rising abruptly and sometimes piercing the surface to become islands. Abyssal hills are probably Earth's most common topographic feature. They are found over 50% of the Atlantic sea floor and about 80% of the Pacific floor; they are also abundant in the Indian Ocean. Most abyssal hills are probably volcanic, but some may have been formed by other movements of the sea floor. Submerged, flat-topped seamounts, known as **guyots,** are found most often in the Pacific Ocean. The tops of Pacific guyots are 1000–1700 m (3300–5600 ft) below the surface; many are at the 1300 m (4300 ft) depth. Many guyots show the remains of shallow-water coral reefs and evidence of wave erosion at their summits. These features indicate that at one time they were warm-water surface features and that their flat tops are the result of wave erosion. They have since subsided owing to their weight, the accumulated rock load bearing down on the oceanic crust, and the natural subsidence of the sea floor with increasing age as the crust cools and grows in density as it moves farther away from the ridge where it formed. They have also been submerged by rising sea level during periods when glacial ice melted on land.

In the warm waters of the Atlantic, Pacific, and Indian Oceans, coral reefs and coral islands are formed in association with seamounts. Reef-building corals are warm-water animals that require

a place of attachment and grow in intimate association with a single-cell, plantlike organism; reef-building corals are confined to sunlit, shallow tropical waters. When a seamount pierces the sea surface to form an island, it provides a base on which the coral can grow. The coral grows to form a **fringing reef** around the island. If the seamount sinks or subsides slowly enough, the coral continues to grow upward, and a **barrier reef** with a lagoon between the reef and the island is formed. If the process continues, eventually the volcanic portion of the seamount disappears below the surface and the coral reef is left as a ring, or **atoll.** This process is illustrated in figure 3.10.

On the basis of observations he made during the voyage of the *Beagle* from 1831–36, Charles Darwin suggested that these were the steps necessary to form an atoll. Darwin's ideas have been proved to be substantially correct by more recent expeditions when drilling through the debris on a lagoon floor found the basalt peak of a seamount that once protruded above the sea surface.

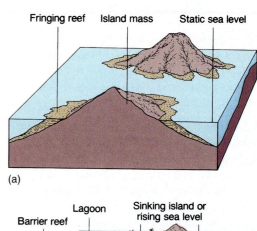

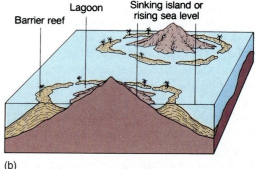

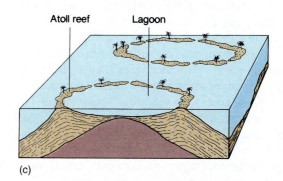

Figure 3.10 Types of coral reefs and the steps in the formation of a coral atoll shown in profile. (a) Fringing reef. (b) Barrier reef. (c) Atoll reef.

Ridges, Rises, and Trenches

The most notable features of the ocean floor are the mid-ocean ridge and rise systems stretching for 65,000 km (40,000 mi) around the world and running through every ocean. Their origin and role in plate tectonics were discussed in chapter 2 (see fig. 2.10). Review their distribution using figure 3.11. Recall also the roles played by the rift valleys and transform faults.

The relationship of the deep-sea trenches to plate tectonics was also discussed in chapter 2 (see fig. 2.10). Use figure 3.12 to trace the Japan-Kuril Trench, the Aleutian Trench, the Philippine Trench, and the deepest ocean trench, the Mariana Trench. All these trenches are associated with **island arc systems.** The Challenger Deep, a portion of the Mariana Trench, has a depth of 11,020 m (36,150 ft), making it the deepest known spot in all the oceans. The longest of the trenches is the Peru-Chile Trench,

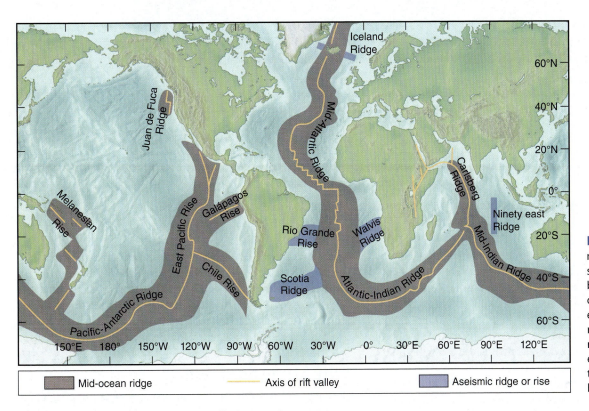

Figure 3.11 The mid-ocean ridge and rise system of divergent plate boundaries. Locations of major aseismic (no earthquakes) ridges and rises are added. Aseismic ridges and rises are elevated linear features thought to be created by hot-spot activity.

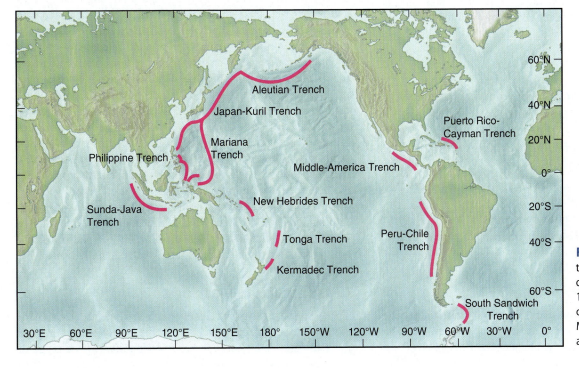

Figure 3.12 Major ocean trenches of the world. The deepest ocean depth is 11,020 m (36,150 ft), east of the Philippines in the Mariana Trench. It is known as the Challenger Deep.

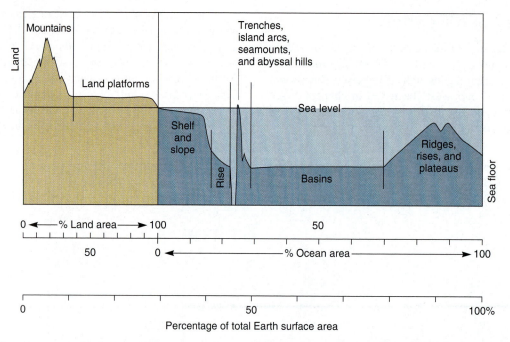

Figure 3.13 Earth's main topographic features shown as percentages of Earth's total surface and as percentages of the land and of the oceans.

stretching 5900 km (3700 mi) along the western side of South America. To the north, the Middle-America Trench borders Central America. The Peru-Chile and Middle-America Trenches are associated with volcanic chains on land. In the Indian Ocean, the great Sunda-Java Trench runs for 4500 km (2800 mi) along Indonesia. In the Atlantic, there are only two comparatively short trenches: the Puerto Rico-Cayman Trench and the South Sandwich Trench, both associated with chains of volcanic islands.

In figure 3.13, the topography of the land and the bathymetry of the sea floor are summarized as percentages of Earth's area. Compare the tectonically active areas of trenches and ridges, as well as the area of low-lying land platforms with the area of the ocean basins.

QUICK REVIEW

1. Identify the different parts of the continental margin.
2. Describe the difference between passive and active continental margins.
3. List the main provinces and features of the ocean basin floor.
4. Explain the formation of an atoll.
5. How are islands, seamounts, abyssal hills, and guyots the same? How are they different?
6. Distinguish between a ridge and a rise.
7. Where are the major deep-sea trenches?

3.3 Sediments

The margins of the continents and the ocean basin floors receive a continuous supply of particles from many sources. Whether these particles have as their origin living organisms, the land, the

atmosphere, or the sea itself, they are called sediment when they accumulate on the sea floor. The thickest deposits of sediment are generally found near the continental margins, where sediment is deposited relatively rapidly; in contrast, the deep-sea floor receives a constant but slow supply of sediment that produces a thinner layer that varies in thickness with the age of the oceanic crust.

Why Study Sediments?

Oceanographers study the rate at which sediments accumulate, the distribution of sediments over the sea bottom, their sources and abundance, their chemistry, and the history they record in layer after layer as they slowly but continuously accumulate on the ocean floor. Marine sediments can provide valuable information about how Earth and its environmental systems function on long time scales. They can provide valuable information concerning past climate change that can be used in predicting possible future environmental change. More immediately, sediments can help us understand how seafloor habitats impact fisheries and other biological communities. Knowledge of the characteristics of marine sediments is important in locating offshore mineral resources, including sand for beach replenishment. Sediment studies can help evaluate the possible impact of offshore waste disposal and map offshore pollution patterns, thus helping us sustain healthy coasts. Information about sediments is also critical in identifying sites for seabed communications cables, offshore drilling platforms, and coastal structures such as piers, breakwaters, and jetties.

Classification Methods

There are different ways to classify marine sediments. One classification method is based on the size of the particles that comprise the sediment. Sediments can be described both by the range of particle sizes found in a sample and by the dominant particle size, if there is one. Another way of classifying sediments is by their geographical location—where they are found in relation to distance from the coast, for instance. The rate of deposition of sediment often varies significantly for different locations, particularly with distance from the coast. Perhaps the most common classification method for marine sediments is one that is based on the origin of the particles and their chemical composition: Where did the individual particles come from and what are they made of? Each of these methods provides specific information about sediments. The importance of that information will depend on the problem you are trying to solve by studying the sediment.

Particle Size

Sediment particles are classified by size, as indicated in table 3.1. Familiar terms such as *gravel, sand,* and *mud* are used to identify broad size ranges of large, intermediate, and small particles, respectively. Within each of these ranges, particles are further ranked to produce a more detailed scale from boulders to the very smallest clay-sized particles, which can only be seen with a microscope.

When a sediment sample is collected, it can be dried and shaken through a series of woven-mesh sieves of decreasing opening size. Material that passes through one sieve but not the next is classified by one of the sizes listed in table 3.1.

A sample is said to be "well sorted" if it is nearly uniform in particle size and "poorly sorted" if it is made up of many different particle sizes. Size influences the horizontal distance a particle is transported before settling out of the water and the rate at which it sinks. In general, it takes more energy to transport large particles than it does small particles. In the coastal environment, when poorly sorted sediment is transported by wave or current action, the larger particles will settle out and be deposited first, while the smaller particles may be carried farther away from the coast and deposited elsewhere. In the open ocean, the variation in sinking rate between large and small particles has a tremendous influence on how long it takes for a particle to sink to the deep-sea floor and, hence, how far the settling particle may be transported by deep horizontal currents (table 3.2). A very fine sand-sized particle may settle to the deep-sea floor in a matter of days, where it could come to rest a short horizontal distance away from the point at the surface where it began its journey. In contrast, it may take clay-sized particles over 125 years (nearly 50,000 days) to make the same journey (see Stokes Law for small-particle settling velocity in appendix C). The speed of deep horizontal currents in the oceans is generally quite slow, but even at a speed of 5 cm (2 in) per second, a clay-sized particle could theoretically be transported around the world five times before it reached the deep-sea floor. Smaller soluble particles also have time to dissolve as they slowly sink in the deep ocean. Settling rate is also influenced by particle shape. Stokes Law assumes that the sediment particles are spherical, and the resulting calculated settling rates tend to be maximum rates. Angular grains generate small turbulent eddies that slow their rate of descent. Relatively flat particles such as clays also settle more slowly than spheres of the same density.

Scientists have puzzled over the close correlation between the particle types found in surface waters and those found almost directly below on the sea floor. This observation seems to contradict the large horizontal displacement of very slowly sinking particles due to currents in the water. Some mechanisms must be working to aggregate the tiny particles into larger particles. Scientists have observed that small particles often attract each other owing to their electrical charges. This attraction forms larger particles, which sink more rapidly. The remains of microscopic organisms known as **phytoplankton** and **zooplankton** can also be found on the sea floor. When zooplankton or phytoplankton are consumed by larger organisms, their inorganic cell walls or **tests** are expelled by these organisms in large fecal pellets that

Table 3.1 Sediment Size Classifications

Descriptive Name		Diameter (mm)
Gravel	Boulder	> 256
	Cobble	64–256
	Pebble	4–64
	Granule	2–4
Sand	Very coarse	1–2
	Coarse	0.5–1
	Medium	0.25–0.5
	Fine	0.125–0.25
	Very fine	0.0625–0.125
Mud	Silt	0.0039–0.0625
	Clay	< 0.0039

Table 3.2 Sediment Sinking Rate and Distance Traveled

Sediment Size	Approximate Sinking Rate (m/s)	Time for a Vertical Fall of 4 km (days)	Horizontal Distance Traveled in a 5 cm/s Current (km)
Very fine sand	9.8×10^{-3}	4.7	20.4
Silt	9.8×10^{-5}	470	2040
Clay	9.8×10^{-7}	47,000	204,000

Note: The sinking rate of a particle depends on its density, shape, and diameter. These rates are based on the assumption that the particles are spherical and have a density similar to that of quartz. Estimates of the speed of deep currents vary. A conservative estimate of 5 cm/s is chosen for purposes of illustration. See appendix C for the formula for small-particle settling velocity.

sink rapidly. It is estimated that as many as 100,000 tests can be packaged into a single, large fecal pellet. This packaging of small particles into larger particles decreases the time for the remains of plankton to sink to the sea floor from years to just ten to fifteen days, minimizing their horizontal displacement by water movements. Once the fecal pellets are deposited on the bottom, breakdown of the remaining organic portion of the pellets liberates the small inorganic particles.

Location and Rates of Deposition

Marine sediments are classified as either neritic (*neritos* = of the coast) or pelagic (*pelagios* = of the sea) based on where they are found (fig. 3.14).

Neritic sediments are found near continental margins and islands and have a wide range of particle sizes. Most neritic sediments are eroded from rocks on land and transported to the coast by rivers. Once they enter the ocean, they are spread across the continental shelf and down the slope by waves, currents, and turbidity currents. The largest particles are left near coastal beaches, whereas smaller particles are transported farther from shore. Accumulation rates of neritic sediments are highly variable. In river estuaries, the rate may be more than 800,000 cm (315,000 in) per 1000 years, or 8 m (over 26 ft) per year. Each year the rivers of Asia, such as the Ganges, the Yangtse, the Yellow, and the Brahmaputra, contribute more than one-quarter of the world's land-derived marine sediments. In quiet bays, the rate may be 500 cm (about 200 in) per 1000 years, and on the continental shelves and slopes, values of 10–40 cm (about 4–16 in) per 1000 years are typical, with the flat continental shelves receiving the larger amounts. Many sediments covering the continental shelves away from river mouths were deposited thousands of years ago when sea level was lower and the shoreline was located on the shelf. Such sediments are called **relict sediments.**

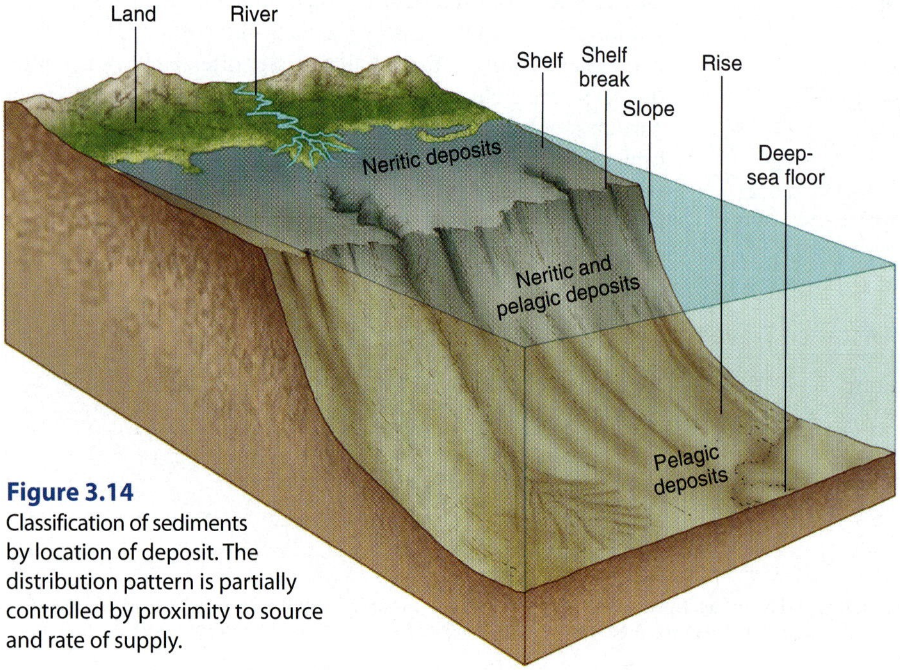

Figure 3.14
Classification of sediments by location of deposit. The distribution pattern is partially controlled by proximity to source and rate of supply.

Pelagic sediments are fine-grained and collect slowly on the deep-sea floor. The thickness of pelagic sediments is related to the length of time they have been accumulating or the age of the sea floor they cover. Consequently, their thickness tends to increase with increasing distance from mid-ocean ridges (see fig. 2.18). Accumulation rates for pelagic sediments are much slower than those of typical neritic sediments. An average accumulation rate for deep-ocean pelagic sediment is 0.5–1.0 cm (0.2–0.4 in) per 1000 years. Although deep-sea sedimentation rates are extremely slow, there has been plenty of time during Earth history for them to accumulate. The average thickness of pelagic sediments on the older areas of sea floor and the continental rises is about 500–600 m (1600–2000 ft). At a rate of 0.5 cm (0.2 in) per 1000 years, it takes only 100 million years to accumulate 500 m (1600 ft) of sediment, and the oldest sea floor is known to be roughly 200 million years old.

The actual thickness of marine sediments is controlled by a variety of factors including the age of the sea floor and its tectonic history, the nature and location of the sediment sources, and the nature of the processes that delivered the sediment to a particular location. Generally speaking, sediment thickness increases away from mid-ocean ridges as the age of the sea floor increases and maximum sediment thickness is found along continental margins where major rivers empty into the oceans (fig. 3.15).

Source and Chemistry

Marine sediments are also classified by the source of the particles that make up the sediment and may be further subdivided by their chemistry. Sedimentary particles may come from one of four different sources: preexisting rocks, marine organisms, seawater, or space.

Sediments derived from preexisting rocks are classified as **lithogenous** (*lithos* = stone, *generare* = to produce) **sediments.** These are also commonly called **terrigenous** (*terri* = land, *generare* = to produce) **sediments.** While terrigenous sediment technically includes any type of material coming off the land, such as rock fragments, wood chips, and sewage sludge, the majority of terrigenous material consists of lithogenous particles. Active volcanic islands in the ocean basins are also an important source of lithogenous sediment. Rocks on land are weathered and broken down into smaller particles by wind, water, and seasonal changes in temperature that result in freezing and thawing. The resulting particles are transported to the oceans by water, wind, ice, and gravity. Windblown dust from the continents, ash from active volcanoes, and rocks picked up by glaciers and embedded in icebergs are additional sources of lithogenous materials.

Lithogenous material can be found everywhere in the oceans. It is the dominant neritic sediment because the supply

Total Sediment Thickness of the World's Oceans & Marginal Seas

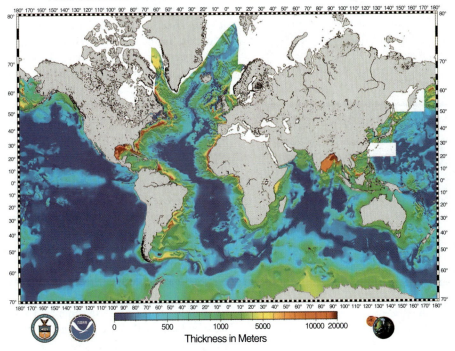

Thickness in Meters

Figure 3.15 Estimated thickness of marine sediment in meters. The average thickness of deep-ocean sediments is approximately 500 m. Thinner deposits of sediments are found along the mid-ocean ridge system. Sediment thickness generally increases closer to continental margins. The thickest deposits of sediment are found near the mouths of major rivers.   Source of Data: National Oceanic and Atmospheric Administration (NOAA).

sediment can provide important information concerning changes in wind patterns and intensity through time.

Clays are abundant because they are produced by chemical weathering. Four clay minerals make up the deep-sea clays: chlorite, illite, kaolinite, and montmorillonite. The distribution of these four clays reflects different climatic and geologic conditions in the areas where and when they originated as well as along the paths they traveled before settling on the sea floor. These conditions often have a strong dependence on latitude. The warm, moist climate of low latitudes supports strong chemical weathering on land. Mechanical weathering tends to be dominant in the cold, dry climate typical of high latitudes. Chlorite is highly susceptible to chemical weathering and can be altered to form kaolinite. Consequently, chlorite is abundant in deep-sea clays at high latitudes, where chemical weathering is less effective. Kaolinite is produced in the strong chemical

of lithogenous particles from land simply overwhelms all other types of material. Pelagic lithogenous sediments on the deep-sea floor, called **abyssal clay,** are composed of at least 70% by weight clay-sized particles. Abyssal clay accumulates very slowly at rates that are generally less than 0.1 cm (0.04 in) per 1000 years. Because the accumulation rate is so slow, even a thin deposit represents a very long period of time. It is important to understand that where abyssal clay is the dominant pelagic sediment, it is only because of the lack of other types of material, not because of an increase in the supply of clay-size particles. This is generally the case in regions where there is little marine life in the surface waters above. This fine rock powder, blown out to sea by wind and swept out of the atmosphere by rain, may remain suspended in the water for many years. These clays are often rich in iron, which oxidizes in the water and turns a reddish brown color; hence, they are frequently called **red clay** (fig. 3.16*a*). The distribution of red clay is illustrated in figure 3.17.

The composition of lithogenous sediments, generally various clays and quartz, is controlled by the chemistry of the rocks they came from and their response to chemical and mechanical weathering. Most lithogenous sediments have quartz because it is one of the most abundant and stable minerals in continental rocks. Quartz is very resistant to both chemical and mechanical weathering, so it can easily be transported long distances from its source. The distribution pattern of quartz grains in the

(a)

(b)

Figure 3.16 (a) Manganese nodules resting on red clay photographed on deck in natural light. Nodules are 1–10 cm in diameter. (b) A cross section of a manganese nodule showing concentric layers of formation.

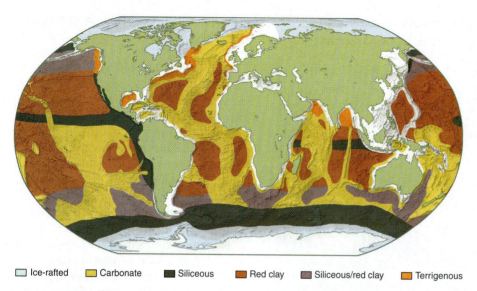

Ice-rafted ☐ | Carbonate ☐ | Siliceous ☐ | Red clay ☐ | Siliceous/red clay ☐ | Terrigenous ☐

Figure 3.17 Global distribution of surficial sediments in the world ocean. Equatorial upwelling in the Pacific and ice-edge processes in the Antarctic contribute to the productivity of diatoms and, hence, the accumulation of siliceous sediments. Carbonate sediments are generally confined to shallower regions of the world ocean. Terrigenous sediments dominate near the mouths of major rivers.

a clear hemispheric rather than climatic distribution. In the Southern Hemisphere, it comprises up to 20%–50% of the clay minerals; in the Northern Hemisphere, it usually accounts for more than 50% of the clay minerals. Illite forms under a variety of conditions that are not dependent on latitude, so its abundance in marine sediment depends on the degree of dilution by other clay minerals. Montmorillonite is produced by the weathering of volcanic material on land and on the sea floor. It is common in regions of low sedimentation near sources of volcanic ash. It is more abundant in the Pacific and Indian Oceans than in the Atlantic Ocean, where there is little volcanic activity along the surrounding coastlines.

weathering of minerals to form soil. It is ten times as abundant in the tropics as in polar regions, where soil-forming processes are very slow. Illite is the most widespread clay mineral. It has

Sediments derived from organisms are classified as **biogenous** (*bio* = life, *generare* = to produce) **sediments.** These may include shell and coral fragments as well as the hard skeletal parts of single-celled phytoplankton and zooplankton that live in the surface waters. Pelagic biogenous sediments are composed almost entirely of the shells, or tests, of plankton (fig. 3.18). The chemical composition of these tests is either

(a)

(b)

(c)

Figure 3.18 Scanning electron micrographs of biogenous sediments: (a) Diatoms and coccolithophorids. The small disks are detached coccoliths. (b) Radiolarians. (c) Foraminifera.

calcareous (calcium carbonate: CaCO$_3$, as in most seashell material) or siliceous (silicon dioxide: SiO$_2$, clear and hard). If pelagic sediments are more than 30% biogenous material by weight, the sediment is called an **ooze;** specifically, either a **calcareous ooze** or **siliceous ooze,** depending on the chemical composition of the majority of the tests. The distribution of calcareous and siliceous oozes on the sea floor is related to the supply of organisms in the overlying water, the rate at which the tests dissolve as they descend, the depth at which they are deposited, and dilution with other sediment types (see fig. 3.17).

Calcareous tests are created by a group of phytoplankton called **coccolithophorids** (covered with calcareous plates called **coccoliths**), snails called **pteropods,** and amoeba-like animals called **foraminifera** (fig. 3.18a and c). Most coccoliths are smaller than 20 μm. Pteropod tests range from a few millimeters to 1 cm in size, while foraminifera tests range from about 30 μm to 1 mm. These deposits are often named for their principal constituent: coccolithophorid ooze, pteropod ooze, or foraminiferan ooze. Calcareous oozes are the dominant pelagic sediments (see fig. 3.17). The dissolution, or destruction, rate of calcium carbonate varies with depth and temperature and is different in different ocean basins. Calcium carbonate generally dissolves more rapidly in cold, deep water, which characteristically has a higher concentration of CO$_2$ and is slightly more acidic (this is discussed in detail in the sections on the pH of seawater and dissolved gas in chapter 5). The depth at which calcareous skeletal material first begins to dissolve is called the **lysocline**. Beneath the lysocline, seawater becomes undersaturated in dissolved calcium carbonate and there is a progressive decrease in the amount of calcareous material preserved in the sediment (fig. 3.19). The depth at which the amount of calcareous material preserved falls below 20% of the total sediment is called the **carbonate compensation depth (CCD)**. The CCD is also commonly defined as the depth at which the rate of accumulation of calcium carbonate is equal to the rate at which it is dissolved. The lysocline and the CCD will differ in depth depending on the rate of biological productivity, and the production of calcareous

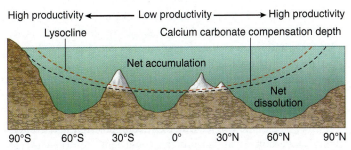

Figure 3.20 The actual depths of the lysocline and carbonate compensation depth vary regionally. The depths may be roughly equal beneath areas of low surface biological productivity where relatively little calcareous skeletal material is produced. Beneath areas of high biological productivity, the amount of calcareous material added to the water can be high, depressing the depth of the CCD below the lysocline.

material, near the surface (fig. 3.20). Beneath areas of low surface productivity, there is a small supply of sinking calcium carbonate particles and the depth of the CCD will not be substantially different from the lysocline. Beneath areas of high productivity, the large supply of sinking calcium carbonate particles will delay the point at which the supply of calcium carbonate equals the rate of dissolution. This will deepen the CCD below the lysocline. Calcareous ooze tends to accumulate on the sea floor at depths above the CCD and is generally absent at depths below the CCD. The CCD has an average depth of about 4500 m (14,800 ft), or roughly midway between the depth of the crests of ocean ridges and the deepest regions of the abyssal plains. In the Pacific, the CCD is generally at depths of about 4200–4500 m (13,800–14,800 ft). An exception to this is the deepening of the CCD to about 5000 m (16,400 ft) in the equatorial Pacific, where high rates of biological productivity result in a large supply of calcareous material. In the North Atlantic and parts of the South Atlantic, it is at or just below depths of 5000 m (16,400 ft). Calcareous oozes are found at temperate and tropical latitudes in shallower areas of the sea floor such as the Caribbean Sea, on elevated ridge systems, and in coastal regions.

Siliceous tests are created by another group of phytoplankton called diatoms and a type of zooplankton called radiolaria (fig. 3.18a and b). Their skeletal remains are the dominant components of diatomaceous and radiolarian ooze, respectively. The pattern of dissolution of siliceous tests is opposite to that of calcareous tests. The oceans are undersaturated in silica everywhere, so siliceous material will dissolve at all depths, but it dissolves most rapidly in shallow, warm water. Siliceous oozes are only preserved below areas of very high biological productivity in the surface waters (see fig. 3.17). Even in these areas, an estimated 90% or more of the siliceous tests are dissolved, either in the water or on the sea floor.

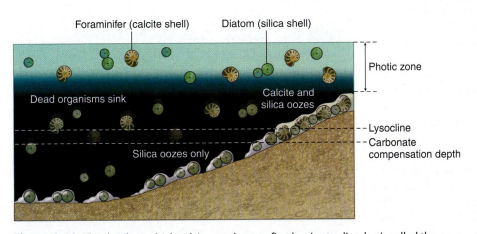

Figure 3.19 The depth at which calcium carbonate first begins to dissolve is called the lysocline. Below the lysocline, less and less calcium carbonate material is preserved. The carbonate compensation depth is the depth at which the rate of supply of calcium carbonate equals the rate of dissolution.

Diatomaceous ooze is found at cold and temperate latitudes around Antarctica and in a band across the North Pacific. Because diatoms are photosynthetic, they require sunlight and inorganic **nutrients,** such as nitrate, phosphate, and silicate (similar to what is found in fertilizers) for growth. The sunlight is available at the ocean's surface; the nutrients are produced by the decomposition of organisms in the ocean, and these nutrients are liberated in the deeper water as decomposition takes place. Only at certain locations are these nutrients returned to the surface by the large-scale upward flow of deeper water. Where this upward flow occurs, sunlight combines with the nutrients to produce the conditions needed for high levels of phytoplankton production. Large numbers of diatoms are found in the areas that combine suitable light, nutrients, and the correct temperature.

Radiolarian ooze is found beneath the warm waters of equatorial latitudes. Radiolaria thrive in warm water, producing siliceous outer shells that are often covered with long spines. Figure 3.18*b* shows radiolaria tests at high magnification.

Sediments derived from the water are classified as **hydrogenous** (*hydro* = water, *generare* = to produce) **sediments.** Hydrogenous sediments are produced in the water by chemical reactions. Most are formed by the slow precipitation of minerals onto the sea floor, but some are created by the precipitation of minerals in the water column in plumes of recirculated water at hydrothermal vents along the ocean ridge system. Hydrogenous sediments include some **carbonates** (limestone-type deposits), **phosphorites** (phosphorus in the form of phosphate in crusts and nodules), **salts,** and **manganese nodules.** In addition, hydrothermally generated sulfides rich in iron and other metals form along the axis of spreading centers on young sea floor, and carbonates and magnesium-rich minerals form off the axis of spreading centers on older sea floor, as discussed in chapter 2.

Hydrogenous carbonates are known to form by direct precipitation in some shallow, warm-water environments as a result of an increase in water temperature or a slight decrease in the acidity of the water. In shallow, warm water with high biological productivity, photosynthetic organisms can remove enough dissolved carbon dioxide in the water to decrease the acidity of the water and trigger the precipitation of calcium carbonate (see the discussion of carbon dioxide as a buffer in chapter 5). The calcium carbonate often precipitates in small pellets called **ooliths** (*oon* = egg) about 0.5–1.0 mm (0.02–0.04 in) in diameter. In the present oceans, there are relatively few places where this is known to be occurring. The largest modern deposits of hydrogenous carbonates are currently forming on the Bahama Banks. Additional deposits are forming on Australia's Great Barrier Reef and in the Persian Gulf.

Phosphorites contain phosphorus in the form of phosphate and are most abundant on the continental shelf and upper part of the continental slope. They are occasionally found as nodules as much as 25 cm (10 in) in diameter or in beds of sand-size grains, but more often they form thick crusts. Most phosphate deposits on continental margins do not appear to be actively accumulating. Phosphorite deposits are currently forming in regions of high biological productivity off the coasts of southwestern Africa and Peru.

Salt deposits occur when a high rate of evaporation removes most of the water and leaves a very salty brine in shallow areas. Chemical reactions occur in the brine, and salts are precipitated or separated from solution and then deposited on the bottom. In such processes, carbonate salts are formed first, followed by sulfate salts, and then chlorides, including sodium chloride. Studies of precipitated material on the floor of the Mediterranean Sea have provided clues to its past isolation from the Atlantic Ocean.

Manganese nodules are composed primarily of manganese and iron oxides but also contain significant amounts of copper, cobalt, and nickel. They were first recovered from the ocean floor in 1873 during the *Challenger* expedition. They are found in a variety of marine environments, including the abyssal sea floor, on seamounts, along active ridges, and on continental margins. Their chemistry is related to the ocean basin they are found in as well as the specific marine environment where they have grown (tables 3.3 and 3.4). Nodules from the Pacific Ocean tend to have the highest concentrations of metals, with the exception of iron. Nodules in the Atlantic Ocean generally have the highest iron concentration. The average weight percent

Table 3.3   Average Chemistry of Manganese Nodules from the Three Ocean Basins

Element	Atlantic	Pacific	Indian	Average for All Three Oceans
Mn	16.18	19.75	18.03	17.99
Fe	21.2	14.29	16.25	17.25
Ni	0.297	0.722	0.510	0.509
Co	0.309	0.381	0.279	0.323
Cu	0.109	0.366	0.223	0.233

Note: The average abundances of manganese (Mn), iron (Fe), nickel (Ni), cobalt (Co), and copper (Cu) in manganese nodules from the Atlantic, Pacific, and Indian Ocean Basins. Numbers are weight % of each metal.

Table 3.4   Average Chemistry of Manganese Nodules from Different Environments

Element	Seamounts	Active Ridges	Continental	Abyssal Depths
Mn	14.62	15.51	38.69	17.99
Fe	15.81	19.15	1.34	17.25
Ni	0.351	0.306	0.121	0.509
Co	1.15	0.400	0.011	0.323
Cu	0.058	0.081	0.082	0.233
Mn/Fe	0.92	0.81	28.9	1.04

Note: The average abundances of manganese (Mn), iron (Fe), nickel (Ni), cobalt (Co), and copper (Cu), and the manganese-to-iron ratio in manganese nodules, from different environments. Numbers are weight % of each metal.

of manganese and iron in nodules is about 18% and 17%, respectively, while the average weight percent of nickel, cobalt, and copper varies from about 0.5% down to 0.2%. Nodules that form on continental margins are very distinct chemically. They have very high manganese concentrations combined with very low iron concentrations. The chemistry of nodules can also be influenced by their position on the sea floor with respect to other sediments. Manganese nodules may lie on top of the other sediment (see fig. 3.16a) or be buried at shallow depth in the sediment. Nodules lying on top of the sediment react chemically with the seawater and can become enriched in iron and cobalt. Those that are buried react with both the seawater and the sediment and can become enriched in manganese and copper. The concentric layers in a nodule typically have slightly different chemistries (see fig. 3.16b). This chemical layering is the result of changes in the chemistry of the seawater as the nodule grew.

On the deep-sea floor, manganese nodules form black or brown rounded masses typically 1–10 cm (0.5–4 in) in diameter, roughly the size of a golf ball or a little larger. Continental margin manganese and iron oxide deposits can take a variety of forms, from nodules similar to those found on the deep-sea floor to extensive slabs, or crusts. Most manganese nodules grow very slowly: 1–10 mm (0.004–0.04 in) per million years for deep-sea nodules, roughly 1000 times slower than accumulation rates of other pelagic sediments. Nodules grow layer upon layer, often around a hard skeletal piece such as a shark's tooth, rock fragment, or fish bone that acts as a seed, much as a pearl grows around a grain of sand. They generally form in areas of very little sediment supply from other sources or where rapid bottom currents prevent them from being deeply buried. Manganese nodules on continental margins are unique in their rapid growth, having growth rates on the order of 0.01–1 mm per year—from 1000 to 1 million times faster than their deep-sea counterparts. Manganese nodules have been mapped in all oceans except the Arctic. They are most abundant in the central Pacific north and south of the biogenous oozes along the equator (see fig. 3.17). In the Atlantic and Indian Oceans, there are higher rates of lithogenous and biogenous sedimentation and consequently fewer deposits of manganese nodules.

Sediments derived from space are classified as **cosmogenous** (*cosmos* = universe, *generare* = to produce) **sediments.** Particles from space constantly bombard Earth. Most of these particles burn up as they pass through the atmosphere, but roughly 10% of the material reaches the surface of Earth. Cosmogenous particles are generally small, and those that survive the passage through the atmosphere and fall in the ocean stay in suspension in the water long enough usually to dissolve before they reach the sea floor. These iron-rich sediments are found in small amounts in all oceans, mixed in with the other sediments. The pattern of related cosmic materials can indicate the direction of the particle shower that supplied them. The particles become very hot as they pass through Earth's atmosphere and partially melt; this melting gives the particles a characteristic rounded or teardrop shape. Cosmic bodies can disintegrate and melt surface materials as they strike Earth. Their impact can cause a splash of melted particles that spray outward and produce splash-form **tektites** (fig. 3.21). Microtektites are found on the ocean floor and on land.

A brief summary of the major sediment types is given in table 3.5.

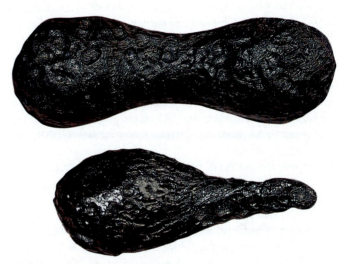

Figure 3.21 Examples of splash-form tektites. These splash-form tektites were collected in Southeast Asia.

Table 3.5   Sediment Summary

Type	Source	Areas of Significant Deposit	Examples	Percent of Ocean Floor Area Covered
Lithogenous (terrigenous)	Eroded rock, volcanoes, airborne dust	Dominantly neritic, pelagic in areas of low productivity	Coarse beach and shelf deposits, turbidites, red clay	~45%
Biogenous	Living organisms	Regions of high surface productivity, areas of upwelling, dominantly pelagic, some beaches, shallow warm water	Calcareous ooze (above the CCD), siliceous ooze (below the CCD), coral	~55%
Hydrogenous	Chemical precipitation from seawater	Mid-ocean ridges, areas starved of other sediment types, neritic and pelagic	Metal sulfides, manganese nodules, phosphates, some carbonates	~1%
Cosmogenous	Space	Everywhere but in very low concentration	Meteorites, space dust	Trace

Patterns of Deposit on the Sea Floor

The patterns formed by the sediments on the sea floor reflect both distance from their source and processes that control the rates at which they are produced, transported, and deposited. Seventy-five percent of marine sediments are terrigenous. The majority of terrigenous sediments are initially deposited on the continental margins but are moved seaward by the waves, currents, and turbidity flows that move across the continental shelves and down the continental slopes. The terrigenous sediments of coastal regions are primarily lithogenous, supplied by rivers and wave erosion along the coasts. Worldwide river sediment transport is about $12–15 \times 10^9$ metric tons per year. The majority of this sediment enters the tropical and subtropical oceans.

Coarse sediments are concentrated close to their sources in high-energy environments—for example, beaches with swift currents and breaking waves. The waves and currents move quite large rock particles in the shore zone, but these larger particles settle out quickly. Finer particles are held in suspension and are carried farther away from their source. This pattern results in a gradation by particle size: coarse particles close to shore and to their source, with finer and finer particles predominating as the distance from the source increases.

Finer sediments are deposited in low-energy environments, offshore away from the currents and waves or in quiet bays and estuaries. In higher latitudes, deposits of rock and gravel carried along by glaciers are found in coastal environments, whereas in low latitudes, fine sediments predominate and are considered to be products of large rivers, heavy rainfall, and loose surface soils.

At the present time, most of the land-derived sediments are accumulating off the world's river mouths and in estuaries. Estuaries and river deltas serve as sediment traps, preventing terrigenous sediments from reaching the deep-sea floor in such places as the Chesapeake and Delaware Bay systems along the North Atlantic coast, in the Georgia Strait of British Columbia, and in California's San Francisco Bay along the North Pacific coast. If sediments are supplied to a delta faster than they can be retained, the sediments will move across the shelf into the deeper water environments. This is currently the case with the sediments of the Mississippi River. Much of the thick sediment layer on the outer continental shelf was laid down during the ice ages, when the sea level was lower—for example, at Georges Bank southeast of Cape Cod. Little is currently being added to these outer regions of the continental shelf.

The accumulation of sediments on the passive shelves of continents results in unstable, steep-sided deposits that may slump, sending a flow of terrigenous sediment moving rapidly down the continental slope in a turbidity current. Turbidity currents move coarse terrigenous materials farther out to sea; in doing so, they distort the general deep-sea sediment pattern and reduce the abundance of pelagic deposits. Near shore, the spring flooding of rivers alternates with periods of low river discharge in summer and fall. The floods bring large quantities of sediment to the coastal waters, and the contributions of this flooding are recorded in the layering of the sediments. Sudden

Figure 3.22 A deep-sea sediment core obtained by the drilling ship *Glomar Challenger*. Note the layering of the sediments.

masses of sediment from the collapse of a cliff or the eruption of a volcano are seen in the sediment pattern as specific additions of large quantities of sand or ash.

Oceanic sediments form visually distinct layers characterized by color, particle size, type of particle, and supply rate (fig. 3.22). Seasonal variations and the patterns of long and short growing seasons for marine life can also be determined from the properties and thicknesses of the layers of biogenous material. Over long periods of geologic time, climatic changes such as the ice ages have altered the biological populations that produce sediment and have left a record in the sediment layers.

In shallow coastal areas, cycles of climate change cause variation in rates of sediment production. Along passive continental margins, biogenous sediments may also be diluted by large amounts of lithogenous sediment washing from the land. In coastal areas where marine life is very abundant and river deposits are sparse, biogenous sediments are formed from both shell fragments and broken corals. In the more homogeneous environment of the deep sea, biogenous sediments make up the majority of the pelagic deposits. There is less dilution with terrigenous materials, and few environmental changes disturb the deep bottom deposits, allowing them to remain relatively unchanged for long periods of time. Calcareous oozes are found where the production of organisms is high, dilution by other sediments is small, and depths are less than 4000 m (13,000 ft). (See the areas including the mid-ocean ridges and the warmer shallower areas of the South Pacific in fig. 3.17.) Siliceous oozes cover the deep-sea floor beneath the colder surface waters of 50°–60°N and S latitude and in equatorial regions where cold, deeper water is brought to the surface by vertical circulation. Deep basin areas of the Pacific have extensive deposits of red clay (again, see fig. 3.17).

Large rock particles of land origin are also moved out to sea by a process known as **rafting.** Glaciers carry sand, gravel, and rocks embedded in the ice. When the glacier reaches the sea, parts break off and fall into the water as icebergs. The icebergs are carried away from land by the currents and winds, taking the terrigenous materials far from their original sources. As the ice melts,

Figure 3.23 Satellite image taken October 2, 2007, showing wind-blown dust from the western Sahara Desert moving over the Atlantic Ocean.

rocks and gravel that were frozen in the ice sink to the sea floor. In addition, sea ice formed in shallow water along the shore can incorporate material from the sea floor and transport it out to sea. Figure 3.17 indicates areas of terrigenous deposits that are affected by ice rafting. It is estimated that ice-rafted material can be found over about 20% of the sea floor. Sometimes large, brown seaweeds known as kelp, which grow attached to rocks in coastal areas, are dislodged by storm waves. The kelp may have enough buoyancy to float away, carrying the attached rock. When the seaweed dies or sinks, the rock is deposited on the ocean floor at some distance from its origin. The deposition of larger rocks by this rafting process is infrequent and irregular.

The wind is an effective agent for moving lithogenous materials out to sea in some parts of the world. Winds blowing offshore from the Sahara Desert or other arid regions transfer sand particles directly from land to sea, sometimes 1000 km (600 mi) or more offshore (fig. 3.23). A similar process can occur between sand dunes and coastal waters. In the open ocean, airborne dust probably supplies much of the deep-sea red clay material. Figure 3.24 indicates the frequency with which winds carry dust, or haze, out to sea. The world's volcanoes are another source of airborne particles. Volcanic ash is present in seafloor sediments and can be found in layers of significant thickness associated with past volcanic events.

Formation of Rock

Loose sediments on the sea floor are transformed into **sedimentary rock** in a process known as **lithification.** Lithification can occur through burial, compaction, recrystallization, and cementation. As one layer of sediment covers another, the weight of the sediments puts pressure on the lower sediment layers, and the sediment particles are squeezed more and more tightly together. The particles begin to stick to each other, and the pore water between the sediment particles, with its dissolved solids, moves through the sediments. As it does so, minerals precipitate on the surfaces of the particles and, in time, act to cement the sediment particles together into a mass of sedimentary rock. The sediments in these processes are also exposed to increasing temperature with increasing depth of burial. Chemical changes also occur in sedimentary particles through interaction with seawater and pore water in a process called **diagenesis.** One example of diagenesis is the gradual lithification of calcareous ooze to form chalk or limestone. In this process, calcite particles in the sediments are cemented by calcite precipitated from the pore waters. The transformation of calcareous ooze to chalk occurs at a sediment depth of a few hundred meters, and the further transformation to limestone occurs with additional cementation under about 1 km burial. Siliceous oozes can be lithified to form a very hard rock called chert.

Sedimentary rock may preserve the layering of the sediments in visually distinct features and strata. Ripple marks from the motion of waves and currents may be seen, and fossils may also be present. Sedimentary rocks are found beneath the sediments of the deep-sea floor, along the passive margins of continents, and on land where they have been thrust upward along active margins or formed in ancient inland seas. Sedimentary rocks include sandstone, shale, and limestone.

If sediments are subjected to greater changes in temperature, pressure, and chemistry, **metamorphic rock** results. Slate is a metamorphic rock derived from shale, and marble is recrystallized metamorphosed limestone.

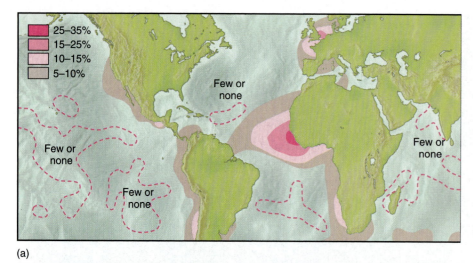

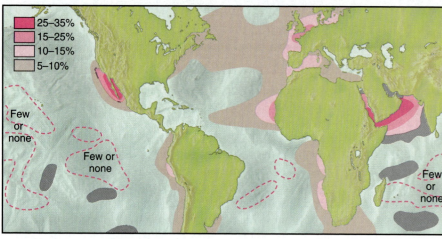

Figure 3.24 Frequency of haze as a result of airborne dust during the Northern Hemisphere's (a) winter and (b) summer. Values are given in percentages of total observations.

Sampling Methods

To analyze sediments, the geological oceanographer must have an actual bottom sample to examine. A variety of devices have been developed to take a sample from the sea floor and return it to the laboratory for analysis. **Dredges** are net or wire baskets that are dragged across a bottom to collect loose bulk material, surface rocks, and shells in a somewhat haphazard manner (fig. 3.25). **Grab samplers** are hinged devices that are spring- or weight-loaded to snap shut when the sampler strikes the bottom. See figure 3.26 for examples of this device. Grab samplers sample surface sediments from a fixed area of the sea floor at a single known location.

A **corer** is essentially a hollow pipe with a sharp cutting end. The free-falling pipe is forced down into the sediments by its weight or, for longer cores, by a piston device that enables water pressure to help drive the core barrel into the sediment; coring devices are shown in figure 3.27*a–e*. The product is a cylinder of mud, usually 1–20 m (3.3–65.6 ft) long, that contains undisturbed sediment layers (see fig. 3.22). Box corers (fig. 3.27*e*) are used when a large and nearly undisturbed sample

of surface sediment is needed. These corers drive a rectangular metal box into the sediment; they have doors that close over the bottom before the sample is retrieved. Long cores that penetrate the thick sediment overlying older sea floor and reach the older sediment layers nearer the oceanic basalt may be obtained by drilling through both loose sediments and rock. The highly sophisticated drilling techniques used by the research vessel *JOIDES Resolution* are discussed in chapter 2.

Geologic oceanographers and geophysicists also study sediment distribution and seafloor structure with high-intensity sound, a technique known as **acoustic profiling.** Bursts of sound are directed toward the sea floor, where the sound waves either reflect from or penetrate into the sediments. Sound waves that penetrate the sediments are refracted and change speed as they pass through the different layers of sediments. A surface vessel tows an array of underwater microphones, or hydrophones, to sense the returning sound waves, and a recorder plots the returning sound energy to produce a profile of the sediment structure. This technique details the structure of the continental margin and finds buried faults, filled submarine canyons, and clues to oil and gas deposits.

Today's ocean scientists are searching for records of Earth's history in the sediment and rock layers of the ocean floor. These layers hold evidence for understanding the formation of the ocean basins and continents, changing climate, periods of unusual volcanism, the presence and absence of various life-forms, and much more. The information is there, but it requires a combination of sophisticated technical know-how at sea and increasingly detailed scientific research in the laboratory to discern and understand it.

Sediments as Historical Records

Marine sediments and the skeletal materials in them provide important information about processes that have shaped the planet and its ocean basins over the past 200 million years. The study of the oceans through an analysis of sediments is called **paleoceanography.** Two examples of the use of marine sediments to unravel history are (1) the study of the distribution of skeletal remains of marine organisms to date the initiation of the Antarctic Circumpolar Current (ACC) and (2) the study of the relative abundance of different oxygen isotopes in foraminifera tests preserved in the sediment to determine variations in climate and seawater temperature.

Prevailing westerly winds at high southern latitudes cause the ACC to flow continuously from west to east around Antarctica. The ACC is a very deep current, extending to depths of 3000–4000 m (9800–13,000 ft), and it is able to flow unimpeded around the

Figure 3.26 Grab samplers: Van Veen (*left*) and orange peel (*right*), both in open positions. Grabs take surface sediment samples.

(a)

(b)

Figure 3.25 (a) Rocks can be recovered from the sea floor with a dredge having a chain basket. Sediments and other fine material escape through the chains. (b) Basalt dredged from a depth of about 8 km (5 mi) near the Tonga Trench in the western Pacific Ocean. The dredge is in the *foreground*.

globe because there are no shallow seafloor features to block its path. This situation has not always existed, however. The southern continents began to break apart at different times. About 135 million years ago, Africa and India first began to separate from Antarctica, South America, and Australia (see fig. 2.39). As recently as 80 million years ago, South America, Antarctica, and Australia were still effectively one landmass. Sometime around 55 million years ago, some sea floor existed between Australia and Antarctica, and by 35 million years ago, they had separated sufficiently to create a narrow expanse of water called the Austral Gulf. South America had not yet separated from Antarctica. Marine sediments deposited at this time indicate that a small, single-celled, shallow-water marine foraminiferan called *Guembelitria* lived in the restricted waters of the Austral Gulf. The absence of its remains in other Southern Hemisphere sediments of the same age indicates that the organism had not been spread to other areas by ocean currents. Skeletal remains of *Guembelitria* appear quite suddenly in sediments deposited all around Antarctica about 30 million years ago. Even though the Drake Passage between South America and Antarctica did not fully open before 20 million years ago, there must have been a shallow channel a few hundred meters deep as early as 30 million years ago that allowed the ACC to first flow around the continent, carrying *Guembelitria* with it.

Calcite tests found in successive layers of sediment can provide information about changes in climate and seawater temperature over time through a careful analysis of the relative abundance of different oxygen isotopes in the calcite. **Isotopes** are atoms of the same element that have different numbers of neutrons in the nucleus; thus, they have different atomic masses but behave identically chemically. Some marine organisms remove oxygen from water molecules in the ocean to construct calcareous hard parts. Water contains the two main isotopes of oxygen: the common ^{16}O and the rarer ^{18}O. These isotopes are stable and do not decay radioactively, so once they have been incorporated into an organism's skeletal material, their relative proportion ($^{18}O{:}^{16}O$) remains constant even after the organism dies. The $^{18}O{:}^{16}O$ ratio in a skeletal fragment depends in part on the relative abundance of the isotopes in the seawater at the time the organism formed it. Thus, calcareous biogenous remains record changes in the isotopic chemistry of seawater that are related to changes in global temperature.

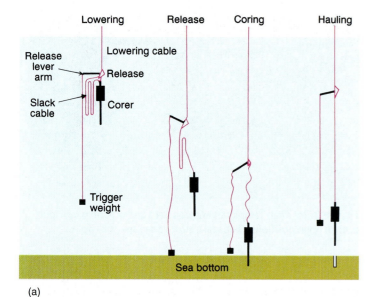

(a)

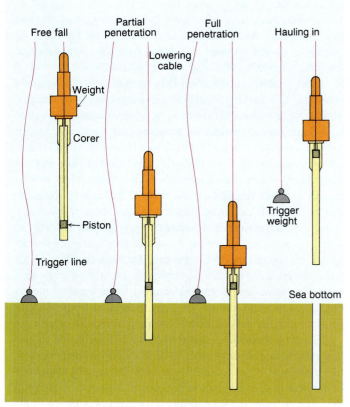

(b)

(c)

(d)

(e)

Figure 3.27 (a) The Phleger corer is a free-fall gravity corer. The weights help to drive the core barrel into the soft sediments. Inside the corer is a plastic liner. The sediment core is removed from the corer by removing the plastic tube, which is capped to form a storage container for the core. (b) A sketch of a piston corer in operation. The corer is allowed to fall freely to the sea bottom. The action of the piston moving up the core barrel owing to the tension on the cable allows water pressure to force the core barrel into the sediments. (c) Recovering a piston corer. The barrel of a gravity corer is in the left foreground. (d) A gravity corer ready to be lowered. (e) A box corer is used to obtain large, undisturbed seafloor surface samples.

Water molecules containing ^{16}O are lighter than molecules containing ^{18}O, so they are more easily removed from the oceans by evaporation. During glacial periods, the water evaporated from the sea surface is trapped in ice sheets; the sea level is lowered, and ^{16}O is removed from the ocean system. This process increases the $^{18}O{:}^{16}O$ ratio in the seawater and in skeletal parts that organisms are forming at that time. When these organisms die, their skeletal parts sink to the sea floor and are incorporated into the sediment. During warmer, interglacial periods, the melting of ice sheets causes a rise in sea level and returns ^{16}O-enriched fresh water to the oceans. The result is a drop in the $^{18}O{:}^{16}O$ ratio in the seawater and in the skeletal parts that are being formed. The isotopic composition of skeletal parts is also influenced by seawater temperature. As temperature decreases, organisms preferentially take up more ^{18}O than ^{16}O in their skeletons. The actual $^{18}O{:}^{16}O$ ratio preserved in a skeletal fragment is primarily due to changes in seawater composition related to the growth and decay of global ice sheets and consequent fall and rise of sea level.

QUICK REVIEW

1. Describe three different ways to classify marine sediments.
2. Relate the distribution of calcareous ooze to seafloor features.
3. Explain the factors that contribute to the accumulation of siliceous oozes and red clays.
4. What is the difference between the lysocline and the CCD?
5. How can small particles sink quickly to the sea floor?

3.4 Seabed Resources

Long ago, people began to exploit the materials of the seabed. The ancient Greeks extended their lead and zinc mines under the sea, medieval Scottish miners followed seams of coal under the Firth of Forth, and, more recently, coal has been mined from undersea strata off Japan, Turkey, and Canada. As technology has developed and as people have become concerned about the depletion of onshore mineral reserves, interest in seabed minerals and mining has grown. At present, the United States is showing little interest in new seabed resources, but international interest remains strong; research continues in exploration, technology development, and environmental studies, especially in Japan, India, China, and South Korea. Keep in mind that each potential deep-sea source is in competition with an onshore supply. Whether the seabed source will be developed depends largely on international markets, needs for strategic materials, and whether offshore production costs can compete with onshore costs.

Sand and Gravel

The largest superficial seafloor mining operation is for sand and gravel, widely used in construction. The technology and cost required to mine sand and gravel in shallow water differ very little from land operations. This is a high-bulk, low-cost material tied to the economics of transport and the distance to market.

Sand and gravel mining is the only significant seabed mining done by the United States at this time. It is estimated that the United States has a reserve of 450 billion tons of sand off its northeastern coast; there are large deposits of gravel along Georges Bank off New England and in the area off New York City. Along the coasts of Louisiana, Texas, and Florida, shell deposits are mined for use in the lime and cement industries, as a source of calcium oxide used to remove magnesium from seawater as part of the process of making magnesium metal, and, when crushed, as a gravel substitute for roads and highways.

Sands are mined as a source of calcium carbonate throughout the Bahamas, which have an estimated reserve of 100 billion metric tons. Coral sands are mined in Fiji, in Hawaii, and along the U.S. Gulf Coast. Other coastal sands contain iron, tin, uranium, platinum, gold, and diamonds. The "tin belt" stretches for 3000 km (1800 mi) from northern Thailand and western Malaysia to Indonesia. Here, sediments rich in tin have been dredged for hundreds of years and supply more than 1% of the world's market. Iron-rich sediments are dredged in Japan, where the reserve of iron in shallow coastal waters is estimated at 36 million tons. The United States, Australia, and South Africa recover platinum from some sands, and gold is found in river delta sediments along Alaska, Oregon, Chile, South Africa, and Australia. Diamonds, like gold, are found in sediments washed down the rivers in some areas of Africa and Australia. Muds bearing copper, zinc, lead, and silver also occur on the continental slopes, but they lie too deep for exploitation, considering the present demand and their market value.

Phosphorite

Phosphorite, which can be mined to produce phosphate fertilizers, is found in shallow waters as phosphorite muds and sands containing 12%–18% phosphate and as nodules on the continental shelf and slope. The nodules contain about 30% phosphate, and large deposits are known to exist off Florida, California, Mexico, Peru, Australia, Japan, and northwestern and southern Africa. Recently, a substantial source of phosphorite was located in Onslow Bay, North Carolina. Eight beds have been found, and five are thought to be economically valuable; they have been estimated to contain 3 billion metric tons of phosphate concentrates.

The world's ocean reserve of phosphorite is estimated at about 50 billion tons. Readily available land reserves are not in short supply, but most of the world's land reserves are controlled by relatively few nations. Therefore, political considerations may make these marine deposits attractive as mining ventures for some countries.

Oil and Gas

Oil and gas represent more than 95% of the value of all resources extracted from the sea floor or below. Oil and gas deposits are almost always associated with marine sedimentary rocks and are believed to be produced by the slow conversion of marine plant and animal organic matter to hydrocarbons. Conditions must be just right for marine organic material to eventually be converted to oil and gas. It must first accumulate in relatively shallow, quiet water with low oxygen content. Anaerobic bacteria can then utilize the organic matter to produce methane and other light hydrocarbons. As these simple hydrocarbons are buried beneath deeper layers of sediment, they are subjected to higher pressure and temperature. Over a period of millions of years, they can be converted to oil or gas. Oil forms if the depth of burial is on the order of about 2 km (1.2 mi). If the organic material is buried even deeper or cooked for a longer period of time at higher temperature, gas is produced. Oil deposits are generally found at depths less than 3 km (1.9 mi), and below 7 km (4.3 mi) only gas is found.

Because oil and gas are very light, they migrate upward over time, moving slowly out of the source rock and into porous rocks above. This upward migration continues until the fluids reach an impermeable layer of rock. The oil and gas then stop their ascent and fill the pore spaces of the reservoir rock below this impermeable layer.

Petroleum-rich marine sediments are more likely to accumulate during periods of geologic time when sea level is unusually high and the oceans flood extensive low-lying continental regions to create large shallow basins. Much oil and gas are found in marine rocks that formed from sediments deposited during a relatively short period of time during the Jurassic and Cretaceous, between about 85 million and 180 million years ago, when sea level was high.

Major offshore oil fields are found in the Gulf of Mexico, the Persian Gulf, and the North Sea, and off the northern coast of Australia, the southern coast of California, and the coasts of the Arctic Ocean.

Bringing the offshore oil fields into production has required the development of massive drilling platforms and specialized equipment to withstand heavy seas and fierce storms and to allow drilling and well development at great depth. Although the cost of drilling and equipping an offshore well is three to four times greater than that of a similar venture on land, the large size of the deposits allows offshore ventures to compete successfully. The gas and oil potential in even deeper offshore waters is still unknown, but the deeper the water in which the drilling must be done, the higher the cost.

The new methods and equipment developed and used for deep-sea oceanographic drilling and research have provided the prototypes for new generations of deep-sea commercial drilling systems. Even though legal restraints, environmental concerns, and worldwide political uncertainties will continue to contribute to the slow development of offshore deposits, petroleum exploration and development will undoubtedly continue to be the main focus of ocean mining in the near future.

Gas Hydrates

In recent years, interest has been growing in gas hydrates trapped in marine sediments (fig. 3.28). Gas hydrates are a combination

(a)

(b)

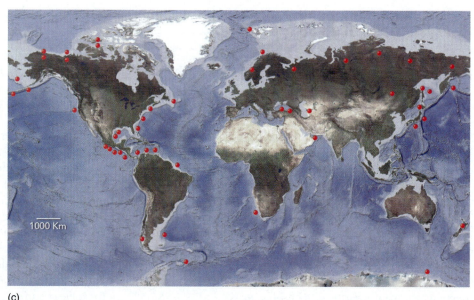
1000 Km
(c)

Figure 3.28 (a) The cream-colored, icy material in this underwater ledge is gas hydrate. Frozen water can trap other molecules within its cage-like structure, including molecules of methane gas. (b) Gas hydrate is called the *ice that burns*. Here, the methane being released from the ice is burning, not the water ice, which is incombustible. (c) Gas hydrates occur within ocean-floor sediments, especially in parts of the ocean that are cold and deeper than 500 m (1,600 ft). Common settings are along passive margins and trenches. Gas hydrates also occur on land, beneath the frozen arctic tundra.

of natural gas, primarily methane (CH_4), and water, which forms a solid, icelike structure under pressure at low temperatures. Drill cores of marine sediment have recovered samples of gas hydrates that melt and bubble as the natural gas escapes. These melting samples burn if lit. Gas hydrates are a subject of intense interest for three reasons: they are a potential source of energy, they may contribute to slumping along continental margins, and they may play a role in climate change.

When 1 cubic foot of gas hydrate melts, it releases about 160 cubic feet of gas. A gas hydrate accumulation thus can contain a huge amount of natural gas. Estimates of the amount of natural gas contained in the world's gas hydrate accumulations are speculative and range over three orders of magnitude, from about 2800 to 8,000,000 trillion (2.8×10^{15} to 8×10^{18}) cubic meters of gas. By comparison, in 2000, the U.S. Geological Survey estimated conventional natural gas accumulations for the world at approximately 440 trillion (4.4×10^{14}) cubic meters. Despite the enormous range in the estimated amount of natural gas contained in gas hydrates, even the lowest estimates suggest that gas hydrates are a much greater resource of natural gas than conventional accumulations and may be a substantial source of energy in the future (table 3.6). It is important to note, however, that none of these assessments have predicted how much gas could actually be produced from the world's gas hydrate accumulations given present technology and their location.

A second reason gas hydrates are significant is their effect on seafloor stability. Along the southeastern coast of the United States, a number of submarine landslides, or slumps, have been identified that may be related to the presence of gas hydrates. The hydrates may inhibit normal sediment consolidation and cementation processes, creating a weak zone in the sediments. Alternately, the lowering of sea level during the last glacial period may have reduced the pressure on the sea floor enough to allow some of the gas to escape from the hydrates and accumulate in the sediment, decreasing its strength.

A final reason for studying gas hydrates is their potential link to climate changes. The amount of methane stored in hydrates is believed to be about 3000 times the amount currently present in the atmosphere. Since methane is a greenhouse gas, its release from hydrates could affect global climate.

Manganese Nodules

Manganese nodules are found scattered across the world's deep-ocean floors, with particular concentrations in the red clay regions of the northeastern Pacific (see figs. 3.16 and 3.17).

The nodule chemistry varies from place to place, but the nodules in some areas contain 30% manganese, 1% copper, 1.25% nickel, and 0.25% cobalt: these are much higher concentrations than are usually found in land ores. Cobalt is of particular interest, since it is classified as being of "strategic" importance to the United States and hence essential to the national security. Cobalt is an important component in the manufacture of strong alloys used in tools and aircraft engines. The nodules grow very slowly, but they are present in huge quantities. An estimated 16 million additional tons of nodules accumulate each year.

Cobalt-enriched manganese crusts, or hard coatings on other rocks, were discovered in relatively shallow water on the slopes of seamounts and islands within U.S. territorial waters in the 1980s. The concentration of cobalt in these deposits is roughly twice that found in typical pelagic manganese nodules and about one and one-half times that found in known continental deposits. These crusts are not being actively mined because of the relatively low cost and continued availability of continental sources.

Sulfide Mineral Deposits

Expeditions to the rift valleys of the East Pacific Rise near the Gulf of California, the Galápagos Ridge off Ecuador, and the Juan de Fuca and Gorda Ridges off the northwestern United States have found sulfides of zinc, iron, copper, and possibly silver, molybdenum, lead, chromium, gold, and platinum. Molten material from beneath Earth's crust rises along the rift valleys, fracturing and heating the rock. Seawater percolates into and through the fractured rock, forming metal-rich hot solutions. When these solutions rise from the cracks and cool, the metallic sulfides precipitate to the sea floor. Deposits may be tens of meters thick and hundreds of meters long. Too little is presently known about these deposits to determine whether they might be of economic importance at some future date. No practical technology exists to sample or retrieve them at this time, and, like the manganese nodules, these deposits are found outside national economic zones, so there are ownership problems.

In the 1960s, metallic sulfide muds were discovered in the Red Sea. Deposits of mud 100 m (330 ft) thick were found in small basins at depths of 1900–2200 m (6200–7200 ft). High amounts of iron, zinc, and copper and smaller amounts of silver and gold were found. The salty brines over these muds contained hundreds of times more of these metals than normal seawater.

Table 3.6 Potential Significance of Gas Hydrates

Estimated Volume of Gas Hydrates, EVGH (10^{12} m³)	Ratio of EVGH to World Supply of Natural Gas	Ratio of EVGH to Natural Gas Consumption in the: United States in 2000	World in 1999
Low of 2800	6.4:1	4375:1	1175:1
High of 8,000,000	18,200:1	12,500,000:1	3,355,700:1

QUICK REVIEW

1. How are sand and gravel used as valuable resources both directly and indirectly?
2. What seafloor resource is used in the production of fertilizer?
3. What resources account for the vast majority of the monetary value of all resources extracted from the ocean?
4. List multiple reasons why gas hydrates are important.
5. What important metals are found in manganese nodules?
6. Hydrothermal vents along mid-ocean ridges are often the site of what natural resource?

Summary

Ocean-depth measurements were made first with a hand line, then with wire, and, since the 1920s, with echo sounders. Today, they are made with precision depth recorders. Seafloor features can also be sensed by satellites that measure the distance between the satellite and the sea surface.

The bathymetric features of the ocean floor are as rugged as the topographic features of the land but erode more slowly. The continental margin includes the continental shelf, slope, and rise. The continental shelf break is located at the change in steepness between the continental shelf and the continental slope. Submarine canyons are major features of the continental slope and, in some cases, the continental shelf. Some canyons are associated with rivers; others are believed to have been cut by turbidity currents. Turbidity currents deposit graded sediments known as turbidites.

The ocean basin floor is a flat abyssal plain, but it is interrupted by scattered abyssal hills, volcanic seamounts, and flat-topped guyots. In warm, shallow water, corals have grown up around the seamounts to form fringing reefs. A barrier reef is formed when a seamount subsides while the coral grows. An atoll results when the seamount's peak is fully submerged. The mid-ocean ridges and rises extend through all the oceans; trenches are associated with island arcs and are found mainly in the Pacific Ocean.

Sediment classifications are based on their size, location, origin, and chemistry. Sediment particles are broadly categorized in order of decreasing size as gravel, sand, and mud. Within each of these categories, particles can be further subdivided by size. The sinking rate and distance traveled in the water column are related to sediment size, shape, and currents. Small particles sink more slowly than large particles. The very smallest particle sizes, silts and clays, sink so slowly that they may be transported large distances while falling to the sea floor. The sinking rate of particles is increased by clumping and incorporation into fecal pellets.

Sediments that accumulate on continental margins and the slopes of islands are called neritic sediments. Sediments of the deep-sea floor are pelagic sediments. In general, pelagic sediments accumulate very slowly and neritic sediments accumulate more rapidly.

Sediments formed from particles of preexisting rocks are called lithogenous sediments. These sediments are also sometimes called *terrigenous sediments.* Since lithogenous sediments are typically derived from the land, they are also known as terrigenous sediments. Pelagic lithogenous sediment is dominated by red clay. Red clay dominates marine sediments only in regions that are starved of other sources of sediment. Biogenous sediments come from living organisms. Sediments composed of at least 30% biogenous material are called oozes; this material accumulates in regions of high biological productivity. Siliceous sediments are subjected to dissolution everywhere in the oceans, while calcareous sediments dissolve rapidly in deep, cold water below the CCD. Sediments that precipitate directly from the water are called hydrogenous sediments. These include manganese nodules on the deep-sea floor and metal sulfides along mid-ocean ridges. Sediments containing particles that originate in space are called cosmogenous sediments.

Patterns of sediment deposit result from the distance from the source area, the abundance of living organisms contributing remains, the seasonal variations in river flow, waves and currents including turbidity currents, the variability in land sources, the prevailing winds, and sometimes rafting.

Coarse sediments are concentrated close to shore; finer sediments are found in quiet offshore or nearshore environments. Terrigenous sediments are found mainly along coastal margins; most deep-sea sediments come from biogenous sources. The distributions of particle sizes reveal the processes that formed the deposit, and the sediment layers provide clues to ancient climate patterns.

In general, sedimentation rates are slowest in the deep sea and greatest near the continents. In some areas, relict sediments were deposited under conditions that no longer exist. Mechanisms that increase the sinking rates of particles include clumping and incorporation of sediment particles into larger fecal pellets of small marine organisms. Loose sediments are transformed into sedimentary rock in which the layering of the sediments may be preserved.

Sediments are sampled with dredges, grabs, and corers; deep-sea drilling takes samples through the sediments and from the seafloor rock below.

Calcareous biogenous sediments preserve records of changes in the oxygen isotopic composition of seawater that are related directly to water temperature and hence can be used to study changes in global climate. Consequently, variations in ^{18}O:^{16}O isotopic ratios in calcareous skeletal remains record fluctuations in global coverage by ice sheets and in sea level.

Seabed resources include sand and gravel used in construction and landfills. Sands and muds that are rich in mineral ores are mined. Phosphorite nodules are the raw material of fertilizer. Oil and gas are the most valuable of all seabed resources. Manganese nodules are rich in copper, nickel, and cobalt; they are present on the ocean floor in huge numbers. Retrieval of sea-floor mineral resources is slowed by disputes over international law, high mining costs, and low market prices. Sulfide mineral deposits have been discovered along rift valleys; their economic importance is unknown.

Large deposits of gas hydrates are being studied to determine their potential as economically important sources of methane gas. These deposits are icelike accumulations of natural gas and water that form at low temperature and high pressure on the sea floor. Scientists are also studying their possible role in submarine landslides and global climate change.

Key Terms

soundings, 84	seamount, 90	biogenous sediment, 96	manganese nodule, 98
fathom, 84	guyot, 90	ooze, 97	oolith, 98
echo sounder, 84	fringing reef, 90	calcareous ooze, 97	cosmogenous sediment, 99
depth recorder, 84	barrier reef, 90	siliceous ooze, 97	tektite, 99
continental margin, 86	atoll, 90	coccolithophorids, 97	rafting, 101
passive margin, 86	island arc system, 91	coccolith, 97	sedimentary rock, 101
active margin, 86	phytoplankton, 93	pteropod, 97	lithification, 101
continental shelf, 86	zooplankton, 93	foraminifera, 97	diagenesis, 101
continental shelf break, 88	test, 93	lysocline, 97	metamorphic rock, 101
continental slope, 88	neritic sediment, 94	carbonate compensation	dredge, 102
submarine canyon, 88	relict sediment, 94	depth (CCD), 97	grab sampler, 102
turbidity current, 89	pelagic sediments, 94	nutrient, 98	corer, 102
turbidite, 89	lithogenous sediment, 94	hydrogenous sediment, 98	acoustic profiling, 102
continental rise, 90	terrigenous sediment, 94	carbonate, 98	paleoceanography, 102
abyssal plain, 90	abyssal clay, 95	phosphorite, 98	isotope, 103
abyssal hill, 90	red clay, 95	salt, 98	

Study Problems

1. Assume an accumulation rate of 0.8 cm per 1000 years for deep-ocean pelagic sediment. How long would it take to accumulate 500 m of sediment?

2. Assume an accumulation rate of 30 cm per 1000 years for continental shelf sediment. How long would it take to accumulate 500 m of sediment?

3. If underwater cables are spaced 14 km apart on the sea floor, and if monitoring equipment shows that they break in sequence from the shallowest to the deepest at 15-minute intervals, what can you determine about the event that caused the breaks?

4. If the average concentration of suspended sediment in the water of a harbor is 1 g/m^3, the volume of water in the harbor is 158 km^3, and the daily sediment supply rate averages 1×10^7 kg, what is the average residence time of sediment in the water? If the harbor has an average depth of 15.8 m, what is the surface area of the harbor?

The Physical Properties of Water

Learning Outcomes

After studying the information in this chapter students should be able to:

1. *review* the physical properties of water listed in table 4.1,

2. *distinguish* between temperature and heat,

3. *construct* a plot of temperature vs. heat gain in 1 gram of water as it goes from a temperature of −10°C to 110°C,

4. *construct* a plot of density vs. temperature for pure water from a temperature of −2°C to 10°C,

5. *contrast* the change in density of pure water and average salinity seawater as they cool from a temperature of 10°C to −2°C,

6. *illustrate* the attenuation of light in open ocean water and coastal water, and

7. *theorize* how submarines could evade detection by surface ships using acoustic equipment.

Breaking wave along the coast at La Jolla, California.

Water is one of the most common substances on Earth, yet it is uncommon in many of its properties. Water is a unique compound. It makes life possible, and its properties largely determine the characteristics of the oceans, the atmosphere, and the land. To understand the oceans, one must examine water as a substance and learn something of its physical and chemical characteristics. In this chapter, we learn about the structure of the water molecule and explore the properties of water.

4.1 The Water Molecule

The properties of water have excited scientists for over 2000 years. The early Greek philosophers (500 B.C.) counted four basic elements from which they believed all else was made: fire, earth, air, and water. In 1783, more than 2200 years later, English scientist Henry Cavendish determined that water was not a simple element but a substance made up of hydrogen and oxygen. Shortly afterward, another Englishman, Sir Humphrey Davey, discovered that the correct formula for water was two parts hydrogen to one part oxygen, or H_2O.

The chemical properties that make water such a special, useful, and essential substance result from its molecular structure. These properties are the subject of this chapter and are summarized in table 4.1.

The water molecule is deceptively simple, made up of three atoms: two hydrogen atoms and one oxygen atom. An atom is the smallest unit of matter that retains the properties of an element. We can envision an atom as consisting of three distinct types of particles located in two different regions. At the center is the nucleus. The nucleus contains positively charged particles called protons tightly packed together with electrically neutral particles called neutrons. A characteristic number of protons are present in the nucleus of every atom of a given element. For instance, every atom with one proton in the nucleus is an atom of hydrogen, and every atom with eight protons in the nucleus is an atom of oxygen. Negatively charged particles called electrons orbit the nucleus in a series of energy levels. The different energy levels can hold different numbers of electrons. Electrically neutral atoms have the same number of electrons as protons. In the case of hydrogen, there is one electron in the first energy level—an energy level that can hold a maximum of two electrons. In the case of oxygen, two electrons fill the first energy level, and six additional electrons are in the second energy level—a level that can hold a total of eight electrons when it is full. Thus, a hydrogen atom's outermost energy level is one electron short of being full and an oxygen atom's outermost energy level is two electrons short of being full. When two hydrogen atoms and one oxygen atom combine to form a water molecule, each hydrogen atom shares its single electron with the oxygen atom, and the oxygen atom shares one of its electrons with each hydrogen atom. Shared pairs of electrons

form **covalent bonds.** The formation of covalent bonds in the water molecule has the effect of filling the outer energy levels of all three atoms in the molecule (fig. 4.1a).

The angle between the hydrogen atoms in the water molecule is about 105° (fig. 4.1b). A molecule of water is electrically neutral, but the negatively charged electrons within the molecule are distributed unequally. The shared electrons spend more time around the oxygen nucleus than they do around the hydrogen nuclei, giving the oxygen end of the molecule a slightly negative charge. The hydrogen end of the molecule carries a slightly positive charge. As a consequence, the opposite ends of the water molecule have opposite charges, and the molecule is an electrically unbalanced, or **polar, molecule.**

When one end of a water molecule comes close to the oppositely charged end of another water molecule, a bond forms between their positively and negatively charged ends (fig. 4.1c). These bonds are known as **hydrogen bonds,** and each water molecule can establish hydrogen bonds with neighboring water molecules. Any single hydrogen bond is weak (less than one-tenth the strength of the covalent bonds between the hydrogen and oxygen atoms), but as one hydrogen bond is broken, another is formed. Consequently, water is characterized by an extensive but ever-changing three-dimensional network of hydrogen-bonded molecules; the result is an atypical liquid with the extraordinary properties that are the subject of this chapter.

QUICK REVIEW

1. Describe the shape of the water molecule.
2. Why is the water molecule called a "polar" molecule?
3. How does the shape of the water molecule affect its behavior?

4.2 Temperature and Heat

The atoms and molecules in any gas, liquid, or solid are always in motion. Thus, each individual atom or molecule has a certain amount of kinetic energy that is equal to one-half its mass times its velocity squared (kinetic energy = $1/2 \ mv^2$). Even in solids, where atoms and molecules are tightly packed and fixed in place, unable to move from one location to another as in a gas or liquid, those atoms and molecules will vibrate around their average position; they will have kinetic energy. The **temperature** of a substance is a measure of the average kinetic energy of the atoms and molecules in the substance. For a homogeneous material consisting of identical atoms or molecules all having the same mass, temperature is simply related to the average velocity of the atoms or molecules. The colder a substance is, the slower the motion of its atoms and molecules. Conversely, the warmer a substance is, the faster the motion of its atoms and molecules. Temperature is measured in **degrees** using one of three different scales (fig. 4.2). The two most commonly used scales are the Fahrenheit (°F) and Celsius (°C) scales (see "Temperature," appendix B). The third scale is the Kelvin (K) scale. It is used to measure extremes of hot and cold and is constructed in such a manner that 0 K

Table 4.1 Properties of Water

Definition	Comparison	Effects
PHYSICAL STATES		
Gas, Liquid, Solid. Addition or loss of heat breaks or forms bonds between molecules to change from one state to another.	The only substance that occurs naturally in three states on Earth's surface.	Important for the hydrologic cycle and the transfer of heat between the oceans and atmosphere.
SPECIFIC HEAT		
One calorie per gram of water per °C.	Highest of all common solids and liquids.	Prevents large variations of surface temperature in the oceans and atmosphere.
SURFACE TENSION		
Elastic property of water surface.	Highest of all common liquids.	Important in cell physiology, water surface processes, and drop formation.
LATENT HEAT OF FUSION		
Heat required to change a unit mass from a solid to a liquid without changing temperature.	Highest of all common liquids and higher than most solids.	Results in the release of heat during freezing and the absorption of heat during melting. Moderates temperature of polar seas.
LATENT HEAT OF VAPORIZATION		
Heat required to change a unit mass from a liquid to a gas without changing temperature.	Highest of all common substances.	Results in the release of heat during condensation and the absorption of heat during vaporization important in controlling sea surface temperature and the transfer of heat in the atmosphere.
COMPRESSIBILITY		
Average pressure on total ocean volume 200 atmospheres; ocean depth decreased by 37 m (121 ft).	Seawater is only slightly compressible. $4–4.6 \times 10^{-5}$ cm³/g for an increase of 1 atmosphere of pressure.	Density changes only slightly with pressure. Sinking water can warm slightly due to its compressibility.
DENSITY		
Mass per unit volume: grams per cubic centimeter, g/cm³.	Density of seawater is controlled by temperature, salinity, and pressure.	Controls the ocean's vertical circulation and layering. Affects ocean temperature distribution.
VISCOSITY		
Liquid property that resists flow. Internal friction of a fluid.	Decreases with increasing temperature. Salt and pressure have little effect. Water has a low viscosity.	Some motions of water are considered friction free. Low friction dampens motion; retards sinking rate of single-celled organisms.
DISSOLVING ABILITY		
Dissolves solids, gases, and liquids.	Dissolves more substances than any other solvent.	Determines the physical and chemical properties of seawater and the biological processes of life-forms.
HEAT TRANSMISSION		
Heat energy transmitted by conduction, convection, and radiation.	Molecular conduction slow; convection effective. Transparency to light allows radiant energy to penetrate seawater.	Affects density; related to vertical circulation and layering.
LIGHT TRANSPARENCY		
Transmits light energy.	Relatively transparent for visible wavelength light.	Allows plant life to grow in the upper layer of the sea.
SOUND TRANSMISSION		
Transmits sound waves.	Transmits sound very well compared to other fluids and gases.	Used to determine water depth and to locate objects.
REFRACTION		
The bending of light and sound waves by density changes that affect the speed of light and sound.	Refraction increases with increasing salt content and decreases with increasing temperature.	Makes objects appear displaced when viewed by light and sound.

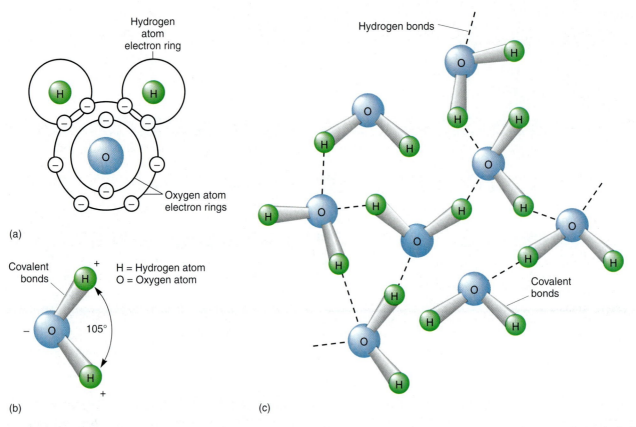

(a)

(b)

(c)

Figure 4.1 The water molecule. (a) The hydrogen atoms share electrons with the outer ring of the oxygen atoms. (b) The angle at which the hydrogen atoms form covalent bonds with the oxygen atom results in a polar molecule. (c) The positive and negative charges allow each water molecule to form hydrogen bonds with other water molecules.

corresponds to absolute zero, the temperature at which all atomic and molecular motion ceases. Absolute zero, 0 K, is equal to −273.2°C or −459.7°F; it is a temperature that can never be reached. **Heat** is a measure of the total kinetic energy of the atoms and molecules in a substance. The amount of heat in a substance is determined by the sum of the product of one-half the mass of every atom or molecule in the substance and its velocity squared. Heat is measured in **calories.** One calorie is the amount of heat needed to raise the temperature of 1 g of water by 1°C from 14.5°–15.5°C (see "Energy," appendix B). One thousand calories is equivalent to 1 Calorie, kilocalorie (kcal), or food calorie.

The difference between heat and temperature can be readily understood with a simple example. Imagine comparing boiling water in a pot on your stove to the water in a swimming pool on a warm day. The boiling water in the pot clearly has a higher temperature than the water in the pool; the average kinetic energy, or velocity, of the water molecules in the pot is much faster than in the pool. However, there is far more heat in the pool water. Even though the average kinetic energy, or velocity, of the water molecules in the pool is relatively small, there are so many of them that the total kinetic energy, or heat, is greater.

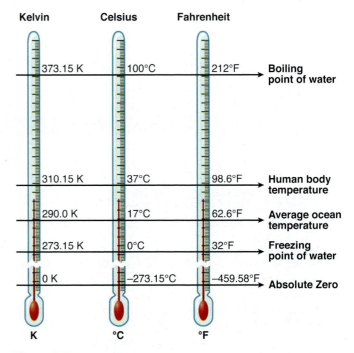

Figure 4.2 A comparison of the Farenheit (F), Celsius (C), and Kelvin (K) scales for measuring temperature.

QUICK REVIEW

1. What is heat, and how is it measured?
2. How does heat differ from temperature?
3. Explain what a calorie is.

4.3 Changes of State

Water exists on Earth in three physical states: solid, liquid, and gas. When it is a solid, we refer to it as ice, and when it is a gas, we call it water vapor. Pure water ice melts at 0°C and liquid pure water boils at 100°C at standard atmospheric pressure. Pure water is defined as fresh water without suspended particles or dissolved substances, including gases.

Because of the hydrogen bonding between the water molecules, it takes energy (heat) to separate them from each other: that is, for liquid water to evaporate or for ice to melt. In the natural environment, this heat is supplied by the Sun.

When pure water makes any of the changes among liquid, solid, and gas, it is said to change its state. Changes of state are due to the addition or loss of heat. When enough heat is added to solid water or ice, the hydrogen bonds break and the ice melts, forming water. When heat is added to liquid water, the temperature of the water rises and some of the water molecules escape from the liquid or evaporate to form water vapor. When heat is removed from water vapor and the temperature falls below the **dew point,** or temperature of water vapor saturation, the water vapor condenses to liquid. When liquid water loses heat and its temperature is lowered to its freezing point, ice is formed as the water molecules form a crystalline lattice.

To change pure water from its solid state (ice) to liquid water at 0°C requires the addition of 80 calories for each gram of ice (fig. 4.3). There is no change in temperature; there is a change

in the physical state of the water as hydrogen bonds break. The reverse of this process is required to change liquid water to ice. For each gram of liquid water that becomes ice, 80 calories of heat must be removed at 0°C. The heat necessary to change the state of water between solid and liquid is known as the **latent heat of fusion.** This addition or loss of heat takes time in nature. A lake does not freeze immediately, even though the surface water temperature is 0°C, nor does the ice thaw on the first warm day. Time is required to remove or add the heat needed for the change of state.

One gram of liquid water requires 1 calorie of heat to raise its temperature 1°C. Therefore, 100 calories are needed to raise the temperature of 1 g of water from 0° to 100°C. In comparison, the 80 calories of heat required per gram to convert ice to and from liquid water at 0°C with no change in temperature is relatively large. Ice is a very stable form of water; a large addition of heat is required to melt ice, and a large removal of heat is required to freeze water.

The change of state between liquid water and water vapor requires 540 calories of heat to convert 1 g of water to water vapor at 100°C (fig. 4.3). When 1 g of water vapor condenses to the liquid state, 540 calories of heat are liberated. Again, there is no change in temperature; there is only a change in the water's physical state. The heat needed for a change between the liquid and vapor states is the **latent heat of vaporization.**

When heat energy is added to water it can result in a change in state or an increase in the temperature of the water (fig. 4.4). As long as water is in a constant state (solid, liquid, or water vapor) the addition of heat energy will raise the temperature of the water. However, if heat energy is added during a change in state, the temperature of the water will remain constant. This is because all of the heat energy added is going into changing the state. As you can see in figure 4.4, it takes much more energy to change the state of water from liquid to water vapor than it does from solid to liquid.

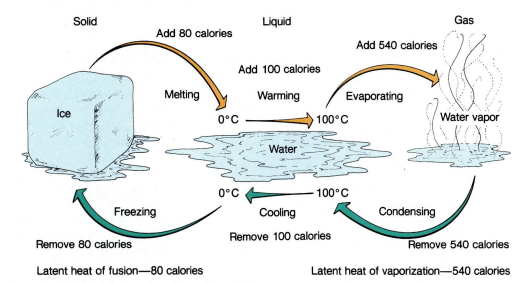

Figure 4.3 Heat energy must be added to convert a gram of ice to liquid water and to convert liquid water to water vapor. The same quantity of heat must be removed to reverse the process.

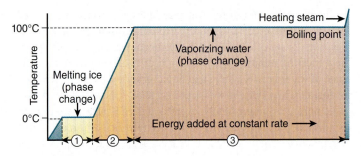

1. Latent heat of fusion = 80 cal/g
2. Energy required to heat/cool 1 g liquid water between
   0 and 100°C = 100 cal/g
3. Latent heat of vaporization = 540 cal/g

Figure 4.4 The addition of heat energy to water can result in a change in temperature or a change in state. For a given state (solid, liquid, or gas), the addition of 1 calorie of heat to 1 gram of water will raise water temperature by 1°C. During a change in state, or phase transition, water temperature remains constant because the heat energy added is used entirely to change the state.

It is possible to change the state of water from liquid to water vapor at temperatures other than 100°C; for example, rain puddles evaporate and clothes dry on the clothesline. This change requires slightly more heat to convert the liquid water to a gas at lower temperatures (fig. 4.5). It is also possible for water to remain in the liquid state at temperatures greater than 100°C if you increase pressure. This is why hot vent water along mid-ocean ridges can reach temperatures above 300°C without turning into water vapor, or steam.

As water is evaporated from the world's lakes, streams, and oceans and returned as precipitation, heat is being removed from Earth's surface and liberated into the atmosphere, where condensation occurs to form clouds. This heat is a major source of the energy used to power Earth's weather systems (see chapter 6).

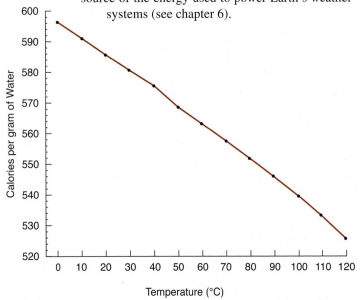

Figure 4.5 The amount of energy required to convert liquid water to water vapor as a function of temperature (the latent heat of vaporization).

Under certain conditions it is possible to change ice directly to a gas, a process known as **sublimation.** Sublimation is seen in nature when snow or ice evaporates directly under very cold and dry conditions.

The behavior of water, as described, is explained by considering the processes occurring between the water molecules. At the molecular level, the addition of heat energy increases the speed of the molecules, while the loss of heat energy decreases their rate of motion. Heat energy is required to break hydrogen bonds between water molecules, and heat is released when hydrogen bonds are formed. The addition of heat to water causes a relatively small change in the temperature because much of the heat energy is used to disrupt the hydrogen bonds. These bonds must be broken before the molecules are able to move more rapidly. When heat is removed from water, the molecules slow, many additional hydrogen bonds are formed, and considerable energy is released as heat; this process prevents any rapid drop in temperature.

Water molecules stay close together because of their polarity. If the molecules are moving fast enough, they overcome the attractions between them and leave the liquid, entering the air as a gas. Relatively large amounts of heat are needed to evaporate water because hydrogen bonds must first be broken. The greater the addition of heat, the more hydrogen bonds are disrupted and the greater the average energy of motion of the molecules. Under these conditions, more water molecules leave the liquid more quickly. In the same way, large amounts of heat must be extracted from water to form the hydrogen bonds required to freeze water. On Earth, natural temperatures required for boiling are rare and those for freezing are frequent but geographically limited; the average Earth temperature of about 16°C ensures that liquid water is abundant.

Dissolved salts in water change its boiling and freezing points. The boiling temperature is raised and the freezing temperature is lowered. The amount of change is controlled by the amount of salt dissolved. The rise in the boiling temperature is of little consequence to oceanography because seawater does not normally reach such high temperatures in nature, but the lowering of the freezing point is important in the formation of sea ice. Seawater freezes at about −2°C.

QUICK REVIEW

1. Compare the amount of energy required to change the state of water from solid to liquid to the amount required to change the state from liquid to water vapor.
2. Why is there such a big difference between the latent heat of fusion and the latent heat of vaporization?

4.4 Specific Heat

Of all the naturally occurring earth materials, water changes its temperature the least for the addition or removal of a given amount of heat. The ability of a substance to give up or take in a given amount of heat and undergo large or small changes in temperature is a measure of the substance's **specific heat.** The specific heat of water is very high compared to that of soil, rock, and air (table 4.2). For example, summer temperatures in the

Libyan Desert reach 50°C and temperatures in the Antarctic drop to −50°C, for a worldwide temperature range of 100°C on land. Ocean temperatures vary from a high of approximately 28°C in the equatorial areas to a low of −2°C in Antarctic waters, for a worldwide water temperature range of 30°C. The water in a lake changes its temperature very little between noon and midnight, but the adjacent land and air temperature changes are large during this same time. The high specific heat of water and the ability of water to redistribute heat over depth allow the world's lakes and oceans to change temperature slowly, helping to keep Earth's surface temperature stable.

Specific heat of a material is the quantity of heat required to produce a unit change of temperature in a unit mass of that material. The specific heat of water is 1.0 calorie per gram per degree Celsius (cal/g/°C). This is much higher than specific heats of most other liquids because of water's extensive hydrogen bonding. Energy that is used to break hydrogen bonds between water molecules is used in other liquids to directly increase the molecular motion and temperature. The high specific heat of water allows water to gain or lose large quantities of heat with little change in temperature. When salt is added to pure water, the changes in heat capacity, latent heat of fusion, and latent heat of vaporization are small.

The **heat capacity** of a material is the quantity of heat required to produce a unit change of temperature in the material. A material's heat capacity depends on the material's specific heat and the mass of the material. For instance, the heat capacity of 1 kilogram of water is 1000 calories/°C, and the heat capacity of 1 kilogram of sandy soil is 240 calories/°C (see table 4.2). The specific heat of a material is numerically the same as the heat capacity of 1 gram of the material.

QUICK REVIEW

1. Describe the difference between specific heat and heat capacity.
2. Which would have more heat energy—a pot of boiling water or a cold lake in winter? Explain your choice.
3. Why are larger and more rapid temperature changes seen on land than in the ocean?

Table 4.2   Specific Heat of Earth Materials

Material	Specific Heat (Calories/g/°C)
Water	1.00
Air	0.25
Sandstone	0.47
Shale	0.39
Sandy soil	0.24
Basalt	0.20
Limestone	0.17

4.5 Cohesion, Surface Tension, and Viscosity

In the liquid state, the bonds between water molecules form, break, and re-form with great frequency. Each bond lasts only a few trillionths of a second. However, at any instant, a substantial percentage of all water molecules are bonded to their neighbors. Therefore, water has more structure than other liquids. Collectively, the hydrogen bonds hold water together; this property is known as **cohesion.**

Cohesion is related to **surface tension,** which is a measure of how difficult it is to stretch or penetrate the surface of a liquid. At the surface between air and water, water molecules arrange themselves in an ordered system, hydrogen-bonded to each other laterally and to the water molecules beneath. This arrangement forms a weak elastic membrane that can be demonstrated by filling a water glass carefully; the water can be made to brim above the top of the glass but not overflow. A steel needle can be floated on water; insects such as the water strider walk about on the surface of lakes and streams. These things are possible because water has a high surface tension. This property is important in the early formation of waves. A gentle breeze stretches and wrinkles the smooth water surface, enabling the wind to get a better grip on the water and to add more energy to the sea surface.

The addition of salt to pure water increases the surface tension. Decreasing the temperature also increases the surface tension, and increasing the temperature decreases it.

Liquid water pours and stirs easily. It has little resistance to motion or internal friction. This property is called **viscosity.** Water has a low viscosity when compared to motor oil, paint, or syrup. Viscosity is affected by temperature. Consider pancake syrup. When it is stored in the refrigerator, it becomes more viscous and slow to pour. It has a high viscosity. When the syrup is returned to room temperature or heated, it becomes more runny; it has a low viscosity. The same is true of water, but the change in viscosity is much less, and it is not noticeable with normal temperature variations. Surface water at the equator is warmer and therefore less viscous than surface water in the Arctic. Minute microscopic organisms find it easier to float in the more viscous polar waters; some of their tropical-water cousins have adapted to the less-viscous water by developing spines and frilly appendages to help keep them afloat. The addition of salt to water increases the viscosity of the water, but the change is small.

QUICK REVIEW

1. How does water's high surface tension aid in the generation of waves?
2. What kinds of marine organisms are affected by changes in seawater viscosity with temperature?

4.6 Density

Density is defined as mass per unit volume of a substance. Water density is usually measured in grams per cubic centimeter (g/cm³). The density of pure water is often determined at 3.98°C, the temperature of maximum density, or approximately 4°C, and is 1 g/cm³. Therefore, a cube of pure water with 1 cm long edges has a mass of 1 g or a density of 1 g/cm³. The density of seawater is greater than the density of pure water at the same temperature because seawater contains dissolved salts. At 4°C, the density of seawater of average salinity is 1.0278 g/cm³. The densities of other substances may be considered in the same way (table 4.3). Less-dense substances will float on denser liquids (for example, oil on water, dry pine wood on water, and alcohol on oil). Ocean water is denser than fresh water; therefore, fresh water floats on salt water.

The Effect of Pressure

Pure water is nearly incompressible, and so is seawater. Pressure in the oceans increases with increasing depth. For every 10 m (33 ft) in depth, the pressure increases by about 1 atmosphere, or 14.7 lb/in². See figure 4.6 for the effect of water pressure at 2000 m (6560 ft). Pressure in the deepest ocean trench, 11,000 m (36,000 ft) deep, is about 1100 atm. These great pressures have only a small effect on the volume of the oceans because of the slight compressibility of the water. A cubic centimeter of seawater at the surface will lose only 1.7% of its volume if it is lowered to 4000 m (13,000 ft), where the pressure is 400 atm. The average pressure acting on the total world's ocean volume results in the reduction of ocean depth by about 37 m (121 ft). In other words, if the ocean water were truly incompressible, sea level would stand about 37 m higher than it does at present. The pressure effect is small enough to be ignored in most instances except when very accurate determination of seawater density is required.

The Effect of Temperature

Water density is very sensitive to temperature changes. When water is heated, energy is added and the water molecules speed up and move apart; therefore, the mass per cubic centimeter

Figure 4.6 These oceanography students hold research cruise souvenirs—polystyrene coffee cups that were attached to a sampler and lowered 2000 m (6560 ft) into the sea. The water pressure compressed the cups to the size of thimbles.

becomes less because there are fewer molecules per cubic centimeter. For this reason, the density of warm water is less than that of cold water, and warm water floats on cold water. When water is cooled, it loses heat energy, and the water molecules slow down and come closer together; there are then more water molecules, or a greater mass, per cubic centimeter. Because cold water is denser than warm water, it sinks below the warm water. To see these changes, try the following experiment. Chill a small quantity of water in the refrigerator; mix it with a little dye such as ink or food coloring to help see the effect. Allow the water from the tap to run hot and fill a glass half full of the hot water. Now slowly and carefully pour the colored water down the side of the glass on top of the hot water and watch what happens. You should see the cold, colored water sink below the hot water to create two distinct layers: the higher-density cold water will be on the bottom, and the lower-density hot water will be on top.

In pure fresh water, molecules move more slowly and come closer and closer together as water is cooled to 4°C. At this temperature, the water molecules are so close together and moving so slowly that each molecule can form hydrogen bonds with four other molecules. Unlike other fluids that continuously increase in density as they cool, water reaches its greatest density, 1 g/cm³, at 3.98°C, or about 4°C. As the temperature of the water falls below 4°C, the molecules move slightly apart, and at 0°C, they form an open latticework that is the stable structure for ice. Water as a solid takes up more space than water as a liquid; there are fewer water molecules per cubic centimeter, and therefore, ice is less dense than water and floats on water (fig. 4.7). Frozen water's increase in volume is dramatically and often disastrously demonstrated in the winter when water pipes freeze and then burst.

When ice absorbs enough heat, the hydrogen bonds between molecules are broken, the lattice starts to collapse, and the molecules move closer together. Above 4°C, the molecules move apart as they increase their motion. If enough heat is added, the energy level of the molecules increases until they overcome the attractive forces in the liquid and evaporate into the air as water

Table 4.3   Densities of Common Materials

Material	Density (g/cm³)
Ice (pure) 0°C	0.917
Water (pure) 0°C	0.99987
Water (pure) 3.98°C	1.0000
Water (pure) 20°C	0.99823
White pine wood	0.35–0.50
Olive oil 15°C	0.918
Ethyl alcohol 0°C	0.791
Seawater 4°C (salt 35 g/kg)	1.0278
Steel	7.60–7.80
Lead	11.347
Mercury	13.6

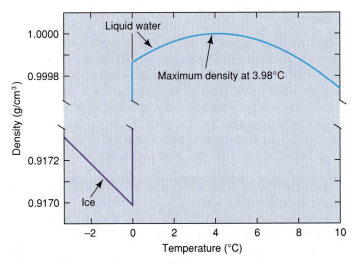

Figure 4.7 The density of pure water reaches its maximum at 3.98°C. Pure water is free of dissolved gases.

vapor. Because water vapor is less dense than the mixture of gases that form the atmosphere, a mixture of water vapor and dry air is less dense than dry air alone at the same temperature and pressure.

The Effect of Salt

When salts are dissolved in water, the density of the water increases because the salts have a greater density than water. In other words, they have more mass per cubic centimeter. The

average density of seawater at 4°C is 1.0278 g/cm³, compared with 1 g/cm³ for fresh water. Therefore, fresh water floats on salt water. To see this happen, take two small quantities of fresh water; add dye to one and some salt to the other. Carefully and slowly add the salty water to the dyed fresh water. You should see the salty water sink below the dyed fresh water to create two layers: the lower-density fresh water on top and the higher-density salty water on the bottom.

If seawater contains less than 24.7 grams of salt per kilogram, then the seawater, although denser, will behave much like fresh water. Seawater of low salt concentration will reach maximum density before freezing at a temperature less than 0°C. Cooling surface water with a salt content of less than 24.7 g/kg causes the water to increase in density and sink. This process continues until the temperature of maximum density is reached. Further cooling causes this low-salinity surface water to become less dense and remain at the surface.

At a salt content of 24.7 g/kg, the freezing point and temperature of maximum density of seawater coincide at −1.332°C. If the salt content is greater than 24.7 g/kg, freezing occurs before a maximum density is reached. This relationship is shown in figure 4.8. Open-ocean water generally has an average salt content of 36 g/kg; therefore, the density of seawater increases, and the seawater sinks continuously as it is cooled to its freezing point. The effect of temperature and salt content on the density of water is shown in figure 4.9. Notice in this figure that (1) if temperature remains constant, density increases with increasing salt content, and (2) if the salt content remains constant and is greater than 24.7 g/kg (review fig. 4.8), density increases as temperature decreases.

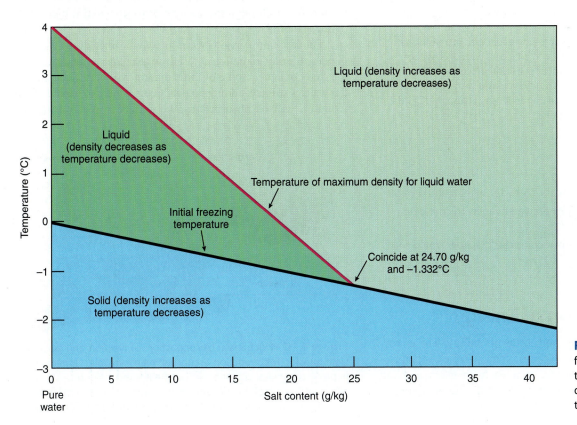

Figure 4.8 The initial freezing point and the temperature of maximum density of water decrease as the salt content increases.

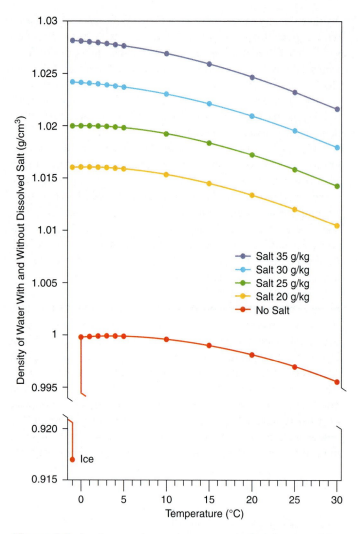

Figure 4.9 The density of seawater increases with increasing salt content at constant temperature and increases with decreasing temperature to the freezing point at constant salt content. The density of pure water with zero salinity increases with decreasing temperature until it reaches a maximum density at about 4°C and then decreases to the freezing point.

QUICK REVIEW

1. Why are ice and water vapor less dense than liquid water?
2. How does the density of seawater vary with changing temperature and salinity?
3. If water was completely incompressible, how would sea level be different?

4.7 Transmission of Energy

Fresh water and salt water both transmit energy. The energy may be in the form of heat, light, or sound. Because the principles of transmission are the same in fresh water and salt water, our discussion will center on energy transmission in the oceans.

Heat

There are three ways in which heat energy may be transmitted within a material: by **conduction,** by **convection,** and by **radiation** (fig. 4.10). Conduction is a molecular process. When heat is applied at one location, the molecules move faster because of the addition of energy; gradually, this more rapid molecular motion passes on to the adjacent molecules and the motion spreads. For example, if a metal spoon is placed in a hot liquid, the handle soon becomes hot. Heat has been conducted from the part of the spoon that is in contact with the heat source to the handle. Metals are excellent conductors; water is a poor conductor and transmits heat slowly in this way.

Convection is a density-driven process in which heated material moves and carries its heat to a new location. In older home heating systems, hot air is supplied through vents at floor level; the hot air rises because it is less dense than the cooler air above it, and the rising air carries the heat with it. When this air is cooled, it sinks toward the floor because it has become denser. In these energy-conscious days, ceiling fans are used to force the less dense hot air down, to keep the room warm at all levels rather than accumulating excess heat at the ceiling. Water behaves in the same way; it rises when heated and its density decreases; it sinks when cooled and its density increases. Remember that convection cells were discussed in chapter 2.

Radiation is the direct transmission of heat from its energy source. Switch on a heat lamp and feel, from a foot away, the heat released. This is radiant energy. The Sun provides Earth with radiant energy that penetrates and warms the surface waters of the planet. Unlike conduction and convection, which require a medium for the transfer of heat, radiated heat can be transferred through the vacuum of space and through transparent materials.

Think about trying to warm the water in each of two containers. If solar radiation is used to heat the surface water of the first container, some of the radiation will be reflected from the water's surface, adding no heat to the water, and some of the radiation will be absorbed by the surface water, raising the

Figure 4.10 Heat transfer by conduction, convection, and radiation. Heat transfer by conduction occurs between two objects that are in physical contact with one another. Heat transfer by convection occurs as a result of the movement of molecules. Heat transfer by radiation occurs through the radiation of electromagnetic energy.

water's temperature. Because warm water is less dense, it remains at the surface. A small amount of heat may be transmitted slowly downward by molecular conduction, but unless the container is stirred by mechanical energy from another source, the heat remains at the water's surface. If the second container of water is heated from below, the water at the bottom of the container gains heat, decreases in density, and rises, taking heat with it and distributing the heat through the water volume. As the warmed water rises, it is replaced by colder water from the surface, which is warmed in turn. Convection is a much more rapid and efficient method of distributing heat in water than are radiation and conduction.

The oceans are heated from above by solar radiation absorbed in the upper surface layers of water. The heat gained at the surface is transmitted slowly downward by the natural turbulent stirring action of the wind and the currents, as well as by the slower molecular conduction. In addition, the ocean's surface water may lose heat to the overlying atmosphere as a result of two processes: (1) direct transfer of heat from warm water to colder air by conduction and convection processes, and (2) the transfer of water vapor to the atmosphere by evaporation and atmospheric condensation. Heat transfers that warm the atmosphere from below create atmospheric convection that enhances this heat transfer.

Light

The light striking the ocean surface is one of the many forms of **electromagnetic radiation** Earth receives from the Sun. The full range of this radiation may be seen in the **electromagnetic spectrum,** shown in figure 4.11. Note that visible light occupies a very narrow segment of the spectrum at wavelengths from about 390–760 nm (390–760 $\times$ 10^{-7} cm). Wavelengths shorter than visible light include ultraviolet light, X-rays, and gamma rays, whereas infrared (heat) waves, microwaves, TV, and FM and AM radio waves all have wavelengths longer than visible light. Visible light may be broken down into the familiar spectrum of the rainbow: red, orange, yellow, green, blue, and violet. Each color represents a range of wavelengths; the longest wavelengths are at the red end of the spectrum, and the shortest are at the blue-violet end. When combined, these wavelengths produce white light.

As light passes through the water, it is **absorbed** and **scattered** by water molecules, ions, and suspended particles, including silt and microorganisms. It is also absorbed by organisms for photosynthesis, to be used in their life processes. This decrease in the intensity of light over distance is known as **attenuation.**

Seawater transmits only a small portion of the electromagnetic spectrum, primarily in the visible band. Light energy is attenuated very rapidly with depth, particularly the longer infrared wavelengths. The intensity of light at any depth can be calculated using the following equation (known as Beer's Law):

$$I_z = I_0 e^{-kz}$$

where I_0 is the intensity of the light at the surface, I_z is the intensity of the light at a depth of z meters, and k is the attenuation

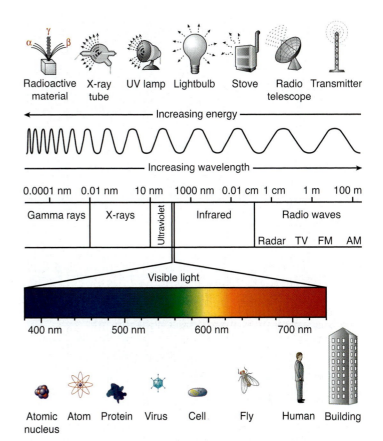

Figure 4.11 The elecromagnetic spectrum. Short wavelength radiation, like ultraviolet radiation, has important implications for the health of humans and organisms. Long wavelength radiation, like infared radiation, is involved in heat. Microwaves and radio waves provide important tools for satellite observations of the world ocean. Visible light occupies only a narrow section in the middle part of the spectrum.

coefficient. Beer's Law can be used to calculate the depth at which a given light-intensity ratio (I_z/I_0) occurs in the water by rearranging terms:

$$z = -\ln(I_z/I_0)/k$$

The attenuation coefficient k varies with the clarity of the water. The clearer the water, the smaller the attenuation coefficient and the greater the light penetration (fig 4.12 and table 4.4). In typical open ocean water, about 50% of the entering light energy is attenuated in the first 10 m (33 ft) ($I_z/I_0 = 0.5$), and almost all of the light is attenuated 100 m (330 ft) beneath the surface ($I_z/I_0 = 0.001$). In coastal water, where the water is not as clear, about 96% of the light energy is attenuated in the first 10 m (33 ft), and almost all the light is attenuated only 20 m (66 ft) beneath the surface.

The rate of light attenuation also depends heavily on the wavelength of the light (table 4.5). The short wavelengths in the ultraviolet regions and the long wavelengths in the red region and beyond are attenuated very rapidly (fig. 4.13). Light in the blue-green to blue-violet range is attenuated less rapidly and penetrates to the greatest depth (fig. 4.14).

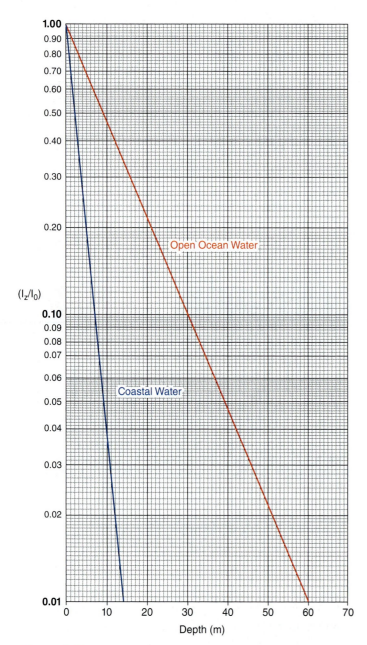

Figure 4.12 Ratio of the intensity of visible light at a depth of "z" meters (I_z) to the intensity at the surface (I_0) in open-ocean water (*red*) and coastal water (*blue*). The slope of each line is equal to the negative of the attenuation coefficient for blue light (see tabular representation of this information in table 4.4).

Table 4.4   Attenuation of Visible Light in Open Ocean and Coastal Water*

Depth (m)	% of Light Attenuated Open Ocean Water	% of Light Attenuated Coastal Water
0	0	0
0.1	0.8	3.3
1	7.3	28.4
10	53.2	96.5
20	78.1	99.9
30	89.8	~100
40	95.2	~100
50	97.8	~100
100	99.9	~100
150	~100	~100

* Values used for attenuation coefficient (k) are average values for blue light (480 nm); $k = 0.076$ m^{-1} in the open ocean and $k = 0.334$ m^{-1} in coastal water.

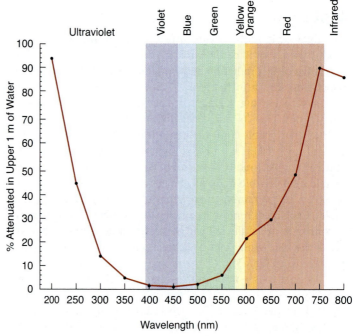

Figure 4.13 Percent of light attenuated in the upper 1 meter of clear ocean water as a function of wavelength. Light outside of the visible band is attenuated very rapidly. Within the visible band, light at blue and violet wavelengths experiences the least attenuation.

Color perception is due to the reflection back to our eyes of wavelengths of a particular color. Because all wavelengths, or colors, are present to illuminate objects in shallow water, objects just below the surface are seen in their natural colors. Objects in deeper waters usually appear dark because they are illuminated mainly by blue light. Ocean water usually appears blue-green because wavelengths of this color, being attenuated the least, are most available to be reflected and scattered back to an observer. Coastal waters vary in color, appearing green, yellow, brown, or red; these waters usually contain silt from rivers as well as large numbers of microscopic organisms and organic substances. The presence of these inclusions is revealed by the light reflecting from depth in the water.

Table 4.5 Attenuation of Light at Different Wavelengths[*]

Color	Wavelength (nm)	Attenuation k (m^{-1})	% Attenuated in Upper 1 m of Water	Depth at Which 99% Is Attenuated (m)
Ultraviolet (10–390 nm)	200	3.14	95.7	1
	250	0.588	44.5	8
	300	0.154	14.3	30
	350	0.0530	5.2	87
	400	0.0209	2.1	220
Violet (390–455 nm)	450	0.0168	1.7	274
Blue (455–492 nm)	500	0.0271	2.7	170
Green (492–577 nm)	550	0.0648	6.3	71
Yellow (577–597 nm)	600	0.245	21.7	19
Orange (597–622 nm)	650	0.350	29.5	13
	700	0.650	47.8	7
Red (622–760 nm)	750	2.47	91.5	2
Infrared (760 nm–1 mm)	800	2.07	87.4	2

[*] Attenuation coefficients (k) are values for clearest ocean water.

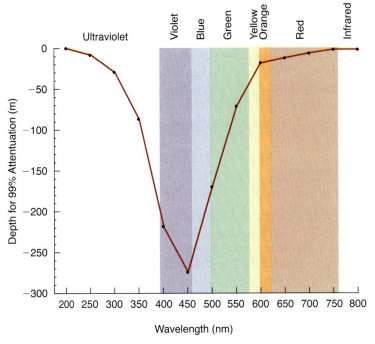

Figure 4.14 Depth (m) at which 99% of the incident light at the surface has been attenuated in the clearest ocean water. Light in the blue and blue-violet wavelengths penetrates to the greatest depth.

Open-ocean water beyond the influence of land contains little suspended matter. Light penetrates deeper in water that contains less particles, and open-ocean water appears bluer than coastal waters.

When light passes from air into water it is bent, or **refracted,** because the speed of light is faster in less dense air than in denser water. Because of the light refraction, objects seen through the water's surface are not where they appear to be (fig. 4.15). Refraction is affected slightly by changes in salinity, temperature, and pressure.

The simplest way of measuring light attenuation in surface water is to use a **Secchi disk** (fig. 4.16). This is a disk, about 30 cm (12 in) in diameter, that is lowered on a line to the depth at which it just disappears from view. This depth can be used to estimate the average attenuation coefficient for light using the following relationship:

$$k = (1.44/z)$$

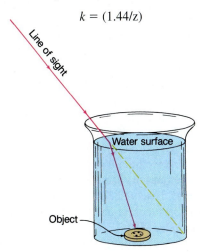

Figure 4.15 Objects that are not directly in the individual's line of sight can be seen in water because of the refraction of the light rays. The refraction is caused by the decreased speed of light in water.

Figure 4.16 A Secchi disk measures the transparency of water.

Figure 4.17 The Marine Optical Buoy (MOBY) is lowered into the sea. MOBY measures radiant energy entering and exiting the sea surface at three different depths. The data are used to adjust ocean color in satellite images.

where *k* is the average attenuation coefficient and z is the depth in meters at which the disk can no longer be seen. Secchi disks are either entirely white, as shown in figure 4.16, or divided into alternating black and white quadrants. In water rich with living organisms or suspended silt, the Secchi disk may disappear from view at depths of 1–3 m (3–10 ft); in the open ocean, visibility usually extends down to 20–30 m (65–100 ft), although a Secchi disk reading of 79 m (260 ft) was reported from Antarctica's Weddell Sea in 1986. Although crude, a Secchi disk has the advantages of low cost, never needing adjustment, never leaking, and never requiring replacement of electronic components. However, this technique explains little about how seawater affects light.

Changes in light attenuation are measured by devices that project a beam of light over a fixed distance to an electronic photoreceptor. Such devices are lowered meter by meter into the sea, and measurements are taken at chosen intervals. The cumulative effect of attenuation may also be calculated to any depth. Different wavelengths of light are used to separate the effects of absorption and scattering by inorganic and organic particles as well as absorption by dissolved substances and phytoplankton in the water. The role of natural fluorescence of material exposed to sunlight is also under investigation. Most light is produced when atoms become excited and give off radiation in the form of light and heat. In fluorescence, only the atom's electrons are agitated; they give off radiation in the form of light but practically no heat. Fluorescence is sometimes described as "cold light." Multichannel color sensors are used to observe both the changing properties of solar radiation as it penetrates the ocean and the radiance emitted by the fluorescence of dissolved compounds. Light transmission data are used to adjust color quality and clarity of underwater photographs and video.

The Marine Optical Buoy (MOBY; fig. 4.17) was launched in 1997; it carries instruments to measure light and color at the surface and at three different depths. Measurements are transmitted by fiber optics from the buoy to instruments onboard ship.

These measurements are used to improve satellites' sensing of sea surface color. Satellites monitor sea surface color to determine the abundance of phytoplankton at the sea surface, which are used as indicators of the biological productivity of the oceans.

Sound

The sea is a noisy place; waves break, fish grunt and blow bubbles, crabs snap their claws, and whales whistle and sing. Sound travels farther and faster in seawater than it does in air; the average velocity of sound in seawater is 1500 m/s (5000 ft/s) compared with 334 m/s (1100 ft/s) in dry air at 20°C.

The speed of sound in seawater is a function of the axial modulus and density of the water:

$$\text{speed of sound} = \sqrt{\frac{\text{axial modulus}}{\text{density}}}$$

The axial modulus of a material is a measure of how it compresses. A material with a high axial modulus is more difficult to compress than a material with a low axial modulus; a golf ball has a higher axial modulus than a tennis ball. Both the axial modulus and the density of seawater depend on temperature, salinity, and pressure (or depth). The speed of sound in seawater increases with increasing temperature, salinity, or depth and decreases with decreasing temperature, salinity, or depth. Sound speed increases with increasing temperature because the density decreases faster than the axial modulus as the water becomes warmer. Sound speed increases with increasing salinity and depth, despite the fact that both these changes increase seawater's density, because the axial modulus of the water increases faster than its density.

Water dissipates the energy of high-frequency sounds faster than that of low-frequency sounds. Therefore, high-frequency sounds do not travel as far as the lower-frequency sounds.

Sound is reflected after striking an object; therefore, sound can be used to find objects, sense their shape, and determine their distance from the sound's source. If a sound signal is sent into the water and the time required for the return of the reflected sound, or echo, is measured accurately, the distance to the object may be determined. For example, if six seconds elapse between the outgoing sound pulse and its return, the sound has taken three seconds to travel to the object and three seconds to return. Because sound travels at an average speed of 1500 m/s in water, the object is 4500 m away, as shown in figure 4.18.

Water depth is measured by directing a narrow sound beam vertically to the sea floor. The sound beam passes through the nearly horizontal layers of water, in which salt content, temperature, and pressure vary. The sound speed continuously changes as it passes from layer to layer, until it reaches the sea floor and is reflected back to the ship. Little refraction, or bending, of the sound beam occurs, because its path is perpendicular to the water layers. Echo sounders, or depth recorders, are used by all modern vessels to measure the depth of water beneath the ship. Oceanographic vessels record depths on a chart, producing a continuous profile of depth as the vessel moves along its course. The precision depth recorder (PDR) on an oceanographic research vessel uses a very narrow sound beam to give detailed and continuous traces of the bottom (fig. 4.19).

Geologists can detect the properties of the sea floor by studying echo charts because some seafloor materials reflect stronger signals than others. In general, basalt reflects a stronger signal than sediment. The frequency of the sound will also determine the depth of penetration of the sound energy beneath the sea floor. Sound frequencies used by PDRs are generally 5–30 kHz (1 kHz = 1000 cycles per second). High-frequency signals do not penetrate far beneath the sea floor and are used

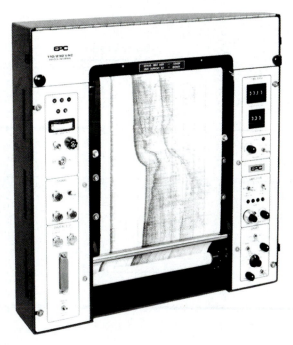

Figure 4.19 A precision depth recorder displaying bottom and subbottom profiles.

simply to measure water depth. Lower-frequency sound energy can penetrate the seafloor sediments and reflect from boundaries between sediment layers. It is used to study sediment thickness and layering. An example of a PDR record illustrating sediment thickness and layering is seen in figure 4.19.

Depth recorders can also record large numbers of small organisms, including fish, that move toward the surface during the night and sink to greater depths during the day. These organisms form a layer known as the **deep scattering layer (DSL).** This layer reflects a portion of the sound beam energy and creates the image of a false bottom on the depth recorder trace.

Other echoes are returned from mid-depths by fish swimming in schools or by large individual fish. Echo sounders are designed and marketed as "fish finders," and persons who fish often learn to recognize fish school reflections. The echoes provide information on the depth to set nets, while the location of the school and the appearance of the echo pattern give information on the fish species.

Porpoises and whales use sound in water in the same way that bats use sound in air. The animal produces a sound, which travels outward until it reaches an object, from which the sound is reflected. The animal is able to judge the direction from which the sound returns, the distance to the reflecting object, and the properties of that object. Humans have produced an underwater location technology called **sonar** (sound navigation and ranging) that uses sound in a similar way. Sonar technicians send directional pulses through the water, searching for targets that return echoes. They are then able to determine the distance and direction of the target.

The target, however, may not be at the depth, distance, and angle indicated, because the sound beam may change speed and direction as it passes through water layers of differing densities

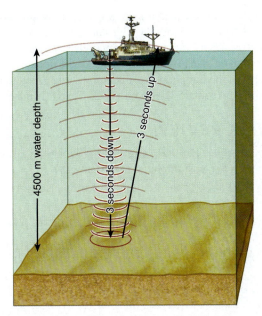

Figure 4.18 Traveling at an average speed of 1500 m per second, a sound pulse leaves the ship, travels downward, strikes the bottom, and returns. In 4500 m of water, the sound requires three seconds to reach the bottom and three seconds to return.

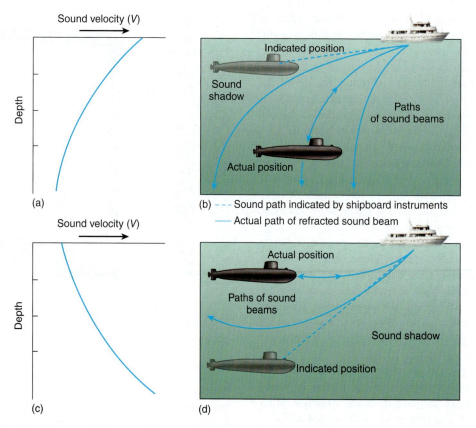

Figure 4.20 Sound waves change velocity and refract as they travel at an angle through water layers of different densities (a and c). The angle at which the sound beam leaves the ship indicates a target in the indicated, or ghost, position (b and d).

(fig. 4.20*a* and *c*). Figure 4.20*b* and *d* illustrate the refraction of sound beams and the formation of **sound shadow zones,** areas of the ocean into which sound does not penetrate. Sound beams bend toward regions in which sound travels more slowly and away from regions in which sound waves travel more rapidly.

To interpret the returning echo and to determine distance and depth correctly, the sonar operator must have information about the properties of the water through which the sound passes. Governments and their navies have conducted intensive research in the field of underwater sound. Survival for a surface vessel depends on its accuracy in locating a submarine's position, while the submarine must remain at the correct depth and distance from the sonar detector to remain invisible in the shadow zone.

At about 1000 m (3280 ft), the combination of salt content, temperature, and pressure creates a zone of minimum velocity for sound, the **sofar** (sound fixing and ranging) **channel** (fig. 4.21*a*). Sound waves produced in the sofar channel do not escape from it unless they are directed outward at a sharp angle. Instead, the majority of the sound energy is refracted back and forth along the channel for great distances (fig. 4.21*b*). Test

explosions in the channel near Australia have produced sound heard as far off as Bermuda. This channel was used by a project known as ATOC (acoustic thermometry of ocean climate) to send sound pulses over long distances to look for long-term changes in the temperature of ocean waters. (See the Diving In box titled "Acoustic Thermometry of Ocean Climate.")

QUICK REVIEW

1. What part of the electromagnetic spectrum is transmitted with the least attenuation in seawater?
2. What affects the attenuation of light in seawater?
3. Why does the sea generally appear blue?
4. Why can sound be used to measure depth whereas light cannot?
5. How does sound refraction affect sonar signals?
6. Explain how the sofar channel is formed and its effect on sound.

Diving
in

Acoustic Thermometry of Ocean Climate

Based on the principle that sound's speed in seawater is determined primarily by the temperature of the water, oceanographers Carl Wunsch of the Massachusetts Institute of Technology and Walter Munk of the Scripps Institution of Oceanography in California envisioned a transoceanic experiment to follow ocean temperature response to global warming. Because sound waves travel faster in warmer water, if the oceans are warming, the amount of time required for sound to travel from one location to another will decrease.

In 1991, the travel time of low-frequency sound pulses transmitted through the sofar channel from a site near Heard Island in the southern Indian Ocean was repeatedly measured at special listening stations around the world. The precision of sound travel time measurements from source to receiver was about 1 millisecond over a path 1000 km (660 mi) long, so temperature changes of a few thousandths of a degree could be detected. Test results were promising, and a new series of tests called acoustic thermometry of ocean climate (ATOC) began in the Pacific Ocean in 1995.

ATOC broadcasts sound from underwater sources near San Francisco and Hawaii. The sound is picked up by arrays of sensitive hydrophones as far away as Christmas Island and New Zealand (box fig. 1). After eighteen months of testing, temperature readings of the Pacific Ocean were found to be even more precise than had been projected. Scientists can detect variations as small as 20 milliseconds in the hour-long travel time of the pulses. This precision enables researchers to calculate the average ocean temperature along the sound pulse path to within 0.006°C. Repeating the measurements will allow long-term temperature changes at mid-ocean depths to be measured before they can be deduced by any other method.

Questions concerning the effect of sound signals on marine mammals began with the Heard Island tests and delayed the start of the ATOC experiment. Marine mammalogists have monitored the behavior of whale and elephant seal populations near the sound source off California. They report having seen no changes in the animals' swimming activity or distribution. Mammal researchers have also released deep-diving elephant seals farther out to sea and used satellite tags to track the paths of the animals as they returned to shore. The seals made no attempt to avoid the sound source. Additional experiments are being conducted to see whether whale vocalizations are affected.

A similar experiment that measured water temperatures in the Arctic Ocean was performed in the spring of 1994 by a joint U.S.-Russian-Canadian project, the Transarctic Acoustic Propagation Experiment. Sound signals were sent from an ice camp north of Spitsbergen to a camp 900 km (540 mi) in the Lincoln Sea and to another camp 2600 km (1600 mi) away in the Beaufort Sea. Travel times were predicted by using water temperatures from earlier research, but the measured travel times were shorter, implying that the mid-depth Atlantic water that penetrates the Arctic Ocean had warmed by 0.2°–0.4°C since the mid-1980s. The U.S.-Canada Arctic Ocean Section cruise had also measured such a warming trend, but scientists emphasize that it is too early to say whether the change is due to global warming or whether it is a part of some other natural cycle.

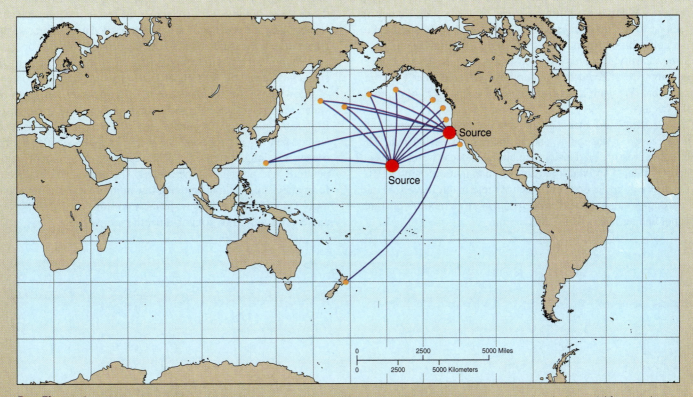

Box Figure 1 Acoustic thermometry of ocean climate (ATOC) takes the temperature of the Pacific Ocean using two sound sources, California and Hawaii, and twelve receivers.

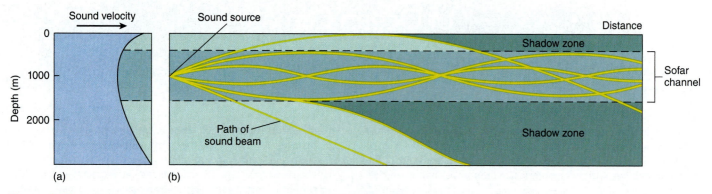

Figure 4.21 (a) The temperature, salinity, and pressure variation with depth combine to produce a minimum sound velocity at about 1000 m (3280 ft). (b) Sound generated at this depth is trapped in a layer known as the sofar channel.

Summary

A water molecule is made up of two positively charged atoms of hydrogen and one negatively charged atom of oxygen. The molecule has a specific shape with oppositely charged sides; it is a polar molecule. Because of this distribution of electrostatic charges, water molecules interact with each other by forming hydrogen bonds between molecules. The water molecule is very stable. The structure of the water molecule is responsible for the properties of water.

Water exists as a solid, a liquid, and a gas. Changes from one state to another require the addition or extraction of heat energy. Water has a high heat capacity; it is able to take in or give up large quantities of heat with a small change in temperature. The surface tension of water, which is related to the cohesion between water molecules at the surface, is high. The viscosity of water is a measure of its internal friction; it is primarily affected by temperature. Water is nearly incompressible. Pressure increases 1 atm for every 10 m of depth in the oceans.

The density, or mass per unit volume, of water increases with a decrease in temperature and an increase in salt content. The effect of pressure on density is small. Less-dense water floats on denser water. Pure water reaches its maximum density at 4°C. Open-ocean water does not reach its maximum density before freezing. Water vapor is less dense than air; a mixture of water vapor and air is less dense than dry air.

The ability of water to dissolve substances is exceptionally good. River water dissolves salts from the land and carries them to the sea.

Seawater transmits energy as heat, light, and sound. The sea surface layer is heated by solar radiation, and heat is transmitted downward by conduction. This is an inefficient process compared to convection.

The long red wavelengths of light are lost primarily in the first 10 m (33 ft) of seawater; only the shorter wavelengths of blue-green to blue-violet light penetrate to depths of 150 m (500 ft) or more. Light passing into water is refracted. Attenuation, or the decrease in light over distance, is the result of absorption and scattering by the water and particles suspended in the water. Light attenuation is measured by a Secchi disk and photoreceptors.

Sound travels farther and faster in water than in air. Its speed is affected by the temperature, pressure, and salt content of the water. Echo sounders are used to measure the depth of water, and sonar is used to locate objects. Sound is refracted as it passes at an angle through water of different densities, and sound shadows are formed. The sofar channel, in which sound travels for long distances, is the result of salt content, temperature, and pressure in the oceans. The deep scattering layer, formed by small animals moving toward the surface at night and away from it during the day, reflects portions of a sound beam and creates a false bottom on a depth recorder trace.

Key Terms

covalent bond, 112
polar molecule, 112
hydrogen bond, 112
temperature, 112
degree, 112
heat, 114
calorie, 114
dew point, 115

latent heat of fusion, 115
latent heat of vaporization, 115
sublimation, 116
specific heat, 116
heat capacity, 117
cohesion, 117
surface tension, 117
viscosity, 117

conduction, 120
convection, 120
radiation, 120
electromagnetic radiation, 121
electromagnetic spectrum, 121
absorption, 121
scattering, 121
attenuation, 121

refraction, 123
Secchi disk, 123
deep scattering layer
   (DSL), 125
sonar, 125
sound shadow zone, 126
sofar channel, 126

Study Problems

1. Use figure 4.12 to determine the ratio of the intensity of light at depths of 10, 30, and 50 m to the intensity of light at the surface in open-ocean water.

2. If you start with 1 g of ice at $-2°C$, how many calories of heat would you have to add to end up with 1 g of liquid water at a temperature of $+2°C$?

3. Convert $32°F$ to $°C$ and K.

4. If a PDR measures the depth of the water as 3500 m, but the instrument is known to have a timing error of ±0.001 second over the total time period of the measurement, what is the error in the depth measurement?

The Chemistry of Seawater

Learning Outcomes

After studying the information in this chapter students should be able to:

1. *sketch* the pattern of high and low sea surface salinity on a map of the world's oceans,

2. *explain* how sea surface salinity is modified by evaporation, precipitation, and runoff from the continents,

3. *review* the sources of major constituent ions in seawater,

4. *rank* the six most abundant constituent ions in seawater in order of their concentration,

5. *calculate* the residence time of an ion given its concentration and rate of supply,

6. *diagram* the distribution of oxygen and carbon dioxide with depth,

7. *describe* the pH scale and *explain* the role of carbon dioxide in buffering seawater pH,

8. *identify* the three ions considered important marine nutrients, and

9. *compare* and *contrast* two different methods of desalination.

An evaporate salt pond in Bonaire, Netherlands Antilles.

Seawater is salt water, and historically, seawater has been valued for its salt. Until recently, salt was enormously important as a food preservative, and at one time, salt formed the basis for a major commercial trade. Today, although salt is still extracted from seawater, it is the water that has become increasingly valuable in many areas of the world. Seawater is also much more than salt water. Seawater is a complex solution containing dissolved gases, nutrient substances, and organic molecules as well as salts.

In this chapter, we investigate seawater, and we explore physical, chemical, and biological processes that regulate its composition. We also review the commercial extraction of salts from seawater and the possibilities of increasing our supply of fresh water by desalination.

5.1 Salts

Dissolving Ability of Water

When immersed in water, compounds can break apart into individual atoms or groups of atoms that have opposite electrical charges. A charged atom or group of atoms is called an **ion**. An ion with a positive charge is a **cation**; an atom with a negative charge is an **anion**. The salts in seawater are present in dissolved form as cations and anions. A good example is the salt sodium chloride (NaCl). Sodium chloride is held together by **ionic bonds** in which electrons are transferred from a metal atom (in this case, the metal sodium) to a nonmetal atom (in this case, the chlorine atom), creating ions of opposite charge that attract each other. Ionic bonds are easily broken in water because of the polar nature of the water molecule. Thus, when salts are added to water, the salts dissolve, or dissociate (break apart) into ions. This process can be written as

$$\text{sodium chloride} \rightarrow \text{sodium ion} + \text{chloride ion}$$

or it can be written with chemical abbreviations as

$$\text{NaCl} \rightarrow \text{Na}^+ + \text{Cl}^-$$

In a glass of seawater, you would find individual sodium cations (Na^+) and chloride anions (Cl^-). If the water was left out in the sun to evaporate, these ions would combine to form a solid precipitate, the salt compound sodium chloride (NaCl).

Because of their polarity, water molecules can surround both cations and anions in solution and prevent them from combining into molecules. Figure 5.1a shows water molecules clustered around a sodium cation (Na^+) with their negative (oxygen) ends pointing toward it. Around the chloride anion (Cl^-), the orientation of the water molecules is reversed (fig. 5.1b), with the positive (hydrogen) ends of the water molecules pointing toward the anion.

For millions of years, volcanism and rain washing over the land have supplied the oceans with dissolved salts. Once these reach the ocean basins, most of the dissolved salts stay in the ocean. Water is recycled to the land by oceanic evaporation, but salts remain in the sea and the bottom sediments. Some land salt deposits resulted from geologic uplift processes that bring seafloor deposits above sea level. Other salt deposits on land are the remnants of ancient shallow seas that became isolated over geologic time—the water evaporated, leaving the salt deposits behind. Salts in the seawater and marine sediments are recycled as a result of subduction and subsequent volcanism, as discussed in chapter 2.

Units of Concentration

The concentration of dissolved constituents in seawater can be expressed by weight, by volume, or in molar terms. Concentrations of different constituents vary by several orders of magnitude.

When **expressed by weight**, concentrations are given as g/kg (parts per thousand), mg/kg (parts per million), or even μg/kg (parts per billion), depending on the abundance of the constitutent. Thus,

$$\begin{aligned} 1 \text{ g/kg (1 part per thousand)} &= 1000 \text{ mg/kg} \\ &\quad (10^3 \text{ parts per million}) \\ &= 1 \text{ million μg/kg} \\ &\quad (10^6 \text{ parts per billion}) \end{aligned}$$

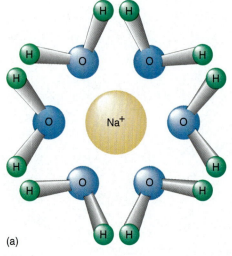

(a)

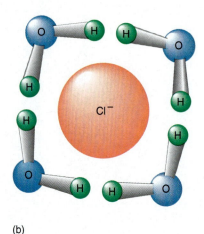

(b)

Figure 5.1 Salts dissolve in water because the polarity of the water molecule keeps positive ions separated from negative ions. (a) Sodium ions are surrounded by water molecules with their negatively charged portion attracted to the positive ion. (b) Chloride ions are surrounded by water molecules with their positively charged portion attracted to the negative ion.

132

or:

$$1 \,\mu g/kg \,(1 \text{ part per billion}) = 0.001 \text{ mg/kg}$$
$$(10^{-3} \text{ parts per million})$$
$$= 0.000001 \text{ g/kg}$$
$$(10^{-6} \text{ parts per thousand})$$

When **expressed by volume**, concentrations are given as g/l, mg/l, or µg/l. Since 1 liter of seawater weighs very nearly 1 kilogram (the actual weight is approximately 1.027 kg), concentrations measured by volume are numerically similar to concentrations measured by weight. For example, the concentration of chloride in seawater is approximately 19.87 g/l, or 19.35 g/kg.

For some purposes, it is useful to **express concentrations in molar terms**. The mass of 1 mole of a substance, expressed in grams, is exactly equal to the substance's mean atomic or molecular weight (table 5.1). For example, the mean molecular weight of water (H_2O) is about $(1 \text{ g} \times 2) + 16 \text{ g} = 18 \text{ g}$, so 1 mole of water is about 18 grams. A mole of sodium (Na^+) contains 23 g of sodium; a mole of sulfate (SO_4^{2-}) contains $32 \text{ g} + (16 \text{ g} \times 4) = 96 \text{ g}$ of sulfate; and a mole of bicarbonate (HCO_3^-) contains $1 \text{ g} + 12 \text{ g} + (16 \text{ g} \times 3) = 61 \text{ g}$ of bicarbonate. Once again, because 1 liter of seawater weighs nearly 1 kilogram, molar concentrations of a given dissolved constituent expressed as moles/kg and moles/l are numerically similar.

Ocean Salinities

In the major ocean basins, 3.5% of the weight of seawater is, on the average, dissolved salt and 96.5% is water, so a typical 1000 g or 1 kg sample of seawater is made up of 965 g of water and 35 g of salt. Oceanographers measure the salt content of ocean water in grams of salt per kilogram of seawater (g/kg), or parts per thousand (‰). The total quantity of dissolved salt in seawater is known as **salinity,** and the average ocean salinity is approximately 35‰.

The salinity of ocean surface water is associated with latitude. Latitudinal variations in evaporation and precipitation, as well as freezing, thawing,

Table 5.1 Approximate Atomic Weight of Selected Elements

Element	Atomic Weight (g)
Hydrogen (H)	1
Carbon (C)	12
Oxygen (O)	16
Sodium (Na)	23
Sulfur (S)	32

and freshwater runoff from the land, affect the amount of salt in seawater. The relationship among evaporation, precipitation, and mid-ocean surface salinity with latitude is shown in figure 5.2. Notice the low surface salinities in the cool and rainy 40°–50°N and S latitude belts, high evaporation rates and high surface salinities in the desert belts centered on 25°N and S, and low surface salinities again in the warm but rainy tropics centered at 5°N. Sea surface salinities during the Northern Hemisphere summer are shown in figure 5.3.

In coastal areas of high precipitation and river inflow, surface salinities fall below the average. For example, during periods of high flow, the water of the Columbia River lowers the Pacific Ocean's surface salinity to less than 25‰ as far as 35 km (20 mi) at sea. Also, sailors have dipped up water fresh enough to drink from the ocean surface 85 km (50 mi) from the mouth of the

(a)

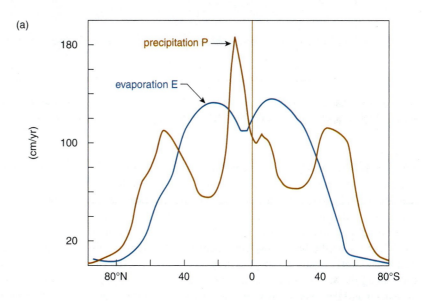

(b)

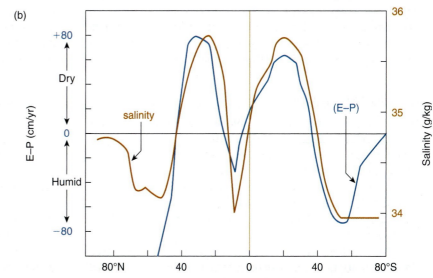

Figure 5.2 (a) Mid-ocean precipitation (*red*) and evaporation (*blue*) values as a function of latitude. (b) Mid-ocean average surface salinity values (*red*) match the average variation in [evaporation minus precipitation] values (*blue*) that occur with latitude.

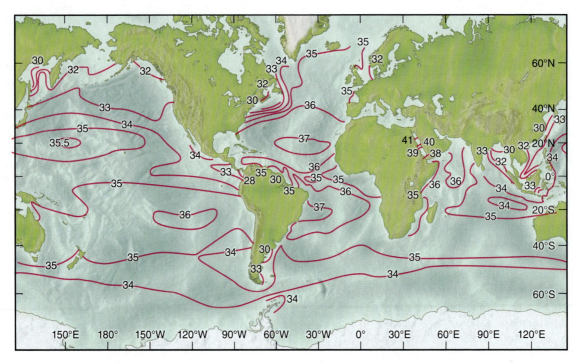

Figure 5.3 Average sea surface salinities in the Northern Hemisphere summer, given in parts per thousand (‰). (High salinities are found in areas of high evaporation; low salinities are common in coastal areas and regions of high precipitation.)

Amazon River. In subtropic regions of high evaporation and low freshwater input, the surface salinities of nearly landlocked seas are well above the average: 40–42‰ in the Red Sea and the Persian Gulf and 38–39‰ in the Mediterranean Sea. In the open ocean at these same latitudes, the surface salinity is closer to 36.5‰. Surface salinities change seasonally in polar areas, where the surface water forms sea ice in winter, leaving behind the salt and raising the salinity of the water under the ice. In summer, a freshwater surface layer forms when the sea ice melts. Deep-water samples from the mid-latitudes are usually slightly less salty than the surface waters in part because the deep water is formed at the surface in high latitudes with high precipitation. The formation of these deep-water types is discussed in chapter 7.

Dissolved Salts

Six ions make up more than 99% of the salts dissolved in seawater. Four of these are cations: sodium (Na^+), magnesium (Mg^{2+}), calcium (Ca^{2+}), and potassium (K^+); two are anions: chloride (Cl^-) and sulfate (SO_4^{2-}). Table 5.2 lists these six ions and five more, arranged in order of their abundance in seawater. The ions listed in table 5.2 are known as the **major constituents** of seawater. Note that sodium and chloride ions account for 86% of the salt ions in seawater.

All the other elements dissolved in seawater are present in concentrations of less than one part per million and are called **trace elements** (table 5.3). Most trace elements are present in such small concentrations that it is common to report their concentrations at the parts per billion level. Some of these elements are important to organisms that are able to concentrate the ions.

For example, long before iodine could be determined chemically as a trace element of seawater, it was known that shellfish and seaweeds were rich sources for this element, and seaweed was harvested for commercial iodine extraction.

Because the ratios of the major constituents of seawater do not change with changes in total salt content and because these constituents are not generally removed or added by living organisms, the major constituents are termed **conservative constituents.** Certain of the ions present in much smaller quantities, some dissolved gases, and assorted organic molecules and complexes do change in concentrations because of biological and chemical processes that occur in some areas of the oceans; these are known as **nonconservative constituents.**

Sources of Salt

The original sources of sea salts include the crust and the interior of Earth. The chemical composition of Earth's rocky crust can account for most of the positively charged ions found in seawater. Large quantities of cations are present in rocks that are formed by the crystallization of molten magma from volcanic processes. The physical and chemical weathering of rock over time breaks it into small pieces and the rain dissolves out ions, which are carried to the sea by rivers. Anions are present in Earth's interior and may have been present in Earth's early atmosphere. Some anions may have been washed from the atmosphere by long periods of rainfall, but the more likely source of most of the anions is thought to have been Earth's mantle. During the formation of Earth, gases from the mantle are believed to have supplied anions to the newly forming oceans.

Table 5.2   Major Constituents of Seawater[1]

Constituent		Symbol	Concentration in Seawater			Percentage by Weight
			g/kg	g/l	mole/l	
Chloride	The six most abundant ions	Cl^-	19.35	19.87	0.560	55.07
Sodium		Na^+	10.76	11.05	0.481	30.62
Sulfate		SO_4^{2-}	2.71 (34.91)	2.78 (35.84)	0.029 (1.145)	7.72 (99.36)
Magnesium		Mg^{2+}	1.29	1.32	0.054	3.68
Calcium		Ca^{2+}	0.41	0.42	0.0105	1.17
Potassium		K^+	0.39	0.40	0.0102	1.10
Bicarbonate		HCO_3^-	0.14	0.144	0.0024	0.40
Bromide		Br^-	0.067	0.069	0.00086	0.19
Strontium		Sr^{2+}	0.008	0.008	0.00009	0.02
Boron		B^{3+}	0.004	0.004	0.00037	0.01
Fluoride		F^-	0.001	0.001	0.00005	0.02
Total			35.13	36.07	1.1485	99.99

[1] Nutrients and dissolved gases are not included.

Acidic gases released during volcanic eruptions (for example, hydrogen sulfide, sulfur dioxide, and chlorine) dissolve in rainwater or river water and are carried to the oceans as Cl^- (chloride) and SO_4^{2-} (sulfate).

Tests show that the most abundant ions in today's rivers (table 5.4) are the least abundant ions in ocean water because the rivers have previously removed the most easily dissolved land salts and are now carrying the less soluble salts. Exceptions to this pattern are found in rivers used for irrigation. These rivers are flowing through arid soils that have not lost much of their salt content. The water is frequently used several times and passes through a number of irrigation projects on its way downriver, causing the water to become increasingly salty and unfit for irrigation purposes. This situation has produced years of continuous conflict between the United States and Mexico over the waters of the Colorado River and Rio Grande, which irrigate much of the agricultural land of the U.S. desert Southwest before becoming available to the farmlands of Mexico.

In addition, we know that hot-water vents located on the sea floor supply chemicals to the ocean water and also remove them. Hydrothermal activity found at the mid-ocean ridges and associated with hotspots and ridge formation may play an important role in stabilizing the ocean's salt composition. When hot magma is introduced, the cold crust cracks and becomes permeable. Pressure from the water above the sea floor is high (1 atm for every 10 m of water depth), and it forces water into cracks and voids, where it is heated to extremely high temperatures. The salinity of seawater entering the hydrothermal system is relatively constant worldwide, but the water emerging from hydrothermal vents has a salinity that may be more than double its original value. The processes controlling this salinity change and their role in the ocean's total salt balance are unclear at this time.

Regulating the Salt Balance

We know from the age of older marine sedimentary rocks that the oceans have been present on Earth for about 3.5 billion years. Chemical and geologic evidence from rocks and salt deposits leads researchers to believe that the salt composition of the oceans has been the same for about the last 1.5 billion years. The total amount of dissolved material in the world's oceans is calculated to be 5×10^{22} g for ocean water of 36‰ salinity. Each year, the runoff from the land adds another 2.5×10^{15} g, or 0.000005% of total ocean salt. If we assume that the rivers have been flowing to the sea at the same rate over the last 3.5 billion years, more dissolved material has been added by this process than is presently found in the sea. For the oceans to remain at the same salinity, the rate of addition of salt by rivers must be balanced by the removal of salt; input must balance output.

Salt ions are removed from seawater in a number of ways; some are shown in figure 5.4. Sea spray from the waves is blown ashore, depositing a film of salt on land. This salt is later returned to the oceans by runoff from the land. Over geologic time, shallow arms of the sea have become isolated, the water has evaporated, and the salts have been left behind as sedimentary deposits called **evaporites.** Salt ions can also react with each other to form insoluble products that precipitate on the ocean floor. Biological processes concentrate salts, which are removed from the water if the organisms are harvested or if the organisms become part of the sediments. Organisms' excretion products trap ions, which are transferred to the sediments or returned to the seawater. Other biological processes remove Ca^{2+} (calcium) by incorporating it into shells, and silica is used to form the hard parts of diatoms and radiolarians. These accumulate in the sediments when the organisms die. Chapter 3 discussed the sediments produced by biological processes.

Table 5.3   Concentrations of Trace Elements in Seawater[1]

Element	Symbol	Concentration[2]	Element	Symbol	Concentration[2]
Aluminum	Al	5.4×10^{-1}	Manganese	Mn	3×10^{-2}
Antimony	Sb	1.5×10^{-1}	Mercury	Hg	1×10^{-3}
Arsenic	As	1.7	Molybdenum	Mo	1.1×10^{1}
Barium	Ba	1.37×10^{1}	Nickel	Ni	5×10^{-1}
Bismuth	Bi	$\leq 4.2 \times 10^{-5}$	Niobium	Nb	$\leq (4.6 \times 10^{-3})$
Cadmium	Cd	8×10^{-2}	Protactinium	Pa	5×10^{-8}
Cerium	Ce	2.8×10^{-3}	Radium	Ra	7×10^{-8}
Cesium	Cs	2.9×10^{-1}	Rubidium	Rb	1.2×10^{2}
Chromium	Cr	2×10^{-1}	Scandium	Sc	6.7×10^{-4}
Cobalt	Co	1×10^{-3}	Selenium	Se	1.3×10^{-1}
Copper	Cu	2.5×10^{-1}	Silver	Ag	2.7×10^{-3}
Gallium	Ga	2×10^{-2}	Thallium	Tl	1.2×10^{-2}
Germanium	Ge	5.1×10^{-3}	Thorium	Th	(1×10^{-2})
Gold	Au	4.9×10^{-3}	Tin	Sn	5×10^{-4}
Indium	In	1×10^{-4}	Titanium	Ti	$< (9.6 \times 10^{-1})$
Iodine	I	5×10^{1}	Tungsten	W	9×10^{-2}
Iron	Fe	6×10^{-2}	Uranium	U	3.2
Lanthanum	La	4.2×10^{-3}	Vanadium	V	1.58
Lead	Pb	2.1×10^{-3}	Yttrium	Y	1.3×10^{-2}
Lithium	Li	1.7×10^{2}	Zinc	Zn	4×10^{-1}
			Rare earths		$(0.5\text{–}3.0) \times 10^{-3}$

[1] Nutrients and dissolved gases are not included.
[2] Parts per billion, or μ g/kg.
Note: Parentheses indicate uncertainty about concentration.

Table 5.4   Dissolved Salts in River Water

Ion	Symbol	Percentage by Weight
Bicarbonate	HCO_3^-	48.7
Calcium	Ca^{2+}	12.5
Silicon dioxide (nonionic)	SiO_2	10.9
Sulfate	SO_4^{2-}	9.3
Chloride	Cl^-	6.5
Sodium	Na^+	5.2
Magnesium	Mg^{2+}	3.4
Potassium	K^+	1.9
Oxides (nonionic)	$(Fe, Al)_2O_3$	0.8
Nitrate	NO_3^-	0.8
Total		100.0

Note: Average river salt concentration is 0.120‰.

A chemical process known as **adsorption,** the adherence of ions and molecules onto a particle's surface, removes other ions and molecules from seawater. In this process, tiny clay mineral particles, weathered from rock and brought to the oceans by winds and rivers, bind ions such as K^+ (potassium) and trace metals to their surfaces. The ions sink with the clay particles and are eventually incorporated into the sediments. Strongly adsorbable ions replace weakly adsorbable ions in a process known as **ion exchange.** If the clay minerals and sediments adsorb and exchange one ion more easily than another, there is a lower concentration of the more easily adsorbed ion in the seawater. For example, potassium is more easily adsorbed than sodium, so K^+ is less concentrated than Na^+ in seawater. Nickel, cobalt, zinc, and copper are adsorbed on nodules that form on the ocean floor. The fecal pellets and skeletal remains of small organisms also act as adsorption surfaces. The settling of this organic debris transports metallic ions to the sediment, where they can be adsorbed on nodules. In these cases, the ions are removed from the water and are transferred to the sediments.

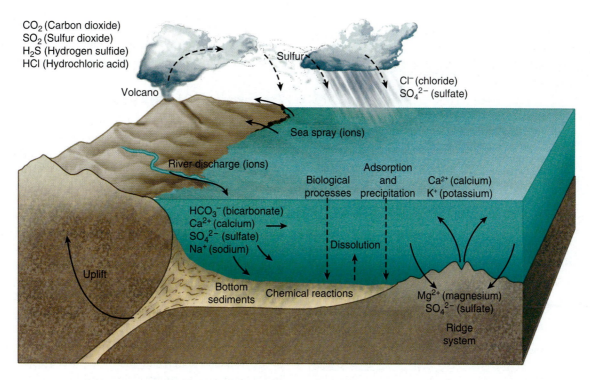

Figure 5.4 Processes that distribute and regulate the major constituents in seawater. Salt ions are added to seawater from rivers, volcanic events, ridge systems, and decay processes. Salt ions are removed from seawater by adsorption and ion exchange, spray, chemical precipitation, biological uptake, and addition to crustal rocks at ridge systems.

The process of forming the new crust at the ridge system of the deep-ocean floor (see chapter 2) participates in the input and output of the ions in seawater. Where molten rock rises from the mantle into the crust, magma chambers are formed. These chambers are found mainly along plate boundaries but also above volcanic hotspots in the middle of plates. Cold water seeps down several kilometers through the fractured crust at a spreading center. The seawater becomes heated by flowing near the magma chamber. By convection, the heated water rises through the crust and reacts chemically with the rocks. Magnesium ions (Mg^{2+}) are transferred from the water to form minerals in the crust. At the same time, hydrogen ions (H^+) are released and the seawater solution becomes more acidic. Chemical reactions change sulfate (SO_4^{2-}) to sulfur and then to hydrogen sulfide (H_2S). The hot saline water dissolves metals from the crust, including copper, iron, manganese, and zinc, and releases potassium and calcium. It has been estimated that a volume of water equivalent to the entire mass of the oceans circulates through the crust at oceanic ridges every 10 million years.

The most important process for the removal of most elements from seawater is still adsorption of ions onto fine particles and their removal to the sediments. Ions deposited in the sediments are trapped there. They are not soon redissolved in the seawater, but geologic uplift elevates some marine sediments to positions above sea level. Erosion then works to dissolve and wash these deposits back to the sea.

Residence Time

The relative abundances of the salts in the sea are due in part to the ease with which they are introduced from Earth's crust and in part to the rate at which they are removed from the seawater. Sodium is moderately abundant in fresh water but is so highly soluble in seawater that it remains dissolved in the ocean. Calcium is removed rapidly from seawater to form limestone and the shells of marine organisms, but it is also replaced rapidly by calcium ions moving down rivers and in hot springs of the ridge systems on the sea floor (fig. 5.4).

The average, or mean, time that a substance remains in solution in the ocean is called its **residence time** (table 5.5). Aluminum, iron, and chromium ions have short residence times, in the hundreds of years. They react with other substances quickly and form insoluble mineral solids in the sediments. Sodium, potassium, and magnesium are very soluble and have long residence times, in the millions of years. If the concentration of an ion in seawater is constant, the rates of supply and removal are equal. If the total amount of an ion present in the ocean is divided by either its rate of supply or its rate of removal, the residence time for that ion is known. For example, there are roughly 5.74×10^{20} g of Ca^{2+} in the oceans. It is estimated that Ca^{2+} is added to the oceans at a rate of about 5.4×10^{14} g per year. Therefore

$$\text{residence time } (Ca^{2+}) = \left(\frac{5.74 \times 10^{20} \text{ g}}{5.55 \times 10^{14} \text{ g/yr}} \right)$$

$$= 1.06 \times 10^6 \text{ years}$$

Table 5.5   Approximate Residence Time of Constituents in the Ocean

Constituent	Residence Time (years)
Chloride (Cl⁻)	100,000,000
Sodium (Na⁺)	68,000,000
Magnesium (Mg²⁺)	13,000,000
Potassium (K⁺)	12,000,000
Sulfate (SO₄²⁻)	11,000,000
Calcium (Ca²⁺)	1,000,000
Carbonate (CO₃²⁻)	110,000
Silicon (Si)	20,000
Water (H₂O)	4100
Manganese (Mn)	1300
Aluminum (Al)	600
Iron (Fe)	200

Constant Proportions

Seawater is a well-mixed solution; currents and eddies in surface and deep water, vertical mixing processes, and wave and tidal action have all helped to stir the oceans through geologic time. Because of this thorough mixing, the ionic composition of open-ocean seawater is the same from place to place and from depth to depth. That is, the ratio of one major ion or seawater constituent to another remains the same. Whether the salinity is 40‰ or 28‰, the major ions exist in the same proportions. This concept was first suggested by the chemist Alexander Marcet in 1819. In 1865, another chemist, Georg Forchhammer, analyzed several hundred seawater samples and found that these constant proportions did hold true. During the world cruise of the *Challenger* expedition (1872–76), seventy-seven water samples were collected from different depths and locations. When chemist William Dittmar analyzed these samples, he verified Forchhammer's findings and Marcet's suggestion. These analyses led to the **principle of constant proportion** (or **constant composition**) of seawater, which states that regardless of variations in salinity, the ratios between the amounts of major ions in open-ocean water are constant. For example

$$Cl^-/Na^+ = 1.796 \text{ everywhere in the open ocean, and}$$
$$Na^+/SO_4^{2-} = 3.972 \text{ everywhere in the open ocean}$$

Note that the principle applies to major conservative ions in the open ocean; it does not apply along the shores, where rivers may bring in large quantities of dissolved substances or may reduce the salinity to very low values.

The ratios of abundance of minor nonconservative constituents vary because some of these are closely related to the life cycles of living organisms. Populations of organisms remove ions during periods of growth and reproduction, reducing the amounts in solution. Later, the population declines, and decay processes return these ions to the seawater.

Determining Salinity

Seawater transmits, or conducts, electricity because it contains dissolved ionic salts; the more ions in solution, the greater the conductance. Therefore, the salt content, or salinity, can be determined by using an instrument, called a **salinometer,** that measures electrical conductivity. Salinity readings in parts per thousand (S‰) are made quickly and directly on the water sample with an electrical probe. Because conductivity of seawater is affected by salinity and temperature, a conductivity instrument must correct for the temperature if the instrument reads directly in salinity units. This correction is also necessary if the instrument reads conductance only. If conductance and temperature are measured separately, a computer program calculates the salinity.

To be sure that all salinity determinations are comparable, the world's oceanographic laboratories use a standard method of analysis and a standard seawater reference. At present, the Institute of Oceanographic Services in Wormly, England, is responsible for the production of standard seawater adjusted to both constant chloride content and electrical conductance to ensure standard calibration of laboratory instruments.

Historically, the quantity of chloride ions in a water sample was measured to establish the salinity of the sample. To do so, silver nitrate was added because the silver combines with the chloride ions. If the amount of silver required to react with all the chloride ions in a sample is known, the amount of chloride is known. The chloride concentration measured in this way is termed **chlorinity (Cl‰)** and is measured in parts per thousand or grams of chloride per kilogram of seawater. When the chlorinity of a sample is known, the concentration of any other major constituent can be calculated by using the principle of constant proportion.

Chlorinity and salinity are related by the equation

$$\text{salinity (‰)} = 1.80655 \times \text{chlorinity (‰)}$$

or

$$S‰ = 1.80655 \times Cl‰$$

QUICK REVIEW

1. What is the difference between major constituents and trace elements in seawater?

2. How does the relative abundance of major constituent ions change in open-ocean water when the salinity changes?

3. Why can salinity be calculated from chlorinity?

4. Relate changes in open-ocean salinity to variations in latitude and explain these changes.

5. Why do coastal sea surface salinities differ from open-ocean sea surface salinities at the same latitude?

6. How are the abundance, solubility, and residence time of ions related?

5.2 Gases

Gases move between the sea and the atmosphere at the sea surface. Atmospheric gases dissolve in seawater and are distributed to all depths by mixing processes and currents. Abundant gases in the atmosphere and in the oceans are nitrogen (N_2), oxygen (O_2), and carbon dioxide (CO_2). The percentages of each of these gases in the atmosphere and in seawater are given in table 5.6. Oxygen and carbon dioxide play important roles in the ocean because they are necessary to all life, and biological activities modify their concentrations at various depths. Although nitrogen gas is used directly only by bacteria, its use also plays an important role in ocean processes. Gases such as argon, helium, and neon are present in small amounts, but they are chemically inert and do not interact with the ocean water or its inhabitants.

The maximum amount of any gas that can be held in solution is the **saturation concentration.** The saturation concentration changes because it depends on the temperature, salinity, and pressure of the water. If the temperature or salinity decreases, the saturation concentration for the gas increases. If the pressure decreases, the saturation concentration decreases. In other words, colder water holds more dissolved gas than warmer water, less-salty water holds more gas than more-salty water, and water under more pressure holds more gas than water under less pressure.

Distribution with Depth

During **photosynthesis,** plants, seaweeds, and phytoplankton use carbon dioxide, sunlight, and inorganic nutrients (see section 5.4) to produce organic matter. In the process of photosynthesis, oxygen is generated. Phytoplankton and seaweeds grow in surface waters where there is sufficient sunlight to carry out photosynthesis. This lighted portion of the ocean is referred to as the **euphotic zone** (euphotic means "well lit" in Greek). In coastal waters, the euphotic zone is relatively shallow and may extend to only about 15 to 20 m (49–66 ft). In the open ocean where there are less suspended particles, the euphotic zone extends much deeper to about 150 to 200 m (492–656 ft). Photosynthetic organisms produce oxygen and use carbon dioxide

in surface waters. In contrast, heterotrophic organisms consume organic compounds as food and use respiration to derive energy from the consumed materials. During respiration, the organic matter is **oxidized** using oxygen to produce carbon dioxide. Thus, heterotrophic organisms consume oxygen and produce carbon dioxide. All living organisms, regardless of whether they are photosynthetic or heterotrophic, carry out respiration. As a consequence, respiration occurs at all depths within the ocean. Bacteria also respire (although they, of course, do not breathe like humans). Bacterial respiration of organic matter becomes the most important factor in the removal of oxygen from seawater at depth because bacteria are the most abundant organisms deep in the water column.

The depth at which the rate of photosynthesis balances the rate of respiration is called the **compensation depth.** Above the compensation depth, photosynthetic organisms produce oxygen at the expense of carbon dioxide; below it, carbon dioxide is produced at the expense of oxygen. Oxygen can be added to the oceans only at the surface, from exchange with the atmosphere or as a waste product of photosynthesis. Carbon dioxide also enters from the atmosphere at the surface, but it is also produced at all depths from respiration.

Dissolved oxygen concentrations vary from 0–10 ml/l of seawater. Very low or zero concentrations occur in the bottom waters of isolated deep basins, which have little or no mixing with surface water. Such an area can occur at the bottom of a trench, in a deep basin behind a shallow entrance sill (as in the Black Sea), or at the bottom of a deep fjord (300–400 m; 900–1300 ft). If the deep water is only slowly flushed, respiration can use up the oxygen faster than the slow circulation to this depth is able to replace it. The bottom water becomes **anoxic,** or stripped of dissolved oxygen; **anaerobic** (or nonoxygen-using) bacteria live in such water. Because oxygen is more soluble in cold water than in warm water, more oxygen is found in surface waters at high latitudes than at lower latitudes. If the water is quiet, the nutrients and sunlight are abundant, and a large population of photosynthetic organisms is present, oxygen values at the surface can rise above the equilibrium (or saturation) value to 150% or more. This water is **supersaturated.** Wave action tends to liberate oxygen to the atmosphere and return the water to its 100% saturation state.

Table 5.6   Abundance of Gases in Air and Seawater

Gas	Symbol	Percentage by Volume in Atmosphere	Percentage by Volume in Surface Seawater[1]	Percentage by Volume in Total Oceans
Nitrogen	N_2	78.03	48	11
Oxygen	O_2	20.99	36	6
Carbon dioxide	CO_2	0.03	15	83
Argon, helium, neon, etc.	Ar, He, Ne	0.95	1	
Totals		100.00	100	100

[1] Salinity = 36‰, temperature = 20°C.

Figure 5.5 shows typical oxygen and carbon dioxide concentrations with depth. The concentration of both oxygen and carbon dioxide is influenced by biology. The concentration of oxygen is high and the concentration of carbon dioxide is low in surface waters because of photosynthesis. Below the euphotic zone, oxygen decreases as respiration of organic material removes the oxygen. The oxygen minimum occurs at about 800 m (2600 ft). Below this depth, the rate of removal of oxygen decreases because the population density of animals and the abundance of organic matter have decreased. The slow supply of oxygen to greater depths by water sinking from the surface gradually increases the concentration above that found at the oxygen minimum.

Carbon dioxide levels range between 45 and 54 ml/l throughout the oceans. The carbon dioxide concentration at the surface is low because it is used in photosynthesis. Below the surface layer, the concentration increases with depth as respiration continually produces carbon dioxide and adds it to the water. The deep water is able to hold high concentrations of CO_2 because the saturation value is high at low temperatures and high pressures. This is why calcareous oozes are preserved in the warmer, shallower water above the CCD and dissolved below it (review the discussion of biogenous sediments in chapter 3). At

shallow depths, the concentration of CO_2 in the oceans is relatively low, in part because of high rates of photosynthesis and in part because the warm, shallow water has a low saturation value.

The Carbon Dioxide Cycle

At present, the net annual ocean uptake of carbon as carbon dioxide (CO_2) from the atmosphere is estimated to be 2 billion–3 billion metric tons per year. The rate at which the oceans absorb CO_2 is controlled by water temperature, pH (discussed in section 5.3), salinity, the chemistry of the ions (presence of calcium and carbonate ions), and biological processes, as well as mixing and circulation patterns (fig. 5.6).

The transfer of carbon from CO_2 to organic molecules by photosynthesis results in the addition of CO_2 to the intermediate and deep-ocean water when the organic material sinks and decays. This process is often called the **biological pump.** Phytoplankton are responsible for about 40% of Earth's total production of organic material by photosynthesis. These organisms inhabit the shallow surface water, where sufficient sunlight is available for photosynthesis. About 90% of the phytoplankton organic matter is recycled in the euphotic zone as a result of consumption and respiration. Most of the remaining 10% (between 70% and 90%) is recycled before it reaches the sea floor, where the remainder is consumed by bottom-dwelling animals, decomposed by bacteria, or preserved in the sediment. This biochemical pump works along with the chemical solubility pump discussed previously to concentrate carbon in the deep ocean.

The Oxygen Balance

The oceans also play a large role in regulating the oxygen balance in Earth's atmosphere. Photosynthesis in the oceans releases oxygen, which is consumed by respiration and decay processes in the same way as on land. However, some organic matter is incorporated into the seafloor sediments, preventing decay and decomposition. Therefore, oxygen is not consumed to balance the oxygen produced in photosynthesis, and 300 million metric tons of excess oxygen are produced each year. This excess amount of oxygen is not released into the atmosphere because the marine sediments are formed into rocks by Earth's geologic processes. Some of these rocks are uplifted onto land, and oxygen is eventually consumed in the weathering and oxidation of the materials in the rocks. This process balances the atmosphere's oxygen budget. The mechanisms that link and control the process are not well understood.

Measuring the Gases

The amount of dissolved oxygen in seawater samples can be measured by traditional chemical techniques in the laboratory. It is also possible to directly measure oxygen in the oceans by using specialized probes that send an electronic signal back to the ship or store the information in the testing unit. The concentration of dissolved carbon dioxide in seawater is very small. Nearly all of the carbon dioxide in seawater reacts with water to form carbonic acid and its dissociation products (see

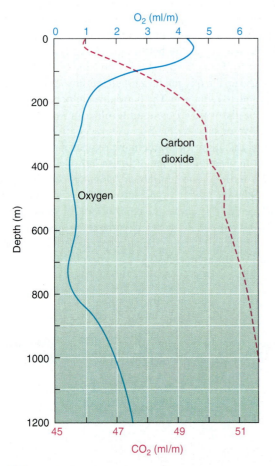

Figure 5.5 The distribution of O_2 and CO_2 with depth. Changes in O_2 concentration may exceed 400% from the surface to depth, whereas CO_2 concentrations change by less than 15%.

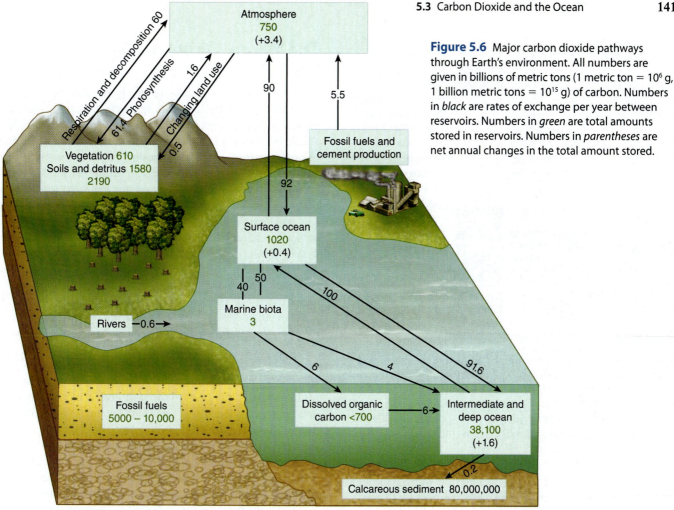

Figure 5.6 Major carbon dioxide pathways through Earth's environment. All numbers are given in billions of metric tons (1 metric ton = 10^6 g, 1 billion metric tons = 10^{15} g) of carbon. Numbers in *black* are rates of exchange per year between reservoirs. Numbers in *green* are total amounts stored in reservoirs. Numbers in *parentheses* are net annual changes in the total amount stored.

section 5.3). Consequently, carbon dioxide concentrations can either be measured directly or determined by measuring the pH of the water.

QUICK REVIEW

1. Which two gases are biologically important in the ocean?
2. Describe how the concentrations of these two gases change as depth increases in the ocean.
3. How is it possible for seawater to become supersaturated with oxygen?

5.3 Carbon Dioxide and the Ocean

The pH of Seawater

The water molecule, H_2O, can dissociate (break apart) to form a hydrogen cation, H^+, and a hydroxide anion, OH^-. Consequently, in any water solution, there will always be a combination of H_2O molecules, H^+ ions, and OH^- ions. The concentration of H_2O molecules always greatly exceeds the concentrations of the two ions. In a pure water solution (one in which there is only water molecules) at 25°C, a very small fraction of the water molecules, about 10^{-7}, will spontaneously dissociate into H^+ and OH^- ions.

In other words, the concentration of both H^+ and OH^- ions will be 10^{-7}, as one in every 10 million (10^7) water molecules breaks apart. Solutions in which the concentrations of these two ions are equal are called neutral solutions.

In solutions that are not pure water, chemical reactions can remove or release hydrogen ions, making the concentrations of H^+ and OH^- unequal. The concentrations of H^+ and OH^- in a water solution are inversely proportional to each other. In other words, a tenfold increase in the concentration of one ion results in a tenfold decrease in the concentration of the other. An imbalance in the relative concentrations of these ions results in either an acidic solution (if there are more H^+ cations than OH^- anions) or an alkaline, also called basic, solution (if there are more OH^- anions than H^+ cations).

The acidity or alkalinity of a solution is measured using the **pH** scale, which ranges from a low of 0 to high of 14 (fig. 5.7). The pH scale is a logarithmic scale that measures the concentration of the hydrogen ion (written [H^+]) in a solution. The formal definition of pH is:

$$pH = -\log_{10}[H^+]$$

In pure water, where the concentrations of H^+ and OH^- are both 10^{-7}, the pH is equal to 7,

$$pH = -\log_{10}[10^{-7}] = -(-7) = 7,$$

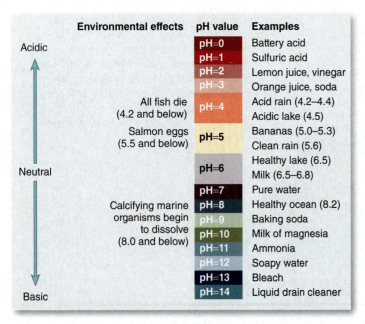

Figure 5.7 The pH scale with examples and environmental effects for aquatic organisms. The term pH stands for the "power" (in terms of exponents) of the negative logarithm (log) of the hydrogen concentration of a substance, or formally: pH = −log [H$^+$]. The pH is given as the negative log of the concentration of the hydrogen ion because it simplifies the scale to values between zero (0) and fourteen (14).

and the solution is neutral. If the concentration of the hydrogen ion is increased by a factor of ten to 10^{-6}, or one part in 1 million instead of one part in 10 million, the pH drops to 6 and the solution is slightly acidic. Solutions with pH less than 7 (high H$^+$ concentrations) are acidic and those with pH greater than 7 (low H$^+$ concentrations) are alkaline, or basic.

Solutions with lower pH are more acidic than solutions with higher pH. For example, the following statements are all equivalent:

pH = 1 is more acidic than pH = 3, or

[H$^+$] = 10^{-1} is more acidic than [H$^+$] = 10^{-3}, or

[H$^+$] = 0.1 is more acidic than [H$^+$] = 0.001

Similarly, solutions with higher pH are more alkaline than solutions with lower pH.

Because the hydrogen ion is very reactive, acidic water (water with a relatively high concentration of H$^+$ and pH less than 7) is an effective chemical weathering agent capable of decomposing and dissolving rock. The pH of rainwater is normally about 5.0–5.6 (slightly acidic), but in some heavily industrialized regions where emissions combine with water droplets to form acid rain, it can be much lower. Typical rain in the eastern United States has a pH of about 4.3, roughly ten times more acidic than normal, and can in some cases drop as low as 3, or 100 times more acidic than normal.

Seawater is slightly alkaline with a pH between 7.5 and 8.5. The pH of the world's ocean averaged over all depths is approximately 7.8. Surface water currently has an average pH of about 8.2.

The Marine Carbonate System and Buffering pH

The pH of seawater remains relatively constant because of the buffering action of the carbonate system in the water. A **buffer** is a substance that prevents sudden, or large, changes in the acidity or alkalinity of a solution. If some process changes the concentration of hydrogen ions in seawater, causing the pH to rise above or fall below its average, or mean value, the buffer becomes involved in chemical reactions that release or capture hydrogen ions, returning the pH to normal. When carbon dioxide dissolves in seawater, the CO_2 combines with the water to form carbonic acid (H_2CO_3). The carbonic acid rapidly dissociates into bicarbonate (HCO_3^-) and a hydrogen ion (H$^+$), or carbonate (CO_3^{2-}) and two hydrogen ions (2 H$^+$). The CO_2, H_2CO_3, HCO_3^-, and CO_3^{2-} exist in equilibrium with each other and with H$^+$, as shown in the following equation and in figure 5.8. The double arrows indicate that the reactions can move in either direction, either producing or removing hydrogen ions as is necessary to maintain a relatively constant pH.

$$CO_2 + H_2O \Leftrightarrow H_2CO_3 \Leftrightarrow HCO_3^- + H^+ \text{ or } CO_3^{2-} + 2H^+$$

If seawater becomes too alkaline, or basic, then the reactions in this equation progress to the right, releasing hydrogen ions and decreasing the pH. If seawater becomes too acidic, then the reactions progress to the left, removing free hydrogen ions from the water and increasing the pH. This buffering capacity of carbon dioxide in seawater is important to organisms requiring a relatively constant pH for their life processes and to the chemistry of seawater, which is controlled, in part, by its pH.

From the reactions described, it is clear that the pH of seawater strongly depends on the concentration of CO_2 in the water. The lower the concentration of CO_2 in the water, the higher its

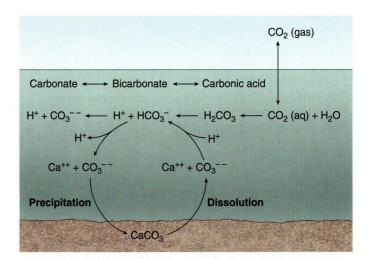

Figure 5.8 The carbonate system in seawater. The concentrations of any one of the "species" in the marine carbonate system are pH dependent. At high pH (above 9), the equilibrium shifts toward the right, favoring carbonate ions. At low pH (below 6), the equilibrium favors the left-hand products, namely CO_2. At the intermediate pH of the world ocean (~8.2), most of the dissolved carbon dioxide exists as a bicarbonate.

pH and the more alkaline it becomes. In general, the pH of surface water tends to be higher, or more alkaline, than average because of lower levels of CO_2 in the water due to the consumption of CO_2 in the process of photosynthesis. The pH at the ocean surface may be as high as 8.5 if the water is warm and if the rate of primary production, or photosynthesis, is high. Raising the pH of the water releases carbonate ions, CO_3^{2-}, that bond with the abundant calcium ions, Ca^{2+}, in solution to form calcium carbonate $CaCO_3$. In cold, deep water, where the concentration of CO_2 is high, the pH drops, making the water more acidic and dissolving calcium carbonate shells.

Anthropogenic Carbon Dioxide and Ocean Acidification

Burning of fossil fuels and deforestation has led to a dramatic increase in the concentration of CO_2 in the atmosphere. Carbon dioxide produced by human activities is known as **anthropogenic carbon dioxide**. Since 1850 and the start of the Industrial Revolution, the concentration of CO_2 in the atmosphere has increased from 280 parts per million (ppm) to over 390 ppm. Recently, the average rate of increase has been 1.5–2 ppm per year. A continuous measurement by David Keeling of Scripps Institution of Oceanography, from the top of Mauna Loa on the Big Island of Hawaii, of atmospheric CO_2 concentration began in 1958. The **Keeling Curve** is the most widely recognized measurement of human impact on the environment in existence (fig. 5.9). The short-term variations in CO_2 concentration seen in the data are due to the natural seasonal variation in plant photosynthesis in the Northern Hemisphere (fig. 5.10). A reduction in the concentration of CO_2 in the late spring and summer occurs as plants increase active photosynthesis and extract CO_2 from the atmosphere. The concentration of CO_2 increases in the fall and winter because of a decrease in photosynthesis when plants lose their leaves, and also because of the release of CO_2 to the atmosphere by decay processes.

The increase in CO_2 in the atmosphere has resulted in a corresponding increase in the concentration of the gas in the ocean as CO_2 is absorbed by seawater at the sea surface. The increasing concentration of CO_2 in the water is causing a decrease in the pH of the water, an effect that is referred to as **ocean acidification** (fig. 5.11). The average pH of the ocean is predicted to fall by up to 0.5 units by 2100 if global emissions of CO_2 continue to rise at present rates. This increase in ocean acidity could have a major impact on shallow-water marine organisms that build shells of calcium carbonate, which could dissolve rapidly in more acidic water (fig. 5.12).

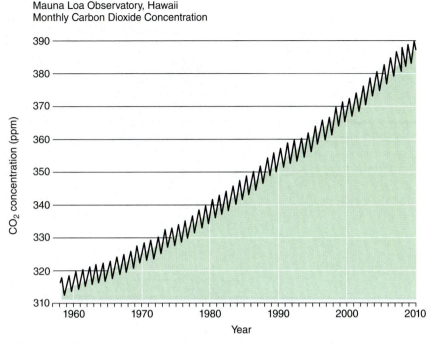

Mauna Loa Observatory, Hawaii
Monthly Carbon Dioxide Concentration

Figure 5.9 The Keeling Curve. Concentration of atmospheric carbon dioxide in parts per million (ppm) observed at Mauna Loa Observatory, Hawaii.

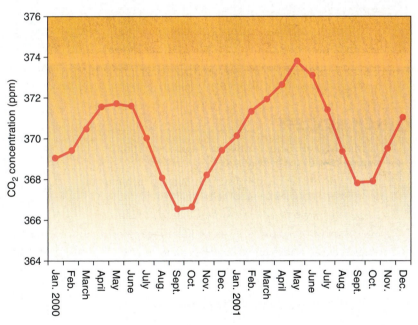

Figure 5.10 The concentration of carbon dioxide in parts per million (ppm) for the years 2000 and 2001 measured at Mauna Loa Observatory, Hawaii. Season variations in concentration reflect seasonal variations in plant photosynthesis. Low concentrations in spring and summer are due to high rates of photosynthesis, which takes carbon dioxide out of the atmosphere. High concentrations in fall and winter are due to low rates of photosynthesis and the release of carbon dioxide to the atmosphere by decay.

The most important sources of CO_2 in seawater are direct transfer of the gas from the atmosphere, the respiration of marine organisms, and the oxidation of organic matter during decay.

Figure 5.11 Estimated change in annual mean sea surface pH between the pre-industrial period (1700s) and the present day. Rising levels of carbon dioxide in the atmosphere result in more carbon dioxide being absorbed in ocean surface water, thus producing ocean acidification though the lowering of surface water pH. Increasing acidity of seawater makes it more difficult for marine organisms to construct hard calcium carbonate forms such as shells and coral.

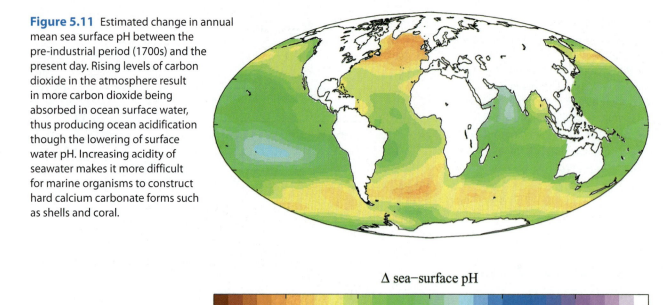

Δ sea–surface pH

-0.12   -0.1   -0.08   -0.06   -0.04   -0.02   0

Figure 5.12 A calcium carbonate pteropod (zooplankton) shell being progressively dissolved in acidic seawater.

QUICK REVIEW

1. Explain how pH varies relative to the concentration of carbon dioxide in the water.
2. Would you expect the pH of surface water to be higher or lower than the average pH of seawater? Why?
3. Why is the range of pH in seawater relatively small?
4. How does carbon dioxide act as a buffer in seawater?
5. What is ocean acidification and what is causing it?

5.4 Nutrients and Organics

Nutrients

Ions required for plant or phytoplankton growth are known as nutrients; these are the fertilizers of the oceans. As on land, phytoplankton require nitrogen and phosphorus in the form of nitrate (NO_3^-) and phosphate (PO_4^{3-}) ions. A third nutrient required in the oceans is the silicate ion (SiO_4^{4-}), which is needed to form silica (SiO_2), the hard outer wall of the single-celled diatoms and the skeletal parts of some protozoans. These three nutrients are among the dissolved substances brought to the sea by the rivers and land runoff. Despite their importance, they are present in very low concentrations (table 5.7).

The concentrations of nutrient ions vary because some of these ions are closely related to the life cycles of organisms. The relative molar abundance of carbon, nitrogen, and phosphorus in marine phytoplankton is C:N:P = 106:16:1. This relationship is called the **Redfield Ratio.** Analysis of the composition of siliceous marine organisms makes it possible to calculate a Redfield Ratio for silicon as well: C:Si:N:P = 106:40:16:1. The consumption, decomposition, and recycling of organic matter as it sinks through the water column result in an increase in carbon:nutrient ratios as nutrients are released to the water. Nutrients are removed from the water as the plant populations grow and reproduce, temporarily reducing the amounts in solution. Later, when the populations decline, death and

Table 5.7   Nutrients in Seawater

Element	Concentration μg/kg[1]	Relative Molar Abundance
Nitrogen (N)	500	16
Phosphorus (P)	70	1
Silicon (Si)	3000	40

[1] Parts per billion.

decay return the ions to the seawater. Nutrients are cycled to different consumers as zooplankton feed on phytoplankton. The zooplankton are in turn eaten by other consumers, and eventually the nutrients are returned to the oceans by death and bacterial decomposition. Excretory products from zooplankton and larger animals are also added to the seawater, broken down, and used by a new generation of zooplankton and phytoplankton. Nutrients are nonconservative; they do not maintain constant ratios in the way most major salt ions do.

Organics

A wide variety of organic substances are present in seawater. Proteins, carbohydrates, lipids (or fats), vitamins, hormones, and their breakdown products are all present. Some are eventually oxidized or broken down into smaller molecules; others are used directly by organisms and are incorporated into their systems. Another portion of the organic matter accumulates in the sediments, where over geologic time it may slowly provide hydrocarbon molecules to form deposits of oil and gas. In the areas of the ocean that are high in plant and animal life, the surface layer may take on a green-yellow color owing to the presence of organic decay products. The incorporation of soluble organics into glacial ice at the Antarctic ice shelves is related to the formation of green ice.

QUICK REVIEW

1. What are nutrients and why are they important?
2. Why is the concentration of some nutrients so variable?
3. What is the Redfield Ratio?
4. Why are nutrients considered to be nonconservative materials?
5. Silicate is a nonconservative constituent of seawater and does not obey the principle of constant proportions. Explain why.

5.5 Practical Considerations: Salt and Water

Chemical Resources

About 30% of the world's table salt is extracted from seawater. The industrially produced energy required to remove the water is kept to a minimum to keep extraction costs low. In warm, dry climates, seawater is allowed to flow into shallow ponds and evaporate down to a concentrated brine solution. More seawater is added, and the process is repeated several times, until a dense brine is produced. Evaporation continues until a thick, white salt deposit is left on the bottom of the pond. Several different salts are in the deposit. These salts form in the order listed in table 5.8. The salt deposit is collected and refined to separate out sodium chloride (halite), or table salt. This technique is used in southern France, Puerto Rico, and California (fig. 5.13).

Table 5.8 Sequence of Salts Formed from Evaporation of Seawater

Order of Precipitation	Solid	% of Total Solid
1	$CaCO_3 + MgCO_3$	1
2	$CaSO_4$ (gypsum)	3
3	NaCl (halite)	70
4	Na-Mg-K-SO_4 and KCl, $MgCl_2$	26

In cold climate areas, salt has been recovered by freezing the seawater in similar ponds. The ice that forms is nearly fresh; the salts are concentrated in the brine beneath the ice. The brine is removed and heated to remove the last of the water.

Of the world's supply of magnesium, 60% comes from the sea, and so does 70% of the bromine. There are vast amounts of dissolved constituents in the world's seawater, but their concentrations are typically very low (1 part per billion or less), making their extraction more costly than their current value.

Figure 5.13 The southern end of San Francisco Bay was diked into shallow ponds, where seawater is evaporated to obtain salt. Most of these ponds are being converted back to "natural" conditions.

Desalination

Desalination is the process of obtaining fresh water from salt water. There are three main desalination methods:

1. processes involving a change of state of the water: liquid to solid or liquid to vapor;
2. processes requiring ion exchange columns; and
3. processes using a semipermeable membrane: electrodialysis and reverse osmosis.

The simplest process involving a change of state is a solar still (fig. 5.14). In this process, a pond of seawater is capped by a low plastic dome. Solar radiation penetrates the dome and evaporates the seawater. The evaporated water condenses on the undersurface of the dome and trickles down to be caught in a trough, where it accumulates and flows to a freshwater reservoir. The rate of production is slow, and a very large system is needed to supply the water requirements of even a small community.

When water is distilled by boiling, evaporation proceeds at a rapid rate and large quantities of fresh water are produced, but the energy requirement is high. If water is introduced to a chamber with a reduced air pressure, the boiling occurs at a much lower temperature and therefore uses less energy. Change of state by freezing can also be used to recover fresh water from seawater. The energy requirement is approximately one-sixth of that needed for evaporation, but the mechanical separation of the freshwater ice from the salt brine remains difficult.

Columns containing ion exchange resins that extract ions from salt water work well with water of low salinity, but the resins need to be replaced periodically (fig. 5.15b). Small ion exchange units are manufactured for household use to improve drinking water quality.

Electrodialysis uses an electrical field to transport ions out of solution and through **semipermeable membranes;** this technique also works best in low-salinity (or brackish) water (fig. 5.15a).

Osmosis is the movement of water across a semipermeable membrane; the water moves from the side with the higher concentration of water molecules (or low salinity) to the side with the lower concentration of water molecules (or high salinity); this movement creates a higher pressure on the low-water concentration (or high-salinity) side of the membrane (fig. 5.16a). **Reverse osmosis** produces fresh water from seawater by applying pressure to seawater and forcing the water molecules through a semipermeable membrane, leaving behind the salt ions and other impurities (fig. 5.16b). The pressure applied to the seawater must exceed 24.5 atm, or 25.84×10^6 dynes/cm^2. A pressure of about 101.5 atm, or $10^3 \times 10^6$ dynes/cm^2, is required to achieve a reasonable rate of freshwater production. The energy requirement is about one-half that needed for the evaporative process.

Reverse osmosis is the most popular and rapidly growing form of desalination technology. As older evaporative plants wear out, reverse osmosis plants are replacing them. The advantages of reverse osmosis include no energy requirement for heating the water, no thermal pollution from the discharge, and removal of unwanted contaminants—including pesticides, bacteria, and some chemical compounds. High-salinity wastewater returning to the coastal environment can be a disadvantage. The cost of the energy required to pump the water under pressure makes the cost of this type of desalinated water very high. In southern California, desalinated water is about fifteen times more expensive than local groundwater and five times more expensive than water imported from the northern part of the state.

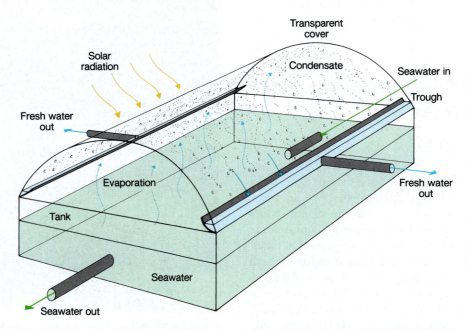

Figure 5.14 Solar energy is used to evaporate fresh water from seawater. Solar radiation penetrates the transparent cover over the seawater contained in the tank.

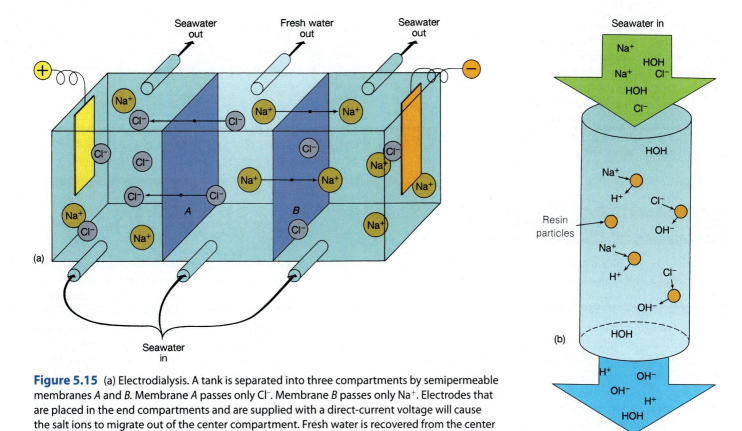

Figure 5.15 (a) Electrodialysis. A tank is separated into three compartments by semipermeable membranes *A* and *B*. Membrane *A* passes only Cl^-. Membrane *B* passes only Na^+. Electrodes that are placed in the end compartments and are supplied with a direct-current voltage will cause the salt ions to migrate out of the center compartment. Fresh water is recovered from the center compartment; excess salt water is removed from each side compartment. (b) Ion-exchange column. Seawater passes through a column of resin particles that exchange H^+ for Na^+ and Cl^- for OH^- to produce HOH, or fresh water (H_2O).

Because of the high price involved, desalination plants in southern California are typically operated only during periods of drought, when reservoirs and groundwater levels are low. During years when normal or high rainfall provides sufficient fresh water, the plants are shut down and only maintenence level work is done to ensure that they will be operational when needed.

In areas such as Kuwait, Saudi Arabia, Morocco, Malta, Israel, the West Indies, California, and the Florida Keys, water is a limiting factor for population and industrial growth. The greatest drawback to the production of fresh water from seawater is the high cost, which is linked to the energy required. In Persian Gulf countries, water costs are low because fuel costs are low.

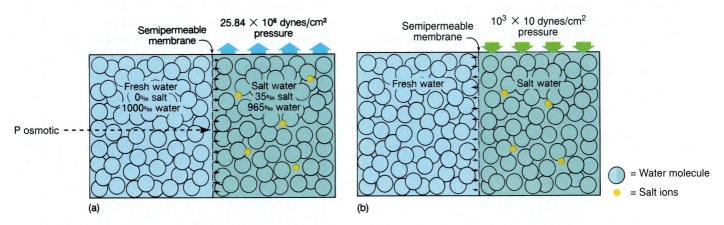

Figure 5.16 (a) Osmosis. Water molecules move from the freshwater side to the saltwater side through a semipermeable membrane. (b) Reverse osmosis. When pressure on the salt water exceeds 25.84×10^6 dynes/cm^2, water molecules move from the saltwater side to the freshwater side of a semipermeable membrane.

Large floating masses of naturally desalinated ice have attracted interest as sources of fresh water. The first of a series of meetings on this possibility was hosted in 1978 by Arab interests. Discussions centered on the feasibility of towing icebergs from the Antarctic to the Red Sea. Although the expense would be enormous and much of the ice would melt as it was towed into equatorial latitudes, the need for water is so great that it is believed that sufficient ice would remain after the journey to make the project worthwhile.

QUICK REVIEW

1. Explain the operation of a solar still.
2. Where are most of the world's desalination plants located and why?
3. What is the process of reverse osmosis?

Summary

Seawater is slightly alkaline with a pH between 7.5 and 8.5. The average pH value for all of the oceans over all depths is about 7.8. Seawater pH remains fairly constant because of the buffering action of carbon dioxide in the water.

The average salinity of ocean water is 35‰. The salinity of the surface water changes with latitude and is affected by evaporation, precipitation, and the freezing and thawing of sea ice. Soluble salts are present as ions in seawater. Positive ions are cations; negative ions are anions. Six major constituent ions make up 99% of the salt in seawater. Trace elements that are present in very small quantities are particularly important to living organisms.

Most of the positively charged ions come from the weathering and erosion of Earth's crust and are added to the sea by rivers. Gases from volcanic eruptions are dissolved in river water as anions. Because the average salinity of the oceans remains constant, the salt gain must be balanced by the removal of salt; input must equal output. Salts are removed as sea spray, evaporites, and insoluble precipitates, as well as by biological reactions, adsorption, chemical reactions, and uplift processes. Seawater circulating near magma chambers in Earth's crust deposits dissolved metals and releases other chemicals in solution. The time that salts remain in solution, known as residence time, depends on their reactivity.

The proportion of one major ion to another remains the same for all open-ocean salinities. Ratios may vary in coastal areas and in association with biological processes.

Salinity is determined by measuring a sample's electrical conductivity. Historically, salinity has been determined chemically by measuring the quantity of chloride ions in a sample.

The saturation value of gases dissolved in seawater varies with salinity, temperature, and pressure. Carbon dioxide is added to seawater from the atmosphere and by respiration and decay processes at all depths; it is removed at the surface by photosynthesis. Oxygen is added only at the surface from the atmosphere and the photosynthetic process; it is depleted at all depths by respiration and decay. Seawater may become supersaturated with oxygen, or it may become anoxic. Carbon dioxide levels tend to change little over depth. Carbon dioxide has the additional role of buffer in keeping the pH range of ocean water between 7.5 and 8.5. Large quantities of CO_2 are absorbed by the oceans. Biological processes pump carbon as carbon dioxide into deep water, where it is fixed in the marine sediments as calcium carbonate. Atmospheric oxygen is regulated by oceanic processes. The amount of oxygen present in seawater is measured chemically and electronically. Carbon dioxide content is determined from the pH of the water.

Nutrients include the nitrates, phosphates, and silicates required for plant growth. A wide variety of organic products are also present.

Salt, magnesium, and bromine are currently being commercially extracted from seawater. Direct extraction of other chemicals is neither economic nor practical at present. Fresh water is an important product of seawater. Desalination methods include change-of-state processes, movement of ions across semipermeable membranes, and ion exchange. The practicality of desalination is determined by cost and need. Reverse osmosis has become the most popular option; it is nonpolluting but costly because of its energy requirements.

Key Terms

ion, 132
cation, 132
anion, 132
ionic bond, 132
salinity, 133
major constituent, 134
trace element, 134
conservative constituent, 134
nonconservative
    constituent, 134

evaporite, 135
adsorption, 136
ion exchange, 136
residence time, 137
principle of constant
    proportion (constant
    composition), 138
salinometer, 138
chlorinity (Cl‰), 138
saturation concentration, 139

photosynthesis, 139
euphotic zone, 139
oxidized, 139
compensation depth, 139
anoxic, 139
anaerobic, 139
supersaturation, 139
biological pump, 140
pH, 141
buffer, 142

anthropogenic carbon
    dioxide, 143
Keeling Curve, 143
ocean acidification, 143
Redfield Ratio, 144
desalination, 146
electrodialysis, 146
semipermeable membrane, 146
osmosis, 146
reverse osmosis, 146

Study Problems

1. If there is 1.4×10^{21} kg of water in the oceans, what is the potential mass of NaCl (sodium chloride) in the oceans? Use table 5.2.

2. If the chloride ion (Cl⁻) content of a seawater sample is 18.5 ppt, what is the concentration of magnesium in the same sample? Express your answer in g/kg.

3. Determine the residence time of calcium using the following information:

   calcium ion concentration in seawater = 0.41 g/kg

   seawater in the oceans = 1.4×10^{21} kg
   calcium ion concentration by weight in
   river water = 12.5 g/kg
   average salt content of river water = 0.12 g/kg
   annual river runoff into the oceans = 3.6×10^{16} kg/yr

4. How many kilograms of seawater would have to be processed to obtain 1 kg of gold? Use table 5.3.

The Atmosphere and the Oceans

Learning Outcomes

After studying the information in this chapter students should be able to:

1. *outline* and *discuss* Earth's heat budget,

2. *distinguish* between specific heat and heat capacity,

3. *list* the layers of the atmosphere in order of ascending height and *sketch* a plot of temperature vs. elevation in each layer,

4. *argue* that global warming is enhanced by the accumulation of greenhouse gases, such as carbon dioxide, in the atmosphere,

5. *explain* the characteristics of the Antarctic ozone hole and *relate* its size to cloud formation and temperature,

6. *explain* the Coriolis effect and *describe* its direction and magnitude as a function of latitude,

7. *sketch* the pattern of major wind systems and regions of vertical motion in the atmosphere,

8. *list* three names for large, intense low-pressure systems,

9. *relate* ENSO events to global weather patterns and variations in sea surface temperature, and

10. *review* the major historical and physical details of Hurricane Katrina.

Hurricane Linda off Baja California Sur, 1999.

The Sun's energy reaches Earth's surface through the atmosphere, a thin shell of mixed gases we call air. The atmosphere and the ocean are in contact over 71% of Earth's surface; their interaction is continuous and dynamic. Processes that occur in the atmosphere are closely related to processes that occur in the oceans, and together they form much of what we call weather and climate. Clouds, winds, storms, rain, and fog are all the result of interplay among the Sun's energy, the atmosphere, and the oceans. This complex of interactions provides Earth's average climate and its daily weather, sometimes pleasant and stable, at other times severe and turbulent. Some of these interactions and processes are more predictable than others; some are better understood than others. Understanding the oceans requires an understanding of the atmosphere's influence on them. This chapter presents an overview of these self-adjusting relationships as well as specific examples of their combined effects.

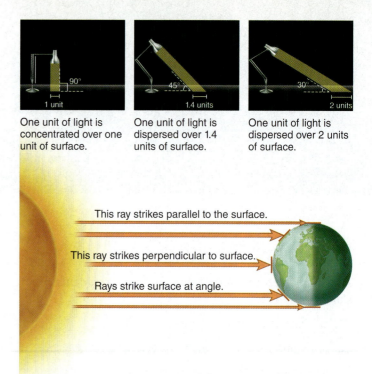

One unit of light is concentrated over one unit of surface.

One unit of light is dispersed over 1.4 units of surface.

One unit of light is dispersed over 2 units of surface.

This ray strikes parallel to the surface.

This ray strikes perpendicular to surface.

Rays strike surface at angle.

Figure 6.1 Areas of Earth's surface that are equal in size receive different levels of solar radiation as they become more oblique to the Sun's rays. As latitude increases, the Sun's rays and Earth's surface become more nearly parallel, and the solar radiation received on equal surface areas decreases. With increasing latitude, the Sun's rays also have to travel an increasing distance through the atmosphere before reaching Earth's surface.

6.1 Heating and Cooling Earth's Surface

The heating and cooling of Earth's surface are accomplished by energy exchanges that act to alter the densities of two fluid envelopes surrounding Earth: the atmosphere and the oceans. The initial source of the energy is solar radiation, which varies in both time and location on Earth's surface. This energy is absorbed, reflected, reradiated, converted into other forms of energy, and redistributed over Earth to create not only the structure but also the dynamics of the atmosphere and ocean.

Distribution of Solar Radiation

Instantaneous solar radiation per unit area of Earth surface has its greatest intensity at the equator, moderate intensity in the middle latitudes, and least intensity at the poles. If Earth had no atmosphere, the intensity of solar radiation available on a surface at right angles to the Sun's rays would be 2 calories per square centimeter per minute ($cal/cm^2/min$); this value is called the **solar constant.** The solar constant is approached only at latitudes between 23½°N (the Tropic of Cancer) and 23½°S (the Tropic of Capricorn), because only between those latitudes does sunlight strike Earth at a right angle. Because of Earth's spherical shape, at all other latitudes, Earth's surface is inclined to the Sun's rays at an angle other than 90°. Compare the angles at which the Sun's rays strike Earth in figure 6.1. Because the Sun is so far away from Earth, its rays are parallel when they reach Earth. Where the rays strike Earth at right angles, the same amount of radiant energy strikes each unit area. But as the angle the Sun's rays make with Earth decreases, the rays and the surface become nearly parallel, and the unit areas receive less energy. This difference is also shown in figure 6.1.

When the Sun stands directly above the equator at noon, during either the vernal or the autumnal equinox, the radiation value is about 1.6 $cal/cm^2/min$. This value is less than the solar constant because Earth's atmosphere stands between Earth's surface and the incoming solar radiation, and the atmosphere both absorbs and reflects portions of the Sun's energy. Atmospheric interference causes the solar radiation per unit of surface area to decrease with increasing latitude in each hemisphere, because the greater the latitude, the longer the distance through the atmosphere the Sun's rays must travel. The combined effects of atmospheric path length and inclination of Earth's axis cause Earth to receive more solar heat in the tropics, less at the temperate latitudes, and the least at the poles. The intensity of solar radiation also varies as points on the turning Earth move from darkness to light and return to darkness and as the distance between the Sun and Earth changes seasonally.

Heat Budget

To maintain its long-term mean surface temperature of 16°C, Earth must lose heat as well as gain it. To maintain a constant average temperature, Earth and its atmosphere must reradiate as much heat back to space as they receive from the Sun. These gains and losses in heat are represented by a **heat budget.** If less heat was returned to space than is gained, Earth would become hotter, and if more heat was returned than is gained, Earth would become colder. In both cases, the planet would change dramatically.

To follow the long-term average gains and losses in Earth's heat budget, we will first assume that the total available energy to heat Earth's surface is incoming short-wave solar radiation (light) (fig. 6.2). Let's imagine that the total incoming solar radiation can be expressed as 100 units of energy (fig. 6.2). This will make it easy to visualize different parts of the heat budget as percentages of the whole. To balance the heat budget, the amount of incoming energy has to equal the amount of outgoing energy. Thus, we can express our heat budget as:

100 units of incoming energy = 100 units of outgoing energy

First, let's look at what happens to the 100 units of incoming energy. Roughly 30 units of incoming energy are backscattered, or reflected, as short-wave energy without being changed. This energy is effectively lost and does not contribute to heating Earth. This leaves roughly 70 units of incoming energy that are absorbed, heating Earth.

(1) 100 units of incoming energy →

+70 units absorbed (gained) by Earth

−30 units reflected (lost) to space

Now, let's account for the 100 units of outgoing energy. We already know that 30 units of outgoing energy are the short-wave incoming energy that is immediately reflected and lost. The remaining 70 units are lost by radiation of long-wave (infrared)

energy from the atmosphere, clouds, and the ground (land, water, and ice).

(2) 100 units of outgoing energy →

−70 units infrared, radiated (lost) to space

−30 units light, reflected (lost) to space

(−6 units atmosphere)

(−20 units clouds)

(−4 units ground)

In our accounting of incoming and outgoing energy, expressed in relationships (1) and (2), it is clear that Earth's heat budget is a balance between the 70 units of incoming short-wave energy absorbed that provide heating and the 70 units of radiated long-wave energy lost that provide cooling. So, what happens to the absorbed and radiated energy? What processes allow them to balance each other? First we'll track the incoming energy that is absorbed and used for heating. Of the 70 units of incoming energy, 51 units will be absorbed by the ground. The remaining 19 units will be absorbed by the atmosphere.

Heating

(3) +70 units absorbed (gained) by Earth →

+51 units absorbed by the ground

+19 units absorbed by the atmosphere

(+3 units by clouds)

(+16 units by water vapor, CO_2, dust)

+70 absorbed (gained) by Earth

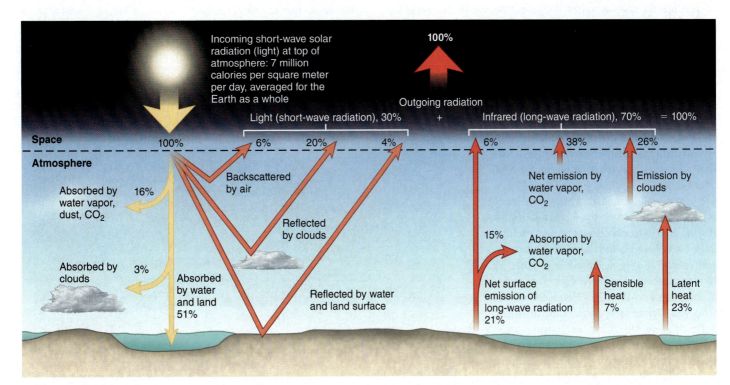

Figure 6.2 Earth's heat budget. Incoming solar energy is balanced by reflected and reradiated energy. The atmosphere's loss of heat is balanced by heat transferred from the ground (land, water, and ice) to the atmosphere by evaporation, conduction, and reradiation.

Now we'll track the radiated energy that is lost and used for cooling. Of the 70 units of outgoing energy, 6 units will be radiated by the ground. The remaining 64 units will be radiated by the atmosphere.

Cooling

(4) −70 units radiated (lost) to space →

> −6 units radiated by the ground
> −64 units radiated by the atmosphere
> (−26 units by clouds)
> (−38 units by water vapor, CO_2)
> _____
> −70 units radiated (lost) to space

Looking carefully at relationships (3) and (4), it appears that the ground absorbs more energy than it radiates, thus having a net gain of heat (+51 − 6 = +45), whereas the atmosphere radiates more energy than it absorbs, thus having a net loss of heat (−64 + 19 = −45). Global long-term average temperatures of the ground and atmosphere are very stable because the excess heat absorbed by the ground is transferred to the atmosphere, replacing heat lost by the atmosphere.

(5) +45 units transferred from ground →

> +15 units radiated and absorbed by water vapor, CO_2
> +7 units sensible heat (heated rising air)
> +23 units latent heat (evaporation and condensation)
> _____
> +45 units transferred to atmosphere

Thus, over long time periods, the amount of heat gained by Earth is balanced by the amount of heat lost.

Earth's surface, including the oceans, is heated from above. If we consider the short-term heat budget of only a small portion of the oceans rather than the total, the following must be taken into consideration: total energy absorbed at the sea surface in that area, loss of energy due to evaporation, transfer of heat into and out of the area by currents, vertical redistribution, warming or cooling of the overlying atmosphere by heat from the sea surface, and heat reradiated to space from the sea surface. These processes vary with time, showing daily and seasonal variations.

Measurements made over Earth's surface show that, on the average, more heat over the annual cycle is gained than lost at the equatorial latitudes, while more heat is lost than gained at the higher latitudes (fig. 6.3). Winds and ocean currents remove the excess heat accumulated in the tropics and release it at high latitudes to maintain Earth's present surface temperature patterns. Figure 6.4 shows Northern Hemisphere winter ocean surface temperatures. North-south deflections in the colors of constant temperature indicate the displacement of surface water by currents carrying warm water from low to high latitudes and cold water from high to low latitudes, redistributing heat energy over Earth's surface.

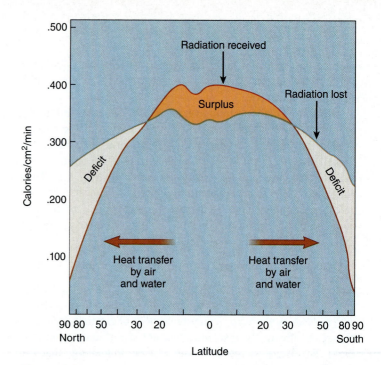

Figure 6.3 Comparison of incoming solar radiation and outgoing long-wave radiation with latitude. A transfer of energy is required to maintain a balance.

Annual Cycles of Solar Radiation

The total Earth long-term heat budget and the average distribution of incoming and outgoing radiant energy with latitude do not include the annual cycle of radiant energy changes related to the seasonal north-south migration of the Sun. When these variations are included, the annual cycle of seasonal variation in average daily solar radiation is most pronounced at the middle and higher latitudes. Here, the angle at which the Sun's rays strike Earth and the length of daylight change dramatically from summer to winter. The seasonal variation is illustrated in figure 6.5.

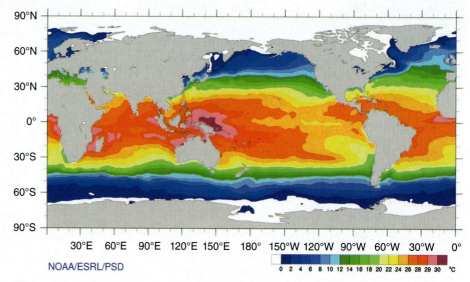

Figure 6.4 Weekly average sea surface temperature during the first week of February 2011.
Source of Data: National Oceanic and Atmospheric Administration (NOAA).

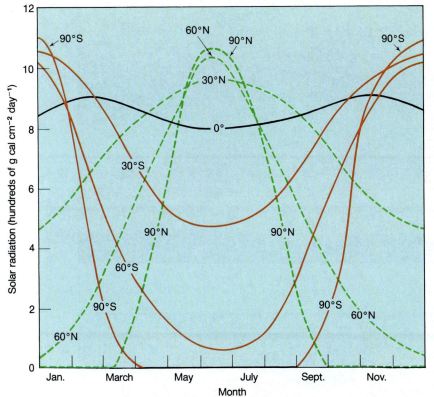

the land can undergo large temperature changes as heat is gained or lost between day and night or summer and winter. The oceans have a very high heat capacity due to the high specific heat of water. As a result, the oceans can absorb and release large amounts of heat with little change in temperature.

The average annual range of surface temperatures for land and ocean are given in figure 6.6. Note that the seasonal temperature ranges at high latitudes are greater for land than for the oceans and that the annual range of ocean surface temperatures is greatest at the middle latitudes. Because of the unequal distribution of land between the two hemispheres, the summer-to-winter variation in temperature for land at the middle latitudes is much greater in the Northern Hemisphere than in the Southern Hemisphere. Because of the lack of land in the Southern Hemisphere, the oceans, with their high heat capacity, control the annual temperature range in southern middle latitudes. These differences between land and

Figure 6.5 Average daily solar radiation values at different latitudes during the year. When it is summer in the Northern Hemisphere, it is winter in the Southern Hemisphere; therefore, the higher values in the Northern Hemisphere coincide with the lower values in the Southern Hemisphere. Peak values of solar radiation at the higher latitudes occur during summer in each hemisphere as a result of more hours of sunlight.

Changes in incident radiation produce seasonal variations in land and sea surface temperatures because of heat losses or gains. The intensity of solar radiation remains fairly constant at tropical latitudes (between 23½°N and 23½°S) over the year, because the Sun's noontime rays are always received at an angle approaching 90° and the length of the daylight period is nearly constant. During the Sun's annual migration between 23½°N and 23½°S, its rays perpendicular to Earth's surface cross the intervening latitudes twice. This twice-a-year crossing produces a small-amplitude, semiannual variation in the intensity of tropical solar radiation. The effect is most evident at the equator and can be seen in figure 6.5. Between 23½°N and 90°N and between 23½°S and 90°S, the Sun's noontime rays always strike Earth at an oblique angle. The long duration of daylight hours in summer at polar latitudes produces a high level of incident solar radiation averaged over the twenty-four-hour day. However, the intensity of radiation per unit surface area per minute of daylight and the annual average radiation level are much lower than those found at lower latitudes (see fig. 6.1).

Specific Heat and Heat Capacity

Land and sea respond differently to solar radiation because of the difference in the specific heat of land materials and water (review table 4.2). The land has relatively low heat capacity because of the low specific heat of rock and soil. Consequently,

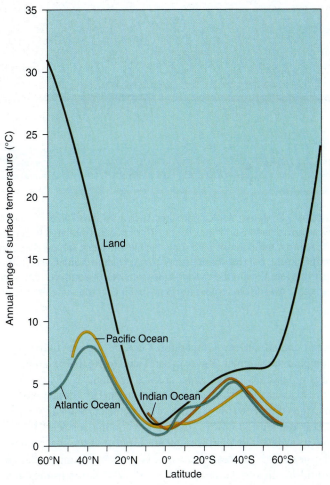

Figure 6.6 The annual range of mid-ocean sea surface temperatures is considerably less than the annual range of land surface temperatures at the highest latitudes. The maximum annual range of sea surface temperatures occurs at the middle latitudes.

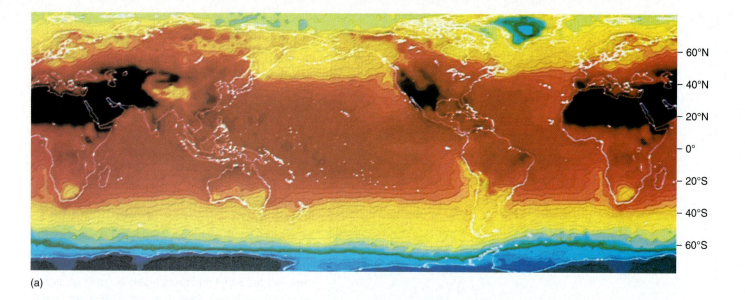

(a)

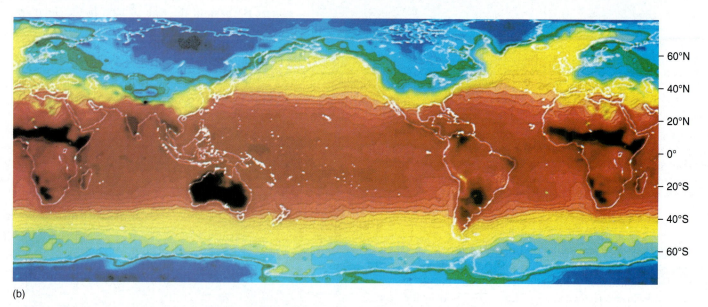

(b)

Figure 6.7 Meteorological satellites, such as NOAA's *TIROS*, carry high-resolution infrared sensors that measure long-wave radiation emitted from Earth's surface and atmosphere. This radiation is related to Earth's surface temperatures. *Green* and *blue* indicate temperatures below 0°C; warmer temperatures are shown in *red* and *black*. (a) July data show the Northern Hemisphere landmasses with considerably warmer temperatures, but the ocean waters do not change dramatically from winter to summer. The oceans' surface temperature distribution moves north and south with the change in seasons. North-south currents along the coasts of continents are also visible. (b) In January, Siberia and Canada show surface temperatures near −30°C; at the same time, latitudes between 30°S and 50°S show warm summer temperatures.

ocean annual surface temperatures can be seen in figure 6.7. Compare the summer (fig. 6.7*a*) and winter (fig. 6.7*b*) temperatures of landmasses and water areas at about 60°N.

Heat that is absorbed at the ocean surface in summer is transferred downward by winds, waves, and currents. In winter, heat is transferred upward toward the cooling surface. Heat is also transferred to and from the atmosphere at the sea surface. The net effect of these processes is the small annual change in mid-ocean surface temperatures: 0°–2°C in the tropics and 5°–8°C at the middle latitudes. The smaller 2°–4°C change in temperature at polar latitudes results from the heat transferred locally in the formation and melting of sea ice.

QUICK REVIEW

1. Why does the intensity of radiation per square meter of ground vary with latitude?

2. Why are daily solar radiation values high at the poles in summer?

3. Why do Southern Hemisphere summer radiation values exceed the summer values in the Northern Hemisphere?

4. How is Earth's atmosphere heated?

5. Compare the heat capacity of land and water.

6. How efficient is Earth in capturing available solar radiation for heating?

6.2 Sea Ice and Icebergs

Sea Ice

As seawater begins to freeze in the polar winters, a layer of slush forms, covering the ocean with a thin sheet of ice. Sheets of new **sea ice** can be broken into **pancakes** (fig. 6.8) by waves and wind; then as the freezing continues, the pancakes move about, unite, and form floes. An **ice floe** is a floating chunk of ice that is less than 10 kilometers (6 miles) in its greatest dimension. Ice floes move with the currents and the wind, collide with each other, and form ridges and hummocks (fig. 6.9). Some floes shift constantly, breaking apart and freezing together; others remain anchored to a landmass. Sea ice that is anchored to a landmass or shallow parts of the continental shelf is called **fast ice** whereas sea ice that floats freely is called **drift ice**. Continuous, or nearly continuous, sea ice is also called **pack ice**. As sea ice forms, the heat of fusion is transferred to the cold atmosphere, and the seawater temperature remains at the freezing point.

Ice about 2 m (6 ft) thick can be formed in one season. The thickness of the ice is limited because the latent heat of fusion from the underlying water must be extracted through the ice by conduction, a slow process even at polar temperatures. Snow that falls on the ice surface acts as an insulator, further retarding ice formation.

As sea ice is formed, some seawater is trapped in the voids between the ice crystals. If the ice forms slowly, most of the trapped seawater drains out and escapes; if the ice forms quickly, more salt water is trapped. As time passes, the salt water slowly escapes through the ice, and eventually the ice becomes fresh enough to drink when melted. The formation of the fresh sea ice concentrates salt in the underlying seawater, increasing its salinity and density and causing it to sink.

Figure 6.9 A pressure ridge formed by colliding ice floes.

Icebergs

Icebergs are massive, irregular in shape, and float with about 12% of their mass above the sea surface (fig. 6.10). They are formed by glaciers—large masses of freshwater ice—that begin inland in the snows of central Greenland, Antarctica, and Alaska, and inch their way downhill toward the sea under the influence of gravity. The forward movement, melting at the base of the glacier where it meets the ocean, and wave and wind action cause blocks of ice to break off and float out to sea. The castle bergs produced in the Arctic (fig. 6.10a) drift south with currents as far as New England and the busy shipping lanes of the North Atlantic. It was one of these icebergs, probably from a glacier in Greenland, that sank the *Titanic* with a loss of 1517 lives on its maiden voyage in 1912. Alaskan icebergs are usually released in narrow channels and bays; they do not easily escape into the open ocean. Flat, tabular icebergs (fig. 6.10b) produced from the broad continental ice sheets of Antarctica tend to stay close to the polar continent, caught by the circling currents, although they have been known to reach latitudes of 40°S. In late 1987, a large tabular berg, 155 km (96 mi) long and 230 m (755 ft) thick, with a surface area about the size of Delaware, broke from Antarctica and drifted 2000 km (1250 mi) along the Antarctic coast. Two years later, it grounded and broke into three pieces. The volume of ice in this berg was estimated to have been enough to provide everyone on Earth with two glasses of water daily for about 2000 years, or enough water for the city of Los Angeles for 100 years.

QUICK REVIEW

1. How is sea ice formed?
2. What limits the thickness of seasonal sea ice?
3. How are icebergs formed, and where are they most commonly found?
4. What is the difference between fast ice and drift ice?
5. What is pancake sea ice?

Figure 6.8 Pancake sea ice in the Ross Sea, Antarctica.

(a)

(b)

Figure 6.10 (a) Castle berg near Cape York, Greenland. The hole was caused by weathering effects due to waves, wind, and melting. (b) A tabular iceberg that broke off Antarctica in 2002. The dimensions of the berg in early June were 49.9 by 23.4 kilometers, giving it an area of 622 square kilometers, or seven times the area of Manhattan Island.

6.3 Structure and Composition of the Atmosphere

The absorption, reflection, and transmission of solar energy in the atmosphere depend on the gaseous composition of the atmosphere, suspended particles, and the abundance and types of clouds. The transfer of heat energy from Earth's surface acts to heat the atmosphere from below and set this gaseous fluid into convective motion. This convective motion produces the winds, which redistribute heat over Earth's surface and produce waves and currents in the oceans.

Structure of the Atmosphere

The atmosphere is a nearly homogeneous mixture of gases extending 90 km (54 mi) above Earth. Ninety-nine percent of the mass of atmospheric gases is contained in a layer extending upward 30 km (18 mi), and 90% is within a layer extending only 15 km (9 mi) above Earth's surface. The lowest layer of the atmosphere is the **troposphere;** here, the temperature decreases with altitude, changing from a mean Earth surface value of 16°C to −60°C at an altitude of 12 km (7 mi). The tropopause marks the minimum temperature zone between the troposphere and the layer above it, the **stratosphere.** In the stratosphere, the temperature increases with increasing altitude until the stratopause is reached at 50 km (31 mi) (fig. 6.11).

The troposphere is warmed from below by heat energy reradiated and conducted from Earth's surface and by condensation of water vapor in the upper troposphere (see fig. 6.2). Precipitation, evaporation, convective circulation, wind systems, and clouds are all found within the troposphere. **Ozone,** a highly reactive form

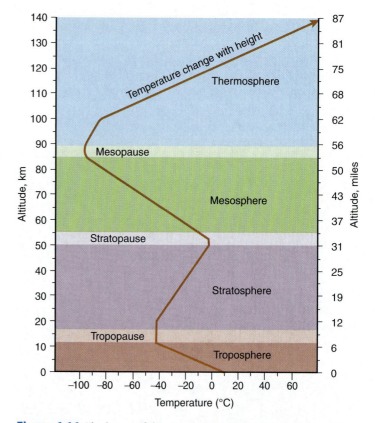

Figure 6.11 The layers of the atmosphere. Temperature decreases with altitude in the troposphere and mesosphere. Temperature increases with altitude in the stratosphere and thermosphere. Ninety-nine percent of the mass of the atmosphere is below a height of 30 km and 90% of the mass of the atmosphere is below a height of 15 km.

of oxygen, occurs principally in the stratosphere. Each ozone molecule, O_3, is made up of three atoms of oxygen instead of two, as found in oxygen, O_2. Ozone absorbs ultraviolet radiation from sunlight and therefore raises the temperature of the stratosphere. By absorbing ultraviolet radiation, the ozone lowers the incidence of ultraviolet light at Earth's surface, protecting living organisms from harmful high-intensity ultraviolet radiation.

At altitudes higher than 50 km (31 mi), there is little absorption of solar radiation, so the temperature again decreases with height in the layer known as the **mesosphere.** Here, the number of molecules per cubic centimeter is reduced by 1000, and the pressure is only 1/1000 of the atmospheric pressure at Earth's surface. The mesosphere extends upward to 90 km (54 mi), and above the mesosphere, the **thermosphere** extends upward to a height of about 500 km (310 mi). (See again fig. 6.11.)

The emphasis here is on the troposphere, for within this layer, heat and water move between Earth's surface and the atmosphere, causing the motions that produce the winds, the weather, the ocean's waves, and ocean surface currents. The tropopause is also of interest because this is the region of the high-altitude winds called jet streams that play a role in wind systems and storm tracks.

Composition of Air

The atmosphere is composed of gases, suspended microscopic particles, and water droplets. This mixture is commonly and simply called air. Atmospheric gases are typically categorized as being permanent or variable. The permanent gases are present in a constant relative percentage of the atmosphere's total volume, while the concentration of the variable gases changes with time and location (table 6.1 and figure 6.12).

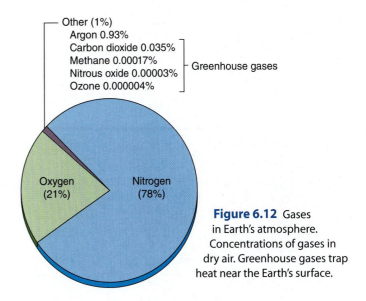

Figure 6.12 Gases in Earth's atmosphere. Concentrations of gases in dry air. Greenhouse gases trap heat near the Earth's surface.

Other (1%)
Argon 0.93%
Carbon dioxide 0.035% ⎤
Methane 0.00017%   ⎥ Greenhouse gases
Nitrous oxide 0.00003% ⎥
Ozone 0.000004%   ⎦
Oxygen (21%)
Nitrogen (78%)

Table 6.1   Composition of the Atmosphere

PERMANENT GASES			
Gas	Formula	Percent by Volume	Molecular Weight
Nitrogen	N_2	78.08	28.01
Oxygen	O_2	20.95	32.00
Argon	Ar	0.93	39.95
Neon	Ne	1.8×10^{-3}	20.18
Helium	He	5.0×10^{-4}	4.00
Hydrogen	H_2	5.0×10^{-5}	2.02
Xenon	Xe	9.0×10^{-6}	131.30
VARIABLE GASES			
Gas	Formula	Percent by Volume	Molecular Weight
Water vapor	H_2O	0 to 4	18.02
Carbon dioxide	CO_2	3.5×10^{-2}	44.01
Methane	CH_4	1.7×10^{-4}	16.04
Nitrous oxide	N_2O	3.0×10^{-5}	44.01
Ozone	O_3	4.0×10^{-6}	48.00

The density of air is controlled by three variables: temperature, the amount of water vapor in the air, and altitude. The density of air decreases with increasing temperature and increases with decreasing temperature. In other words, warm air is generally lighter than cold air. Air density decreases if its humidity increases or the concentration of water vapor increases, and it increases if the humidity decreases. In general, moist air is lighter than dry air. When water vapor is added to the atmosphere, the relatively low molecular-weight water molecules replace higher molecular-weight permanent gases (compare the molecular weight of water to the molecular weights of the permanent gases listed in table 6.1). Thus, cold, dry air is more dense than warm, moist air. Finally, the density of air decreases with increasing altitude. The air at any given altitude is compressed by the weight of the column of air above it. The greater the compression, the greater the density. Changes in the density allow air to move vertically and cause atmospheric convective motion.

Carbon Dioxide and the Greenhouse Effect

There are three active reservoirs for carbon dioxide (CO_2): the atmosphere, the oceans, and the terrestrial system; in addition, there is the geologic reservoir of Earth's crust. The oceans store the largest amount of CO_2, and the atmosphere has the smallest amount (fig. 6.13). The atmosphere is the link with the other reservoirs, and the ocean plays a major part in determining the atmosphere's concentration of CO_2 through physical (mixing and circulation), chemical, and biological processes.

Atmospheric CO_2 is transparent to incoming short-wave solar radiation, but at the same time, it reduces by absorption the amount of outgoing long-wave radiation from Earth. In this way, Earth and its atmosphere are warmed by what is commonly known as the **greenhouse effect.** In the Northern Hemisphere, the natural cycle of carbon dioxide shows decreasing atmospheric CO_2 in the late spring and summer (fig. 5.10). At this

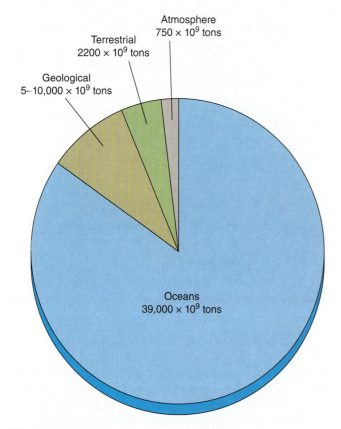

Terrestrial
2200 × 10⁹ tons

Geological
5–10,000 × 10⁹ tons

Atmosphere
750 × 10⁹ tons

Oceans
39,000 × 10⁹ tons

Figure 6.13 World carbon dioxide distribution.

time, plants increase active photosynthesis and remove more CO_2 than is contributed by respiration and decay. In the fall and winter, the photosynthetic activity is reduced, plants lose their leaves, and decay processes release CO_2; atmospheric CO_2 increases. The effects of deforestation and conversion of forest land to agriculture, the burning of fossil fuels, and the growth of human populations have been superimposed on this natural cycle. In preindustrial times, the human impact on this seasonal cycle was small and the cycle was in balance.

Since 1850 and the start of the Industrial Revolution, however, the concentration of CO_2 in the atmosphere has increased from 280 parts per million (ppm) to over 390 ppm. These are the highest values observed for the last 420,000 years. Recently, the average rate of increase has been 1.5–2 ppm per year. For over fifty years, scientists have been recording a steady increase in the CO_2 concentration in the atmosphere due to the burning of coal, oil, and other fossil fuels (fig. 5.9). If this trend continues, the concentration of CO_2 will double its 1850 value sometime before the end of this century. This increase will warm Earth and alter its average heat budget by reducing the surface heat loss to space by long-wave radiation. Instead, more long-wave radiation will be absorbed into the atmosphere, increasing its temperature and forcing it to lose more long-wave radiation to space in order to maintain the heat budget of Earth and atmosphere systems.

Based on this trend of increasing CO_2, climate researchers have predicted global warming of 2°–4°C in the next hundred years. With global warming, several scenarios are possible. It is expected that such warming would affect the higher latitudes, causing melting of polar land ice and raising sea level about 1 m (3 ft) by the year 2100. A decrease in sea ice could provide more open water for marine phytoplankton populations with a corresponding increase of photosynthesis, leading to increased carbon storage in the ocean reservoir. Other possibilities are related to the interaction between clouds, the oceans, and Earth's surface (see the box titled "Diving In: Clouds and Climate"). Changes in sea surface temperature could also affect the oceanic circulation and the surface wind systems that drive the ocean currents. Changes in the ocean current patterns would modify the transfer of heat and water vapor from low to high latitudes and alter Earth's climate patterns. Also, increases in available atmospheric CO_2 have the potential to stimulate photosynthesis both on the land and in the oceans with unknown effects.

Ozone

Depletion of the stratospheric ozone layer that screens Earth from much of the Sun's ultraviolet radiation was first reported in 1985 by members of the British Antarctic Survey, who discovered that significant ozone loss had been occurring over Antarctica since the late 1970s. The Antarctic "ozone hole" is the result of a large-scale destruction of the ozone layer over Antarctica that occurs when temperatures in the ozone layer drop low enough for clouds to form. These clouds form in the stratosphere, at heights of between 10 and 30 km, when the temperature there falls below −80°C. In these clouds, chemical reactions take place that lead to the destruction of ozone. Without the clouds, there is little or no ozone destruction. Only during the Antarctic winter does the atmosphere get cold enough for these clouds to form widely through the center of the ozone layer. The Antarctic ozone hole is usually largest in early September (spring season in the Southern Hemisphere) after the ozone has been decreasing through the Antarctic winter, and deepest in late September to early October (fig. 6.14). The size and shape of the hole vary from year to year with natural variations in the temperature of the stratosphere (fig. 6.15). The word *hole* is a misnomer; the hole is really an area over the pole that experiences a significant reduction (up to 70%) in the ozone concentrations normally found over Antarctica.

The Northern and Southern Hemispheres have different temperature conditions in the ozone layer. The temperature of the Arctic ozone layer during winter is normally some 10 degrees warmer than that of the Antarctic. This means that clouds in the stratosphere over the Arctic are rare, but sometimes the temperature is lower than normal and they do form. Under these circumstances significant ozone depletion can take place over the Arctic, but it is usually for a much shorter period of time and covers a smaller area than in the Antarctic.

The most widely accepted theory of ozone destruction is related to the release of chlorine into the atmosphere. Chlorine is commonly released as a component of chlorofluorocarbons (CFCs). CFCs are used as coolants for refrigeration and air conditioning, as solvents, and in the production of insulating foams. CFCs are distributed throughout the troposphere by the winds and gradually leak into the stratosphere, where the ultraviolet light

Diving in

Clouds and Climate

Clouds are beautiful, always changing, ever moving; we see them white and fluffy, high and wispy, black and threatening. Clouds form from liquid water droplets with temperatures above freezing, supercooled water droplets with temperatures below freezing, and solid particles. In the lower troposphere, condensation nuclei—small particles of dust, salt, or other matter—serve as the cores for condensing water droplets. Clouds of the upper troposphere are composed of ice crystals.

In 1803, Luke Howard, an English pharmacist, proposed the first useful classification of clouds: stratus (layered), cumulus (puffy), cirrus (wispy), and nimbus (clouds releasing snow or rain that travels to the ground) (box fig. 1). This system is the basis for the cloud types recognized by today's World Meteorological Organization (WMO) (box fig. 2). Different cloud types absorb radiant energy at different rates and affect the heat budget and climate of Earth.

Clouds have a dual role in the atmosphere: they simultaneously heat and cool Earth. The tops and sides of clouds appear white because they reflect and scatter short-wave radiation. If clouds are dense enough, they absorb, reflect, and scatter sufficient incoming radiation to produce dark shadows on their undersides and on Earth's surface. These dense clouds help to cool Earth. Clouds absorb long-wave radiation from Earth's surface because of their water content, and because they are generally cooler than Earth, they reradiate only part of this absorbed long-wave radiation and absorbed solar radiation to space. The net effect is that clouds play a significant part in warming Earth and its atmosphere.

Recent research indicates that on a global scale, Earth's present cloud cover provides more shielding from incoming solar radiation than

Box Figure 1   Subtropical cumulus clouds over the Florida Keys.

trapping of long-wave radiation. At present, the result is negative, a total net reduction in radiation to Earth of about 14–21 watts per square meter per month. It is estimated that if there were no clouds, average Earth surface temperature would warm by about 10°C.

Not only do the clouds affect Earth's climate; they are also affected by it. This feedback mechanism could impose a new cloud-controlled radiation balance if Earth's climate changed. If the present negative radiation balance became less negative, Earth's surface would warm; if cloud feedback produced a more negative radiation balance, Earth's surface would become cooler than it is at present. If Earth warmed because of the greenhouse effect, there could be an increase in evaporation and therefore an increase in low-level water-droplet clouds over the ocean; these clouds would absorb and reflect incoming short-wave radiation

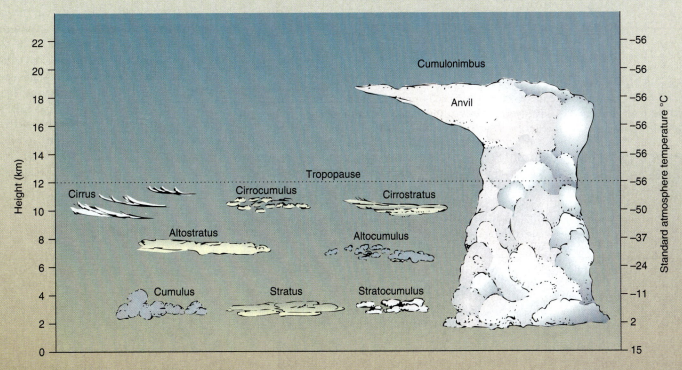

Box Figure 2   Cloud types recognized by the World Meteorological Organization.

Continued next page—

and help to cool Earth. However, an increase in high-altitude ice clouds would warm Earth because ice clouds pass incoming short-wave radiation but reduce Earth's long-wave radiation loss.

The formation of clouds is related to the availability of condensation nuclei. If nuclei are sparse, the droplets are fewer and larger; the change of precipitation is increased, and the amount of nuclei is further decreased. Many small condensation nuclei produce abundant small droplets, increasing the cloud reflectivity and decreasing incoming radiation. In oceanic areas where precipitation is frequent, condensation particles are removed by precipitation, and their low supply may hinder cloud formation. In these areas, satellite photos show passing ships leaving cloud trails as condensation occurs on particles from the ships' exhausts. Dimethyl sulfide is a source of condensation nuclei over ocean areas, and its availability is controlled by its own feedback system. Volcanic activity contributes particles that act as condensation nuclei,

as do severe dust storms and major fires such as those that occurred in Brazil in 1997 and in Indonesia in 1998. The carbon particles in smoke increase the cloud absorption of solar energy while, at the same time, they reduce the size of water droplets.

Although clouds are among the most common of atmospheric phenomena, our understanding of cloud formation and distribution and of climate change and feedback mechanisms is presently incomplete. Scientists use large computer models known as general circulation models (GCMs) to test their ideas, modifying them as new information is acquired. A concerted effort is ongoing to directly measure radiation levels above and below clouds to better understand the complexities of cloud-climate systems. Early results indicate that the solar radiation shielding by clouds has been underestimated. By using improved GCMs to more clearly define the role of clouds in climate changes, scientists hope to better understand the consequences of modifying our atmosphere.

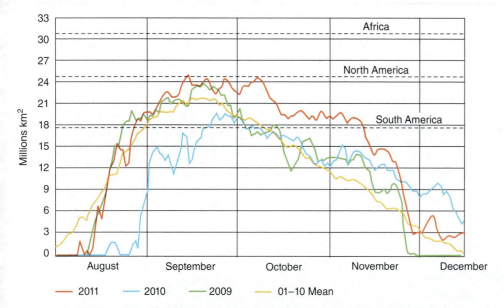

Figure 6.14 Variation in the area of the Antarctic ozone hole from late winter to early spring. Areas are in millions of square kilometers for 2009 (green), 2010 (blue), 2011 (red), and the mean value for 2001–2010 (yellow). Areas of three continents are included for comparison. Source of Data: National Oceanic and Atmospheric Administration (NOAA).

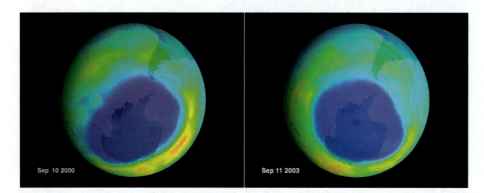

Figure 6.15 Comparison of the Antarctic ozone hole on September 10, 2000 (*left*) and September 11, 2003 (*right*). The maximum area in 2000 was 29.8×10^6 km² and in 2003 was 28.3×10^6 km².

breaks them apart. Gases in the atmosphere react with the chlorine and trap it as inert molecules in stratospheric clouds, which are common during the polar winter. In the presence of sunlight, chlorine is liberated and free to attack ozone molecules.

CFCs appear to have reached their maximum level in the troposphere and are expected to decline gradually as efforts to reduce their production continue. Because it takes many years for CFCs in the troposphere to work their way into the stratosphere, they will continue to destroy ozone for many years to come.

At ground level, ozone is a pollutant and a health hazard; in the stratosphere, it absorbs most of the ultraviolet radiation from the Sun, protecting life-forms on land and at the sea surface. The loss of ozone is a significant concern because a 50% decrease in ozone is estimated to cause a 350% increase in ultraviolet radiation reaching Earth's surface. Increased ultraviolet radiation is responsible for increases in skin cancers and has been shown to affect the growth and reproduction of some organisms.

QUICK REVIEW

1. What are the four major layers of the atmosphere? How does temperature vary with elevation in each of them?

2. How have carbon dioxide levels contributed to the greenhouse effect?

3. Describe the effect of temperature and cloud formation in the concentration of ozone in the atmosphere.

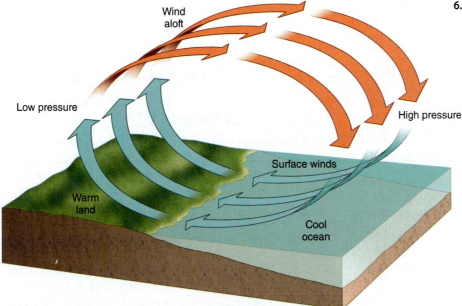

forms a convection cell based on vertical air movements due to changes in air's density. Less-dense, typically relatively warm and moist, air rises whereas denser, often cold and dry, air sinks. Areas of rising air are associated with low pressure and areas of descending air are associated with high pressure.

Atmospheric Pressure

Atmospheric pressure is the force with which a column of overlying air presses on an area of Earth's surface. The average atmospheric pressure at sea level is 1013.25 millibars (1 bar = 1×10^6 dynes/cm^2), or 14.7 lb/in^2. This standard atmospheric pressure is also equal to the pressure produced by a column of mercury standing 760 mm (29.92 in) high. Barometers can measure atmospheric pressure in millibars, or in millimeters or inches of mercury. Pressure can also be recorded in torrs, where 1 torr equals the pressure of a column of mercury 1 mm high. Where the density of air is less than average, atmospheric pressure is below average, and a **low-pressure zone** of rising air is formed. Regions of air with a density greater than average are known as **high-pressure zones** of descending air. Average global atmospheric pressures are strongly influenced by both latitude and the presence of continents and ocean basins (fig. 6.17).

Figure 6.16 A convection cell is formed in the atmosphere when air is warmed at one location and cooled at another.

6.4 The Atmosphere in Motion

In a simple sense, we can say that air moves because at one place, less-dense air rises, while in another place, denser air sinks toward Earth. Between these areas, the air that flows horizontally along Earth's surface is the wind. This process is shown in figure 6.16. There are really two horizontal airflows, or wind levels, moving in opposite directions: one at Earth's surface and one aloft. Air circulating in this manner

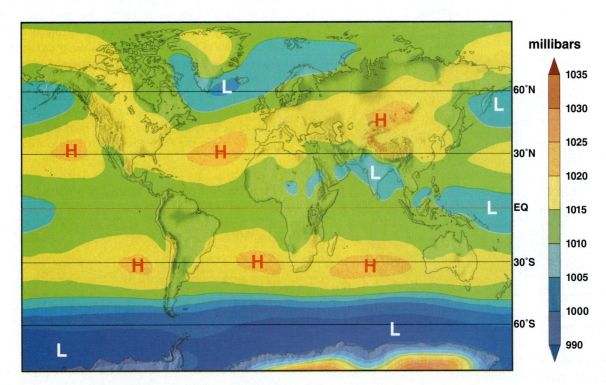

Figure 6.17 Annual mean sea level atmospheric pressure (millibars) averaged from 1979–95. L = low-pressure zone, H = high-pressure zone. In general, in temperate latitudes regions of high pressure are found over ocean basins and regions of low pressure are found over continents.

Winds on a Nonrotating Earth

The heating and cooling of air and the gains and losses of water vapor in air are related to the unequal distribution of the Sun's energy over Earth's surface, the presence or absence of water, and the variation in temperature of Earth's surface materials in response to heating. These act to affect the air's density. Imagine a model Earth with no continents and with no rotation but heated like the real Earth. On this stationary model covered with uniform layers of atmosphere and water, the wind pattern is very simple. Around the equator, the air, warmed from below, rises. Once aloft, the air flows toward the poles, where it is cooled and sinks to flow back toward the equator. Because of the unequal distribution of the Sun's heat over the model's surface, large amounts of heat and water vapor are transferred to the atmosphere around the equator. This less-dense air rises, and as it rises, it cools; condensation exceeds evaporation, clouds form, and it rains. Equatorial regions are known for their warm, wet climate. The cool, dry air remains aloft and flows toward the poles, where it sinks, producing high evaporation, dense air, and a zone of high atmospheric pressure. Such air movement is shown in figure 6.18; note each hemisphere's two large convective circulation cells, each extending from a pole to the equator. In this model, the Northern Hemisphere surface winds blow from north to south, and the upper winds blow from south to north; in the Southern Hemisphere, the reverse is true, with the surface winds blowing from south to north and the upper winds blowing from north to south.

It is important to remember that winds are named for the direction *from which they blow*. A north wind blows from north to south; a south wind blows from south to north. In this model, the Northern Hemisphere surface winds are north winds, and the Southern Hemisphere surface winds are south winds.

The Effects of Rotation

Now let's consider the same model of a water-covered Earth but with rotation added. This will dramatically change atmospheric circulation. Each of us, standing on Earth's surface, is moving with Earth in its daily eastward rotation. The speed of this motion varies with latitude because the circumference of a circle of latitude decreases with distance from the equator (table 6.2). Standing on Earth's surface we are unaware of this motion, but the movement of air masses is affected by this change in rotational speed because these air masses are not attached to Earth's surface. When we observe the movement of air masses, we are making those observations from a rotating reference frame. Since the air masses themselves are effectively independent of this reference frame, this creates apparent forces that appear to deflect moving air masses from their intended direction of motion. One way of visualizing this is to consider a plane traveling along a north-south line, or along a single meridian of longitude, with the intent of traveling from point A at latitude 30°N to point B at latitude 45°N (fig. 6.19). The plane takes one hour to fly from 30°N to 45°N. When the plane is on the ground at point A, ready to begin the flight, it is moving eastward with Earth's rotation at 1450 km/hr. It will continue to have this eastward speed once it takes off and is airborne. As it flies directly north, Earth's surface beneath it is moving to the east at a progressively slower speed. After one hour of flight the plane arrives at 45°N. During the hour-long flight it traveled 1450 km east (as well as a distance of 45° − 30° = 15° north). Over the same period of time, point B only traveled 1184 km east. The plane arrives at 45°N, 266 km east of point B. As viewed from the ground, the plane's flight path appears to have veered to the right (fig. 6.19*a*). Now imagine reversing the route, so the plane flies from point B at latitude 45°N with the intent of arriving at point A at latitude 30°N (fig. 6.19*b*). Once again, the plane takes one hour to fly from 45°N to 30°N. When the plane is on the ground at point B, ready to begin the flight, it is moving eastward with Earth's

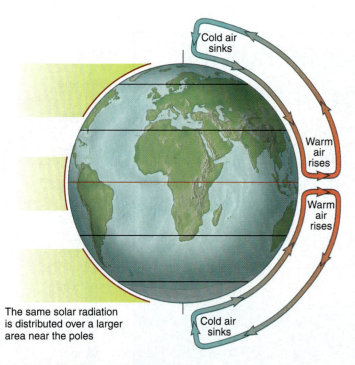

Cold air sinks

Warm air rises

Warm air rises

The same solar radiation is distributed over a larger area near the poles

Cold air sinks

Figure 6.18 Non-rotating Earth model of atmospheric circulation. Heating at the equator and cooling at the poles produce a single, large convection cell in each hemisphere. Air rises at the equator and sinks at the poles.

Table 6.2   Eastward Speed of Earth's Surface with Latitude

Latitude	Speed (km/h)	Speed (mi/h)
90°N and S	0	0
75°N and S	433	269
60°N and S	837	520
45°N and S	1184	735
30°N and S	1450	900
15°N and S	1617	1004
0°	1674	1040

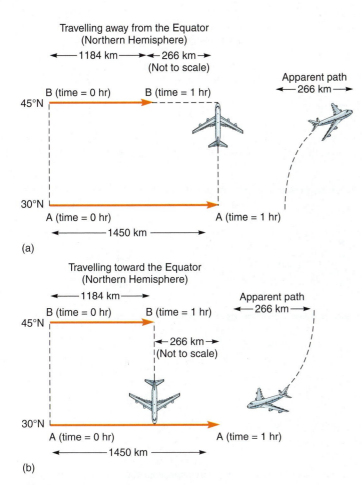

Figure 6.19 Moving parcels of air in the atmosphere, or water in the oceans, appear to be deflected from straight-line movement as a result of the Coriolis effect. The Coriolis effect is a consequence of the variation in eastward speed of the surface of the rotating Earth with latitude. The apparent deflection is to the right in the Northern Hemisphere. This can be seen by imagining the path of a plane traveling north or south. (a) A plane traveling from point A at 30°N to point B at 45°N will arrive at 45°N east of point B because the plane's eastward velocity is greater than the eastward speed of the surface of the rotating Earth at 45°N. Hence, its path will appear to veer to the right. (b) A plane traveling in the opposite direction from point B at 45°N to point A at 30°N will arrive at 30°N west of point A because the plane's eastward velocity is less than the eastward speed of the surface of the rotating Earth at 30°N. Once again, its path will appear to veer to the right. If you perform this same thought experiment with the plane in the Southern Hemisphere, you will see that the Coriolis effect causes an apparent deflection to the left.

rotation at 1184 km/hr. It will continue to have this eastward speed once it takes off and is airborne. As it flies directly south, Earth's surface beneath it is moving to the east at a progressively faster speed. After one hour of flight the plane arrives at 30°N. During the hour-long flight it traveled 1184 km east (as well as a distance of 45° − 30° = 15° south). Over the same period of time, point A traveled 1450 km east. The plane arrives at 30°N, 266 km west of point A. As viewed from the ground, the plane's flight path again appears to have veered to the right (fig. 6.19*b*).

Moving air masses, or the wind, are affected in the same way. The deflection of moving air relative to Earth's surface is called the **Coriolis effect**, after Gaspard Gustave de Coriolis (1792–1843), who mathematically solved the problem of deflection in frictionless motion when that motion occurs relative to a rotating reference frame. If the same thought experiment with the plane is performed in the Southern Hemisphere, the deflection would be to the left of the intended direction of motion.

The magnitude of the Coriolis effect increases with increasing latitude, increases with the speed of the moving air, and is dependent on the variation in rotation rate of Earth's surface with latitude. The magnitude of the Coriolis effect is zero at the equator.

The Coriolis effect is equally important in determining the relative motion of ocean currents and water masses in the oceans. Both moving air and moving water are deflected relative to their intended direction of motion—to the right in the Northern Hemisphere and to the left in the Southern Hemisphere.

Wind Bands

Applying rotation, and therefore the Coriolis effect, to the non-rotating model Earth modifies the winds considerably. Refer to figure 6.20 as you read the following description. The air rises at the equator and flows aloft to the north and the south, but it cannot continue to move northward and southward without being deflected to the right in the Northern Hemisphere and to the left in the Southern Hemisphere. This deflection short-circuits the large, hemispheric, atmospheric convection cells of the stationary model Earth. The deflected air aloft sinks at 30°N and 30°S; it moves along the water-covered surface, either back toward the equator or toward 60°N and 60°S. The upper-level air that reaches the poles cools, sinks, and moves to lower latitudes, warming and picking up water vapor; at 60°N and 60°S, it rises again. The result is three convection cells in each hemisphere wrapped around the rotating Earth.

Consider the flow of surface air in the three-cell system. Between 0° and 30°N and 30°S, the surface winds are deflected relative to Earth, blowing from the north and east in the Northern Hemisphere and from the south and east in the Southern Hemisphere. This deflection creates bands of moving air known as the **trade winds:** the northeast trade winds, north of the equator, and the southeast trade winds, south of the equator. Between 30°N and 60°N, the deflected surface flow produces winds that blow from the south and west, while between 30°S and 60°S, they blow from the north and west. In both hemispheres, these winds are called the **westerlies.** Between 60°N and the North Pole, the winds blow from the north and east, while between 60°S and the South Pole, they blow from the south and east. In both cases, they are called the **polar easterlies.** The six surface wind bands are shown in figure 6.20.

At 0° and 60°N and 60°S, moist, low-density air rises in areas of low atmospheric pressure; these are zones of clouds and rain. Zones of high-density descending air at 30°N and 30°S and 90°N and 90°S are areas of high atmospheric pressure, dry air with low precipitation, and clear skies.

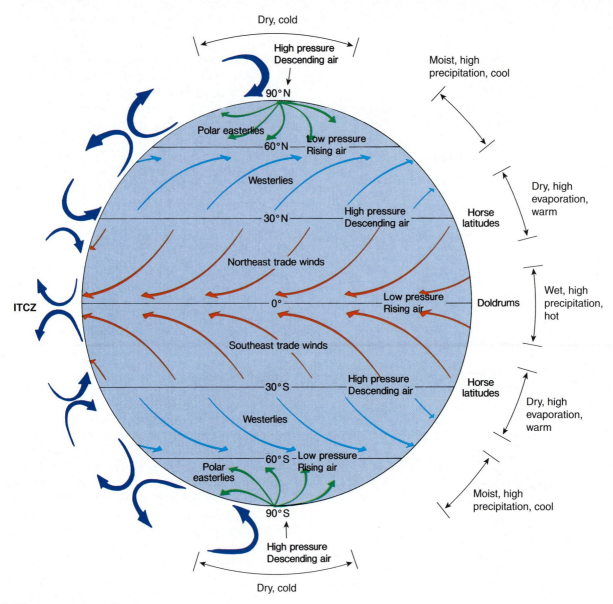

Figure 6.20 The circulation of Earth's atmosphere results in a six-band surface wind system.

The salinity of mid-ocean surface water is controlled by the distribution of the world's evaporation and precipitation zones (see figs. 5.2 and 5.3). Salinity of ocean water is measured in grams of salt per kilogram of seawater and is expressed in parts per thousand (‰). The average ocean salinity is appoximately 35‰. In the tropics, the precipitation is heavy on land and at sea. On land, this produces tropical rain forests; at sea, the surface water has a low salinity, around 34.5‰. At approximately 30°N and 30°S, the evaporation rates are high. These are the latitudes of the world's deserts and of increased salinity in the surface waters, around 36.7‰. Farther north and south, from 50°–60°N and from 50°–60°S, precipitation is again heavy, producing cool but less-salty surface water, around 34.0‰, and the heavily forested areas of the Northern Hemisphere. At the polar latitudes, sea ice is formed in the winter. During the freezing process, the salinity of the water beneath the ice increases, and in the summer, the thawing of the ice reduces the surface salinity once more (see fig. 6.20).

Wind belts are formed when air flows over Earth's surface from regions of higher atmospheric pressure to areas of lower atmospheric pressure. In the zones of vertical motion, between the wind belts, the surface winds are unsteady. Such areas were troublesome to the early sailors, who depended on steady winds for propulsion. The area of rising air at the equator is known as the **doldrums,** and the high-pressure areas at 30°N and 30°S are known as the **horse latitudes.** In all these areas, sailing ships could find themselves becalmed for days. The origin of the word *doldrum* is obscure, but it is likely related to its meaning of a period of low spirits, listlessness, or despondency, feelings often felt by sailors becalmed in this region. The formal name for the zone of low pressure and rising air near the equator is the **Intertropical Convergence Zone (ITCZ)** (fig. 6.20). It is along the ITCZ that the wind systems of the Northern and Southern Hemispheres converge to produce rising air, low pressure, and high precipitation. The horse latitudes are said to have gotten their name from the stories of ships carrying horses that were thrown overboard when the ships were becalmed and the freshwater supply became too low to support both the sailors and the animals.

6.5 Modifying the Wind Bands

Moving from the rotating, water-covered model of Earth to the real Earth requires the consideration of three factors: (1) seasonal changes in Earth's surface temperature due to solar heating, (2) the addition of the large continental land blocks, and (3) the difference in heat capacity of land and water. Both ocean and land surfaces remain warm at the equatorial latitudes and cold at the polar latitudes all through the year, but the middle latitudes have seasonal temperature changes, ranging from warm in summer to cold in winter. Keep in mind that land surface temperatures have a greater seasonal fluctuation than ocean surface temperatures because the heat capacity of the ocean water is greater than the heat capacity of land and because ocean water has the ability to transfer heat from the surface to depth in summer and from depth to the surface in winter. Land does not transfer heat in this way (review fig. 6.6).

Seasonal Changes

The presence of land and water in near-equal amounts at the middle latitudes in the Northern Hemisphere produces average seasonal patterns of atmospheric pressure. During the warm summer months, the land is warmer than the ocean (see fig. 6.7). The air over the land is heated from below and rises, creating a low-pressure area, while the air over the sea cools and sinks, producing a high-pressure area over the water. In the summer, low-pressure zones at 60°N and 0° tend to combine over the land, cutting through the high-pressure belt along 30°N latitude. This breaks the high-pressure belt into several high-pressure cells over the oceans rather than maintaining the latitudinal zones of pressure stretching continuously around Earth. In the winter, the reverse is true at the middle latitudes (see again fig. 6.7). The land becomes colder than the water. The air over the land is cooled, increasing its density and producing a region of descending air and high atmospheric pressure over the land. The air over the relatively warm water is heated from below, decreasing its density and producing a region of rising air and low atmospheric pressure over the water.

Over the land, the polar high-pressure zone spreads toward the high-pressure zone at 30°N, breaking the low-pressure belt

centered about 60°N into discrete low-pressure cells that are centered over the warmer ocean water. This middle-latitude, seasonal alternation of high- and low-pressure cells breaks up the latitudinal pressure zones and wind belts that are seen in the water-covered model to produce the air-pressure distribution shown in figure 6.21.

During the Northern Hemisphere's summer, the air in the high-pressure cells over the central portion of the North Atlantic and North Pacific Oceans descends and flows outward toward the continental low-pressure areas. As the descending air moves outward, it is deflected to its right, producing winds that spiral in a clockwise direction about the high-pressure cells. The wind circulation about a high-pressure cell in the Northern Hemisphere is always clockwise. On the northern side of these high-pressure cells are the westerlies, and on the southern side are the northeasterlies; the eastern side of the cell has northerly winds, and the western side has southerly ones. In the winter, the air circulates counterclockwise about the low-pressure cell over the northern oceans, and the prevailing wind directions reverse. The wind circulation pattern about a low-pressure cell in the Northern Hemisphere is always counterclockwise. The wind directions related to these pressure cells are shown in figure 6.22. Along the Pacific Coast of the United States, the northerly winds cool the coastal areas in the summer, and the southerly winds warm them in the winter. The eastern United States receives warm, moist air from the low latitudes in the summer, and cold air moves down from the high latitudes in the winter. Although these seasonal changes modify the wind and pressure belts that were developed on the water-covered model, the generalized wind and pressure belts are still identifiable over Earth in the northern latitudes when the atmospheric pressures are averaged over the annual cycle.

Rotation of the airflow around high- and low-pressure cells is reversed in the Southern Hemisphere because the Coriolis effect is opposite to that in the Northern Hemisphere. At the middle latitudes in the Southern Hemisphere, there is little land, and the water temperature predominates. There is little seasonal effect; the atmospheric pressure and wind patterns created by surface temperatures change little over the annual cycle and are very similar to those developed for the water-covered model (see figs. 6.20 and 6.21).

Seasonal changes in the surface temperature of the oceans and continents also influence the position of the ITCZ (fig. 6.21). Its actual location is generally associated with the zone of highest surface temperature. During the Northern Hemisphere summer, it is deflected north of the equator over the continents as the land heats up faster than the water. The ITCZ may be deflected as far as 10°–20°N over Africa and southern Asia. During the Southern Hemisphere summer, the ITCZ will be deflected as far as 10°–20°S over Africa and South America. The ITCZ is the rainiest latitude zone in the world, with many locations experiencing rainfall more than 200 days each year. The weather at many areas along the equator is dominated by the ITCZ year-round with almost no dry season. Areas near the extreme northern and southern swings of the ITCZ, however, are subject to brief dry seasons when the zone shifts away. For example, Iquitos, Peru (3°S), is close enough to the equator that

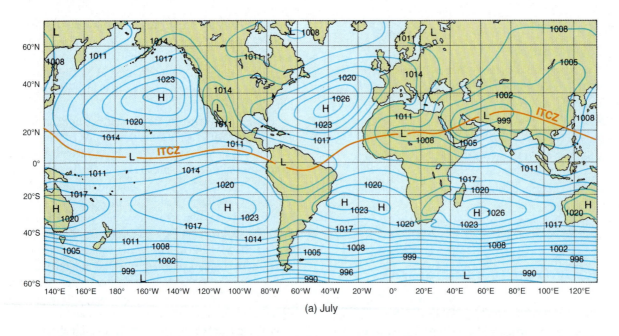

(a) July

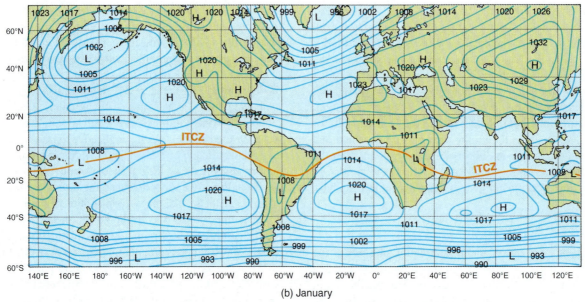

(b) January

Figure 6.21 Average sea-level atmospheric pressures expressed in millibars for (a) July and (b) January. Large landmasses at mid-latitudes in the Northern Hemisphere cause high and low atmospheric pressure cells to change their positions with the seasons. In the Southern Hemisphere, large landmasses do not exist at mid-latitudes, and average air-pressure distribution changes little with the seasons. The location of the Intertropical Convergence Zone (ITCZ) shifts with the seasons.

it is always influenced by the ITCZ, while San José, Costa Rica (10°N), has a relatively dry season from January to March, when the ITCZ is displaced to the south.

The Monsoon Effect

The differences in temperature between land and water produce large-scale and small-scale effects in coastal areas. In the summer along the west coast of India and in Southeast Asia, the air rises over the hot land, creating a low-pressure system (fig. 6.23a). The rising air is replaced by warm, moist air carried on the southwest winds from the Indian Ocean. As this onshore airflow rises over the land, it cools, causing condensation. This produces clouds and a time of heavy rainfall. This is the wet, or summer, **monsoon.** In the winter, a high-pressure cell forms over the land, and northeast winds carry the dry, cool air southward from central Asia and out over the Indian Ocean (fig. 6.23b). This movement produces cool, dry weather over the land, known as the dry, or winter, monsoon. For years, coastal traders in the Indian Ocean who were dependent on sailing craft planned their voyages so that they sailed with the wind to their destination on one phase of the monsoon and returned on the next.

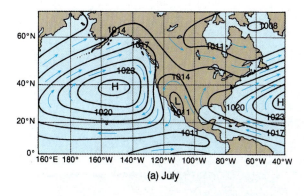

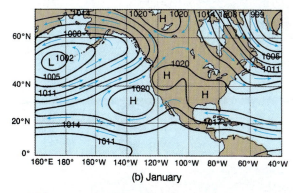

Figure 6.22 Atmospheric pressure cells control the direction of the prevailing winds. Atmospheric pressures are expressed in millibars. In the Northern Hemisphere, air flows outward and clockwise around a region of high pressure and inward and counterclockwise around a low-pressure area. (a) Average wind conditions for summer. (b) Average wind conditions for winter.

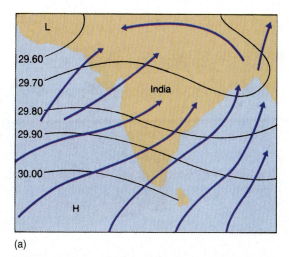

(a)

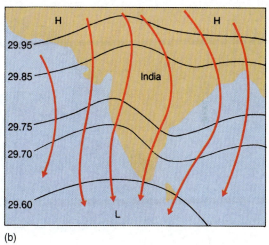

(b)

Figure 6.23 The seasonal reversal in wind patterns associated with (a) the summer (wet) monsoon and (b) the winter (dry) monsoon. The isobars of pressure are given in inches of mercury. Land topography increases friction between winds and land, thereby decreasing the Coriolis effect. The winds blow more directly from high- to low-pressure areas across the isobars.

The monsoon effect is seen on a smaller local scale along a coastal area or along the shore of a large lake. During daylight, the land is warmed faster than the water, and the air rises over the land. The air over the water moves inland to replace it, creating an **onshore** breeze. At night, the land cools rapidly, and the water becomes warmer than the land. Air rises over the water, and the air from the land replaces it, this time creating an **offshore** breeze. Such a local diurnal (once-a-day) wind shift is referred to as a land-sea breeze (fig. 6.24). The onshore breeze reaches its peak in the afternoon, when the temperature difference between the land and the water is at its maximum. The offshore breeze is strongest in the late night and early morning hours; sometimes one can smell the land 30 km (20 mi) at sea, its odors carried by the offshore winds. This daily wind cycle helps fishing boats that depend only on their sails to leave the harbor early in the morning and return late in the afternoon or early evening. This effect brings the summer fogs to San Francisco, California. The area east of San Francisco Bay heats up during the day, and the warm air over the land rises. When the warm marine air reaches the cold coastal waters, a fog is formed that pours in through the Golden Gate, obscuring first the Golden Gate Bridge and then the city (fig. 6.25). When the air reaches the east side of the bay, it warms, and the fog dissipates. At night, the flow is reversed, and the city may be swept

free of the fog. Sometimes San Francisco will remain foggy all night and the fog will not disperse until the morning, when it is burned off by solar heating. This happens when the land and air are close to the same temperature.

The Topographic Effect

Because the continents rise high above the sea surface, they affect the winds in another way. As the winds sweep across the ocean, they reach the land and are forced to rise as they encounter the land, as shown in figure 6.26. The upward deflection cools the air, causing rain on the windward side of the islands and mountains; on the leeward (or sheltered) side, there is a low-precipitation zone, sometimes called a **rain shadow.** For example, the windward sides of the Hawaiian mountains have high precipitation and lush vegetation, whereas the leeward sides are much drier and require irrigation. On the west

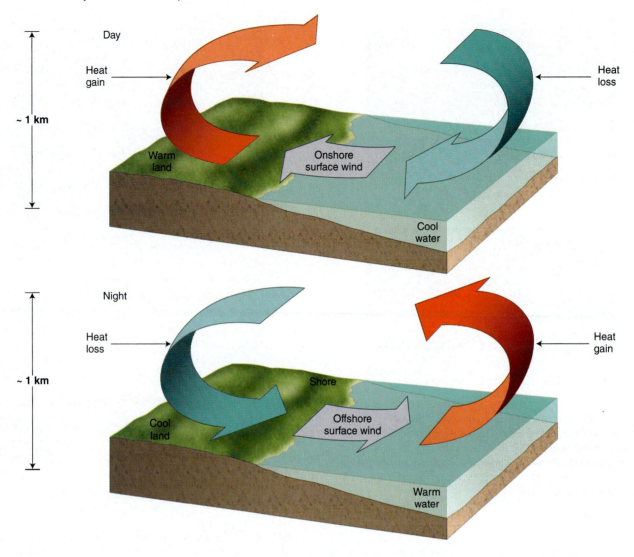

Day

Heat gain →

~ 1 km

Warm land

Onshore surface wind

Cool water

Heat loss

Night

Heat loss →

~ 1 km

Cool land

Shore

Offshore surface wind

Warm water

Heat gain

Figure 6.24 Differences between day and night land-sea temperatures produce an onshore breeze during the day and an offshore breeze at night.

Figure 6.25 Fog obscures the Golden Gate Bridge at the entrance to San Francisco Bay.

coast of Washington, the westerlies moving across the North Pacific produce the Olympic Rain Forest as the air rises to clear the Olympic Mountains. On the western side of the Olympic Mountains, the rainfall is as much as 5 m (200 in) per year; 100 km (60 mi) away, in the rain shadow on the leeward side of the mountains, it is 40–50 cm (16–20 in) per year. The west side of the mountains of Vancouver Island in British Columbia has a high rainfall; the eastern side of the island is known for its sunshine and scenic cruising. The southeast trade winds on the east coast of South America sweep up and across the lowlands and then rise to cross the Andes Mountains, producing rainfall, large river systems, and lush vegetation on the eastern side of the Andes and a desert on their western slope. Over the Indian subcontinent in summer, the air approaching the Himalayas rises, intensifying the wet monsoon; air flows down from these mountains during the winter, the time of the dry monsoon. The control of precipitation patterns due to elevation changes is called the **orographic effect.**

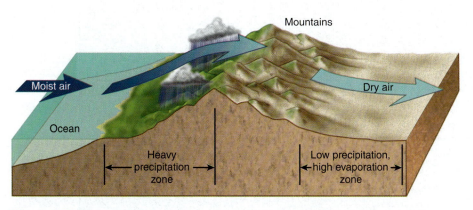

Figure 6.26 Moist air rising over the land expands, cools, and loses its moisture on the mountains' windward side. Descending air compresses, warms, and becomes drier, creating a rain shadow on the leeward side.

QUICK REVIEW

1. Why do high- and low-pressure zones alternate through the temperate latitudes of the Northern Hemisphere?

2. Why do winds circulate clockwise around high-pressure zones in the Northern Hemisphere?

3. How does the lack of land in the Southern Hemisphere affect the seasonal wind patterns?

4. What produces a monsoon?

5. Why do local coastal winds frequently change direction from day to night?

6. Explain the effect of land elevation on precipitation patterns.

6.6 Hurricanes and Coastal Flooding

Hurricanes

Intense atmospheric storms, known as **hurricanes**, are born over tropical oceans when surface-water temperature exceeds 27.8°C (82°F) (fig. 6.27). On either side of the equator very strong, low-pressure zones pump large amounts of energy from the sea surface into the atmosphere as warm, moist marine air rises rapidly and condenses to form clouds and precipitation. These storms bring strong winds and high precipitation because of the very high rate of condensation associated with the rising air. The large amounts of heat energy liberated from the condensing water vapor fuel the storm, raising wind speeds to destructive levels, up to 300 kilometers per hour (160 knots). A major hurricane contains energy exceeding that of a large nuclear explosion; fortunately, the energy is released much more slowly and over a greater area. When these storms move over colder water or over land, the hurricane is robbed of its energy source and begins to dissipate.

Hurricanes can form on either side of the equator but not at the equator because the Coriolis effect is necessary to create their spiraling winds. When a storm of this type is formed in the west-ern Pacific Ocean it is called a **typhoon** or **cyclone** instead of a hurricane. Areas that give rise to hurricanes and typhoons and their storm tracks are shown in figure 6.28.

Coastal Flooding

Strong storms at sea, such as hurricanes or typhoons, are centered about intense low-pressure systems. Under the center of such a storm, the sea surface rises up, forming a mound, or hill, called **storm surge**. The drop in atmospheric pressure from the outer edge of the storm to its center can be as large as 7.5 cm (3 in) of mercury. Because mercury is 13.6 times denser than water, this pressure change would produce a 97 cm (38 in) rise in elevation of the water beneath the storm's center. In addition, the strong hurricane winds spiraling toward the center of the storm drive water into the mound, further elevating the sea surface. Roughly 5% of the elevation of storm surge is caused by the drop in atmospheric pressure and 95% is wind-driven. Storm surge elevation is typically larger on the side of the storm approaching land where the winds are blowing onshore because the coastline acts like a dam (fig. 6.29).

Maximum sea surface elevation occurs when storm surge coincides with high tide; this is called **storm tide**. Storm tides can raise the water level by up to 6 m (20 ft) or more. Storm surge and storm tide are illustrated in figure 6.30. In coastal regions, storm surge, and particularly storm tide, are often the greatest threat to life and property from a hurricane.

Along shallow areas of the East and Gulf Coasts of the United States, storm surges have caused considerable damage. In 1900, a storm surge caused "the great Galveston flood." It

Figure 6.27 National Oceanic and Atmospheric Administration (NOAA) satellite image of Hurricane Lili taken at 4:45 P.M EDT on October 2, 2002.

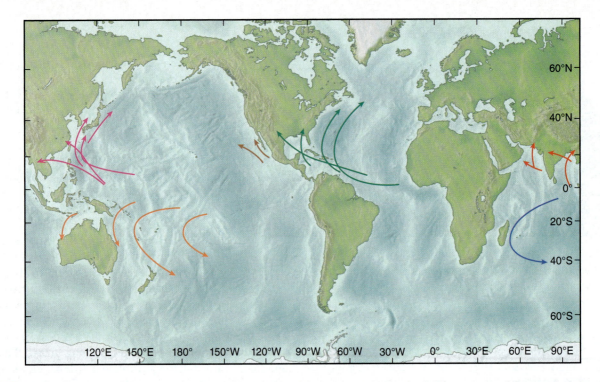

Figure 6.28 Hurricanes, typhoons, and cyclones form on either side of the equator in tropical seas. These storms follow preferred paths in different areas of the oceans.

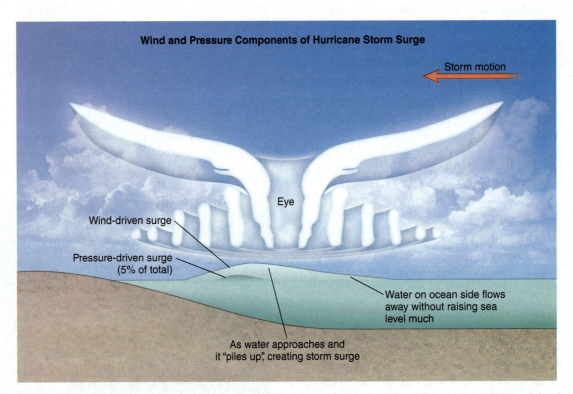

Figure 6.29 Storm surge is produced by a combination of water being pushed toward the center of the storm and toward the shore by the force of the winds and the elevation of the water due to the low pressure of the storm. The impact on the surge of the low pressure associated with intense storms is minimal in comparison to the effect of the wind.   Source of Data: National Oceanic and Atmospheric Administration (NOAA).

produced water depths of 4 m (13 ft) over the island of Galveston, Texas, destroying the city and killing 5000 people. Hurricane Camille, in 1969, was one of the strongest storms ever to hit the Gulf Coast. The high water levels caused severe property damage, and several hundred people were killed. In 1989, a storm surge of 5 m (16.5 ft) came ashore with Hurricane Hugo in South Carolina, destroying much property along the coast and in the city of Charleston. Great loss of life was averted

Figure 6.30 Storm surge is an elevation of the sea surface above tide level caused by the intense low pressure of a hurricane or typhoon. Storm tide is the elevation of the sea surface due to the combination of storm surge and high tide. Illustration from the National Hurricane Center, NOAA. Source of Data: National Oceanic and Atmospheric Administration (NOAA).

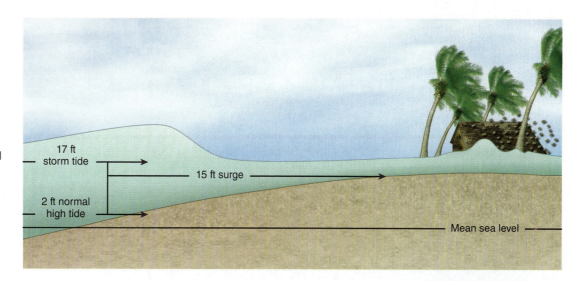

by early warnings and mass evacuation of low-lying coastal areas. In August 1992, Hurricane Andrew struck Florida with the same intensity that Hurricane Hugo brought to the Carolina coast, but the coastal water damage in Florida was much less. Although the winds were high (220 km/h, or 138 mi/h), the storm surge was only 2.4 m (8 ft). The storm traveled a short distance between the Bahamas and Florida at a fast rate of speed; it did not have enough time to build up a large mass of onshore moving water. In addition, some of Andrew's wave energy was absorbed by offshore reefs and coastal mangrove swamps. On shore, however, the wind damage was very severe.

The costliest and most destructive natural disaster in United States history was Hurricane Katrina with damage estimated to be in excess of $200 billion, over a million people displaced, roughly 5 million people without power, and over 1200 deaths. Katrina first made landfall just north of Miami, Florida, on August 25, 2005. As it moved across Florida it weakened but then wind speeds increased as it moved into the Gulf of Mexico over warm water (fig. 6.31). Warm water enters the Gulf through the Yucatan Strait as the Yucatan Current. It then flows clockwise in the Loop Current before exiting the Gulf as the Florida Current. The Loop Current occasionally pinches off to form clockwise rotating eddies or "gyres" of warm water that can slowly drift westward at speeds of 2–5 km/day (1.2–3 mi/day) (the formation of eddies

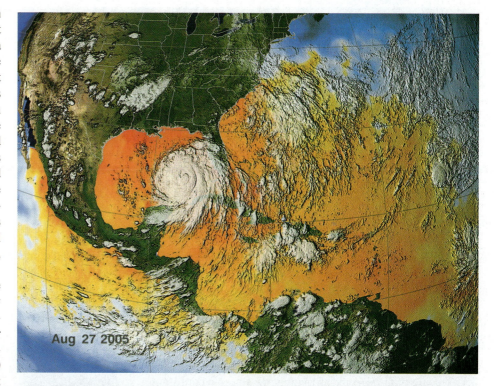

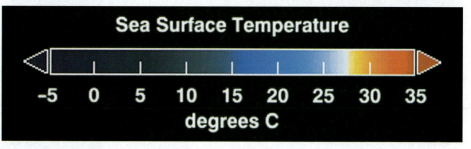

Figure 6.31 Hurricane Katrina gained strength as it moved over warm ocean water in the Gulf of Mexico. This image depicts a three-day average of actual sea surface temperatures (SSTs) for the Caribbean Sea and the Atlantic Ocean, from August 25–27, 2005. Areas in yellow, orange, or red represent 27.8°C (82°F) or above. Hurricanes need SSTs of 27.8°C (82°F) or warmer to strengthen.

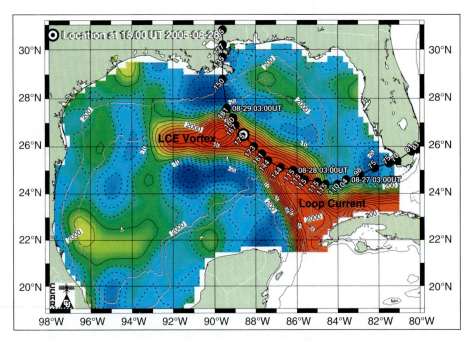

Figure 6.32 An overlay of Hurricane Katrina's track and maximum sustained wind speeds (mph) on the August 28, 2005, sea surface height (SSH) map. The units for SSH are dynamic cm and the contour interval is 5 cm. The SSH map approximates the total sea surface height signal associated with the general ocean circulation in the Gulf and is *false* colored over the range from –30 to +30 cm to highlight cold (*blue*) and warm (*red*) circulation features. At the time of the passage of Katrina, the Loop Current was extended far into the northern and western Gulf and was in the process of shedding a large, warm Loop Current Eddy (LCE) that was named "Vortex." The warm waters associated with the Loop Current and LCE Vortex contributed to the remarkable size and intensity of Hurricane Katrina.

is discussed in detail in chapter 8). As Katrina moved into the Gulf it passed over the warm (roughly 0.5°C or 1.0°F warmer than average) Loop Current and quickly strengthened, having maximum sustained winds of 280 km/hr (175 mi/hr) with gusts as high as 344 km/hr (215 mi/hr) (fig. 6.32). After moving into the Gulf of Mexico, it made landfall again on August 29 along the central Gulf Coast near New Orleans, Louisiana, with sustained winds of 235 km/hr (145 mi/hr). It then moved east along the Louisiana coastline and a few hours later made landfall a third time near the Louisiana/Mississippi border. Because of the size and strength of the storm, record storm surges inundated the entire Mississippi Gulf Coast. A region roughly the size of the United Kingdom covering 233,000 km² (90,000 mi²) was declared a Federal Disaster Area.

New Orleans was spared the worst of the winds when Katrina turned slightly to the east just before landfall. However, torrential rain and storm surge raised the level of Lake Pontchartrain, and on August 30 at 1:30 A.M. the 17th Street Canal barriers failed, and an estimated 224 billion gallons of water poured in, flooding roughly 80% of New Orleans (fig. 6.33).

Hurricane Katrina produced a 3 to 10 m (10 to 30 ft) storm surge along over 200 continuous miles of coastline from southeast Louisiana through the Florida panhandle. The 11.3 m (37 ft) storm surge that struck Pass Christian, Mississippi, is the highest ever recorded in the United States (figs. 6.34 and 6.35). Record

storm surges that had not occurred in at least the last 150 years flooded the entire Mississippi coast.

Following the devastation of Katrina, Hurricane Rita moved into the Gulf of Mexico as the strongest measured hurricane to ever enter the Gulf. Moving into the northern Gulf, it traveled over cooler water and lost energy as it made landfall near the Texas/Louisiana border on September 24, 2005, with sustained wind speeds of 190 km/hr (120 mi/hr). The effects of Hurricane Rita were not as severe as those of Katrina. Storm surges were generally less than 3 m (10 ft), rising to 4.5 to 6 m (15 to 20 ft) in some areas of southwestern Louisiana. Over 2 million people were left without power, and property damages were estimated to be $8 to 11 billion.

Severe storms over the shallow Bay of Bengal spawned storm surges that took an estimated 300,000 lives in 1970. Another 10,000 people were killed in 1985, and in 1991, yet another storm surge struck this same area of Bangladesh, killing an estimated 139,000 people. The expanding population of this area needs land for homes and farms, and as fast as new land appears in the delta of the Ganges River, the people move seaward. Although our ability to predict the severity and the route over which a

Figure 6.33 A Texas Army National Guard Blackhawk helicopter deposits a 6000 pound-plus bag of sand and gravel on-target, Sunday, September 4, as work progresses to close the breach in the 17th Street Canal.

Figure 6.34 What remained of the Pass Christian Middle School, Mississippi following Hurricane Katrina.

Figure 6.35 Aerial photograph of Mississippi Gulf Coast Highway I-90 destroyed as a result of winds and tidal surge from Hurricane Katrina. This section of bridge connects Pass Christian, near Gulfport, to Bay St. Louis.

hurricane will pass is improving, other factors work against the potential life-saving results of this ability. When a storm comes to a heavily populated, low-lying coast, the water rises rapidly and covers hundreds of square miles; such conditions make the evacuation of thousands of people extremely difficult.

The Netherlands has constructed barrier dams to protect its coast from the storm tides of the North Sea. England has installed gates across the River Thames to bar a storm tide moving up the river. The great cost of engineering projects such as these may be justified in heavily populated industrial areas, but storm tides cannot be prevented, and along many stretches of shore, there is no defense. The costs to taxpayers and their governments for emergency services and cleanup of storm tide damage are enormous. Would it be wiser to vacate areas historically prone to storm surges and high-water damage and use them only in nonpermanent ways?

QUICK REVIEW

1. What is the source of energy for intense tropical storms?
2. Where do hurricanes begin and in what direction are they most likely to travel?
3. What is the difference between storm surge and storm tide?
4. How is storm surge generated?

6.7 El Niño–Southern Oscillation

An El Niño is a period of anomalous climatic conditions centered in the tropical Pacific. El Niño events occur approximately every three to seven years. Once developed, they tend to last for roughly a year, although occasionally they may persist for eighteen months or more. El Niño events are perturbations of the coupled ocean-atmosphere system. We do not yet know whether the perturbations begin in the atmosphere or the ocean. We will consider what happens in the atmosphere first and then look at what happens in the ocean.

Under normal conditions, a low-pressure area in the atmosphere, the Indonesian Low, is located north of Australia over Indonesia in the western equatorial Pacific Ocean, and a corresponding high-pressure area, the South Pacific High, is located between Tahiti and Easter Island in the southeastern Pacific Ocean (fig. 6.36a). The prevailing surface winds over the equatorial Pacific from the South Pacific High to the Indonesian Low are the southeast trade winds. The strength of the trade winds depends on the difference in surface atmospheric pressure between the South Pacific High, typically measured at Tahiti, and the Indonesian Low, typically measured at Darwin, Australia. There is a corresponding return flow of air from the Indonesian Low back to the South Pacific High in the upper atmosphere that completes an atmospheric circulation cell known as the Walker circulation (fig. 6.36b). The resulting trade winds drive water away from the west coast of Central and South America. This causes a thick accumulation of warm water with high sea surface temperatures and elevated sea surface in the western equatorial Pacific Ocean (fig. 6.37a). Sea surface temperatures are about 8°C greater and sea surface elevation is about 50 cm (20 in) higher in the western Pacific compared to the eastern Pacific. As water is driven away from the west coast of Central and South America, upwelling brings cold, deep water to the surface along the west coast of Peru. This results in relatively low sea surface temperatures in the eastern equatorial Pacific Ocean.

During an El Niño event, the surface pressure in the area of the Indonesian Low is unusually high, while the surface pressure in the area of the South Pacific High is unusually low. As a result, the Indonesian Low moves eastward into the central tropical Pacific. A region of high pressure moves in to replace it, thus reversing the normal surface pressure gradient across the Pacific. As a result, the trade winds weaken or even reverse direction. This periodic reversal in position of the high- and low-pressure regions on either side of the tropical Pacific Ocean is called the **Southern Oscillation.** As the Indonesian Low moves into the central Pacific, the ITCZ also changes position, moving southward and eastward. The collapse of the trade winds

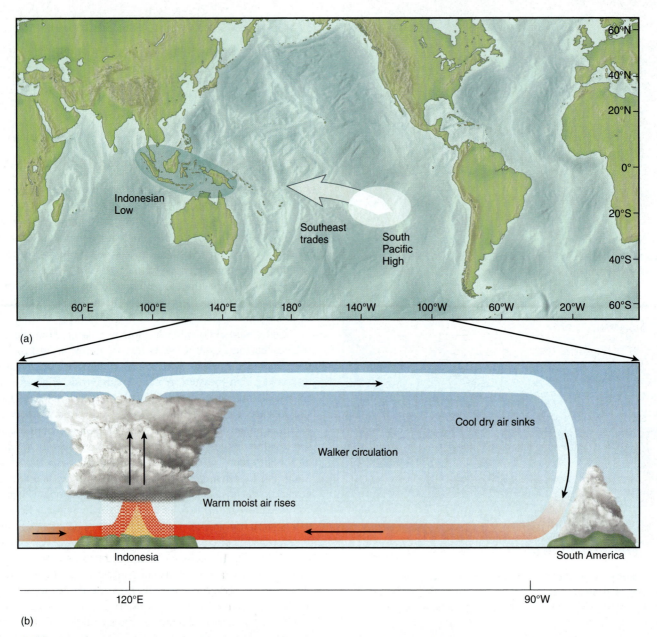

Figure 6.36 (a) Schematic representation of normal (not an El Niño event) atmospheric conditions in the tropical Pacific Ocean. (b) Walker circulation between the South Pacific High and the Indonesian Low. Surface winds are the southeast trade winds.

occurs abruptly. As a result of the Southern Oscillation, the deep, warm pool of surface water that normally occupies the western equatorial Pacific moves toward the eastern equatorial Pacific over a period of two to three months to accumulate along the coast of the Americas (fig. 6.37*b*). This effectively shuts down the upwelling of deep water off the Peruvian coast and raises the sea surface temperature by several degrees centigrade. The flow of water from the west also raises the sea surface elevation by 30 cm (12 in) or more (the change in elevation in the eastern equatorial Pacific between January 1997 and November 1997 shown in figure 6.37 was approximately 34 cm). This sequence of events is known as **El Niño,** or Christ Child, named for its frequent occurrence around the Christmas season. Eventually,

the normal atmospheric pressure distribution is reestablished, the trade winds again blow to the west, and the equatorial Pacific Ocean again has warm water in the west and cool water in the east, signaling the end of the El Niño. Figure 6.37*c* illustrates the transition to normal conditions following the end of the 1997–98 El Niño. Because of the relationship between El Niño and the Southern Oscillation, the two events together are known as ENSO (an acronym for El Niño–Southern Oscillation).

The climatic effects of El Niño are highly variable and appear to depend on the size of the warm pool of water and its temperature. The same region may experience higher-than-normal rainfall and flooding during one event and drought conditions during the next event. However, there do appear to be some very regular

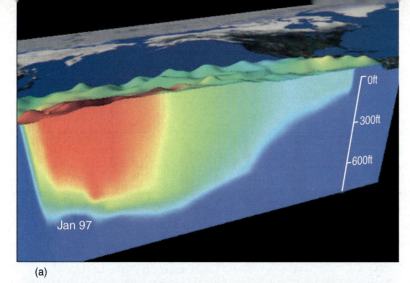

(a)

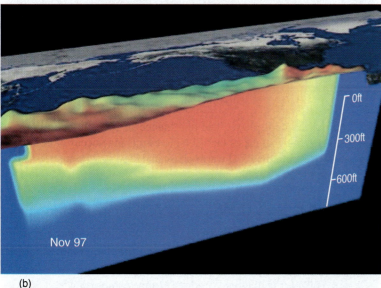

(b)

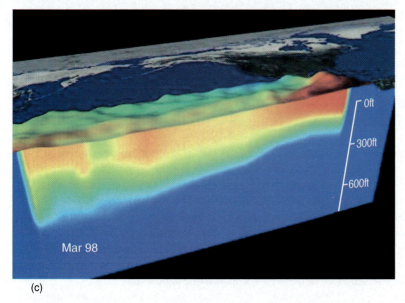

(c)

Figure 6.37 These images show sea surface topography from NASA's *TOPEX* satellite, sea surface temperatures from NOAA's *AVHRR* satellite sensor, and sea temperature below the surface as measured by NOAA's network of TAO moored buoys in the equatorial Pacific Ocean. *Red* is 30°C, and *blue* is 8°C. (a) Normal conditions as seen in January 1997, (b) El Niño conditions during November 1997, (c) the end of El Niño conditions and the beginning of a return to normal conditions in March 1998.

consequences of El Niño events. The northern United States and Canada generally experience warmer-than-normal winters. The eastern United States and normally dry regions of Peru and Ecuador typically have high rainfall, whereas Indonesia, Australia, and the Philippines experience drought. In addition, El Niño years are associated with less-intense hurricane seasons in the Atlantic Ocean.

The strongest El Niño on record occurred in 1997–98. The greatest warming occurred in the eastern tropical Pacific, where surface water temperatures were as much as 8°C above normal near the Galápagos Islands and off the coast of Peru. Normally dry regions in Ecuador and Peru that usually receive only 10–13 cm (4–5 in) of rain annually had as much as 350 cm (138 in, or 11.5 ft). Severe drought struck areas of the west Pacific Ocean in the Philippines, Indonesia, and Australia. Because El Niño tends to reduce precipitation in the wet monsoon areas of Asia, a weak rainy season was predicted, but India's rains were 2% above normal. A series of strong storms caused severe beach erosion, flooding, and landslides along the California coast. At the same time, the Pacific Northwest and Midwest winters were mild and the southern portion of the United States experienced wetter-than-normal conditions. The jet stream was diverted far to the south over North America, inhibiting the growth of hurricanes in the Atlantic Ocean. By the spring of 1998, Pacific Ocean sea surface temperatures were returning to normal and signaling the end of this El Niño event.

Between El Niño events, surface temperatures off Peru may drop below normal; an event of this type is known as **La Niña,** "the girl child." These colder-than-normal years also produce wide-scale meteorological effects. The trade winds strengthen, and surface-water temperatures of the eastern tropical Pacific are colder, whereas those to the west are warmer than normal. These changes help establish very dry conditions over the coastal areas of Peru and Chile, whereas rainfall and flooding increase in India, Myanmar, and Thailand.

Computerized ENSO models attempt to forecast El Niño events by means of the Multivariate ENSO Index (MEI) (fig. 6.38). The MEI is calculated from measurements of sea surface temperature, sea-level pressure, surface-air temperature, the east-west and north-south velocity components of the trade winds, and the total amount of cloudiness. Positive values of the MEI, also known as the ENSO warm phase, are associated with conditions occurring during El Niño events, while negative values of the MEI, known as the ENSO cool phase, are associated with conditions at the onset of La Niña events. Normal conditions correspond to an MEI at or near zero. Computer models successfully predicted the onset of the 1991–92 El Niño event, but the polar jet stream split over the eastern Pacific, and not all predictions came to pass. The 1997–98 episode was predicted six months in advance, although distinguishing the signs of the approaching El Niño from the unusually warm conditions that persisted throughout the 1990s was difficult. The advance warnings of the 1997–98 El Niño are estimated to have saved $1 billion to $2 billion in property damage in the United States.

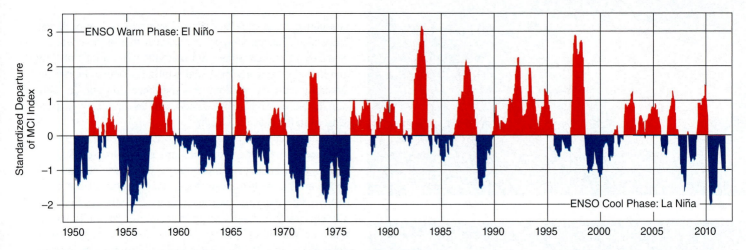

Figure 6.38 The Multivariate ENSO Index (MEI) from 1950 to February 2012. ENSO warm phases (*red*) correspond to El Niño events and cool phases (*blue*) correspond to La Niña events. The larger the positive value of the index, the stronger the El Niño; the larger the negative value, the stronger the La Niña.   Source of Data: National Oceanic and Atmospheric Administration (NOAA).

The cyclic alternation between these two events has been quite regular for the last hundred years, except for the periods between 1880 and 1900, when La Niña conditions prevailed, and the 1975–97 period of El Niños. In 2000–2001, a La Niña prevailed. This led to decreased rainfall in the Pacific Northwest and increased rain in California.

In addition to ENSO events, another natural cycle in the ocean-atmosphere system in the Pacific has been identified with a period of roughly seventeen to twenty-six years. This cycle is known as the Pacific Decadal Oscillation (PDO) (see chapter 8). The PDO appears to have a direct influence on sea surface temperatures. Satellite data indicate that the Pacific Ocean was in the warm phase of the PDO from 1977–99 and that it has now entered the cool phase. The dominance of El Niño conditions from 1975–97 suggests that El Niño events may occur more frequently during the PDO warm phase and they may be suppressed during the PDO cool phase.

QUICK REVIEW

1. How does El Niño begin?
2. How does El Niño change surface-water temperatures in the tropical Pacific Ocean?
3. Why is the ability to predict El Niño so important?
4. What variables are used to develop the Multivariate ENSO Index?

Summary

The intensity of solar radiation over Earth's surface varies with latitude. Incoming radiation and outgoing radiation are equal when averaged with time over the whole Earth. Reflection, re-radiation, evaporation, conduction, and absorption keep the heat budget in balance with the incoming solar radiation. In any local area of the oceans, there is daily and seasonal variation. Winds and ocean currents move heat from one ocean area to another to maintain the surface temperature patterns.

The oceans gain and lose large quantities of heat, but their temperature changes very little. The heat capacity of the oceans is high compared to that of the land and the atmosphere. Mixing and evaporative cooling work to reduce the temperature contrast between the sea surface and deeper water.

Clouds and weather occur in the troposphere, where temperature decreases with altitude. In the stratosphere, temperature increases with altitude because of the ozone and its ability to absorb ultraviolet radiation. The temperature of the higher mesosphere decreases with height.

The atmosphere is a mixture of gases, including water vapor. Atmospheric pressure is the force with which air presses on Earth's surface; high-density air creates high-pressure zones, and low-density air forms low-pressure zones.

Concern that Earth's climate may be changing is related to changes in the balance of the gases in the atmosphere. Increases in CO_2 concentration due to burning and use of fossil fuels are leading to the prediction of global warming because carbon dioxide traps outgoing long-wave radiation by the greenhouse effect. The ozone layer has been significantly depleted, most probably due to the release of chlorine into the atmosphere.

The density of air is controlled by temperature, pressure, and water vapor content. The winds are the horizontal air motion in convection cells produced by heating the atmosphere from below. Winds are named for the direction from which they blow.

Because of the Coriolis effect, winds are deflected to their right in the Northern Hemisphere and to their left in the Southern Hemisphere. This action produces a three-celled wind system in each hemisphere that results in the surface wind bands of the trade winds, the westerlies, and the polar easterlies. Zones of rising air occur at 0° and at 60°N and S; these are low-pressure areas of clouds and rain. Zones of descending air at 30° and 90°N and S are high-pressure areas of clear skies and low precipitation. Surface winds are unsteady and unreliable at the zones of rising and sinking air, producing the doldrums and the horse latitudes. The doldrum belt is displaced north of Earth's geographic equator.

Seasonal atmospheric pressure changes modify these wind bands and cause coastal winds to change direction seasonally in the Northern Hemisphere. Differences in temperature between land and water produce the monsoon effect. The seasonal reversal in wind pattern causes the wet and dry monsoons of the Indian Ocean; a similar daily reversal causes the onshore and offshore winds of any coastal area. Winds from the ocean rising to cross the land also produce heavy rainfall. Intense storms known as hurricanes, typhoons, and cyclones, develop over warm water in the tropical oceans. The sea surface beneath the center of such a storm is elevated because of a combination of the strong winds driving water toward the storm center and the low pressure beneath the storm center. The elevated mound of water is called storm surge. When storm surge combines with high tide it is called storm tide. Most of the damage done by severe storms that reach coastal areas is due to very high storm surge.

In some years, warm tropical surface water moves eastward across the Pacific and accumulates along the west coast of the Americas, blocking the normal upwelling. This phenomenon is El Niño; it is associated with changes in atmospheric pressure and wind direction. Colder-than-normal surface-water temperatures off coastal Peru and associated weather phenomena are known as La Niña; La Niña episodes appear to alternate with El Niño events.

Key Terms

solar constant, 152
heat budget, 152
sea ice, 157
pancakes, 157
ice floe, 157
fast ice, 157
drift ice, 157
pack ice, 157
icebergs, 157
troposphere, 158

stratosphere, 158
ozone, 158
mesosphere, 159
thermosphere, 159
greenhouse effect, 159
atmospheric pressure, 163
low-pressure zone, 163
high-pressure zone, 163
Coriolis effect, 165
trade winds, 165

westerlies, 165
polar easterlies, 165
doldrums, 166
horse latitudes, 166
Intertropical Convergence
    Zone (ITCZ), 166
monsoon, 168
onshore, 169
offshore, 169
rain shadow, 169

orographic effect, 170
hurricane, 171
typhoon, 171
cyclone, 171
storm surge, 171
storm tide, 171
Southern Oscillation, 175
El Niño, 176
La Niña, 177

Study Problems

1. Two people stand on the prime meridian—John at 15°N and Susan at 75°N. As Earth rotates, how far does each person travel in one hour in:
   a. degrees of longitude?
   b. kilometers?
   c. miles?

2. If Earth had no atmosphere, the intensity of solar radiation available on Earth's surface at the equator on the vernal or autumnal equinox would be 2 cal/cm^2/min. What would be the intensity of the solar radiation at 45° latitude and at 60° latitude on these days?

3. The atmospheric pressure at the center of a hurricane is 985 millibars. What would be the height of the pressure-induced storm surge below the center of the storm if the atmospheric pressure at the periphery of the storm is equal to the average atmospheric pressure at sea level?

CHAPTER

7

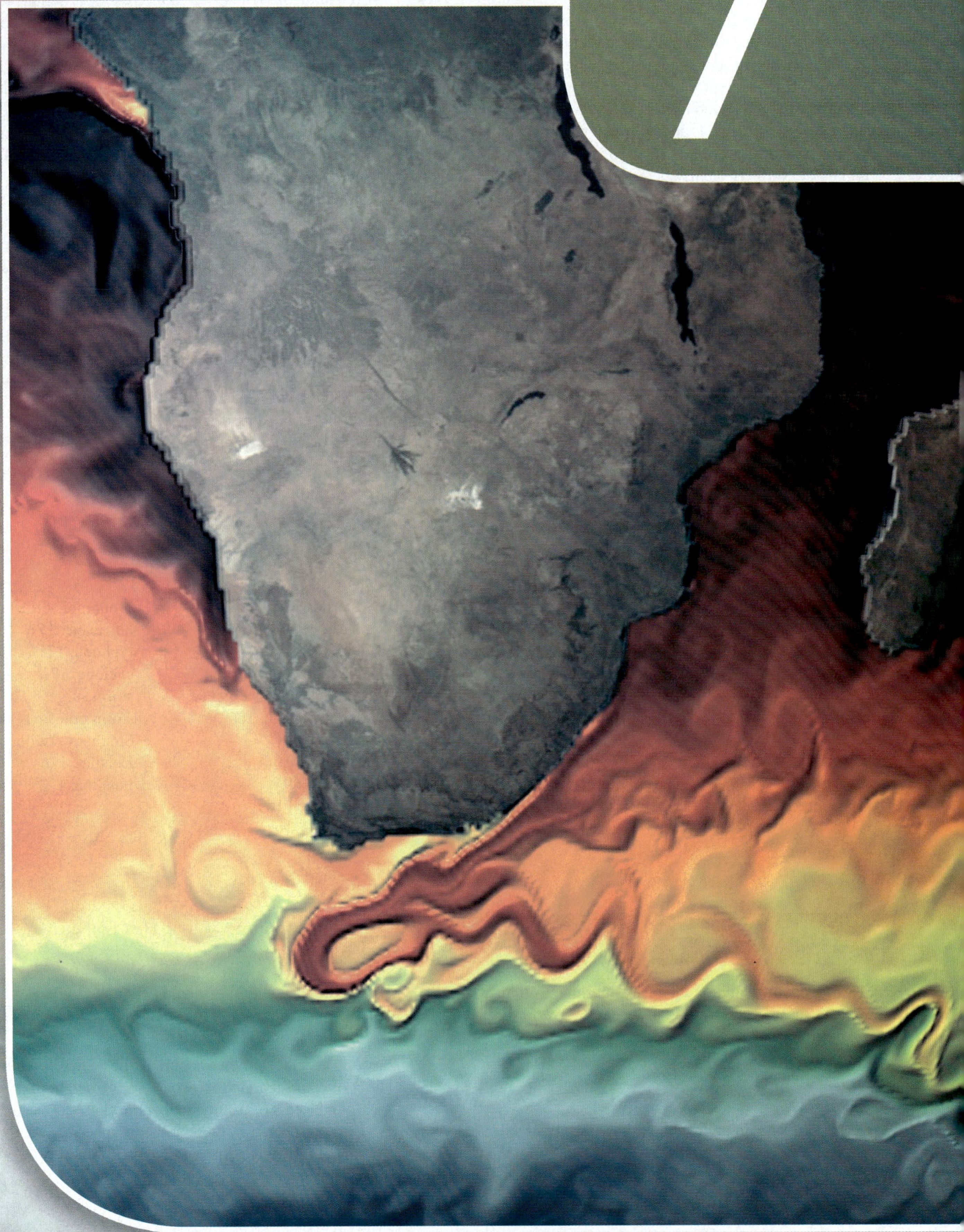

Ocean Structure and Circulation

Learning Outcomes

After studying the information in this chapter students should be able to:

1. *estimate* the density of a mixture of two samples of seawater that have the same density but different temperatures and salinities,

2. *describe* and *sketch* changes in the seasonal thermocline at mid-latitudes through the year,

3. *plot* temperature and salinity as a function of depth and *identify* the thermocline and halocline,

4. *list* five different water masses and *describe* how they form,

5. *relate* surface convergence and divergence to downwelling and upwelling,

6. *describe* the properties of water masses in each ocean basin,

7. *describe* and *sketch* the motion of water in the Ekman layer,

8. *diagram* the formation of surface current gyres,

9. *locate* the major surface currents on a map of the ocean basins,

10. *explain* the process of western intensification,

11. *relate* patterns of surface convergence and divergence to downwelling and upwelling, and

12. *sketch* the Great Ocean Conveyor Belt.

Sea surface temperature (SST) simulation created by scientists at the National Oceanic and Atmospheric Administration's Geophysical Fluid Dynamics Laboratory using a coupled atmosphere-ocean model. Currents and eddies off the southern tip of Africa are evident.

Earth is surrounded by two great oceans: an ocean of air and an ocean of water. Both are in constant motion, driven by the energy of the Sun and the gravity of Earth. Hidden below the ocean's surface is its structure. If we could remove a slice of ocean water in the same way we might cut a slice of cake, we would find that, like a cake, the ocean is a layered system. The layers are invisible to us, but they can be detected by measuring the changing temperature and salt content, and by calculating the density of the water from the surface to the ocean floor. This layered structure is a dynamic response to processes that occur at the surface: the gain and loss of heat, the evaporation and addition of water, the freezing and thawing of ice, and the movement of water in response to wind. These surface processes produce a series of horizontally moving layers of water, as well as local areas of vertical motion. Surface currents carry heat from one location to another, altering Earth's surface temperature patterns and modifying the air above. The interaction between the atmosphere and the ocean is dynamic; as one system drives the other, the driven system acts to alter the properties of the driving system.

In this chapter, we will study both the surface processes and their below-the-surface results in order to understand why the ocean is structured and how its structure is maintained. We will also explore the formation of the ocean's surface currents. We follow these currents as they flow, merge, and move away from each other. We examine both horizontal and vertical circulation, and consider ways in which they are linked to the overall interaction between the atmosphere and the ocean.

- a surface layer tens to a few hundreds of meters thick, called the **mixed layer**;
- a region called the **thermocline**, extending from the bottom of the mixed layer to a depth of about 1000 m (3280 ft); and
- the region from the base of the thermocline to the sea floor.

The mixed layer is an **isothermal** layer—a layer of constant temperature. The thickness of the mixed layer is variable. It depends on the depth to which the surface water is mixed by turbulence caused by waves and wind. The mixed layer can be as thick as 200–300 m (~650–1000 ft) at mid-latitudes in the open ocean, whereas in protected coastal waters in the summer, it can be as little as 10 m (~33 ft) thick. Between about 200–300 m and 1000 m depth is the thermocline, where temperature decreases rapidly throughout much of the ocean. This layer is also frequently known as the "permanent thermocline" because seasonal changes in climate at the surface do not influence water temperature at these depths. Below 1000 m depth is the third, and largest in volume, layer, which extends to the sea floor. This deep water is nearly isothermal, temperature decreases very slowly with depth, and it is uniformly cold everywhere. Roughly 75% of the water in the ocean has a temperature between 0 and 4°C (table 7.1 and fig. 7.3). The discovery that the deep water of the ocean, even in tropical regions, is very cold was made in the eighteenth century. The obvious conclusion that followed was that deep seawater must originate in polar regions, where cold, dense surface water sinks and flows toward the Equator along the ocean floor.

The details of actual temperature-versus-depth profiles vary considerably depending on latitude and season of the year (fig. 7.4). At mid-latitudes, the temperature and depth of the mixed layer undergo seasonal changes with the formation of a shallow seasonal thermocline in the summer and its disappearance in the winter (fig. 7.4a). Throughout the winter, when strong

7.1 Ocean Structure

Variation of Temperature with Depth

With very few exceptions, the temperature of seawater decreases with depth. Effectively all of the energy available to heat the ocean comes from incoming solar radiation. Consequently, only a thin surface layer of the ocean is heated directly because of how rapidly solar radiation is absorbed with depth (fig. 7.1). Nearly half of the total solar energy at the sea surface is absorbed within 10 cm of the surface, and all of the infrared energy is absorbed within about a meter of the surface. A typical seawater temperature-versus-depth profile consists of three "temperature layers" (fig. 7.2):

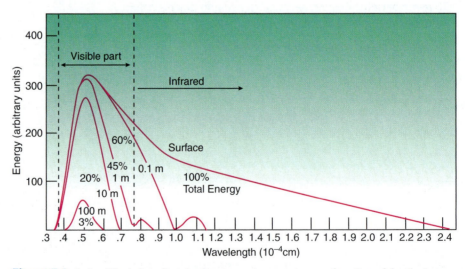

Figure 7.1 A simplified plot of total solar energy in seawater as a function of depth. Area below each curve is representative of the percent of available solar energy at the surface that reaches that depth. A little over half of the total solar energy at the surface is absorbed in the upper 1 m of water (only 45% penetrates to a depth of 1 m). Infrared energy is absorbed particularly rapidly, with nearly all of it absorbed in the upper 1 m of water.

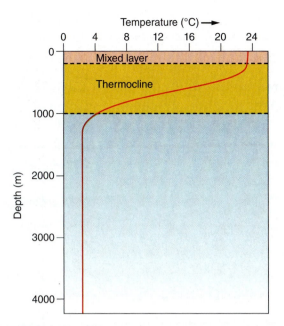

Figure 7.2 Simplified profile of temperature with depth in the ocean. A shallow mixed layer with relatively constant temperature overlies the thermocline, where temperature decreases rapidly with depth. Beneath the thermocline, below a depth of about 1000 m, temperature is fairly constant and cold.

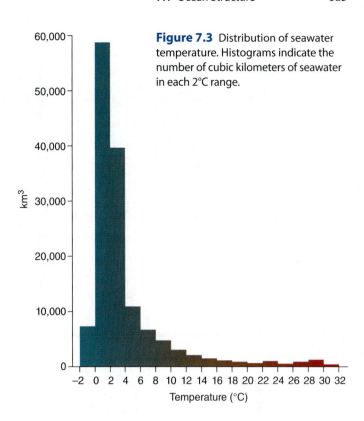

Figure 7.3 Distribution of seawater temperature. Histograms indicate the number of cubic kilometers of seawater in each 2°C range.

Table 7.1 Distribution of Seawater Temperature

Temperature Range (°C)	Seawater Volume (km³)	Percent of Ocean Volume
−2–0	6409	4.7
0–2	57,990	42.4
2–4	40,065	29.3
4–6	11,849	8.7
6–8	6059	4.4
8–10	4222	3.1
10–12	2632	1.9
12–14	2358	1.7
14–16	1340	1.0
16–18	985	0.7
18–20	685	0.5
20–22	638	0.5
22–24	475	0.3
24–26	426	0.3
26–28	489	0.4
28–30	269	0.2
30–32	2	0.001

winds produce deep mixing and surface-water temperatures are cold, the mixed layer may extend all the way to the top of the permanent thermocline, producing an essentially vertical temperature profile in the upper 200–300 m (fig. 7.5, *March*). Throughout the summer, as surface temperatures rise and winds decrease, the mixed layer will become more shallow and a strong (steep temperature gradient) seasonal thermocline can develop above the permanent thermocline (fig. 7.5, *August*). At low latitudes, surface temperatures are warm and constant throughout the year. Consequently, there is no development of seasonal thermoclines and the three distinct temperature layers are quite stable (fig. 7.4*b*). At high latitudes above about 60°, there is no permanent thermocline (fig. 7.4*c*). However, weak seasonal thermoclines can develop in the summer. At high latitudes there is often a layer of cold water 50–100 m below the surface. Overall, the presence of three temperature layers in the ocean is illustrated in the plot of seawater temperature in the Pacific Ocean versus depth and latitude shown in figure 7.6.

Variation of Salinity with Depth

As discussed in chapter 5, the salinity of surface seawater varies as a function of latitude in a relatively stable pattern (review figs. 5.2 and 5.3). However, it is more difficult to draw general conclusions about vertical profiles of salinity versus depth. Depending on latitude, salinity may be relatively constant, decrease, or increase with depth down to about 1000 m. Below about 1000 m depth, the influence of surface processes is minor and salinity is fairly constant. The total salinity range of 75% of the ocean is between 34.5 and 35.0 ppt (table 7.2 and

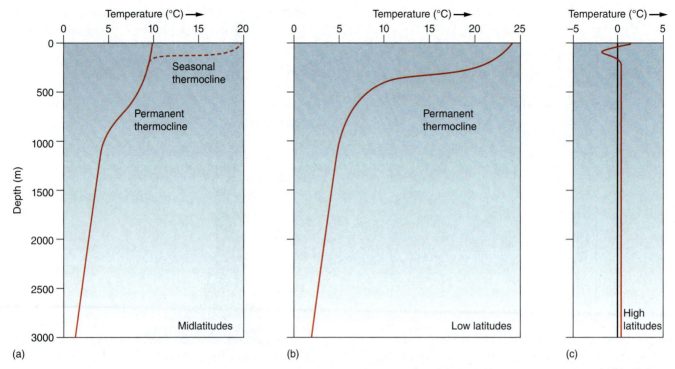

Figure 7.4 Simple temperature-versus-depth profiles for three latitude zones: (a) mid-latitudes with significant seasonal variation, (b) low latitudes where climate tends to be uniformly warm through the year, and (c) high latitudes where climate tends to be uniformly cold through the year.

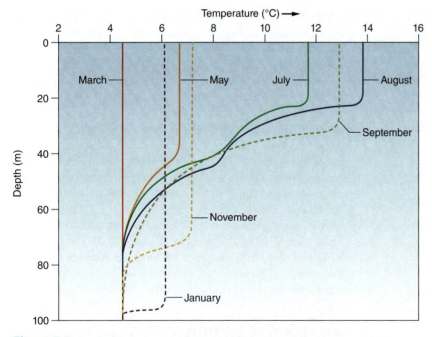

Figure 7.5 Detailed variation in the depth of the seasonal thermocline and the temperature of the surface mixed layer at mid-latitudes. Solid lines indicate the growth of the seasonal thermocline as it increases in strength and shoals during the summer, and dashed lines indicate its decay as it becomes deeper and weaker in winter. Note the different scales compared to figure 7.4.

found at high latitudes. In either case, this marked change in salinity with depth is called the **halocline** (fig. 7.8). The general distribution of seawater salinity in the Pacific Ocean is illustrated in the plot of salinity versus depth and latitude shown in figure 7.9.

Variation of Density with Depth

Variations in temperature, salinity, and pressure (depth) combine to control the density of seawater. Seawater density is inversely proportional to temperature and directly proportional to salinity and pressure. This can be expressed as ($\uparrow$ = increases, and $\downarrow$ = decreases):

$$\text{Density} \uparrow \text{ as temperature } \downarrow$$
$$\text{Density} \uparrow \text{ as salinity } \uparrow$$
$$\text{Density} \uparrow \text{ as pressure } \uparrow$$

In general, variations in temperature and salinity are more influential in determining seawater density than are variations in pressure (or depth). Consequently, we can consider density to be a function of temperature and salinity. The stratification of the ocean by temperature and salinity that we have just discussed results in a stratification of the ocean into density layers also. A typical seawater density-versus-depth profile consists of three "density layers" (fig. 7.10):

• the mixed layer, a surface layer tens to a few hundreds of meters thick;

fig. 7.7). In the upper 1000 m, areas where salinity decreases with depth are typically found at low and middle latitudes whereas areas where salinity increases with depth are typically

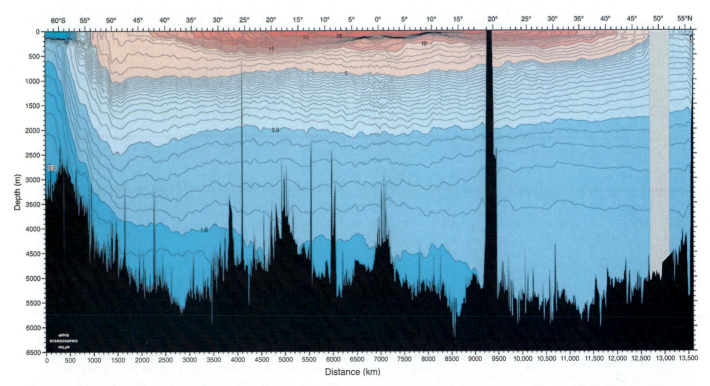

Figure 7.6 Ocean temperature versus depth and latitude. North-south profile though the Pacific Ocean along the 150°W meridian. Temperature decreases with depth. The depth of the surface mixed layer and the permanent thermocline are less at low latitudes than at mid-latitudes because winds are generally weaker and seasonal temperature contrasts are less at low latitudes. The black represents seafloor bathymetry.

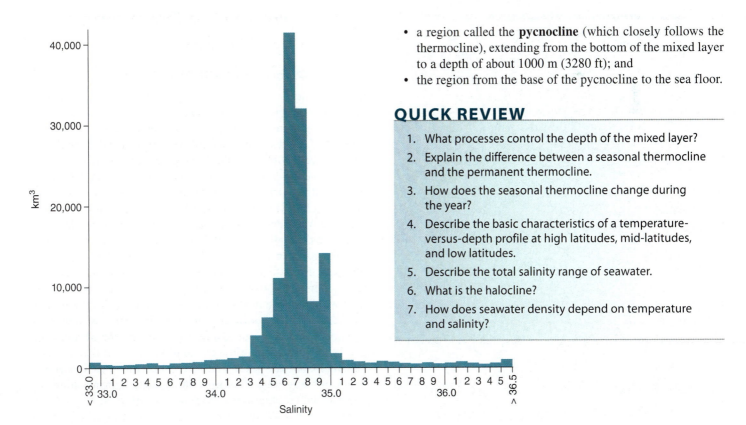

- a region called the **pycnocline** (which closely follows the thermocline), extending from the bottom of the mixed layer to a depth of about 1000 m (3280 ft); and
- the region from the base of the pycnocline to the sea floor.

QUICK REVIEW

1. What processes control the depth of the mixed layer?
2. Explain the difference between a seasonal thermocline and the permanent thermocline.
3. How does the seasonal thermocline change during the year?
4. Describe the basic characteristics of a temperature-versus-depth profile at high latitudes, mid-latitudes, and low latitudes.
5. Describe the total salinity range of seawater.
6. What is the halocline?
7. How does seawater density depend on temperature and salinity?

Figure 7.7 Distribution of seawater salinity. Histograms indicate the number of cubic kilometers of seawater in each 0.1 ppt range.

Table 7.2   Distribution of Seawater Salinity

Salinity Range (ppt)	Seawater Volume (km³)	Percent of Ocean Volume
<33.0	226	0.17
33.0–33.1	44	0.03
33.1–33.2	28	0.02
33.2–33.3	28	0.02
33.3–33.4	44	0.03
33.4–33.5	80	0.06
33.5–33.6	40	0.03
33.6–33.7	86	0.06
33.7–33.8	97	0.07
33.8–33.9	266	0.19
33.9–34.0	483	0.35
34.0–34.1	1058	0.77
34.1–34.2	1152	0.84
34.2–34.3	1896	1.38
34.3–34.4	4112	3.00
34.4–34.5	6515	4.76
34.5–34.6	12,301	8.98
34.6–34.7	44,235	32.31
34.7–34.8	32,761	23.93
34.8–34.9	7989	5.83
34.9–35.0	14,211	10.38
35.0–35.1	2396	1.75
35.1–35.2	1094	0.80
35.2–35.3	870	0.64
35.3–35.4	668	0.49
35.4–35.5	765	0.56
35.5–35.6	539	0.39
35.6–35.7	434	0.32
35.7–35.8	325	0.24
35.8–35.9	362	0.26
35.9–36.0	144	0.11
36.0–36.1	178	0.13
36.1–36.2	199	0.15
36.2–36.3	173	0.13
36.3–36.4	148	0.11
36.4–36.5	192	0.14
>36.5	786	0.57

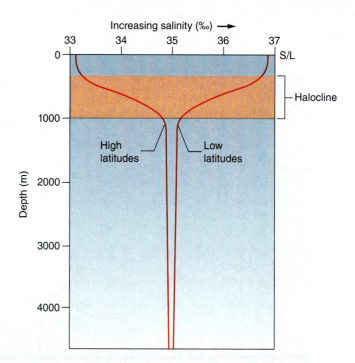

Figure 7.8 Simplified profiles of salinity with depth in the ocean for high and low latitudes. In each case, there is a surface mixed layer of relatively constant salinity. Beneath the mixed layer at high latitudes, salinity decreases rapidly to a depth of about 1000 m. Beneath the mixed layer at low latitudes, salinity increases rapidly to a depth of about 1000 m. The region of rapid change in salinity is called the halocline.

7.2 Thermohaline Circulation and Water Masses

Thermohaline Circulation

If the density of the water increases with depth, the water column from the surface to that depth is **stable**. If there is higher-density water on top of lower-density water, the water column is **unstable**. An unstable water column cannot persist; the denser surface water sinks and the less-dense water at depth rises to replace the water above it. When dense water from the surface sinks and reaches a level at which it is denser than the water above but less dense than the water below, it spreads horizontally as more water descends. At the surface, water moves horizontally into the region where sinking is occurring. The dense water that has descended displaces deeper water upward, completing the cycle. Because water is a fixed quantity in the oceans, it cannot be accumulated at one location or removed at another location without movement of water between those locations. This concept is called **continuity of flow**. This motion, caused by variations in density due to differences in temperature and salinity, is called **thermohaline circulation**. Areas of thermohaline circulation where water sinks are called **downwelling zones**; areas of rising waters are **upwelling zones**. Downwelling is a mechanism that transports oxygen-rich surface water to depth, where it is needed for deep-living animals. Upwelling returns low oxygen-content water with dissolved, decay-produced nutrients to the surface, where the nutrients act as fertilizers to

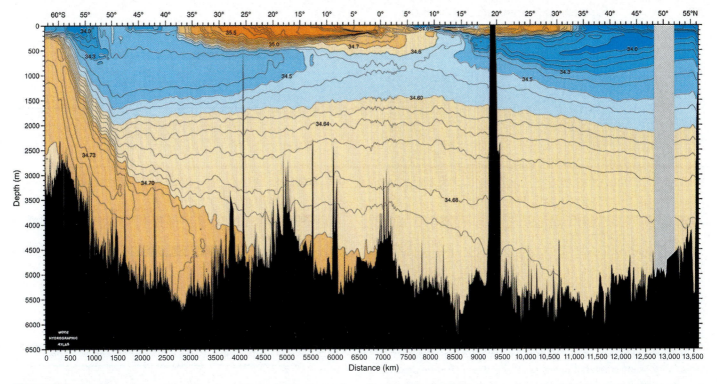

Figure 7.9 Seawater salinity versus depth and latitude. North-south profile through the Pacific Ocean along the 150°W meridian. Most of the variation in salinity is found in surface water with high values at low and middle latitudes and low values at high latitudes. Beneath about 1000–1500 m, salinity shows only minor variations. The black represents seafloor bathymetry.

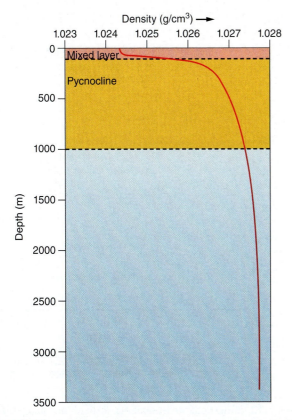

Figure 7.10 Simplified profile of density with depth in the ocean. A shallow mixed layer with relatively constant density overlies the pycnocline, where density increases rapidly with depth (this layer corresponds closely to the thermocline shown in figure 7.2). Beneath the pycnocline, density increases slowly to the sea floor.

promote photosynthesis and the production of more oxygen in the sunlit surface waters. Upwelling and downwelling can also be caused by wind-driven surface currents. When the surface waters are driven together by the wind or against a coast, a surface **convergence** is formed. Water at a surface convergence sinks, or downwells. When the wind blows surface waters away from an area or a coast, a surface **divergence** occurs and water upwells from below. Surface convergences and divergences are discussed further later in this chapter.

The speed of upwelling and downwelling water is about 0.1–1.5 m (0.3–5 ft) per day. Compare these flows to ocean surface currents, which reach speeds of 1.5 m (5 ft) per second. Horizontal movement at mid-ocean depth due to thermohaline flow is about 0.01 cm (0.004 in) per second. Water caught in this slow but relentlessly moving cycle can spend 1000 years at the greater ocean depths before it again reaches the surface.

If the water column has the same density over depth, it has neutral stability and is termed **isopycnal**. A neutrally stable water column can easily be mixed vertically by wind, wave action, and currents. If the water temperature is unchanging over depth, the water column is isothermal; if the salinity is constant over depth, it is **isohaline**.

Water Masses

There are many different combinations of temperature and salinity that produce the same seawater density. This can be illustrated in a temperature-salinity diagram, or T-S diagram, where lines of equal density, or isopycnals, are drawn for many combinations of temperature and salinity (fig 7.11). As the salinity

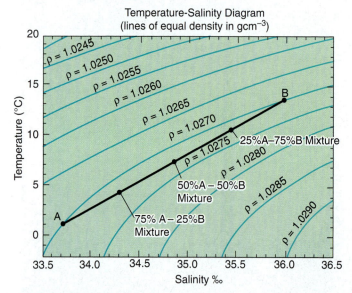

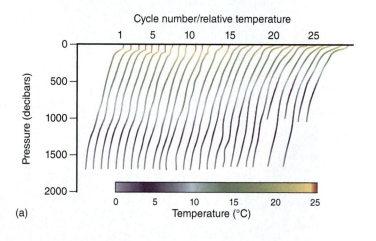

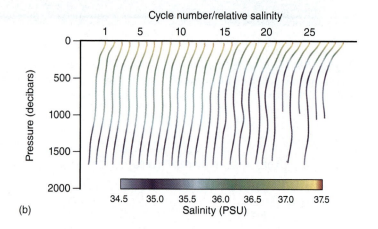

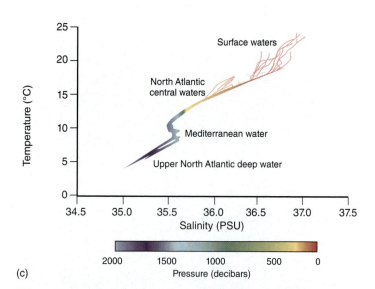

Figure 7.11 The density of seawater, measured in grams per cubic centimeter, is abbreviated as ρ (rho) and varies with temperature and salinity. Many combinations of salinity and temperature produce the same density. Low densities are at the *upper left* and high densities are at the *lower right*. The *straight line* is the mixing line for waters *A* and *B*, both with the same density. A mixture of *A* and *B* lies on the mixing line and is more dense than either *A* or *B*.

increases, the density increases; as the temperature increases, the density decreases. Salinity can be increased by evaporation or by the formation of sea ice; it can be decreased by precipitation, the inflow of river water, the melting of ice, or a combination of these factors.

Any seawater sample can be plotted on a T-S diagram as a single point and the density of the sample can be determined from the isopycnals. Seawater samples from a common source that all plot very near each other on a T-S diagram, with a narrow range of temperature and salinity, define a **water type**. When two water types (e.g., *A* and *B* in fig. 7.11) having the same density, but different values of salinity and temperature, are mixed, they form a new water type that lies on a mixing line (a straight line drawn from point *A* to point *B*). The location of the new water type on the mixing line will depend on the relative amounts of the two original water types that were combined to form the mix. Notice that the isopycnals are curved. Consequently, the mixture of two water types with the same density will produce a new water type with a density greater than either of the two original water types, and the newly created mixed water will sink. This mixing and sinking process is known as **caballing** and it occurs wherever surface waters converge and mix.

A **water mass** is a large body of water that has similar values of temperature and salinity throughout. Water masses can be identified by measuring water temperature and salinity along vertical profiles in the ocean and plotting these data on a T-S diagram (fig. 7.12*a* and *b*). A water mass will be represented on a T-S diagram by closely clustered data points (fig. 7.12*c*). Surface-water masses typically have greater variability in temperature and salinity than do deeper water masses.

Figure 7.12 ARGO float measurements north of the Canary Islands in the North Atlantic (about 31–37°N, 25–27°W) along vertical profiles of (a) temperature and (b) salinity. (c) T-S plot of the data collected can be used to identify specific water masses. Notice the greater range of temperature and salinity found in surface waters (*red*) compared to deep waters (*violet*).

QUICK REVIEW

1. Describe the concept of continuity of flow.
2. How do stable and unstable water columns differ?
3. What is the relationship between upwelling and downwelling? Between surface divergence and convergence?
4. What is the difference between a water type and a water mass?
5. Explain how water types with different temperature and salinity can have the same density.

7.3 The Layered Oceans

Oceanographers have taken salinity and temperature measurements with depth at many locations and for many years. Gradually, they accumulated sufficient data to identify the layers of water that make up each ocean and the surface source of the water forming each layer. The structure of an ocean is determined by the properties of these layers. Each layer received its characteristic salinity, temperature, and density at the surface. The water's density controls the depth to which the water sinks; the thickness and horizontal extent of each layer are related to the rate of its formation and the size of the surface source region. Water that sinks from the surface to spread out at depth and slowly mixes with adjacent layers eventually rises at another location. In all cases, water that sinks displaces an equivalent volume of water upward toward the surface at some other location so that the oceans' vertical circulation is continuous.

The layers of water just described are associated with depth zones: surface, intermediate, deep, and bottom. The surface zone extends to 200 m (650 ft), and the intermediate zone lies between 200 and 2000 m (1000 and 6500 ft). Deep water is found between 2000 and 4000 m (6500 and 13,000 ft), and bottom water is below 4000 m (13,000 ft).

The Atlantic Ocean

The properties of the layers of water making up the Atlantic Ocean are shown in figure 7.13. At the surface in the North Atlantic, water from high northern latitudes moves southward, while water from low latitudes moves northward along the coast of North America and then east across the North Atlantic. These waters converge in areas of cool temperatures and high precipitation at approximately 50°–60°N in the Norwegian Sea and at the boundaries of the Gulf Stream and the Greenland and Labrador Currents. The resulting mixed water has a salinity of about 34.9‰ and a temperature of 2°–4°C. This water, known as **North Atlantic deep water (NADW),** sinks and moves southward. North Atlantic deep water from the Norwegian Sea moves south along the east side of the Atlantic, while water formed at the boundary of the Labrador Current and the Gulf Stream flows along the western side. Above this water at 30°N, a low-density lens of very salty (36.5‰) but very warm (25°C) surface water remains trapped by the circular movement of the major oceanic surface currents. Between this surface water and the North Atlantic deep water lies water of intermediate temperature (10°C) and salinity (35.5‰). This water is a mixture of surface water and the upwelled colder, saltier water from the subtropical regions. It moves northward to reappear at the surface south of the convergence in the North Atlantic.

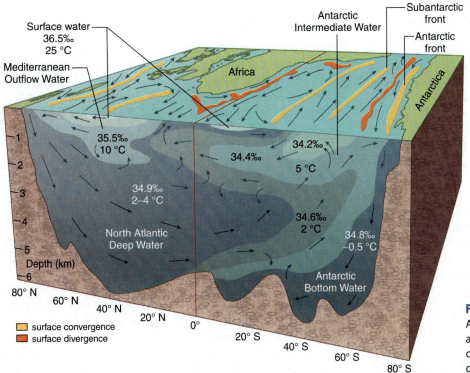

Figure 7.13 North-south cross section of the Atlantic Ocean illustrating major water masses and their general movement. Areas of surface convergence and divergence associated with downwelling and upwelling are also shown.

Near the equator, the upper boundary of the North Atlantic deep water is formed by water produced at the convergence centered about 40°S. This is **Antarctic intermediate water (AAIW).** Because it is warmer (5°C) and less salty (34.4‰) than the North Atlantic deep water, it is less dense and remains above the denser and saltier water below. Along the edge of Antarctica, very cold (−0.5°C), salty (34.8‰), and dense water is produced at the surface by sea ice formation during the Southern Hemisphere's winter. This is **Antarctic bottom water (AABW),** the densest water in the oceans. This water sinks to the ocean floor and flows slowly northward, creeping beneath North Atlantic deep water, as it continues on through the deep South Atlantic ocean basins west of the Mid-Atlantic Ridge. Antarctic bottom water does not accumulate enough thickness to be able to flow over the mid-ocean ridge system into the basins on the African side of the ridge; it is confined to the deep basins on the west side of the South Atlantic and has been found as far north as the equator.

At the same time, the North Atlantic deep water between the Antarctic bottom and intermediate waters rises to the ocean's surface in the area of the 60°S divergence. As it reaches the surface, it splits; part moves northward as **South Atlantic surface water** and Antarctic intermediate water; part moves southward toward Antarctica, to be cooled and modified to form Antarctic bottom water. A mixture of North Atlantic deep water and Antarctic bottom water becomes the circumpolar water for the Southern Ocean that flows around Antarctica. The Antarctic circumpolar water becomes the source of the deep water found in the Indian and Pacific Oceans. In this way, the Atlantic Ocean and its circulation play a defining role in the structure and circulation of all the oceans.

Warm (25°C), salty (36.5‰) surface water in the South Atlantic is also caught by the circular current pattern at the surface and is centered about 30°S. South of the southern tips of South America and Africa, the water flows eastward, driven by the prevailing westerly winds, which move the water around and around Antarctica.

Water from the Mediterranean Sea has a temperature of about 13°C and a salinity of 37.3‰ as it leaves the Strait of Gibraltar. This water, mixing with Atlantic Ocean water, forms an intermediate density water, **Mediterranean Intermediate Water (MIW),** also known as Mediterranean Outflow Water (MOW), that sinks in the North Atlantic to a depth of approximately 1000 m (3300 ft). The influence of Mediterranean water can be traced 2500 km (1500 mi) from the Strait of Gibraltar before it is lost through modification and mixing.

Because the Atlantic Ocean is a narrow, confined ocean of relatively small volume but great north-south extent, the water types are readily identifiable and the movement of the layers can be followed quite easily. In addition, the bordering nations of the Atlantic have had a long-standing interest in oceanography, so the vertical circulation and layering of the Atlantic are the most studied and the best understood of all the oceans.

The Pacific Ocean

In the vast Pacific Ocean, waters that sink from relatively small areas of surface convergences lose their identity rapidly, making the layers difficult to distinguish. Antarctic bottom water forms in small amounts along the Pacific rim of Antarctica, but it is quickly lost in the great volume of the Pacific Ocean. The deeper water of the South Pacific Ocean is the water of the Antarctic circumpolar flow. Because the North Pacific is isolated from the Arctic Ocean, only a small amount of water comparable to North Atlantic deep water can be formed. In the extreme western North Pacific, convergence of the southward-flowing cold water from the Bering Sea and the Sea of Okhotsk and the northward-moving water from the lower latitudes produces only a small volume of water that sinks to mid-depths. There is no large source of deep water similar to that found in the North Atlantic. Warm, salty surface water occurs at subtropical latitudes (30°N and 30°S) in each hemisphere, and Antarctic intermediate water is produced in small quantities, but its influence is small. Deep-water flows in the Pacific are sluggish, and conditions are very uniform below 2000 m (6600 ft). The slow circulation of the Pacific means that it has the oldest water at depths where age is measured as time from the water's last contact with the surface. Residence time for deep water in the Pacific is about twice that of deep water in the Atlantic.

The Indian Ocean

The Indian Ocean is principally an ocean of the Southern Hemisphere and has no counterpart of the North Atlantic deep water. Small amounts of Antarctic bottom water are soon mixed with the deeper waters to form a fairly uniform mixture of Antarctic circumpolar water brought into the Indian Ocean by the Antarctic circumpolar current. There is a small amount of Antarctic intermediate water, and in the subtropics, a lens of warm, salty water occurs at the surface.

The Arctic Ocean

The Arctic Ocean basin is unique. About one-third of its area is covered by extensive continental shelves, the widest of any ocean basin. Two basins, the Eurasian to the east and the Canadian to the west, occupy the central portion of the ocean; they are separated by the Lomonosov Ridge extending due north from Greenland.

The density of Arctic Ocean water is controlled more by salinity than by temperature. Its surface layer is formed from low-salinity water entering from the Bering Sea, fresh water from Siberian and Canadian rivers, and seasonal melting of sea ice. The surface layer from these combined sources is about 80 m (250 ft) deep and has a low salinity (32.5‰) and a low temperature (−1.5°C). Below the surface layer, salinity increases with depth in the halocline layer, 200 m (650 ft) thick, to reach 34.5‰ at its base. The cold, salty water of the halocline layer is produced by annual freezing and formation of sea ice over the continental shelves bordering the ocean. This water sinks and moves across the shelves to spread out in the central ocean basins. West of Spitsbergen, Norway, North Atlantic water (2°C and 35‰) enters the Arctic Ocean and is cooled as it flows under the halocline and fills the Arctic Ocean basins. This water upwells along the edge of the continental shelves, mixing with the water formed during freezing, and exits the Arctic as water of 0.5°C temperature and 34.9‰ salinity along the edge of the shelf adjacent to Greenland. This exiting water moves south along the coast of Greenland and enters the North Atlantic south of Greenland and Iceland, where it combines with Gulf Stream water to form North Atlantic deep water.

Internal Mixing

Mixing between waters in the ocean is most active when turbulence and energy of motion are available to stir the waters and blend their properties. At the sea surface, wind-driven waves and currents supply energy for mixing, and the tides create currents at all depths. The large eddies that may form at the boundaries of currents also stir together dissimilar waters, acting to homogenize them. When surface currents coverage, mixing at current boundaries may produce caballing of the mixed water. When currents and their associated turbulence are weak, mixing is reduced. Mixing by diffusion occurs continually at the molecular level, but diffusion is much weaker than mixing by turbulent processes.

If a parcel of water is displaced vertically by turbulence, buoyancy forces tend to return the parcel to its original density level. Therefore, vertical mixing between the water types that form the oceans' internal layers is weak. Horizontal mixing is more efficient because it requires less energy than vertical mixing. A parcel of water displaced horizontally along a surface of constant density remains at its new position and shares its properties with the surrounding water.

In areas under warm, high-salinity surface water with an appreciable salinity and temperature decrease with depth, internal vertical mixing processes occur despite the stability of the water column. Vertical columnar flows, approximately 3 cm (5 in) in diameter, are called salt fingers; they develop and mix the water vertically, causing a stair-step salinity and temperature change with depth. This phenomenon is caused by the ability of seawater to gain or lose heat faster by conduction than it gains or loses salt by diffusion. This causes the density of the vertically moving water to change relative to that of the surrounding water, and the moving water is propelled either up or down. Salt fingers mix water over limited depths, creating homogeneous layers 30 m (100 ft) thick. These layers exist from about 150–700 m (500–2300 ft) deep and are estimated to occur over large areas of the oceans when the required conditions are present.

QUICK REVIEW

1. Identify the water masses of the Atlantic Ocean. What is the origin of each? In which direction does each flow?
2. Why is the layering of the Pacific Ocean less dramatic than the layering of the Atlantic Ocean?
3. In what ways is the water of the Indian Ocean similar to the water of the South Atlantic Ocean?
4. What is the origin of Arctic Ocean deep water?

7.4 What Drives the Surface Currents?

When the winds blow over the oceans, they set the surface water in motion, driving the large-scale surface currents in nearly constant patterns. The density of water is about 1000 times greater than the density of air, and once in motion, the mass of the moving water is so great that its inertia keeps it flowing. The currents flow more in response to the average atmospheric circulation than to the daily weather and its short-term changes; however, the major currents do shift slightly in response to seasonal changes in the winds. The currents are further modified by interactions between the currents and along zones of converging and diverging water. The major surface currents have been called the rivers of the sea; they have no banks to contain them, but they maintain their average course.

Because the frictional coupling between the ocean water and Earth's surface is small, the moving water is deflected by the Coriolis effect in the same way that moving air is deflected (see chapter 6). But because water moves more slowly than air, it takes longer to move water the same distance as air. During this longer time period, Earth rotates farther out from under the water than from under the wind. Therefore, the slower-moving water appears to be deflected to a greater degree than the overlying air. The surface-current acted upon by the Coriolis effect is deflected to the right of the driving wind direction in the Northern Hemisphere and to the left in the Southern Hemisphere. In the open sea, the surface flow is deflected at a 45° angle from the wind direction, as shown in figure 7.14.

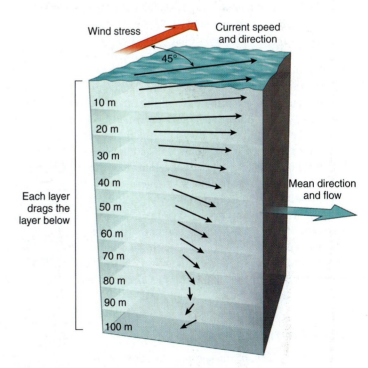

Figure 7.14 The Ekman spiral and Ekman transport. Water motion in the surface "Ekman Layer" is due to wind stress.

The Ekman Spiral and Ekman Transport

Wind-driven surface water sets the water immediately below it in motion. But because of low-friction coupling in the water, this next deeper layer moves more slowly than the surface layer and is deflected to the right (Northern Hemisphere) or left (Southern Hemisphere) of the surface-layer direction. The same is true for the next layer down and the next. The result is a spiral in which each deeper layer moves more slowly and with a greater angle of deflection to the surface flow (fig. 7.14). This current spiral is called the **Ekman spiral**, after the physicist V. Walfrid Ekman, who developed its mathematical relationship. The spiral extends to a depth of approximately 100–150 m (330–500 ft), where the much-reduced current will be moving in the opposite direction to the surface current. The surface layer of water corresponding to the Ekman spiral is known as the **Ekman layer**. Over the depth of the spiral, the average flow of the water set in motion by the wind, or the net flow, moves 90° to the right (Northern Hemisphere) or left (Southern Hemisphere) of the surface wind. This is known as **Ekman transport**. This relationship is in contrast to the surface water, which moves at an angle of 45° to the wind direction.

Ocean Gyres

The major surface currents in the ocean are driven primarily by the trade winds, blowing in a westerly direction toward the equator, and the westerlies, blowing in an easterly direction away from the equator (review figure 6.20). Surface water is driven at a 45° angle to the direction of these winds (fig. 7.15). Thus, the trade winds drive surface currents that flow from east to west on either side of the equator. When these currents reach the western boundaries of ocean basins, they are deflected away from the equator and move to higher latitudes where they enter the region of the westerlies. Once under the influence of the westerlies, the surface currents are then driven back across the ocean basins from west to east. When these eastward-flowing currents reach the eastern boundaries of the ocean basins, they are largely deflected back toward the equator where they again come under the influence of the trade winds. This completes a full cycle of surface currents that rotate clockwise in the Northern Hemisphere and counterclockwise in the Southern Hemisphere. These large, circular-motion, wind-driven current systems are known as **gyres** (fig. 7.16). In high southern latitudes, no land separates the Atlantic, Pacific, and Indian Oceans; here, the surface currents, driven by the westerlies, continue around Earth in a circumpolar flow around Antarctica.

Geostrophic Flow

Because of Ekman transport, a portion of the wind-driven surface water is deflected toward the center of each of the large, circular current gyres

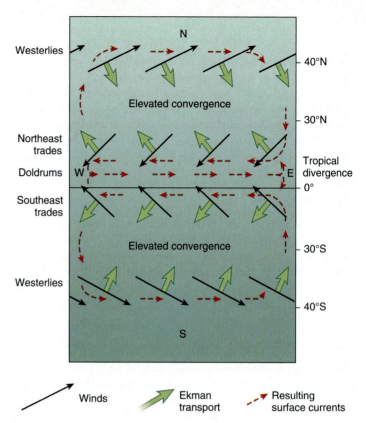

Figure 7.15 Wind-driven transport and resulting surface currents in an ocean bounded by land to the east and to the west. The currents form large oceanic gyres that rotate clockwise in the Northern Hemisphere and counterclockwise in the Southern Hemisphere.

just described (fig. 7.16). A convergent lens of surface water is elevated as much as 2 m (6.5 ft) above the equilibrium sea level, and this lens depresses the underlying denser water.

Figure 7.16 There are five major ocean gyres: (a) Indian Ocean, (b) North Pacific, (c) South Pacific, (d) North Atlantic, and (e) South Atlantic gyres. Each has a strong and narrow "western boundary current," and a weak and broad "eastern boundary current."

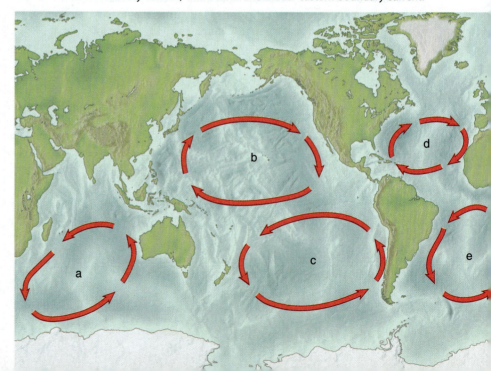

Figure 7.17 Geostrophic flow (*V*) exists around a gyre when *Fc*, the inward deflection force due to the Coriolis effect, is balanced by *Fg*, the outward-acting pressure force created by the elevated water and gravity. This example is of a clockwise gyre in the Northern Hemisphere.

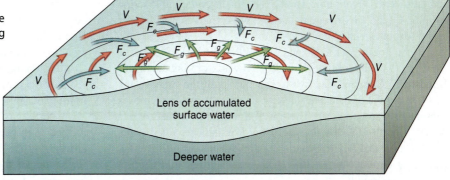

The thickness of the surface lens is about 1000 times greater than the elevation of the lens above sea level. This is because the difference in density between the surface water and the deeper water is only about 1/1000 of the density difference between air and water at the sea surface. The surface slope of the mound increases as deflected water moves inward until the outward pressure driving the water away from the gyre center equals the Coriolis effect, acting to deflect the moving water into the raised central mound. At this balance point, **geostrophic flow** is said to exist, and no further deflection of the moving water occurs. Instead, the currents flow smoothly around the gyre parallel to its elevation contours. See figure 7.17 for a diagram of this process.

QUICK REVIEW

1. If a north wind blows across the sea surface, what direction does the surface current flow in the Northern Hemisphere? In the Southern Hemisphere?

2. What is the direction of Ekman transport in question 1?

3. Why is the sea surface elevated in the interior of the major current gyres?

4. Why do gyres rotate in opposite directions in the two hemispheres?

7.5 Ocean Surface Currents

The currents that make up the large oceanic gyre systems and other major currents have been given names and descriptions based on their average positions. These are presented here ocean by ocean and can be followed on figure 7.18. As you follow these current paths, review their associations with the large gyre systems and their overlying wind belts.

Pacific Ocean Currents

In the North Pacific Ocean, the northeast trade winds push the water toward the west and northwest; this is the **North Equatorial Current.** The westerlies create the **North Pacific Current,** or **North Pacific Drift,** moving from west to east. Note that the trade winds move the water away from Central and South America and pile it up against Asia, whereas the westerlies move the water away from Asia and push it against the west coast of North America. The water that accumulates in one area must flow toward areas from which the water has been removed. This movement forms two currents: the **California Current,** moving from north to south along the western coast of North America, and the **Kuroshio Current,** moving from south to north along the east coast of Japan. The Kuroshio and California Currents are not wind-driven currents; they provide continuity of flow and complete a circular motion centered around 30°N latitude. This circular, clockwise flow of water is called the North Pacific gyre. Other major North Pacific currents include the **Oyashio Current,** driven by the polar easterlies, and the **Alaska Current,** fed by water from the North Pacific Current and moving in a counterclockwise gyre in the Gulf of Alaska. Little exchange of water occurs through the Bering Strait between the North Pacific

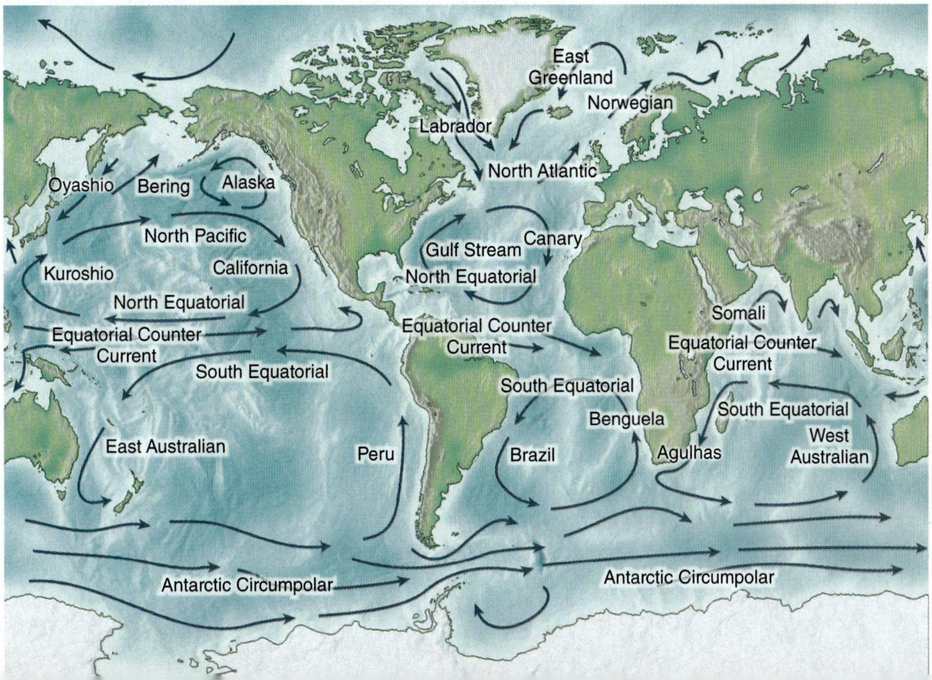

Figure 7.18 The long-term average flow of the major wind-driven surface currents.

and the Arctic Ocean; no current exists that is comparable to the Atlantic Ocean's Norwegian Current, which moves warm water to the Arctic Ocean.

In the South Pacific Ocean, the southeast trade winds move the water to the left of the wind and westward, forming the **South Equatorial Current.** The westerly winds push the water to the east. The current formed, the Antarctic Circumpolar Current, moves continuously around Earth. The tips of South America and Africa deflect a portion of this flow northward on the east sides of the South Pacific and South Atlantic Oceans. As in the North Pacific, continuity currents form between the South Equatorial Current and the Antarctic Circumpolar Current. The **Peru Current,** or **Humbolt Current,** flows from south to north along the coast of South America, while the **East Australia Current** can be seen moving weakly from north to south on the west side of the ocean. These four currents form the counterclockwise South Pacific gyre.

The North Pacific and South Pacific gyres form on either side of 5°N because the meteorological equator, or doldrums belt, is displaced northward from the geographic equator (0°), owing to the unequal heating of the Northern and Southern Hemispheres. Also between the North and South Equatorial Currents, in the zone of the doldrums is a current moving in the opposite direction, from west to east. This is a continuity current known as the **Equatorial Countercurrent,** which helps to return accumulated surface water eastward across the Pacific. Under the South Equatorial Current is a subsurface current flowing from west to east called the **Cromwell Current.** This cold-water continuity current also returns water accumulated in the western Pacific.

Atlantic Ocean Currents

The North Atlantic westerly winds move the water eastward as the **North Atlantic Current,** or **North Atlantic Drift.** The northeast trade winds push the water to the west, forming the **North Equatorial Current.** The north-south continuity currents are the **Gulf Stream,** flowing northward along the coast of North America, and the **Canary Current,** moving to the south on the eastern side of the North Atlantic. The Gulf Stream is fed by the **Florida Current** and the North Equatorial Current. The North Atlantic gyre rotates clockwise. The polar easterlies provide the driving force for the **Labrador** and **East Greenland Currents,** which balance water flowing into the Arctic Ocean from the **Norwegian Current.**

In the South Atlantic, the westerlies continue the West Wind Drift. The southeast trade winds move the water to the west, but the bulge of Brazil deflects part of the **South Equatorial Current** northward into the Caribbean Sea and eventually into the Gulf of Mexico, where it exits as the Florida Current and joins the Gulf Stream. A portion of the South Equatorial Current moves south of the Brazilian bulge along the western side of the South Atlantic to form the **Brazil Current.** The **Benguela Current** moves northward along the African coast. The South Atlantic gyre is complete, and it rotates counterclockwise.

Because much of the South Equatorial Current is deflected across the equator, the Equatorial Countercurrent appears only weakly in the eastern portion of the mid-Atlantic. The north-

ward movement of South Atlantic surface water across the equator results in a net flow of surface water from the Southern Hemisphere to the Northern Hemisphere. This flow is balanced by a flow of water at depth from the Northern Hemisphere to the Southern Hemisphere. This deep-water return flow is the North Atlantic deep water. Again, the equatorial currents are displaced northward, although not as markedly as in the Pacific Ocean.

The Sargasso Sea marks the middle of an ocean gyre. It is located in the central North Atlantic Ocean, and its boundaries are the Gulf Stream on the west, the North Atlantic Current to the north, the Canary Current on the east, and the North Equatorial Current to the south. The circular motion of the gyre currents isolates a lens of clear, warm, downwelling water 1000 m (3000 ft) deep. The region is famous for the floating mats of *Sargassum,* a brown seaweed, stretching across its surface. The extent of the floating seaweed frightened early sailors, who told stories of ships imprisoned by the weed and sea monsters lurking below the surface. Except for the floating *Sargassum,* with its rich and specialized ecological community, the clear water is nearly a biological desert.

Indian Ocean Currents

The Indian Ocean is mainly a Southern Hemisphere ocean. The southeast trade winds push the water to the west, creating the **South Equatorial Current.** The Southern Hemisphere westerlies still move the water eastward in the West Wind Drift. The gyre is completed by the **West Australia Current** moving northward and the **Agulhas Current** moving southward along the east coast of Africa. Because this is a Southern Hemisphere ocean, the currents are deflected left of the wind direction, and the gyre rotates counterclockwise. The northeast trade winds in winter drive the **North Equatorial Current** to the west, and the **Equatorial Countercurrent** returns water eastward toward Australia. Again, these equatorial currents are displaced approximately 5°N. With the coming of the wet monsoon season and its west winds, these currents are reduced. The strong seasonal monsoon effect controls the surface flow of the Northern Hemisphere portion of the Indian Ocean. In the summer, the winds blow the surface water eastward, and in the winter, they blow it westward (fig. 7.19). This strong seasonal shift is unlike anything found in the Atlantic or the Pacific Ocean.

Arctic Ocean Currents

The relentless drift of water and ice in the Arctic Ocean moves in a large clockwise gyre driven by the polar easterly winds. This gyre is centered not on the North Pole, as early explorers expected, but is offset over the Canadian basin at 150°W and 80°N (fig. 7.20). Although the currents and the winds move the ice slowly at 0.1 knot (2 mi/day), Arctic explorers trying to reach the North Pole found that they traveled south with the drifting ice and water at speeds almost equal to their difficult progress north.

The Arctic Ocean is supplied from the North Atlantic by the Norwegian Current; some of this flow enters west of

Northeast Monsoon (January)

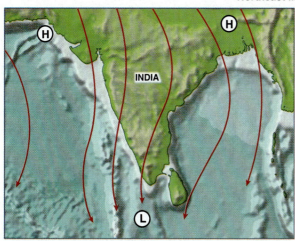

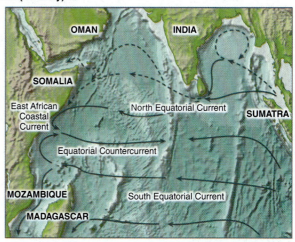

Southwest Monsoon (July)

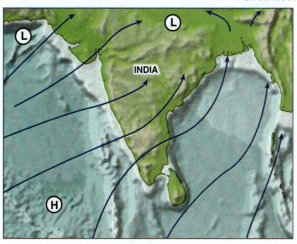

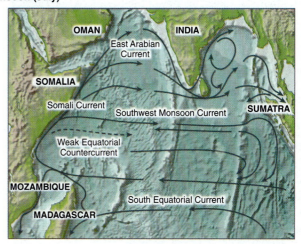

Figure 7.19 Indian Ocean monsoonal circulation. In the winter, high pressure over the continent creates dry monsoon winds roughly out of the northeast that drive water in the Northern Hemisphere portion of the ocean to the west. In the summer, low pressure over the continent creates wet monsoon winds roughly out of the southwest that reverse the surface currents in the North Hemisphere portion of the ocean.

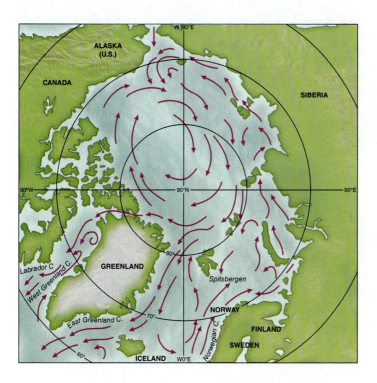

Spitsbergen, but most flows along the coast of Norway and moves eastward along the Siberian coast into the Chukchi Sea. A small inflow of water entering the Arctic through the Bering Strait brings water from the Bering Sea to join the eastward flow along Siberia and the large Arctic gyre. The western side of the gyre crosses the center of the Arctic Ocean to split north of Greenland. Here, the larger flow forms the East Greenland Current flowing south and taking Arctic Ocean water into the North Atlantic. The lesser flow moves along the west side of Greenland to join the Labrador Current and move south along the Canadian coast.

Outflow from Siberian rivers is caught in the eastward flow of water and ice along Siberia. Eventually, this discharge joins the gyre, distributing sediments and pollutants throughout the Arctic.

Figure 7.20 The circulation in the Arctic Ocean is driven by the polar easterlies, which produce a large, clockwise gyre. Water enters the Arctic Ocean from the North Atlantic by way of the Norwegian Current and exits to the Atlantic by the East Greenland Current and the Labrador Current.

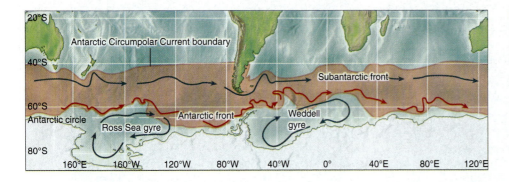

Figure 7.21 The Antarctic Circumpolar Current, also known as the West Wind Drift, provides exchange of water among the Pacific, Atlantic, and Indian Ocean basins.

Antarctic Currents

At high southern latitudes, the tips of South America and Africa deflect some of the eastward surface flow back toward the equator, completing the Southern Hemisphere gyres. However, much of this eastward-moving water continues its journey, circling Antarctica as the **Antarctic Circumpolar Current**, also known as the **West Wind Drift** (fig. 7.21). This current is unique because it is the only current that flows completely around the globe without interruption. It is also an important current because it provides a mechanism for sharing, or mixing, water among the Pacific, Atlantic, and Indian Ocean basins. In spite of its great length, about 24,000 km (15,000 mi), the Antarctic Circumpolar Current has very consistent characteristics. Generally, its path doesn't differ much from that shown in figure 7.21, but in the central and western Pacific, it can shift by more than 10° of latitude between summer and winter. The Antarctic Circumpolar Current extends all the way to the sea floor. It is the largest surface current on Earth. Two clockwise gyres are found in the Ross Sea and the Weddell Sea.

The Indonesian Throughflow

The western border of the South Pacific Ocean basin is not a solid landmass. Rather, it consists of a number of islands of varying size that comprise the Indonesian archipelago. When the Pacific Ocean's westward-moving Equatorial Currents reach the western side of the ocean basin, some of their water is deflected away from the equator as the Kuroshio and East Australia Currents. The remainder, known as the **Indonesian Throughflow**, continues westward as a complicated series of currents winding through the Indonesian archipelago. (fig. 7.22). Because the water in the western equatorial Pacific Ocean has a higher temperature and lower salinity than the water in the Indian Ocean, the Indonesian

Throughflow transports large amounts of relatively warm and fresh water to the Indian Ocean.

The volume of water moving in the Indonesian Throughflow can vary with changing conditions in the Pacific. In particular, the amount of flow of Pacific water through the Indonesian archipelago is less during an El Niño event.

QUICK REVIEW

1. Follow the major ocean gyres and identify the major currents in each.
2. Why are the equatorial countercurrents located under the doldrums?
3. Where does water enter and exit the Arctic Ocean?
4. Why is it possible for the Antarctic Circumpolar Current to flow all the way around Antarctica without interruption?

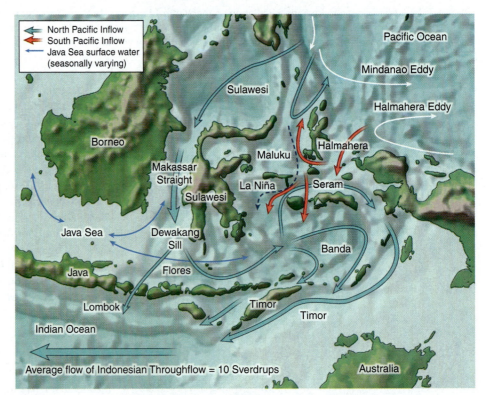

Figure 7.22 The complicated current system through the Indonesian archipelago that comprises the Indonesian Throughflow, carrying Pacific water into the Indian Ocean.

7.6 Current Characteristics

Current Speed

Wind-driven open-ocean surface currents move at speeds that are about 1/100 of the wind speed measured 10 m (30 ft) above the sea surface. The water moves between 0.25 and 1.0 knot, or 0.1–0.5 m (0.3–1.5 ft) per second. Currents flow faster when a large volume of water is forced to flow through a narrow gap. For example, the North and part of the South Atlantic Equatorial Currents flow into the Caribbean Sea, then into the Gulf of Mexico, and finally exit to the North Atlantic as the Florida Current through the narrow gap between Florida and Cuba. The Florida Current's speed may exceed 3 knots, or 1.5 m (5 ft) per second. Once into the Atlantic Ocean this current turns north and becomes the Gulf Stream.

The flow is distributed over the width and depth of the current. When the cross-sectional area of the current expands, the current slows down; when the cross-sectional area decreases, the current speeds up. Speed of flow may not be directly related to surface wind speed but can be affected by the depth and width of the current as determined by land barriers, by the presence of another current, or by the rotation of Earth, as explained in the "Western Intensification" section of this chapter.

Current Volume Transport

Major ocean currents transport enormous volumes of water. A convenient unit to report transport volume is the Sverdrup (Sv) (named after Harald Sverdrup, a leading oceanographer of the last century and former Director of the Scripps Institution of Oceanography). A Sverdrup equals 1 million cubic meters ($\sim3.5 \times 10^7$ ft³) per second. The transport rate of fresh water in all of the world's rivers into the ocean is about 1 Sv. Transport rates of ocean currents are difficult to measure accurately and can vary by both location in the current and time of the year. The Gulf Stream transports about 30 Sv passing through the Strait of Florida as the Florida Current. This increases steadily as it moves north along the coast until it transports about 80 Sv near Cape Hatteras. The transport of the Gulf Stream continues to increase downstream of Cape Hatteras at a rate of 8 Sv every 100 km, reaching a maximum transport of about 150 Sv at 55°W. The downstream increase in transport between Cape Hatteras and 55°W is thought to be caused by increased velocities in the deep waters of the Gulf Stream. The current transports a maximum amount of water in the fall and a minimum in the spring.

The largest current in the oceans, the Antarctic Circumpolar Current, transports an estimated 125 Sv of water.

Western Intensification

When we look at the shape of the actual mounds of water created by Ekman transport in the major subtropical gyres (review figs. 7.16 and 7.17), it is clear that the peaks of the mounds are not in the center of the gyres. The highest point of the mounds is displaced to the western side of the gyres (fig. 7.23). This creates a more gently sloping sea surface, over a longer distance, on the eastern side of the mound and a more steeply sloping surface,

over a shorter distance, on the western side (fig. 7.23a). The displacement of the mounds of water in the North Pacific and North Atlantic gyres is particularly easy to see in the satellite sea surface elevation data shown in figure 7.24. The geostrophic currents that flow around the mound have very different properties on the two sides. Currents flowing on the western side of these gyres, where the slope is steeper, tend to be much faster, deeper, and narrower. Currents flowing on the eastern side of these gyres, where the slope is gentler, tend to be slower, shallower, and broader. This phenomenon is known as the **western intensification** of currents. The Gulf Stream and Kuroshio Currents are faster and narrower than the Canary and California Currents, although both the eastern and western boundary currents transport about the same amount of water to preserve continuity of flow around the gyres. Western intensification of currents traveling from low to high latitudes is related to (1) the eastward turning of Earth, (2) the increase in the Coriolis effect with increasing latitude, (3) the changing strength and direction of the east-west wind field (trade winds and westerlies) with latitude, and (4) the friction between land masses and ocean water currents. These factors cause a compression of the currents on the western side of the oceans, where water is moving from lower to higher latitudes. This compression requires that the current speed increase to transport the water circulating about the gyre. On the eastern side of the gyre, where currents are moving from higher to lower latitudes, the currents are stretched in the east-west direction. Here, the current's speed is reduced, but it still transports the required volume of water. The changing speed of flow around the gyre causes the Coriolis effect to vary. Where the current speed is high, the Coriolis effect is large, and a steeper surface slope is required to create a geostrophic flow balance.

Fast-flowing, western-boundary currents move warm equatorial surface water to higher latitudes. Both the Gulf Stream and the Kuroshio Current bring heat from equatorial latitudes to moderate the climates of Japan and northern Asia (in the case of the Kuroshio) and the British Isles and northern Europe (in the case of the Gulf Stream, via the North Atlantic and Norwegian Currents). Western intensification is obscured in the South Pacific and South Atlantic, because both Africa and South America deflect portions of the West Wind Drift and create strong currents on the eastern side of these oceans. The deflection of water from the Atlantic's South Equatorial Current to the Northern Hemisphere removes water from the South Atlantic gyre and strengthens the Gulf Stream. The flow of surface water from the Pacific to the Indian Ocean through the islands of Indonesia also helps to prevent the development of strongly flowing currents on the west side of Southern Hemisphere oceans.

QUICK REVIEW

1. Explain in general terms differences in the characteristics of the Kuroshio and California Currents.
2. What forces balance to create geostrophic currents?
3. What do surface currents transport?
4. How does changing the width of a current affect its speed?
5. What is western intensification?

Northern Hemisphere Subtropical Gyre

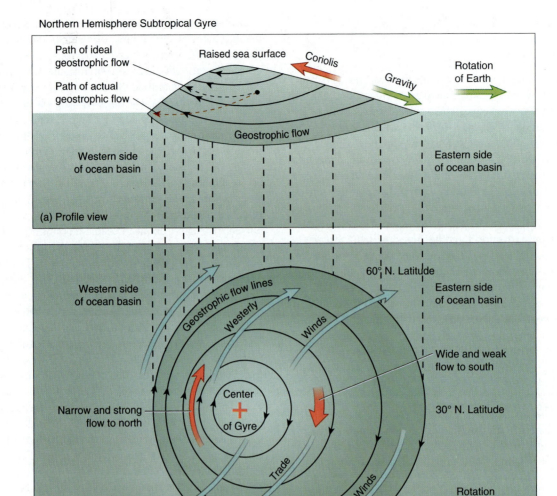

(a) Profile view

(b) Map view

Figure 7.23 (a) A cross-sectional view of a subtropical gyre illustrating the accumulation of water in the center, elevating the sea surface by as much as 2 m (6.6 ft). The peak of the mound is displaced to the west. Gravity and the Coriolis effect balance to create geostrophic flow around the hill. Friction makes the current gradually spiral downslope. (b) A map view of the same gyre showing the concentration of the currents on the western side of the gyre compared to the currents on the eastern side, creating the phenomenon called western intensification.

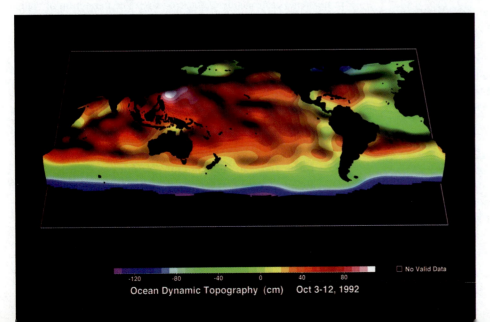

Figure 7.24 Variations in the average height of the sea surface are shown in this image constructed from data taken in 1992 by the *TOPEX/Poseidon* satellite. Red indicates the highest elevations; green and blue the lowest. Note that the highest elevations within the subtropical gyres are found on the western sides of the ocean basins. These are particularly easy to see in the western Pacific and western Atlantic in the Northern Hemisphere. Source of Data: National Oceanic and Atmospheric Administration (NOAA).

7.7 Eddies

When a narrow, fast-moving current moves into or through slower-moving water, the force of its flow displaces the quieter water and captures additional water as it does so. The current oscillates and develops waves along its boundary that are known as meanders. These meanders break off to form **eddies,** or pockets of water moving with a circular motion; eddies take with them energy of motion from the main flow and gradually dissipate this energy through friction. Eddies also act to mix and blend water.

As the Gulf Stream moves away from the North American coast, it is likely to develop a meandering path. The western edge of the Gulf Stream develops oscillations, and the indentations are filled by cold water from the Labrador Current side. When these indentations pinch off, they become counterclockwise-rotating, cold-water eddies that are displaced eastward through the Gulf Stream and into the warm-water core of the gyre. Bulges at the western edge of the Gulf Stream are filled with warm Gulf Stream water. When these bulges are cut off, they become warm-water, clockwise-rotating eddies drifting into cold water to the west and north of the Gulf Stream. Follow these processes in figures 7.25 and 7.26. These eddies may maintain their physical identity for weeks as they wander about the oceans; they are especially numerous in the area north of the Sargasso Sea. Current meanders and eddy formation produce surface flow patterns that differ markedly from the uniform current flows shown on current charts. These charts show average current flow, not daily or weekly variations.

Large and small eddies generated by horizontal flows or currents exist in all parts of the oceans; these eddies are of varying sizes, ranging from tens to several hundred kilometers in diameter. Each eddy contains water with specific chemical and physical properties and maintains its identity and rotational inertia as it wanders through the oceans. Eddies may appear at the sea surface or be embedded in waters at any depth.

Eddies rotate in a clockwise or counterclockwise direction. They stir the ocean until they gradually dissipate because of fluid friction, losing their chemical and thermal identity and their energy of motion. By testing the water properties of an eddy, oceanographers are able to determine the eddy's place of origin. Small surface eddies encountered 800 km (500 mi) southeast of Cape Hatteras in the North Atlantic have been found with water properties of the eastern Atlantic near Gibraltar, more than 4000 km (2500 mi) away. Eddies from the Strait of Gibraltar are formed in the salty water of the Mediterranean as it sinks and spreads out into the Atlantic 500–1000 m (1600–3300 ft) down; these eddies have been nicknamed "Meddies." Deep-water eddies near Cape Hatteras may come from the eastern and western Atlantic, the Caribbean, or Iceland. Researchers estimate that some of these eddies are several years old; age determination is based on drift rates, distance from source, and biological consumption of oxygen.

The rotational water speed in the large eddies that form at the western boundary of the Gulf Stream is about 0.51 m/s (1 knot), but because of the water's density, the force of the flow is

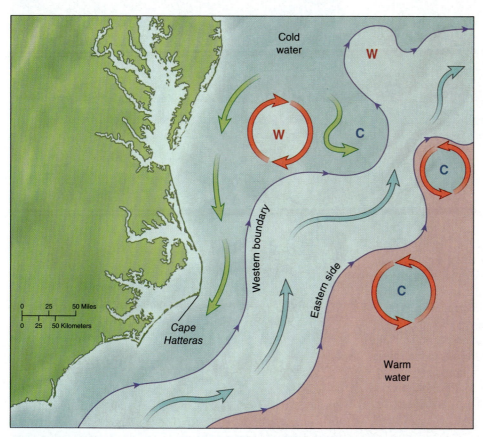

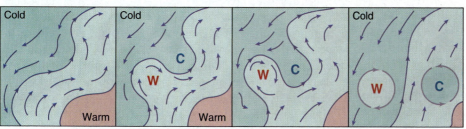

Figure 7.25 The western boundary of the Gulf Stream is defined by sharp changes in current velocity and direction. Meanders form at this boundary after the Gulf Stream leaves the U.S. coast at Cape Hatteras. The amplitude of the meanders increases as they move downstream (a and b). In time, the current flow pinches off the meander (c). The current boundary re-forms, and isolated rotating cells of warm water (*W*) wander into the cold water, whereas cells of cold water (*C*) drift through the Gulf Stream into the warm water (d).

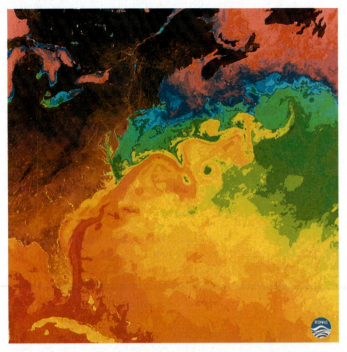

Figure 7.26 A composite satellite image of the sea surface reveals the warm (*orange* and *yellow*) and cold (*green* and *blue*) eddies that form along the Gulf Stream. (*Reddish blue* areas at the top are the coldest waters.) These eddies may stir the water column right down to the ocean floor, kicking up blizzards of sediment.

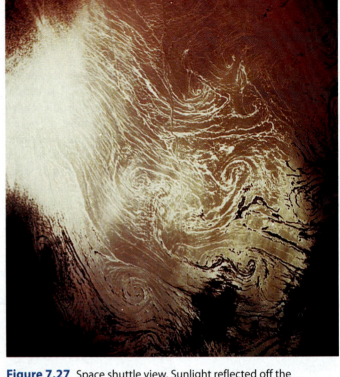

Figure 7.27 Space shuttle view. Sunlight reflected off the Mediterranean reveals spiral eddies; their effects on climate are being monitored. Dimensions of this image are 500 km × 500 km (310 mi × 310 mi).

similar to that generated by a 35-knot wind. The diameter of the eddies may be as much as 325 km (200 mi), and their effect may reach to the sea floor. At the sea floor, the rotation rate is zero; therefore, a few meters above the bottom the speed of rotation diminishes very rapidly, and considerable turbulence is generated as the energy of the eddy is dissipated. These eddies are similar in many ways to the winds rotating about atmospheric pressure cells, and they are sometimes called abyssal storms. As the eddies wander through the oceans, they stir up bottom sediments, producing ripples and sand waves in their wakes; they also mix the water, creating homogeneous water properties over large areas. Eventually, the eddies lose their energy to turbulence and blend into the surrounding water.

Eddies constantly form, migrate, and dissipate at all depths. Eddy motion is superimposed on the mean flow of the oceans. To understand the role of eddies in mixing the oceans, we need more data and better tracking of eddy size, position, and rate of dissipation. Satellites are important tools for detecting surface eddies because they can precisely measure temperature, increased elevation, and light reflection of the sea surface (fig. 7.27). Deep-water eddies are monitored by using instruments designed to float at a mid-depth density layer. The instruments are caught up in the eddies, moving with them and sending out acoustic signals that are monitored through the sofar channel. In this way, the rate of deep-water eddy formation, the numbers of major eddies, their movements, and their life spans can be observed.

QUICK REVIEW

1. What are eddies and where do they form?
2. How are parcels of warm water transferred across the Gulf Stream and embedded in regions of cold water?
3. How can we monitor the movement of deep-water eddies?

7.8 Convergence and Divergence

Changes in density and the accompanying concepts of upwelling, downwelling, convergence, divergence, and continuity of flow were introduced in section 7.2. The convergence and divergence zones discussed here are the product of wind-driven surface currents that produce large-scale areas of convergence and divergence at the sea surface. For example, convergence zones are at the centers of the large oceanic gyres, and when wind-driven surface currents collide or are forced against landmasses, they produce convergences. When surface currents move away from each other or away from a landmass, they produce a surface divergence. Upwellings and downwellings of this type are nearly permanent but do react to seasonal changes in Earth's surface winds.

Langmuir Cells

A strong wind blowing across the sea surface often causes streaks of foam and surface debris that are seen trailing off in the direction the wind is blowing. These streaks are called

The vertical extent of these Langmuir cells is 4–10 m (12–30 ft). The distance between rows becomes larger at higher wind speeds. Langmuir cells are not long-lasting but help to mix the surface waters and organize the distribution of suspended organic matter in convergence and divergence zones. Sinking particles and organisms tend to congregate at depth in the regions of rising currents, while floating particles and organisms accumulate at the surface where currents converge and descend.

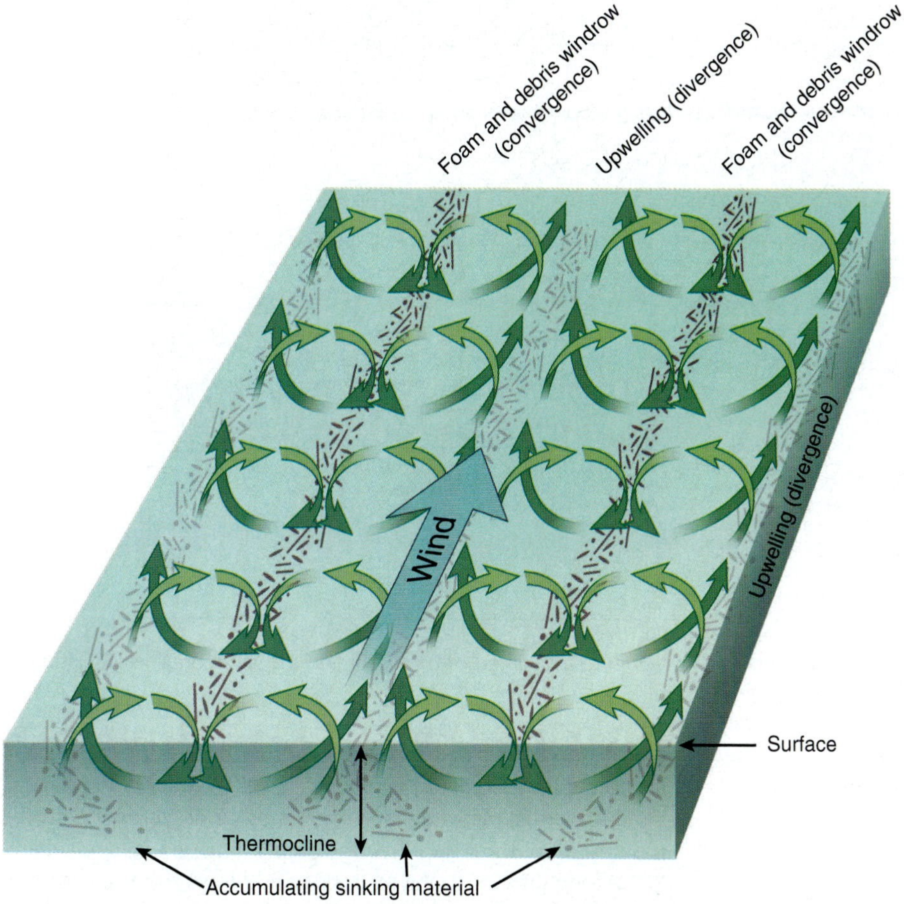

Permanent Zones

Areas of surface-water convergence and divergence occur in association with high- and low-pressure systems over the ocean. In the Northern Hemisphere, a low-pressure system produces cyclonic winds that rotate counterclockwise. These winds result in Ekman transport of water away from the center of the system, creating an area of surface divergence. This surface divergence produces upwelling of deep water (fig. 7.30a). High-pressure systems in the Northern Hemisphere produce anticyclonic winds that rotate clockwise. These winds drive Ekman transport of water toward the center of the system, creating an area of surface convergence. This surface convergence produces downwelling of surface water (fig. 7.30b).

There are also zones of permanent convergence and divergence, and associated downwelling and upwelling, related to the large ocean gyres and the Ekman transport occurring in them (figs. 7.13 and 7.31). There are five major zones of convergence: the **tropical convergence** at the equator and the two **subtropical convergences** at approximately 30°–40°N and S. These convergences mark the centers of the large ocean gyres. The **Arctic** and **Antarctic convergences** are found at about 50°N and S. Surface convergence zones are regions of downwelling. These areas are low in nutrients and biological productivity. There are three major divergence zones: the two **tropical divergences** and the **Antarctic divergence.** Upwelling associated with divergences delivers nutrients to the surface waters to supply the food chains that support the anchovy and tropical tuna fisheries and the richly productive waters of Antarctica.

Figure 7.32 illustrates global patterns of wind-induced upwelling and downwelling. The equatorial upwelling is particularly easy to see in this image as well as slower rates of upwelling around Antarctica and along the eastern coasts of continents.

In coastal areas where trade winds move the surface waters away from the western side of continents, upwelling occurs nearly continuously throughout the

Figure 7.28 Windrows in the water of Rodeo Lagoon, San Francisco, CA, created by Langmuir circulation.

windrows and may be 100 m (330 ft) in length. Windrows mark the convergence zones of shallow circulation cells known as **Langmuir cells** (fig. 7.28). These cells are composed of paired right- and left-handed helixes (fig. 7.29). The spacing between the windrows varies between 5 and 50 m (16 and 160 ft), and the rows are closer together if the thermocline depth is shallow.

Foam and debris windrow (convergence)

Upwelling (divergence)

Foam and debris windrow (convergence)

Wind

Upwelling (divergence)

Surface

Thermocline

Accumulating sinking material

Figure 7.29 Langmuir cells are helixes of near-surface, wind-driven water motion. The cells extend downwind to form windrows of surface debris along convergences. At the same time, sinking materials are swept into zones of upwelling water.

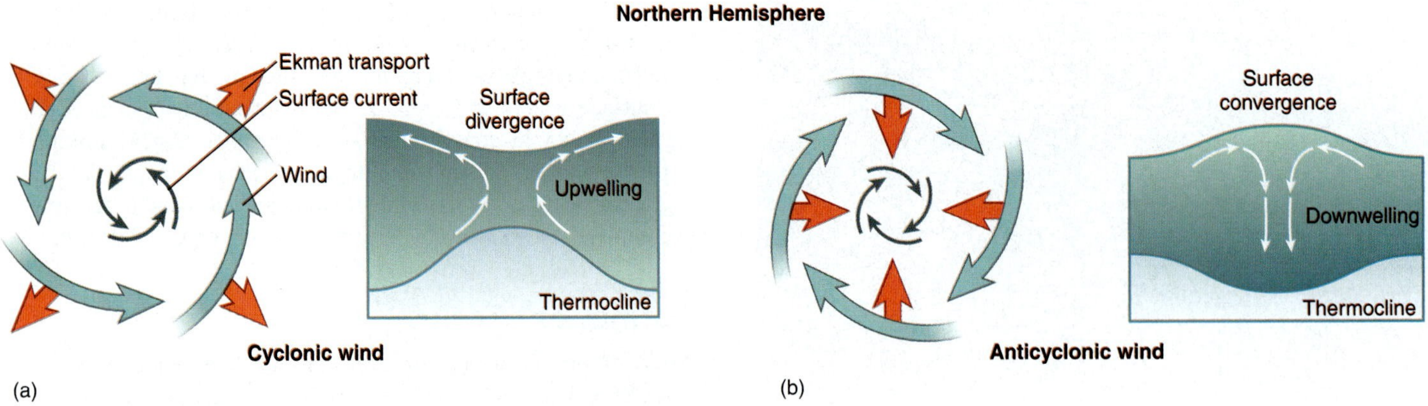

Northern Hemisphere

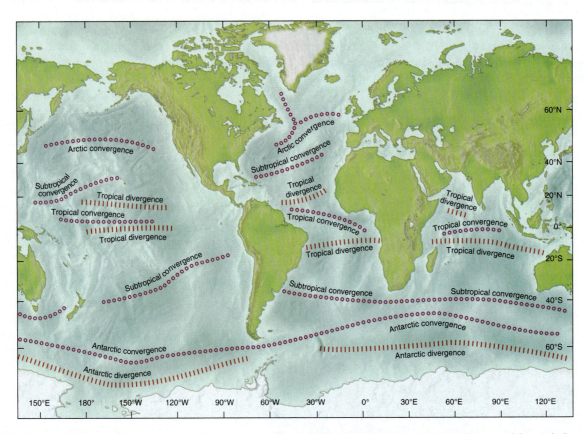

(a)   (b)

Figure 7.30 Surface convergence and surface divergence occur in association with low- and high-pressure systems. (a) In the Northern Hemisphere, cyclonic winds (low-pressure systems) result in Ekman transport of water away from the center of the low pressure, producing surface divergence and upwelling. (b) Anticyclonic winds (high-pressure systems) result in Ekman transport of water toward the center of the high pressure, producing surface convergence and downwelling.

Figure 7.31 The principal zones of open-ocean surface convergence and divergence associated with wind-driven and thermohaline circulation.

year. For example, the trade winds drive upwellings off the west coasts of Africa and South America that are very productive and yield large fish catches.

Seasonal Zones

Off the west coast of North America, downwelling and upwelling occur seasonally as the Northern Hemisphere temperate wind pattern changes from southerly in winter to northerly in summer.

The downwelling and upwelling occur because of the change in direction of Ekman transport (fig. 7.33). Remember that the wind-driven Ekman transport moves at an angle of 90° to the right or left of the wind direction, depending on the hemisphere.

Along this coast, the average wind blows from the north in the summer, and the net movement of the water is to the west, or 90° to the right of the wind. This action results in the offshore transport of the surface water and the upwelling of deeper water along the coast to replace it. The upwelling zone is evident from

July Global Wind-induced Upwelling (cm/day)

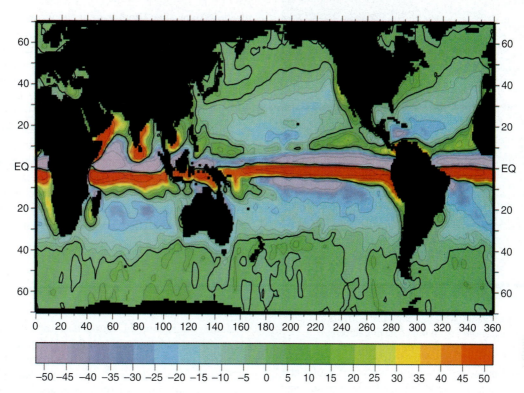

Figure 7.32 Equatorial upwelling occurs in all of the ocean basins but is particularly well developed in the Pacific. Positive values indicate upwelling associated with surface divergence. Negative values indicate downwelling associated with surface divergence.

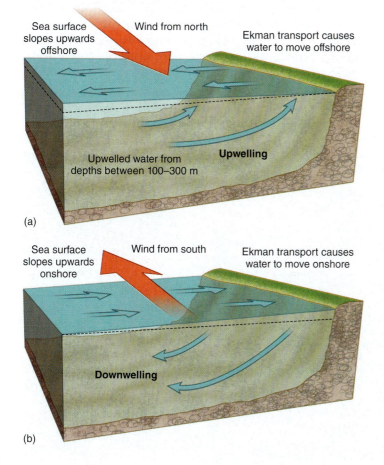

central California to Vancouver Island (fig. 7.34). In winter, the winds blow from the south; the wind-driven surface waters move to the east, onshore against the coast; and downwelling occurs. The summer upwelling pattern is what produces the band of cold coastal water at San Francisco that helps cause the frequent summer fogs. Because of the lack of land at the middle latitudes in the Southern Hemisphere, this type of seasonal upwelling is less common.

Convergence and divergence of surface currents result not only in upwellings and downwellings but also in the mixing of water from different geographic areas. Waters carried by the surface currents converge, share properties due to mixing, and form new water mixtures with specific ranges of temperature and salinity, which then sink to their appropriate density levels. After sinking, the water moves horizontally, blending and sharing its properties with adjacent water, and eventually, it rises to the surface at a new location. This process, known as caballing, is discussed as thermohaline circulation in section 7.2. Thermohaline circulation and wind-driven surface currents are closely

Figure 7.33 Winds blowing along the coastline can generate Ekman transport of water toward, or away from, the coast. Along the northwest coast of North America (a) northerly summer winds transport water away from the coast, depressing the sea surface and producing upwelling, and (b) southerly winter winds transport water toward the coast, elevating the sea surface and producing downwelling.

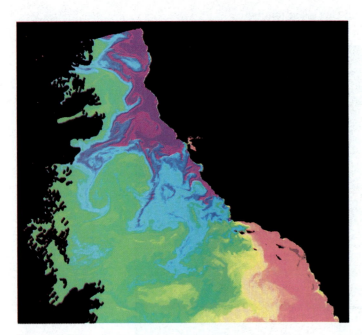

Figure 7.34 Coastal upwelling (shades of *purple*) appears as patches of cold water extending from the coast of California.

related. The connections are so close among formation of water mixtures, upwelling and downwelling, and convergence and divergence that it is difficult to assign a priority of importance to one process over another. Changes in these flows and events that might cause such changes are discussed in the next section.

7.9 The Great Ocean Conveyor Belt

It was first suggested in the 1980s that the movement of heat around the globe in ocean currents and water masses could be modeled as a thermohaline conveyor belt (fig. 7.35). This model is not meant to be a literal map of actual warm and cold currents—it is a representation of the overall effect of the mass movement of warm and cold water on the vertical circulation in the ocean. We can think of the overall driving mechanism of this conveyor belt as the formation and sinking of dense water at polar latitudes. The formation of deep-water masses depends on the production of dense surface water, either by significant cooling and/or an increase in salinity. In the Northern Hemisphere, deep water is created in the North Atlantic; this is the North Atlantic Deep Water (NADW) discussed earlier in the chapter. NADW is dense because it is both cold and has a relatively high salinity. NADW sinks and moves slowly southward, extending all the

way to the southern tip of Africa where it joins the deep water circling Antarctica. From there it can move into the Atlantic and Indian Ocean basins. Deep water is also created in the Southern Hemisphere around Antarctica; this is the Antarctic Bottom Water (AABW) discussed earlier in the chapter. AABW is the densest water in the oceans. Deep water eventually wells up in the centers of the ocean basins and moves toward regions of dense, sinking surface water. This transport of water through the entire conveyor belt can take as long as 1000 years or more. As water is transported, so is heat. Tropical water, moving near the surface toward the poles, carries heat with it. As this water moves to higher latitudes it releases its heat to the environment. Seawater temperature beneath the thermocline is uniformly cold everywhere in the oceans because all deep water has its origin at polar latitudes where low environmental temperatures contribute to high seawater density.

7.10 Changing Circulation Patterns

North Pacific Oscillations

W. James Ingraham, Jr., an oceanographer at NOAA's Seattle laboratory, developed a North Pacific Current model, or North Pacific Ocean Surface Current Simulation. When the model was used to map current patterns for North Pacific surface water between 1902 and 1997, it showed a north-south current oscillation associated with changes in atmospheric pressure and climate shifts. Cold and wet conditions are associated with a southerly current flow, and warm and dry conditions predominate with a northerly flow (fig. 7.36). This evidence was compared to climate-sensitive tree-ring data from western juniper trees in eastern Oregon. These trees show wide rings during periods when currents were displaced to the north and narrow rings when currents were displaced to the south.

Figure 7.35 The Great Ocean Conveyor Belt of thermohaline circulation within and between oceans circulates at the surface (*red arrows*) and at depth (*blue arrows*). *Dark-blue dots* mark regions of deep-water formation.

surface flow
deep flow

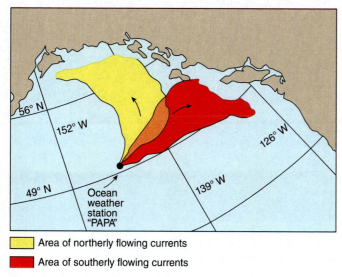

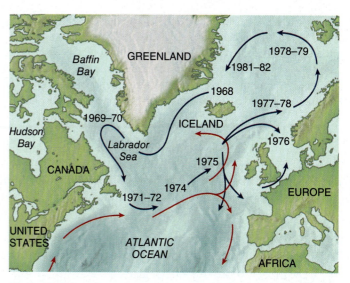

Figure 7.36 North-south current shifts in the Northeast Pacific are associated with changes in atmospheric pressure, winds, precipitation, and water temperature. This phenomenon is known as the Pacific Decadal Oscillation (PDO).

Figure 7.37 *Purple arrows* follow the path of the North Atlantic cool pool. *Red arrows* show warm-water flow from the Gulf Stream.

Tree-ring data cover many more years than oceanographic and meteorological data, and from the tree rings, it is calculated that thirty-four north-south oscillations have occurred since the time of Columbus. The most common time period between fast and slow growth is seventeen years, with twenty-three- and twenty-six-year periods common. Tree-ring data and current oscillations agree.

The climate pattern is presently warm and dry with a northerly current flow that has not changed since 1967. This is one of the longest periods without a reversal that has been found during the last 500 years. Following the wet conditions of the 1998–99 winter and the 1999–2000 La Niña, the winter of 2000–2001 in the Pacific Northwest was exceptionally warm and dry, and the winter of 2002 again brought wet El Niño conditions. These current shifts are associated with changes in atmospheric pressure, winds, precipitation, and water temperature; together, they are known as the Pacific Decadal Oscillation, or PDO. Because the PDO affects coastal surface temperatures from California to Alaska, it is thought to affect the survival of fish stocks of the area, especially salmon.

North Atlantic Oscillations

A large pool of cold, low-salinity surface water appeared off Greenland, north of Iceland, in 1968. It was about 0.5‰ less salty and 1° and 2°C colder than usual. Within two years, this cool pool had moved west into the Labrador Sea off eastern Canada; then it crossed the Atlantic and, in the mid-1970s, moved north into the Norwegian Sea. It had returned to its place of origin by the early 1980s. Follow the path of this pool in figure 7.37. During this period, harsh winters plagued Europe, and the entire Northern Hemisphere had cooler-than-average temperatures for more than ten years.

Recent studies have further investigated the North Atlantic, seeking to improve our ability to predict climate. The new analysis shows pools of warm and cool surface water that circle the North Atlantic; the pools seem to have life spans of four to ten years. Because of the alternation of climatic conditions, this is referred to as the North Atlantic Oscillation (NAO). Investigators are finding pieces of the puzzle that are associated with NAO but have not yet found the factors that control the system. A counterclockwise wind circulation centered over Iceland and a high-pressure clockwise circulation residing near the Azores are the usual situation. If the air-pressure differences between these two locations are large, then strong westerly winds supply Europe with heat from the North Atlantic Current. If the air-pressure difference decreases, then weaker-than-normal westerlies drive less warm water into the Norwegian Current and less heat is delivered to Europe. The periods of 1950–71 and 1976–80 were recognized as prolonged cool periods for Europe. The winds may also drive cold, low-salinity water from the Arctic into the area where North Atlantic deep water is formed. The influx of this water may cause a small-scale reduction in the formation of North Atlantic deep water and an accompanying reduction of warm surface water moving northward. How much freshening is required to change ocean circulation? This question is unanswered, but if the current melting of Arctic sea ice continues, more information may soon be available.

Natural cycles operate at a variety of time scales and magnitudes, and they interact and react in ways we do not fully comprehend. In the enormously complex interactions that they are observing, oceanographers and meteorologists are beginning to identify the strands that united the ocean-atmosphere and the world current systems. Scientists use mathematical equations assembled into a model to describe interactions between the oceans and the atmosphere. Supercomputers are needed to make the millions of calculations that a model requires to predict changes in environmental conditions over a given time period.

This process is repeated many times to predict conditions. Models are verified and tuned by adjusting the equations so that the predicted model changes closely agree with changes that have been observed in the past. Modeling is nearly as much an art as a science. Sometimes predictions work; sometimes they don't. It depends on how well the model has been conceived, how much data are available, and the capability of the computers to process the data. The model may be constructed to approximate the whole ocean-atmosphere system or to apply only to part of Earth. Eventually, long-term predictions concerning the oceans and the atmosphere will become possible; how we use such information and whether it will benefit ocean resources are not yet clear.

QUICK REVIEW

1. Describe the Pacific Decadal Oscillation and its causes.
2. Describe the North Atlantic Oscillation.
3. What effect does the North Atlantic Oscillation have on Europe's climate?

7.11 Measuring the Currents

Direct measurements of currents fall into two groups: (1) those that follow a parcel of the moving water and (2) those that measure the speed and direction of the water as it passes a fixed point. Moving waters may be followed with buoys designed to float at predetermined depths. These buoys signal their positions acoustically to a research vessel or shore station; their paths are followed and their speed and displacement due to the current are calculated. Autonomous profilers also measure currents. Surface water may be labeled with buoys or with dye that can be photographed from the air. Buoy positions may also be tracked by satellites using GPS. A series of pictures or position fixes may be used to calculate the speed and direction of a current from the buoys' drift rates. Buoys can be instrumented to measure other water properties such as temperature and salinity (fig. 7.38).

Figure 7.38 Surface buoys (*red*) carrying instruments are deployed from the research vessel *Thomas G. Thompson.*

A variation of this technique uses **drift bottles.** Thousands of sealed bottles, each containing a postcard, are released at a known position. When the bottles are washed ashore, the finders are requested to record the time and location of the find and return the card. In this case, only the release point, the recovery points, and the elapsed time are known; the actual path of motion is assumed. See the Diving In box titled "Ocean Drifters."

Sensors used to measure current speed and direction at fixed locations are called **current meters.** The current meters used by oceanographers over the last twenty years include a rotor to measure speed and a vane to measure direction of flow (fig. 7.39*a*). If the current meter is lowered from a stationary vessel, the measurements can be returned to the ship by a cable or stored by the meter for reading upon retrieval. If the current meter is attached to an independent, bottom-moored buoy system, the signals can be transmitted to the ship as radio signals or stored on tape in the meter, to be removed when the buoy and meter are retrieved.

Figure 7.39 (a) An internally recording Aanderaa current meter. The vane orients the meter to the current while the rotor determines current speed. (b) A Doppler current meter sends out sound pulses in four directions. The frequency shift of the returning echoes allows the detection of the current. These meters are also equipped with salinity-temperature-depth sensors as well as instruments for measuring water turbidity and oxygen. The data may be stored internally and collected at another time.

(a)   (b)

Diving in

Ocean Drifters

In May 1990, a severe storm in the North Pacific caused the Korean container ship. *Hansa Carrier*, to lose overboard twenty-one deck-cargo containers, each approximately 40 feet long. Among the items lost were 39,466 pairs of NIKE brand athletic shoes on their way to the United States. Six months to a year later these shoes began washing up along the beaches of Washington, Oregon, and British Columbia (box fig. 1). They were wearable after washing and having the barnacles and the oil removed, but the two shoes of a pair had not been tied together for shipping, and pairs did not come ashore together. As beach residents recovered the shoes (some with a retail value of $100 a pair), swap meets were held in coastal communities to match the pairs.

In May of 1991, oceanographer Curtis Ebbesmeyer of Seattle, Washington, read a news article on the beached NIKEs. He was intrigued and realized that 78,932 shoes was a very large number of drifting objects compared to the 33,869 drift bottles used in a 1956–59 study of North Pacific currents. He contacted Steve McLeod, an Oregon artist and shoe collector, who had information on locations and dates for some 1600 shoes that had been found between northern California and the Queen Charlotte Islands in British Columbia. Additional beachcombers were asked for information, and Ebbesmeyer mapped the times and locations where batches of 100 or more shoes had been found (box fig. 2). Next Ebbesmeyer visited Jim Ingraham at NOAA's National Marine Fisheries Service's offices in Seattle to study his computer model of Pacific Ocean currents and wind systems north of 30°N latitude. Using the spill date (May 27, 1990), the spill location (161°W, 48°N), and the dates of the first shoe landings on Vancouver Island and Washington State beaches between Thanksgiving and Christmas 1990, Ebbesmeyer found that the shoe drift rates agreed with the computer model's predicted currents.

News of Ebbesmeyer and Ingraham's interest in the shoe spill reached an Oregon news reporter and was then circulated nationally. Readers sent letters describing their own shoe finds. Reports of single shoes were valuable, because each shoe had within it a NIKE purchase order number that could be traced to a specific cargo container. Ebbesmeyer was able to determine from these numbers that only four of the five shoe containers broke open, so that only 61,820 shoes were set afloat.

The computer model and previous experiments with satellite-tracked drifters showed that there would have been little scattering of the shoes as the ocean currents carried them eastward and approximately 1500 miles from the spill site to shore, but the shoes were found scattered from California to northern British Columbia. The north-south scattering is related to coastal currents that flow northward in winter, carrying the shoes to the Queen Charlotte Islands, and southward in spring and summer, bringing the shoes to Oregon and California. In the spring of 1992, three sneakers from one of the containers were found at Pololu, at the north end of the island of Hawaii, indicating that they were making their way across the Pacific.

Because it takes about four and a half years for an object to drift completely around the North Pacific Current gyre, the shoes were expected to arrive on Japanese beaches during 1994–95, but no shoes have been reported on Japanese beaches.

Box Figure 1 Shoes and toys after their rescues from Pacific cargo spills.

Other Scenarios

Ebbesmeyer was interested in seeing where the shoes might have gone if they had been lost on the same date but under different conditions in other years. The computer allowed simulations for May 27 of each year from 1946 to 1991. Box figure 3 shows the wide variation in model-predicted drift routes. If the shoes had been lost in 1951, they would have traveled in the loop of the Alaska Current. If they had been lost in 1982, they would have been carried far to the north during the very strong El Niño of 1982–83, and if lost in 1973, they would have come ashore at the Columbia River.

On December 15, 2002, a ship carrying cargo containers between Los Angeles and Tacoma, Washington, ran into 25-foot seas off Cape Mendocino in northern California. Several containers fell overboard, one containing 33,000 NIKE athletic shoes. By mid-January 2003, the northward-flowing Davidson Current had brought the shoes 833 km (450 nautical miles) to the Washington coast. Once again, the shoes were not tied together but notices were posted and more swap meets occurred.

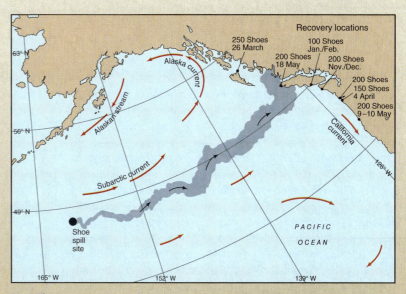

Box Figure 2 Site where 80,000 Nike shoes washed overboard on May 27, 1990, and dates and locations where 1300 shoes were discovered by beachcombers (dots at upper right). Drift of the shoes is simulated with a computer model (colored plume).

Continued next page—

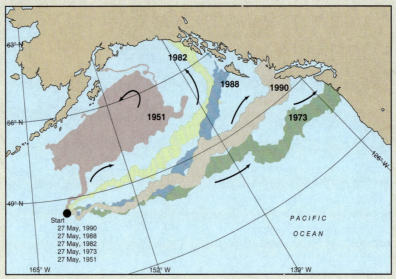

Box Figure 3 Projected drift tracks for the sneakers for the years 1951, 1973, 1982, 1988, and 1990 based on computer modeling of ocean currents and weather.

Tub Toys and Hockey Gloves Come Ashore

A similar situation occurred in January 1992 when twelve cargo containers were lost from another vessel in the North Pacific at 180°W, 45°N. One of these containers held 29,000 small, floatable, bathtub toys. Plastic blue turtles, yellow ducks, red beavers, and green frogs began arriving on beaches near Sitka, Alaska in November 1992. Advertisements asking for news of toy strandings were placed in local newspapers and the Canadian lighthouse-keepers newsletter. Beachcombers reported a total of about 400 of these toys.

None of the toys were found south of 60°N latitude, suggesting that once the toys reached the vicinity of Sitka, they drifted to the north. This time the computer model showed that if the toys had continued to float with the Alaska Current, they would have moved with the Alaska stream on through the Aleutian Islands into the Bering Sea. Some could have continued through the Bering Sea into the Arctic Ocean to the vicinity of Point Barrow, and from there drifted north of Siberia with the Arctic pack ice. Eventually some plastic turtles, ducks, beavers, or frogs may have come to rest on the coast of western Europe. If the toys turned south, they merged with the Kuroshio Current and were carried past the location where they were spilled.

After a December 1994 fire, 2500 cases of hockey gloves (34,300 gloves) were lost from a container ship in the North Pacific. In August 1995, a fishing vessel found seven gloves 800 miles west of the Oregon coast, and by January 1996, the barnacle-covered gloves began to arrive on Washington State beaches. The most northerly glove sighting came from Prince William Sound, Alaska in August 1996. The gloves were expected to follow the tub toys along the coast of Alaska and into the Arctic.

To measure a current at a location, a current meter must not move. Although a vessel can be moored in shallow water so that it does not move, it is very difficult, if not impossible, to moor a ship or surface platform in the open sea so that it will not move and thus move the current meter. The solution is to attach the current meter to a buoy system that is entirely submerged and not affected by winds or waves. See figure 7.40 for a diagram of this taut-wire moorage system. The string of anchors, current meters, wire, and floats is preassembled on deck and is launched, surface float first, over the stern of the slowly moving ship. As the ship moves away, the float, meters, and cable are stretched out on the surface. When the vessel reaches the sampling site, the anchor is pushed overboard to pull the entire string down into the water. The floats and meters are retrieved by grappling from the surface for the ground wire or by sending a sound signal to a special acoustical link (fig. 7.40b), which detaches the wire from the anchor. The anchor is discarded, and the buoyed equipment returns to the surface. The technique is

straightforward, but many problems can occur in launching, finding, and retrieving instruments from the heaving deck of a ship at sea. Whenever oceanographers send their increasingly sophisticated equipment over the side, they must cross their fingers and hope to see it again.

A new technique for measuring currents does not need the energy of the moving water to run the rotor of a current meter. This technique uses sound pulses and takes advantage of the change in the pitch, or frequency, of sound as it is reflected from particles suspended in the moving water. When sound is reflected off particles moving toward the meter, the pitch increases; when the sound is reflected off particles moving away from the meter, the pitch decreases. This is the **Doppler effect;** the same effect increases the pitch of the horn or siren of an approaching vehicle and decreases the pitch as the vehicle passes and moves away. To use this effect, the sound source is mounted on a vessel or a buoy system, or placed on the sea floor. A seafloor-mounted Doppler current meter is shown

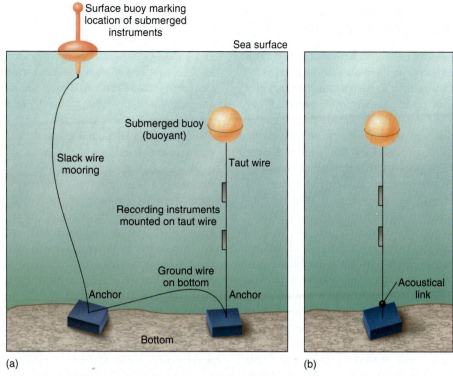

Figure 7.40 Taut-wire moorage. (a) Recovery is accomplished by retrieving the surface buoy and hauling in the wire. If the surface buoy is lost, it is possible to grapple for the ground wire. (b) In this system, a sound signal disconnects the anchor, and the equipment floats to the surface.

in figure 7.41. Four beams of sound pulses, set at a precisely known frequency, are sent out at right angles to each other. The change in pitch of the returning echoes provides the speed and direction of the water moving along each sound pulse path, and the direction of the resulting current is computed by comparison to an internal compass.

Using satellite altimeter data, scientists are able to map the topography of the sea surface on a global scale. Surface elevations and depressions are analyzed to determine the roles of gravity, periodic tidal motion, air pressure, and geostrophic flow in producing sea surface topography. Large amounts of data are required if researchers are to distinguish between the assortment of interacting currents. Oceanwide measurements of this type were not possible until satellite coverage of the oceans became available.

QUICK REVIEW

1. What measurements are necessary to define a current?
2. What are the advantages of electronic current meters? Are there any disadvantages?

Figure 7.41 Retrieving a Doppler current meter from the sea floor. The sound signals from the meter are analyzed to determine the speed and direction of the current.

Summary

The absorption and exchange of energy at the sea surface control the properties of surface seawater. Sea surface exchanges of heat, radiant energy, and water alter the temperature and salinity of the surface water and affect the density of the water. Many combinations of salinity and temperature can produce seawater of the same density. When waters with different properties but the same density are mixed, the resulting water has a greater density than either of its components. The density-driven vertical circulation that results is known as caballing. The waters of different densities produced at the sea surface, and the resulting vertical circulation, create a layered ocean that is primarily stably stratified.

The geographical distribution of surface salinities reflects Earth's latitudinal and seasonal patterns of evaporation, precipitation, and sea-ice formation. Cooling, evaporation, and freezing increase the density of the sea surface water. Heating, precipitation, and ice melt decrease its density. The surface water changes its density with changes in salinity and temperature that are keyed to latitude. The densities at depth are more homogeneous. Thermoclines and haloclines form where the temperature and salt concentrations change rapidly with depth. If the density increases with depth, the water column is stable; unstable water columns overturn and return to a stable distribution. Neutrally stable water columns are easily mixed vertically by winds and waves.

Vertical circulation driven by changes in surface density is known as thermohaline circulation. In the open ocean, temperature is generally more important than salinity in determining the surface density. Salinity is the more important factor close to shore and in areas of large seasonal ice melt.

Water sinks at downwellings and rises at upwellings. A downwelling occurs at the convergence of surface currents and transfers oxygen to depth. Upwellings bring nutrients to the surface and occur at zones of surface current divergence.

The oceans are layered systems. The layers (or water types) are identified by specific ranges of temperature and salinity. The water types of the Atlantic Ocean are formed at the surface at different latitudes; they sink and flow northward or southward. The water types of the Pacific Ocean lose their identity in the large volume of this ocean; their movements are sluggish. The water types of the Indian Ocean are less distinct than those of the Atlantic. Mediterranean Sea and Red Sea waters enter the Atlantic and Indian Oceans at depth as discrete water types that can be tracked for long distances. Water enters and exits the Arctic Ocean from the North Atlantic. The density of Arctic water is controlled more by salinity than by temperature.

Winds push the surface water 45° to the right of their direction in the Northern Hemisphere and 45° to the left in the Southern Hemisphere. Wind moves the water in layers that are deflected by the Coriolis effect to form the Ekman spiral; net flow over the depth of the spiral is deflected 90°. Geostrophic flow is produced when the force of gravity balances the Coriolis effect. Large surface gyres are observed in each ocean. Northern Hemisphere gyres rotate clockwise, and Southern Hemisphere gyres rotate counterclockwise. The currents of the northern Indian Ocean change with the seasonal monsoons.

Large oceanic current systems have names and descriptions based on their average locations. The water transport and speed of a current are affected by the current's cross-sectional area, by other currents, by westward intensification, and by wind speed. Eddies are formed at the surface when a fast-moving current develops meanders along its boundary that break off from the parent current. Eddies occur at all depths, wander long distances, and gradually lose their identity.

Downwelling is produced by converging surface currents, and upwelling is produced by diverging surface currents. Upwelling and downwelling may be shallow and short-lived, as in Langmuir cells, or these processes may involve large volumes and large areas of the oceans. Upwelling occurs nearly continuously along the western sides of the continents in the trade-wind belts, where surface water diverges from the coast. Seasonal upwellings and downwellings occur in coastal areas that have changing wind patterns and an alternating coastal flow of water onshore and offshore due to the Ekman transport.

Cyclic global circulation changes are part of Earth's normal dynamic system. Sudden changes in global ocean circulation may lead to major climate changes and are thought to be triggered by localized events in the North Atlantic. Both the North Pacific and the North Atlantic show decadal oscillations in current flow and climate.

A variety of techniques are available to measure currents: by following the water, by measuring the water's speed and direction as it moves past a fixed point, or by using changes in the frequency of sound.

Key Terms

mixed layer, 182
thermocline, 182
isothermal, 182
halocline, 184
pycnocline, 185
stable water column, 186
unstable water column, 186
continuity of flow, 186
thermohaline circulation, 186
downwelling zone, 186
upwelling zone, 186
convergence, 187
divergence, 187
isopycnal, 187
isohaline, 187
water type, 188
caballing, 188
water mass, 188
North Atlantic deep water
    (NADW), 189

Antarctic intermediate water
    (AAIW), 190
Antarctic bottom water
    (AABW), 190
South Atlantic surface
    water, 190
Mediterranean Intermediate
    Water (MIW), 190
Ekman spiral, 192
Ekman layer, 192
Ekman transport, 192
gyre, 192
geostrophic flow, 193
North Equatorial Current, 193
North Pacific Current, 193
North Pacific Drift, 193
California Current, 193
Kuroshio Current, 193
Oyashio Current, 193
Alaska Current, 193

South Equatorial Current, 194
Peru Current, 194
Humboldt Current, 194
East Australia Current, 194
Equatorial Countercurrent, 194
Cromwell Current, 194
North Atlantic Current, 194
North Atlantic Drift, 194
North Equatorial Current, 194
Gulf Stream, 194
Canary Current, 194
Florida Current, 194
Labrador Current, 194
East Greenland Current, 194
Norwegian Current, 194
South Equatorial Current, 194
Brazil Current, 194
Benguela Current, 194
South Equatorial Current, 194
West Australia Current, 194

Agulhas Current, 194
North Equatorial Current, 194
Equatorial Countercurrent, 194
Antarctic Circumpolar
    Current, 196
West Wind Drift, 196
Indonesian Throughflow, 196
western intensification, 197
eddy, 199
Langmuir cell, 201
tropical convergence, 201
subtropical convergence, 201
Arctic convergence, 201
Antarctic convergence, 201
tropical divergence, 201
Antarctic divergence, 201
drift bottle, 206
current meter, 206
Doppler effect, 208

Study Problems

1. During wet seasons, the Mississippi River discharges fresh water at a rate of as much as 20,000 m³/s while the Amazon River can discharge fresh water at a rate as high as 200,000 m³/s. Compare the average flow rate of the Gulf Stream off Cape Hatteras to these river discharge rates.

2. The following data were taken from a sampling station located at 79°N, 145°W:

Depth (m)	Temp. (°C)	Salinity (‰)	Density (g/cm³)
0	1.28	33.29	1.02659
50	1.29	33.30	1.02659
100	1.36	33.35	1.02669
150	1.39	33.55	1.02694
200	2.73	33.76	1.02701
300	3.07	33.87	1.02708
400	3.12	34.03	1.02713
500	3.14	34.13	1.02721

a. In what ocean region is this station?
b. Is the water column stable or unstable?
c. Does the temperature or the salt content control the density?
d. How deep is the mixed layer?
e. At what time of year were these data obtained?

Oceanography from SPACE

The View from Space

It is perhaps inconceivable today to imagine a world where we cannot visualize the surface of the Earth from space. We rely daily on satellite data for our weather forecasts, just as we rely on another satellite technology, the global positioning satellite system, to provide precise locations of where we are on the Earth, and communications satellites to let us talk with friends and colleagues around the globe. Satellites provide us with an unprecedented ability to monitor the surface ocean and today are routinely used in all aspects of oceanography.

Despite the ubiquity of these systems and data, it was not until 1978 that oceanographers had a dedicated satellite, SEASAT (Fig. OS.1). It included the first **synthetic aperture radar** and was designed to measure sea surface winds, temperature, sea ice, wave heights, and ocean topography. Shortly after that oceanographers got their first glimpse of the extraordinary beauty and complexity of the oceans, when NIMBUS-7 carried the **Coastal Zone Color Scanner (CZCS)** into space. To put this in perspective, oceanographers discovered that hydrothermal vents support biological communities in 1977; so within a few years, we discovered the rich diversity of biogeochemical processes occurring in the vast open oceans and deep sea. These discoveries fundamentally changed our view of the oceans and led to the development of remote sensing technologies.

The first two oceanographic satellite sensors illustrate many of the technologies we use today. Remote sensing via satellite or airborne platforms relies on interpreting information from the **electromagnetic spectrum** (see Figure 4.11). This includes both visible light (such as the human eye sees, from approximately 400–700 nm) and both shorter (UV light) and longer (thermal infrared, x-rays, microwaves, radar) radiation. SEASAT included both **passive sensors** that simply measure the reflected light from the surface of the ocean and **active sensors** that send a pulse of electromagnetic radiation down and then measure the return signal. Synthetic aperture radar is an example of an active sensor. Microwave pulses are directed at the Earth's surface, and a receiver composed of multiple detectors (designed to create a virtual, or synthetic aperture, receiver) captures these returned pulses to measure properties such as surface roughness, ice extent, and wave height. Today we continue to use both types of sensors, but the data we are most familiar with, such as images of sea surface temperature, ocean color, and cloud cover rely on passive sensors that are similar to a digital camera but with multiple bands of information at discrete wavelengths.

SEASAT lasted for only 108 days but proved that satellites could provide valuable information from space. CZCS lasted much longer (1979–1986) and highlighted the difficult challenge of measuring ocean processes from a fixed platform such as a ship (see Figure P.20, Prologue). Many sensors have been launched since those first images were collected and today multiple countries have active space programs for oceanography. The **Sea-Viewing Wide Field-of-view Sensor** (Sea-WiFS) was operated by NASA from 1986–2010 as a replacement for CZCS, with the **Moderate Resolution**

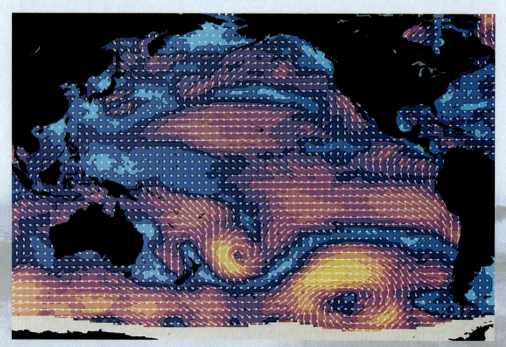

Figure OS.1 SEASAT satellite image of the first global measurements of the Pacific Ocean's wind distribution in 1978.

Figure OS.2 NASA, NOAA, and EUMETSAT satellite data sets (left to right, top to bottom): biosphere (SeaStar/SeaWiFS), water vapor (GOES 9 and 10, Meteosat, and GMS-5), temperature (Globe), fires (AVHRR), clouds (GOES 9 and 10, Meteosat, and GMS-5), methane (UARS), aerosols (TOMS), radiant energy (Globe), vegetation index anomalies (NDVI).

Imaging Spectrometer (MODIS) improving our capabilities in 1999 (MODIS Terra) and 2002 (MODIS Aqua). The two MODIS sensors are complementary, with Terra crossing the equator from north to south in the morning and Aqua crossing from south to north in the afternoon. Together these sensors provide an image of the entire Earth's surface every 1–2 days (Fig. OS.2).

All of these satellites followed a **polar orbit** meaning that they fly around the Earth either north to south or south to north. In contrast, satellite sensors such as the Geostationary Operational Environmental Satellites (GOES) are in **geosynchronous orbit**, meaning that they orbit the Earth along the equatorial plane, matching speed with the rotation of the planet so that they constantly view a fixed region. NOAA launched the first GOES satellite in 1975, and continues to maintain two satellites (GOES-15 and GOES-13) imaging the Pacific and Atlantic oceans respectively.

Following on from the original SEASAT mission, NASA launched the satellite *TOPEX/Poseidon* in 1992 in collaboration with France. This satellite lasted until 2005 and was an active

sensor system that used a form of radar to measure sea surface topography. A major follow-on mission to continue these measurements began in 2001, with the launch of *Jason-1*. This mission was in turn followed by the launch of the *Jason-2* satellite in 2008. Planning for a *Jason-3* mission is ongoing with an expected launch date in 2014.

Many other satellites are currently in orbit or are scheduled for launch and provide both new and improved measurements. For example **Aquarius** launched in 2011, and produced the first monthly maps of global salinity in September–December 2011 (Fig. OS.3). This is a joint effort between NASA and the Space Agency of Argentina (Comisión Nacional de Actividades Espaciales) and incorporates both active and passive L-band sensors.

Although we take for granted the ability to view the Earth nearly instantaneously, satellite oceanography is still a developing field and there are both successes and failures. While CZCS and SeaWiFS lasted many years beyond their expected lifetimes, satellite oceanography can be a risky business. In 2009 the NASA **Orbiting Carbon Observatory** failed to reach

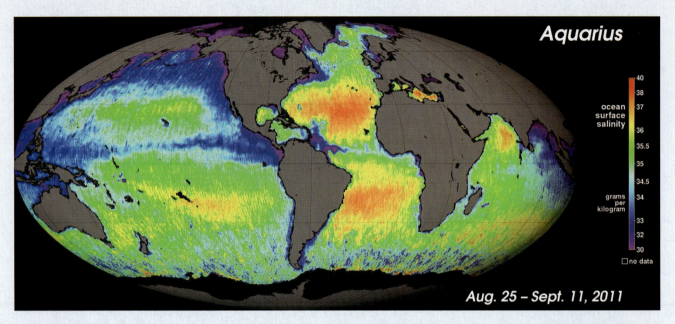

Figure OS.3 The first global map of ocean salinity measured by the Aquarius satellite. The map is a composite of the first two and a half weeks of data. Yellow and red colors represent areas of higher salinity, blues and purples indicate areas of lower salinity.

orbit due to a rocket malfunction. Its replacement, **OCO-2**, is expected to launch in 2014. When it does launch OCO-2 will provide high-resolution measurements of carbon dioxide in our atmosphere over large swaths of the planet, helping to improve predictions of future atmospheric CO_2 increases and linkages between CO_2 and Earth's climate.

Oceanographic Applications for Geology

Geological oceanographers are interested in the structure, topography, and processes (such as plate tectonics) of the ocean floor. Prior to the availability of satellite topography oceanographers had a very imprecise view of the large-scale structures under the ocean's surface, since traditional measurements such as soundings and side scan sonar, while very precise, can only be used over small spatial regions. With the introduction of **satellite altimetry**, which measures the height of the surface ocean relative to a **geoid**, the expected equipotential surface of the surface ocean if the oceans and atmosphere were at equilibrium on a non-rotating Earth, oceanographers can map the major bathymetric features of the ocean basins. Although the surface of the ocean can look smooth, it actually closely follows the bottom topography, creating ridges and bumps in the ocean's surface over ocean ridges and seamounts, and troughs over canyons and trenches. Satellites such as Topex/Poseidon and Jason measure the small differences in sea surface height with an accuracy of 2.5 cm (1 in), turning sea surface height into a map of the underlying bathymetry.

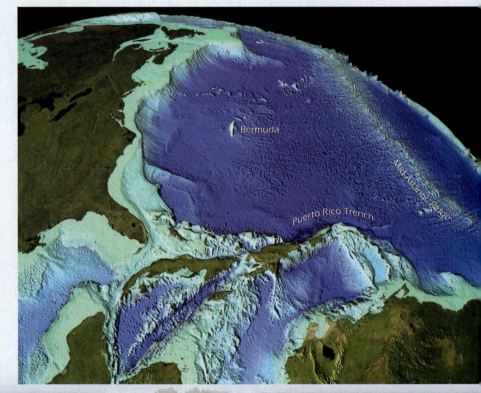

Figure OS.4 Bathymetric features of the Northwest Atlantic using data from the ERS-1 and GEOSAT satellites, analyzed by NOAA's Laboratory for Satellite Altimetry.

These observations can be used not only to provide detailed bathymetric charts of the entire world's oceans (Fig. OS.4) but are also used to track dynamic changes in the Earth's surface. Since these altimetry-based maps can be continuously updated with new information it is possible to track

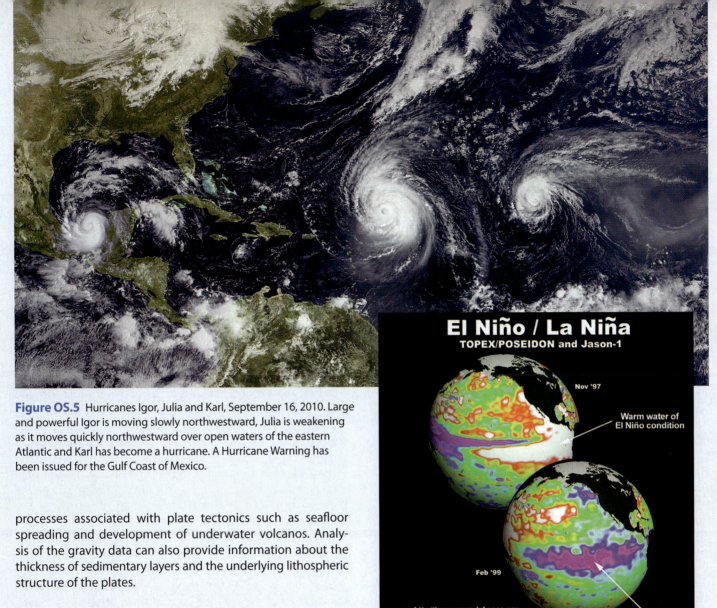

Figure OS.5 Hurricanes Igor, Julia and Karl, September 16, 2010. Large and powerful Igor is moving slowly northwestward, Julia is weakening as it moves quickly northwestward over open waters of the eastern Atlantic and Karl has become a hurricane. A Hurricane Warning has been issued for the Gulf Coast of Mexico.

processes associated with plate tectonics such as seafloor spreading and development of underwater volcanos. Analysis of the gravity data can also provide information about the thickness of sedimentary layers and the underlying lithospheric structure of the plates.

Oceanographic Applications for Physics

Satellite altimetry can provide information beyond bathymetric maps. In addition to the bumps and dips caused by the gravitational pull of the underlying topography, the surface ocean is strongly influenced by ocean currents. A common use for sea surface height measurements is to calculate the **sea surface height anomaly** (SSHa) which over short time scales provides a map of ocean current speed and direction. Until the 1960s oceanographers assumed that ocean circulation was dominated by slow-moving gyres and faster moving boundary currents. Satellite altimetry clearly demonstrated the ubiquity and importance of **mesoscale processes** (~100 km) dominated by eddies. Satellites also provide both ocean sea surface temperature and with the launch of Aquarius, salinity, giving us a global view of these two fundamental properties.

Following the initial development of SEASAT, there are currently several versions of synthetic aperture radar used to track processes such as sea-ice extent, glacier movement, and internal wave propagation. Finally, special radar systems called **scatterometers** are used to measure surface wind speed and direction. Taken together these data provide valuable information

Figure OS.6 The TOPEX/Poseidon satellite measures changes in sea surface height. During an El Niño, there are strong positive anomalies in the eastern Pacific (red and white colors) while during La Niña, unusually strong trade winds result in lower than normal sea surface height and enhanced upwelling (blue and purple colors) along the equatorial Pacific.

about ocean circulation, physical properties (temperature and salinity), mixing, and air-sea interactions, and are often used to track large storms such as hurricanes and typhoons (Fig. OS.5).

Satellite oceanography has strongly influenced our understanding of global phenomena such as the **El Niño/La Niña cycle** (Fig. OS.6). During an El Niño, the western warm pool in the Pacific, normally maintained by the trade winds, spreads eastward, reducing upwelling of cold, nutrient-rich waters in the eastern Pacific. An El Niño event is preceded by a series of **Kelvin waves** that propagate from west to east, and is accompanied by large changes in sea surface height across much of the Pacific. During a La Niña, the trade winds return with stronger than usual intensity, resulting in unusually strong upwelling in the eastern Pacific. Satellites provide an ability to monitor changes in temperature, sea surface height, and winds across the entire Pacific, helping scientists to better understand the causes and consequences of these oceanographic oscillations in the ocean and atmosphere.

Oceanographic Applications for Chemistry and Biology

Until recently chemical oceanographers have only used satellite data indirectly, since there are very few direct measurements of ocean chemistry that can be made with the electromagnetic spectrum. The launch of OCO-2 will provide our first opportunity to directly measure an important chemical variable, carbon dioxide, over much of the world's oceans. Other chemical properties can be inferred by biological changes in the ocean, which in turn can be measured using ocean color. For example, many types of algae produce **Dimethyl Sulfide (DMS)**, a gas that can provide the source for cloud condensation nuclei, and thereby cool the surface of the planet. One particular phytoplankton group, coccolithophores, produce large quantities of DMS and are easily observed from space because these algae use calcium carbonate (chalk) to make a plate-like "armor" to protect their cells. Because of their unique optical properties, these large blooms show up as milky white areas in satellite imagery.

Satellites have also been used to track episodic (natural and man-made) injections of iron into the surface ocean. Large sections of the world's oceans, predominantly in the Pacific and Antarctic, are chronically limited by the availability of iron. These areas are sometimes called **High Nutrient Low Chlorophyll (HNLC)** regions because there are adequate concentrations of macronutrients such as N and P, but less chlorophyll (phyto-plankton biomass) than expected due to the lack of iron. When iron is added, there can be explosive growth of phytoplankton, which show up as blooms from space. There have been multiple iron fertilization experiments conducted by scientists, many of which were documented using ocean color. There are also natural fertilization events when iron-rich dust falls on the surface of the ocean; these events can be tracked by combining satellite data monitoring the amount of dust in the atmosphere and ocean color images of the surface ocean.

These indirect measurements of ocean chemistry often rely on tracking the biological response of phytoplankton. We can do this easily with satellites because chlorophyll provides a direct estimate of phytoplankton biomass in the oceans. Pigments such as chlorophyll both absorb and scatter light, resulting in a change in ocean color. Scientists have been taking advantage of this shift in color since the launch of CZCS to estimate chlorophyll in the ocean. Indeed, some of the first CZCS imagery showed us just how dynamic the surface ocean could be, and emphasized the need for synoptic views from space (Figure P.20).

Today we produce global maps of both terrestrial and land biomass almost daily. When combined with other satellite data including the amount of sunlight and the temperature of the surface ocean, we can also predict **primary productivity**, or the growth rate for the phytoplankton in the ocean. These data in turn allow us to monitor the waxing and waning of phytoplankton on daily, seasonal, and interannual scales as the biology responds to ocean physics and chemistry. For example during an El Niño, the nutrient-rich water that is normally upwelled along the west coast of North and South America is greatly reduced, resulting in a direct decrease in phytoplankton biomass and primary productivity. When the El Niño ends, productivity increases again, driven in part along the equatorial Pacific by the propagation of **tropical instability waves** that move from east to west, responding to the resumption of the trade winds (Fig. OS.7).

Figure OS.7 Following an El Niño, tropical instability waves propagate from east (North and South America) to west along the equator. As the wave passes, it stimulates phytoplankton productivity, producing a "green ribbon" of enhanced chlorophyll (red/orange colors) relative to the low-chlorophyll (blue) tropical waters as seen by the SeaWiFS ocean color satellite (July 28–August 4, 1998). Black indicates missing data.

Our Wired World

It is difficult to imagine conducting modern ocean-ography without the use of satellites. Satellite sensors provide geological, physical, chemical, and biological data that are used to interpret nearly all aspects of the oceans, from the movement of tectonic plates to track-ing of marine mammals. One example of how all this information can be integrated to give us a holistic view of the environment is the **Tagging of Pacific Predators (TOPP)** project, part of the **Census of Marine Life**. In these projects, scientists are seeking to understand how marine organisms utilize ocean habitat. The TOPP program has put electronic tracking tags on more than 4000 individuals from 23 species, including birds, mam-mals (such as seals and whales), fish (including tuna and sharks), squid, and turtles. Using a combination of the tag data and information from satellites, scientists have identi-fied oceanic "hotspots" and migration corridors that span large swaths of the Pacific (see Figure 13.1). Animals take advantage of features such as the California Current and the North Pacific Transition Zone where warm and cool waters mix, and where phytoplankton productivity is also increased (Fig. OS.8). Many of the animals take advantage of other mesoscale features such as the presence of sea-mounts and eddies as part of their migrations. By over-laying the animal tracks with satellite imagery including bathymetric maps, temperature, chlorophyll, sea surface currents, and winds, scientists are better able to under-stand how these animals interact with the ocean environ-ment and how these interactions change through time. This is just one example of how satellite data has revolu-tionized our understanding of the oceans, and how scien-tists increasingly live in a data rich, wired world (Fig. OS.9).

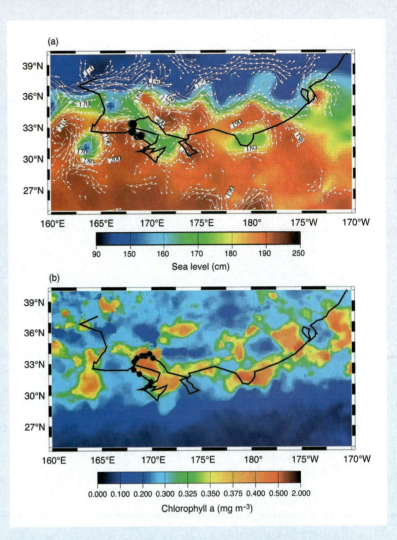

Figure OS.8 Loggerhead turtle tracks along the Transition Zone Chlorophyll Front during Feb. 2001, overlaid on satellite sea surface height (SSH) and chloro-phyll. Scientists can tag animals and track them to see how they respond to and use oceanographic features such as chlorophyll fronts and currents.

Figure OS.9 This graphic (not to scale) depicts the satellites that make up the Afternoon Constellation—The "A-Train." Each satellite in the A-Train crosses the equator within a few minutes of each another at around 1:30 P.M. local time. By combining the different sets of nearly simultaneous observations, scientists are able to gain a better understanding of important parameters related to climate change. OCO-2 will join the A-Train in 2014.

CHAPTER 8

The Waves

Learning Outcomes

After studying the information in this chapter students should be able to:

1. *describe* the process of wave formation, including wave generating and restoring forces,

2. *label* the basic characteristics of a wave, including wave crest, trough, height, and wavelength,

3. *define* wavelength, wave period, wave frequency, and wave steepness,

4. *diagram* wave-induced water motion as a function of depth for both deep- and shallow-water waves,

5. *characterize* deep- and shallow-water waves,

6. *calculate* the speed of deep- and shallow-water waves given wave period, wavelength, wave frequency, and water depth,

7. *review* the factors controlling maximum potential wave height,

8. *diagram* the processes of wave refraction, diffraction, and reflection,

9. *discuss* the formation of tsunamis and *report* the history and characteristics of the 2011 Japan tsunami, and

10. *explain* the formation and properties of internal and standing waves.

A wind-blown wave breaking as it nears the shore.

We have all seen water waves. This up-and-down motion occurs on the surface of oceans, seas, lakes, and ponds, and we speak of waves, ripples, swells, breakers, whitecaps, and surf. Sometimes the waves are related to the wind, or to a passing ship, or even to a stone thrown into the water. Waves may swamp a small boat or, when large, smash and twist the bow of a supertanker. The storm waves of winter crash against our coasts, eating away the shore and damaging structures, such as docks and breakwaters. Heavy wave action may force commercial vessels to slow their speed and lengthen their sailing time between ports, to the inconvenience and increased cost of the shippers. Special waves associated with seismic disturbances have killed thousands of coastal inhabitants and severely damaged their cities and towns. Surfers search for the perfect wave, and ancient peoples navigated by the patterns that waves form.

In this chapter, we describe different kinds of waves, investigate their origin and behavior, and study their major characteristics. Our study is somewhat superficial and simplified, because waves are among the most complex of the ocean's phenomena and because no two waves are exactly the same. Books are written about waves; mathematical models have been devised to explain waves; wave tanks are built to study waves. However, there is much we can learn if we follow the sequence of events beginning at sea where a wave is born and ending at last in the spray and surf of a faraway shore. A summary of many of the characteristics of waves is discussed in this chapter and the next, and given in table 8.1.

Figure 8.1 Short-wavelength capillary waves in front of ordinary gravity waves on a beach. The gravity waves are the larger waves and are propagating "down" toward a region of shallow water. The capillary waves are the very short waves propagating in front of the larger waves. (Photo taken by Fabrice Neyret.)

darken the surface of the water and move quickly, keeping pace with the gusts of wind. Sailors call these fast-moving patches **cat's-paws**. These waves die out rapidly while new ripples form constantly in front of each moving wind gust. When the wind blows, energy is transferred to the water over large areas, for varying lengths of time, and at different rates. As waves form, the surface becomes rougher, and it is easier for the wind to grip the roughened water surface and add energy. As wind speeds increase or winds blow for longer times over the water, the waves become larger, and the restoring force changes from surface tension to gravity. The vast majority of water waves that you see on the ocean, or in a lake, are wind-generated **gravity waves** (fig. 8.2).

8.1 How a Wave Begins

Forces Influencing Waves

There are two "families" of forces that influence waves: **generating forces**—those responsible for creating a disturbance on the water surface, and **restoring forces**—those responsible for returning the water surface to an undisturbed state. While there are a number of different generating forces, there are only two restoring forces: gravity, which is by far the most common, and surface tension, which has a greater influence than gravity on very small waves.

The most common generating force for water waves is wind. As wind blows across a water surface, the friction, or drag, between the air and the water tends to stretch the surface, resulting in wrinkles. These wrinkles are small waves called **ripples**, or **capillary waves** (fig. 8.1). The restoring force for these small capillary waves is the surface tension of the water. Patches of these very small waves can be seen forming, moving, and disappearing as they are driven by pulses of wind. These patches

Figure 8.2 Wind-generated gravity waves at sea.

Table 8.1 Wave Summary

Type	Cause	Period	Class of Wave	Open-Ocean Height	Coastal Height	Wave Speed	Group Speed	Orbit Type
Wind wave	Wind	0.1–20 s	Deep-water $D > L/2$	Up to 30 m (100 ft)	Up to 6 m (20 ft)	$C = 1.56T$ $C^2 = 1.56L$ 0–111 km/hr (0–60 kts)	$V = C/2$	Circle
			Shallow-water $D < L/20$	—	Up to 6 m (20 ft)	$C = \sqrt{gD}$	$V = C$	Ellipse
Tsunami	Seismic upheaval	10–30 min	Shallow-water $D < L/20$	Small	10–18 m (30–60 ft)	$C = \sqrt{gD}$ 740 km/hr (400 kts)	$V = C$	Ellipse
Standing	Wind, tsunamis, tides, pressure changes	10 min–30 hrs	Normally shallow-water $D < L/20$	0–2.5 m (0–8 ft)	0–15 m (0–50 ft)	None	None	None
Tide	Gravity of Sun and Moon	12 hrs 25 min and 24 hrs 50 min	Shallow-water $D < L/20$	0–1.5 m (0–4 ft)	0–15 m (0–50 ft)	$C = \sqrt{gD}$ >740 km/hr (>400 kts)	None	Ellipse

Another important generating force is earthquakes that produce vertical displacement of the sea floor. If the sea floor is disturbed, the water column above it will be displaced also, creating a series of waves that move outward away from the location of the earthquake. These waves are called tsunamis. While tsunamis are relatively uncommon, they can be devastatingly destructive. Tsunamis are discussed in more detail later in this chapter.

Intense low-pressure systems can produce large differences in atmospheric pressure that raise the sea surface and generate very long waves. This can result in storm surge along coastlines, as discussed in chapter 6.

When water is displaced by something like a landslide, or ice breaking off a glacier, it can produce significant waves. The process is something like throwing a stone into water. If a stone is thrown into water it will displace, or push aside, the water. As the stone sinks, the displaced water flows from all sides into the space left behind, and as the water rushes back, the water at the center is forced upward. The elevated water falls back, causing a depression of the surface that is refilled, starting another cycle. This process sets up a series of waves, or oscillations, that move outward away from the point of the disturbance.

The largest waves in the ocean are the tides. These are generated by the combined influence of lunar and solar gravitational attraction and the rotation of Earth. The tides are the subject of chapter 9.

Two Types of Wind-Generated Waves

Common wind-generated gravity waves can be divided into two types based on the depth of the water they travel in compared to their wavelength: deep-water waves or shallow-water waves. Deep-water waves travel in water deeper than one-half their wavelength. Shallow-water waves travel in water shallower than one-twentieth their wavelength. Which category they fall into will influence particle motion in the water column, the speed at which they travel, their shape, and how they may change direction as they travel. Generally speaking, most waves in the open ocean are deep-water waves. These deep-water waves convert into shallow-water waves when they approach a coastline and eventually break on the shore. We will discuss these two different types of wind-generated gravity waves in detail in this chapter.

QUICK REVIEW

1. What two kinds of forces influence waves?
2. Surface tension is the restoring force for what type of wave?
3. How do ripples help large waves to develop?
4. Name four different generating forces.

8.2 Anatomy of a Wave

In any discussion of waves, certain terms are used to describe them; refer to figure 8.3 for help in understanding these terms.

- **crest** The crest is the part of the wave that is elevated the highest above the undisturbed sea surface (also known as the **equilibrium surface**).
- **trough** The trough is the part of the wave that is depressed the lowest below the undisturbed sea surface. The trough has a flatter shape than the crest.
- **wavelength** The wavelength of a wave is the shortest part of the waveform that, if repeated multiple times, will reproduce the wave shape. In practical terms, the wavelength is typically measured as the distance between two successive wave crests or two successive wave troughs.
- **wave height** The height of a wave is the vertical distance from the elevation of the crest to the depth of the trough. Sometimes the term *amplitude* is used in describing waves. The **amplitude** of the wave is equal to one-half the wave height, or the vertical distance from either the height of the crest or the depth of the trough to the undisturbed water level.
- **wave steepness** The steepness of the wave is equal to the wave height divided by the wavelength. This is a dimensionless number.
- **wave period** The period of a wave is the time required for two successive crests or troughs, or one cycle of the wave, to pass a stationary point in space. If you stand on the end of a dock and start a stopwatch as the crest of one wave passes by and then stop the stopwatch as the crest of the next wave passes, the measured time is the period of the wave. Unlike other properties of waves, the period of a wave will not change once the wave has formed.
- **wave frequency** The wave frequency is the reciprocal of the wave period. Frequency is a measure of how many cycles of the wave pass a stationary point in space in a unit length of time. So, if the period of the wave is equal to 10 seconds/cycle, then the frequency is 0.1 cycles/second.

QUICK REVIEW

1. Sketch a wave and label its different parts.
2. Sketch a wave and identify three different segments on the sketch that all correspond to the wavelength of the wave.
3. How is the steepness of a wave related to the slope of the wave from the crest to the trough?
4. If you measure either the period or the frequency of a wave, is it necessary to measure the other parameter? Why?

8.3 Wave Speed

A wave's speed across the sea surface is related to its wavelength and wave period. The speed of any surface wave (*C*) is equal to the length of the wave (*L*) divided by the wave period (*T*):

$$\text{speed} = \frac{\text{length of wave}}{\text{wave period}}$$

$$\text{or}$$

$$C = \frac{L}{T} \qquad \textit{Equation 8.1}$$

Once a wave is created, the speed at which the wave moves may change, but *its period remains the same* (period is determined by the generating force).

In wave studies, *C* stands for *celerity*, a term traditionally used to identify wave speed and to distinguish wave speed from the group speed of waves. Group speed is usually identified by *V* (see section 8.4, Group Speed).

The speed of a wave can be predicted from the equation:

$$C = \sqrt{\frac{gL}{2\pi} \tanh \frac{2\pi D}{L}} \qquad \textit{Equation 8.2}$$

where *g* is the acceleration of gravity (9.8 m/s² or 32 ft/s²), *L* is the wavelength, π is the constant pi (3.14), and *D* is the water depth. The hyperbolic tangent function (tanh) describes a curve that initially rises almost linearly and then gradually levels

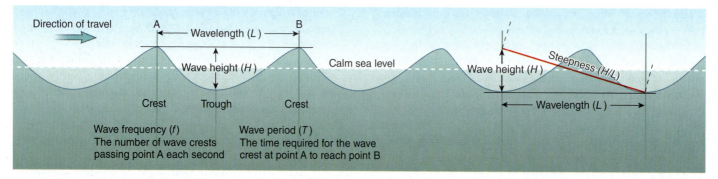

Figure 8.3 Terms used in describing a water wave.

off. In discussing wave speed, all you need to know about the hyperbolic tangent function (tanh x) is that if x is small, say less than 0.05, then tanh $x \approx x$. If x is larger than π, then tanh $x \approx 1$. Several curves of equation 8.2 are plotted for different water depths in figure 8.4. Looking at the curves plotted in figure 8.4, it is clear that the speed of a wave behaves differently in deep water compared to shallow water. This is one reason why wind-generated gravity waves are divided into the two groups: deep-water waves and shallow-water waves.

Deep-water waves are defined as waves that travel in water that is deeper than one-half of their wavelength.

$$D > L/2$$

Shallow-water waves are defined as waves that travel in water that is shallower than one-twentieth of their wavelength.

$$D < L/20$$

At depths between one-half and one-twentieth the wavelength, waves are in a transitional stage where their characteristics progressively change from deep-water to shallow-water waves.

In deep water, the speed of a wave depends primarily on the wavelength of the wave, not on the depth of the water. In shallow water, the opposite it true—wave speed depends on water depth, not on wavelength.

8.4 Deep-Water Waves in Detail

Deep-Water Wave Motion

As a wave moves across the water surface, particles of water are set in motion. Seaward, beyond the surf and breaker zone, where the surface undulates quietly in water deeper than half the wavelength of the waves, the water is not moving toward the shore. Such an ocean wave does not represent a flow of water but instead represents a flow of motion or energy from its origin to its eventual dissipation at sea or against the land. To understand what is happening during the passing of a deep-water wave, let us follow the motion of the water particles as the wave moves through the water.

As the wave crest approaches, the surface-water particles rise and move forward. Immediately under the crest the particles have stopped rising and are moving forward at the speed of the crest. When the crest passes, the particles begin to fall and to slow in their forward motion, reaching a maximum falling speed and a zero forward speed when the midpoint between crest and trough passes. As the trough advances, the particles slow their falling rate and start to move backward, until at the bottom of the trough they have attained their maximum backward speed and neither rise nor fall. As the remainder of the trough passes, the water particles begin to slow their backward speed and start to rise again, until the midpoint between trough and crest passes. At this point, the water particles start their forward motion again and continue to rise with the advancing crest. This motion (rising, moving forward, falling, reversing direction, and rising again) creates a circular path, or **orbit,** for the water particles as the wave passes. Follow this motion in figure 8.5.

It is the orbital motion of the water particles that causes a floating object to bob, or move up and down, forward and backward, as the waves pass. This motion affects a fishing boat, swimmer, seagull, or any other floating object on the surface seaward of the surf zone. The surface-water particles trace an orbit with a diameter equal to the height of the wave. This same type of motion is transferred to the water particles below the surface, but less energy of motion is found at each succeeding depth. The diameter of the orbits becomes smaller and smaller as depth increases. At a depth equal to one-half the wavelength,

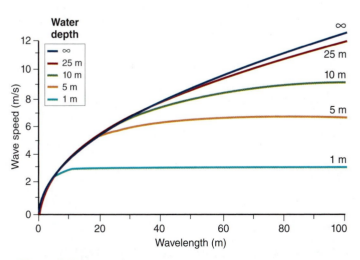

Figure 8.4 Wave speed as a function of wavelength, calculated using equation 8.2, for different water depths. In shallow water, wave speed is effectively independent of wavelength and is controlled by water depth. In deep water, wave speed is dependent on wavelength; speed increases as wavelength increases.

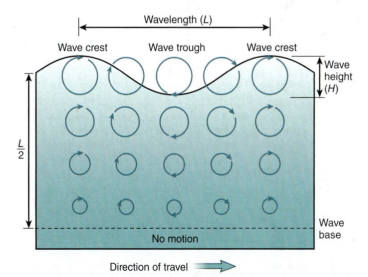

Figure 8.5 A deep-water wave sets water particles in circular motion. At the surface, the diameter of the particle orbit is equal to the wave height. Orbit diameter decreases with increasing depth until there is essentially no motion at a depth of $L/2$. This depth is known as the wave base.

the orbital motion has decreased to almost zero; notice how the orbit decreases in diameter in figure 8.5. Submarines dive during rough weather for a quiet ride because the wave motion does not extend far below the surface.

The orbits just described are based on the mathematical sine wave, not on real sea waves. The water particles of actual sea waves move in orbits whose forward speed at the top of the orbit is slightly greater than the reverse speed at the bottom of the orbit. Therefore, each orbit made by a water particle does move the water slightly forward in the direction the waves travel. This movement is due to the shape of the real waves, whose crests are sharper than their troughs. This difference means that a very slow transport of water in the direction of the waves occurs in nature, but this motion is ignored in calculations based on simple wave models.

Deep-Water Wave Speed

Let's consider the general wave speed equation (equation 8.2) once again. When waves travel in water deeper than one-half their wavelength, the value of:

$$\tanh \frac{2\pi D}{L}$$

in equation 8.2 is approximately equal to one. Consequently, for deep-water waves, the general wave speed equation reduces to:

$$C_{deep} = \sqrt{\frac{gL}{2\pi}} \quad \text{or} \quad C_{deep} = 1.25 \sqrt{L} \qquad Equation\ 8.3$$

where speed is in meters per second and wavelength is in meters. This simple relationship between wave speed and wavelength is often difficult to use because it can be hard to measure wavelength accurately. It is much easier to measure wave period accurately. Fortunately, wave period and wavelength are related to each other by:

$$L = \frac{g}{2\pi}T^2 \quad \text{or} \quad L = 1.56T^2 \qquad Equation\ 8.4$$

where wavelength is in meters and period is in seconds. If we use equation 8.4 to substitute wave period for wavelength in equation 8.3, we obtain the more easily used equation for the speed of deep-water waves:

$$C_{deep} = 1.56\ T \qquad Equation\ 8.5$$

where speed is in meters per second and period is in seconds.

Storm Centers

Most deep-water waves observed at sea are **progressive wind waves.** They are generated by the wind, are restored by gravity, and progress in a particular direction. These waves are formed in local storm centers or by the steady

winds of the trade and westerly wind belts. In an active storm area covering thousands of square kilometers, the winds are not steady but vary in strength and direction. Storm-area winds flow in a circular pattern about the low-pressure **storm center,** creating waves that move outward and away from the storm in all directions. When a storm center moves across the sea surface in a direction following the waves, wave heights are increased because the winds supply energy for a longer time and over a longer distance.

In a storm area, the sea surface appears as a jumble and confusion of waves of all heights, lengths, and periods; there are no regular patterns. Capillary waves ride the backs of small gravity waves, which in turn are superimposed on still higher and longer gravity waves. This turmoil of mixed waves is called a "sea," and sailors use the expression "There is a sea building" to refer to the growth of these waves under storm conditions. When waves are being generated, they are forced to increase in size and speed by the continuing input of energy; these are known as **forced waves.** Because of variations in the winds of the storm area, energy at different intensities is transferred to the sea surface at different pulse rates, resulting in waves with a variety of periods and heights. Remember, wave periods are a function of the generating force; the speed at which a wave moves away from a storm may change, but its period remains the same.

Dispersion

When the waves move away from the storm, they are no longer wind-driven forced waves but become **free waves** moving at speeds controlled by their periods and wavelengths. Waves with long periods and long wavelengths have a greater speed than waves with short periods and short wavelengths. The faster, longer waves gradually move through and ahead of the shorter, slower waves; this process is called **sorting,** or **dispersion.** Groups of these faster waves move as **wave trains,** or packets of similar waves with approximately the same period and speed. Because of dispersion, the distribution of observed waves from any single storm changes with time. Near the storm center, the waves are not yet sorted, while farther away, the faster, longer-period waves are out ahead of the slower, shorter-period waves. This process is shown in figure 8.6.

Away from the storm, the faster, longer-period waves appear as a regular pattern of crests and troughs moving across the sea surface. These uniform, free waves are called **swell;** they carry

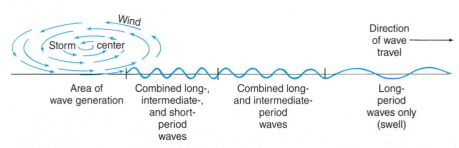

Figure 8.6 Dispersion. The longer waves travel faster than the shorter waves. Waves are shown here moving in only one direction.

Figure 8.7 Waves of uniform wavelength and period, known as swell, approaching the coast at the entrance to Grays Harbor, Washington.

considerable energy, which they lose very slowly (fig. 8.7). The distribution of the waves from a given storm and the energy associated with particular wave periods change predictably with time, allowing the oceanographer to follow wave trains from a single storm over long distances. Groups of large, long-period waves created by storms between 40°S and 50°S in the Pacific Ocean have been traced across the entire length of that ocean, until they die on the shores and beaches of Alaska.

Group Speed

Consider again the waves formed by a stone thrown into the water (fig. 8.8). The wave group, or train, is seen as a ring of waves moving outward from the point of disturbance. Careful observation shows that waves constantly form on the inside of the ring as it moves across the water. As each new wave joins the train on the inside of the ring, a wave is lost from the leading edge, or outside, of the ring, and the number of waves remains the same. The outside wave's energy is lost in advancing the wave form into undisturbed water. Therefore, the speed of each individual wave (C) in the group is greater than the speed of the leading edge of the wave train, and the wave ring moves outward at a speed one-half that of the individual waves. This speed is known as the **group speed,** the speed at which wave energy is transported away from its source under deep-water conditions:

group speed = ½ wave speed = speed of energy transport

or

$$V = \frac{C}{2}$$

Wave Interaction

Waves that escape a storm and are no longer receiving energy from the storm winds tend to flatten out slightly, and their crests become more rounded. These waves moving across the ocean surface as swell are likely to meet other trains of swell moving away from other storm centers. When the two wave trains meet, they pass through each other and continue on. Wave trains may

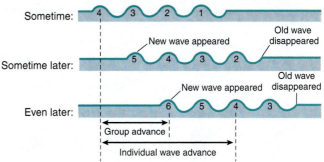

Figure 8.8 The phenomenon of dispersion can be observed by dropping a pebble in a still pond. The ripples created by the pebble become more distant from each other as they progress from the site of initial disturbance. The ripples with longer wavelengths move out faster than the waves with the shorter wavelengths and they disperse. However, these front-running waves soon disappear and a new wave appears at the back of the group of ripples. if you watch carefully, this new wave will travel through the group until it becomes the front wave. Once again, this wave disappears and another wave forms to the rear.

intersect at any angle, and many possible interference patterns may result. If the two wave trains intersect each other sharply, as at a right angle, then a checkerboard pattern is formed (fig. 8.9).

If the waves have similar lengths and heights and approach from opposite directions (fig. 8.10a) and if the crests of one wave train coincide with the crests of another train, the wave trains reinforce each other by constructive interference. If the crests of a wave train coincide with the troughs of another wave train, the waves are canceled through destructive interference (fig. 8.10b). In this way, two or more similar wave trains traveling in the same or opposite direction and passing through each other can join together in phase and suddenly develop large-amplitude waves unrelated to local storms. If these waves become too high, they may break, lose some of their energy, and create new, smaller waves. If such high waves overtake a vessel, they can cause severe damage.

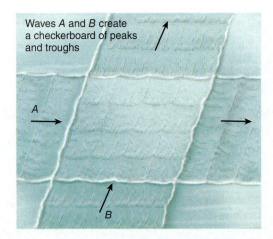

Figure 8.9 Waves meeting at right angles create a checkerboard pattern.

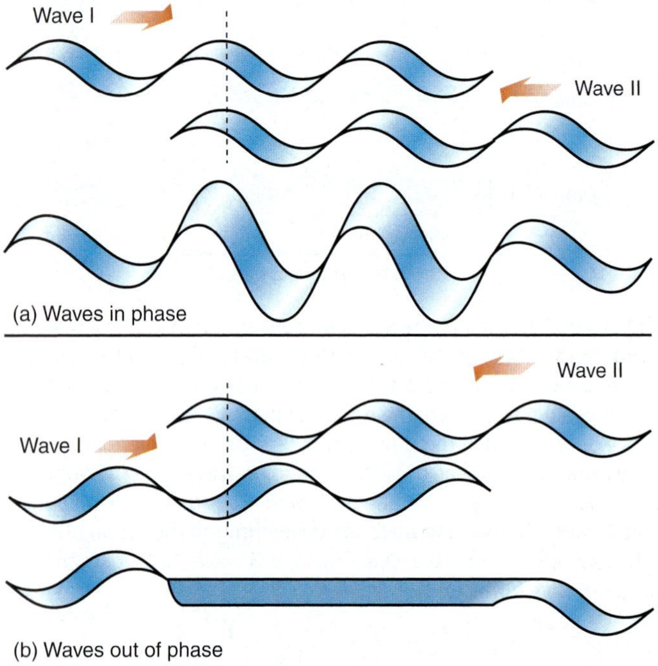

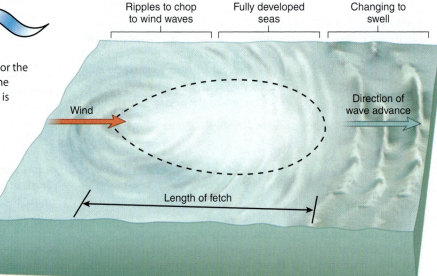

Wave I

Wave II

(a) Waves in phase

Wave II

Wave I

(b) Waves out of phase

Figure 8.10 Waves approach each other. (a) If the crests or the troughs of the approaching waves coincide, the height of the combined waves increases, and the waves are in phase; this is constructive interference. (b) If the crests of one wave and the troughs of the other wave coincide, the waves cancel each other, and the waves are out of phase; this is destructive interference.

Wave Height

The height of wind waves is controlled by the interaction of several factors. The three most important factors are (1) wind speed (how fast the wind is blowing), (2) wind duration (how long the wind blows), and (3) **fetch** (the distance over water that the wind blows in the same

direction) (fig. 8.11). The wave height may be limited by any one of these factors. If the wind speed is very low, large waves are not produced, no matter how long the wind blows over an unlimited fetch. If the wind's speed is great but it blows for only a few minutes, no high waves are produced despite unlimited wind strength and fetch. Also, if very strong winds blow for a long period over a very short fetch, no high waves form. When no single one of these three factors is limiting, spectacular wind waves are formed at sea.

Table 8.2 lists the maximum and significant wave heights possible for certain average wind speeds when fetch and wind duration are not limiting. The significant wave height is defined as the average wave height of the highest one-third of the waves in a long record of measured wave heights. For example, if a wave-height record of 1200 successive storm waves is made and the individual wave heights are arranged in order of height, the average height of the 400 highest waves defines the significant wave height. Significant wave heights are forecast from wind data, and the maximum wave heights are related to the significant wave heights by calculation.

The area of the ocean in the vicinity of 40°–50°S latitude is ideal for the production of high waves (fig. 8.12). Here, in an area noted for high-intensity storms and strong winds of long duration (what sailors have called the "roaring forties and furious fifties"), there are no landmasses to interfere and limit the fetch length. The westerly winds blow almost continuously around Earth, adding energy to the sea surface for long periods of time and over great distances, resulting in waves that move in the same direction as the wind. Although this area is ideal for the production of high waves, such waves can occur anywhere in the open sea, given the proper storm conditions.

A typical maximum fetch for a local storm over the ocean is approximately 920 km (500 nautical mi). Because storm winds circulate around a low-pressure disturbance, the winds continue

Ripples to chop to wind waves

Fully developed seas

Changing to swell

Wind

Direction of wave advance

Length of fetch

Figure 8.11 Wave height is determined in part by the fetch, the distance the wind blows at a relatively constant speed and direction.

Table 8.2 The Relationship Between Wind Speed and Wave Height

Average Wind Speed		Significant Wave Height	Significant Wave Period	Significant Wave Speed	Maximum Wave Height	Minimum Fetch[1]	Minimum Wind Duration[1]
(knots)	(m/s)	(m)	(s)	(m/s)	(m)	(km)	(h)
10	5.1	1.22	5.5	8.58	2.19	16	2.4
20	10.2	2.44	7.3	11.39	4.39	110	10
30	15.3	5.79	12.5	19.50	10.43	450	23
40	20.4	14.33	18.0	28.00	25.79	1136	42
50	25.5	16.77	21.0	32.76	30.19	2272	69

[1] Minimum fetch and minimum wind duration are distances and times required when wind speed is the only limiting factor in wave development.

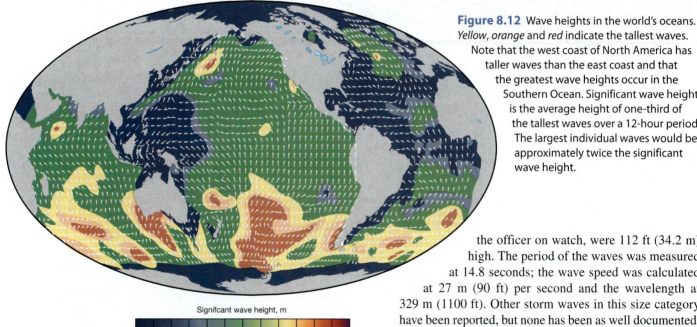

Signifcant wave height, m

1 2 3 4 5 6 7 8 9 10 11

Figure 8.12 Wave heights in the world's oceans. *Yellow, orange* and *red* indicate the tallest waves. Note that the west coast of North America has taller waves than the east coast and that the greatest wave heights occur in the Southern Ocean. Significant wave height is the average height of one-third of the tallest waves over a 12-hour period. The largest individual waves would be approximately twice the significant wave height.

the officer on watch, were 112 ft (34.2 m) high. The period of the waves was measured at 14.8 seconds; the wave speed was calculated at 27 m (90 ft) per second and the wavelength at 329 m (1100 ft). Other storm waves in this size category have been reported, but none has been as well documented. It is also probable that crews confronted with such waves do not always survive to report the incidents.

Episodic Waves

Large waves, or **episodic waves,** can suddenly appear unrelated to local sea conditions. An episodic wave is an abnormally high wave that occurs because of a combination of intersecting wave trains, changing depths, and currents. We do not know a great deal about these waves, as they do not exist for long and they can and do swamp ships, often eliminating any witnesses. They occur most frequently near the edge of the continental shelf, in water about 200 m (660 ft) deep, and in certain geographic areas with particular prevailing wind, wave, and current patterns.

The area where the Agulhas Current sweeps south along the east coast of South Africa and meets the storm waves arising in the Southern Ocean is noted for such waves. Storm waves from more than one storm may combine constructively and run into the current and against the continental shelf, producing occasional episodic waves. This area is also one of the world's

to follow the waves on the side of the storm, along which wave direction is the same as the storm direction. This increases both the fetch and the duration of time over which the wind adds energy to the waves. If the waves move fast enough, their speeds exceed the speed of the moving storm center; the waves escape the generating wind, become free waves, and do not grow larger. Waves 10–15 m (33–49 ft) high are not uncommon under severe storm conditions; such waves are typically between 100 and 200 m (330 and 660 ft) long. This length is about the same as the length of some modern ships, and a vessel of this length encounters hazardous sailing conditions, because the ship may become suspended between the crests of two waves and break its back.

Giant waves over 30.5 m (100 ft) high are rare. In 1933, the USS *Ramapo,* a Navy tanker en route from Manila to San Diego, encountered a severe storm, or typhoon. As the ship was running downwind to ease the ride, it was overtaken by waves that, as measured against the ship's superstructure by

Figure 8.13 A giant wave breaking over the bow of the *ESSO Nederland* southbound in the Agulhas Current. The bow of this supertanker is about 25 m (80 ft) above the water.

busiest sea routes, as supertankers carrying oil from the Middle East ride the Agulhas Current on their trip southward to round the Cape of Good Hope en route to Europe and America. The situation is an invitation to trouble, and tankers have been damaged and lost in this area (fig. 8.13). In the North Atlantic, strong northeasterly gales send large storm waves into the edge of the northward-moving Gulf Stream near the border of the continental shelf, resulting in the formation of large waves. The shallow North Sea also seems to provide suitable conditions for extremely high episodic waves during its severe winter storms.

Researchers studying these waves describe them as having a height equal to a seven- or eight-story building (20–30 m, or 70–100 ft) and moving at a speed of 50 knots, with a wavelength approaching a half mile (0.9 km). If such a wave topples onto a vessel that has dropped its bow into the preceding trough, there is no escape from the thousands of tons of water crashing on the deck. In the North Sea, maximum potential wave height for an episodic wave is calculated as 33.8 m (111 ft), but 22.9 m (75 ft) is the highest that has been observed. In the Agulhas Current area, researchers searched the storm records for a twenty-year period and calculated a possible maximum wave height of about 57.9 m (190 ft).

There are many disappearances of vessels for which episodic waves are now suspected of being the chief cause. Many of the casualties are tankers or bulk carriers of ore, grain, and the like. These vessels are susceptible not only because so many are at sea at any one time but also because of their design. Bulk carriers comprise a bow section and an aft (or rear) section that includes the engine and the crew accommodations. These sections are separated by a series of flat-bottomed boxes, or storage tanks, which make up the majority of the vessel's length. Because these vessels are about 300 m (1000 ft) long, they are subject to great wrenching forces if a large portion of the hull is left unsupported or is only partially supported while suspended between wave

crests. More-traditional and smaller hull forms are stronger, ride more easily, and are less likely to be destroyed by these severe waves.

Wave Energy

A wave's energy is present as **potential energy,** due to the change in elevation of the water surface, and as **kinetic energy,** due to the motion of the water particles in their orbits. The higher the wave, the larger the diameter of the water particle orbit and the greater the speed of the orbiting particle; therefore, the greater the kinetic and potential energy. The energy in a deep-water wave is nearly equally divided between kinetic and potential energy. The energy in a deep-water wave can be described as the total energy distributed over one wavelength of the wave, from the sea surface to a depth of $L/2$, per 1 meter of distance measured along the crest of the wave. Wave energy is a function of the square of the wave height, and can be expressed as:

$$E = 1/8 \, \rho \, g \, H^2 \qquad \textit{Equation 8.6}$$

where ρ is water density, g is the acceleration of gravity, and H is wave height. This relationship is demonstrated in figure 8.14. Clearly, wave energy increases rapidly with increasing height. If you double the wave height, wave energy increases by a factor of four.

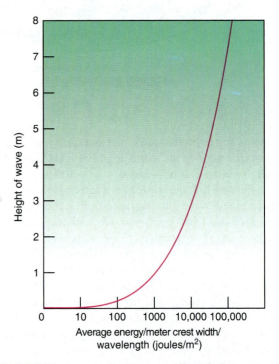

Figure 8.14 Wave energy increases rapidly with the square of the wave height. Average wave energy is calculated per unit width of the crest and averaged over the wavelength *(L)* and the depth *(L/2)*.

Wave Steepness

There is a maximum possible height for any given wavelength. This maximum value is determined by the ratio of the wave's height to the wavelength, and it is the measure of the **steepness** of the wave:

$$\text{steepness} = \frac{\text{height}}{\text{length}} \quad \text{or} \quad S = \frac{H}{L}$$

If the ratio of the height to the length exceeds 1:7, the wave becomes too steep and the wave breaks. For example, if the wavelength is 70 m (230 ft), the wave will break when the wave height reaches 10 m (33 ft). The angle formed at the wave crest approaches 120°, and the wave becomes unstable. Under this condition, the wave cannot maintain its shape; it collapses and breaks (fig. 8.15).

Small, unstable breaking waves are quite common. When wind speeds reach 8–9 m/s (16–18 knots), waves known as whitecaps can be observed. These waves have short wavelengths (about 2 m), and the wind increases their height rapidly. As each wave reaches the critical steepness and crest angle, it breaks and is replaced by another wave produced by the rising wind.

Long waves at sea usually have a height well below their maximum value. Sufficient wind energy to force them to their maximum height rarely occurs. If a long wave does attain maximum height and breaks in deep water, tons of water are sent crashing to the surface. The energy of the wave is lost in turbulence and in the production of smaller waves. Rather than breaking, under such conditions it is more likely that the top of a large wave will be torn off by the wind and cascade down the wave face. This action does not completely destroy the wave and is not considered to be the true collapse, or breaking, of such a wave.

In addition, waves break and dissipate if intersecting wave trains pass through each other in proper phase to form a combined wave with sufficient height to exceed the critical steepness. Waves sometimes run into a strong opposing current, forcing the waves to slow down. Remember that the speed of all waves equals the wavelength divided by the period ($C = L/T$) and that a wave's period does not change. If the speed of a wave is reduced by an opposing current, its wavelength must shorten. In such a case, the wave's energy is confined to a shorter length, so the wave increases in height to satisfy the height-energy relationship. If the increase in height exceeds the maximum allowable height-to-length ratio for the shorter wavelength, the wave breaks. Crossing a sandbar into a harbor or river mouth during an outgoing or falling tide is dangerous because the waves moving over the bar and against the tidal current steepen and break. Entering a harbor or river should be done at the change of the tide or on the rising tide, when the tidal current moves with the waves, stretching their wavelengths and decreasing their heights.

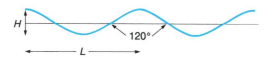

Figure 8.15 Wave steepness. When *H/L* approaches 1:7, the wave's crest angle approaches 120° and the wave breaks.

Universal Sea State Code

In 1806, Admiral Sir Francis Beaufort of the British navy adapted a wind-estimation system from land to sea use. On land, the clues to wind speed included smoke drift, the rustle of leaves, the flapping of flags, slates blowing from roofs, and uprooting of trees. Admiral Beaufort related observations of the sea surface state to wind speed and designed a 0–12 (calm to hurricane) wind scale with typical wave descriptions for each level of wind speed. This Beaufort Scale was adopted by the U.S. Navy in 1838, and the scale was extended from 0–17. At present, a Universal Sea State Code of 0–9, based on the Beaufort Scale, is in international use for wind speeds and related sea surface conditions (table 8.3).

QUICK REVIEW

1. What is the maximum diameter of a wave's water particle orbit?

2. What is the depth at which particle motion will essentially stop in a deep-water wave?

3. Why do waves appear more regular when they are far away from a storm center?

4. Describe the effects of constructive and destructive interference of waves.

5. Compare the energy found in a wave 2 feet high with the energy in a wave 16 feet high.

6. What is the difference between wave speed and group speed? What is the relationship between them?

8.5 Shallow-Water Waves in Detail

Shallow-Water Wave Motion

As a deep-water wave approaches the shore and moves into shallow water, motion within the wave will eventually reach the sea floor and the wave will go through a transitional, or intermediate, phase before becoming a true shallow-water wave. Interaction between the wave and the bottom will begin to affect the shape of the orbits made by the water particles. The orbits gradually become flattened circles, or ellipses (fig. 8.16). The wave begins to "feel" the bottom, and the resulting friction and compression of the orbits reduce the forward speed of the wave.

Remember that (1) the speed of all waves is equal to the wavelength divided by period, and (2) the period of a wave does not change. Therefore, when the wave "feels bottom," it slows, and the accompanying reduction in the wavelength and speed results in increased height and steepness as the wave's energy is condensed in a smaller water volume.

When the wave enters water with a depth of less than one-twentieth the wavelength ($D < L/20$), the wave becomes a **shallow-water wave** (fig. 8.17). Here the wavelength and speed are determined by the square root of the product of Earth's gravitational acceleration (g, 9.81 m/s²) and depth (*D*):

$$C = \sqrt{gD} \quad \text{or} \quad C = 3.13\sqrt{D}$$
$$\text{or} \quad L = 3.13\,T\sqrt{D}$$

Table 8.3   Universal Sea State Code

Sea State Code	Description	Average Wave Height
SS0	Sea like a mirror; wind less than 1 knot	0
SS1	A smooth sea; ripples, no foam; very light winds, 1–3 knots, not felt on face	0–0.3 m 0–1 ft
SS2	A slight sea; small wavelets; winds light to gentle, 4–6 knots, felt on face; light flags wave	0.3–0.6 m 1–2 ft
SS3	A moderate sea; large wavelets, crests begin to break; winds gentle to moderate, 7–10 knots; light flags fully extend	0.6–1.2 m 2–4 ft
SS4	A rough sea; moderate waves, many crests break, whitecaps, some wind-blown spray; winds moderate to strong breeze, 11–27 knots; wind whistles in the rigging	1.2–2.4 m 4–8 ft
SS5	A very rough sea; waves heap up, forming foam streaks and spindrift; winds moderate to fresh gale, 28–40 knots; wind affects walking	2.4–4.0 m 8–13 ft
SS6	A high sea; sea begins to roll, forming very definite foam streaks and considerable spray; winds at strong gale, 41–47 knots; loose gear and light canvas may be blown about or ripped	4.0–6.1 m 13–20 ft
SS7	A very high sea; very high, steep waves with wind-driven overhanging crests; sea surface whitens due to dense coverage with foam; visibility reduced due to wind-blown spray; winds at whole gale force, 48–55 knots	6.1–9.1 m 20–30 ft
SS8	Mountainous seas; very high-rolling breaking waves; sea surface foam-covered; very poor visibility; winds at storm level, 56–63 knots	9.1–13.7 m 30–45 ft
SS9	Air filled with foam; sea surface white with spray; winds 64 knots and above	13.7 m and above 45 ft and above

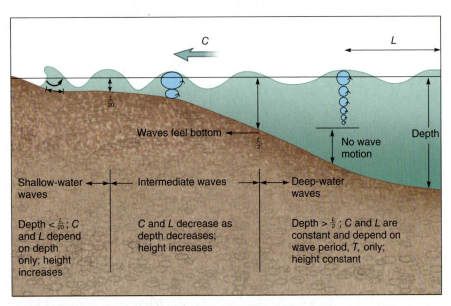

Figure 8.16 Deep-water waves become intermediate waves and then shallow-water waves as depth decreases and wave motion interacts with the sea floor.

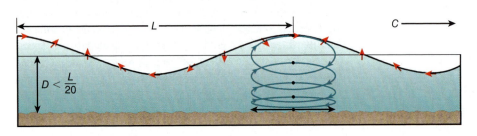

When the condition for a shallow-water wave is met, the orbits of the water particles are elliptical; they become flatter with depth until, at the sea floor, only a back-and-forth oscillatory motion remains (fig. 8.17). Note that the horizontal dimension of the orbit remains unchanged in shallow water.

When the water depth is between $L/2$ and $L/20$, the speed of the wave is also slowed. Waves in this depth range are called intermediate waves. No simple algebraic equation exists to determine the speed of these intermediate waves.

Shallow-Water Wave Speed

Let's consider the general wave speed equation once again.

$$C = \sqrt{\frac{gL}{2\pi} \tanh \frac{2\pi D}{L}} \qquad \textit{Equation 8.2}$$

Figure 8.17 A shallow-water wave sets water particles in elliptical motion. The height of the ellipse decreases with depth until, at the bottom, the water simply moves back and forth, perpendicular to the direction the wave is traveling.

When waves travel in water shallower than one-twentieth their wavelength:

$$\tanh \frac{2\pi D}{L} \approx \frac{2\pi D}{L}$$

consequently, for shallow-water waves, the general wave speed equation reduces to:

$$C_{shallow} = \sqrt{gD} \quad \text{or} \quad C_{shallow} = 3.13\sqrt{D}$$
$$\text{or} \quad L = 3.13 \, T \sqrt{D} \qquad \textit{Equation 8.7}$$

where speed is in meters per second and depth is in meters. The speed of shallow-water waves depends only on water depth. All shallow-water waves travel at the same speed for a given depth of water. The group speed, V, of shallow-water waves is equal to the speed, C, of each wave in the group. In other words, shallow-water waves are not dispersive waves.

Refraction

Waves are refracted, or bent, as they move from deep to shallow water, begin to feel the bottom, and change wavelength and wave speed. When waves from a distant storm center approach the shore, they are likely to approach the beach at an angle. One end of the wave crest comes into shallow water and begins to feel bottom, while the other end is in deeper water. The shallow-water end moves more slowly than that portion of the wave in deeper water. The result is that the wave crests bend, or refract, and tend to become oriented parallel to the shore. This pattern is shown in figure 8.18. The **wave rays** drawn perpendicular to the crests show the direction of motion of the wave crests. Note that the refraction of water waves is similar to the refraction of light and sound waves, described in chapter 4.

When approaching an irregular coastline with headlands jutting into the ocean and bays set back into the land, waves may encounter a submerged ridge seaward from a headland or a depression in front of a bay. As waves approach such a coastline and feel the bottom, the portion of the wave crest over the ridge slows down more than the wave crests on either side. Therefore, the crests wrap around the headland in the pattern shown in figure 8.19. This refraction pattern focuses wave energy on the headland. Waves must gain in height as their wavelength is shortened and the total wave energy is crowded into a smaller volume of water; this increases the energy per unit of surface area. Therefore, more energy is expended on a unit length of shore at the point of the headland than on a unit length of shore elsewhere.

The central area at the mouth of the bay is usually deeper than the areas to each side, so the advancing waves slow down more on the sides than in the center. Because the wavelengths remain long in the center and shorten on each side, the wave crests bulge toward the center of the bay. The waves in the center of the bay do not shorten as much; therefore, they have less height

and less energy to expend per length of shoreline. This pattern is shown in figures 8.20 and 8.21. The result is an environment of low wave energy, providing sheltered water in the bay. Overall,

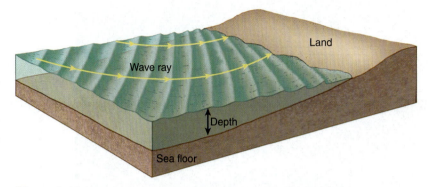

Figure 8.18 Waves moving inshore at an oblique angle to the depth contours are refracted. One end of the wave reaches a depth of $L/2$ or less and slows, while the other end of the wave maintains its speed in deeper water. Wave rays drawn perpendicular to the crests show the direction of wave travel and the bending of the wave crests.

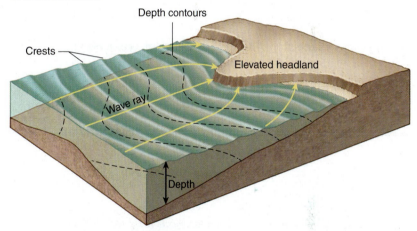

Figure 8.19 The energy of waves refracted over a shallow, submerged ridge is focused on the headland. The converging wave rays show the wave energy being crowded into a small volume of water, increasing the energy per unit length of wave crest as the height of the wave increases.

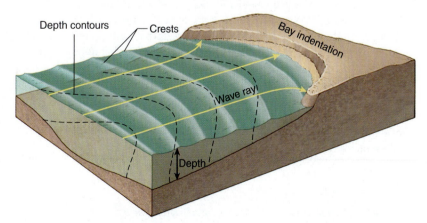

Figure 8.20 Waves refracted by the shallow depths on each side of the bay deliver lower levels of energy inside the bay. The diverging wave rays show the spreading of energy over a larger volume of water, decreasing the energy per unit length of wave crest as the wave height decreases.

Figure 8.21 Wave refraction in an embayment.

this unequal distribution of wave energy along the coast results in a wearing down of the headlands and a filling in of the bays, as the sand and mud settle out in the quieter water. If all coastal materials had the same resistance to wave erosion, this process would lead in time to a straightening of the coastline; but the rocky structure of many headlands resists wave erosion; therefore, the cliffs remain.

Reflection

A straight, smooth, vertical barrier in water deep enough to prevent waves from breaking reflects the waves (fig. 8.22). The barrier may be a cliff, steep beach, breakwater, bulkhead, or other structure. The reflected waves pass through the incoming waves to produce an interference pattern, and steep, choppy seas often result. If the waves reflect directly back on themselves, the resulting waves appear to stand still, rising and falling in place. The behavior of waves reflected from a curved vertical surface depends on the type of curvature. If the curvature is convex, the reflected wave rays spread and disperse the wave energy, but if the curved surface is concave, the reflected wave rays converge and the energy is focused. This situation is similar to the reflection of light from curved mirrors. Great care must be taken in designing walls and barriers to protect an area from waves to be sure that the energy of reflected waves is not focused in a way that will result in another area being damaged.

Diffraction

Another phenomenon that is associated with waves as they approach the shore or other obstacles is **diffraction.** Diffraction is caused by the spread of wave energy sideways to the direction of wave travel. If waves move toward a barrier with a small opening (fig. 8.23a), some wave energy passes through the small opening

(a)

(b)

(c)

Figure 8.22 (a) A wave approaches the coast at an angle. Arrows are behind the wave crest, pointing in the direction of propagation. (b) The wave hits the coastline. (c) The reflected wave moves away from the coast at the same angle that it initially hit the coast. Arrows are again behind the wave crest, pointing in the direction of propagation.

to the other side. Once the waves have passed through the opening, their crests decrease in height and radiate out and away from the gap. A portion of the wave energy is diffracted, transported sideways from its original direction. If more than one gap is open to the waves, the patterns produced by the spreading waves from the openings may intersect and form interference patterns as the waves move through each other (fig. 8.23*b*).

If the waves approach a barrier without an opening, diffraction can still occur. Energy will be transported at right angles to the wave crests as the waves pass the end of the barrier (fig. 8.24). Note that the pattern produced is one-half of the pattern observed when the waves pass through a narrow opening, as in figure 8.23*a*. Energy is transported behind the sheltered (or lee) side of the barrier. This effect is an important consideration

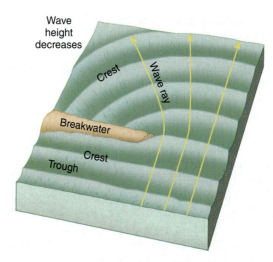

Figure 8.24 Diffraction occurs behind the breakwater.

in planning the construction of breakwaters and other coastal barriers intended to protect vessels in harbors from wave action and possible damage.

QUICK REVIEW

1. How does a deep-water wave change when it converts to a shallow-water wave?
2. What property of waves does not change when they travel from deep water into shallow water?
3. Why do waves change direction as they move into shallow water?
4. On what does the speed of shallow-water waves depend?

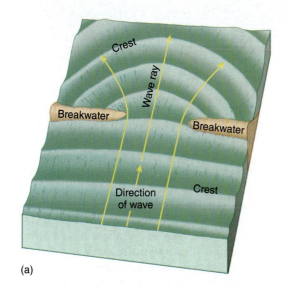

(a)

8.6 The Surf Zone

The surf zone is the shallow area along the coast in which the waves slow rapidly, steepen, break, and disappear in the turbulence and spray of expended energy. The width of this zone is variable and is related to both the wavelength and height of the arriving waves and the changing depth pattern. Longer, higher waves, which feel the bottom before shorter waves, become unstable and break farther offshore in deeper water. If shallow depths extend offshore for some distance, the surf zone is wider than it is over a sharply sloping shore.

Breakers

Breakers form in the surf zone because the water particle motion at depth is affected by the bottom. Orbital motion is slowed and compressed vertically, but the orbit speed of water particles near the crest of

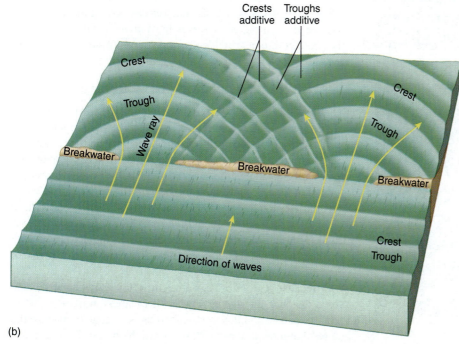

(b)

Figure 8.23 Diffraction patterns produced by waves passing through narrow openings (a, b). Note the interference pattern produced when diffracted waves interact (b).

the wave is not slowed as much. The particles at the wave crest move faster toward the shore than the rest of the wave form, resulting in the curling of the crest and the eventual breaking of the wave. The two most common types of breakers are **plungers** and **spillers** (fig. 8.25).

Plunging breakers form on narrow, steep beach slopes. The curling crest outruns the rest of the wave, curves over the air below it, and breaks with a sudden loss of energy and a splash. The more common spilling breaker is found over wider, flatter beaches, where the energy is extracted more gradually as the wave moves over the shallow bottom. This action results in the less dramatic wave form, consisting of turbulent water and bubbles flowing down the collapsing wave face. The spilling breakers last longer than the plungers because they lose energy more gradually. Therefore, spillers give surfers a longer ride, but plungers give them a more exciting one.

The slow curling over of the crest observed on some breakers begins at a point on the crest and then moves lengthwise along the wave crest as the wave approaches shore. This movement

Figure 8.26 A surfer rides the tube of a large curling wave in Hawaii.

of the curl along the crest occurs because waves are seldom exactly parallel to a beach. The curl begins at the point on the crest that is in the shallowest water or the point at which the crest height is slightly greater, and it moves along the crest as the rest of the wave approaches the beach. The result is the "tube" so sought after by surfers (fig. 8.26).

If the waves approaching the beach are uniform in length, period, and height, they are the swells from some far-distant storm, which have had time and distance to sort into uniform groups. For example, the long surfing waves of the California beaches in summer begin in the winter storms of the South Pacific and Antarctic Oceans. If the waves are of different heights, lengths, and periods and break at varying distances from the beach, then unsorted waves have arrived and are probably the product of a nearby local storm superimposed on the swell.

Water Transport and Rip Currents

The small net drift of water in the direction the waves are traveling is intensified in the surf zone, as the shoreward motion of the water particles at the crest becomes greater than the return particle motion at the trough. Because the crests usually approach the beach at an angle, the surf zone transport of water flows both toward the beach and along the beach. The result is that water accumulates against the beach and flows along the beach until it can flow seaward again and return to the area beyond the surf zone. This return flow generally occurs in quieter water with smaller wave heights, for example, in areas with troughs or depressions in the sea floor.

Because regions of seaward return flow may be narrow and some distance apart, the flow in these areas must be swift in order to carry enough water beyond the surf zone to balance the slower but more extensive flow toward the beach. These regions of rapid seaward flow are called **rip currents.** Rip currents can be a major hazard to surf swimmers.

(a)

(b)

Figure 8.25 Breaking waves. A plunger (a) loses energy more quickly than a spiller (b).

Swimmers who unknowingly venture into a rip current will find themselves carried seaward and unable to swim back to shore against the flow; they must swim parallel to the beach, or across the rip current, and then return to shore. Because of the danger associated with rip currents, swimmers should be on the lookout for indicators of their presence, including (1) turbid water and floating debris moving seaward through the surf zone, (2) areas of reduced wave heights in the surf zone, and (3) depressions in the beach running perpendicular to the shore.

Wave action on the beach stirs up the sand particles and temporarily suspends them in the water. The sand is carried along the beach parallel to the shore until the rip current is reached, and the sand is transported seaward. Viewed from a height, such as a high cliff or a low-flying airplane, rip currents are seen as streaks of discolored turbid water extending seaward through the clearer water of the outer surf zone (fig. 8.27).

Energy Release

Watching the heavy surf pounding a beach from a safe vantage point is an exciting and exhilarating experience; the trick is to determine at what point one is safe. In a narrow surf zone during a period of very large waves, the wave energy must be expended rapidly over a short distance. Under these conditions, the height of the waves and the forward motion of the water particles combine to send the water high up on the beach. The accompanying release of energy is explosive and can result in rocks and debris from the water's edge being hurled high up on the beach by the force of the water. Minot's Ledge Light on the south side of Massachusetts Bay is 30 m (100 ft) high, but it is regularly engulfed in spray (fig. 8.28). Lonely Tillamook Light off the

Figure 8.28 Minot's Ledge Light hit by waves in a winter storm.

Oregon coast had to have steel gratings installed to protect the glass that shields the light 40 m (130 ft) above sea level, after the glass had been broken several times by wave-thrown rocks. In every winter storm, waves displace boulders weighing many tons from breakwaters along the world's coasts.

Waves do not always expend their energy on the shore. Some break farther seaward on sandbars such as those associated with river mouths and estuaries. The famous bar at the mouth of the Columbia River is responsible for extremely hazardous conditions for both fishing and commercial vessels, and in winter it often produces casualties. On an ebbing (or falling) tide, waves approaching the bar against the current rise and break at heights up to 20 m (66 ft). The Coast Guard uses the Columbia River entrance to train its crews to operate roll-over, or self-righting, rescue boats, which are used during storms and heavy surf conditions.

QUICK REVIEW

1. What is the difference between a spiller and a plunger?
2. What can you say about the distance that regular swell waves have traveled when they arrive along a coast?
3. What should you do if you are caught in a rip current?
4. Why are rip currents often a different color from the surrounding water?
5. Why does a gently sloping beach offer protection for the shore from heavy storm waves?

Figure 8.27 Small rip currents carry turbid water seaward through the surf zone.

8.7 Tsunami

Sudden movements of Earth's crust may produce a **seismic sea wave,** or **tsunami.** These waves are often incorrectly called tidal waves. Because a seismic sea wave has nothing to do with tides, oceanographers have adopted the Japanese word *tsunami,* meaning harbor wave, to replace the misleading term *tidal wave.* Tsunami is a synonym for seismic sea wave.

If a large area, maybe several hundred square kilometers, of Earth's crust below the sea surface is suddenly displaced, it may cause a sudden rise or fall in the overlying sea surface. In the case of a rise, gravity causes the suddenly elevated water to return to the equilibrium surface level; if a depression is produced, gravity causes the surrounding water to flow into it. Both events result in the production of waves with extremely long wavelengths (100–200 km, or 60–120 mi) and long periods as well (10–20 minutes). The average depth of the oceans is about 4000 m (4 km, or 13,000 ft); this depth is less than one-twentieth the wavelength of these waves, so tsunamis are shallow-water waves. These waves radiate from the point of the seismic disturbance at a speed determined by the ocean's depth ($C = \sqrt{gD}$) and move across the oceans at about 200 m/s (400 mph). Because they are shallow-water waves, tsunamis may be refracted, diffracted, or reflected in mid-ocean by changes in seafloor topography and by mid-ocean islands.

When a tsunami leaves its point of origin, it may have a height of several meters, but this height is distributed over its many-kilometer wavelength. It is not easily seen or felt when superimposed on the other motions of the sea's surface, and a vessel in the open ocean is in little or no danger if a tsunami passes. The danger occurs only if the vessel has the misfortune to be directly above the area of the original seismic disturbance.

The energy of a tsunami is distributed from the ocean surface to the ocean floor and over the length of the wave. When the path of the wave is blocked by a coast or an island, the wave behaves like any other shallow-water wave; it slows (to approximately 80 km [50 mi] per hour), its wavelength decreases, and its energy is compressed into a smaller water volume as the depth rapidly decreases. This sudden confinement of energy to a smaller volume increases the energy density and causes the wave height to build rapidly, and the loss of energy is equally rapid when the wave breaks. A tremendous surge of moving water races up over the land, flooding the coast for a period that lasts 5–10 minutes before the water flows seaward, exposing the nearshore sea floor. The surge destroys buildings and docks; large ships may be left far up on the shore. The receding water carries the debris from the surge, battering and buffeting everything in its path. In bays and inlets, the rising and falling of the water may not be as destructive as the extreme currents that form at the entrance of a bay or an inlet where large volumes of water oscillate in and out, producing dangerous and destructive current surges. If these oscillating flows occur at the natural oscillation period of the bay, large changes in water level are created.

The leading edge of the tsunami wave group may be either a crest or a trough. If the initial crustal disturbance is an upward motion, a crest arrives first; if the crustal motion is downward, a trough precedes the crest. If a trough arrives at the shore first, the water level drops by as much as 3–4 m (10–13 ft) within 2–3 minutes. People who follow the receding water to inspect exposed sea life may drown, for in another 4–5 minutes, the water rises 6–8 m (20–26 ft), and they will not be able to out-run the wall of advancing water. In still another 4–5 minutes, a trough with a water-level drop of 6–8 m (20–26 ft) arrives, and the water flowing away from the beach will carry debris out to sea.

Tsunamis are most likely to occur in ocean basins that are tectonically active. The Pacific Ocean, ringed by crustal faults and volcanic activity, is the birthplace of most tsunamis. They also appear in the Indian Ocean, in the Caribbean Sea, which is bounded by an active island arc system, and in the Mediterranean Sea. The spectacular havoc and destruction caused by these waves are well recorded. In August 1883, the Indonesian island of Krakatoa erupted and was nearly destroyed in a gigantic volcanic explosion that hurled several cubic miles of material into the air. A series of tsunamis followed; these waves had unusually long periods of one to two hours. The town of Merak, 53 km (33 mi) away and on another island, was inundated by waves over 30.5 m (or 100 ft) high, and a large ship was carried nearly 3 km (2 mi) inland and left stranded 9.2 m (30 ft) above sea level. More than 35,000 people died from these enormous waves, and as the waves moved across the oceans, water-level recorders were affected as far away as Cape Horn (12,500 km, or 7800 mi) and Panama (18,200 km, or 11,400 mi). The waves were traveling with a speed calculated at about 200 m/s, or 400 mi/h.

In recent history there have been a number of destructive tsunamis. On April 1, 1946, an earthquake in the area of the Aleutian Trench, off the coast of Alaska, produced a series of tsunamis that heavily damaged Hilo, Hawaii, killing more than 150 people. On September 1, 1992, a magnitude 7.0 earthquake 100 km (60 mi) off the coast of Nicaragua produced tsunamis that killed 170 people. Eyewitnesses recall a single wave, preceded by a trough that lowered the harbor depth by 7 m (23 ft). On December 12, 1992, a magnitude 7.8 earthquake in the western Pacific generated a tsunami that struck Flores Island in Indonesia. The earthquake and tsunami reported killed 2500 people in and around the town of Maumere. On July 12, 1993, a tsunami struck Okushiri Island, Japan, claiming 185 lives and causing an estimated $600 million in property damage. A very large tsunami struck the northwest coast of Papua, New Guinea, on July 17, 1998. The 7–12 m (22–50 ft) high tsunami destroyed four fishing villages. More than 2000 people were killed or reported missing.

The most destructive tsunami in history in terms of loss of life was the December 26, 2004, Sumatra tsunami that devastated coastlines in the Indian Ocean basin. Shortly before eight o'clock in the morning, a massive earthquake ruptured the sea floor off the northwest coast of Sumatra, resulting in as much as 10 m (33 ft) of vertical displacement. This earthquake was the result of the subduction of the Indian Plate beneath the Eurasian Plate. An international team of scientists studying the effects of the tsunami in Sumatra documented wave heights of 20–30 m (65–100 ft) at the island's northwest end and found evidence suggesting that wave heights may have ranged from 15–30 m

(a)   (b)   (c)

Figure 8.29 (a) A mosque is left standing amid the rubble in Banda Aceh, Sumatra. Several mosques survived and may have been saved by the open ground floor that is part of their design. The tsunami waves reached the middle of the second floor. (b) Boat carried inland by the tsunami. (c) Typical tsunami damage in Banda Aceh. (Photos courtesy of the United States Geological Survey.)

(50–100 ft) along at least a 100 km (62 mi) stretch of the northwest coast. The worst damage is thought to have occurred in the province of Aceh, roughly 100 km (62 mi) from the epicenter of the earthquake (fig. 8.29a–c). About one-third of the 320,000 residents of Aceh's capital, Banda Aceh, are presumed to have been killed.

The tragic loss of life caused by the Sumatra tsunami was compounded by the lack of tsunami detection instruments in the Indian Ocean basin, similar to those in the Pacific Ocean basin,

and a coordinated tsunami alert plan in the region. Tsunami Prediction and Warning Centers were first located in Hawaii and Alaska in 1946, after the April 1, 1946, tsunamis that struck Hawaii. Tsunami detection in the Pacific Ocean basin is now aided by the National Oceanic and Atmospheric Administration's (NOAA) Deep-Ocean Assessment and Reporting of Tsunamis (DART) program. The DART program operates forty open-ocean tsunameters for detecting tsunamis (fig. 8.30). Additional tsunameters are operated by the Australian, Chilean,

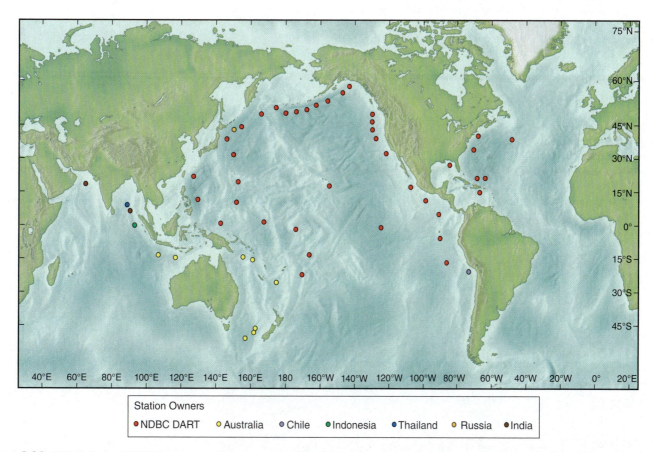

Station Owners

● NDBC DART   ○ Australia   ● Chile   ● Indonesia   ● Thailand   ○ Russia   ● India

Figure 8.30 NOAA's Project DART (Deep-ocean Assessment and Reporting of Tsunamis). Tsunameters are deployed near regions with a history of tsunami generation, to ensure measurement of the waves as they propogate toward coastal communities and to acquire data critical to real-time forecasts. Locations of the forty tsunameters comprising the network (*red* circles) are shown on this map along with additional tsunameters operated by the Australian, Chilean, Indonesian, Thai, Russian, and Indian governments.

Diving in

The March 11, 2011 Japanese Tsunami

BY DR. EDDIE BERNARD

Dr. Eddie Bernard is an expert in the study of tsunamis. He has published over eighty-five scientific papers and three books. He served as Director of NOAA's Pacific Tsunami Warning Center and led the U.S. team that surveyed the damage caused by the 1993 Sea of Japan tsunami. He served as founding Chairman of the National Tsunami Hazard Mitigation Program, a joint federal/state effort.

Tsunamis rank high on the scale of natural disasters. Since 1850 alone, tsunamis have been responsible for the loss of over 450,000 lives and over $311 billion in damage to coastal structures and habitats. Most of these casualties were caused by local tsunamis that occur about once per year somewhere in the world. Predicting when and where the next tsunami will strike is currently not possible. Once a tsunami is generated, however, forecasting its arrival and the extent of flooding is possible through recently developed tsunami modeling and measurement technologies.

Tsunamis can be characterized through three distinct stages of evolution: **generation**, **propagation**, and **Inundation**, or **flooding** (Box fig. 1). These terms are used in describing the March 11, 2011 Japanese tsunami that killed about 20,000 people and caused over $300 billion in economic loss to Japan, *making it the world's most costly natural disaster*. Although the earthquake was powerful, the tsunami was responsible for the huge scale of the catastrophe. A preliminary report released in April 2011 summarizing autopsy results showed 96% of the deaths were tsunami related. For the survivors, two months after the tsunami, 130,229 evacuees were still in 2,559 shelters.

Generation

About 80% of historical tsunamis have been generated by earthquakes in marine and coastal regions. When a powerful earthquake (magnitude 9.0) struck the coastal region of Japan at 2:46 P.M. on March 11, 2011, movements of the sea floor produced a tsunami of about 40 meters along the adjacent coastline. This tsunami was measured in real time by four deep-ocean tsunami detectors, providing an estimate of the ocean surface disturbance within 1 hour of generation (Box fig. 2).

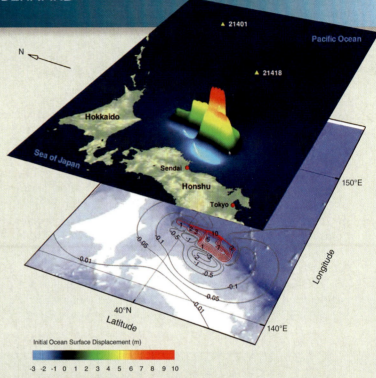

Initial Ocean Surface Displacement (m)

-3 -2 -1 0 1 2 3 4 5 6 7 8 9 10

Box Figure 2 Ocean surface disturbance as derived from deep-ocean tsunami measurements approximately 1 hour after the March 11, 2011 Japanese earthquake began rupturing.   Source: National Oceanic and Atmospheric Administration (NOAA).

Propagation

Because earth movements associated with large earthquakes are thousands of square kilometers in area, any movement of the sea floor immediately changes the sea surface. The resulting tsunami propagates as a set of waves whose energy is concentrated at wavelengths corresponding to the Earth movements (~100 km), at wave heights determined by vertical and horizontal displacement (~10 m), and in wave directions determined by the adjacent coastline geometry and bathymetry. Box figure 2 illustrates the initial sea surface distortion associated with the 2011 tsunami off the coast of northern Japan: The depression rapidly reached the nearby Japanese coast, causing a drawdown of water before flooding, whereas the seaward elevation propagated across the basin and struck the coasts of Hawaii, California, and Chile hours later. Box figure 3 illustrates the sea surface conditions, derived from a propagation model, at one hour propagation time from the source area. Basin-wide distribution of tsunami wave energy is illustrated in box figure 4. The tsunami spread from the generation location near Japan (Box fig. 2) and proceeded to flood Kakului, Hawaii 6 hours later, and damage Crescent City, California harbor 4 hours after flooding Hawaii. As shown in box figure 4, the tsunami energy was steered and amplified by the Emperor Seamounts and Mendocino Escarpment lying along 40°N latitude.

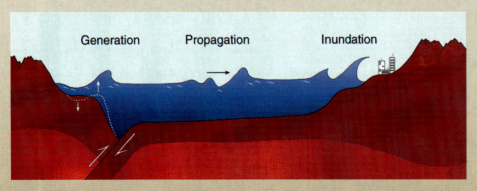

Box Figure 1   Stages of tsunami evolution: generation, propagation, and flooding.

Box Figure 3 Propagation of the 2011 Japanese tsunami away from source depicted in Box figure 2 at 1 hour after generation. The leading edge of the tsunami passed over the first tsunami detector (21418) at 0.5 hour and had reached two detectors within 1 hour. Source: National Oceanic and Atmospheric Administration (NOAA).

Box Figure 5 Tsunami damage at Minami-Sanriku about 40 km south of Rikuzen-Takata. This town had population of 17,000, of which about 1,000 were killed or still missing.

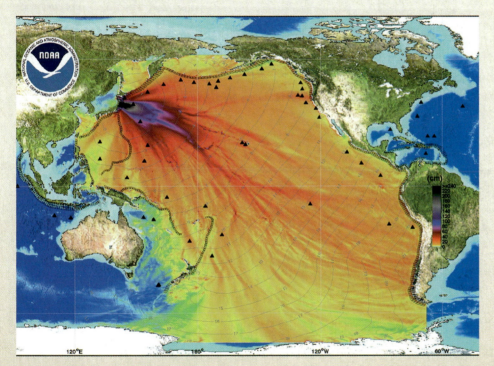

Box Figure 4 Distribution of tsunami energy from the source of the March 11, 2011 tsunami as depicted in Box figure 2. Pacific Ocean bathymetry scatters and focuses tsunami wave energy from the source to distant shorelines. *Warm colors* represent high-energy pathways. *Black* triangles are locations of deep-ocean tsunami detectors that recorded this powerful tsunami. Source: National Oceanic and Atmospheric Administration (NOAA).

Inundation

As a tsunami approaches the coastline, the wave energy is concentrated in smaller volumes of ocean, amplifying the tsunami heights. Box figure 5 illustrates the devastation caused by inundation. The highest water levels (38.9 m) at Aneyoshi Bay south of Miyako City in Iwate Prefecture were the maximum ever measured in a Japan tsunami. Water heights were close to or exceeded 20 meters in most populated coastal communities in Iwate and northern Miyagi prefectures. On the broad plain that characterizes the coast of Miyagi Prefecture south of Sendai, peak water heights averaged 8–10 meters. There were significant tsunami impacts as far south as Chiba Prefecture. The dense city and town centers were

especially vulnerable, even though much of the town or city land area was outside of the inundation zone on the hill slopes and farther inland. The tide gauges show the first tsunami wave arriving 36 minutes after the earthquake at Hachinohe and 29 minutes in Okai Town in Chiba Prefecture. Eyewitnesses in Northern Miyagi and Southern Iwate Prefectures generally reported 25–30 minutes between the earthquake and the tsunami.

Preliminary findings indicate that Japan, the most tsunami-prepared nation in the world, had underestimated tsunami impacts for evacuation planning and coastal structures design. For example, Minami-Sanriku Town (population 17,000) had gained an international reputation for tsunami preparedness before the tsunami and was a featured field-trip stop for tsunami experts. The three rivers flowing through the town featured tsunami gates that could be shut in 15 minutes to keep the tsunami from flooding inland through the river channels. The tsunami extended nearly 3 km up the Hachiman River and nearly 2 km up the adjacent river valleys (see Box fig. 5). Officials successfully lowered the gates on March 11, but the tsunami overtopped the adjacent sea walls and flooded the city. At the Disaster Management Center (Box fig. 6), more than thirty officials, including the town mayor, gathered on the roof-top during the tsunami event, and twenty of those died. ***This tragedy was repeated throughout the region where an estimated thirty-one of eighty designated tsunami evacuation buildings were destroyed (Japan Times, 2011).***

These dramatic accounts certainly bring into focus the importance of the accuracy of hazard assessment. Other reports on public response to the 2011 Japanese tsunami concluded that the warning messages, which upgraded the danger level *four* times, may have contributed to

Continued next page—

233

Diving in *Continued—*

Box Figure 6 The disaster management headquarters for the town of Minamisanriku. About thirty officials gathered on the upper floor and roof on March 11. The tsunami completely flooded the structure and only eleven people survived. Note the location of high ground in the background.

the underestimate of the size of the tsunami because some evacuees may not have received later warning messages of greater danger. The Japan tsunami in 2011 also demonstrated that existing tsunami warning products are confusing to the public. The confusion arises, in part, from using text messages to convey complex information under stressful conditions. An evaluation report of tsunami warning performance during the 2010 Chile tsunami recommended the use of graphic products to reduce public confusion. With smartphone technology, it is now possible to disseminate easy-to-understand graphical flooding products, as shown in box figure 7.

Box Figure 7 Future tsunami warning product delivered via smartphone.

Indonesian, Thai, Russian, and Indian governments. Each instrument consists of a surface buoy (fig. 8.31*a*) for real-time data transmission connected to an anchored seafloor bottom package that includes a bottom pressure recorder (fig. 8.31*b*). Each instrument is designed to be deployed for twenty-four months at depths of up to 6000 m (19,700 ft). These instruments have measured tsunamis characterized by amplitudes less than 1 cm in the deep ocean. Data are transmitted from the bottom pressure recorder on the sea floor to the surface buoy and then relayed via satellite to ground stations.

The tsunameters operate in two modes: standard and event. In standard mode, they measure water pressure due to the height of the water column above them every fifteen seconds to obtain an average value over a fifteen minute period of time. In this mode, four measurements are transmitted each hour. When a computer on the instrument detects a possible tsunami, the instrument goes into event mode. In event mode, the instrument transmits fifteen-second values during the initial few minutes, followed by one-minute averages. The system returns to standard mode after four hours of one-minute transmissions if no additional events are detected.

The March 11, 2011 Japanese tsunami killed about 20,000 people and caused over $300 billion in economic loss to Japan. This event is discussed in detail in the Diving In box entitled *"The March 11, 2011 Japanese Tsunami."* Preliminary findings indicate that Japan, the most tsunami-prepared nation in the world, had underestimated tsunami impacts for evacuation planning and the design of coastal structures. The powerful tsunami was

recorded at thirty tsunameter stations. Data from four of these detectors (fig. 8.31) near Japan were used to accurately forecast tsunami flooding for Kahului, Hawaii, six hours before the tsunami struck Hawaii (fig. 8.32). An evacuation of the area was ordered, saving lives from drowning.

QUICK REVIEW

1. Are tsunamis deep-water waves or shallow-water waves? Describe your reasoning.
2. Why are tsunamis unnoticed by sailors in the open ocean?
3. Describe what you would see along a coastline if the trough of a tsunami was the first part of the wave to arrive. What about if the crest was the first to arrive?
4. Why do tsunamis have variable speed as they cross ocean basins?

8.8 Internal Waves

The waves discussed to this point have all formed at the interface of the atmosphere and the ocean. This interface marks the common boundary between two fluids of different densities, air and water. Another interface between two fluids lies below the ocean surface at the pycnocline that separates the shallow mixed layer from the denser underlying water. In this case, the boundary is less abrupt and the density difference is not as great as it is at the

(a)

(b)

Figure 8.31 (a) A DART tsunameter surface buoy. (b) A bottom pressure recorder used to detect variations in sea surface elevation by measuring changes in water pressure. Data obtained by the pressure recorder are transmitted to the buoy and then on to a satellite which relays the data to tsunami warning centers.

Figure 8.32 Tsunami flooding forecast for Kahului, Hawaii resulting from the March 11, 2011 Japanese tsunami. This forecast was made 6 hours before the tsunami arrived. Comparison with field measurements and video of the tsunami show 80% agreement between the model and observations.

the density boundary (pycnocline or thermocline) depth and decreases downward as well as upward from this depth. When wave heights are large, the crests of the internal waves may show at the sea surface as moving bands. The water over the crests of the internal waves often shows ripples. If the amplitude of the wave approaches the thickness of the surface layer, the deeper water may also be seen breaking the sea surface. The water over the trough of the internal wave is generally smooth. Sometimes, instead of visible bands, elevations and depressions of the sea surface occur as the internal waves pass (fig. 8.33*b*).

Many processes are responsible for internal waves. A low-pressure storm system may elevate the sea surface and depress the pycnocline. When the storm moves away, the displaced pycnocline will oscillate as it returns to its equilibrium level. If the speed of a surface current changes abruptly at the pycnocline, internal waves may be generated. Currents moving over rough bottom topography may also produce internal waves. If a thin layer of low-density surface water allows a ship's propeller to reach the pycnocline, the energy from the propeller creates internal waves; under this condition, the ship's propeller becomes inefficient because internal waves created at the pycnocline carry energy away from the vessel instead of driving the vessel forward. This results in a loss of speed that mariners call the "dead water effect."

The relationship of wavelength and depth to wave speed is similar for both internal and surface waves. When the internal waves are short relative to the water depth, the density (rho = ρ) of both layers must be included in the equation. Wave speed squared (C^2) is equal to Earth's acceleration due to gravity (*g*) divided by 2π times the wavelength (*L*) times a ratio of densities, where ρ (rho) is the density of the lower layer and ρ' is the density of the upper layer:

$$C^2 = \frac{g}{2\pi} L \left[\frac{\rho - \rho'}{\rho + \rho'} \right]$$

When the internal waves are long relative to the water depth, it is necessary to include other relationships. For this equation, see appendix C.

air-water boundary. The waves that form along this boundary are known as **internal waves** (fig. 8.33). These internal waves cause the boundary to oscillate as the wave form progresses between the water layers.

Internal waves are slower than surface waves. They typically have wavelengths from hundreds of meters to tens of kilometers and periods from tens of minutes to several hours. Their height often exceeds 50 m and may be limited by the thickness of the surface layer. The orbital motion generated by internal waves of water particles is illustrated in fig. 8.33*a*. The radius of the circular motion of the water particles is largest at

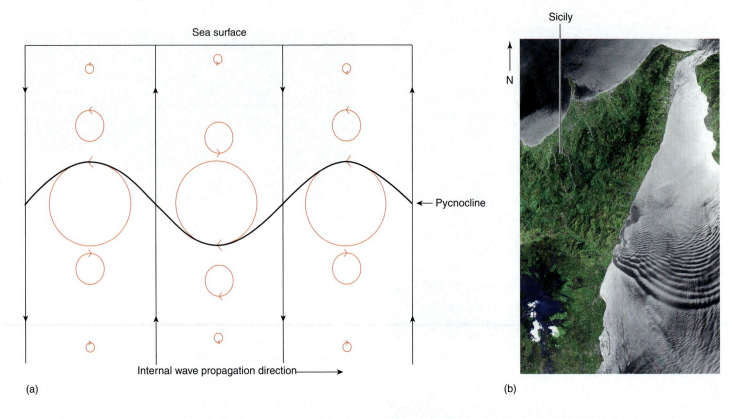

(a)

(b)

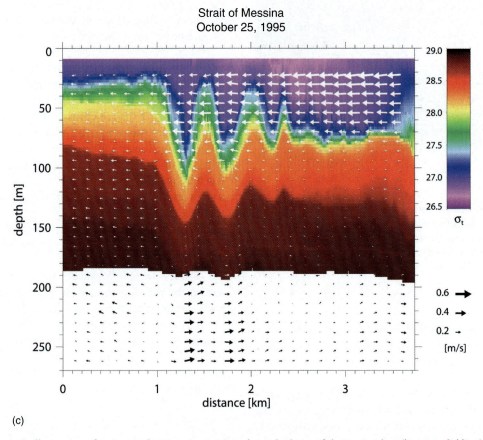

(c)

Figure 8.33 (a) Schematic illustration of an internal wave propagating along the base of the pycnocline (heavy solid line). The orbital motion of water particles is indicated by dashed lines. (b) Satellite image (58 km × 90 km, 36 mi × 56 mi) of internal waves moving through the Strait of Messina separating the island of Sicily from the Italian peninsula. (c) Color-coded density variations with depth (as measured by a CTD) revealing the movement of internal waves in the Strait of Messina. The speed of the waves (m/s) is indicated by the arrows.

QUICK REVIEW

1. Where is the diameter of the particle motion in an internal wave largest?
2. Why are internal waves often larger and slower than surface waves?

8.9 Standing Waves

Deep-water waves, shallow-water waves, and internal waves are all progressive waves; they have a speed and move in a direction. **Standing waves** do not progress; they are progressive waves reflected back on themselves and appear as an alternation between a trough and a crest at a fixed position. They occur in ocean basins, partly enclosed bays and seas, and estuaries. A standing

wave can be demonstrated by slowly lifting one end of a container partially filled with water and then rapidly but gently returning it to a level position. If this is done, the surface alternately rises at one end and falls at the other end of the container. The surface oscillates about a point at the center of the container, the **node;** the alternations of low and high water at each end are the **antinodes** (fig. 8.34a). A standing wave is a progressive wave reflected back on itself; the reflection cancels out the forward motions of the initial and reflected waves. If different-sized containers are treated the same way, the period of oscillation increases as the length of the container increases or its depth decreases.

Notice that the single-node standing wave contains one-half of a wave form (fig. 8.34a). The crest is at one end of the container, and the trough is at the other end. As the wave oscillates, a trough replaces the crest, and a crest replaces the trough. One

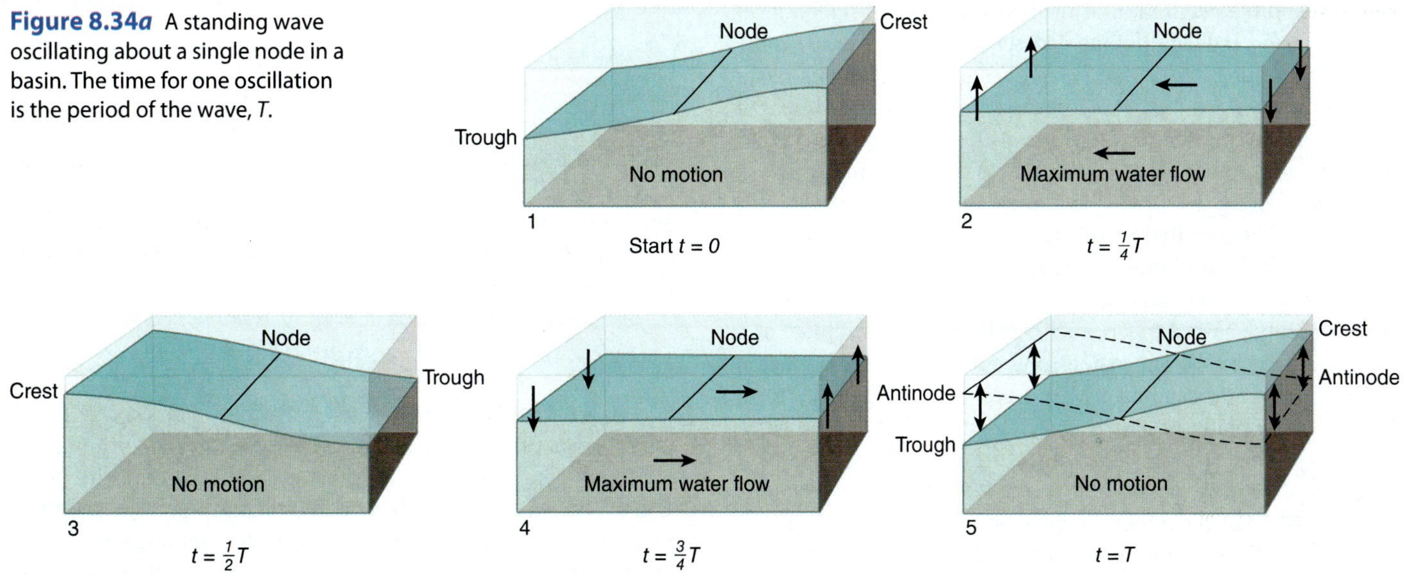

Figure 8.34a A standing wave oscillating about a single node in a basin. The time for one oscillation is the period of the wave, T.

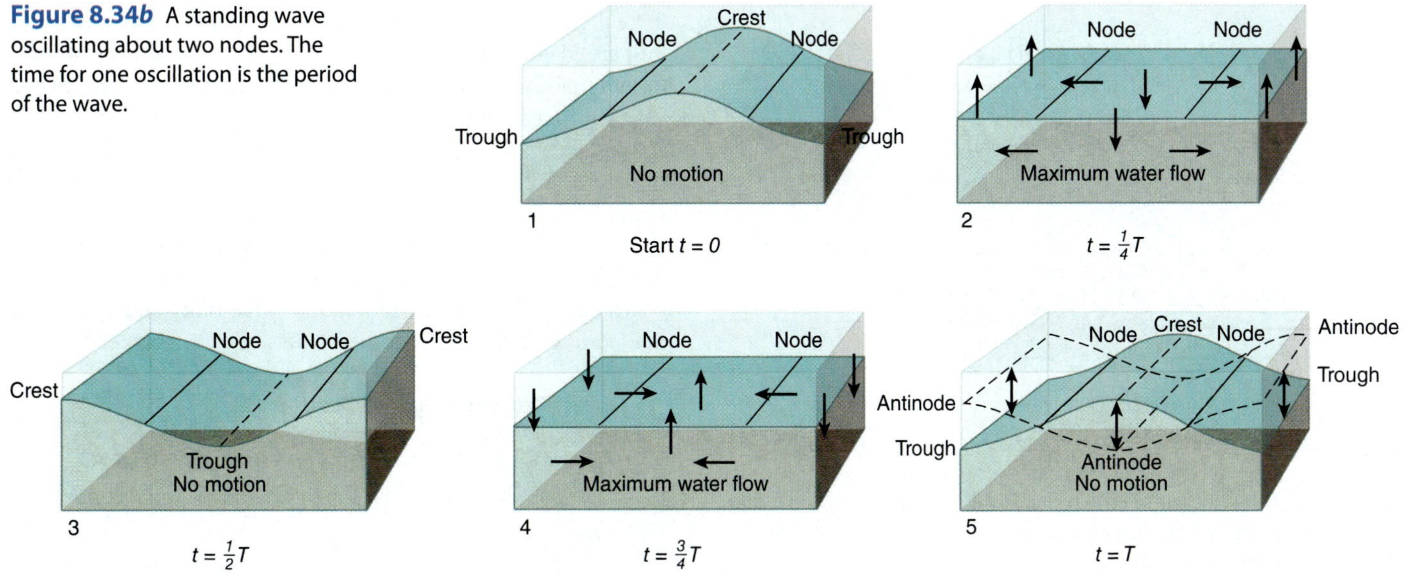

Figure 8.34b A standing wave oscillating about two nodes. The time for one oscillation is the period of the wave.

wavelength, the distance from crest to crest or from trough to trough, is twice the length of the container. By rapidly tilting the basin back and forth at the correct rate, one can produce a wave with more than one node (fig. 8.34b). In the case of two nodes, there is a crest at either end of the container and a trough in the center; this configuration alternates with a trough at each end and the crest in the center. The two nodes are one-quarter of the basin's length from each end. In this case, note that the wavelength is equal to the basin's length. The oscillation period of the wave with two nodes is one-half that of the wave with a single node.

Standing waves in bays or inlets with an open end behave somewhat differently than standing waves in closed basins. A node is usually located at the entrance to the open-ended bay, so only one-quarter of the wavelength is inside the bay. There is little or no rise and fall of the water surface at the entrance, but a large rise and fall occurs at the closed end of the bay (fig. 8.35). Multiple nodes may also be present in open-ended basins.

Standing waves that occur in natural basins are called **seiches,** and the oscillation of the surface is called seiching. In natural basins, the length dimension usually greatly exceeds the depth. Therefore, a standing wave of one node in such a basin behaves as a reflecting shallow-water wave, with the wavelength determined by the length or width of the basin. In water with distinct layers having sharp density boundaries, standing waves may occur along the fluid boundaries as well as at the air-sea boundary. The oscillation of the internal standing waves is slower than the oscillation of the sea surface.

Standing waves may be triggered by tectonic movements that suddenly shake a basin, causing the water to oscillate at a period defined by the dimensions of the basin. This phenomenon occurs during an earthquake when water sloshes back and forth in swimming pools. If storm winds create a change in surface level to produce storm surges, the surface may oscillate as a standing wave in the act of returning to its normal level when the wind ceases. The movement of an air-pressure disturbance over a lake may also cause periodic water-level changes, reaching a

meter or more in height. Tidal currents moving through an area with a sharp pycnocline and an irregular bottom topography may create internal waves that sometimes produce seiches.

If the period of the disturbing force is a multiple of the natural period of oscillation of the basin (an ocean basin or a smaller coastal basin), the height of the standing wave is greatly increased. For example, if a child is riding on a swing, a gentle push timed with each swing period forces the swing higher and higher. The push may be delivered each time the swing passes, every other time, or every third time; all are multiples of the natural period of the swing. In chapter 9, we will learn that repeating tidal forces at the entrance to a bay can produce standing waves in those basins that have natural periods of oscillation approximating the tidal period. In the open ocean, large oceanic basins sometimes have natural periods of oscillation that promote standing wave tides (chapter 9).

A standing wave in a basin is like a water pendulum. The wave's natural period of oscillation is

$$T = \left(\frac{1}{n}\right)\left(\frac{L}{\sqrt{gD}}\right)$$

where n is the number of nodes present, D is the depth of water in the basin, g is Earth's acceleration due to gravity, and L is the wavelength. L equals twice the basin length, l, in a closed basin and four times the basin length in a basin with an open end. This equation is related to the shallow-water wave equation when the number of nodes is equal to 1:

$$\frac{L}{T} = n\sqrt{gD}$$

A progressive wave directly reflected back on itself produces a standing wave, because the two waves—original and reflected—are moving at the same speed but in opposite directions. The checkerboard interference pattern produced by two matched wave systems approaching each other at an angle also creates standing waves, with crests and troughs alternating with each other in fixed positions (see fig. 8.9).

Figure 8.36 shows the relationship between the distribution of total oceanic wave energy and wave period. The energy of ordinary wind waves is high because these waves are always present and well distributed through all the oceans. Storm waves are larger and carry more energy, but they do not occur as frequently and are present over much less of the ocean area. Therefore, storm waves have less total energy than ordinary wind waves. Tsunami-type waves contain a large amount of energy, but they are infrequent and confined to fewer areas of the oceans. The tides, when considered as waves, concentrate their energy in two narrow bands centered on the twice-daily and once-daily tidal periods. Tide wave forms are discussed in chapter 9.

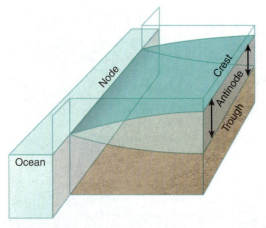

Figure 8.35 A standing wave oscillates about the node located at the opening to a basin. The antinodes produce the rise and fall of water at the closed end of the basin. This type of oscillation is produced by alternating water inflow and outflow at a period equal to the natural period of the basin.

QUICK REVIEW

1. How is a standing wave different from a wind wave?
2. Compare standing waves in open and closed basins.
3. What is a seiche?

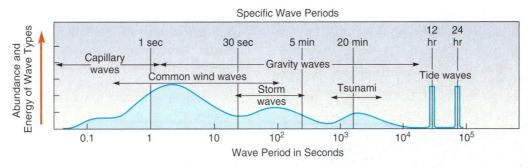

Figure 8.36 The distribution of wave energy with wave period.

8.10 Practical Considerations: Energy from Waves

A tremendous amount of energy exists in ocean waves. The power of all waves is estimated at 2.7×10^{12} watts, which is about equal to 3000 times the power-generating capacity of Hoover Dam. Unfortunately for human needs, this energy is widely dispersed and not constant at any given location or time. It is, therefore, difficult to tap this supply to produce power, except in small quantities.

Wave energy can be harnessed in three basic ways: (1) using the changing level of the water to lift an object, which can then do useful work because of its potential energy; (2) using the orbital motion of the water particles or the changing tilt of the sea surface to rock an object to and fro; and (3) using rising water to compress air or water in a chamber. A combination of these may also be used. If the wave motion is used directly or indirectly to turn a generator, electrical energy may be produced.

Consider a large surface float with a hollow cylinder extending down into the sea (fig. 8.37). Inside the cylinder is a piston, and the up-and-down motion of the surface float causes the cylinder to move up and down over the piston, while the large drag plate restricts the motion of the piston. The system takes in water as the surface buoy rises on the crest of the waves and squirts water out as the surface buoy drops with the passing of the trough. The pumped water can be used to turn a turbine, but because wave energy is distributed over a volume of water, this mechanism does not withdraw much of the passing wave's energy. This system can be adapted to pump air rather than water. An air-compression system called *Sperboy* has been developed and tested in Great Britain. Air displaced by the oscillating water column is passed through turbine generators to produce energy. *Sperboy* is designed to be deployed in large arrays 8 to 12 miles offshore providing large-scale energy generation at a competitive cost.

Another system constructs a tapered channel perpendicular to the shore. Incoming waves force the water up 2–3 m (6–10 ft) in the narrow end of the channel, where it spills into an elevated storage tank, then down through a turbine. This system is used to generate power by a 75-kilowatt plant on Scotland's Isle of Islay, a 350-kilowatt plant at Toftestalen, Norway, and two 1500-kilowatt plants, one in Java and the other in Australia. Both the surface float and the tapered channel are examples of changing the level of an object or the water itself to create potential energy.

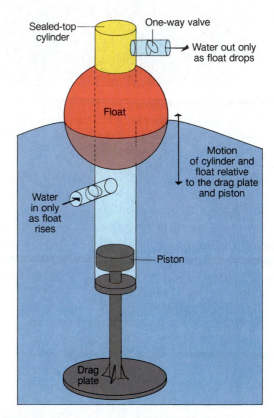

Figure 8.37 The vertical rise and fall of the waves can be used to power a pump.

Wave power systems that use the orbital or rocking motion of the waves have been successfully developed. Long strings of mechanical power units are moored in water where waves are abundant. Each passing wave makes the power units move relative to each other, causing pumps to move oil that passes through a turbine in a closed system. One such system is called *Pelamis*, named after the genus of sea snakes it resembles (fig. 8.38). *Pelamis* consists of a series of semi-submerged, cylindrical sections linked by hinged joints; it is oriented perpendicular to the predominant wave direction. As waves pass along the length of the machine, the sections move relative to one another. This motion causes hydraulic cylinders in the joints to pump high-pressure oil through hydraulic motors. These motors drive electrical generators to produce electricity.

Figure 8.38 The *Pelamis* machine is an attenuating wave converter. It is 492 feet long and has a diameter of 11.5 feet. Waves make the machine oscillate, or bend, at several joints that connect the tubes.

In Western Australia, the Azores, and Japan, other systems using wave energy to compress air are being developed (fig. 8.39). Air traps can be installed along a wave-exposed coast so that the crest of a wave moving into the trap compresses air, forcing it through a one-way valve; the air traps can also be constructed to pass air in either direction. This compressed air powers a turbine. The trough of the wave allows more air to enter the trap, readying it for compression by the following wave crest.

Shores that are continually pounded by large-amplitude waves are most likely to be developed for wave power. Great Britain has a coastline with frequent high-energy waves and an average wave power of about 5.5×10^4 watts (or 55 kilowatts) per meter of coastline. If the wave energy could be completely harnessed along 1000 km (620 mi) of coast, it would generate enough power to supply 50% of Great Britain's present power needs. Along the northern California coast, waves are estimated to expend 23×10^6 kilowatts of power annually; it is thought that 4.6×10^6 kilowatts, or 20%, could be harvested to generate electrical power. The Pacific Gas & Electric Company, a northern California utility, has considered installing a generating device in a breakwater planned for Fort Bragg, California.

When we think about wave energy systems, thoughtful consideration needs to be given to items other than cost. If all the energy were extracted from the waves in a coastal area, what effect would this action have on the shore area? If the nearshore areas are covered with wave energy absorbers 5–10 m (15–33 ft) apart, what will the effect be on other ocean uses? Since the individual units collect energy at a slow rate, can they collect enough energy over their projected life span to exceed the energy used to fabricate and maintain them? Answers to these questions will help us understand that the harvesting of wave energy is not without an effect on the environment, that it may not be either cost- or energy-effective, and that its location may present enormous problems for installation, maintenance, and transport of energy to sites of energy use.

QUICK REVIEW

1. What is the source of power in waves?
2. Where are some good places for wave-generated power efforts?
3. Why can't wave power be harnessed anywhere?

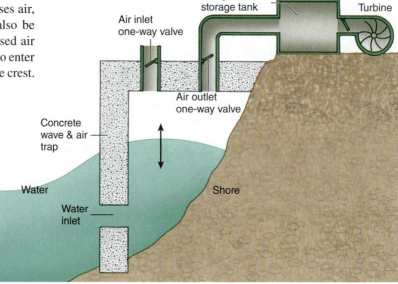

Figure 8.39 Each rise and fall of the waves pumps pulses of compressed air into a storage tank. A smooth flow of compressed air from the storage tank turns a turbine that generates electricity.

Summary

When the water's surface is disturbed, a wave is formed by the interaction between generating and restoring forces. The wind produces capillary waves, which grow to form gravity waves. The elevated portion of a wave is the crest; the depressed portion is the trough. The wavelength is the distance between two successive crests or troughs. The wave height is the difference in elevation between the crest and the trough. Wave period measures the time required for two successive crests or troughs to pass a location. The moving wave form causes water particles to move in orbits. The wave's speed is related to wavelength and period.

Deep-water waves occur in water deeper than one-half the wavelength. Wind waves generated in storm centers are deep-water waves. The period of a wave is a function of its generating force and does not change. Long-period waves move out from the storm center, forming long, regular waves, or swell. The faster waves move through the slower waves and form groups, or trains, of waves. The longer waves are followed by the shorter waves. This process is known as sorting, or dispersion. The speed of a group of waves is half the speed of the individual waves in deep water. Swells from different storms cross, cancel, and combine with each other as they move across the ocean.

Wave height depends on wind speed, wind duration, and fetch. Single, large waves unrelated to local conditions are called episodic waves. The energy of a wave is related to its height. When the ratio of the height to the length of a wave, or its steepness, exceeds 1:7, the wave breaks. The Universal Sea State Code relates wind speeds and sea surface conditions.

Shallow-water waves occur when the depth is less than one-twentieth the wavelength. The speed of a shallow-water wave depends on the depth of the water. As the wave moves toward shore and decreasing depth, it slows, shortens, and increases in height. Waves coming into shore are refracted, reflected, and diffracted. The patterns produced by these processes helped people in ancient times to navigate from island to island.

In the surf zone, breaking waves produce a water movement toward the shore. Breaking waves are classified as plungers or spillers. Water moves along the beach as well as toward it; it is returned seaward through the surf zone by rip currents.

Tsunamis are seismic sea waves. They behave as shallow-water waves, producing severe coastal destruction and flooding.

Internal waves occur between water layers of different densities. Standing waves, or seiches, occur in basins as the sea surface oscillates about a node. Alternate troughs and crests occur at the antinodes.

The energy of the waves can be harnessed by using either the water-level changes or the changing surface angle associated with them. Difficulties include cost, location, environmental effects, and lack of wave regularity.

Key Terms

generating force, 214	amplitude, 216	swell, 218	plunger, 228
restoring force, 214	wave steepness, 216	group speed, 219	spiller, 228
gravity wave, 214	wave period, 216	fetch, 220	rip current, 228
ripple, 214	wave frequency, 216	episodic wave, 221	seismic sea wave, 230
capillary wave, 214	orbit, 217	potential energy, 222	tsunami, 230
cat's-paws, 214	progressive wind wave, 218	kinetic energy, 222	internal wave, 235
crest, 216	storm center, 218	wave steepness, 223	standing wave, 237
equilibrium surface, 216	forced wave, 218	shallow-water wave, 223	node, 237
trough, 216	free wave, 218	wave ray, 225	antinode, 237
wavelength, 216	sorting/dispersion, 218	diffraction, 226	seiche, 238
wave height, 216	wave train, 218	breaker, 227	

Study Problems

1. What is the speed (in m/s) of a deep-water wave with a wavelength of 20 meters?

2. What is the speed (in m/s) of a deep-water wave with a period of 12 seconds?

3. How fast does a tsunami travel (in m/s and miles/hr) in the open ocean if the water depth is 5000 meters?

4. Using the equations $C = L/T$ and $L = (g/2\pi)T^2$, show that wave speed can be determined from (a) wave period only and (b) wavelength only.

5. A submarine earthquake produces a tsunami in the Gulf of Alaska. How long will it take the tsunami to reach Hawaii if the average depth of the ocean over which the tsunami travels is 3.8 km and the distance is 4600 kilometers?

The Tides

Learning Outcomes

After studying the information in this chapter students should be able to:

1. *compare* and *contrast* diurnal, semidiurnal, and semidiurnal mixed tides,

2. *label* the basic characteristics of the three tidal patterns listed above,

3. *explain* why the Moon's tide-raising force is greater than the Sun's despite the much larger mass and gravitational attraction of the Sun,

4. *diagram* the Earth-Moon-Sun system during spring and neap tides,

5. *calculate* the time of the next high tide for diurnal and semidiurnal tides, given the time of the last high tide,

6. *sketch* the Earth-Moon system that leads to declinational tides,

7. *describe* the effect that distance from the amphidromic point has on tidal range,

8. *illustrate* the motion of the ocean surface in a rotary standing tide, and

9. *discuss* the prospects for capturing energy from the tides.

Fishing boats at low tide, Belliveau Cove, Bay of Fundy, Nova Scotia.

Best known as the rise and fall of the sea around the edge of the land, the tides are caused by the gravitational attraction between Earth and the Sun and between Earth and the Moon. Far out at sea, tidal changes go unnoticed, but along the shores and beaches the tides govern many of our water-related activities, both commercial and recreational. Early sailors from the Mediterranean Sea, where the daily tidal range is less than 1 m (3 ft), ventured out into the Atlantic and sailed northward to the British Isles; to their amazement, they found a tidal range in excess of 10 m (30 ft). The movement of the tide into and out of bays and harbors has been helpful to sailors beaching their boats and to food-gatherers searching the shore for edible plants and animals, but it is also recognized as a hazard by navigators and can produce some spectacular effects when rushing through narrow channels.

In this chapter, we survey tide patterns around the world and explore the tides in two ways: one is a theoretical consideration of the tides on an Earth with no land; the other is a study of the natural situation. We also show how to use available tide data to predict water-level changes and coastal tidal currents.

9.1 Tide Patterns

Measurements of tidal movements around the world show us that the tides behave differently in different places (fig. 9.1). In some coastal areas, a regular pattern occurs of one high tide and one low tide each day; this is a **diurnal tide.** In other areas, a cyclic high water–low water sequence is repeated twice in one day; this is a **semidiurnal tide.** In a semidiurnal tidal pattern, the two high tides reach about the same height and the two low tides drop to about the same level. A tide in which the high tides regularly reach different heights and the low tides drop regularly to different levels is called a **semidiurnal mixed tide.** This type of tide has a diurnal (or daily) inequality, created by combining diurnal and semidiurnal tide patterns. The tide curves in figure 9.2 show each type of tide.

9.2 Tide Levels

In a uniform diurnal or semidiurnal tidal system, the greatest height to which the tide rises on any day is known as **high water**, and the lowest point to which it drops is called **low water**. In a mixed-tide system, it is necessary to refer to **higher high water** and **lower high water**, as well as **higher low water** and **lower low water** (see fig. 9.2).

Tide measurements taken over many years are used to calculate the **average** (or **mean**) **tide** levels. Averaging all water levels over many years gives the local mean tide level. Averages are also calculated for the high-water and low-water levels, as

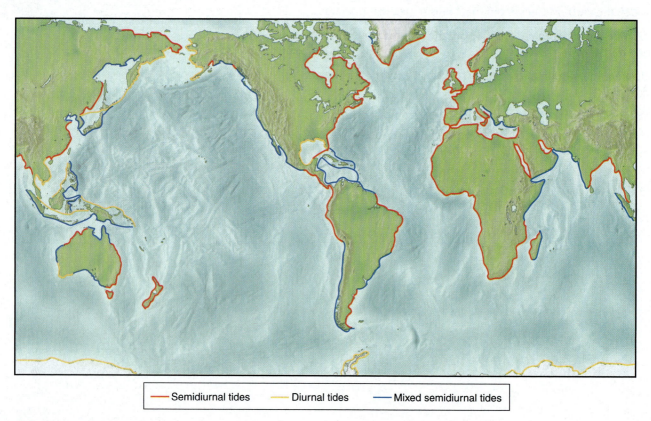

Semidiurnal tides — Diurnal tides — Mixed semidiurnal tides

Figure 9.1 The geographic distribution of different tidal cycles. Areas experiencing diurnal tides are marked in *yellow*, areas experiencing semidiurnal tides are drawn in *red*, and regions with mixed semidiurnal tides are outlined in *blue*. Source of Data: National Oceanic and Atmospheric Administration (NOAA).

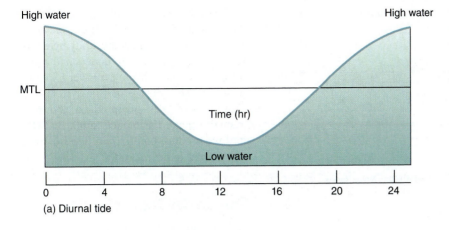

(a) Diurnal tide

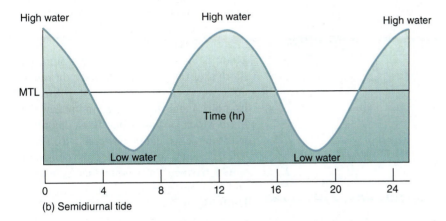

(b) Semidiurnal tide

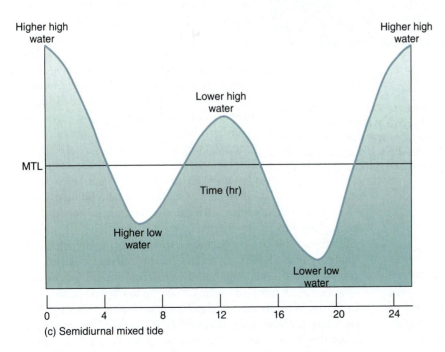

(c) Semidiurnal mixed tide

Figure 9.2 The three basic types of tides: (a) a once-daily, diurnal, tide; (b) a twice-daily, semidiurnal, tide; and (c) a semidiurnal mixed tide with diurnal inequality. MTL equals mean tide level.

mean high water and mean low water. For mixed tides, mean higher high water, mean lower high water, mean higher low water, and mean lower low water are calculated.

Because the depth of coastal water is important to safe navigation, an average low-water reference level is established; depths are measured from this level for navigational charts. The tide level is added to this charted depth to find the true depth of water under a vessel at any particular time. In areas of uniform diurnal or semidiurnal tide patterns, the zero depth reference, or **tidal datum,** is usually equal to mean low water. The use of mean low water assures the sailor that the actual depth of the water is, in general, greater than that on the chart. In regions with mixed tides, mean lower low water is used as the tidal datum for the same reason. When the low-tide level falls below the mean value used as the tidal datum, a **minus tide** results. A minus tide can be a hazard to boaters, but it is cherished by clam diggers and students of marine biology, because it exposes shoreline usually covered by the sea.

As the water level along the shore increases in height, the tide is said to be rising or flooding; a rising tide is a **flood tide**. When the water level drops, the tide is falling or ebbing; a falling tide is an **ebb tide**.

Curves for typical tides at some U.S. coastal cities are shown in figure 9.3.

9.3 Tidal Currents

Currents are associated with the rising and falling of the tide in coastal waters. These **tidal currents** may be extremely swift and dangerous as they move water into a region on the flood tide and out of the region on the ebb tide. When the tide turns, or changes from an ebb to a flood or vice versa, a period of **slack water** occurs during which the tidal currents slow and then reverse. Slack water may be the only time that a vessel can safely navigate a narrow channel with swiftly moving tidal currents, sometimes in excess of 5 m/s (10 knots). The relationships of tidal currents to standing wave tides, to progressive tides, and to tidal current prediction are discussed in this chapter.

QUICK REVIEW

1. Draw and label the water levels in diurnal, semidiurnal, and mixed-semidiurnal tides.

2. What is the difference between a minus tide and an ebb tide?

3. What is the tidal datum?

4. How many times does slack water occur during a semidiurnal tide?

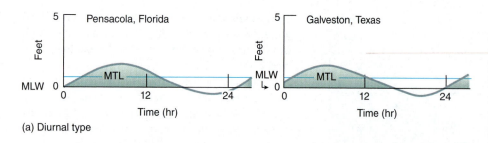

(a) Diurnal type

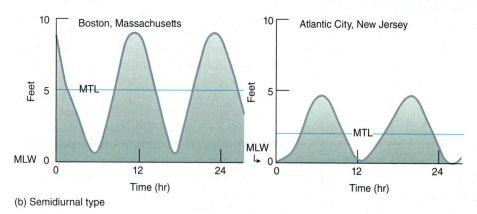

(b) Semidiurnal type

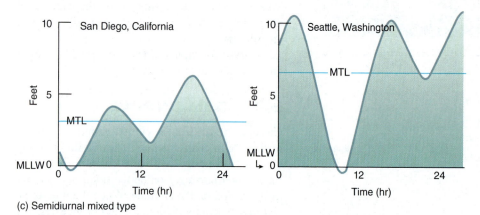

(c) Semidiurnal mixed type

Figure 9.3 Tide types and tidal ranges vary from one coastal area to another. The zero tide level equals mean low water (MLW) or mean lower low water (MLLW), as appropriate. MTL equals mean tide level. All tide curves are for the same date. Tide types at any one location can change with time. These tide curves are intended to illustrate the characteristics of different tide types and should not be interpreted as indicating that these locations experience only one tide type.

9.4 Modeling the Tides

Oceanographers analyze tides in two ways. The tides are studied as mathematically ideal wave forms behaving uniformly in response to the laws of physics. This method is called **equilibrium tidal theory.** It is based on an Earth covered with a uniform layer of water, in order to simplify the relationships between the oceans and the tide-rising bodies, the Moon and the Sun. The tides are also studied as they occur naturally; this method is called **dynamic tidal analysis.** It studies the oceans' tides as they occur, modified by the landmasses, the geometry of the ocean basins, and Earth's rotation.

The effects of the Sun's and the Moon's gravity and the rotation of Earth on tides are most easily explained by studying equilibrium tides. One way to understand equilibrium tides is to begin with very simple models of the Earth-Moon-Sun system, see what kinds of tides they produce, and then gradually make the models more realistic until we produce equilibrium tides that are very similar to the real tides.

The Earth-Moon System

Let's begin with a model that consists only of Earth and the Moon. We'll make the following assumptions: (1) Earth is perfectly smooth and covered by a single ocean of constant depth; (2) the Moon is directly above Earth's equator; and (3) the Moon is always above the same point on Earth's surface (i.e., Earth does not rotate on its axis and the Moon is not moving in its orbit). In this model, we can imagine the Earth-Moon system being like a dumbbell with a large weight on one end and a small weight on the other end. This dumbbell would balance on a point that coincides with the common center of mass of the Earth-Moon system. Because the Earth is so much larger than the Moon, this center of mass is about 1710 km (1062 mi) beneath Earth's surface (fig. 9.4). The commonly held belief that the Moon rotates around Earth's center is actually false. Rather, the Earth-Moon system rotates around its center of mass, making one complete revolution roughly every month (27.3 days) (fig. 9.5). The average distance between the Earth and the Moon remains constant because of a balance of two forces acting on them as they rotate: gravity and inertia. The gravitational attraction between Earth and the Moon helps keep these bodies in their orbital relation to each other. However, without some counterbalancing force to gravity, Earth and the Moon would move closer and closer. This counterbalancing force is inertia, which is the tendency of a moving object to continue moving in a straight line. It is this force (sometimes called **centrifugal force**) that holds water in a bucket when you swing the bucket in an overhead arc. Without inertia, Earth and the Moon would quickly be drawn together and collide. Without gravity, they would fly apart deeper into space. Because these forces are in balance, Earth and the Moon have maintained their orbital relationship for billions of years. However, it is very important to understand that this is an overall balance for the Earth-Moon system as a

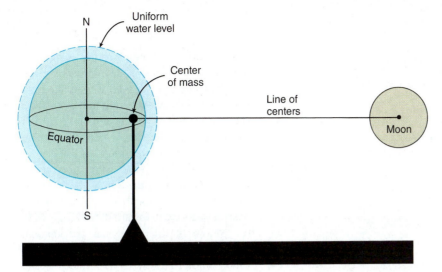

Figure 9.4 The Earth-Moon system balances on its common center of mass. Because Earth's mass is 81 times the Moon's mass, this center of mass is located beneath Earth's surface.

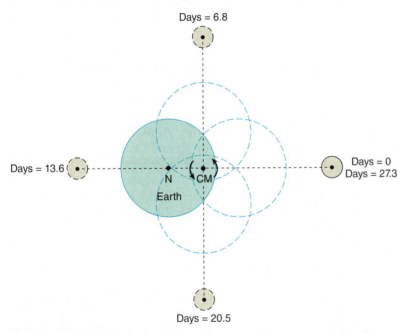

Figure 9.5 The Earth-Moon system rotates around its common center of mass (CM). It makes one complete revolution every 27.3 days. The rotation of the system is in the same direction as Earth's rotation; counterclockwise looking down on the North Pole.

whole. In the Earth-Moon system, the two forces, gravity and inertia, are only balanced (i.e., equal in strength but opposite in direction) at Earth's center. At all other points, and specifically on Earth's surface where the ocean lies, they will not be completely balanced. It is the difference between these two forces that produces the tide-generating force at any particular location (fig. 9.6).

Every point on Earth experiences the same inertia, or centrifugal force. This is constant regardless of where you are. The gravitational force between points on Earth's surface and the Moon is not constant. Sir Isaac Newton's universal law of gravitation tells us that the force of attraction between any two bodies

is proportional to the product of the two masses divided by the square of the distance between the centers of the masses:

$$F = G\left(\frac{m_1 m_2}{R_2}\right)$$

where

G = universal gravitational constant, 6.67×10^{-8} cm³/g/s²
m_1 = mass of body 1 in grams
m_2 = mass of body 2 in grams
R = distance between centers of masses in centimeters

Imagine that body 1 is Earth and body 2 is the Moon. Now let's say that we measure mass in units of "Earth masses." Then $m_1 = 1$ and we can say that the gravitational attraction between the Moon and Earth is *proportional* to $G(M/R^2)$ where M is the mass of the Moon. Since gravity and inertia balance, inertia must also be *proportional* to $G(M/R^2)$.

The gravitational attraction of the Moon is strongest on the side of Earth that happens to be facing the Moon, simply because it is closer. On this "near side" of Earth, the gravitational attraction is greater than inertia, and gravity causes water to be pulled toward the Moon, creating a bulge of water. On the opposite side of Earth, the "far side," the gravitational attraction of the Moon is less because of the greater distance. Here, inertia is greater than gravity. As the Earth-Moon system rotates, water on this side tries to keep moving in a straight line, thus moving away from Earth and creating a second bulge. These two bulges are of equal size and shape but opposite in direction. It is the difference between the two forces, gravity and inertia, that causes the tide-generating force. The tide-generating force is *proportional* to $G(M/R^3)$. The calculation of this tide-generating force is shown in appendix C.

Earth and Moon Rotation

We started this discussion with some simple assumptions, including the idea that Earth does not rotate on its axis. We know that is false, so if we include Earth's rotation, how does that influence the pattern of the tides? Let's begin by assuming that there is no friction between the rotating Earth and the ocean surrounding it. Because water is fluid, the two bulges created in this single, globe-encircling ocean will stay aligned with the Moon—one pointing toward it and the other pointing away. As Earth rotates on its axis, a point on the surface will pass through two tidal bulges and experience two tidal highs during each rotation. Similarly, it will experience two tidal lows (fig. 9.7a). The two tidal highs will be of the same height and the two tidal lows of the same height, creating a semidiurnal tide pattern (fig. 9.7b).

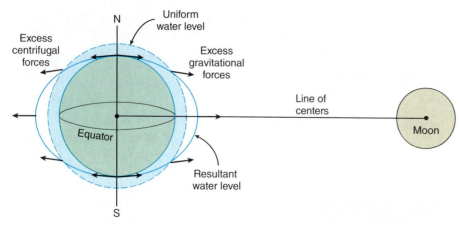

Figure 9.6 Distribution of tide-raising forces on Earth. Excess lunar gravitational and centrifugal forces distort the Earth model's water envelope to produce bulges and depressions. View is roughly in Earth's equatorial plane.

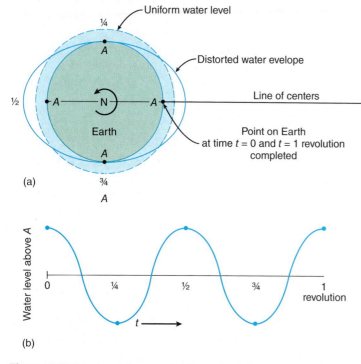

Figure 9.7 The change in water level at point *A* during one Earth rotation through the distorted water envelope [see fig. 9.6]. Fractions indicate portions of a revolution. View is down on Earth's North Pole, perpendicular to Earth's equatorial plane.

What about the Moon's movement? While Earth turns eastward on its axis, the Moon is moving in the same direction in its orbit around Earth. After twenty-four hours, a point on Earth's surface that began directly under the Moon is no longer directly under the Moon. Earth must turn for an additional fifty minutes, about 12°, to bring the starting point on Earth back in line with the Moon. Therefore, a **tidal day** is not twenty-four hours long, but twenty-four hours and fifty minutes long. This explains why the two daily tides in a semidiurnal tide pattern occur twelve hours and twenty-five minutes apart (fig. 9.8). We can think of

this tide pattern as being a **tide wave**. The crest of the tide wave is the high-water level, and the trough of the tide wave is the low-water level. The wavelength in the case of a semidiurnal tide is half the circumference of Earth, and the tide wave's period is twelve hours and twenty-five minutes.

A second way to understand tide-raising forces is to think of the Sun or the Moon as exerting a gravitational force that continuously attracts Earth toward the Sun or toward the Moon. In this case, the gravitational force, or **centripetal force,** of the Sun or the Moon holds Earth in orbit. The centripetal force is constant and equal to the average gravitational force of the Sun or the Moon acting at Earth's center. In this case, the difference between the average gravitational attraction and the gravitational attraction at individual points on Earth produces the tide-raising force. The force difference per unit mass between a surface point and Earth's center is proportional to $G(M/R^3)$ in this second case as well. The two approaches yield the same results. The derivation of $G(M/R^3)$ is found in appendix C.

In the Earth-Moon-Sun system, the mass of the Sun is very great, but the Sun is very far away. By contrast, the Moon is small, but it is close to Earth. Calculating the distribution of these forces for each water particle at Earth's surface shows that the Moon has a greater attractive effect on the water particles than the Sun.

The Sun Tide

Although the Moon plays the greater role in producing the tides, the Sun produces its own tide wave. Despite the Sun's large mass, it is so far away from Earth that its tide-raising force is only 46% that of the Moon. The time required for Earth to revolve on its axis with respect to the Sun is on average twenty-four hours, not twenty-four hours and fifty minutes, as in the case of the Moon tide. For this reason, the tide wave produced by the Moon is not only of greater magnitude than that produced by the Sun, but it also continually moves eastward relative to the tide wave produced by the Sun. Because the tidal forces of the Moon are greater than those of the Sun, the tidal period of the Moon is more important, and the tidal day is considered to be twenty-four hours and fifty minutes.

Spring Tides and Neap Tides

The Moon's orbit requires 29½ days relative to a point on Earth. During this period the Sun, Earth, and the Moon move in and out of phase with each other. At the new Moon, the Moon and Sun are on the same side of Earth, so the high tides, or bulges,

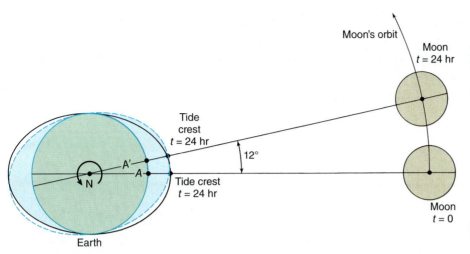

Figure 9.8 Point *A* requires twenty-four hours to complete one Earth rotation. During this time, the Moon moves 12° east along its orbit, carrying with it the tide crest. To move from *A* to *A'* requires an additional fifty minutes to complete a tidal day. View is down on Earth's North Pole, perpendicular to Earth's equatorial plane.

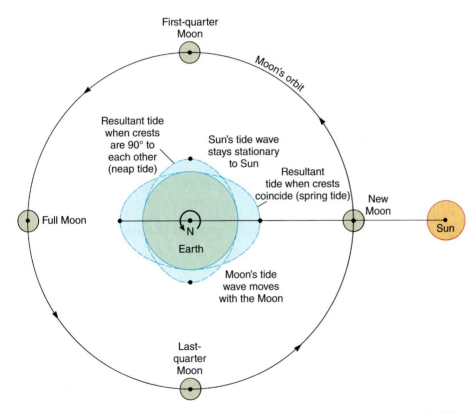

Figure 9.9 Spring tides result from the alignment of Earth, Sun, and Moon during the full Moon and the new Moon. During the Moon's first and last quarters neap tides are produced. The tidal range or vertical difference between high and low water is reduced during neap tides. View is down on Earth's North Pole, perpendicular to Earth's equatorial plane.

produced independently by the Moon and the Sun coincide (fig. 9.9). Because the water level is the result of adding the two wave forms together, tides of maximum height and depression, or tides with the greatest **range** between high water and low

water, are produced. These tides are known as **spring tides.** The vertical displacement, or amplitude, of the tide is one-half the range— the distance above or below mean tide level.

In a week's time, the Moon is in its first quarter; it has moved eastward along its orbit (about 12° per day) and is located approximately at right angles (or 90°) to the line between the centers of Earth and the Sun. The crest, or bulge, of the Moon tide is at right angles to the tide wave created by the Sun; the crests of the Moon tide will coincide with the troughs of the Sun tide, and the same will be true of the Sun's tide crests and the Moon's tide troughs (fig. 9.9). The crests and troughs tend to cancel each other out, and the range between high water and low water is small, producing low-amplitude **neap tides.**

At the end of another week, the Moon is full, and the Sun, Moon, and Earth are again lined up, producing crests that coincide and tides with the greatest range between high and low waters, or spring tides. These spring tides are followed by another period of neap tides, produced by the Moon in its last quarter when it again stands at right angles to the Sun-Earth line (fig. 9.9). The tides follow a four-week cycle of changing amplitude, with spring tides occurring every two weeks and a period of neap tides occurring in between. This progression can be seen in the portions of the tide records reproduced in figure 9.10. The effect occurs each lunar month and is the result of the Moon's tide wave moving around Earth relative to the Sun's tide wave.

Declinational Tides

If the Moon or the Sun stands north or south of Earth's equator, one bulge, or high water, is in the Northern Hemisphere and the other is in the Southern Hemisphere (fig. 9.11). Under these conditions, a point at the middle latitudes on Earth's surface passes through only one crest, or high tide, and one trough, or low tide, each tidal day. A diurnal tide, often called a **declinational tide,** is formed, because the Moon or the Sun is said to have declination when it stands above or below the equator.

Declinational (or diurnal) tides are influenced by both the Moon and the Sun. The Sun stands above 23½°N at the summer solstice and above 23½°S at the winter solstice. This variation causes the bulge created by the Sun to oscillate north and south of the equator in a regular fashion each year and tends to create more diurnal Sun tides

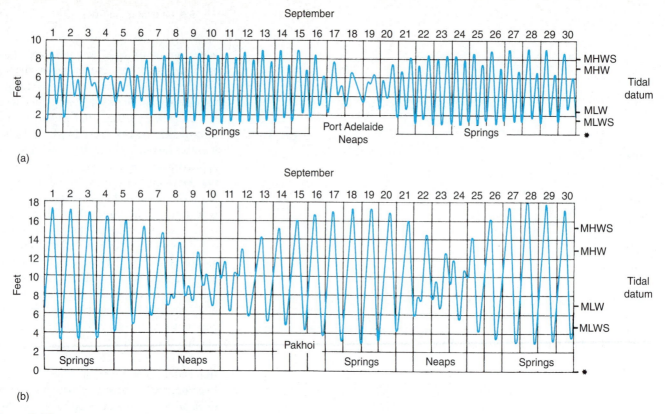

Figure 9.10 Spring and neap tides alternate during the tides' monthly cycle. MHWS is the mean high-water spring tides; MLWS is mean low-water spring tides. (a) A semidiurnal tide from Port Adelaide, Australia. (b) A diurnal tide from Pakhoi, China.

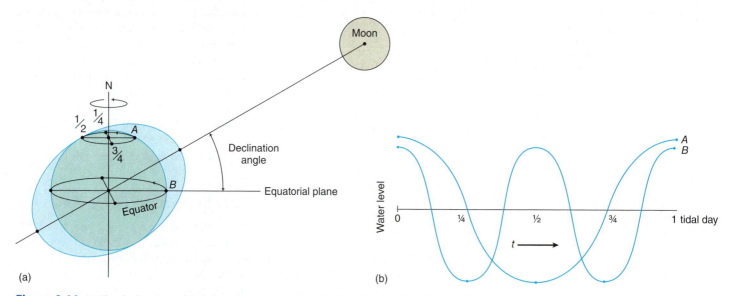

Figure 9.11 (a) The declination of the Moon produces a diurnal tide at latitude *A* and a semidiurnal tide at latitude *B*. (b) Fractions indicate portions of the tidal day. View is roughly in Earth's equatorial plane.

during the winter and summer than during the spring and fall. The Moon's declination varies between 28½°N and 28½°S with reference to Earth. The Moon's orbit is inclined 5° to the Earth-Sun orbit, and it takes 18.6 years for the Moon to complete its cycle of maximum declination. When the Sun's and the Moon's declinations coincide, both tide waves become more diurnal. Each lunar month, the Moon travels from a declination of 5° above the Earth-Sun orbit plane to a declination of 5° below the plane and back.

Elliptical Orbits

The Moon does not move about Earth in a perfectly circular orbit, nor does Earth orbit the Sun at a constant distance. These orbits are elliptical, and therefore, Earth is closer to a certain tide-raising body at some times during an orbit than at other times. During the Northern Hemisphere's winter, Earth is closest to the Sun; therefore, the Sun plays a greater role as a tide producer in winter than in summer.

QUICK REVIEW

1. Why is a tidal day longer than a solar day?
2. Why is a tide at certain times more diurnal or semidiurnal than at other times?
3. If the Sun's gravitational attraction is so much larger than the Moon's, why is the Moon's tide-generating force greater than the Sun's?
4. Explain how gravity and inertia, or centrifugal force, combine to produce the tide-generating force.

9.5 Real Tides in Real Ocean Basins

Equilibrium tidal theory helps us understand the distribution of wave-level changes and tide-raising forces, but it does not explain the tides as observed on the real Earth. Return to figure 9.3 and notice the variety of tidal ranges and tidal periods that appear at different locations on the same date. Refer also to figure 9.10, which shows different tides at different places during the same time period. The Sun-Moon-Earth system for all locations in the two figures is the same, but the equilibrium tidal theory does not explain natural tides at any particular location. Investigating the actual tides requires the dynamic approach, a mathematical study of tide waves as they occur.

The Tide Wave

The behavior of the natural tide wave varies considerably from the tide wave of the water-covered model. Because the continents separate the oceans, the tide wave is discontinuous; the wave starts at the shore, moves across the ocean, and stops at the next shore. Only in the Southern Ocean around Antarctica do the tide waves move continuously around Earth. A tide wave has a long wavelength compared to the depth of the oceans; therefore, it behaves like a shallow-water wave, with its speed controlled by the depth of the water. Because the wave is contained within the ocean basins, it can oscillate in the basin as a standing wave, and it is also reflected from the edge of the continents, refracted by changes in water depth, and diffracted as it passes through gaps between continents. In addition, the persistence of the tidal motion and the scale on which it occurs are so great that the Coriolis effect plays a role in the water's movement. All these factors together produce Earth's real tides. Because these interactions are complex, it is not possible to understand them together until each is first considered separately.

The tide wave's speed as a **free wave** moving across the water's surface is determined by the depth of the water. The tide wave moves as a free wave at about 200 m/s (400 mi/h) in a water depth of 4000 m (13,200 ft), but at the equator, Earth moves eastward under the tide wave at 463 m/s (1044 mi/h). This is more than twice the speed at which the tide can travel freely as a shallow-water wave. Under these conditions, the tide moves as a **forced wave** that is the result of the Moon's attractive force and Earth's rotation. Because Earth turns eastward faster than the tide wave moves freely westward, friction displaces the tide crest to the east of its expected position under the Moon. This eastward displacement continues until the friction force is balanced by a portion of the Moon's attractive force. These two forces, when balanced, hold the tide crest in a position to the east of the Moon rather than directly under it. This process is illustrated in figure 9.12.

Above 60°N or 60°S, the distance around a latitude circle is less than one-half the distance around the equator. Here, the free propagation speed of the tide wave equals the speed at which the sea floor moves under the wave form. Under these conditions, the crest of the tide stays aligned with the Moon, and less friction is generated between the rotating Earth and the moving wave form. Friction between the moving tide wave and the turning Earth also acts to slow the rotation rate of Earth, adding about 1½ milliseconds per 100 years to the length of a day.

Tidal motion extends to all depths in the oceans and can produce internal waves along water density boundaries between the upper and deeper layers in the oceans. Internal waves may become unstable and break, forming smaller waves and generating turbulence, which mixes water and dissipates energy. This tidal turbulence at mid-ocean depths becomes an important mechanism for resupplying nutrients from depth to the upper layers of the oceans.

Progressive Wave Tides

In a large ocean basin, the tide wave moving across the sea surface like a shallow-water wave is a **progressive tide.** Examples of progressive tides are found in the western North Pacific, the

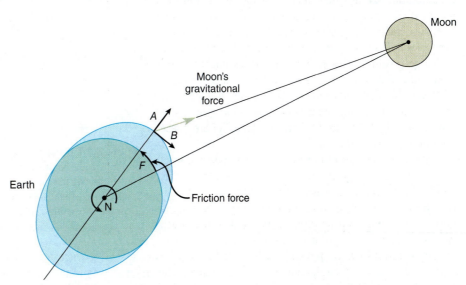

Figure 9.12 The crest of the tide wave is displaced eastward until the *B* component of the gravitational force balances the friction *(F)* between Earth and the tide wave. Component *A* of the gravitational force is the tide-raising force. Component *B* causes the tide wave to move as a forced wave. View is down on Earth's North Pole, perpendicular to Earth's equatorial plane.

eastern South Pacific, and the South Atlantic Oceans. **Cotidal lines** are drawn on charts to mark the location of the tide crest at set time intervals, generally one hour apart. The cotidal lines for the world's ocean tides are shown in figure 9.13 (see also box figure 1 in box titled "Diving In: Measuring Tides from Space" in this chapter).

Because the tide wave is a shallow-water wave, the water particles move in elliptical orbits, and their motion extends to the sea floor. The horizontal component of the motion greatly exceeds the vertical motion. Because the time in which the water particles move in one direction is so long (one-half the tide period), the Coriolis effect becomes important. In the Northern Hemisphere, the water particles are deflected to the right, and in the Southern Hemisphere, they are deflected to the left. This deflection causes a clockwise rotation of the water in the Northern Hemisphere and a counterclockwise rotation in the Southern Hemisphere. This circular (or rotary) movement is the oceanic tidal current described at the end of the "Standing Wave Tides" section.

Standing Wave Tides

In some ocean basins or parts of ocean basins, the tide wave is reflected from the edge of the continents, and a **standing wave tide** is produced (see chapter 8). Remember that if a container of water is tipped so that the water level is high at one end and low at the other, the water flows to the low end, raising the water level at that end as the water level at the high end drops. This movement produces a wave having a wavelength that is twice the length of the container, with antinodes at the ends of the basin and a node at the basin's center. This same process occurs in ocean basins but with some important modifications.

The high-water to low-water change in an ocean basin requires a long time, and the Coriolis effect must be included. The moving water, deflected to the right in the Northern Hemisphere, does not reach the low-tide end but instead is deflected to a position to the right of the initial high-tide position. This movement causes the tide crest to rotate counterclockwise around the basin in which it oscillates, but the tidal current rotates clockwise, because the current is deflected to the right in the Northern Hemisphere (see fig. 9.15). In the Southern Hemisphere, these directions are reversed.

In the **rotary standing tide wave,** the node becomes reduced to a central point, while the tide crest (shown as cotidal lines) progresses around the edges of the basin. (See figure 9.13 for a demonstration of this pattern in the northeastern and southwestern Pacific and in the North Atlantic.) The central point, or node, for a rotary tide is called the **amphidromic point.** The distance between low- and high-water levels, or the tidal range, for a rotary tide is shown on a chart by a series of lines decreasing in value as they approach the amphidromic point. The lines of equal tidal range are called **corange lines** (fig. 9.14). Near the amphidromic point the tidal range is small; the farther from the amphidromic point, the

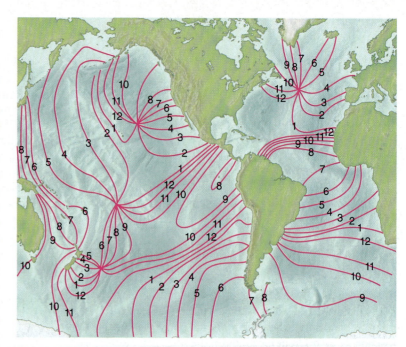

Figure 9.13 Cotidal lines for the world's oceans. The high-tide crest occurs at the same time along each cotidal line. Positions of the high tide are indicated for each hour over a twelve-hour period for semidiurnal tides.

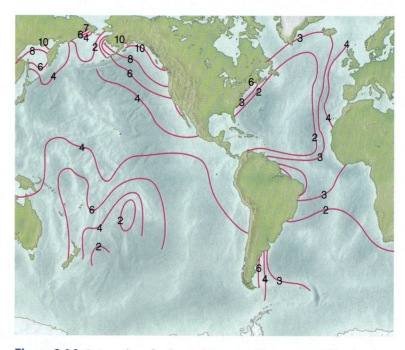

Figure 9.14 Corange lines for the world's oceans. These corange lines connect positions with the same spring tidal range. Open-ocean tidal ranges are less well known than nearshore ranges, where tide-level recorders are commonly used. Satellites now measure water elevation changes in mid-ocean; see the box titled "Diving In: Measuring Tides from Space" in this chapter.

greater the range. Because the amphidromic point is located near the center of an ocean basin, many mid-ocean areas have small tidal ranges, while the shores of the landmasses forming the sides of the basins have larger tidal ranges. The value and position of corange lines in mid-ocean are not as well known as they are near shore.

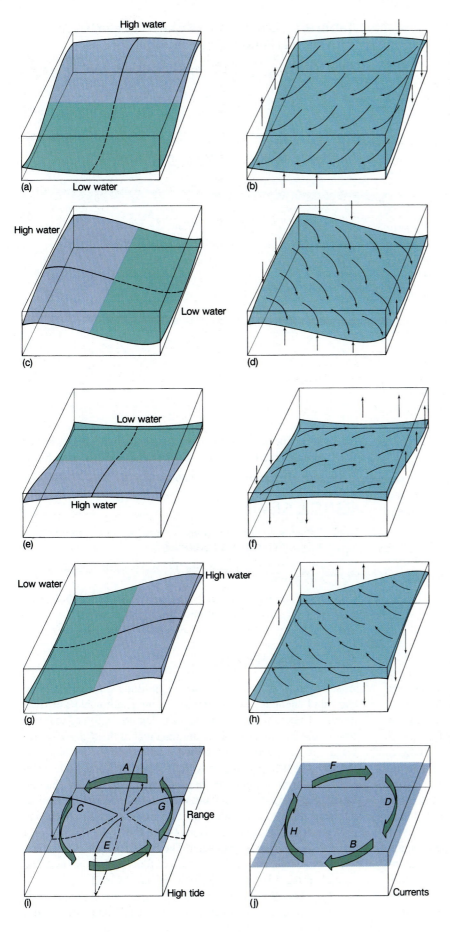

Figure 9.15 Rotary standing tide waves. At the instant high tide occurs at one side of the basin (a), the water begins to flow toward the low-tide side, creating a tidal current (b) that is deflected to its right in the Northern Hemisphere. The current displaces the high tide counterclockwise from (a) by way of (b) to (c). The process continues from (c) by way of (d) to (e), from (e) by way of (f) to (g), and from (g) by way of (h) to (a). This process results in a tide wave that rotates counterclockwise about the amphidromic point (i) and a tidal current that flows clockwise (j).

The flow of water from the high-water side to the low-water side of a standing tide wave produces a rotating tidal current, as shown in figure 9.15. A rotating tidal current is also produced by the orbital motion of water particles in a progressive tide wave. If the tide wave is diurnal, the water particles travel in one complete circle in a tidal day. A semidiurnal tide causes two circles, and a semidiurnal mixed tide produces two circles of unequal size. An example of a rotating semidiurnal mixed tidal current recorded at the Columbia River lightship in the North Pacific is presented in figure 9.16.

Rotary standing tides occur in basins in which the natural period of the basin approximates the tidal period. If the tidal period and the oscillation period of the basin coincide, the tides increase in amplitude. Table 9.1 relates basin depths and lengths or widths that produce natural oscillations equal to tidal periods. Remember that the natural period of oscillation of a standing wave in a basin is

$$T = \left(\frac{1}{n}\right)\left(\frac{L}{\sqrt{gD}}\right)$$

where L, the wavelength, is twice the basin length, l, in closed basins and four times l in open-ended basins. A comparison of the values given in table 9.1 shows that deep-ocean basins must have great length to accommodate standing waves with tidal periods, whereas shallow basins may be much shorter. Most tides are semidiurnal, but dimensions of some basins cause the basin to resonate with a diurnal tidal period rather than a semidiurnal period. See the tide curves in figure 9.3 for Pensacola, Florida, and Galveston, Texas. In an open-ended tidal basin with a mixed tide, the semidiurnal portion of the tide may cause the basin to resonate with a node at the entrance to the basin and another node within the basin. The diurnal portion of the tide may have only a single node at the basin entrance. The result is a diurnal water-level pattern at the second node of the semidiurnal tide and a semidiurnal mixed pattern in all other parts

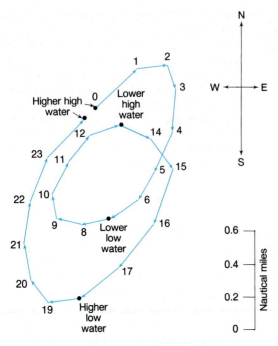

Figure 9.16 Ocean tidal currents are rotary currents. The *arrows* trace the path followed by water particles in a tide wave during a mixed tidal cycle. Two unequal tidal cycles are shown. *Numbers* indicate consecutive hours in each tidal stage. The Coriolis effect deflects the horizontal component of the water particles' orbital motion, causing them to move in a circular path.

Table 9.1   Dimensions of Closed Ocean Basins with Natural Periods Equaling Tidal Periods

Tidal Period	Depth (m)	Length or Width (km)
Semidiurnal	4000	4428
12.42 h	3000	3835
44,712 s	2000	3131
	1000	2214
	500	1566
	100	700
	50	495
Diurnal	4000	8853
24.83 h	3000	7667
89,388 s	2000	6260
	1000	4427
	500	3130
	100	1399
	50	989

coast, the shape of the basin may be such that it decreases rather than increases the tide's range. Every naturally occurring basin is unique in this regard.

QUICK REVIEW

1. Explain why high tide does not appear directly beneath the Moon but instead precedes it.
2. Where, in a rotary standing tide wave, is the tidal range a minimum?
3. How does the tidal range change along a cotidal line as you move from the amphidromic point outward?

of the basin. The harbor of Victoria, British Columbia, Canada, is located near a semidiurnal node so that it registers a diurnal tidal pattern in an inlet system of mixed tides.

Ocean tides are the result of combining progressive and standing wave tides with diurnal and semidiurnal characteristics. Different kinds of tides interact with each other along their boundaries, and the results are exceedingly complex.

Tide Waves in Narrow Basins

Unlike ocean basins, coastal bays and channels are often long and narrow with a length that is considerably greater than their width. These narrow basins have an open end toward the sea, so that the reflection of the tide wave occurs only at the head of the basin. To resonate with a tidal period, the length dimension need be only one-half the length cited for the closed basins in table 9.1. In this type of open basin, the node is at the entrance, and only one-fourth of the tide wave's form is present; an antinode is at the head of the bay. If the open basin is very narrow, oscillation occurs only along the length of the basin; there is no rotary motion because the basin is too narrow. For example, the Bay of Fundy in northeastern Canada has a tidal range near the entrance node of about 2 m (6.6 ft), whereas the range at the head of the bay is 11.7 m (35 ft). This particular bay has a natural oscillation period that is so well matched to the tidal period that every tidal impulse at its entrance creates a large oscillation at the head of the bay (fig. 9.17). In another bay along another

9.6 Tidal Bores

In some areas of the world, large-amplitude tides cause large and rapid changes in water volume along shallow bays or river mouths. Under these conditions, the rising tide is forced to move toward the land at a speed greater than that of the shallow-water wave, whose speed is determined by the depth of the water or the speed of the opposing river flow. When the forced tide wave breaks, it forms a spilling wave front that moves into the shallow water or up into the river. This wave front appears as a wall of turbulent water called a **tidal bore** and produces an abrupt change in water levels as it passes. A single bore may be formed, or a series of bores may be produced. The bores are usually less than a meter in height but can be as much as 8 m (26 ft) high, as in the case of spring tides on the Qiantang River of China.

The Amazon, Trent, and Seine Rivers have bores. Fast-rising tides also send bores across the sand flats surrounding

(a)

Figure 9.18 The tidal bore moving up the Shubenacadie River in Nova Scotia, Canada.

(b)

Figure 9.17 Low tide (a) and high tide (b) at the Hopewell Rocks Ocean Tidal Exploration Site, Hopewell Cape, New Brunswick, Canada. The tidal range at the head of the bay exceeds 14 m (46 ft).

Mont-Saint-Michel in France and into Turnagain Arm of Cook's Inlet in Alaska. The bore in the Bay of Fundy in Canada (fig. 9.18) has been reduced by the construction of a causeway. Towns in areas having tidal bores often post warnings; their turbulence can be a severe hazard because they suddenly flood areas that were open stretches of beach only minutes before.

QUICK REVIEW

1. Explain the hazards of tidal bores.

9.7 Predicting Tides and Tidal Currents

Because of all the natural combinations of progressive and standing tides and the factors that affect them, it is not possible to predict Earth's tides from knowledge of the tide-raising bodies alone; equilibrium tidal theory is not adequate for the task. Accurate, dependable daily tidal predictions are made by combining actual local measurements with astronomical data.

The rise and fall of the tides are measured over a period of years at selected locations. Primary tide stations make these water-level measurements for at least nineteen years, to allow for the 18.6-year declinational period of the Moon. From these data, mean tide levels are calculated. Oceanographers use a technique called **harmonic analysis** to separate the tide record into components with magnitudes and periods that match the tide-raising forces of the Sun and the Moon. They are then able to isolate the effect of the local geography, known as the **local effect.** Tides for any location are predicted by combining the local effect with the predicted astronomical data. Complex and cumbersome mechanical computers or tide machines were once used to predict the tides, but today computers quickly and easily recombine the data and predict the time, date, and elevation of each high-water and low-water level.

Tide Tables

Tide tables give the dates, times, and water levels for high and low water at primary tide stations (table 9.2). There are 196 primary tide stations in the United States, but many more locations require accurate tide predictions. The data for these auxiliary stations are determined by correcting nearby primary station data for time and tide height.

Measuring Tides from Space

Until recently, monitoring and recording of open-ocean tides had been done only by inference from measurements at coastal sites, some mid-ocean islands, and a few seafloor-mounted pressure gauges. Today's satellites, first *SEASAT,* then *GEOSAT,* and now the *ERS* series and *TOPEX/Poseidon,* can measure the absolute elevation of the sea surface by radar altimetry. The satellite uses a radar beam to measure the distance from the satellite to the sea surface, and this distance is compared to the distance between the satellite and Earth's center to obtain an elevation of the sea surface above Earth's center. *TOPEX/Poseidon* provides more accurate sea-level measurements than previous satellites, collecting ten measurements per second and requiring ten days to repeat its ground track between approximately 65°N and 65°S.

The changes in sea surface elevation recorded by the satellite are caused by climate variations, water-density shifts due to temperature change, wind and atmospheric pressure fluctuations, and ocean current meanders, as well as by the passing tide wave.

Tidal elevations can be separated from other height changes because of their definite and recognizable periods, allowing oceanic tidal maps based on satellite data to be drawn with high accuracy.

Oceanic tidal maps for specific tidal components have been derived from the *TOPEX/Poseidon* data. The astronomical tidal component with the largest effect on the tides is the M_2 or principal lunar semidiurnal component. Box figure 1 is a cotidal map of global M_2 tides derived from one year of *TOPEX/Poseidon* sea surface elevation data that have been fitted to a computer model. As longer-term records are collected, it will be possible to clearly separate more of the tidal components, improve computer models, and predict the total ocean tide.

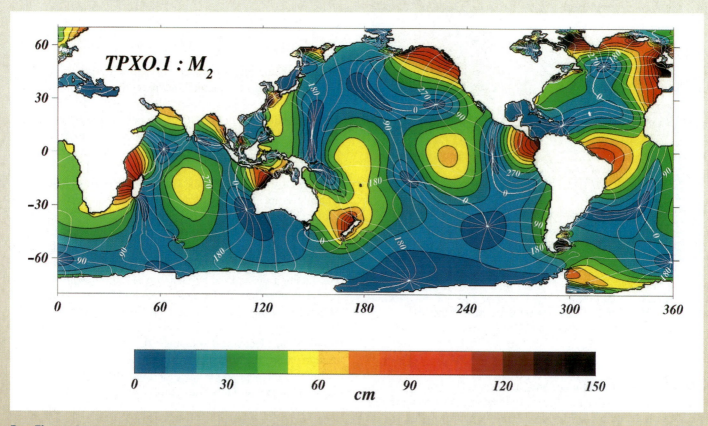

Box Figure 1   *TOPEX/Poseidon* data produced this cotidal chart of the principal lunar semidiurnal tidal component (M_2). The rotation of the tides about amphidromic points and areas of progressive tides are shown. Cotidal lines mark the position of the tidal wave crest, hour by hour; each starting point is a *thick white line* representing 0000 hour GMT. The *numbers* represent the degrees of rotation of the tidal crest about the amphidromic points. Cotidal amplitude (one-half the cotidal range) increases with distance from the amphidromic points. The color scale is in centimeters.

Table 9.2 Predicted Times and Heights of High and Low Waters for July 6–9, 2006, San Diego, California, United States (32.7133° N, 117.1733° W)

Day	Time (h:min)[1]	Height (ft)
6	01:26	0.80 low
	07:42	3.19 high
	11:47	2.46 low
	18:35	5.84 high
7	02:09	0.15 low
	08:39	3.43 high
	12:42	2.55 low
	19:17	6.26 high
8	02:49	−0.94 low
	09:23	3.66 high
	13:31	2.54 low
	19:59	6.69 high
9	03:27	−0.94 low
	10:02	3.87 high
	14:18	2.44 low
	20:42	7.05 high

[1] hours:minutes. Time meridian 120°W. 00:00 is midnight; 12:00 is noon—Pacific Daylight Time. Heights are referred to mean lower low water, which is the chart datum of soundings.

Tidal Current Tables

Tidal currents in the open ocean have been explained as rotary currents formed by the passing tide wave form and the deflection of water particles due to the Coriolis effect. Tidal currents in the deep sea are of scientific interest to oceanographers concerned with removing this circular motion from their data to obtain the net flows of the major ocean currents. Tidal currents in harbors and coastal waters are of major interest to commercial vessels and pleasure boaters because these currents can be very strong and must be taken into account by anyone who wants to navigate in such waters.

Like the tides, the tidal currents are first measured at selected primary locations in important inland waterways and channels. These current data are studied to determine how the speed and direction of the tidal current are related to the predicted tide-level changes. As before, the local effect is determined and is used to predict tidal currents on the basis of the tide tables.

Once tidal currents have been predicted for a location, they can be graphed with time and compared to the tidal height curves for that area. If maximum current times coincide with the times for either low or high water, the tide has a progressive wave form. If maximum current times coincide with mid-tide stages, the tide is a standing wave–type tide.

Tidal current data (table 9.3) are published in a format similar to that of the tide tables. The times of slack water, maximum flood currents, and maximum ebb currents, as well as the speed of the currents in knots and the direction of flow for ebb and flood currents are given for primary channel stations. Auxiliary tidal current stations are keyed to the primary current stations with correction factors to determine current speed, time, and direction at the secondary stations. This information allows the master of a vessel to decide at what time to arrive at a particular channel in order to find the current flowing in the right direction or how long to wait for slack water before choosing to proceed through a particularly swift and turbulent passage.

QUICK REVIEW

1. Why are astronomical data not sufficient to predict the tides for a specific location?
2. Describe the changes in speed and direction of the tidal current from ebb to flood tide.

Table 9.3 Predicted Tidal Currents for August 1–4, 2006, Seymour Narrows, British Columbia, Canada (50.1333°N, 125.3500°W)

Day	Slack Water Time (h:min)[1]	Maximum Current Time (h:min)[1]	Velocity (knots)[2]
1		01:38	−8.15
	04:59 flood tide begins	08:01	6.74
	11:25 ebb tide begins	14:14	−4.85
	17:07 flood tide begins	19:57	4.81
2	22:42 ebb tide begins	02:21	−7.46
	05:46 flood tide begins	08:59	6.83
	12:35 ebb tide begins	15:23	−4.21
	18:19 flood tide begins	20:55	3.56
3	23:23 ebb tide begins	03:14	−6.99
	06:39 flood tide begins	10:02	7.33
	13:49 ebb tide begins	16:43	−4.29
	19:46 flood tide begins	22:03	2.82
4	00:16 ebb tide begins	04:17	−6.91
	07:36 flood tide begins	11:04	8.30
	14:57 ebb tide begins	17:57	−5.20
	21:06 flood tide begins	23:13	2.79

[1] hours:minutes. Time meridian 120°W. 00:00 is midnight; 12:00 is noon—Pacific Daylight Time.

[2] Velocity is positive during flood tide and negative during ebb tide.

9.8 Practical Considerations: Energy from Tides

Tidal energy was used to turn mill wheels in the coastal towns of northern Europe during the nineteenth century. The possibility of obtaining large amounts of energy from the tides still exists where there are large tidal ranges or narrow channels with swift tidal currents. Two systems can be used to extract energy from the rise and fall of the tide. Both require building a dam across a bay or an estuary so that seawater can be held in the bay at high tide. When the tide ebbs, a difference in water-level height is produced between the water behind the dam and the ebbing tide. When the elevation difference becomes sufficient, the seawater behind the dam is released through turbines to produce electrical power. The reservoir behind the dam is refilled on the next rising tide by opening gates in the dam. This single-action system produces power only during a portion of each ebb tide (fig. 9.19*a*). A tidal range of about 7 m (23 ft) is required for this system to produce power.

This same arrangement can be used as a double-action system. In this system, power is produced on both the ebbing and rising tide (fig. 9.19*b*). This system requires manipulating water heights across the dam and two-way flow through turbines.

These systems appear to be simple and cost-effective methods for producing electrical power, but there are few places in the world where the tidal range is sufficient and where natural bays or estuaries can be dammed at their entrances at reasonable cost and effort. Moreover, the appropriate tides and bays are not necessarily located near population centers that need the power. Installation and power-distribution costs, in addition to periodic low-power production because of the changing tidal amplitude over the tide's monthly cycle, make this type of power more expensive than that from other sources.

Since 1966, a 240-megawatt power plant alongside a dam on the Rance River Estuary in France has produced 5.4×10^{10} watt-hours of electricity per year. Present global energy demands could be satisfied by 250,000 plants of this capacity, but only about 255 sites have been identified around the world that have the potential for tidal energy development.

Tidal power was first considered for the Bay of Fundy in 1930. Canadian and American interest was casual because of the expense of the project until the rising cost of fossil fuels focused interest on alternative energy sources. The province of Nova Scotia commissioned a power station in the tidal estuary of the Annapolis River in 1984 (fig. 9.20), the world's largest

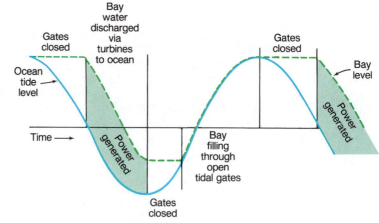

(a) Single-action power cycle; ebb only

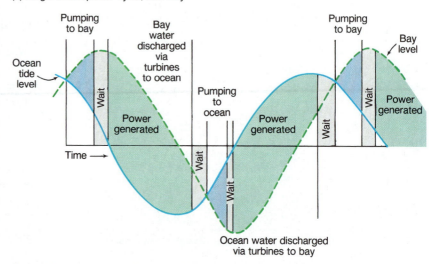

(b) Double-action power cycle; ebb and flood

Figure 9.19 Periods of power generation related to ocean tidal heights and water storage levels. (a) A single-action tidal power system. Power is generated on the ebb tide. (b) A double-action tidal power system. Power is generated on the ebb and flood tides.

Figure 9.20 The Annapolis River tidal power project in Nova Scotia (Bay of Fundy), Canada, is the first tidal power plant in North America.

straight-flow-rim–type single-effect scheme. Tidal ranges at the Annapolis site vary from 8.7 m (29 ft) during spring tides to 4.4 m (14 ft) during neap tides. The unit generates up to 20 megawatts of power from the head of water developed between the upstream basin and sea level downstream at low tide. Initially, this project was intended as a pilot to demonstrate the feasibility of a large-scale, straight-flow turbine in a tidal setting. The station has now been added to the province's principal electrical utility's hydrogenerating system. Annual production is $3–4 \times 10^{10}$ watt-hours. Power availability has been in excess of 95%. Similar plants exist in China and Russia.

Although tidal power does not release pollutants, it is not without environmental consequences. Dams isolate bays from the rivers and estuaries with which they had been connected. At present, the natural period of oscillation of the Bay of Fundy is about thirty minutes longer than the tidal period; these periods are sufficiently alike to cause the tides to resonate. Decreasing a bay's length with a dam shortens its period of oscillation. In the case of the Bay of Fundy, it is estimated that this could increase tidal ranges by 0.5 m (20 in) and tidal currents by 5% along the coast of Maine.

The damming of a bay or an estuary interferes with ship travel and port facilities. The dams are barriers to migratory species and alter the circulation patterns of the isolated basin. As the environmental impacts of dams have become better known, they have become far less popular as possible energy sources.

Swift tidal currents in inshore channels represent another possible energy source. Flowing water has been used for several centuries to turn the equivalent of windmills or water wheels for limited power. Because the tidal currents reverse with the tide, these "water mills" must operate with the current flowing in either direction.

Power generated by windmills depends on the density of the air, the blade diameter, and the cube of the wind speed. Water mills with a similar design depend on the density of the water, the blade diameter, and the cube of the current speed (table 9.4). A windmill in a 20-knot wind produces about the same power as a water mill with blades of the same diameter in a 2-knot current because the density of water is about 1000 times the density of air:

$$0.001 \text{ g/cm}^3 \ (20 \text{ knots})^3 = 1 \text{ g/cm}^3 \times (2 \text{ knots})^3$$

air density $\times$ (speed)^3 = water density $\times$ (speed)^3

Tidal currents, although reversing, are regular and predictable. The currents are a natural, renewable power source that is much easier to harvest than the energy from water waves. British and Norwegian engineering teams have begun looking at these strong currents for producing electrical energy. Propeller-driven turbine generators mounted on tripods have been designed to deliver 300 kilowatts of power per generating unit (fig. 9.21).

Table 9.4 Estimated Power Generation (in kW) for Water Mills at Various Current Speeds

Blade Diameter (m)	Current Speed		
	5 Knots (2.5 m/s)	4.5 Knots (2.25 m/s)	3.5 Knots (1.75 m/s)
2	8.5	5.5	3.7
5	53	35	23
10	210	140	90
15	480	310	205
20	850	550	370
30	1910	1250	820

QUICK REVIEW

1. What are the benefits and disadvantages of generating power from tidal action?
2. What characteristics would you look for in tidal currents and tidal range to determine if a particular site would be good for generating power from the tides?

Figure 9.21 Computer image of a 15–16 m (50–53 ft) blade-driven turbine. This unit sits on the sea floor in an area of fast tidal flow. It is the marine counterpart of a modern windmill.

Summary

Diurnal tides have one high tide and one low tide each tidal day; semidiurnal tides have two high tides and two low tides. A semidiurnal mixed tide has two high tides and two low tides, but the high tides reach different heights and the low tides drop to different levels. For diurnal and semidiurnal tides, the greatest height reached by the water is high water, and the lowest point is low water. Mixed tides have higher high water, lower high water, higher low water, and lower low water. The zero depth on charts is referenced to mean low water or mean lower low water; low tides falling below these levels are minus tides. Rising tides are flood tides; falling tides are ebb tides.

Equilibrium tidal theory is used to explain the tides as a balance between gravitational and centrifugal forces. An equatorial tide is a semidiurnal tide, or a tide wave with two crests and two troughs. Because the Moon moves along its orbit as Earth rotates on its axis, a tidal day is twenty-four hours and fifty minutes long. The period of the semidiurnal tide is therefore twelve hours and twenty-five minutes.

The Sun's effect is less than half that of the Moon's, and the tidal day with respect to the Sun is twenty-four hours. Because the tidal force of the Moon is greater than that of the Sun, the tidal day is still considered to be twenty-four hours and fifty minutes.

Spring tides have the greatest range between high and low water; they occur at the new and full Moons, when Earth, the Sun, and the Moon are in line. Neap tides have the least tidal range; they occur at the Moon's first and last quarters, when the Moon is at right angles to the line of centers of Earth and the Sun.

When the Moon or Sun stands above or below the equator, equatorial tides become more diurnal. Diurnal tides are often called declinational tides. The elliptic orbits of Earth and the Moon also influence the tide.

The dynamic approach to tides investigates the actual tides as they occur in the ocean basins. The tide wave is discontinuous, except in the Southern Ocean. It is a shallow-water wave that oscillates in some ocean basins as a standing wave, and its motions persist long enough to be acted on by the Coriolis effect. The tide wave is reflected, refracted, and diffracted along its route. Because a point on Earth moves eastward faster at the equator than the tide wave progresses westward as a free wave, the tide wave moves as a forced wave, and its crest is displaced to the east of the tide-raising body. Above 60°N and 60°S latitudes, the crest is more nearly in line with the Moon, because the speeds of Earth and the tide wave match more closely.

In large ocean basins, the tide wave can move as a progressive wave. The Coriolis effect causes a rotary tidal current. Standing wave tides can also form in ocean basins; they rotate around an amphidromic point as they oscillate. Cotidal lines mark the progression of the tide crest, and regions of equal tidal range are identified by corange lines. The tidal range increases with the distance from an amphidromic point. Standing wave tides in narrow open-ended basins oscillate about a node at the entrance to the bay or basin; the antinode is at the head of the basin, as in the Bay of Fundy. There is no rotary motion when a bay is very narrow.

A rapidly moving tidal bore is caused when a large-amplitude tide wave moves into a shallow bay or river.

Tidal heights and currents are predicted from astronomical data and actual local measurements. NOAA provides data for annual tide and tidal current tables.

Single- and double-action dam and turbine systems extract energy from the tides. Tidal power plants are in use in France, Russia, and Canada. Few places have large enough tidal ranges and suitable locations for tidal dams. Tidal power has environmental drawbacks as well as high developmental costs. Tidal currents are another energy-producing possibility, but installation and service costs are considered high.

Key Terms

diurnal tide, 244
semidiurnal tide, 244
semidiurnal mixed tide, 244
high water, 244
low water, 244
higher high water, 244
lower high water, 244
higher low water, 244
lower low water, 244
average tide/mean tide, 244

tidal datum, 245
minus tide, 245
flood tide, 245
ebb tide, 245
tidal current, 245
slack water, 245
equilibrium tidal theory, 246
dynamic tidal analysis, 246
centrifugal force, 246
tidal day, 248

tide wave, 248
centripetal force, 248
range, 249
spring tide, 249
neap tide, 249
declinational tide, 249
free wave, 251
forced wave, 251
progressive tide, 251

cotidal line, 252
standing wave tide, 252
rotary standing tide
    wave, 252
amphidromic point, 252
corange line, 252
tidal bore, 254
harmonic analysis, 255
local effect, 255

Study Problems

1. In a simple diurnal tide, if high tide occurs at 1:10 P.M. on a given day, when would you expect the next high tide?

2. In a simple semidiurnal tide, if high tide occurs at 1:10 P.M. on a given day, when would you expect the next high tide?

3. Using the information in table 9.2, plot the San Diego, California, tide curves for July 8 and 9, 2006. What type of tide is the San Diego tide? Label the water levels on the curve.

4. Using the information in table 9.2, what are the maximum and minimum tidal ranges of the San Diego, California, tide during the time July 6 to 9, 2006?

5. Using the information in table 9.3, at what time on August 4, 2006 should you arrive at Seymour Narrows to navigate the narrows at slack water between lunch and dinner?

Coasts, Beaches, and Estuaries

Learning Outcomes

After studying the information in this chapter students should be able to:

1. *recognize* different types of coasts when viewing images of them,

2. *describe* the processes that generate secondary coasts,

3. *label* features of the beach on a diagram of a typical beach profile,

4. *review* the seasonal movement of sand onto and off the beach,

5. *diagram* and explain the movement of sand in a coastal circulation cell,

6. *list* the different types of beaches,

7. *recognize* different types of coastal structures and *know* their intended purpose,

8. *locate* the National Marine Sanctuaries,

9. *characterize* the different types of estuaries, and

10. *sketch* the trend of mean sea level over the past fifteen years and label the axes.

Rocky headlands and a pocket beach along the coast of La Jolla, California. The pier at the Scripps Institution of Oceanography is in the background.

Shores and beaches are the most familiar areas of the ocean for most people, but even the casual visitor senses changes in these areas. On any visit the tide may be high or low, the logs and drift may have changed position since the last visit, and the dunes and sandbars may have shifted since the previous summer. A visit to a beach farther along the coast or along a different ocean presents a different picture. The sand is a different color, or there is no sand at all; the waves break higher or lower across the beach; the slope of the beach is steeper or flatter; and so on. No beach is static, and no two beaches are exactly the same.

On our visits to the coast, we may stop to admire an inlet, photograph a quiet harbor, or visit an estuary. These places are where the salt water from the oceans and the fresh water from the land meet. In some ways, all these areas behave like small oceans, but the frequent addition of fresh water gives them characteristics of their own. These areas share certain features, but, just as beaches differ, one estuary differs from another.

In this chapter, we study the types of coasts and beaches and the natural processes that create and maintain them. We also investigate the estuaries and partially enclosed seas of coastal areas and the circulation patterns that are unique to them. These are complex and sensitive parts of the ocean system; to preserve them we must understand them.

10.1 Major Coastal Zones

The **coasts** of the continents are the areas where the land meets the sea. The terms *coast, coastal area,* and *coastal zone* are used to describe these land edges that border the sea, including areas of cliffs, dunes, beaches, bays, coves, and river mouths. The width of the coast, or the distance to which the coast extends inland, varies and is determined by local geography, climate, and vegetation, as well as by the perception of the limits of marine influence in social customs and culture. However, the coast is most generally described as the land area that is or has been affected by marine processes such as tides, winds, and waves, even though the direct effect of these processes may be felt only under extreme storm conditions. The seaward limit of the coast usually coincides with the beginning of the beach or shore but sometimes includes nearby offshore islands.

The term *coastal zone* includes the open coast as well as the semi-isolated and sheltered bays and estuaries that interrupt it. The coastal zone incorporates land and water areas and has become the standard term used in legal and legislative documents affecting U.S. coastal areas. In this context, the coastal zone's landward boundary is defined by a distance (often 200 ft or about

60 m) from some chosen reference (usually high water); the seaward boundary is defined by state and federal laws.

Coastal areas are regions of change in which the sea acts to alter the shape and configuration of the land. Sometimes these changes are extreme and occur rapidly (for instance, the damage caused by a hurricane). Sometimes the changes are subtle and so slow that they are not perceived by people during their lifetimes but are very impressive when considered over long periods (for example, the formation of the Mississippi River delta or the gradual erosion of Cape Hatteras, North Carolina). In general, coasts composed of soft, unconsolidated materials, such as sand, change more rapidly than coasts composed of rock. Over geologic time, coasts have changed dramatically as they have emerged and submerged with tectonic processes, climatic variations, and changing sea level; some remnants of these changes can still be seen in present coastlines.

The **shore** is a part of the coast; it is that region from the outer limit of wave action on the bottom (seaward of the lowest tide level) to the limit of the waves' direct influence on the land.

This land limit may be marked by a cliff or an elevation of the land above which the sea waves cannot break. Such features act as barriers to the wave-tossed drift of logs, seaweeds, and other debris. The **beach** is an accumulation of sediment (sand or gravel) that occupies a portion of the shore. The beach is not static but moving and dynamic, because the beach sediments are constantly being moved seaward, landward, and along the shore by nearshore wave and current action. Between the high-tide mark and the upper limit of the shore may be dunes or grass flats dotted with drift logs left behind by an exceptionally high tide or a severe storm.

QUICK REVIEW

1. What are the boundaries of a coast?
2. What kinds of features are found along coasts?
3. What factors determine the width of a coast?

10.2 Types of Coasts

Any land form is the product of the processes that have changed it through time, and the study of land forms and the processes that have fashioned them is known as **geomorphology.** Coastal geomorphology considers tectonic processes; wave, wind, and current exposure; tidal range and tidal currents; sediment supply and coastal transport; and climate and climate changes.

Climate change can result in a worldwide, or **eustatic, change** in sea level that may submerge previous coasts or expose previous sea floor. During the last period of glaciation, sea level dropped by an estimated 100 m (328 ft) or more due to the storage of water on land as glacial ice and the cooling and subsequent decrease in volume of the remaining water in the ocean basins. With the global warming that signaled the end of the last ice age, sea level rose to its present height as land ice melted and the increase in average temperature of seawater caused a corresponding increase in volume. Sea level is continuing to rise even now as a result of global warming.

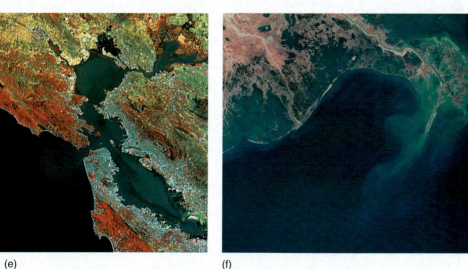

Figure 10.1 Primary coasts: (a) The narrow channel of a fjord, Milford Sound, New Zealand. (b) Ria coasts (drowned river valleys), Delaware Bay (upper right) and Chesapeake Bay (center). (c) Dune coast, Florence, Oregon. (d) A volcanic crater coast, Hanauma Bay, Hawaii, is a crater that has lost the seaward portion of its rim. (e) The San Andreas Fault runs along the California coast to the west of San Fransisco, separating Point Reyes from the mainland. (f) The Mississippi River delta. Because the river is being kept in its channel by levees, river-borne sediments are continuing to the river's mouth and are being deposited in deep water instead of spreading out and replenishing the delta.

There are different ways to classify coasts for the purpose of describing and studying them. One is to classify them as erosional or depositional, depending on whether they predominantly lose or gain sediments, or shoreline. This is often influenced by whether the coast is located at the edge of the plate as part of an active, or leading, continental margin or away from the edge of the plate as part of a passive, or trailing, continental margin (the formation and characteristics of active and passive continental margins are discussed in chapter 2, section 2.4). In the United States, eastern coasts are part of a passive continental margin and are generally considered depositional. Western coasts are located along active continental margins and are generally considered erosional. These categories should not be confused with coastal erosion problems that are prevalent on both coasts.

Coasts of both types may be modified by eustatic sea-level changes. Some coasts (both erosional and depositional) have been eroded and their sediments swept away to be deposited elsewhere, while other coasts have received sediments from rivers and shore currents. In some parts of the world, coasts are modified by seasonal storms; other areas enjoy more benign weather. Coasts of polar regions are modified by their interaction with sea ice; tropical coasts are altered by reef-building corals. Each coast has its own identity, but different types of coasts can be recognized as the results of certain processes.

A system devised by the late Francis P. Shepard of the Scripps Institution of Oceanography organizes coasts into two process categories: (1) coasts that owe their character and appearance to processes that occur at the land-air boundary and (2) coasts that owe their character and appearance to processes that are primarily of marine origin (table 10.1). Further classification depends on whether their large-scale features are the products of tectonic, depositional, erosional, volcanic, or biological processes.

In the first category are coasts that have been formed by (1) erosion of the land by running surface water, wind, or land ice, followed by a sinking of the land or a rise in sea level; (2) deposits of sediments carried by rivers, glaciers, or the wind; (3) volcanic activity, including lava flows; and (4) uplift and subsidence of the land by earthquakes and associated crustal movements. Coasts formed by these processes are **primary coasts,** since there has not yet been time for the sea to substantially alter or modify their original appearance (fig. 10.1). The second category includes coasts formed by (1) erosion

Table 10.1   Coastal Types

Primary Coasts	Features	Examples
Erosional Examples		
Fjord coasts Figure 10.1*a*	Glacially carved drowned channels. Deep and elongate.	Mountainous coasts of previously glaciated temperate and subpolar regions; Alaska, Greenland, New Zealand, Scandinavia.
Drowned river valleys (ria coasts) Figure 10.1*b*	Multibranched shallow areas; sea level has risen or the land has sunk.	Chesapeake or Delaware Bay systems, Rio Plata, Argentina.
Depositional Examples		
Dune coasts Figure 10.1*c*	Sand deposits on coasts by wind from the beach or from the land.	Central and northern Oregon coast. West coast of Sahara Desert and west coast of Kalahari Desert, Africa.
Glacial moraine coasts	Deposits left by glaciers.	Long Island and Cape Cod, Baltic Sea, Sea of Okhotsk.
Deltaic coasts Figure 10.1*f*	Heavy river deposition in shallow coastal regions.	Mouths of rivers such as the Mississippi, Nile, Ganges, and Amazon.
Volcanic Examples		
Lava coasts	Areas of active volcanoes with lava that reaches the sea.	East coast island of Hawaii.
Cratered coasts Figure 10.1*d*	Craters of coastal volcanoes at sea level filled with seawater.	Hawaiian islands, and islands and coastal areas of the Aegean Sea.
Tectonic Examples		
Fault coasts Figure 10.1*e*	Coastal faults filled with seawater.	Tormales Bay, California, coast of Scotland, Gulf of California, Red Sea.

Secondary Coasts	Features	Examples
Erosional Examples		
Regular cliffed coasts Figure 10.2*a*	Irregular coast made more uniform by wave erosion.	Hawaii, southern California, southern Australia, Cliffs of Dover.
Irregular cliffed coasts Figure 10.2*b*	Coast made more irregular by wave erosion.	Sea stack coasts of northern Oregon and Washington, New Zealand, Australia.
Depositional Examples		
Barrier coasts Figure 10.2*c*	Shallow coast with abundant sediments bordered by barrier islands and sand spits.	East Coast of the United States, Gulf of Mexico, Netherlands.
Beach plains	Flat areas of elevated coastal seashore. May be terraced by wave erosion during uplift.	Northern California, Sweden, New Guinea.
Mudflats and salt marshes Figure 10.2*e*	Sediments deposited in wave-protected environments.	Common in many areas.
Built by Marine Organisms		
Coral reef coasts	Shallow tropical waters; corals produce massive coastal structures.	Tropical regions of the Pacific, Indian, and Atlantic Oceans.
Mangrove coasts Figure 10.2*f*	Tropical and subtropical areas with freshwater drainage.	Florida, northern Australia, Bay of Bengal, other tropical areas.
Marsh grass coasts Figure 10.2*g*	Shallow, protected areas of fine sediments where plants root in intertidal land.	Eastern United States, Canada, coast of Europe, temperate climates.

due to waves, currents, or the dissolving action of the seawater; (2) deposition of sediments by waves, tides, and currents; and (3) alteration by marine plants and animals. These coasts are **secondary coasts;** their character, even though it may have been originally land-derived, is now distinctly a result of the sea and its processes (fig. 10.2).

A relatively young coast may be rapidly modified by the sea, while a coast that is old may retain its land-derived characteristics. Keep in mind that absolute age is not really important because the classification is based only on whether the characteristics are derived from the land or from the sea, and it is possible to find both types of features along the same coast.

Primary Coasts

In Shephard's first category are coasts that have been formed by (1) erosion of the land by running surface water, wind, or land ice, followed by sinking of the land or a rise in sea level; (2) deposits of sediments carried by rivers, glaciers, or the wind; (3) volcanic activity, including lava flows; and (4) uplift and subsidence of the land by earthquakes and associated crustal movements. Coasts formed by these processes are described as primary coasts because there has not yet been time for the sea to substantially alter or modify their original appearance (fig. 10.1).

Coasts formed by subaerial erosion followed by sinking of the land or a rise in sea level include those that were covered by glaciers during the ice ages. During these glacial periods, sea level was lower than at present because much of the water was held as ice on the continents. Glaciers moved slowly across the land, scouring out valleys as they inched along to the sea. The weight of the ice caused the land to subside, and in some cases, when the ice began to melt, sea level rose faster than the land could rebound upward to its preglacial elevation. In other cases, glacial troughs were scoured below sea level and filled with seawater as the ice receded. The **fjords** of Norway, Greenland, New Zealand, Chile, and southeastern Alaska are the results of these processes (fig. 10.1a). Fjords are long, deep, narrow channels with a U-shaped cross section. Where the glacier met the sea, there was often a collection of debris that formed a lip, creating a shallow entrance, or **sill.**

When a glacier or ice sheet ceases its forward motion and retreats, it leaves a mound of rubble, called a **moraine,** along the border of its greatest extension. If the glacier has reached the edge of a continent, this material becomes a part of the coastal area. Long Island, off the New York and Connecticut coasts, and Cape Cod, Massachusetts, are moraines. Moraines act as protective barriers to the continental coast. In some areas, land that was deeply covered by ice during the last ice age is still slowly rising; in Scandinavia, the rate of rise is about 1–5 cm (0.4–2 in) per year. Along other coasts, tectonic forces rather than loss of ice have caused uplift. Along the western coasts of North and South America, old wave-cut terraces can be identified now well above sea level.

When sea level was lowered during the ice ages, rivers flowed over the exposed shore to the sea. The rivers cut V-shaped channels into these areas, and in many cases, the channels had numerous side branches formed by feeder streams. As the sea level rose, these channels were filled with seawater, producing areas such as Chesapeake Bay and Delaware Bay on the eastern coast of the United States. A coast of this type is called a **drowned river valley,** or **ria coast,** shown in figure 10.1b.

Rivers carrying extremely heavy sediment loads build deltas on top of the continental shelves. The **delta,** or deposit of river-borne sediment left at a river mouth, produces a flat, fertile coastal area. Examples of these deposits are found at the mouths of the Mississippi, Ganges, Nile, and Amazon Rivers. A similar type of coast is produced when the eroded materials carried down from the hills by surface runoff and many small rivers join together to form an **alluvial plain.** The eastern coast of the United States south of Cape Hatteras was produced in this way.

It is estimated that every second, the rivers of the world carry 530 tons of sediment to the sea. This rate of removal is equal to the erosion of a layer 6 cm (2.4 in) thick from all land above sea level every 1000 years. This sediment helps to form and maintain the world's beaches, and much of it finally finds its way to the deep-ocean floor. All of it passes through the coastal zone and takes part in coastal processes. Refer to chapter 3 for sources of terrigenous material.

A **dune coast** is a wind-modified depositional coast. In Africa, the Western Sahara is gradually growing westward toward the Atlantic Ocean as the prevailing winds move sand from the inland desert to the coast. Along other dune coasts, for example, along central and northern Oregon (fig. 10.1c), the winds move the sands inland to form dunes; elsewhere dunes driven by the wind migrate along the shore.

The Hawaiian Islands are the tops of large seamounts and have excellent examples of coasts formed by volcanic activity. Lava flows extending to the sea form black sand beaches or **lava coasts.** Craters formed by volcanic explosions became concave bays when they lost their rims on the seaward side, forming **cratered coasts** (fig. 10.1d).

When tectonic activity results in faulting and displacement of Earth's crust, the coast is changed in characteristic ways. The California San Andreas Fault system lies along a boundary where crustal plates are moving parallel to each other along a transform fault. (Plate movements and transform faults are described in chapter 2.) Over long periods of time, faults formed in this area filled with seawater. The Gulf of California, also known as the Sea of Cortez, between Baja California and the mainland of Mexico, is found at the southern end of the fault system. At the northern end lie San Francisco and Tomales Bays, where the fault runs out into the Pacific Ocean. Tomales Bay is a particularly good example of a **fault bay.** See the satellite image of this fault system in figure 10.1e. In other parts of the world, the Red Sea and the Scottish coast provide examples of **fault coasts.**

Rivers often carry large amounts of sediment that has been eroded from the land. When rivers reach the coast, they are no longer confined to a narrow river channel and their speed slows down significantly. This reduces their ability to transport the sediment they carry and the sediment is deposited at the mouth of the river. This accumulated sediment is called a delta (fig. 10.1*f*). The shape of the delta depends on the volume of sediment transported and the action of coastal currents that may redistribute the sediment.

Secondary Coasts

In Shephard's second category are coasts that have been formed by (1) erosion due to waves, currents, or the dissolving action of the seawater; (2) deposition of sediments by waves, tides, and currents; and (3) alteration by marine plants and animals. These coasts are secondary coasts; their character, even though it may have been originally land-derived, is now distinctly a result of the sea and its processes (fig. 10.2).

Secondary coasts owe their present-day appearance to marine processes. As waves batter the coast, they constantly erode and grind away the shore; rocks and cliffs are undercut by the wave action and fall into the sea, where they are ground into sand. A coastal area formed of uniform material may have been irregular in shape with headlands and bays at one time, but the concentrated energy of the waves wears down the headlands more rapidly than the straighter shore and cove areas. With time, a more regular coastline results; examples are found in southern California, southern Australia, New Zealand (fig. 10.2*a*), and England's cliffs of Dover. If the original coastline is made up of materials that vary greatly in composition and resistance to wave erosion, the result will be an increasingly irregular coastline of hard rock headlands separated by sandy coves. In some places, such as northern California, Oregon, Washington, Australia, and New Zealand, the headlands erode irregularly, and small rock islands and tall, slender pinnacles of resistant rock are formed; these are known as **sea stacks** (fig. 10.2*b*). Headlands still connected to the mainland are undercut to produce sea caves or are cut through to produce arches or small windows.

In some cases, eroded materials are carried seaward by the waves and currents to areas just off the coast. If sufficient sand material is deposited in offshore shallows paralleling the beach, **bars** are produced. If still greater amounts of material are added, the bars may grow until they break the surface and form **barrier islands** (fig. 10.2*c*). The barrier islands of the southeastern coast of the United States were formed during a period of rising sea level; seawater flooded low coastal areas, isolating the high dunes at the sea's edge and converting a primary coast to a secondary coast. Once a barrier island has formed, plants begin to grow, and the vegetation helps to stabilize the sand and increase the island's elevation by trapping sediments and accumulating organic matter.

A line of barrier islands along a coast protects the shoreline behind the islands from storm waves and erosion, but the islands can sustain heavy damage. One of the greatest natural disasters to strike the United States was the September 8, 1900, Galveston hurricane. Galveston sits on Galveston Island, a barrier island along the Texas coast in the Gulf of Mexico. In 1900, the city had a population of 37,000, and the highest elevation on the island was 2.7 m (8.7 ft). Just before the full force of the hurricane struck, people were still playing on the beach, enjoying the novelty of unusually large waves. No one thought it would be a dangerous storm. By the time the citizens of Galveston realized how strong the hurricane was, it was too late to escape to the mainland. Winds estimated at 140 mph swept over the island, leaving devastation in their wake. The island was pounded by large waves, and a storm surge (chapter 6, section 6.6) 4.8 m high (15.7 ft) swept across it, destroying more than 3600 buildings in the city (fig. 10.3). An estimated 6000 to 8000 of the town's residents died. After the storm, Galveston constructed a seawall and raised the elevation of the island to protect it from future hurricanes. During 1989's Hurricane Hugo, Folly Island, along the coast of South Carolina, suffered extensive storm damage. Eighty-six of 290 oceanfront structures were more than 50% destroyed, and fifty were damaged beyond repair. The U.S. National Flood Insurance Program issued nearly $3 million in damage payments to Folly Island property owners. In 1997, Tropical Storm Josephine moved eastward across the Gulf of Mexico toward Florida. This storm did not develop hurricane-force winds, but it did cause extensive damage to coastal areas of Texas in the vicinity of Galveston. Josephine remained 482 km (300 mi) off the coast, producing steady, onshore winds of approximately 12 m/s (27 mi/h) for over a hundred hours. The winds raised the sea level at Galveston 80 cm (32 in) above normal high-tide levels and caused average wave heights of 3.5 m (11 ft) along the coast. The sustained period of high water and wave action brought by this storm produced severe beach erosion, loss of homes, and loss of the dunes that stood between the remaining homes and the water. Attempts are made to halt storm-caused erosion by building seawalls and beach-holding devices, but these do not always prove successful. Although well intentioned, they often result in aggravating the loss of material from barrier islands rather than halting it.

Sand spits and **hooks** are bars connected to the shore at one end (fig. 10.2*d*). Spits and hooks may grow, shift position, wash away in a storm, or rebuild under more moderate conditions. The area between the mainland and these spits and hooks is protected from open ocean waves and is often the site of beach flats formed of sand or mud (fig. 10.2*e*). If a spit grows sufficiently to close off the mouth of an inlet, a shallow lagoon is formed. Water percolates through the sand, and inside the lagoon, the water rises and falls with the tides.

In the tropical oceans, **reef coasts** result from the activities of sea organisms. Corals grow in the shallow, warm waters surrounding a landmass, and the small animals gradually build a fringing reef, which is attached directly to the landmass. In other places, a lagoon of quiet water may lie between the barrier reef and the land, and, in a few cases, the coral encircles and overlies a submerged seamount to form an atoll.

The Great Barrier Reef, stretching along the northeastern coast of Australia toward New Guinea, is the largest and most famous of the world's coral reefs. Coral atolls in the Pacific include Tarawa, Kwajalein, Enewetak, and Bikini. The reef-encircled islands of Iwo Jima and Okinawa are familiar as the sites of major battles in the Pacific during World War II.

Other marine animals form reeflike structures as their shells are deposited layer on layer, gradually building up a mass of hard material. There are large reef deposits of oyster shells in the Gulf

Figure 10.2 Secondary coasts. (a) A cliff coast made more regular by erosion, New Zealand. (b) Sea stacks are common features along the southern coast of Australia; these stacks are known as the Twleve Apostles. (c) Sea Island, Georgia, is a barrier island that has been extensively developed. The shallow bay between the island and the coast is visible on the left. (d) A spit has formed partway across the entrance to Sequim Bay, Washington. (e) A protected mudflat shore along Puget Sound, Washington. (f) Mangrove trees along the shore of Florida Bay, Florida. (g) Coastal saltwater marsh, Gaspé, Québec, Canada.

Figure 10.3 Residents of the city inspect the damage after the September 8, 1900, Galveston Island, Texas, hurricane.

of Mexico off the coasts of Louisiana and Texas. These reefs are so large that the shells are harvested commercially for lime production. Along the eastern coast of Florida, large populations of shell-bearing animals have contributed their shells directly to the shore; small shell fragments form the sand on the beaches.

Plants as well as animals may modify a coastal area. Along low-lying coasts in warm climates, mangrove trees grow in shallow water. Their great roots form a nearly impenetrable tangle, providing shelter for a unique community of other plants and animals (fig. 10.2*f*). Coastal mangrove swamps are found along the Florida coast, northern Australia, the Bay of Bengal, and in the West Indies. In more temperate climates, low-lying protected coasts with areas of sand and mud are often thickly covered with grasses, forming another type of plant-maintained environment

(fig. 10.2*g*). These **salt marshes** may extend inland a considerable distance if the land is flat enough to permit periodic tidal flooding. Such marshes also form around protected bays and coves that have large changes in tide level. Salt marshes are extremely productive in terms of organic material and extend the shoreline seaward by trapping sediments.

QUICK REVIEW

1. How do primary and secondary coasts differ?
2. Give examples of primary coasts and explain their features.
3. Give examples of secondary coasts and explain their features.

10.3 Anatomy of a Beach

The beach includes the sand and sediment lying along the shore and the sediments carried along the shore by the nearshore waves and currents. Let us consider first the beach where we walk, sunbathe, and relax. A specific terminology for beaches and beach features has developed from the many studies of beaches around the world. These terms allow each area to be described and prevent confusion when one beach is compared to another. A beach profile cuts through a beach perpendicular to the coast; the profile in figure 10.4 shows the major features that might appear on any beach. Use this figure while reading this section, but remember that a given beach may not have all these features, for each beach is the product of its specific environment.

The **backshore** is the dry region of a beach that is submerged only during the highest tides and severest storms. The **foreshore** extends out past the low-tide level, and the **offshore** includes the shallow-water areas seaward of the low-tide terrace to the limit of wave action on the sea floor.

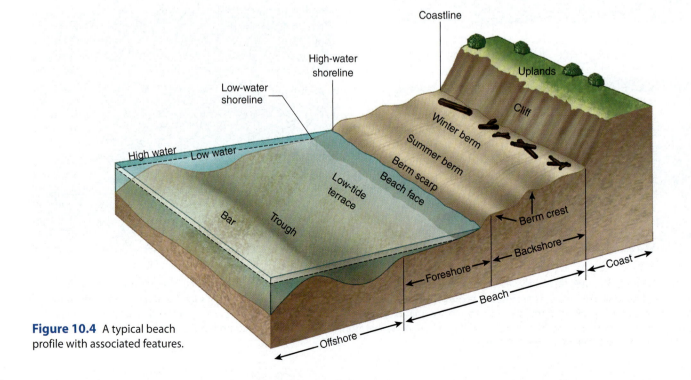

Figure 10.4 A typical beach profile with associated features.

Figure 10.5 A summer berm on a sandy beach. The berm is located at the top of the sloping beach face and shows an accumulation of seaweed.

Figure 10.6 A beach scarp at the intersection of the foreshore and backshore of a sandy beach on Sanibel Island, Florida.

Low terraces called **berms** appear on beaches in the backshore area. Berms are formed by wave-deposited material and have flat tops like the top of a terrace. Berms are recognizable by their slope, or rise in elevation. Some berms have a slight ridge crest that runs parallel to the beach; such a ridge is called the **berm crest.** When two berms are found on a beach (figs. 10.4 and 10.5), the berm higher up the beach, or closer to the coastline, is the **winter,** or **storm, berm.** It is formed during severe winter storms when the waves reach high up the beach and pile up material along the backshore. The seaward berm, the berm closer to the water, is the **summer berm** formed by the gentle waves of spring and summer that do not reach as far up the beach. Once the winter berm is formed, it is not disturbed by the less-intense storms during the year. The waves that produce a winter berm erase the summer berm, but after the storm season is over, a new summer berm is formed by the reduced wave action.

Between the berm and the water may be a wave-cut **scarp** at the high-water level. The scarp is an abrupt change in the beach

slope that is caused by the cutting action of waves at normal high tide (fig. 10.6). Berms and scarps do not form in the foreshore because of the wave action and the continual rise and fall of the water.

The foreshore area is divided into two regions by the water level at low tide on the beach. Seaward of the low-tide level, the foreshore is often flat, forming the shallow, submerged **low-tide terrace.** Landward of the water level at low tide is the part of the beach that is alternately exposed and covered by the water at both low and high tides; this region is called the **beach face.** The beach face is also typically flat but, in some cases, can have low regions called **runnels** bordered by high areas called **ridges** (fig. 10.7). Ridges and runnels are not apparent when the beach face is submerged by high tide, but at low tide, water can be trapped, along with small marine animals, in the runnels. The slope of the beach face is related to the particle size of the loose beach material and to the wave energy. In general, the beach face steepens with larger particles and higher wave energy.

Seaward of the low-tide terrace, in the offshore region, may be **troughs** and bars that run parallel to the beach. These structures change seasonally as beach sediments move seaward, enlarging the bars during winter storms, and then shoreward in summer, diminishing the bars. When a bar accumulates enough sediment to break the surface and is stabilized by vegetation, it may become a barrier island.

Beaches do not occur along all shores. Along rocky cliffs, there may be no beach area between low and high tides. However, there may be small pocket beaches, each separated from the next by rocky headlands or cliffs (fig. 10.8).

QUICK REVIEW

1. Distinguish among the backshore, foreshore, and offshore regions.
2. What is a berm and how is it formed?
3. Distinguish between a summer berm and a winter berm.
4. What is the function of bars along a beach?

Figure 10.7 Ridge and runnel formation on the beach face at Sanibel Island, Florida. Picture is taken at low tide.

Figure 10.8 A pocket beach between rocky headlands along the southern California coast.

periods of late winter and spring river flooding. If a beach annually receives as much sand as it loses, the beach is in equilibrium and does not appear to change from year to year.

Single violent events may alter a beach. A great storm may arrive from such a direction that the usual beach current is reversed. A landslide may pile rubble across the beach and some distance into the water, interfering with the flow of sand along the beach. In either case, the beach changes because of changes in sediment supply, transport, and removal.

Waves moving toward a beach produce a current in the surf zone that moves water onshore and along the beach. This landward motion of water is called the **onshore current,** and this current transports sediment toward the shore in what is called **onshore transport.** Waves usually do not approach with their crests completely parallel to the beach but strike the shore at a slight angle. This pattern sets up a surf-zone **longshore current** that moves along the beach. Both onshore and longshore flow in the surf zone are illustrated in figure 10.10. The turbulence generated as the waves break in the surf zone tumbles the beach material in the water. The wave-produced longshore current moves this sediment parallel to the shore in the surf zone, producing **longshore transport.** Along both coasts of the North American continent, longshore transport generally moves the sediments in a southerly direction.

10.4 Beach Dynamics

A sand beach exists because there is a balance between the supply and the removal of the sand. A sand beach that does not appear to change with time is not necessarily static (no new material supplied, no old material removed) but is more likely to represent a **dynamic equilibrium,** the supply of new sand equaling the loss of sand from the beach. The beach is continually changing, but it appears unchanged and remains in balance.

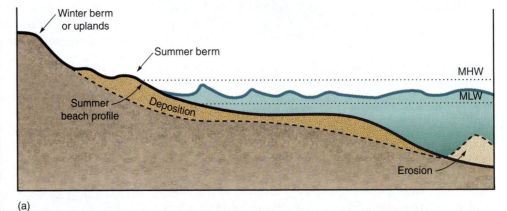

(a)

Natural Processes

The gentler waves of summer move sand shoreward and deposit it, where it remains until the winter. The large storm waves of winter remove the sand from a beach and transport it offshore, to a sandbar. These processes occur alternately, winter and summer, leaving the beach rocky and bare each winter and covered with sand during the summer (fig. 10.9). These seasonal changes are fluctuations about the equilibrium state of the beach each year. Other changes also appear in a cyclic fashion, such as the arrival of more sand during

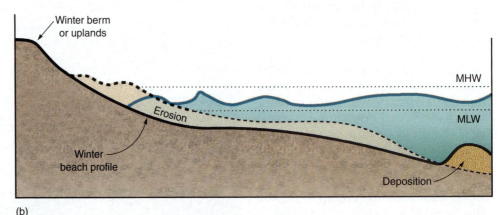

(b)

Figure 10.9 Seasonal beach changes. (a) The small waves and average tides of summer move sand from offshore to the beach and form a summer berm. (b) The high waves and storms of winter erode sand from the beach and store it in offshore bars. Winter conditions remove the summer berm, leaving only the winter berm on the beach.

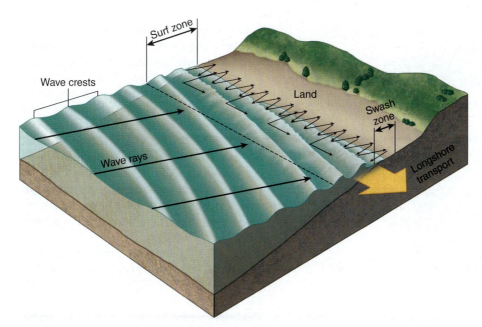

Figure 10.10 Waves in the surf zone produce a longshore current that transports sediments along the beach. *Arrows* indicate onshore and longshore water movement.

Figure 10.11 At Point Reyes National Seashore in California, wave action forms a series of cusps along a shore that has been made regular by wave erosion. How cusps form is not clearly understood.

Evenly spaced crescent-shaped depressions called **cusps** are sometimes found along the sand and cobble beaches of quiet coves and longer, straighter coastlines (fig. 10.11). How cusps are formed is still somewhat of a mystery. They tend to form during neap tide periods, when the tidal range is at a minimum, and are often destroyed during periods of spring tide, when the tidal range is at a maximum. The size of cusps appears to be directly related to the amount of wave energy on the beach. Large cusps are associated with high wave energy, and smaller cusps seem to form where the wave energy is low. Cusp formation may be related to some form of wave interference as waves approach the beach.

Studies of sediment movements along coasts, into navigational channels, and around coastal structures provide important information to property owners, to engineers involved in the design and construction of shoreline installations—for example, piers, breakwaters, and seawalls—and to the military when planning troop and equipment movements in coastal areas.

The path traveled by beach sediments as the longshore current transports them from their source to their area of deposition is called a **drift sector.** Beaches within a drift sector usually remain in dynamic equilibrium and change little in appearance. Although the flow of material along the central portion of a drift sector may be large, the amounts of sediment supplied and lost from the area are usually equal. At either end of the drift sector the beaches change with time. Areas at the sediment source are erosional; they are supplying or losing sediment. Areas at the depositional end of the drift sector are growing, or **accreting.** These processes are shown in figure 10.12. Although the longshore current is the usual transport mechanism within a drift sector, in some places tidal currents and coastal currents associated with large-scale oceanic circulation affect sediment transport.

The up-rush, or **swash,** of water from each breaking wave moves the sand particles diagonally up and along the beach in the direction of the longshore current. The backwash from the receding wave moves downslope toward the surf zone, but the backwash is weaker than the swash because much of the receding water percolates down into the sand. The combined action of the swash and backwash moves particles in a zigzag, or sawtooth, path along the swash zone as part of the longshore transport.

Coastal Circulation

Onshore transport accumulates water as well as sand along a beach. This water must flow back out to sea and will do so one way or another. A headland may deflect the longshore current seaward, or the water flowing along the beach can return to the area beyond the surf zone in quieter water, such as areas over troughs or depressions in the sea floor. Regions of seaward return

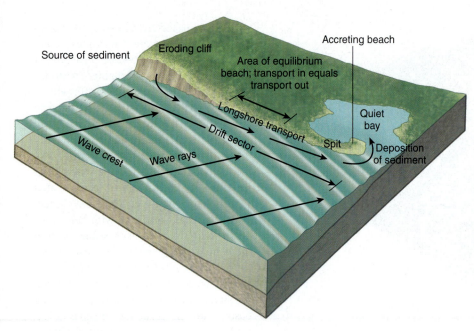

Figure 10.12 A drift sector extends along the coast from the sediment source to the area of deposit.

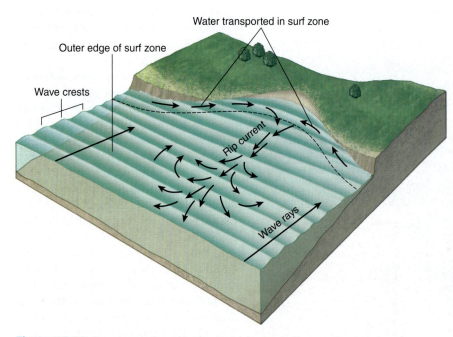

Figure 10.13 Rip currents form along a beach in areas of low surf, reduced onshore flow, and converging longshore currents.

flow are frequently narrow and fast-moving. These are the rip currents that carry sediment through the surf zone (fig. 10.13); rip currents are also discussed in chapter 8 (see fig. 8.27).

Just seaward of the surf, the rip current dissipates into eddies. Some of the sediment is deposited here in quieter, deeper water and is lost from the down-beach flow. However, some is returned to shore with the onshore transport on either side of the rip current. The return of sediments to the beach, transport along the beach, transport back to sea in a rip current, and return again to the beach are processes that occur in a drift sector.

A series of these drift sectors, when linked together along a stretch of coast, forms a **coastal circulation cell.** In a major coastal circulation cell, the seaward transport of sediment results in deposits far enough offshore to prevent the sand from returning to the beach. At the end of a coastal circulation cell, the longshore transport is deflected away from the beach and the sand moves across the continental shelf and down a submarine canyon to the ocean basin floor. This sand has been removed from the coastal system, and sand beaches disappear beyond this point until a new source contributes sand to the next coastal circulation cell.

South of Point Conception in southern California, oceanographers recognize four distinct coastal circulation cells (fig. 10.14). Each cell begins and ends in a region of rocky headlands where beaches are sparse. Submarine canyons are found offshore in each cell. Beaches become wider as rivers contribute their sand to the longshore current, and in each cell, the current deposits the sediment in a submarine canyon, where it cascades to the ocean basin floor. Beaches just south of the canyon are sparse, and a new cell begins. Refer to chapter 3 for a discussion of submarine canyons.

Estimates of sediment transport rates along a beach are made by observing the rate at which sand is deposited, or accreted, on the upstream side of an obstruction or by observing the rate at which a sand spit migrates. Transport of sand along a section of coast varies between zero and several million cubic meters per year. Average values fall between 150,000 and 1,500,000 m³ per year; 150,000 m³ is more than 30,000 dump-truck loads. Once the enormous volume of the naturally moving sand is recognized, it becomes apparent why poorly designed harbors fill very quickly, beaches disappear, and spits migrate a considerable distance during a year. It also becomes obvious that the effort required to keep pace with the supply by dredging or pumping operations is enormously expensive and in many cases impossible.

The energy for the movement of beach materials comes from the waves and the wave-produced currents. Surface winds supply about 10^{14} watts of power to the ocean surface to produce the waves and currents. There are about 440,000 km (264,000 mi) of coastal zone in the world, and approximately half of this is exposed directly to the ocean waves. Ocean waves average 1 m (3.3 ft) in height and produce power that is equivalent to 104 watts for each meter of exposed coastline. Under storm conditions, the wave height averages about 3 m (10 ft), and a meter of coastline receives 10^5 watts. It is this supply of energy that causes beach erosion and the migration of sand along the narrow coastal zone.

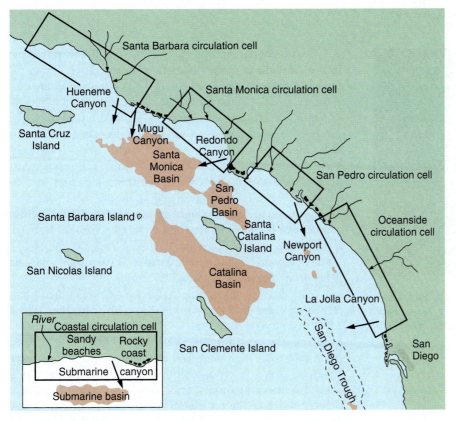

Figure 10.14 Major coastal sediment circulation cells along the California coast. Each cell starts with a sediment source and ends where beach material is transported into a submarine canyon.

QUICK REVIEW

1. How do sand beaches change with the seasons?
2. How is the longshore current formed? How does it function to keep a beach in dynamic equilibrium?
3. Why are rip currents hazardous?
4. What are the components of a coastal circulation cell?

10.5 Beach Types

Beaches are described in terms of (1) shape and structure, (2) composition of beach materials, (3) size of beach materials, and (4) color. In the first category, beaches are described as wide or narrow, steep or flat, and long or discontinuous (pocket beaches). A beach area that extends outward from the main beach and turns and parallels the shore is called a spit (see fig. 10.2*d*). Spits frequently change their shape in response to waves, currents, and storms. In a wave-protected environment, a spit may extend from the shore to an offshore island. If this spit builds until it connects the rock or island to the shore, it forms what is called a **tombolo.** A spit extending offshore in a wide, sweeping arc bending in the direction of the prevailing current is called a hook. The sediment moves around the end of a hook and is deposited in the quiet water behind the point, often producing a broad, rounded, hook-shaped point. See figure 10.15 for an example of a tombolo and figure 10.21 for an example of a hook.

Materials that form beaches include shell, coral, minerals, rock particles, and lava. Shingles are flat, circular, smooth stones

formed from layered rock when the beach slope, wave action, and stone size combine to slide the stones back and forth with the water movement. When stones roll rather than slide, rounded stones or cobbles are formed. The terms *sand, mud, pebble, cobble,* and *boulder* describe the size of beach particles (see chapter 3).

The composition and size of the beach material are related to the source of the material and the forces acting on the beach. Land materials are brought to the coast by rivers or are derived from the local cliffs by wave erosion. The sands of many beaches are the ground-up, eroded products of land materials rich in minerals such as quartz and feldspar. Small particles such as sand, mud, and clay are easily transported and redistributed by waves and currents; larger rocks are usually found close to their source, because they are too large to be moved any distance. The finer particles are carried away by the currents, and the larger rocks are left scattered on the beach. In general, a beach littered with large blocks is an eroded beach, and the pebbles and boulders are called a **lag deposit.** If the lag deposit accumulates in sufficient quantity to protect a beach from further wave and water erosion, the beach is known as an **armored beach** (fig. 10.16).

Some beach materials come from offshore areas. Coral and shell particles broken by the pounding action of the waves are carried to the beaches by the moving water. A beach covered with uniform small particles of sand or mud is a depositional beach.

Some of the world's beaches have distinctive colors. In Hawaii, there are white sand beaches derived from coral, and black sand beaches derived from basaltic lava. Green sands can be found in areas where a specific mineral (such as olivine or glauconite) is available in large enough quantities, and pink sands occur in regions with sufficient shell material.

Figure 10.15 A spit connected to an offshore island forms a tombolo. Tombolos are often formed by converging currents from nearshore eddies; the currents transport sediment toward the tombolo from both sides.

Figure 10.16 An armored beach. Lag deposits of large boulders and cobbles are left on an eroded beach. The rocks on this beach are eroded from glacial deposit cliffs that contain a large assortment of particle sizes, from clays to boulders.

QUICK REVIEW

1. What characteristics can be used to classify beaches?
2. What is the difference between a tombolo and a hook?
3. Name a variety of materials that make up beaches.
4. What kind of materials make up white beaches and black beaches?

10.6 Modifying Beaches

People like beaches and have great desires to own property along these sensitive fringes of land. Approximately 3.8 billion people (60% of the world's population) live within 100 km (60 mi) of a coastline. Population experts expect more than 6.3 billion people to live in coastal areas worldwide within the next fifty years. In the United States today, over one-half of the population lives within 80 km (50 mi) of the coasts (including the Great Lakes). The shore is affected directly and indirectly as increasing numbers of people make their homes in these vulnerable areas.

Coastal zone engineering projects such as **breakwaters** and **jetties** are built to protect harbors and coastal areas from the force of the waves (figs. 10.17 and 10.18). Breakwaters are usually built parallel to the shore; jetties extend seaward to protect or partially enclose a water area or to stabilize a tidal inlet. The protected areas behind jetties and breakwaters are quiet, and the sediments suspended by the wave action and carried by the longshore current settle out in these quiet basins. The result is less material available for beaches farther down the coast, and, as the longshore current continues to flow, the next beach begins to disappear. Small-scale examples of this process are seen where **groins,** rock or timber structures, are placed perpendicular to the beach to trap sand carried in the longshore transport (fig. 10.17); sand is deposited on the up-current side of the groins, and sand is lost from the beach on the down-current side.

Individual owners, trying to stop the erosion of their property, build bulkheads and seawalls along their beaches. These expensive structures are made of timbers, concrete, or large boulders. The land is armored in hopes of preventing erosion by storm waves, high tides, and boat wakes. Many times these structures make the erosion problem worse because the waves expend their energy over a very narrow portion of the beach. If only a portion of the coastline is protected, wave energy may concentrate at the ends of the seawall and move behind it, or severe erosion may occur at the base of the seawall, and the wall may collapse. Seawalls can reflect waves to combine with incoming waves and cause higher, more damaging waves at the seawall or at some other place along the shore.

The natural shore allows waves to expend their energy over a wide area, eroding the land but maintaining the beaches, all a part of nature's processes of give and take. Time and time again, coastal facilities have been built in response to the demands of people, only to trigger a chain of events that results in newly created problems.

Despite our efforts to protect them, beaches are disappearing because of rising sea level and coastal erosion. Rising sea level has caused worldwide coastal land loss. Beaches erode back from the waterline roughly 100 times faster than sea level rises vertically. Every millimeter rise in sea level results in 10 cm of shoreline retreat.

Coastal Structures

People change beaches by damming rivers to control floods and generate power. When a dam is built, a lake is formed behind the dam, and silt, sand, and gravel that once moved down the river to the coast are deposited in the lake behind the dam; an important source of sediment to the beaches has been removed. Although the sediment supply has been reduced, the longshore current continues to flow and carry away beach material. The net result is a loss of sediment and sand from the beaches, and the beaches erode.

Figure 10.17 When groins or jetties perpendicular to the shoreline disrupt longshore currents, deposition occurs up-current, erosion below. Ocean City, New Jersey.

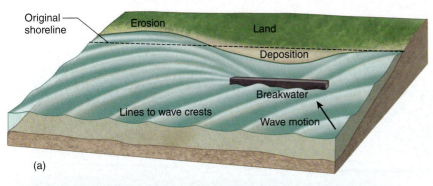

Figure 10.18 Sediment is deposited behind a breakwater; erosion occurs down-current through the action of water that has lost its original sediment load. Breakwaters and groins together combat beach erosion at Presque Isle State Park, Pennsylvania.

Sea-level rise results in beach erosion for several reasons. Deeper water decreases wave refraction, which increases the capacity for longshore transport. Higher water level allows waves to break closer to shore and expend more energy on the beach. In addition, when sea level rises, the effects of waves and currents reach farther up the beach profile, causing a readjustment of the landward portion of the profile.

According to a U.S. Army Corps of Engineers study, more than 40% of the U.S. continental shoreline is losing more sediment than it receives (fig. 10.19).

The Santa Barbara Story

The Santa Barbara, California, harbor project is a classic example of interference with coastal zone processes. At Santa Barbara, a jetty and a breakwater were constructed to form a boat harbor. The jetty at the west side of the harbor juts out into the sea and turns southeastward, forming a breakwater that runs parallel to the coast (fig. 10.20). The longshore current and sediment transport move from west-northwest to east-southeast along this part of the Southern California coast (see fig. 10.14). This jetty-breakwater system creates a wave-sheltered area north and east of the structure and blocks the longshore current. Sand was deposited on the west side, where the longshore current was blocked, and the beach began to grow. The beaches to the southeast began to disappear because they were starved of sand.

When the beach to the west had grown until it reached the seaward limit of the jetty, the longshore current could again move sediments southeastward along the ocean side of the breakwater. When the longshore current and sediment reached the east end of the breakwater and the entrance to the harbor, the current formed an eddy that spiraled into the quiet water of the harbor. The sediment settled out and began to fill the harbor, forming a spit connected to the end of the breakwater. The sediment settling in the harbor deprived the beaches farther southeast of their supply but did not alter the forces acting to remove the sediment from these same beaches.

Today, a dredge pumps the sediments from the harbor through a pipe and back into the longshore current. In this way, the harbor remains open and the beaches to the southeast receive their needed supply of sand. Interference with a natural process requires the expenditure of much time, effort, and money to do the work nature did for no cost.

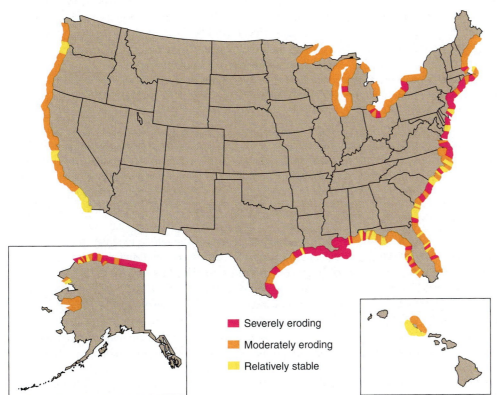

Figure 10.19 Shoreline erosion of the thirty-two coastal and Great Lakes states.

- ■ Severely eroding
- ■ Moderately eroding
- ■ Relatively stable

Figure 10.20 Santa Barbara harbor as seen from the south. In the *foreground,* a dredge removes the sediment that has built up in the protected, quiet water behind the jetty.

The History of Ediz Hook

Twenty-four hundred kilometers (1500 mi) north of Santa Barbara, another harbor has been altered through interference with the supply of sand. The harbor of Port Angeles, on the Strait of Juan de Fuca in Washington, is protected from storm waves by a naturally occurring, 5.6 km (3.5 mi) long curving spit known as Ediz Hook (fig. 10.21). The hook is composed mainly of sand and gravel and protects an area large enough and deep enough to accommodate the largest and most modern commercial vessels. In recent years, the hook has undergone considerable erosion and is in danger of waves breaking through near where the hook is attached to the mainland.

To understand how the hook was formed and how its present condition has come about, one must go back about 124,000 years, to the time when the last glaciers retreated from this area. At that time, sea level was lower than it is now, and the Elwha River west of Port Angeles carried glacial sediments to the sea, forming a delta that spread to the east under the influence of local currents and waves. As sea level rose, the river, currents, and waves continued to carry sediment eastward, and the waves began to erode the cliffs, adding still more sediment to the longshore current. The result was a hook-type spit that built out into the water from where the shoreline makes an abrupt angle with the strait. Ediz Hook was gradually produced and was kept supplied with sand and gravel from cliff erosion and the river.

In 1911, the Elwha Dam was constructed to provide power and a freshwater reservoir. A second dam, the Glines Canyon Dam, was built farther upstream on the river in 1925. The sedi-

ments that had previously moved toward the hook from the river (estimated at 30,000 m³ per year) were now being deposited behind the dams. In 1930, a pipeline to deliver fresh water from the Elwha Reservoir to Port Angeles was constructed around the cliffs at beach level. Bulkheads built along the face of the cliffs to protect the pipeline cut off the hook's supply of cliff-eroded sediments, estimated at 380,000 m³ per year.

Ediz Hook does more than protect the harbor; it is also the site of a large paper mill, a U.S. Coast Guard station, and several small harbor facilities. All are connected by a road running the length of the hook. In the 1950s, the city acted to protect the seaward side of the hook's base by armoring with large boulders and steel bulkheads. Cliff material was blasted into the water in hopes of supplying the required sediments. A constant battle between the people and the sea was fought. Nearly as fast as the armor was applied to the base of the spit, the waves tore away the protective barriers and removed the sediments. Only one-seventh of the total sediments once available to feed the spit remained in the longshore transport drift sector, and studies by the Army Corps of Engineers showed a possible loss of 270,000 m³ of sand each year from the outside of the hook. The hook was sediment-starved.

In the winter of 1973–74, severe storms again damaged the hook, and a Corps of Engineers' program to strengthen the hook began. Over a fifty-year period, annual maintenance costs have been projected at about $30 million compared with revenue from the harbor of $425 million. Since the benefit-to-cost ratio is about 14:1, the project seems to be economically viable.

Since the two dams were built on the Elwha River early in the century, annual salmon runs declined from 380,000 to less

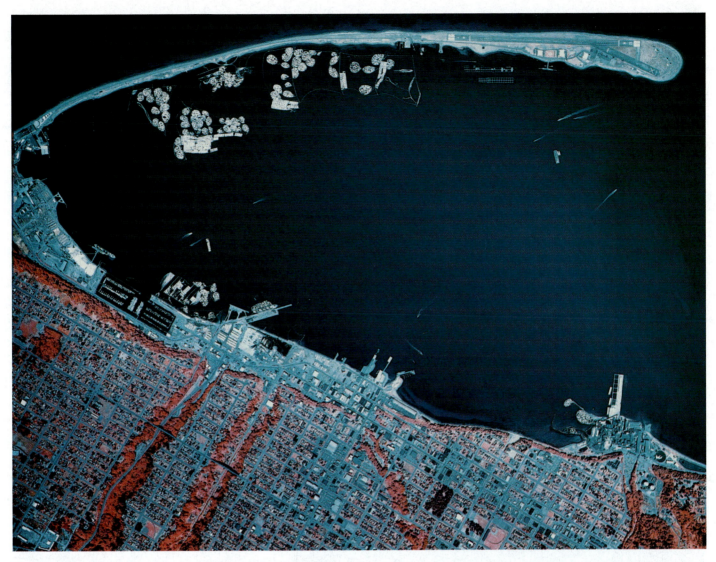

Figure 10.21 Ediz Hook is a long spit forming a natural breakwater at Port Angeles, Washington. The protected harbor behind the spit is deep enough for mooring the largest supertankers (infrared false-color photo).

than 3000 in the 1990s. Congress enacted a law in 1992 that authorized the removal of the dams to help salmon restoration. Once the dams are removed, the natural transport of sediment by the river to the coast will be restored, and Ediz Hook should no longer be as severely starved of sediment.

QUICK REVIEW

1. Why is it important to control the construction of jetties, groins, and breakwaters rather than allow individual homeowners to construct them as they wish?

2. Describe the consequences of constructing a dam on a river that would normally transport a large volume of sediment to the coast.

3. What is the effect of each structure when built along a beach: (a) breakwater, (b) jetty, (c) seawall?

10.7 Estuaries

River mouths, fjords, fault bays, and other semi-enclosed bodies of salt water are all part of the coastal zone. One such body with a free connection with the ocean and fresh water diluting its average salinity to less than that of the adjacent sea is an **estuary.** Estuaries can be described by the geologic processes that formed them, and they can also be described by the dynamics of their freshwater and saltwater circulation. Geomorphology may distinguish between estuary types in the same way that it is used to classify coasts. However, such a system does not consider the complex and dynamic combination of factors that come together when fresh water and salt water meet. In this chapter, estuaries are described mainly in terms of water exchange and salinity variation. Tides, river flow, and the geometry of an inlet are additional factors to be considered. Four basic estuary types are distinguishable.

Types of Estuaries

The simplest type of estuary is the **salt wedge estuary.** Salt wedge estuaries occur within the mouth of a river flowing directly into salt water. The fresh water flows rapidly out to sea at the surface, while the denser seawater flows upstream along the river bottom. The seawater is held back by the flow of the river, and an abrupt boundary forms, separating the intruding wedge of seawater and the fresh water moving downstream. A salt wedge estuary is shown in figure 10.22. The net seaward surface flow is almost entirely fast-moving river water because the large discharge of fresh water is forced into a surface layer above the salt wedge. The salt wedge moves upstream on the rising tide or when the river is at a low flow stage; the salt wedge moves downstream on the falling tide or when the river flow is high. The boundary between the salt water and the overriding river water is kept sharp by the rapidly moving river water. The moving fresh water erodes seawater from the face of the wedge. The waters of contrasting salinity mix and move upward into river water, raising the salinity of the seaward-moving fresh water. This one-way mixing process is called **entrainment;** very little river water is mixed downward into the salt wedge. Seawater from the ocean is continually added to the salt wedge to replace the salt water entrained into the seaward-flowing river water. In a salt wedge estuary, the circulation and mixing are controlled by the rate of river discharge; the influence of tidal currents is generally small compared to the influence of the river flow. Examples of salt wedge estuaries are found in the mouths of the Columbia, Hudson, and Mississippi Rivers. In the Columbia River, the salt wedge moves as far as 25 km (15 mi) upstream at times of high tide and low river flow, maintaining its identity as a sharp boundary between the two water types. Salt wedges also occur where rivers enter other types of estuaries: for example, at the mouth of the Sacramento River in San Francisco Bay.

Estuaries not of the salt wedge type are divided into three additional categories on the basis of their net circulation and the vertical distribution of salinity. These categories are the well-mixed estuary, the partially mixed estuary, and the fjord-type estuary.

Well-mixed estuaries (fig. 10.23) have strong tidal mixing and low river flow, creating a slow net seaward flow of water at all depths. Note that mixing is so complete that the salinity of the water is uniform over depth and decreases from the ocean to the river. There is little or no transport of seawater inward at depth; instead, salt is transferred inward by mixing and diffusion. The nearly vertical lines of constant salinity move seaward on the falling tide or when river flow increases, and they move landward on the rising tide or when river flow decreases. Many shallow estuaries, including the Chesapeake and Delaware Bays, are well-mixed estuaries.

Partially mixed estuaries (fig. 10.24) have a strong net seaward surface flow of fresh water and a strong inflow of seawater at depth. The seawater is mixed upward and combined with the river water by tidal current turbulence and entrainment to produce a seaward surface flow that is larger than that of the river water alone. This two-layered circulation acts to rapidly exchange water between the estuary and the ocean. Salt moves into a partially mixed estuary by diffusion but also, and more important, by **advection,** the inflow of seawater at depth. Examples of partially mixed estuaries include deeper estuaries, such as Puget Sound, San Francisco Bay, and British Columbia's Strait of Georgia.

Fjord-type estuaries (fig. 10.25) are deep, small-surface-area estuaries with moderately high river input and little tidal

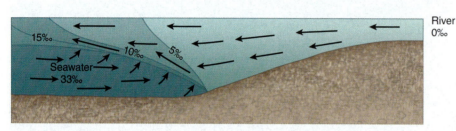

Figure 10.22 The salt wedge estuary. The high flow rate of the river holds back the salt water, which is entrained upward into the fast-moving river flow. Salinity is given in parts per thousand (‰).

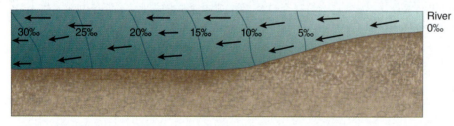

Figure 10.23 The well-mixed estuary. Strong tidal currents and wind-driven circulation distribute and mix the seawater throughout the shallow estuary. The net flow is weak and seaward at all depths. Salinity is given in parts per thousand (‰).

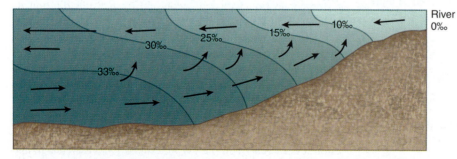

Figure 10.24 The partially mixed estuary. Seawater enters below the mixed water that is flowing seaward at the surface. Seaward surface net flow is larger than river flow alone. Salinity is given in parts per thousand (‰).

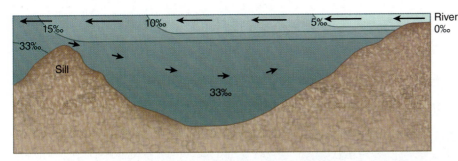

Figure 10.25 The fjord-type estuary. River water flows seaward over the surface of the deeper seawater and gains salt slowly. The deeper layers may become stagnant owing to the slow rate of inflow. Salinity is given in parts per thousand (‰).

mixing. This pattern occurs in the deep and narrow fjords of British Columbia, Alaska, the Scandinavian countries, Greenland, New Zealand, Chile, and other glaciated coasts. In these estuaries, the river water tends to remain at the surface and move seaward without much mixing with the underlying salt water. Most of the net flow is in the surface layer, and there is little influx of seawater at depth. The deeper water may stagnate because the entrance sill isolates it. Little inflow occurs below the surface layer.

The salt wedge estuary is a strongly stratified estuary. These estuaries are never very well mixed vertically, except above the boundary between the seaward-moving river water and the salt wedge. The other estuary types have various degrees of vertical stratification between the highly stratified and poorly mixed, and weakly stratified or well mixed. The processes that govern the degree of vertical mixing and stratification are the strength of the oscillatory tidal currents, the rate of freshwater addition, the roughness of the topography of the estuary over which these currents flow, and the average depth of the estuary. Tidal flows that change direction with the rise and fall of the tide produce energy for mixing; it does not usually result in a net directional flow and should not be confused with the net seaward or landward flow in each estuary.

Not all estuaries fit neatly into one of these categories. There are estuaries that fit between these types, and some estuaries change from one type to another seasonally with changes in river flow or weekly as the tides change from springs to neaps. Studying how an individual estuary compares to examples of the different types helps the oceanographer understand the processes in an estuary and the water exchange with the ocean.

Circulation Patterns

This discussion of estuary circulation is based on a partially mixed estuary, with its inflow from the ocean at depth and its surface outflow of mixed river and seawater. Tidal currents are produced as water flows into the estuary on the rising tide and flows out on the falling tide. Water in these currents moves back and forth in the estuary, but it does not immediately leave the system and enter the open sea. However, the surface water moves farther seaward on the falling tide than it moves inward on the rising tide, and a parcel of surface water is moved progressively farther seaward on each tidal cycle. In the same way, the deeper water

from the sea moves into the estuary on the rising tide farther than it flows seaward on the falling tide. Averaged over many tidal cycles, this pattern produces a net movement seaward at the surface and a net movement landward at depth.

The **net circulation** of an estuary out (or seaward) at the top and in (or landward) at depth is of great importance because the seaward surface circulation carries wastes and accumulated debris seaward and disperses them in the larger oceanic system.

It is also through this circulation that organic materials and juvenile organisms produced in the estuaries and their marsh borderlands are moved seaward, while nutrient-rich water is brought inward at depth and replenishes the estuary's salt water.

Understanding the net circulation of an estuary and evaluating the flows of surface water seaward and the underlying ocean water landward can be a long process. To determine this pattern directly, the oceanographer installs recording current meters at various depths and at several cross-channel locations. Currents are measured over many tidal cycles for both spring and neap tides and at times of low, intermediate, and high freshwater discharge. The current records from each depth and location are averaged to discover the net current distribution for a given cross section in the estuary and the net flow of the estuary. Data are averaged over one-week periods to determine the importance of spring and neap tides, over monthly cycles to find changes in patterns due to seasonal river and climate fluctuations, and over several years to produce an annual pattern. The cost of installing and maintaining equipment, as well as the costs of processing the data, make this direct approach very expensive.

A less expensive, indirect approach is to assume a **water budget** with inflow equal to outflow and a constant estuary volume averaged over time; all processes that add or remove water must be considered. A **salt budget** is also assumed; salt added is equal to salt removed. At the entrance to the estuary, measurements of salinity are made over space and time, establishing the average salinity of the outflowing surface water, $\overline{S}_o$, and the average salinity of the deeper inflowing seawater, $\overline{S}_i$. The ratio

$$\frac{\overline{S}_i - \overline{S}_o}{\overline{S}_i}$$

is used to determine the fraction of river water in the seaward-moving surface layer at the estuary entrance.

If the average rate of river water inflow (R) is known for this same period, the volume rate of the surface layer's seaward flow (T_o) is found by using the following formulas:

$$\frac{\overline{S}_i - \overline{S}_o}{\overline{S}_i}(T_o) = R, \quad \text{or} \quad T_o = \frac{\overline{S}_i}{\overline{S}_i - \overline{S}_o}(R)$$

In a partially mixed estuary, the net seaward flow (T_o) is always larger than the river flow (R) and the river flow and the net saltwater inflow (T_i) combine to produce the seaward flow (T_o) and maintain the water budget. Inflow equals outflow.

$$T_o = T_i + R$$

Both evaporation, E, and precipitation, P, also remove and add water at the surface, as shown in figure 10.26. If both E and P are large, then R in the previous equations becomes ($R - E + P$).

This method assumes that, over the time period for which the calculations are used, the average water volume and the total salt content of the estuary remain constant. For more information on these equations, see Appendix C.

Temperate-Zone Estuaries

In the middle latitudes of North America, most estuaries gain their fresh water from rivers, and evaporation from the estuary surface is minor or nearly balanced by direct precipitation. The salt concentration of the entering seawater, $\overline{S}_i$, is about 33‰, and $\overline{S}_o$ is about 30‰. When we use the preceding equation and the example in figure 10.27, T_o is shown to be about eleven times the rate of river inflow, or R. T_i in this case is ten times R. The inflow rate of seawater at depth decreases as it moves into the estuary, and the surface flow increases seaward because of mixing and the incorporation of seawater from below. While this mixing is occurring, the average salt concentration of the seaward-moving surface flow is also increasing, and T_o is formed when 10 unit volumes of seawater have combined with the 1 unit volume of river water. These relatively large values for T_o and T_i compared to R make these net flows the key to understanding the exchange of estuary water with ocean water.

In some fjord-like estuaries, the surface layer is only as deep as the shallow sill, and the average salt content of the surface layer is very low, owing to weak mixing. In this case, T_o is approximately equal to R; T_i is negligible, and the deep water of the fjord stagnates.

QUICK REVIEW

1. What is an estuary?
2. Describe the salinity patterns of the four types of estuaries.
3. Why is bottom water in fjords sometimes stagnant?

10.8 Regions of High Evaporation

Many bays located near 30°N and 30°S latitudes have low precipitation and high evaporation rates. Although rivers may

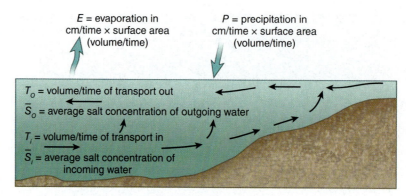

Figure 10.26 Water and salt enter and exit a partially mixed estuary.

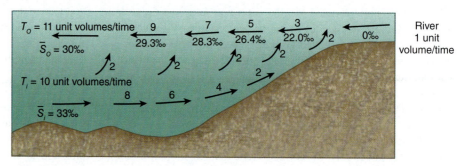

Figure 10.27 The seaward-moving surface water (*odd-numbered arrows*) increases in flow volume as the inflowing seawater (*even-numbered arrows*) moves upward. Upward mixing is indicated by the *vertical arrows*. For every 2 units of seawater mixed upward, the inflow decreases and the outflow increases by the same 2 units. Salinity is given in parts per thousand (‰).

flow into these bays, these bays are not estuaries if they do not experience net dilution. The contribution from rivers is usually minor when compared to the loss of water due to evaporation. The Red Sea and the Mediterranean Sea are examples. In these seas, evaporation increases the surface water salinity and density directly, causing surface water to sink and accumulate at depth, where it flows toward the open ocean. Ocean water flows into the sea at the surface, because it is less dense than the high-salinity outflow of deep water. The T_o and T_i flows have reversed, with T_o at depth and T_i at the surface (fig. 10.28). Because of this reversed circulation pattern, these bays are sometimes called **inverse estuaries.**

Compare figures 10.27 and 10.28. The evaporative bay in figure 10.28 shows a typical $\overline{S}_i$ value (36‰) for ocean surface conditions at latitudes of 30°N and 30°S. The evaporative loss from the surface is 1 unit volume per time, resulting in a T_i flow of 20 units and a T_o flow of 19 units. The resulting $\overline{S}_o$ value is 37.9‰. The estuary in figure 10.27 has a river input of 1 unit volume per time and T_o and T_i flows of 11 and 10 units, respectively. In general, if the evaporation removal rate is similar to the river input of a temperate estuary, an inverse estuary or evaporative bay has a better rate of exchange with the ocean. Not only is T_o larger, but the exchange of water is more complete because the evaporative bay has a continuous overturn as surface water sinks and moves seaward.

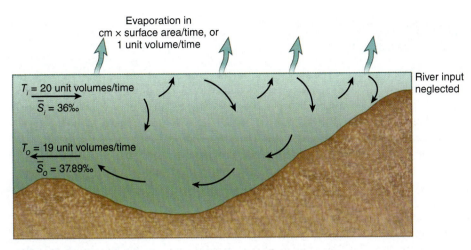

Figure 10.28 The evaporative sea or inverse estuary. Evaporation at the surface removes water (1 unit), and river inflow is negligible. Seawater flows inward at the surface (T_i = 20 units), and the seaward flow (T_o = 19 units) is at depth. Salinity is given in parts per thousand (‰).

QUICK REVIEW

1. Compare the circulation in a bay located in a high-evaporation environment to the circulation in a partially mixed estuary.

10.9 Flushing Time

The mean volume of an estuary divided by T_o gives an estimate of the length of time required for the estuary to exchange its water. This is known as **flushing time.** If the net circulation is rapid and the total volume of an estuary is small, flushing is rapid, and the flushing time is short. A rapidly flushing estuary has a high carrying capacity for wastes because the wastes are moved rapidly out to sea and dispersed. A slowly flushing estuary risks accumulating wastes and building up high concentrations of land-derived pollutants. Population pressures on estuaries are heavy, for these areas are used as seaports, recreational areas, and industrial terminals as well as for their fishing resources. Understanding the circulation of estuary systems is essential to maintaining them as healthy, productive, and useful bodies of water.

If the waste products are associated only with the fresh water entering the estuary, it is possible to determine the ability of the fresh water to carry these products through the estuary without considering the action of the entire estuary. To do so the amount of fresh water in the estuary at any one time must be known. The freshwater volume is estimated from the average salinity of the estuary. For example, an estuary with an average salt concentration of 20‰ is one-third fresh water if the adjacent ocean salinity is 30‰. Dividing the freshwater volume by the rate of river water addition yields the flushing time for fresh water and its pollutants.

Some bays and harbors do not have a freshwater source; these areas are not estuaries and are flushed only by tidal action.

On each change of the tide, a volume of water equal to the area of the bay multiplied by the difference between the high and low tide levels exits and enters the bay. When this volume, known as the **intertidal volume,** leaves the bay and enters the ocean, it may be displaced along the coast by a prevailing coastal current. On the next rising tide, an equivalent volume of different ocean water enters the bay. Under these conditions, the flushing time can be estimated by the number of tidal cycles required to remove and replace the bay volume. This number is found by dividing the mean bay volume by the intertidal volume exchanged per tidal cycle. However, the flushing is often not complete because currents along the coast do not always move the exiting intertidal volume a sufficient distance to prevent some portion of the same water from recycling into the bay. Therefore, the number of tidal cycles required for flushing may be greater than the calculated estimate.

The same conditions occur in estuaries that have a two-layered flow if mixing incorporates some of the exiting surface water into the deeper water flowing in from the sea. This mixing results in partial recycling of the T_o flow and modification of the T_i flow, and increases aeration of the deeper parts of the estuary.

Occasionally, the net circulation in an estuary is altered by ocean conditions. For example, if coastal upwelling occurs, the inflow of cooler, denser water to the estuary increases, and this increased inflow accelerates the estuary's circulation and causes the outflow to increase. When upwelling ceases and downwelling occurs, less-dense seawater is present at the estuary entrance, and the inflow is reduced. In these circumstances, the estuary's circulation may temporarily reverse as the denser, deeper water flows back to sea and the less-dense seawater moves inward at the surface.

Although each estuary has its own distinctive characteristics, the knowledge gained from studying one estuary can be used to understand other similar estuary systems. Understanding the circulation patterns and the processes that control these patterns allows us to judge the degree to which an estuary can be modified and used while still preserving its environmental and economic value. Because estuaries lie at the contact zone between the land and the sea, and because humans are inhabitants of the land, alterations have been made in these areas and probably cannot be entirely avoided in the future. But we owe it to ourselves and to those who come after us to make intelligent and knowledgeable uses of these areas, for they represent a great renewable natural resource if treated with care.

QUICK REVIEW

1. Define *flushing time.*
2. What variables influence the intertidal volume?
3. How is the intertidal volume related to the flushing time?

Summary

The coast is the land area affected by the ocean. The shore extends from the low-tide level to the top of the wave zone. The beach is the accumulation of sediment along the shore.

Primary coasts are formed by nonmarine processes (for example, land erosion; river, glacier, or wind deposition; volcanic activity; and faulting). Primary coasts include fjords, drowned river valleys, deltas, alluvial plains, dune coasts, lava and cratered coasts, and fault coasts. Secondary coasts are modified by ocean processes (for example, ocean erosion; deposition by waves, tides, or currents; and modification by marine plants and animals). Erosion produces regular and irregular coastlines. Deposition of eroded materials creates sandbars, sand spits, and barrier islands. Barrier islands protect the continental coastline from storms, but the islands receive each storm's energy and destructive force.

Beaches exist in a dynamic equilibrium; supply balances removal of beach material. Gentle summer waves move the sand toward shore during onshore transport. In winter, high-energy storm waves scoop the sand off the beach and deposit it in a sandbar during offshore transport. In the surf zone, the breaking waves produce a longshore current. The longshore current moves the sediment along the shore as longshore transport. A beach accumulating as much material as it loses is in equilibrium. Rip currents move water and sediment seaward through the surf zone.

A series of drift sectors forms a coastal circulation cell that defines the paths followed by beach sediments from source to deposition. The energy for the movement comes from waves and the wave-produced currents.

A typical beach has features that include an offshore trough and bar and a beach area comprising a low-tide terrace, beach face, beach scarp, and berms. Winter, or storm, and summer berms are produced by seasonal changes in wave action. Beaches are described by their shape; the size, color, and composition of beach material; and their status as eroded or depositional beaches.

Dams, breakwaters, groins, and jetties interfere with the dynamic processes that create and maintain the oceans' beaches. When beaches are destabilized, they erode, and today, 40% of the U.S. continental shoreline loses more sediment than it receives. Santa Barbara harbor and Ediz Hook are examples of human interference with natural processes.

An estuary is a semi-isolated portion of the ocean that is diluted by fresh water. In a salt wedge estuary, seawater forms a sharply defined wedge moving under the fresh water with the tide; circulation and mixing are controlled by the rate of river discharge. In a shallow, well-mixed estuary, there is a net seaward flow at all depths due to strong mixing and low river flow. Salinity is uniform over depth but varies along the estuary. A partially mixed estuary has good mixing, seaward surface flow of mixed fresh water and seawater, and an inflow of seawater at depth. Salt is transported in this system by advection and mixing. Fjord-type estuaries are deep estuaries in which the fresh water moves out at the surface and there is little mixing or inflow at depth.

In a partially mixed estuary, progressive movement of surface water seaward and deep water inward occurs during each tidal cycle. The circulation of the estuary can be measured directly with current meters, a long and expensive process, or it can be measured indirectly by determining the water and salt budgets of the estuary.

In temperate latitudes, the volume transport of water between the estuary and the sea is much greater than the addition of fresh water. In fjords, the inward flow is small and the deep water tends to stagnate. In semienclosed seas with high evaporation rates, a net removal of fresh water occurs; the seaward flow is at depth and the ocean water enters at the surface.

Estuaries that flush rapidly have a higher capacity for dissipating wastes than those that flush slowly. In bays and harbors that are flushed only by tidal action, resident water and wastes can be recycled. Partial recycling can also happen in estuaries with two-layered flow as exiting surface water is mixed with incoming water.

Key Terms

coast, 264
shore, 264
beach, 264
geomorphology, 264
eustatic change, 264
primary coast, 265
secondary coast, 267
fjord, 267
sill, 267
moraine, 267
drowned river valley, 267
ria coast, 267
delta, 267
alluvial plain, 267
dune coast, 267
lava coast, 267
cratered coast, 267
fault bay, 267

fault coast, 267
sea stack, 268
bar, 268
barrier island, 268
sand spit, 268
hook, 268
reef coast, 268
salt marsh, 270
backshore, 270
foreshore, 270
offshore, 270
berm, 271
berm crest, 271
winter berm, 271
storm berm, 271
summer berm, 271
scarp, 271

low-tide terrace, 271
beach face, 271
runnels, 271
ridges, 271
trough, 271
dynamic equilibrium, 272
onshore current, 272
onshore transport, 272
longshore current, 272
longshore transport, 272
swash, 273
cusp, 273
drift sector, 273
accretion, 273
coastal circulation cell, 274
tombolo, 275
lag deposit, 275

armored beach, 275
breakwater, 276
jetty, 276
groin, 276
estuary, 279
salt wedge estuary, 280
entrainment, 280
well-mixed estuary, 280
partially mixed estuary, 280
advection, 280
fjord-type estuary, 280
net circulation, 281
water budget, 281
salt budget, 281
inverse estuary, 282
flushing time, 283
intertidal volume, 283

Study Problems

1. Determine the flushing time of an estuary in which $T_0 = 9 \times 10^7$ m³/day and the volume of water is 30×10^8 m³.

2. An estuary has a volume of 50×10^9 m³. This water is 5% fresh water, and the fresh water is added at the rate of 6×10^7 m³/day. Why does the rate of addition of fresh water from the estuary to the ocean equal the input of fresh water from the land to the estuary? Consider the average salinity of the estuary as constant. What is the residence time of the fresh water?

3. Water entering an estuary at depth has a salinity of 34.5‰. The water leaving the estuary has a salinity of 29‰. The river inflow is 20×10^5 m³/day. Calculate T_0, the seaward transport.

4. A bay with no freshwater input can flush only by tidal exchange. If on each tidal cycle (ebb and flood) 10% of the bay's water volume is exchanged with ocean water, how much of the original water will still be in the bay after four tidal cycles? Solve for (a) no mixing between bay water and ocean water; and (b) complete mixing between bay water and ocean water.

5. What is the flushing time of the bay in problem 4 for each case? Which case best duplicates natural conditions?

The Living Ocean

Learning Outcomes

After studying the information in this chapter students should be able to:

1. *explain* the relationship among evolution, taxonomy, and phylogeny,

2. *explain* the differences and similarities between chemosynthesis and photosynthesis,

3. *describe* the relationship between organism size and abundance,

4. *explain* how different groups of organisms adapt to different regions of the ocean, and

5. *identify* the major zones in the ocean.

A feeding whale shark (*Rhincodon typus*), the world's largest fish, in West Papua, New Guinea

The previous chapters combined basic principles of geology, physics and chemistry to explain ocean processes. In this chapter and those that follow, we use these same principles to describe how and where organisms live in the oceans, and how they interact with and modify the ocean processes discussed in previous chapters. Marine organisms have much in common with organisms on land, but the ocean environment also presents distinct challenges and survival problems. Scientists that study ocean life are referred to as either biological oceanographers or marine biologists. The distinctions between these two titles are subtle with no obvious boundary between them. Marine ecosystems are composed of an astonishing diversity of organisms, both large and small. Indeed, the oceans include both the largest and smallest varieties of life on Earth, and include representatives from every major group. Consequently, scientists that study ocean biology ultimately have to understand the interactions between different sizes and different groups of organisms. Regardless of whether scientists are referred to as marine biologists or biological oceanographers, they study the abundance and distribution of marine organisms and the relationships between the organisms and their environment.

11.1 Evolution in the Marine Environment

The marine environment represents the single largest habitat on Earth. Every drop of ocean water is teeming with life, although the vast majority of these organisms are impossible to see with the naked eye. Many of these organisms have adapted to what we would consider unusual or harsh conditions, often surviving and thriving under immense pressure, in the cold and dark, at extremes of temperature and salinity. The organisms in the modern ocean represent the offspring of groups or types of living creatures that have adapted to the oceanic environment over geologic time scales. Those organisms that are best adapted to life in a particular environment are more likely to reproduce successfully. This is called **natural selection**—the survival of individuals influenced by environmental factors, leading to more success for that population.

This concept was first proposed by Charles Darwin and Alfred Wallace, two nineteenth-century English naturalists. These ideas led to the **Theory of Evolution**, which states that a population will gradually adapt to take on the traits of the most successful individuals within that population, ultimately leading to the development of new species. The converse is also true—that poorly adapted populations may go extinct. Modern evolutionary theory builds on this basic premise by recognizing that much of the adaptation is driven by mutations in genetic material (DNA), and that these mutations provide variability among individuals, leading to natural selection.

Darwin's Theory of Evolution also led to the recognition that life on Earth probably arose from a **common ancestor** (fig. 11.1) many billions of years ago. This theory provides a basis for grouping similar types of organisms, a process called **classification**. One of the most common methods for classification is based on **taxonomy**, grouping organisms based primarily on shared morphology. Taxonomic systems were developed about 250 years ago by the Swedish botanist Carolus Linnaeus and are still used today to name and classify related groups of organisms using the **Linnaean Classification System**. For example, the genus and species name *Homo sapiens* is accepted in all countries as the name for humans, and all humans share similar, readily recognizable morphological characteristics. The Linnaean system arranges all organisms into categories starting with the least specific grouping (kingdoms) and ending with the most specific groupings (species). Taxonomic categories for some common marine organisms discussed in this chapter are listed in table 11.1.

Another common classification system used today is based on **phylogeny**, the evolutionary relationships between ancestor organisms and their descendents, typically using DNA sequence information to infer these relationships. Importantly, relatedness categories do not necessarily overlap with functional categories. Organisms that are not closely related can carry out similar functions within an ecosystem. It is also possible for organisms that look nearly identical and that exhibit similar lifestyles to be different species (reproductively isolated); these are referred to as **cryptic species**.

The broadest category of relatedness used to group different organisms is based on a combination of cell morphology/structure and DNA sequences. This method of classification recognizes three basic domains of life: Bacteria, Archaea, and Eukarya (fig. 11.1). (Note that these names are capitalized when used to refer to the domain names; the names are not capitalized when used to refer to the common names bacteria, archaea, and eukaryotes.) Bacteria and Archaea belong to the most ancient groups of organisms, the **prokaryotes.** Fossil remnants of prokaryotes date back to about 3.5 billion years ago. Prokaryotes are unicellular, possess a simple cell architecture, and lack membrane-bound internal structures including a cell nucleus. Bacteria and Archaea share a similar morphology but can be readily distinguished from one another based on DNA sequences. Recent estimates suggest that as much as 40% of the prokaryotes in marine waters deeper than ~500 m may be archaea. **Eukaryotes** appear later in the fossil record, dating back to about 2.1 billion years ago. Eukaryotes are characterized by a complex cell architecture. All eukaryotes possess a **nucleus** that houses the majority of a cell's DNA. Most eukaryotes also contain **mitochondria** that are used to derive energy from food, and some eukaryotes contain **chloroplasts** that are used to harvest sunlight by photosynthesis.

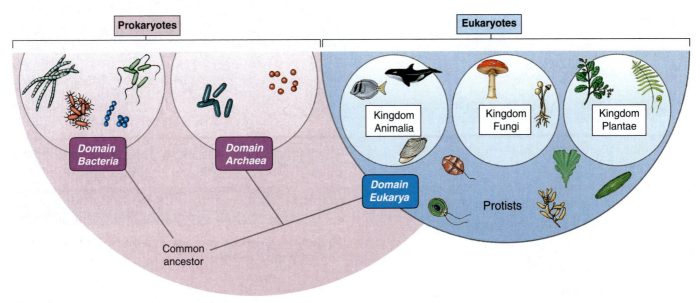

Figure 11.1 The tree of life is divided into three domains: Archaea, Bacteria, and Eukarya. The Bacteria and Eukarya are single cells without nuclei. The Eukarya domain includes everything from single-celled organisms with nuclei to blue whales.

Table 11.1 Taxonomic Categories of Some Marine Organisms

	Sperm Whale	Emperor Penguin	Chinook Salmon	Bay Scallop	Giant Kelp
Phylum	Chordata	Chordata	Chordata	Mollusca	Phaeophyta
Class	Mammalian	Aves	Osteichthyes	Bivalvia	Phaeophycea
Order	Cetacea	Sphenisciformes	Salmoniformes	Ostreoida	Laminariales
Family	Kogiidae	Spheniscidae	Salmonidae	Pectinidae	Lessoniaceae
Genus	*Kogia*	*Aptenodytes*	*Oncorhynchus*	*Argopecten*	*Macrocystis*
Species	*sima*	*forsteri*	*tshawytscha*	*irradians*	*pyrifera*

QUICK REVIEW

1. How many domains and kingdoms are represented in the oceans?
2. Describe the difference between classical taxonomy and molecular phylogeny.
3. Describe how unrelated organisms can serve the same function in an ecosystem.

11.2 The Flow of Energy

In addition to classifying organisms by taxonomy and phylogeny, we can group organisms functionally (how they interact with the environment). The broadest functional division between organisms is based on how they derive energy and carbon for growth (fig. 11.2). This method of classification recognizes two groups of organisms, **autotrophs** and **heterotrophs.** Autotrophs generate organic matter, such as the sugars and proteins needed for growth, from inorganic compounds such as nitrate or ammonium, phosphate, and carbon dioxide. The energy needed to

drive formation of organic matter is derived either from sunlight or from chemical energy, and the carbon is derived from carbon dioxide. The organisms that harness light energy to generate organic matter are termed **photoautotrophs,** and the process they carry out is **photosynthesis.** An important by-product of photosynthesis is the generation of oxygen. The organisms that use chemical energy to drive production of organics are termed **chemoautotrophs,** and the process they carry out is **chemosynthesis.** Until recently, the number of different organisms able to chemosynthesize was believed to be relatively restricted. It now appears that a large percentage of the archaea that live in deep waters may carry out chemosynthesis. Autotrophs are said to occupy the base of the food web because they generate organic compounds from inorganic, non-living matter and because they fuel the growth of organisms that require organic matter as food. Heterotrophs consume the organic matter generated by autotrophs; they use the organic matter as both a carbon source and an energy source. Most heterotrophs use oxygen to oxidize the organic matter to derive energy. Heterotrophs generate carbon dioxide as a by-product of the oxidation of organic matter.

Autotrophs
Plants
Phytoplankton

Heterotrophs
Animals
Bacteria

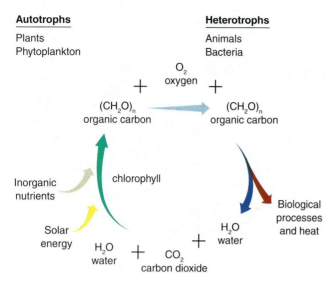

Figure 11.2 Simplified schematic of connections between autotrophic and heterotrophic organisms. Photosynthetic organisms use solar energy to generate organic carbon from carbon dioxide and as a by-product, they generate oxygen. Heterotrophic organisms consume organic matter and breathe oxygen to drive biological processes and as a by-product, they generate carbon dioxide.

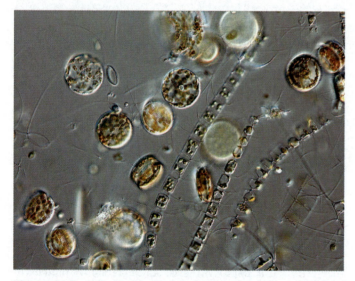

Figure 11.3 Phytoplankton sample collected by pulling a net with a mesh size of 20 μm through surface waters. Both chains and individual cells of different species of diatoms are present.

The functional categories of autotrophs and heterotrophs span across the relatedness categories of prokaryotes and eukaryotes. Prokaryotes can be photoautotrophic (cyanobacteria), chemoautotrophic (for example, the bacteria that live in worms from the hydrothermal vents and derive energy from the oxidation of hydrogen sulfide), or heterotrophic (for example, the bacteria such as *Escherichia coli* that live in our guts and feed on what we leave behind after digestion). Eukaryotes can be photoautotrophic (plants, algae; Fig. 11.3) or heterotrophic (for example, humans, whales, fish, zooplankton; Fig. 11.4). There are no known

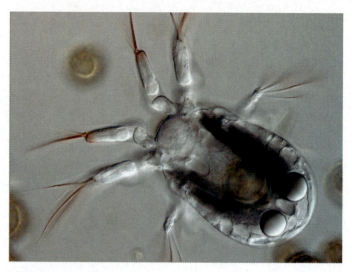

Figure 11.4 Juvenile stage of one of the crustacean zooplankton collected from coastal waters.

examples of eukaryotes that are chemoautotrophic. However, there are examples of eukaryotes, such as the worm *Riftia pachyptila*, that rely solely on the organic material generated by the chemosynthetic bacteria that live in close association with them.

QUICK REVIEW

1. Describe similarities and differences between photosynthesis and chemosynthesis.
2. Explain why chemosynthesis is more common in the deep ocean.
3. Describe what *respiration* means. What organisms respire?

11.3 The Importance of Size

In the previous section we introduced the concept of three domains of life. You may have noted that two of these domains are composed of very small organisms, collectively referred to as **microbes** or **microorganisms**. These are the most abundant organisms in the sea, and the vast majority are generally only a few microns (the micron, μm, is 1 millionth of a meter or 1/1000 of a millimeter). The organisms of the sea that can be seen easily, such as invertebrates, fish, or whales, vary dramatically in size. For example, a small marine worm may be less than an inch in length, and a blue whale may be almost 100 feet (33 m) in length.

The oceans thus contain both the smallest (microbes) and largest (blue whales) organisms on Earth, spanning about ten orders of magnitude. The abundance of organisms in the ocean follows a similar pattern: for every tenfold decrease in size, there is approximately a tenfold increase in abundance (fig. 11.5). So one end of the size spectrum (whales) have relatively small total abundance, whereas the other end (microbes) can be found at concentrations of millions of cells per milliliter of ocean water.

Typically, the individual cells that make up a multicellular organism are more or less the same size (nerve cells are an exception to this rule), and they are comparable to the size of most microbes. Only about three orders of magnitude separate the size of the small 0.5 μm microbes from the size of large

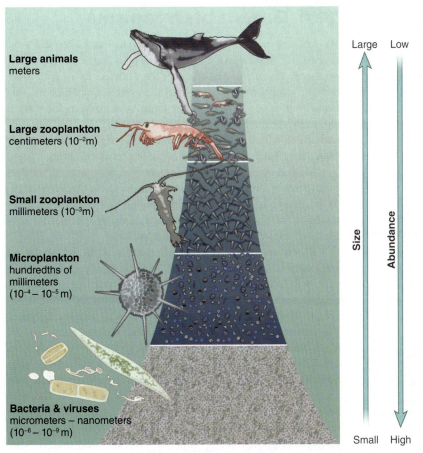

Large animals
meters

Large zooplankton
centimeters (10⁻²m)

Small zooplankton
millimeters (10⁻³m)

Microplankton
hundredths of
millimeters
(10⁻⁴ – 10⁻⁵ m)

Bacteria & viruses
micrometers – nanometers
(10⁻⁶ – 10⁻⁹ m)

Large   Low

Size   Abundance

Small   High

Figure 11.5 The ocean is filled with organisms ranging in size from less than a millionth of a meter to tens of meters. Size and abundance of organisms covaries, with relatively few large animals such as whales and millions of microbes in every drop of seawater.

100 μm cells. Why should this be so? Why aren't enormous organisms composed entirely of enormous cells? What determines the shape that organisms can assume, and how does shape relate to lifestyle? Why do some organisms have elaborate shapes and others have relatively simple shapes? As described below, the answers to these questions are derived from fundamental principles of physics and chemistry such as conservation of mass and energy covered in earlier chapters: organisms must accomplish tasks involving transport and retention of mass and energy, and factors such as size and shape affect how effectively they are able to do so. The field of research that looks for answers to how chemistry and physics affect basic biological characteristics such as size and shape is known as **biomechanics** and will serve as a starting point for thinking about life in the ocean.

Heat content, chemical constituents, and mass of living organisms differ from that of their environment. At the level of a cell, the main underlying mechanism used to maintain these differences is molecular transport. Molecular transport that acts upon mass is referred to as **diffusive,** transport that acts upon heat is referred to as **conductive,** and transport that acts upon momentum is referred to as **viscous.** These transport mechanisms all act across surfaces. Also, they act strongly across shorter

distances and are weaker across longer distances. Think of how increasing the thickness of your clothes by putting on a sweater reduces the rate of heat conduction away from you when you are sitting in a cold room. Conduction alone can be used to dissipate heat from microscopic-sized organisms but cannot be used as the only means of dissipating heat from larger organisms composed of large numbers of closely packed cells. Similarly, diffusion alone can supply nutrients to and remove wastes from organisms the size of typical cells but not organisms the size of whales.

A fundamental difference between small organisms and large organisms is the ratio of their surface area to volume. Let's imagine a shrinking sphere. As the sphere shrinks from having a radius of 1000 units to a radius of only 1 unit, the surface area decreases by the square of the radius (area $= 4\pi r^2$), but the volume decreases by the cube of the radius (volume $= 4/3\ \pi r^3$). Thus the surface area to volume ratio $[(4\pi r^2\ /\ 4/3\ \pi r^3) = (3/r)]$ will increase as the radius decreases. A smaller sphere has a larger surface area to volume ratio than a larger sphere (table 11.2). The same principle holds for organisms: A microscopic phytoplankton has a much larger surface area to volume ratio than a macroscopic whale. Organisms less than about a millimeter in size tend to be more impacted by viscosity, and larger organisms tend to be more impacted by gravity and inertia. The result of the different relative impacts of these forces is that small organisms live in a marine environment dominated by viscous forces and momentum is essentially nonexistent. Larger organisms live in a marine environment dominated by inertia; they leave behind the water they are swimming in, and they tend to continue to move for a period of time after they stop swimming. Gliding due to momentum and streamlining of body types is critical to how large organisms move through water.

Cells are small because of the constraints set by molecular transport; diffusion alone simply cannot act fast enough to supply enough nutrients to an enormous cell. The shapes of small marine organisms can be very different from the shapes of large marine organisms because of the constraints set by the different forces that act upon them. As will be seen, there are apparently few biomechanical "solutions" to some of the "problems"

Table 11.2 Relation Between Size and the Ratio of Surface Area to Volume

Radius (r)	Surface Area (4 πr²)	Volume (4/3 πr³)	Surface Area: Volume (3/r)
1	12.56	4.188	3
10	1256	4188	0.3
100	125,600	4,188,000	0.03
1000	1,256,0000	4,188,000,000	0.003

faced by marine organisms. For example, bivalves, sponges, and microscopic ciliates are unrelated, and yet they all use cilia to filter the water for food (fig. 11.6); the bivalves and sponges use multiple ciliated cells, and the microscopic ciliates are a single-celled organism. The physical and chemical interactions between organisms and their environment and between each other ultimately determine ecosystem structure.

QUICK REVIEW

1. How many microbes are in a typical drop (ml) of ocean water?
2. Explain why large organisms aren't composed of large cells, and why single-celled organisms must be small.
3. Explain the relationship between surface area and volume for a cell.
4. Describe the relationship between size and abundance of organisms in the ocean.

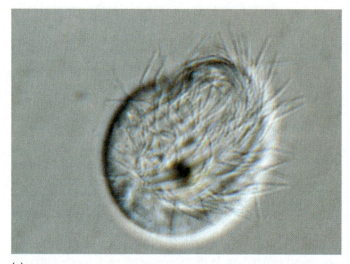

(a)

(b)

Figure 11.6 (a) A single-celled ciliate covered with cilia are the small projections emanating from the cell surface. (b) Flame scallops feed on plankton that they filter from the seawater with cilia.

11.4 Groups of Organisms

We just saw how cell size and the physical and chemical constraints of the environment influence interactions. Grouping organisms together based on either their relatedness or their function provides further insight into the complex interactions that occur within ecosystems. Taxonomic systems were developed about 250 years ago by the Swedish botanist Carolus Linnaeus and are still used today to name and classify related groups of organisms. Taxonomy is based primarily on shared morphology.

Marine ecosystems are composed of diverse communities of interacting organisms. Earlier we learned about various ways of grouping classifying organisms, including by taxonomy, phylogeny, and function. For life in the ocean, we can additionally group organisms based on size and habitat.

Plankton are generally small (less than a few millimeters in size) and are unable to move faster than horizontal currents, so they drift with the water's overall movement. Some large organisms such as *Sargassum* seaweed and some jellyfish are also considered to be planktonic because they also drift with the currents. **Phytoplankton** photosynthesize and thus are photoautotrophic. They can be either prokaryotic or eukaryotic; they are almost always unicellular, although some species form chains of connected cells. **Zooplankton** consume organic matter and thus are heterotrophic. All zooplankton are eukaryotic, but they can be either unicellular or multicellular (fig. 14.6). Heterotrophic prokaryotes are members of the **bacterioplankton.** Some phytoplankton can both photosynthesize and consume organic matter depending on environmental conditions, and they are known as **mixotrophic plankton.** Organisms that are larger than the plankton and can swim faster than the currents are **nekton** (fig. 11.7). Both plankton and nekton live within the pelagic zone, or the water environment. They are found in both neritic

Figure 11.7 All fish are members of the nekton. Here, a diver observes Sargent Major fish at Santo, Vanuatu.

Figure 11.8 The benthic sun star, *Solaster,* typically has ten arms and a diameter of 25 cm (10 in).

and oceanic zones. Organisms that live on, in, or attached to the sea floor are the **benthos** (fig. 11.8), and they live in the benthic zone, which extends from the **supralittoral zone,** or splash zone, to the deepest depths of the ocean, or the **hadal zone**. The habitats of many organisms change as an individual matures during its life cycle. For example, the juvenile stage of a crab is part of the zooplankton, whereas the adult stage of a crab is part of the benthos.

QUICK REVIEW

1. Describe the major groups of organisms in the ocean.
2. Explain the difference between plankton and nekton.

11.5 Facts of Life in the Ocean

Various aspects of the marine environment pose challenges for the organisms that live there. As will be seen in subsequent chapters, there is an enormous diversity of organisms that live in the sea, but as discussed in section 11.3, they all have to fit within the relatively limited number of potential biomechanical solutions to these challenges. What follows is a discussion of broad-scale chemical and physical environmental conditions that influence the distribution and abundance of marine life.

Light

The most abundant source of energy to the ocean is sunlight, which both heats the waters and supplies the light energy necessary for photosynthesis. The **euphotic zone** is the area of the ocean where there is sufficient sunlight for growth of photosynthetic organisms. The **aphotic zone** is the area of the ocean where no light penetrates. In between is the **disphotic**, or twilight, zone, where there is sufficient light to see during the daytime but not enough light energy to support photosynthesis. The depth to which light penetrates varies depending on a number of factors including (1) the angle at which the Sun's rays hit

Earth's surface, (2) absorption of different wavelengths of light by water itself, and (3) the amount of particulate and dissolved matter in water (review the discussion of light transmission in the oceans in chapter 4). The angle at which sunlight hits the ocean is determined by latitude, season, and time of day. Absorption of light by water molecules means that the total amount of light decreases exponentially with depth (fig. 11.9*a*); red wavelengths of light are absorbed at the shallowest depths, and blue wavelengths of light are absorbed at deeper depths (fig. 11.9*b*). The euphotic zone of coastal waters is shallower than the euphotic zone of the open ocean because more particles (and colored dissolved matter both living and nonliving) are found in water closer to land, and these materials absorb light.

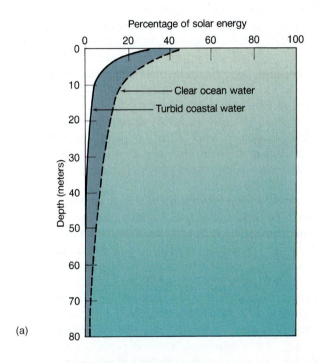

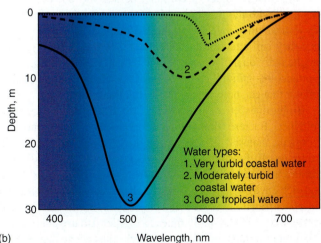

Figure 11.9 (a) The percentage of solar energy available at depth in clear and turbid water. (b) Depth penetration of different wavelengths of light in three different water types. Note that shorter wavelengths of light penetrate to deeper depths in clearer waters.

In the open ocean, the euphotic zone may extend to nearly 200 m (600 ft), whereas, in coastal regions, the euphotic zone may extend to only a few tens of meters or less. The penetration of sunlight is much greater in clear than in turbid seawater (fig. 11.9a).

Phytoplankton are the most abundant photosynthetic organisms in the oceans. Even those phytoplankton that are able to swim cannot move faster than the currents. Depending on the depth of ocean mixing, a community of phytoplankton may be mixed deep into colder, darker waters one day only to be returned to the warmer surface ocean the next day where they are confronted with bright sunlight. Phytoplankton must be able to adapt to wide changes in their physical environment over relatively short time periods.

Another source of light in the ocean is **bioluminescence.** This light is generated by the organisms themselves. On a dark night, the wake of a boat or moving fish or even an oar pulled through surface waters will cause the flashing of certain organisms, primarily a group of phytoplankton known as dinoflagellates, some jellyfish, shrimp, and squid. Bioluminescence is produced by cells when the enzyme luciferase acts on its substrate luciferin to produce light. Bioluminescence in the sea is commonly stimulated when an organism is physically agitated, which is why we see bioluminescence when something moves through the water. Bioluminescence also occurs on land (fireflies use flashes of light to identify mates), but bioluminescence is much more common in the ocean.

The most common function of bioluminescence in the ocean is thought to be predator avoidance. An attacking predator can be startled or distracted by the light produced by its prey, which may allow the prey to escape. For example, the arms of certain brittle stars (relatives of sea stars) will begin to flash in repetitive patterns when the animal is attacked. If the attack continues, the brittle star loses the arm closest to the attacking animal. The discarded arm will continue to flash, holding the attention of the attacker, while the darkened animal crawls away, although with one less arm. The "burglar alarm" hypothesis has been developed to explain why some marine organisms produce light. If movement of water associated with the proximity of a predator causes the potential prey to flash, then repeated flashes act as a sort of burglar alarm, since they call attention to the predator. In this instance, rather than the individual that triggered the burglar alarm being arrested by police, this individual may instead be eaten by an even larger predator that can now see its newly illuminated prey!

Bioluminescence serves other functions besides predator avoidance. Bacteria that colonize marine snow (large particles of organic material that fall through the water column) often bioluminesce. Presumably fish are enticed to eat the glowing marine snow, and the bacteria find themselves once again in the food-rich environment of a fish gut. The use of submersibles and mid-water trawls suggests that a majority of deep-water fish and invertebrates can also bioluminesce. Angler fish use a lure filled with bioluminescent bacteria that dangles just above their mouth to lure their food. Flashlight fish have pouches beneath both eyes that are filled with bioluminescent bacteria that they use to see their prey. Many mid-water fish living in the "twilight zone" produce light on their underside, to avoid producing a shadow.

Figure 11.10 The peacock flounder can change its color and the pattern on its skin to match the sea floor.

In order to effectively mask themselves using this technique, the fish must be able to detect and match the incoming ambient light.

Color (or absence of color), instead of bioluminescence, is used by other organisms to avoid being eaten. Many zooplankton, including jellyfish, are nearly transparent and blend in well with their watery background. Many fish use color rather than transparency to blend in. In the relatively clear waters of the tropics, fish use bright colors to become almost invisible against their backdrops of colorful coral reefs. Other fish conceal themselves with bright color bands and blotches that disrupt the outline of the fish and may draw a predator's eye away from a vital spot. Bright colors are also commonly used for sexual recognition during breeding times. In coastal waters of temperate latitudes, less light penetrates to depth, and organisms use browns and grays to conceal themselves against rocks and in kelp beds. Bottom fish are commonly similar in color to the sea floor or are speckled with neutral colors. Flatfish have skin cells that expand and contract to produce color changes, and they can change their color to match almost any bottom type on which they live (fig. 11.10). Fish that swim near the ocean surface–such as herring, tuna, and mackerel—commonly have dark backs and light undersides. This pattern of coloring is known as countershading, and it allows the fish to blend in with the bottom when seen from above and with the surface when seen from below (fig. 11.11). Deep-water fish are usually small (rarely larger than 10 cm [4 in]), and they typically appear black at depth. Some species of deep-water shrimp and fish are red when seen at the surface. At depths below the penetration of red light, these organisms appear dark and inconspicuous.

Temperature

The temperature of 90% of the ocean is remarkably constant and varies from about −1°C to 4°C (30°F to 40°F). Surface waters are more variable and can range from about −1°C in Arctic and Antarctic waters to over 30°C (86°F) in the surface waters of the tropics and may reach even higher temperatures in shallow tidal

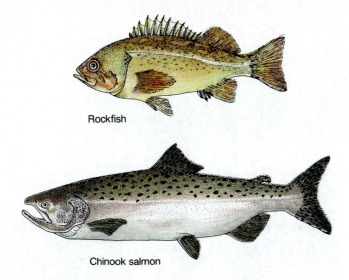

Rockfish

Chinook salmon

Figure 11.11 Viewed from above, the dark dorsal surface of the fish blends with the sea floor; viewed from below, the light ventral surface blends with the sea surface. This type of coloration is known as countershading.

pools. The temperature of surface waters is greatly influenced by season and latitude. Annual changes in open-ocean surface temperatures are small at high and low latitudes, whereas the changes are much larger in the mid-latitudes. Temperature changes in surface waters are more extreme near the coast because this water is shallower and is influenced by temperature changes on land. Northerly winds that blow along the coast in the Northern Hemisphere and southerly winds in the Southern Hemisphere will cause deep, cold water to be upwelled and can make for cold surface waters near the coasts even on a warm, sunny day (review chapter 6).

About 20% of Earth's surface is frozen, including the sea ice in the Arctic and Southern Oceans. The lowest temperatures found anywhere in the oceans occur in the brine channels of winter sea ice. Brine channels are created when salts are expelled into the water as the ice crystals grow during sea ice formation. The temperature of these brine channels can drop to an amazing −35°C (−31°F) because the salinities in these channels can reach levels about six times higher than seawater. Cold-loving organisms that live in very cold environments are known as **psychrophiles,** and they have developed specialized attributes to allow them to exist in these environments. Many psychrophilic microorganisms possess **cryoprotectants,** such as dimethylsulphoniopropionate (DMSP), that lower the freezing point of their internal fluids. Many also possess ice-binding proteins that likely enhance cell survival during the cycling between winter freezes and summer thaws. These organisms also use different fatty acids to maintain flexible membranes even under the very cold and salty conditions of this environment. The hottest temperatures in the ocean occur at the black smokers of deep-sea hydrothermal vents (review the discussion of hydrothermal vents in chapter 2). The fluid exiting the vents is ~350°C (660°F). No organisms survive in 350°C water, but because the water that surrounds the vents is only about 2°C, this extraordinarily hot water is quickly cooled.

Invertebrates and most fishes are **poikilotherms.** They possess no means of metabolically regulating their body temperature, and

instead they rely solely upon heat conductance. Their internal temperature responds to the temperature of surrounding waters, and hence, their physiology is regulated by water temperature. Metabolic processes tend to proceed more rapidly when poikilotherms are in warmer water rather than in colder water. Cold-water species commonly grow more slowly, live longer, and attain a larger size. Fish that live in polar regions must prevent their blood from freezing. Without special adaptations, fish blood would freeze at about −0.8°C (31°F), which means that their blood would freeze in the approximately −2°C (28°F) waters of the poles. Antarctic fish possess anti-freeze proteins that lower the freezing point of their blood and allow them to survive. For some poikilotherms, a change in water temperature triggers spawning or dormancy. The geographical range of poikilotherms is largely restricted by water temperatures.

Seabirds and mammals are **homeotherms.** They can maintain a nearly constant body temperature, often well above the temperature of the surrounding seawater. For example, Weddell seals live in the sea ice around Antarctica, yet their body temperature is a balmy 36°C (remember, our body temperature is 37°C, or 98.6°F) despite the fact that they live in an environment where the air and water temperature is often far below 0°C. They use a thick layer of blubber to reduce heat loss. Since the physiological capabilities of homeotherms are less regulated by water temperature, they often have wider geographical ranges. For example, whales annually migrate between polar and tropical waters. Some fish, notably tunas and lamnid sharks (such as the great white shark), are **endotherms.** They can maintain a higher temperature than the surrounding seawater but do not have the same level of temperature control as homeotherms. Endothermic fish use specialized arrays of blood vessels to prevent heat generated in muscles from being lost when blood circulates through the gills or to their extremities. This is called **countercurrent heat exchange** (fig. 11.12). An excellent example is the bluefin tuna, which can maintain a core body temperature as much as 20°C warmer than the surrounding seawater!

Salinity

The salinity of surface waters from different regions of the ocean differs depending on climate, proximity to freshwater sources, season, temperature, and circulation patterns. The most dramatic differences occur near estuaries. The salinity of deep waters is relatively constant. Organisms must maintain balanced internal salt levels. To maintain a salt content that differs from surrounding waters, an organism actively removes or adds salt. The membranes that surround all cells are permeable only to certain molecules. **Osmosis** is a special type of diffusion in which water moves across a cell membrane from areas of low salinity (high water concentration relative to salt concentration) to areas of high salinity (low water concentration relative to salt concentration).

Different organisms maintain different internal salt concentrations. The body fluids of many bottom-dwelling creatures, such as sea cucumbers and sponges, contain the same salt content as seawater (fig. 11.13). There is no concentration gradient, so the water moves equally in both directions across cell membranes

Diving in

Bioluminescence in the Sea

BY GWYNNE S. RIFE, PhD

Gwynne S. Rife, PhD, is a science educator and aquatic scientist at The University of Findlay. Her interests include invertebrate behavior and ecology of the deep sea.

Box Figure 1 Some dinoflagellates produce intense bioluminescence, as seen in these breaking waves along the shore of Sonoma County, California.

Imagine a dream where you are looking out to sea at night and view a luminescent wave, or that you are sailing on the ocean and see that you are sailing through snow (box fig. 1). Thanks to bioluminescent plankton, you would not have to be dreaming to see either of these scenes. The "milky sea" phenomenon might be the result of blooms of luminous plankton that occur during the monsoon season in the northwest Indian Ocean. In the Arabian Gulf you might be surprised by erupting balls of light as schools of fish drive though luminous organisms and stimulate them as they break the surface. The most impressive accounts of luminous display, however, would have to be the "phosphorescent wheels" described on rare occasions when they are observed, first as lines of glowing bands that rotate toward the horizon and can move faster than 50 km/hr.

These phenomena can be explained in part by the chemical reaction known as chemoluminescence. When this process occurs in living organisms it is called bioluminescence. Bioluminescence should not be confused with phosphorescence, where the luminescent part must absorb radiant light before it can glow, because in chemoluminescence a chemical reaction produces the light energy liberated. You can make a great chemoluminescence demonstration by mixing two solutions, one with the chemical luminol (you would need to purchase) and some common chemistry lab equipment.

Here is one method of making your own bioluminescence to show this effect:

In addition to safety glasses and a laboratory to work in with a balance (that can measure 0.1 g or 0.01 oz.), you need:

- 0.2 g luminol powder
- 4 g sodium carbonate, Na_2CO_3
- 24 g sodium bicarbonate, $NaHCO_3$, aka baking soda
- 0.5 g ammonium carbonate, $(NH_4)_2CO_3$
- 0.4 g copper(II) sulfate pentahydrate, $CuSO_4 \cdot 5H_2O$
- 2 L bottles with caps and 1 gallon distilled water
- 50 ml 3% hydrogen peroxide solution
- 100 ml graduated cylinder or 8 oz measuring cup

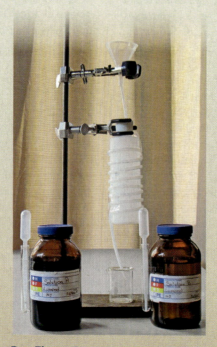

Box Figure 2 A set-up to demonstrate chemoluminescence with easy-to-find materials.

First, weigh out separately on a balance 0.2 g (0.01 oz.) of luminol, 4.0 g (0.14 oz.) of sodium carbonate, 24.0 g (0.85 oz.) of sodium bicarbonate, 0.5 g (0.02 oz.) of ammonium carbonate, and 0.4 g (0.02 oz.) of copper (II) sulfate pentahydrate. Transfer the chemicals to an empty 2 L bottle and fill the bottle about half full with distilled water. Cap the bottle and shake it until all of the compounds have dissolved. Label it solution A. Then, measure out 50 ml (1.7 oz.) of 3% hydrogen peroxide solution and add it to a second empty 2 L bottle. Fill the bottle about half full with distilled water, cap the bottle, and shake it briefly to mix the contents. Label that bottle B. Finally, you are ready to find a darkened room to combine the contents of the two 2 L bottles. The luminol mixture (solution A) will glow green from the luminescence reaction when added to solution B. A fun way to show the reaction is to put together a ring stand with a clear spiral of aquarium tubing to pour the solution in together (see box fig. 2). To show how a bioluminescent plankton might look, you could soak a thin ring of kitchen sponge in solution and then float it in a pan of the second solution.

Luminous bacteria are the smallest plankton found floating free in the ocean but are also sometimes symbionts in deep-sea nekton. Luminous symbionts are housed in the light organs of some fish—for example, as in the ones that hang from the roof of the mouth of Prince Axel's wonderfish, *Thaumatichthys axeli*, (box fig. 3). Deep-sea angler fish of many species use these light organs as lures for prey and many can wiggle them from the end of a structure that makes them appear to be fishing (box fig. 4).

Other common, single-celled types of luminous plankton are dino-flagellates and radiolarians. Luminous dinoflagellates are often the cause of sea surface luminescence—for example, the glowing effect you might see if you were kayaking through a bloom at night. Individual dinoflagellates flash when exposed to turbulence created by other things moving through the water. In daylight, because they live close to the surface, their light would be invisible, so most species have a circadian rhythm that conserves their energy by inhibiting the reaction during the day. Dinoflagellates and radiolarians likely defend themselves against planktonic predators by their flashing, which has the added "burglar alarm" benefit of alerting larger predators to the presence of the original grazer.

Even larger plankton are copepod and ostracods, cnidarians (jellyfish and siphonophores), and comb jellies. All of these groups have members that are bioluminescent. Most copepods do not flash, but have luminous glands on their limbs or bodies from which they exude droplets of light. Ostracods, though less abundant, also produce excretions of light that appear to be squirted. The males of some shallow-water ostracods of the genus *Vargula*, often called the sea-firefly, swim up off the bottom to signal to females. They encode a luminous message in the combination of the frequency of their light puffs, the frequency at which they flash, and direction of movement.

Other crustaceans also have luminous members, including amphipods and mysids (Peracardieans), euphausiids (krill), and other decapods. Many decapods have well-developed structures called photophores that are generally located on the underside of the body and eyestalks and provide a ventral illumination. Predators from below would normally see the shrimp-like form as a silhouette against the dim daylight filtering down through the water, but by emitting light of the same color and intensity as the daylight, they match the background, an adaptation termed counterillumination camouflage. Certain fish that prey on members of the plankton also use counterillumination to avoid becoming prey themselves; hatchet fish are an example (box fig. 5).

Current technology can be used to identify many luminous species by their bioluminescent signatures. Duration, behavior, and intensity of

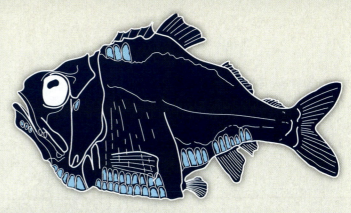

Box Figure 5   Hatchet fish are common midwater fish that use counterillumination camouflage.

the light produced are often species specific. For example, the intense flash from the "sternchaser" light organ on the tail of a lantern fish (box fig. 6) has a characteristic glow and flash sequence that might last many seconds, as in jellyfish, while the vast majority of planktonic organisms such as dinoflagellates, copepods, and ostracods, have much shorter flash durations of between 0.1–1 s.

The mysid *Gnathophausia* will squirt an intense cloud of luminescence into the water if it is startled and then will disappear into the surrounding darkness. Some of the species living in the upper 1000 m have both squirted luminescence and ventral photophores.

A demonstration of chemoluminescence involves a reaction where luciferin from luminol is converted in the presence of oxygen (with the help of a catalyst, luciferinase) to produce the light energy, just as it does in bioluminescence. Although it is similar in appearance, the adaptations for its use by plankton are as different as the fourteen marine phyla that are luminous. From counterillumination to a burglar alarm, we can get a glimpse of the dreamlike vision if we are in the right place at the right time. Bioluminescence is just one more of the many wonderful adaptations we have to observe and investigate as we continue to learn about the plankton community of our world ocean.

To learn more:

Buskey, E. J. 1992. Epipelagic planktonic bioluminescence in the marginal ice zone of the Greenland Sea. *Mar. Biol.* 113: 689–698.

Haddock, S. H. D., M. A. Moline, and J. F. Case. 2010. Bioluminescence in the Sea. *Annu. Rev. Marine. Sci.* 2010.2: 443–93. Downloaded from www.annualreviews.org by University of Findlay on 02/07/12.

Herring, P. J. 1977. Bioluminescence in marine organisms. *Nature*, London 267: 788–93.

Lapota D., M. L. Geiger, A. V. Stiffey, D. E. Rosenberger and D. K. Young 1989. Correlations of planktonic bioluminescence with other oceanographic parameters from a Norwegian fjord. *Mar. Ecol. Progr. Ser.* 55: 217–227.

Widder, E. A. 2006. A look back at quantifying oceanic bioluminescence: Seeing the light, flashes of insight and other bad puns. *Mar Tech Soc J.* 40(2):136–37.

Box Figure 3   Light organs hanging from the upper jaw in *Thaumatichthys axeli*, a deep-sea fish.

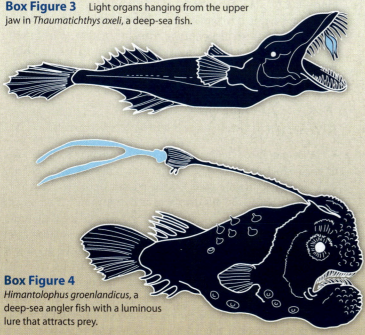

Box Figure 4
Himantolophus groenlandicus, a deep-sea angler fish with a luminous lure that attracts prey.

Box Figure 6   Some lantern fish (myctophids) have very bright light organs near their tail.

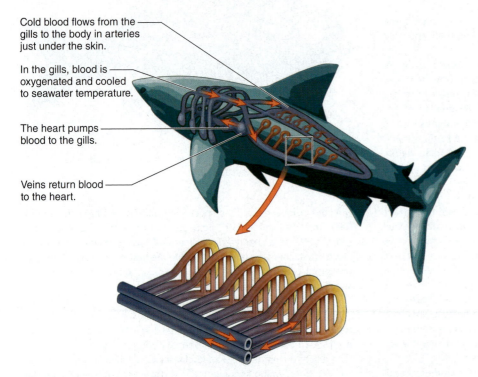

Cold blood flows from the gills to the body in arteries just under the skin.

In the gills, blood is oxygenated and cooled to seawater temperature.

The heart pumps blood to the gills.

Veins return blood to the heart.

Figure 11.12 Countercurrent heat exchange allows some marine organisms to maintain body temperature by passing cold (arterial) blood past warm (venous) blood, minimizing the loss of heat in the extremities and during oxygenation of blood in the gills.

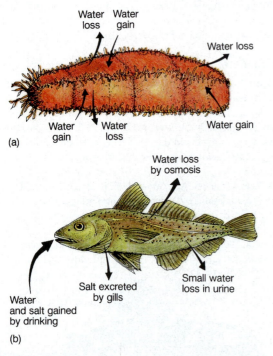

Water loss   Water gain

Water loss

(a)

Water gain   Water loss   Water gain

Water loss by osmosis

Salt excreted by gills

Small water loss in urine

Water and salt gained by drinking

(b)

Figure 11.13 (a) The salt concentration of the seawater is the same as the salt concentration of the sea cucumber's body fluids (35‰). The water diffusing out of the sea cucumber is balanced by the water diffusing into it. (b) The salt concentration in the tissues of the fish is much lower (18‰) than that of the seawater (35‰). To balance the water lost by osmosis, the fish drinks salt water, from which the salt is removed and excreted.

such that the salt content remains the same on both sides. In contrast, the internal salt concentration of most fish is lower than that of seawater, and marine fish tend to lose water from their tissues. Fish expend considerable energy to prevent dehydration, which would increase their salt content. Fish maintain a fluid balance by almost continuously drinking seawater and excreting salt. The outer skin of fishes is not entirely permeable to seawater, so the salt is excreted at the gills (fig. 11.13). Sharks and rays maintain high concentrations of urea in their tissues. The urea acts as a salt and eliminates the osmotic gradient and thus prevents water from leaving cells.

Species distributions are limited by salinity conditions. Organisms adapted to life in fresh waters are rarely able to live in ocean waters, and vice versa, organisms adapted to life in the ocean are rarely adapted to life in fresh water. One slight exception to this rule is that some fish and crustaceans use the low-salinity coastal bays and estuaries as breeding-grounds and nursery areas for their young. The adults then migrate further offshore into higher-salinity waters. Salmon and eels stand in stark contrast to most other animals because they can live in both fresh water and salt water, depending on their life cycle stage. Salmon are **anadromous.** They spawn in fresh water, and after one to two years (depending on the species of salmon), the juvenile fish migrate down rivers to mature as adults in the open ocean. After several years, the adults return to their home streams to spawn, and the cycle begins again. The American and European eels are **catadromous** and display the opposite migratory pattern. They spawn in the open ocean of the Sargasso Sea but mature in fresh water. Juvenile eels are members of the plankton, and they drift north and east with the Gulf Stream. They return to estuaries and rivers where they live for up to ten years before migrating back to the Sargasso Sea to spawn.

Buoyancy

Whether an object floats or sinks in water depends on the difference between the buoyant force, which pushes an object up, and the gravitational force, which pushes an object down. A submerged object always displaces water upwards. If an object weighs more than the weight of the water displaced, then the object will sink. If an object weighs less that the weight of the water displaced, the object will float. Consequently, an object sinks if its density is greater than the density of water.

Organisms regulate their buoyancy through a variety of methods. Most plankton, despite their small size, are denser than seawater (fig. 11.14). Many species of plankton store oil droplets to decrease their density. They also possess various

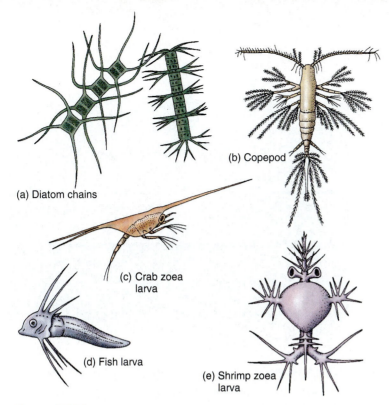

(a) Diatom chains

(b) Copepod

(c) Crab zoea larva

(d) Fish larva

(e) Shrimp zoea larva

Figure 11.14 Despite being very small, many planktonic organisms are still denser than seawater. They often develop long spines or projections to increase surface area (drag), helping to minimize sinking. (a) Diatoms, (b) the copepod *Augaptilus*, (c) zoea larvae of porcellanid crabs, (d) fish larva or *Lophius*, and (e) zoea larva of *Sergestes*, a shrimp.

appendages such as spines and "feathers" that increase their surface area to volume ratio, further enhancing viscous forces and slowing sinking.

Larger organisms can enhance their buoyancy using trapped gases. For example, the Portuguese-man-of-war fills a bell with carbon monoxide, and many seaweeds use trapped gas in their fronds. Many fish maintain neutral buoyancy by using an internal swim bladder filled with gas. Some species fill their bladders by gulping air at the surface; others release gas from their blood through a gas gland to the swim bladder. When a fish changes depth, it adjusts the gas pressure in its swim bladder to compensate for pressure changes. You may have seen the effects of bringing a deep-swimming fish to the surface too quickly: the bulging eyes and distended body result from the uncontrolled expansion of the swim bladder. Whales and seals decrease their density and increase flotation by storing large quantities of blubber. Sharks and other fish store oil in their livers and muscles; the giant squid excludes denser ions from its cells and replaces them with less dense ions. Seabirds float by storing fat, possessing light bones, and using air sacs developed for flight. Their feathers are waterproofed by an oily secretion called preen, which acts to seal air between the feathers and the skin. This helps to keep the birds warm and afloat.

Inorganic Nutrients

Autotrophic organisms require the same inorganic nutrients for growth that the plants in your garden require as fertilizer.

All phytoplankton and seaweeds require the inorganic macro-nutrients phosphate (PO_4^{-3}) and nitrogen, most commonly in the form of nitrate (NO_3^-) or ammonium (NH_4^+). A small number of bacteria, including some prokaryotic phytoplankton known as cyanobacteria, can use N_2 gas as a source of nitrogen when NO_3^- or NH_4^+ concentrations are less than the cells' minimum requirements. One group of eukaryotic phytoplankton, diatoms (which will be covered in more detail in chapter 12), also require silicic acid ($SiOH_4$) for growth. Autotrophs additionally require micro-nutrients such as vitamins, iron, zinc, and manganese. The terms "macro-" and "micro-nutrients" refer to the relative concentrations of the nutrients required by autotrophs. Macro-nutrients are required in higher concentrations than micro-nutrients.

Inorganic nutrients are quickly consumed within the euphotic zone by the phytoplankton to generate nitrogen- and phosphorus-containing organic matter. As this organic matter sinks through the water column, it is consumed by organisms and broken down into its inorganic constituents. Biologically required inorganic nutrients are found in the lowest concentrations in surface waters because phytoplankton use up the nutrients as they grow (fig. 11.15). Inorganic nutrients therefore display the opposite depth profile of sunlight, which has important implications for the distribution of phytoplankton species within the water column; the highest concentrations of nutrients are commonly found deeper in the water column where light levels are low. Nutrients enter surface waters through two main mechanisms: (1) nutrient runoff from land and (2) upwelling of nutrient-rich deep waters. Nitrogen gas (N_2), like CO_2, diffuses

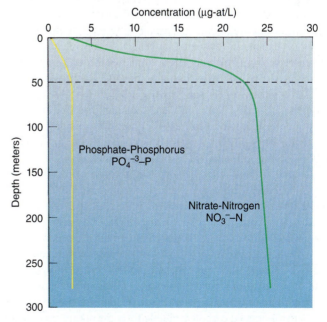

Figure 11.15 Nitrate and phosphate distribution in the main basin of Puget Sound in the late summer. The low surface values are the result of nutrient utilization by phytoplankton. The depth of the euphotic zone is indicated by the *dashed line*.

into seawater from the atmosphere. Until recently, nitrogen fixation by prokaryotic phytoplankton was thought to be limited to primarily the pelagic tropical ocean. New methods, particularly the use of molecular techniques, have challenged this view and it is now thought that direct utilization of N_2 gas by marine prokaryotes is widespread throughout the oceans, and of profound biogeochemical significance. Geographically, the abundance of phytoplankton is greatest where the supply of nutrients is the highest, which occurs in coastal waters or in upwelling regions such as off the coast of Peru. It is important to note that the supply of a nutrient to surface waters can be high due to a process such as upwelling, but the amount of nutrient measured in those same waters can be low because of the rapid uptake of the nutrient by phytoplankton. If the concentration of a required nutrient falls below the minimum required concentration, phytoplankton growth slows, and some species will not be able to grow at all even if other nutrients are present at sufficient concentrations to support growth.

Dissolved Gases

All life on our planet is based on carbon. Autotrophic organisms (both photosynthetic and chemosynthetic) utilize inorganic carbon dioxide (CO_2) as their carbon source to generate the organic carbon needed for growth. Heterotrophic organisms use organic carbon as their carbon source. As described in chapter 5, carbon dioxide is a gas that dissolves in seawater and is stored in large quantities in the ocean. Carbon dioxide also acts as a buffer to maintain the pH of the ocean in a range that is suitable for living organisms (review the discussion of the pH of seawater in chapter 5).

Oxygen enters the ocean in two ways: as a by-product of photosynthesis by marine photoautotrophs and through equilibration of surface waters with the atmosphere. Oxygen concentrations are highest in surface waters where there is efficient gas exchange between the ocean and atmosphere. Oxygen is removed from seawater predominantly through respiration, which is carried out by all living organisms. **Respiration** refers to the oxidation of organic matter to CO_2 to derive energy; oxygen is the most commonly used oxidant (see fig. 11.2). The biologically mediated profiles of O_2 and CO_2 display opposite depth distributions: photosynthesis generates O_2 and removes CO_2; respiration removes O_2 and generates CO_2.

Large regions of the ocean called **oxygen minimum zones** are associated with anoxic and hypoxic conditions; because these environments promote microbes that can utilize these other oxidizing agents, these regions are increasingly recognized as key areas controlling the biogeochemistry of the oceans.

As heterotrophic organisms in deeper waters consume organic matter, they generate CO_2 and remove oxygen. The rate of removal of oxygen slows to a minimum at the oxygen minimum zone (about 800 m depth) because of the decreasing amounts of organic carbon available from the photic zone; below the oxygen minimum zone, oxygen concentration increases because of the vertical circulation of oxygenated waters. The net effect of photosynthesis in the surface waters and the raining

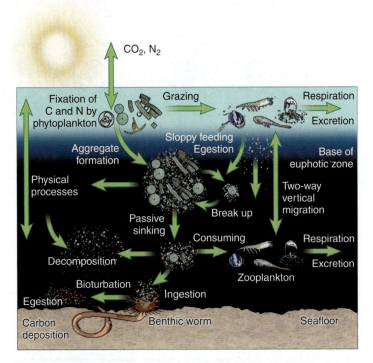

Figure 11.16 A conceptual model of the biological pump showing the transport of carbon and nitrogen from the upper ocean to the sea floor.

down of organic matter to deeper waters is that CO_2 is drawn down from the atmosphere and stored in deeper waters until ocean circulation brings the deep water back to the surface for equilibration with the atmosphere. This process of drawing CO_2 from the atmosphere into the ocean through the activity of biological processes is known as the **biological pump** (fig. 11.16). The biological pump plays a critical role in modulating atmospheric levels of CO_2. Estimates suggest that, if the biological pump was suddenly turned off, atmospheric levels of CO_2 would rise about 200 ppm, almost doubling preindustrial levels of atmospheric CO_2.

If the circulation of deep water is sluggish relative to the rate of removal of oxygen by respiration, oxygen levels in deeper waters may decrease to concentrations that are harmful to organisms. Respiration rates in deep waters may be enhanced when high concentrations of nutrients are introduced into stratified bodies of waters, as is currently being seen in many estuaries of the U.S. East Coast and Gulf coast. When nutrients are added to surface waters, more organic matter is generated via photosynthesis. As the additional organic matter rains down to the sluggishly moving deeper waters, respiration can remove oxygen from the waters faster than it can be replenished via circulation. Fish require at least 5 mg/L of oxygen, and below about 3 mg/L, many marine organisms cannot survive. **Hypoxia** occurs when oxygen concentrations fall below the level necessary to sustain most animal life; generally about 2 mg/L. Fish kills can result when bottom waters become hypoxic. **Anoxia** occurs when oxygen concentrations reach 0 mg/ml. Relatively few ocean environments are anoxic. Only **anaerobes** can survive anoxic conditions because they do not rely on oxygen to oxidize organic matter for energy. They instead use other oxidizing agents such as NO_3^- or SO_4^{2-} that are present in the water.

QUICK REVIEW

1. Describe or draw the change in light intensity with depth. How far does sunlight penetrate in the oceans?

2. Choose a marine organism and give at least three examples of how it is adapted to the marine environment.

3. Explain why red coloration is a useful trait in deep-water organisms.

4. Give examples of adaptations that marine organisms use to increase their buoyancy.

5. Explain why oxygen and carbon dioxide change concentration with increasing depth.

11.6 Environmental Zones

We previously described how marine organisms can be classified by habitat. The marine environment that organisms live within is vast and varied. It can be divided into two general zones: the **pelagic zone**, or water environment, and the **benthic zone**, or seafloor environment. The pelagic zone is further divided into the coastal, or **neritic zone** (above the continental shelf), and the **oceanic zone** (open ocean away from the direct influence of land). Both the benthic and pelagic zones can be divided into additional zones based on different increments of depth, as illustrated in figure 11.17. These depth zones are convenient for describing various habitats in the ocean, but it is important to remember that many organisms routinely cross these depth zones.

The sea floor can also be divided into zones. The **supralittoral zone**, or splash zone, is covered by wave spray only during the highest spring tides. The organisms that exist within this zone must cope with extreme changes in their environment: hot versus cold, wet versus dry, pounding surf versus exposure to air, and coverage by salty ocean water versus fresh rainwater. The **littoral zone**, or intertidal zone, is covered and uncovered with seawater once or twice a day as the water level changes between high and low tides. Conditions within the littoral zone vary greatly from site to site, changing from rock to sand, crashing waves to gentle surf, from high- to low-amplitude tides. Climatic conditions within the littoral zone vary depending on latitude and season. The **sublittoral zone**, or subtidal zone, is below the low-tide level and extends out over the continental shelf. The **bathyal** and **abyssal zones** are areas of complete darkness, without seasonal changes. The **hadal zone** lies beneath 6000 m (19,685 ft) and is associated with ocean trenches.

Benthic organisms of the deep sea rely on the organic material that rains down from the euphotic zone. The organisms that live in the supralittoral and littoral zones are the most familiar to the average beach-goer. In addition to the depth of the bottom environment, the base, or **substrate**, on or in which the organisms live is also critical. Rock, mud, sand, and gravel each provide a different type of food, shelter, or place for attachment that attracts different groups of organisms. The substrate of the sea floor is most variable along shallow coastal areas: sandbars, mud flats, rocky points, and stretches of gravel and cobble are frequently found along the same strip of coastline. At increased depths, the substrate is more uniform with a decrease in the particle size of the sediment.

Benthic organisms either live on or in the sea floor. **Epifauna** live on the surface, and **infauna** live within the sea floor. Epifauna are benthic animals and can be divided into two general categories: sessile and mobile. Suspension feeders such as barnacles, mussels, and some worms live attached to hard surfaces and feed by filtering the water for food particles. Because only a limited amount of food can be gained by filtering the water, all suspension feeders need to filter a maximum amount of water, while expending a minimum amount of energy. As noted earlier, all suspension feeders use cilia to bring water and food to them. The epifauna also include organisms that move across hard surfaces, including crabs, snails, and sea stars. These animals commonly "hunt" other animals as food, graze on algae or microbes growing on the sea floor, or scavenge detritus and dead animals that have sunk to the sea floor.

In addition to classifying the ocean by depth or substrate, scientists increasingly define regions of the oceans based on **biogeochemical provinces**. This is a way of dividing the ocean

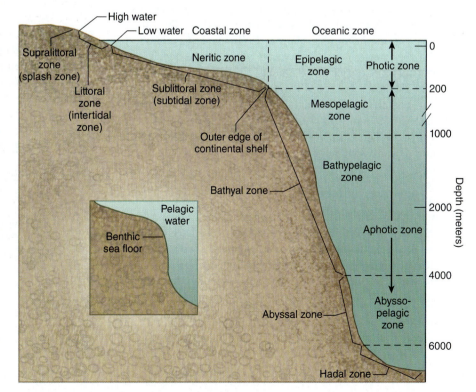

Figure 11.17 Zones of the marine environment.

up into sections, or habitats, based on similar physical, chemical, and geological properties that lead to similar biological ecosystems. While these habitats change from season to season and year to year, these provinces provide another useful way to describe oceanic habitats. A similar set of regions called **large marine ecosystems (LMEs)** has also been established (fig. 11.18); the primary difference is that the LMEs focus on the coastal ocean and marine resource management issues.

QUICK REVIEW

1. Name and describe the major environmental zones of the ocean.
2. Describe how the physical factors (e.g., temperature, light) change for each zone.
3. Explain why benthic organisms are linked to processes occurring in the surface ocean.

11.7 Marine Biodiversity

The richness and variety of life found on our planet are referred to as **biological diversity**, or **biodiversity**. In the ocean, this spans everything from microbes to whales; in fact, the oceans have representatives from twenty-eight phyla compared to just eleven phyla in the terrestrial environment (phyla are major groupings within Kingdoms; see table 11.1). With the introduction of molecular methods, the exact number of species is difficult to determine given the potential discrepancies between taxonomic and phylogenetic classification systems. However, on Earth today (excluding the many species that have gone extinct over geological history), there are approximately 1.6 million eukaryotic species identified, and an estimated 7–100 million total species, with estimates of as many as 1 billion marine microbes.

At the species level, the oceans appear to be much less diverse than terrestrial environments, accounting for perhaps 15% of the total, or 250,000 formally described (excluding bacteria and archaea). Of that number, the vast majority inhabit the benthic environment. Why are there so few marine (and even fewer pelagic) species? There are many possible factors, but one is clearly that the oceans have fewer habitats and greater environmental stability compared to land. New species tend to evolve when populations are separated in space, time, or reproductive capability. Geographical barriers divide the oceans into a series of environments. The arrangement of the continents and oceans combined with latitude, topography, and related climatic zones organizes the oceans into a series of areas with different patterns of circulation and different water properties. Boundaries in this water world exist in both the horizontal and the vertical

and include water mass borders; changes in temperature, salinity, light, and density; and isolating currents, but there are fewer such regions in the ocean compared to land (see fig. 11.8). The oceans are also generally more stable than land; in the deep ocean the temperature, salinity, and density are essentially constant, for example. According to the **Intermediate Disturbance Hypothesis**, species diversity is maximized when ecological disturbance is neither very rare nor very frequent. These disturbances provide opportunities for natural selection to act upon populations, ultimately producing new species. Much of the ocean presumably falls into the category of too rarely disturbed.

Another possibility is that we are simply missing the vast majority of organisms in the ocean. To help address this issue, 2700 marine scientists from eighty nations collaborated on a ten-year project (2000–2010) called the *Census of Marine Life*. The *Census* is creating an online encyclopedia of every marine organism identified, and has the ambitious goal of categorizing every existing form of marine life. Although it is unlikely that the oceans will ever approach the terrestrial environment in number of total species (excluding, perhaps, microbes) the *Census* has provided us with a new view of the richness and diversity of the ocean.

So why do we care how many species exist in the ocean? Biodiversity is one measure of the health of an ecosystem. Over the past few decades, concern has increased over the loss of species and the importance of conserving Earth's biodiversity. The functioning of marine ecosystems is driven by the complex interactions of the abiotic environment and organisms that live in that environment. Biologists do not understand ecosystems well enough to predict what will happen with the loss of individual species, but there is clear evidence that ecosystems generally function better with more species diversity. A diverse ecosystem uses resources more efficiently, produces more food, results in fewer waste products, and is generally more stable than less-diverse ecosystems. More pragmatically, the richer the diversity of life, the greater the opportunity for discovery of valuable drugs and other marine products. For these and other reasons, we are increasingly becoming aware of the importance of measuring and preserving ocean biodiversity.

QUICK REVIEW

1. How are the oceans both more and less diverse than terrestrial environments?
2. Describe biodiversity, and explain why it is important.
3. How could the physical conditions in the deep ocean help to explain, in combination with the intermediate disturbance hypothesis, why there are fewer species in the ocean compared to on land?

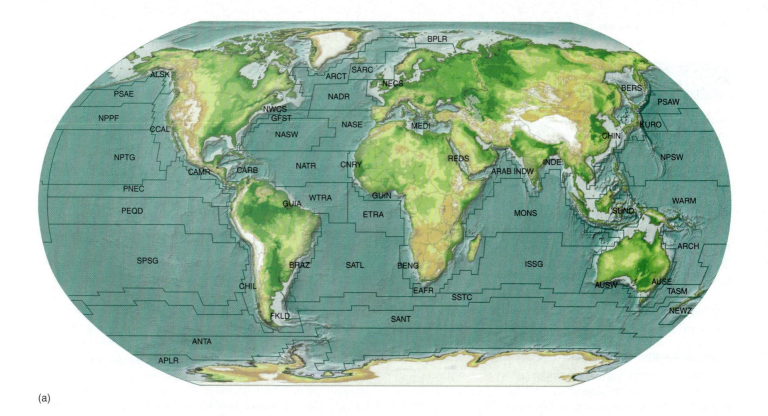

(a)

(b)

Figure 11.18 The marine environment can be classified based on (a) biogeographical provinces—regions that share similar physical and biological characteristics, and (b) large marine ecosystems—coastal regions of elevated productivity and marine resource use. The *letter codes* in (a) refer to specifc provinces, while the *blue-shaded* regions in (b) identify LMEs.

Summary

Marine biologists and biological oceanographers study interactions between marine organisms and their environment. The marine environment is subdivided into different zones. The major environments are the pelagic and benthic zones, both of which have numerous subdivisions based on ocean depth.

Life in the oceans is incredibly diverse. The most abundant organisms in the ocean are microscopic in size, although the largest organisms on Earth are also found in the sea. The dominant physical forces that act on very small organisms differ from those that act on larger organisms. This in turn influences the shapes that organisms can assume.

Marine organisms can be categorized based on their carbon and energy sources and/or their relationships to one another. Unrelated organisms can carry out the same function within an environment. Marine organisms are commonly divided into plankton, nekton, and benthos.

Important physical and chemical factors that influence the distribution of marine organisms include light, temperature, pressure, salinity, and inorganic nutrients. The availability of light and inorganic nutrients limits the growth of phytoplankton populations. Depth profiles of light and biologically required inorganic nutrients display opposite patterns. Light levels are highest in surface waters, whereas inorganic nutrients are highest in deep waters. Phytoplankton produce oxygen and generate the organic matter that serves as food for other organisms. Regions of the ocean where the supply of nutrients to surface waters is high will support more phytoplankton, which in turn will support larger numbers of organisms that consume organic matter.

The vast majority of the ocean is relatively cold and dark, with nearly constant salinity and high pressures. The coldest temperature in the ocean is found in sea ice. The hottest temperature in the ocean is found at hydrothermal vents. The organisms that live under extreme temperatures or high pressures have developed specialized attributes that allow them to exist in these environments. Many organisms that live in the sea have also developed means of generating their own light in a process known as bioluminescence. Bioluminescence in the ocean is primarily used either to attract food or to avoid being eaten.

Ocean regions can be defined based on proximity to land, water column depth, and bottom type. Biogeographical provinces, or regions sharing similar physical, chemical, and biological features, can also be identified. For nearshore habitat, these regions are referred to as large marine ecosystems. The oceans show greater diversity at the level of phyla, but there many fewer species than are found in the terrestrial environment. This is at least partly due to greater environmental stability and fewer biogeographical provinces in the marine environment.

Key Terms

natural selection, 288
Theory of Evolution, 288
common ancestor, 288
classification, 288
taxonomy, 288
Linnaean classification
   system, 288
phylogeny, 288
cryptic species, 288
prokaryote, 288
eukaryote, 288
nucleus, 288
mitochondria, 288
chloroplasts, 288
autotrophs, 289
heterotrophs, 289
photoautotrophs, 289
photosynthesis, 289
chemoautotrophs, 289

chemosynthesis, 289
microbe, 290
microorganism, 290
biomechanics, 291
diffusive, 291
conductive, 291
viscous, 291
plankton, 292
phytoplankton, 292
zooplankton, 292
bacterioplankton, 292
mixotrophic plankton, 292
nekton, 292
benthos, 293
supralittoral zone, 293, 301
hadal zone, 293
euphotic zone, 293
aphotic zone, 293
disphotic zone, 293

bioluminescence, 294
psychrophile, 295
cryoprotectants, 295
poikilotherm, 295
homeotherm, 295
endotherm, 295
countercurrent heat
   exchange, 295
osmosis, 295
anadromous, 298
catadromous, 298
respiration, 300
oxygen minimum zone, 300
biological pumps, 300
hypoxia, 300
anoxia, 300
anaerobes, 300
pelagic zone, 301
benthic zone, 301

neritic zone, 301
oceanic zone, 301
littoral zone, 301
sublittoral zone, 301
bathyal zone, 301
abyssal zone, 301
hadal zone, 301
substrate, 301
epifauna, 301
infauna, 301
biogeochemical provinces, 301
large marine ecosystems
   (LMEs), 302
biological diversity
   (biodiversity), 302
Intermediate Disturbance
   Hypothesis, 302

Study Problems

1. Sperm whales can dive as deep as 3000 m in search of prey. Hydrostatic pressure increases with depth, at a rate of 1 atmosphere of pressure every 10 meters. How many atmospheres of pressure would a sperm whale experience at 3000 m depth? Boyle's law states that volume and pressure are inversely proportional ($V_1 \times P_1 = V_2 \times P_2$) if temperature is held constant. How much would the volume of air in the sperm whale's lungs change if temperature is held constant?

2. The volume and surface area of a cell increase proportionally with size, but at different rates. Assuming a single-celled organism can be represented as a sphere, calculate the surface to volume ratio of the cell (sphere) as the radius increases from 1 μm, to 10 μm, to 100 μm. Explain or graph how surface area to volume ratios change as a function of size.

3. There are many examples of nonlinear relationships in ocean sciences, such as the surface to volume ratio as a function of cell size. Another example is the decrease of light with depth in the oceans. This relationship can be described by the Beer-Lambert law:

$$E = E_o \times e^{-kz}$$

Where E, is the light at some depth z, E_o is the light at the surface, z is the depth (m), and k is the extinction coefficient (inverse meters, or m^{-1}), or a factor describing how quickly light disappears. If the extinction coefficient (k) is $1\ m^{-1}$, calculate the depth where the light is 1% of the surface. Now plot the light levels versus depth from 100% to 1%. (Hint: You can choose various light levels, solve the equation for each, and fit a curve to the points as a function of depth in meters.)

4. The ocean is dominated by microbes, with 1 million bacteria in each milliliter of coastal seawater not uncommon. If we assume bacteria can be described as spheres, and a typical size (diameter) is 0.5 μm, how many bacteria could fit into 1 ml of seawater?

CHAPTER 12

The Plankton, Productivity, and Food Webs

Learning Outcomes

After studying the information in this chapter students should be able to:

1. *describe* the main types of photosynthetic organisms in the ocean and *explain* how they differ from heterotrophs,

2. *explain* under which conditions you would expect to find different groups of phytoplankton (e.g., cyanobacteria, dinoflagellates, diatoms),

3. *understand* the role of bacteria and the microbial loop,

4. *explain* how productivity varies seasonally in tropical, temperate, and polar environments,

5. *compare* and *contrast* food chains and food webs, and

6. *identify* regions of high and low productivity globally.

Garibaldi damselfish (*Hypsypops rubicundus*) swim in a forest of giant kelp (*Macrocystis pyrifera*) that reaches upward toward sunlight.

The focus of this chapter is on photosynthetic autotrophs and the heterotrophic plankton, including bacterioplankton and zooplankton. These groups are intimately linked in the oceans. Photoautotrophs use sunlight and inorganic compounds to generate organic matter that serves as food for life in the sea. Photoautotrophs include marine plants, macroalgae, and phytoplankton. The planktonic heterotrophs, one of the most diverse and abundant groups of organisms in the ocean, consume this organic material, acting as a link between the photosynthetic autotrophs (the "grasses of the sea") and higher trophic levels such as fish, birds, and marine mammals. The bacterioplankton also recycle this material, converting and transforming organic matter back into inorganic compounds. Without the plankton, the ocean would be a very different place, with few (if any) of the more familiar, large marine organisms. In this chapter we will describe how producers and consumers drive the flow of energy and life in the oceans.

Figure 12.1 *Sargassum* seaweed, one of the only large, multicellular planktonic algae.

12.1 The Marine Photoautotrophs

Living on land, we often think of trees, grasses, and moss as the dominant photosynthesizers, or photoautotrophs, on our planet. Perhaps surprisingly, then, about half of the oxygen we breathe comes from microscopic, single-celled marine phytoplankton, which are not true plants (most photosynthetic organisms in the marine environment are not members of the kingdom Plantae). The few specialized marine plants that do exist include **sea grasses**, which resemble grass but are actually part of the lily family, salt marsh plants such as **cordgrasses** and other salt-tolerant plants (**halophytes**), and **mangroves**, which are trees and shrubs adapted to living in tropical and subtropical coastal regions. All of these plants are bottom-dwelling and can only survive at the ocean margins. The primary distinction between sea grasses and halophytes is that sea grasses are permanently submerged by seawater and are therefore true marine plants. Halophytes can tolerate exposure to seawater but cannot survive prolonged periods of submersion. Mangroves are essentially land plants that have developed mechanisms to tolerate salt water to some degree. Salt marshes replace mangroves at higher latitudes where cold winter temperatures limit the survival of the mangroves. Although confined to the ocean margins, these ecosystems serve as nursery grounds for many marine species, stabilize the coast, and provide an important source of primary production.

Seaweeds, or **macroalgae**, are large, plantlike algae. Unlike true plants, they do not produce flowers or seeds, and they have relatively simple tissue structure. Most of the macroalgae are benthic (attached to the bottom). *Sargassum* is one of only a few large, multicellular planktonic alga (fig. 12.1). *Sargassum* can form large mats that provide shelter and food for a wide variety of specialized organisms, including certain fish and crabs, that

occur nowhere else. The Sargasso Sea is part of the large, subtropical gyre in the North Atlantic and is well known for great floating masses of *Sargassum* seaweed. These floating masses frightened early mariners and became part of many sea stories. Lesser quantities of *Sargassum* are distributed throughout other warm-water environments. We will return to the macroalgae in chapter 14. For now, the key point to remember is that both true plants and algae photosynthesize, converting inorganic compounds into organic compounds with the help of pigments such as chlorophyll *a* and sunlight.

QUICK REVIEW

1. Explain the difference between marine plants and algae. Why are marine plants not very common?
2. Why is *Sargassum* considered to be part of the plankton?

12.2 The Plankton

The word *plankton* comes from the Greek *planktos*, meaning to wander. In general, plankton are small organisms, commonly less than a few millimeters in size. Their small size means that they tend to be moved by currents from place to place while suspended in seawater. **Phytoplankton** use solar energy to generate oxygen and the organic food that fuels most of the rest of life in the seas. Phytoplankton form the base of most food webs. Cyanobacteria are the only bacterial members of the phytoplankton. All other phytoplankton are eukaryotic (review section 11.4 for definitions). **Zooplankton** are composed of unicellular as well as multicellular organisms. In general, zooplankton consume other organisms. Many juvenile stages of non-planktonic adults are also members of the zooplankton. Bacteria are not members of the zooplankton. Zooplankton contain the largest members of the plankton. For example, some jellyfish are huge, trailing 15 m (50 ft) tentacles. The **bacterioplankton** are composed of members of the domains Bacteria and Archaea. Bacterioplankton commonly rely upon inorganic and organic compounds that are dissolved in the seawater. Bacterioplankton play incredibly diverse roles in marine ecosystems.

Most plankton are small enough that they can be seen only with the aid of a microscope. Plankton are commonly categorized according to their size. **Picoplankton** are less than 2–3 microns in size. To get a sense of how small a picoplankton is, remember that the average human hair is about 100 microns in diameter. A 2 micron picoplankton is significantly smaller than the width of an average human hair! Picoplankton are composed of phytoplankton, unicellular zooplankton, and bacterioplankton. The most abundant phytoplankton in the world's oceans is a group of cyanobacteria known as *Prochlorococcus*. These organisms are less than 1 micron in size. The extent of diversity of this group of organisms is only just beginning to be uncovered. Within the last few years, there has been an explosion of new information about the diversity of the picoplankton because of the new ability to identify organisms based on their DNA sequences. **Nanoplankton** are between 2 and 20 microns in size. The picoplankton and nanoplankton dominate the plankton of open-ocean environments. **Microplankton,** or **net plankton,** range between 20 and 200 microns in size (fig. 12.2). The microplankton have been well studied over the years because it is relatively straightforward to capture and concentrate these organisms using a plankton net made of fine mesh that can be dragged through the water. **Macroplankton** are 200 microns to 2 mm in size and can also be captured using a plankton net. The larger microplankton and macroplankton are visible members of coastal and upwelling environments. In some locations, their abundance is so great that their vertical movement through the water column can be detected with a depth recorder. This layering of organisms is referred to as the deep scattering layer.

QUICK REVIEW

1. Explain how plankton move from place to place.
2. Describe the typical size of some common plankton groups.

12.3 Phytoplankton

Microscopic phytoplankton are often referred to as the "grasses of the sea." Just as a land without grass could not support the insects, small rodents, and birds that serve as food for larger organisms, a sea without phytoplankton could not support zooplankton and other larger organisms.

Common members of marine phytoplankton are **diatoms**, **dinoflagellates**, **coccolithophorids**, **cyanobacteria**, and **green algae**. These different groups can be readily distinguished from one another based on morphology (such as aspects of their cell walls) and the suite of pigments they possess. Diatoms, dinoflagellates, and coccolithophorids are important members of the larger size classes—the nanoplankton and microplankton. Green algae and cyanobacteria are important members of the picoplankton. They commonly dominate the open-ocean phytoplankton. There are many other groups of marine phytoplankton such as cryptomonads, silicoflagellates, and chrysomonads, but only the dominant players will be discussed here.

Diatoms are estimated to consist of tens of thousands of different species. Diatoms are found in both marine and freshwater environments in both pelagic and benthic realms. They come in a wide variety of sizes and shapes. They are important members of coastal ecosystems, although they are also found in the open ocean. In some regions, diatoms generate as much as 90% of the organic matter produced during phytoplankton blooms. Overall, diatoms are estimated to be responsible for as much as 40% of marine primary productivity.

The cell wall of diatoms is called the **frustule**. A defining feature of diatoms is that their frustule is composed primarily of silica that is embedded with a small amount of organic matter that prevents the silica from dissolving in seawater. The frustule displays elaborate nano-sized structures that are commonly used to define species. Although there are other organisms in the sea that can use silicate, such as sponges and silicoflagellates, diatoms essentially control the cycling of biogenic silicate in the world's oceans.

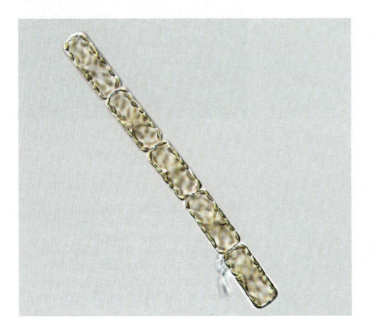

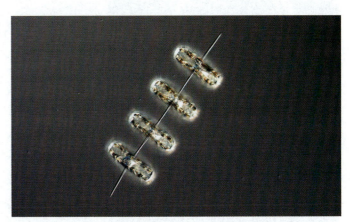

Figure 12.2 Most phytoplankton are unicellular. Some phytoplankton can form chains of connected cells. However, the individual cells within the chains act independently of each other. Two representative chain-forming diatoms are members of the genera *Melosira* (*left*) and *Thalassiosira* (*above*). The *Melosira* chain is composed of five linked cells; the *Thalassiosira* chain is composed of four linked cells.

Figure 12.3 Scanning electron micrograph of a diatom frustule. The organic casing that normally surrounds living diatoms has been removed, so the elaborate patterns of pores in the frustule can be seen.

Diatoms display two general shapes with either **radial** or **bilateral symmetry** (fig. 12.3). Radially symmetrical species are shaped more or less like Petri dishes or hat boxes and are members of the **centric** diatoms. Bilaterally symmetrical species are elongate and are shaped more or less like a cigar box and are members of the **pennate** diatoms. Micrographs of representative members of the centric and pennate diatoms are shown in figure 12.4, and drawings of representative species are shown in figure 12.5. The frustule of both pennate and centric species is composed of two halves (known as valves) held together by a series of bands also made of silica (think of wrapping tape around a Petri dish or hat box to hold the two halves together) (fig. 12.6). When a cell grows, the two halves slide apart due to the addition of new bands; the diameter of a valve cannot increase. All phytoplankton reproduce predominantly via asexual divisions: one cell divides to form two cells. When a diatom cell has grown large enough to undergo division, two new inner halves of the frustule are created, and the cell divides into two daughter cells. By this process, one daughter cell is the same size as the parent cell, and one is slightly smaller (fig. 12.6). Over successive generations, the mean cell size of the population will decrease. The most common means of escaping this cycle of diminishing cell size is through sexual reproduction.

The silica cell wall of diatoms is heavier than seawater, and unless a diatom actively controls its buoyancy, the cell will sink. Diatoms produce spines and other protrusions (fig. 12.4) that increase the surface area to volume ratio, slow sinking, and help maintain cells within surface waters (review section 11.5 on this topic). Eventually, however, either diatoms are eaten or they sink out of the euphotic zone to deeper waters. The zooplankton that consume diatoms produce fecal pellets that contain the heavy frustules, helping these organic matter-rich pellets to sink.

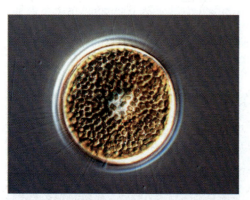
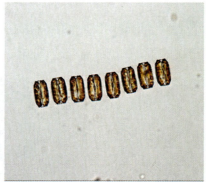
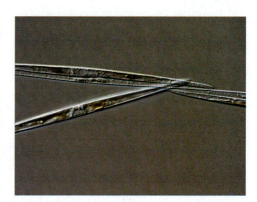

Figure 12.4 Representative members of coastal diatom communities of temperate waters. Four species of centric diatoms (*clockwise beginning upper left: Coscinodiscus, Thalassiosira, Chaetoceros,* and *Ditylum*) are shown in the *left panels* and one species of pennate diatom (*Pseudo-nitzschia*) is shown in the *right panel.*

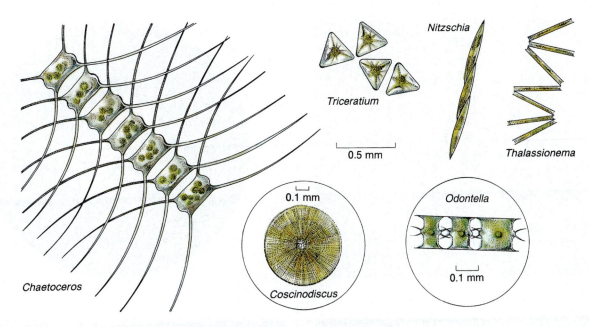

Figure 12.5 Drawings of centric and pennate diatoms.

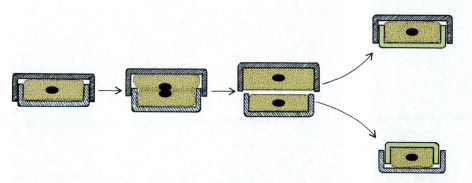

Figure 12.6 The division of a parent centric diatom into two daughter diatoms. The two halves of the pillbox-like cell separate, the cell contents divide, and a new inner half is formed for each pillbox. One daughter cell remains the same size as the parent; the other is smaller.

Dinoflagellates are another important member of the nano- and microplankton. Dinoflagellates tend to do well in relatively calm and well-stratified waters. The cell wall of dinoflagellates is composed of cellulose and some species will form plate-like structures as their cell wall. Dinoflagellates assume a wide variety of shapes. One of the more intriguing aspects of dinoflagellates is that some species are autotrophic and possess pigments necessary for photosynthesis (fig. 12.7a and b); some species are heterotrophic (fig. 12.7c and d) and are incapable of photosynthesis; and some species are both autotrophic and heterotrophic, depending on food availability. The capacity to feed heterotrophically, either exclusively or as a supplement to photosynthesis, means that dinoflagellates are often found in low-nutrient environments, including those of the open ocean.

Dinoflagellates possess two flagella (fig. 12.8). Many species of dinoflagellates can swim toward the surface during the day to photosynthesize, and they can swim toward the more nutrient-rich deeper waters at night. Despite their ability to swim vertically, dinoflagellates cannot swim against the currents and so are members of the plankton.

For two reasons, dinoflagellates are probably the group of phytoplankton best known to the general public. First, some species of dinoflagellates bioluminesce. If you go swimming at night during the summer in temperate waters, for example, the glow you see as you move through the water is most likely due to dinoflagellate bioluminescence (see section 11.5). The second reason that dinoflagellates are well known is because some species produce toxins that are biomagnified through the food web. Consumption of toxin-contaminated food can impact the health of marine mammals and humans.

Coccolithophorids are a third dominant member of the larger phytoplankton. They possess a cell wall embedded with calcium carbonate plates known as coccoliths (fig. 12.9a). Coccoliths are shed as cells grow and divide, and the coccoliths sink slowly through the water column.

Cyanobacteria are the most abundant phytoplankton in low-nutrient open-ocean environments. They also represent the most ancient lineage of oxygen-producing photosynthetic organisms. It is through their photosynthetic activity that early Earth became oxygenated (review chapter 1, "Earth's Age and Time"). The two most common groups of open-ocean cyanobacteria today are *Prochlorococcus* and *Synechococcus*. These organisms are about 1 micron in size and thus are members of the picoplankton. *Prochlorococcus* possesses special forms of the pigments chlorophyll *a* and *b*; *Synechococcus* possesses chlorophyll *a*. *Prochlorococcus* is globally distributed within tropical and sub-tropical waters at latitudes between 40°N and 40°S. *Prochlorococcus* dominates the euphotic zone of these

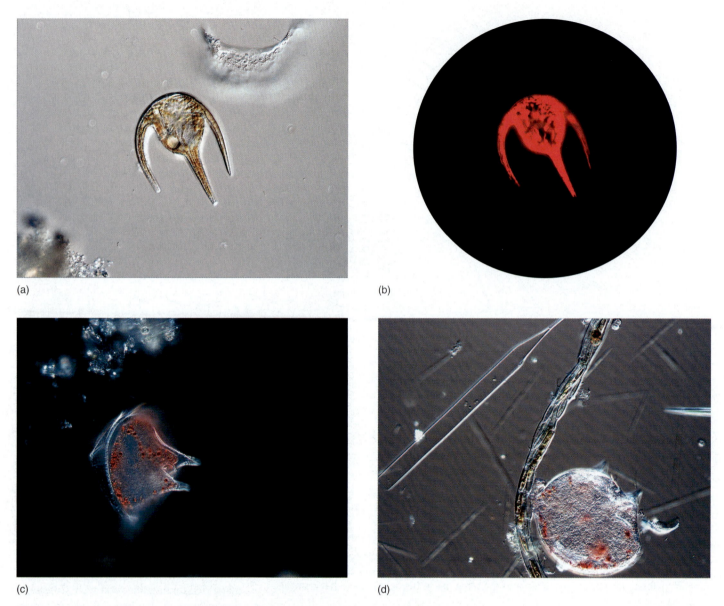

(a)

(b)

(c)

(d)

Figure 12.7 Two representative dinoflagellates from coastal environments. Dinoflagellates such as *Ceratium* (a) can be photosynthetic, and when illuminated with blue light, chlorophyll *a* fluorescence is detected as red light (b). Dinoflagellates such as *Protoperidinium* (c, d) are heterotrophic, and they use elaborate methods to capture their phytoplankton prey. Shown here is a *Protoperidinium* cell preparing to feed on chains of diatoms (d).

open-ocean environments, particularly when surface waters are highly stratified and inorganic nutrients such as nitrate are depleted. *Prochlorococcus* is estimated to be responsible for between 10 and 50% of the net primary production of the open ocean depending on time of year. This single group of organisms plays an incredibly important role in the global carbon cycle. *Synechococcus* has a broader global distribution than *Prochlorococcus*. *Synechococcus* blooms when the water column becomes mixed due to winter storms, and nitrate concentrations are higher.

Cyanobacteria also play an important role in nitrogen fixation—the use of nitrogen gas to generate nitrogen-containing organic matter. Nitrogen gas is used by nitrogen-fixing organisms when the concentrations of other inorganic nitrogen

sources are low. Neither *Prochlorococcus* nor *Synechococcus* can use nitrogen gas. Instead, three general groups of nitrogen-fixing cyanobacteria have been identified. First, one subset of nitrogen-fixing cyanobacteria are symbionts of other organisms. For example, a stable symbiosis exists between the cyanobacterium *Richelia* and the open-ocean diatom *Rhizoselenia*. Second, the filamentous cyanobacterium *Trichodesmium* has long been considered the dominant nitrogen-fixing cyanobacterium in the open ocean. Surface blooms of *Trichodesmium* can be seen from on board the deck of a ship. Water samples containing these cells can be monitored for rates of nitrogen fixation. Recently, a third type of nitrogen-fixing cyanobacterium known as *Crocosphaera* was discovered by scientists using a combination

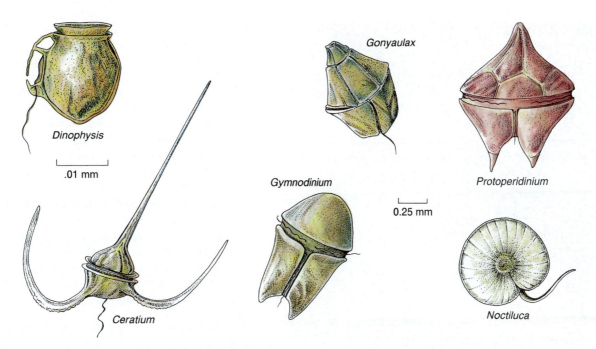

Figure 12.8 Representative species of coastal dinoflagellates.

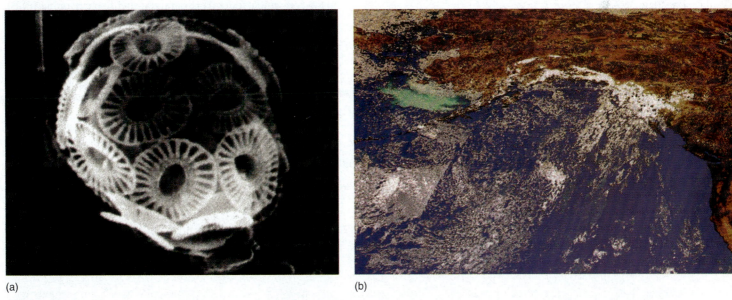

(a)

(b)

Figure 12.9 (a) A scanning electron micrograph of a coccolithophorid. The plates are made of calcium carbonate and are called coccoliths. (b) A true-color image showing the extent of a coccolithophorid bloom over the continental shelf of the eastern Bering Sea in September 1997. Image provided by the SeaWiFS Project, NASA/Goddard Space Flight Center. High concentrations of coccolith plates can accumulate on the sea floor as sediment and can eventually become a rock we call chalk.

of DNA technology and cell isolation. This nanoplankton-sized cyanobacterium likely plays an important role in nitrogen fixation in tropical open oceans.

Prasinophytes are eukaryotic phytoplankton that are closely related to land plants. The smallest known eukaryotic cell—*Ostreococcus*—is a member of the prasinophytes. Another important member of this group is *Micromonas*. These organisms are found in both open-ocean and near-shore environments.

Harmful Algal Blooms

While most of the phytoplankton in the ocean are harmless (indeed, they are the base of the food chain supporting all life!), a small subset can produce toxins, discolor the water, or otherwise negatively impact human and wildlife health. These organisms are collectively referred to as **harmful algal blooms**, or **HABs**.

The term harmful algal blooms is used to describe both toxic and nuisance phytoplankton blooms. HAB, rather than **red tide**,

Figure 12.10 The nontoxic dinoflagellate *Noctiluca scintillans* is the cause of spectacular red tides in many parts of the world, including coastal New Zealand.

has become the preferred scientific term because the blooms have nothing to do with the tides, and they may or may not color the water red (fig. 12.10). During a toxic bloom, particular species of

phytoplankton produce a toxin that can be biomagnified through the food web. The production of toxin is commonly only detected once it has reached measurable levels in an organism of direct interest to humans such as shellfish or marine mammals.

In addition to toxic blooms, other HABs lead to animal mortality due to direct physical contact or irritation of gills, development of low oxygen conditions during bloom decay, and formation of **ecologically disruptive algal blooms**, or **EDABs**. For example, the picoplankton *Aureococcus anophagefferens* can shade out other organisms, including benthic macroalgae, and can cause shellfish to stop feeding; this disrupts the normal functioning of the ecosytem and can lead to economic and ecosystem damage.

The vast majority of HABs are dinoflagellates, but there are also representatives from other groups of algae, including diatoms and cyanobacteria. Only a few species of phytoplankton produce toxins. The toxins associated with HABs (table 12.1) are often powerful nerve poisons that in the most extreme cases can cause paralysis, memory loss, or death. For example, one set of algal toxins known as saxitoxins is fifty times more lethal than strychnine. It is still not clear exactly which factors cause a limited number of species of phytoplankton to produce toxins. The syndromes in humans associated with consumption of toxin-contaminated food are paralytic shellfish poisoning (PSP), neurotoxic shellfish poisoning (NSP), diarrhetic shellfish poisoning (DSP), ciguatera fish poisoning, and amnesiac shellfish poisoning (ASP). These syndromes all have the term "shellfish" or "fish" in their name because consumption of toxin-containing shellfish or fish is the most common route of toxicity in humans.

Table 12.1   Phytoplankton Toxins

Toxin	Condition Produced	Common Phytoplankton Responsible	Characteristics
Azaspiracid	Azaspiracid shellfish poisoning (AZP)	*Protoperidinium;* dinoflagellate	Mode of action unknown; causes gastrointestinal illness in humans, neurotoxic effects in lab animals
Brevetoxin	Neurotoxic shellfish poisoning (NSP)	*Karenia brevis;* dinoflagellate	Affects nervous system; respiratory failure in fish and marine mammals; food poisoning symptoms in humans
Ciguatoxin	Ciguatera fish poisoning (CFP)	*Gambierdiscus toxicus;* dinoflagellate	Affects nervous system; human symptoms variable
Domoic acid	Amnesic shellfish poisoning (ASP)	*Pseudo-nitzschia;* diatom	Acts on vertebrate nervous system
Exotoxins		*Pfisteria piscicida;* dinoflagellate	Mode of action unknown; produces mortality in fish, neurotoxic symptoms in humans
Okadaic acid Dinophysistoxins	Diarrhetic shellfish poisoning (DSP)	*Dinophysis; Prorocentrum;* dinoflagellates	Affects metabolism, membrane transport, cell division; tumor promoter
Saxitoxin	Paralytic shellfish poisoning (PSP)	*Alexandrium; Gonyaulax; Gymnodinium;* dinoflagellates	Causes paralysis and respiratory failure in humans
Yessotoxin		*Protoceratium reticulatum; Gonyaulax spinifera; Lingulodinium polyedrum;* dinoflagellates	Cytotoxic effects in vertebrates; tumor promoter

The shellfish, for example, feed on toxin-producing phytoplankton and concentrate the toxin in their tissues. Humans commonly consume multiple shellfish at a time, further concentrating the toxin. In addition to their impact on human health, the toxins can kill fish, shellfish, marine mammals, birds, and other animals that may consume toxin-contaminated seafood.

The frequency and geographical distribution of HAB incidents have increased over the last few decades. Part of the observed increase is due to enhanced monitoring systems; more toxic events are reported because better detection methods are now in place. However, there are also documented cases of expansion of toxic blooms to previously unaffected areas. Researchers are currently trying to determine the role that human activities may play in the expansion of toxic blooms. Some see the expansion of harmful and toxic algal blooms as an indication of large-scale marine ecologic disturbances.

QUICK REVIEW

1. Describe the typical size and shape of some common phytoplankton groups.
2. Identify similarities and differences between diatoms and dinoflagellates.
3. Why do scientists prefer the term *harmful algal bloom* versus *red tide*?
4. Describe potential impacts from phytoplankton that would make them harmful to humans or wildlife.

12.4 Zooplankton

Zooplankton are heterotrophs, and they consume other organisms. Zooplankton fit within two general categories of plankton: the holoplankton and the meroplankton. **Holoplankton** spend their entire lives as plankton. **Meroplankton** spend only a portion of their lives as plankton. The meroplankton include the eggs and the larval and juvenile stages of many organisms that spend most of their lives as either free swimmers (fish) or bottom dwellers (such as crabs and sea stars). Some feed almost exclusively on phytoplankton and are herbivorous zooplankton. Others feed on other members of the zooplankton and are carnivorous. Those that feed on both phytoplankton and zooplankton are omnivorous. Many zooplankton can swim, and some can even dart rapidly over short distances in pursuit of prey or to escape predators. Even those zooplankton that can swim vertically are still transported by currents and therefore are members of the plankton. As you read about zooplankton, consider the roles that different groups may play in different food webs. We begin this section with representatives of the holoplankton.

The life histories of different types of zooplankton are varied and show many strategies for survival in a world where reproduction rates are high and life spans are short. The unicellular zooplankton or protozoa grow rapidly, sometimes as fast as the phytoplankton because they, too, reproduce by cell division. In contrast, the multicellular zooplankton reproduce sexually and produce juvenile offspring that must mature. In warm waters where food supplies are abundant, some species may produce three to five generations of offspring a year. At high latitudes, where the season for phytoplankton growth is brief, zooplankton may produce only a single generation in a year. The voracious appetites, rapid growth rates, and short life spans of zooplankton lead to rapid liberation of nutrients that can be reused by phytoplankton.

Zooplankton exist in patches of high population density between areas that are much less heavily populated. The high population patches attract predators, and the sparser populations between the denser patches preserve the stock, as fewer predators feed there.

Plankton accumulate at the density boundaries caused by the layering of the surface waters; variation of light with depth and day-night cycles play additional roles. Some zooplankton migrate toward the sea surface each night to feed on the food-rich surface layer and descend during daylight to possibly reduce grazing by their predators or to follow their food resources. This daily migration may be as great as 500 m (1650 ft) or less than 10 m (33 ft) (fig 12.11).

Accumulations of organisms in a thin band extending horizontally along a pycnocline or at a preferred light intensity or food resource level are capable of partially reflecting sound waves from depth sounders. The zooplankton layer is seen on a bathymetric recording as a false bottom or a deep scattering layer, the DSL (see chapters 6 and 11).

Among the most well-studied and widespread zooplankton types worldwide are small **crustaceans** (shrimplike animals), the **copepods**, and **euphausiids** (fig. 12.12). These animals are basically herbivorous and consume more than half their body weight each day. They are found throughout the world. Copepods are a link between the phytoplankton, or producers, and first-level carnivorous consumers. Euphausiids are larger, move more slowly, and live longer than copepods. Euphausiids, because of their size, also eat some of the smaller zooplankton along with the phytoplankton that make up the bulk of their diet. Copepods and euphausiids both reproduce more slowly than diatoms, doubling their populations only three to four times a year. In the Arctic and Antarctic, the euphausiids are the **krill** (fig. 12.13), occurring in such quantities that they provide the main food for the **baleen** (or whalebone) whales. These whales have no teeth; instead, they have netlike strainers of baleen suspended from the roof of their mouth. After the whales gulp the water and plankton, they expel the water through the baleen, leaving the tiny krill behind. It is remarkable to consider that the largest organisms in the sea feed on some of the smallest organisms.

The high iodide content of krill meat prevents marketing for human consumption, and the economic future of the krill fishery is questionable. Krill is sold whole, as peeled tail meat, minced, or as a paste; it is used as livestock and poultry feed in Eastern Europe and as fish feed by the Japanese. Because krill deteriorate rapidly, krill used for consumption must be processed within three hours, limiting the daily harvest. The distance to the fishing grounds is long and the costs of vessels and fuel are high.

The Convention for the Conservation of Antarctic Marine Living Resources was established by treaty in 1981. The convention's goal is to keep any harvested population, including krill, from dropping below levels that ensure the replenishment of the adult population.

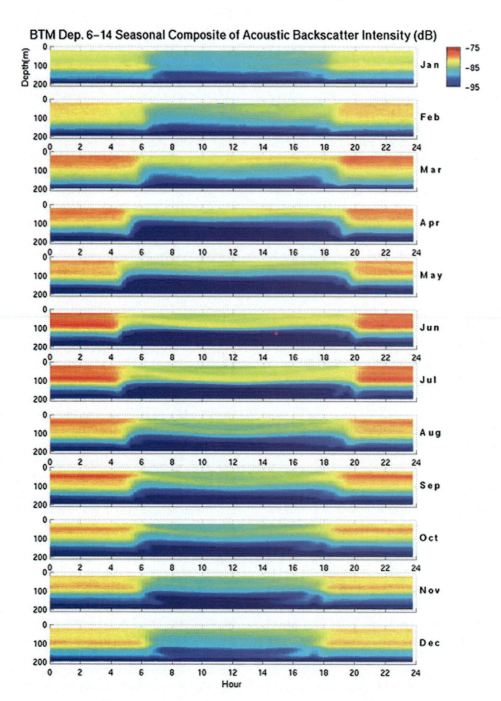

Figure 12.11 A seasonal composite of acoustic backscatter reveals the daily and seasonal patterns of vertical migraters. The weakest signals (*blues* and *greens*) occur during daylight hours when the organisms are absent; the strongest signals (*yellows* and *reds*) are observed at night when the organisms migrate toward the surface. Vertical migraters remain at the surface for less time during the long summer days but their concentrations are highest during spring and summer.

Arrowworms, or **chaetognaths** (see fig. 12.12), are abundant in ocean waters from the surface to great depths. These macroscopic (2–3 cm, or 1 in), nearly transparent, voracious carnivores feed on other members of the zooplankton. These are members of the trophic level corresponding to primary consumers. Several species of arrowworms are found in the sea, and in some cases, a particular species is found only in a certain water mass. The association between organism and water mass is so complete that the species can be used to identify the origin of the water sample in which it is found. In the North Atlantic, the arrowworm *Sagitta setosa* inhabits only the North Sea water mass and *Sagitta elegans* is found only in oceanic waters.

Foraminifera and **radiolarians** are microscopic, single-celled, amoeba-like protozoans; they are shown in figure 12.14. They

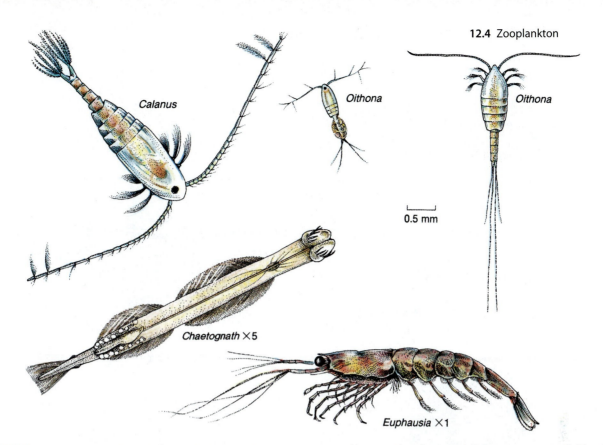

Figure 12.12 Crustacean members of the zooplankton and an arrowworm, or *chaetognath*. *Calanus* and *Oithona* are copepods. The shrimplike *Euphausia* is known as krill.

Figure 12.13 The Antarctic krill, *Euphausia superba*, dominates the zooplankton of the Antarctic Ocean.

feed primarily on phytoplankton but some can also eat small zooplankton. Foraminifera, such as the common *Globigerina*, are encased in a compartmented calcareous covering, or shell. Radiolarians are surrounded by a silica test, or shell. The radiolarian tests are ornately sculptured and covered with delicate spines. Pseudopodia (false feet), many with skeletal elements, radiate out from the cell. Radiolarians feed on diatoms and small protozoans caught in these pseudopodia. Both foraminifera and radiolarians are found in warmer regions of the oceans. After death,

their shells and tests accumulate on the ocean floor, contributing to the sediments. Calcareous foraminifera tests are found in shallow-water sediments; the siliceous radiolarian tests, which are resistant to the dissolving action of the seawater, predominate at greater depths, commonly below 4000 m (13,200 ft) (see chapter 3). **Tintinnids** (fig. 12.14) are tiny protozoans with moving, hairlike structures, or **cilia**. These organisms are often called bell animals and are found in coastal waters and the open ocean.

Pteropods (fig. 12.15) are mollusks; they are related to snails and slugs. They have a foot that is modified into a transparent and gracefully undulating "wing." This grouping includes animals with shells and animals without shells. The shelled group are primarily filter feeders that capture plankton on a mucous net that they then consume. The group without shells specializes in eating soft-bodied animals. Their hard, calcareous remains contribute to the bottom sediments in shallow tropical regions. Some pteropods are herbivores and some are carnivores.

Transparent, gelatinous, and bioluminescent, the **ctenophores**, or comb jellies (fig. 12.16), float in the surface waters. Some have trailing tentacles; all are propelled slowly by eight rows of beating cilia. The small, round forms are familiarly called sea gooseberries or sea walnuts; by contrast, the beautiful, tropical, narrow, flattened Venus' girdle may grow to 30 cm (12 in) or more in length. All ctenophores are carnivores, feeding on other zooplankton.

The tunicate, another transparent member of the zooplankton, is related to the more advanced vertebrate animals (animals with backbones) through its tadpolelike larval form. **Salps**

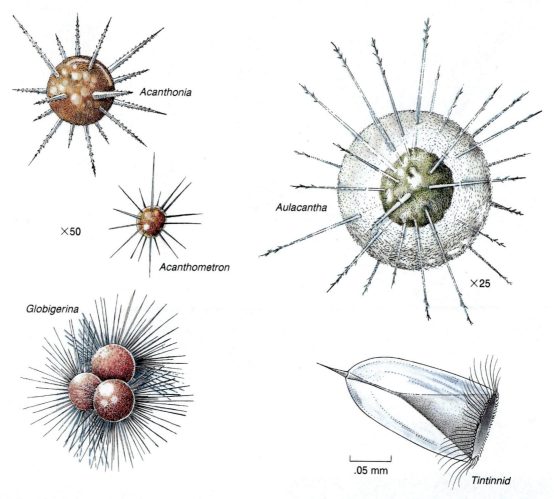

Figure 12.14 Selected radiolarians (*Acanthonia, Acanthometron,* and *Aulacantha*). A foraminiferan (*Globigerina*) is at *bottom left;* a *tintinnid* is at *bottom right.*

Figure 12.15 The pteropods are planktonic mollusks.

(fig. 12.16) are pelagic tunicates that are cylindrical and transparent; they are commonly found in dense patches scattered over many square kilometers of sea surface. Salps feed on phytoplankton and particulate matter.

Both ctenophores and pelagic tunicates, although jellylike and transparent, are not to be confused with jellyfish (fig. 12.17). True jellyfish, or sea jellies, come from another and unrelated group of animals, the **Coelenterata**, also called **Cnidaria**. Another group of unusual jellyfish are **colonial organisms**, including the Portuguese man-of-war, *Physalia*, and the small by-the-wind-sailor, *Velella*. Both are collections of individual but specialized animals. Some gather food, reproduce, or protect the colony with stinging cells, and others form a float.

In contrast to the holoplankton that spend their entire lives as members of the plankton (fig. 12.18), the meroplankton spend only part of their lives as members of the plankton (fig. 12.19). For a few weeks, the **larvae** (or young forms) of oysters, clams, barnacles, crabs, worms, snails, sea stars, and many other organisms are a part of the zooplankton. The currents carry these larvae to new locations, where they find food sources and places to settle. In this way, areas in which species may have died out are repopulated and overcrowding in the home area is reduced. Sea animals produce larvae in enormous numbers, so these meroplankton are an important food source for other members of the zooplankton and other animals. The parent animals may produce millions of spawn, but only small numbers of males and females must survive to adulthood to guarantee survival of the stock.

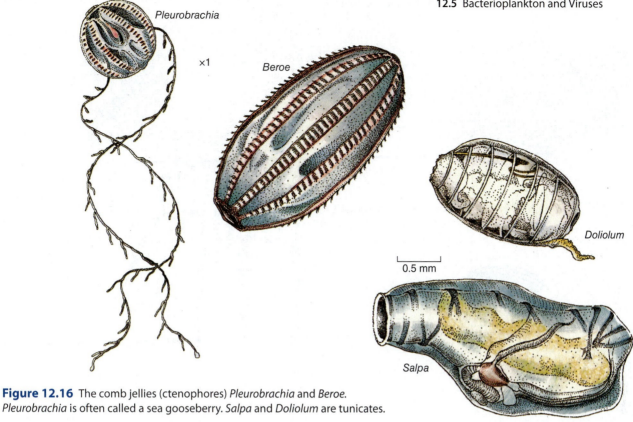

Figure 12.16 The comb jellies (ctenophores) *Pleurobrachia* and *Beroe*. *Pleurobrachia* is often called a sea gooseberry. *Salpa* and *Doliolum* are tunicates.

Other members of the meroplankton include fish eggs, fish larvae, and juvenile fish. The young fish feed on other larvae until they grow large enough to hunt for other foods. Some large seaweeds also release **spores**, or reproductive cells, that drift in the plankton until they are consumed or settle out to grow attached to the sea bottom.

QUICK REVIEW

1. Why are jellyfish considered to be part of the plankton?
2. Describe how zooplankton control the biomass of primary producers in the ocean.
3. Give an example of a meroplanktonic larvae; explain why it is advantageous for this organism to spend part of its life as plankton.

12.5 Bacterioplankton and Viruses

Bacterioplankton are composed of members of the two most ancient domains of life—Bacteria and Archaea. Bacterioplankton share a similar morphology: they are unicellular, they are commonly around 1–2 microns in size, and they possess no internal membrane-bound cell structures. Members of the Bacteria and Archaea can be distinguished from one another based on DNA sequences. It is estimated that about 1×10^{29} bacterioplankton exist within marine environments.

Bacterioplankton have an enormous impact on biogeochemical cycles on our planet. They are directly involved in recycling nutrients, some can use nitrogen gas and they therefore influence nitrogen cycles, and they play important roles in the global carbon cycle.

Most marine bacterioplankton exist either as single cells that float freely in the water column or as colonies of cells attached to sinking particles. Marine bacteria can be a significant food source for planktonic larvae and single-celled zooplankton and mixotrophic algae. A film of bacteria is commonly found on particles of floating organic matter. The attached bacteria help break down the organic matter, regenerating inorganic nutrients. The small size of some particles, with their attached microbial populations, makes them ideal food for small zooplankton. Bacterioplankton also utilize dissolved organic matter (DOM) that is released by phytoplankton and that is also a product of zooplankton feeding and excretion. Overall, about half the organic matter generated by phytoplankton is consumed as DOM by the bacterioplankton. **"The microbial loop"** refers to the processes that convert DOM into biomass that can be consumed by other organisms (fig. 12.20).

With the advent of DNA-based identification technologies, knowledge of bacterioplankton diversity has increased exponentially. However, most members of the Bacteria and Archaea in the sea are still known only by a small segment of DNA sequence (the 16S rRNA gene). The inability to study these DNA-identified microbes in the laboratory means that it is still difficult to know exactly what processes they carry out within the world's oceans. This creates a problem: Scientists know that a particular type of bacteria is present within the water column or sediment sample based on its DNA sequence, but they don't know exactly what processes the organism carries out.

Figure 12.17 Jellyfish belong to Coelenterata, or Cnidaria. *Velella*, the by-the-wind-sailor *(top left)*, and *Physalia*, the Portuguese man-of-war *(top right)*, are colonial forms.

Another breakthrough has come from an increased understanding of Archaea. Members of the Archaea were originally considered to be restricted to extreme environments such as the hydrothermal vents (fig. 12.21) where vent water can be acidic and extremely hot. The use of DNA-based technologies now indicates that Archaea are more widespread than originally realized. One type of Archaea, known as marine group I, comprises about 40% of the bacterioplankton within mesopelagic waters, making them one of the more abundant groups of microorganisms on the planet.

A **virus** is a noncellular particle composed of genetic material surrounded by a protein coat. Viruses are obligate parasites and are only able to carry out their metabolic activities and replicate themselves inside a host. Viruses likely infect all living organisms in the sea, from bacteria to whales. The marine environment is also a potential reservoir for disease-causing viruses. For example, some viruses are thought to cycle between marine and terrestrial mammals. In general, viral abundance is strongly correlated with planktonic biomass in the oceans. Viral abundance is highest where bacteria and phytoplankton abundances

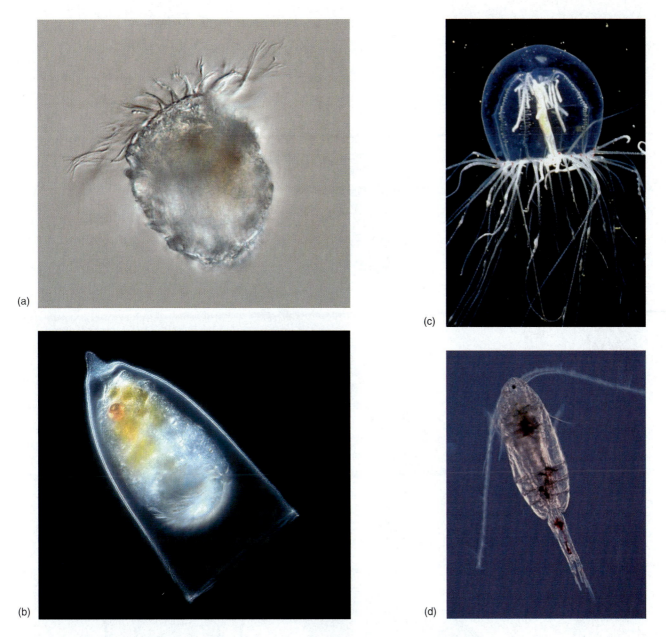

Figure 12.18 Organisms that spend their whole lives in the zooplankton are known as holoplankton and include single-celled protozoa such as (a) a ciliate and (b) a tintinnid, and multicellular oganisms such as (c) the jellyfish *(Aurelia)* and (d) a copepod.

are highest; viral abundance decreases with depth and distance from shore. It is estimated that there are about 3×10^6 viruses in a milliliter of deep seawater and about 1×10^8 viruses in a milliliter of productive coastal waters (fig. 12.22). Viruses are the most abundant biological entity in seawater. Viruses are so abundant that scientists estimate that if they were all laid out end to end they would extend even further into outer space than the nearest sixty galaxies!

Viruses are major pathogens of marine microbes. It is likely that the extent of viral-induced death of microbes varies depending on the environment under study, but it is clear that viruses have a huge impact on microbial communities. When viruses kill a microbe, the cell is lysed, and the cell's organic matter is released as DOM. This DOM is available for utilization by

heterotrophic bacteria. Viruses enhance recycling of nutrients and likely impact each interaction of the microbial loop depicted in figure 12.20.

QUICK REVIEW

1. Why are microbes considered to dominate the ocean's nutrient and carbon cycling?
2. Describe how DNA technology has changed our view of the presence and abundance of microscopic organisms in seawater.
3. Describe the role that viruses play in the microbial food web.
4. Explain the importance of dissolved organic matter (DOM).

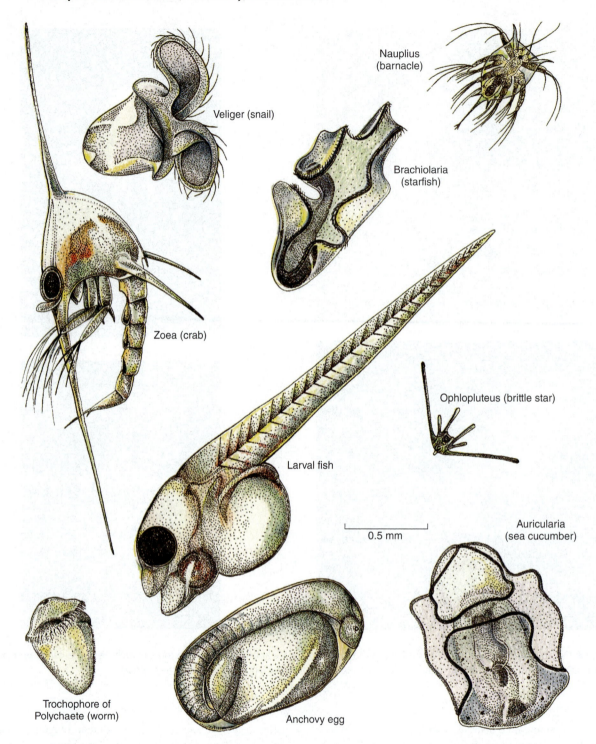

Veliger (snail)

Nauplius
(barnacle)

Brachiolaria
(starfish)

Zoea (crab)

Ophlopluteus (brittle star)

Larval fish

0.5 mm

Auricularia
(sea cucumber)

Trochophore of
Polychaete (worm)

Anchovy egg

Figure 12.19 Members of the meroplankton. All are larval forms of nonplanktonic adults.

12.6 Primary Production

A common factor linking plants, macroalgae, and phytoplankton is that they carry out photosynthesis. The oxygen and organic carbon that are generated through photosynthesis are critical for most life in the ocean. The absolute requirements of phytoplankton are relatively simple: sunlight, inorganic nutrients, and carbon dioxide. The required nutrients and carbon dioxide are dissolved in seawater. Growth of phytoplankton can be limited by temperature or by a lack of adequate amounts of either inorganic nutrients or sunlight; carbon dioxide is rarely limiting in the world's oceans. For the remainder of the discussion, we will focus on phytoplankton.

Phytoplankton use **pigments** to absorb energy from sunlight. Photosynthetic organisms that produce oxygen possess the pigment **chlorophyll *a*,** regardless of whether they live on

Extremophiles are microorganisms that thrive under conditions that would be fatal to other life-forms: extreme temperatures (hot and cold), high levels of acid or salt, no oxygen, no sunlight. Extremophiles not only flourish under these severe conditions but may actually require them to reproduce. Although some extremophiles have been known for forty years or more, scientists have been discovering more and more of these organisms in environments that were once thought lifeless.

These single-celled microorganisms resemble bacteria, for they have no membrane-bounded nucleus. However, when their genes were compared to the genes of bacteria, it was discovered that they are distinctly different. In fact, they appear to share a common ancestor with the eukaryotes, organisms such as ourselves with a membrane-bounded nucleus and cellular functional units, or organelles. This discovery opened for review the basic categories of all living organisms, and the result is a reorganization of life categories into three major domains: Bacteria, Archaea (the extremophiles), and Eukarya (all nuclei-containing organisms) (see fig. 11.1). This discovery was made by Professor Carl Woese at the University of Illinois; it has completely revised the way we think about microbial biology and the organization of life on Earth.

In the oceans, heat-loving members of the Archaea require temperatures in excess of 80°C (176°F) for maximum growth. *Pyrolobus fumarii* is an extreme example. It was found at a depth of 3650 m (12,000 ft) in a hot vent in the mid-Atlantic Ridge southwest of the Azores. Its name means "fire lobe of the chimney" from its shape and the black-smoker vent where it was found. *P. fumarii* stops growing below 90°C (194°F) and reproduces at temperatures up to 113°C (235°F). It uses hydrogen and sulfur compounds as sources of energy and can also use nitrogen gas. It is able to live with or without oxygen. Other Archaea are commonly found in the plumes of hot water that occur after an undersea eruption; it is unclear how deep into Earth's crust these microorganisms are able to exist (box fig. 1).

Another surprise was finding that close relatives of the hot-vent Archaea are common and abundant components of the marine plankton, living in the cold, oxygenated waters off both coasts of North America. They have since been found at all latitudes in water below 100 m (330 ft), in the guts of deep-sea cucumbers, as well as in marine sediments. A microorganism found in Antarctic sea ice grows best at 4°C (39°F) and does not reproduce at temperatures above 12°C (54°F). And still other extremophiles have been found living in the salt ponds constructed for the evaporation of seawater.

Scientists in the United States, Japan, Germany, and other countries have a particular interest in the enzymes of extremophiles. Enzymes are required in all living cells to speed up chemical reactions without being altered themselves. Standard enzymes stop working when they are exposed to heat or other extremes, so those used in research and industrial processes must be protected during reactions and while in storage. Heat-loving extremophile enzymes are already being used in biological

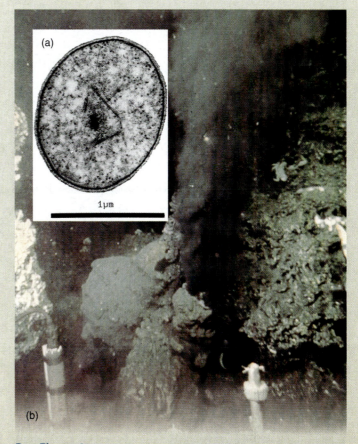

Box Figure 1 The extremophile *Pyrococcus endeavorii* (a) was isolated from an East Pacific Rise black smoker (b). This microorganism grows at temperatures that exceed 100°C (212°F).

and genetic research, forensic DNA testing, and medical diagnosis and screening procedures for genetic susceptibility to some diseases. In industry, these enzymes have increased the efficiency of compounds that stabilize food flavorings and reduce unpleasant odors in medicines. Enzymes that work at low temperatures may be useful in food processing where products must be kept cold to prevent spoilage. An important result of research with enzymes from extremophiles is learning how to redesign conventional enzymes to work under harsher conditions.

The discovery of microorganisms where none was assumed to exist and the recognition of a new branch of life are astonishing. It is likely that more discoveries and new technologies based on such discoveries will continue for some time to come.

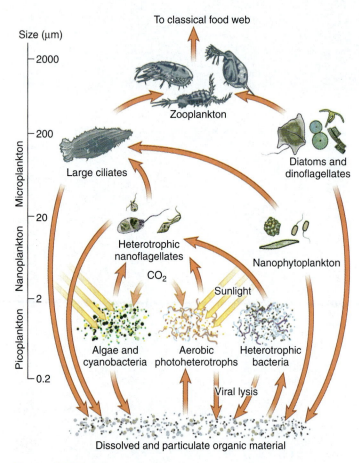

Figure 12.20 The microbial food web.

Figure 12.22 An environmental sample collected from 5 m depth in San Pedro Channel near Santa Catalina Island, California, observed under a microscope. The *small green dots* are viruses, the *larger green and orange dots* are bacteria, and the *largest objects* are a diatom and an unidentified protist. The sample was stained with a fluorescent dye to make the organisms visible.

used to drive chemical reactions such as the splitting of water to produce oxygen and the generation of organic carbon from carbon dioxide. The process of generating organic carbon from carbon dioxide is commonly referred to as **carbon fixation.**

Photosynthesis can be represented by the overall equation

$$6\,CO_2 \;+\; 12\,H_2O \;\xrightarrow[\text{chlorophyll }a]{\text{Solar energy}}\; C_6H_{12}O_6 \;+\; 6\,O_2 \;+\; 6\,H_2O$$

| 6 molecules carbon dioxide | + | 12 molecules water | | 1 molecule sugar | + | 6 molecules oxygen | + | 6 molecules water |

Different wavelengths of light penetrate to different depths in the water column. Phytoplankton have evolved different types of pigments to allow them to utilize a broader spectrum of light than can be captured by chlorophyll *a* alone. For example, certain phytoplankton, such as diatoms, contain the pigment fucoxanthin (pronounced fu′ko′zan′thin) in addition to chlorophyll *a*; other phytoplankton, such as cyanobacteria, contain the pigment phycoerythrin (pronounced fi′ko′e′rith′rin) in addition to chlorophyll *a* (fig. 12.23). Possession of additional pigments expands the spectrum of light that a phytoplankton species is able to absorb and thus extends the depths at which different phytoplankton are able to grow. Some absorbed solar energy is lost as heat, and some is lost as **fluorescence.** Chlorophyll *a* fluorescence from individual phytoplankton cells can be viewed using a fluorescence-detecting microscope. An example of a fluorescent cell is shown in figure 12.24.

Production of organic material from inorganic nutrients using light energy is termed **primary production.** The total amount of organic material produced through photosynthesis is **gross primary production.** However, not all the carbon fixed by phytoplankton is available for consumption by other organisms. The

Figure 12.21 The perforated probe was driven into the seafloor crust by the ROV *Jason II*. The probe allows the sampling of fluids circulating in the upper oceanic crust. *Photo courtesy of Dr. H. Paul Johnson, School of Oceanography, University of Washington.*

land or in the sea. Chlorophyll *a* absorbs light primarily in the blue and red regions of the spectrum. In the remarkable series of chemical reactions that occur during photosynthesis, absorbed light energy is converted into chemical energy, which in turn is

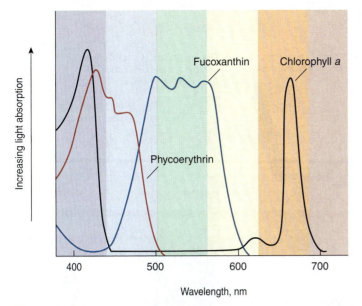

Figure 12.23 Light absorption by three photosynthetic pigments: chlorophyll *a*, fucoxanthin, and phycoerythrin. Note that possession of either phycoerythrin or fucoxanthin in addition to chlorophyll *a* expands the spectrum of light that can be absorbed.

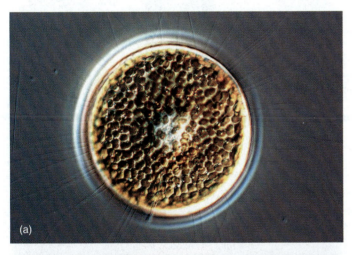

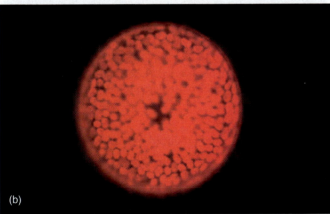

Figure 12.24 Micrographs of a diatom, a type of phytoplankton. The left image (a) was obtained by illuminating the cell with white light. The right image (b) was obtained by illuminating the cell with high-intensity blue light and the resulting red fluorescence from the chlorophyll *a* was photographed. In these types of cells, chlorophyll *a* is located within small structures known as chloroplasts.

phytoplankton consume some of their organic matter through respiration because all organisms respire, regardless of whether they are able to breathe or not. As shown in the equation below, respiration produces chemical energy and generates carbon dioxide from organic carbon. Autotrophs respire the organic carbon obtained from fixation of carbon dioxide. Heterotrophs respire the organic carbon obtained from consumption of organic matter, commonly in the form of whole organisms (review section 11.2). Respiration is represented by the net overall equation

$$1\ C_6H_{12}O_6\ +\ 6\ O_2\ \longrightarrow\ 6\ CO_2\ +\ 6\ H_2O\ +\ \text{chemical energy}$$

| 1 molecule sugar | + | 6 molecules oxygen | | 6 molecules carbon dioxide | + | 6 molecules water | + | chemical energy |

Net primary production is the gain in organic matter from photosynthesis by phytoplankton minus the reduction in organic matter due to respiration by phytoplankton. Net primary production reflects the gain in phytoplankton biomass that is available for consumption by heterotrophs. In marine ecosystems, the amount of net primary production determines the amount of food available for large organisms such as fish. Primary productivity is typically expressed in units of carbon because this is the element that organic substances are based upon. It is also expressed in units of time because productivity is the rate at which phytoplankton biomass is produced due to photosynthesis. It can be expressed relative to area (per m^2) or relative to volume (per m^3). The amount of net primary productivity in coastal waters can be about 10 g C/m^2/day; in surface waters of the open ocean such as the North Pacific, for example, net primary productivity is commonly less than 1 g C/m^2/day. Note that, although the terms "primary productivity" and "primary production" seem similar, they are not the same thing. "Primary productivity" refers to a rate and "primary production" refers to an amount.

QUICK REVIEW

1. Explain the difference between gross and net production.
2. Why do phytoplankton have pigments other than chlorophyll *a*?
3. Compare photosynthesis and respiration. How are they related?
4. Explain why photoautotrophs produce fluorescence.

12.7 Measuring Primary Productivity

Almost all photosynthetic organisms on Earth use chlorophyll *a* as a catalyst to drive photosynthesis. If you review the overall reactions from the previous section, you will notice that oxygen production is directly related to the amount of carbon dioxide consumed. These relationships provide several straightforward ways to measure primary production. First, if we are interested in net primary production, one can determine the amount of biomass at a given time, and repeat the measurement some time later. Any change in biomass (positive or negative) is related to the change in primary production. Since all photosynthetic organisms use chlorophyll, we can, in theory, measure the amount of chlorophyll in some volume or region of the ocean at two different time points to estimate primary production. In practice, with the exception of satellite data (where we can "see" most of the surface ocean in a very short period of time), this is impractical because the ocean is constantly moving (unlike, for example, a forest) and it is nearly impossible to measure the same region or volume of water.

The most common methods oceanographers use to measure primary production instead involve keeping track of the amount of oxygen produced or carbon dioxide consumed. To measure primary productivity at a given site, we can collect bottles of water at different depths and either add the radioactive isotope carbon-14 (fig. 12.25) or measure the change in oxygen over time (fig. 12.26). Carbon-14 acts as a tracer that is incorporated into primary producers proportionally with the amount of carbon dioxide consumed, whereas oxygen is produced in proportion to carbon dioxide consumption. After filling the bottles and either adding carbon-14 or measuring the oxygen concentration, the bottles are held for some period of time (usually twenty-four hours) at the same light and temperature from which the water was collected. At the end of the twenty-four-hour period, either the carbon-14 incorporated into the organic matter or the oxygen produced (or consumed) is measured.

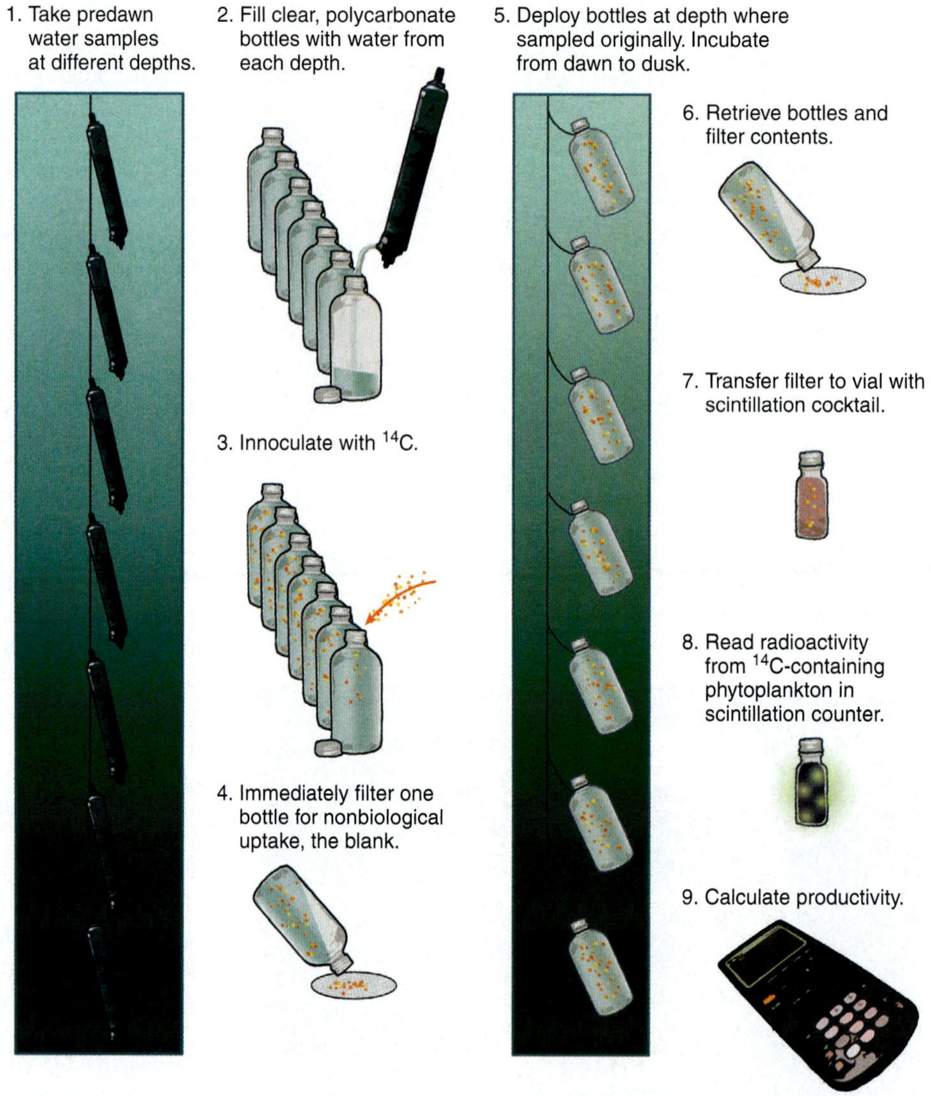

By using matched pairs of light and dark bottles, we can separate primary production (which requires sunlight) and respiration (which is independent of the amount of light). The carbon-14 or oxygen data from the light bottles represent net production because respiration has consumed some fraction of the oxygen and resulted in respiration of organic carbon back to carbon dioxide. In the dark bottles, only respiration can occur, so by comparing the two sets of data we can calculate the gross (total), and net production, and respiration.

The amount of carbon dioxide taken up and the amount of oxygen produced during photosynthesis reflect a chemical balance between starting material (carbon dioxide) and product (oxygen). A similar relationship exists between the amounts of nitrogen and phosphorus removed from the water and the

1. Take predawn water samples at different depths.

2. Fill clear, polycarbonate bottles with water from each depth.

3. Innoculate with ^{14}C.

4. Immediately filter one bottle for nonbiological uptake, the blank.

5. Deploy bottles at depth where sampled originally. Incubate from dawn to dusk.

6. Retrieve bottles and filter contents.

7. Transfer filter to vial with scintillation cocktail.

8. Read radioactivity from ^{14}C-containing phytoplankton in scintillation counter.

9. Calculate productivity.

$$\text{Carbon uptake} = \frac{^{14}\text{C in photoplankton on filter} \times \text{available carbon} \times 1.05 \text{ (fractionation factor)}}{\text{Total } ^{14}\text{C added}}$$

Figure 12.25 The carbon-14 method for determination of primary productivity. Water is collected from different depths, and carbon-14 is added as a tracer. After some period of time (typically twenty-four hours or less, to avoid running out of tracer), the sample is filtered, and the filter is counted to determine the amount of carbon-14 incorporated into organic matter. Unlike the oxygen method, dark bottles are not typically used. The carbon-14 method is extremely sensitive, allowing it to be used in even the most oligotrophic waters.

amount of organic matter produced. When all nutrients required by phytoplankton are abundant and the phytoplankton are growing optimally, the fixed ratios of the different elements by weight are (on average)

$$O_2 : C : N : P = 109 : 41 : 7.2 : 1$$

This is known as the **Redfield ratio** after the researcher who first reported the observation. It is often defined as molar quantities of 106:16:1 (C:N:P). These ratios can be used to estimate primary productivity by measuring either the rate of nutrient uptake by phytoplankton or the rate of oxygen production. If the

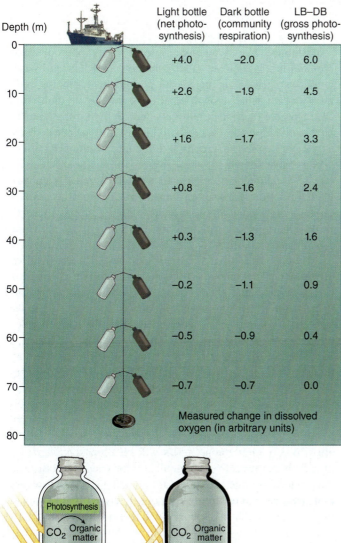

Depth (m)	Light bottle (net photosynthesis)	Dark bottle (community respiration)	LB–DB (gross photosynthesis)
0	+4.0	−2.0	6.0
10	+2.6	−1.9	4.5
20	+1.6	−1.7	3.3
30	+0.8	−1.6	2.4
40	+0.3	−1.3	1.6
50	−0.2	−1.1	0.9
60	−0.5	−0.9	0.4
70	−0.7	−0.7	0.0
80			

Measured change in dissolved oxygen (in arbitrary units)

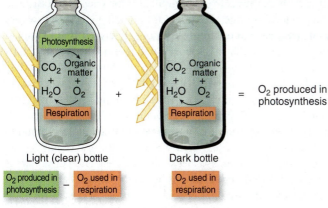

Light (clear) bottle $\quad$ Dark bottle

$$O_2 \text{ produced in photosynthesis} - O_2 \text{ used in respiration} \qquad O_2 \text{ used in respiration}$$

Figure 12.26 The light-dark bottle method of determination of primary productivity using oxygen. The decrease in oxygen in the *dark bottle* (DB) equates to respiration of the phytoplankton, bacteria, and other organisms. The change in oxygen in the *light bottle* (LB) represents the net difference between rates of photosynthesis and respiration.

rate of delivery of nitrogen (N) or phosphorus into a region by upwelling and currents and the rate at which these compounds are removed by other currents are known, then the rate at which the nutrient is incorporated into phytoplankton biomass can be calculated and primary productivity can be determined. This approach allows oceanographers to estimate primary productivity over large areas of the ocean and to associate it with large-scale water movement and chemical cycles.

QUICK REVIEW

1. Why do oceanographers use the carbon-14 method more often than the oxygen bottle method?

2. Explain how we can convert to another element (such as N or P) given an estimate of primary productivity in carbon units.

3. Why do the light bottles in the oxygen method measure only photoautotrophic oxygen production, but the dark bottles measure community respiration?

12.8 Phytoplankton Biomass

The total biomass of the phytoplankton community at any instant in time is referred to as the phytoplankton **standing stock,** and is a product of two main processes. Growth and reproduction increase phytoplankton numbers and are dependent upon net primary productivity. Death and grazing of phytoplankton reduce phytoplankton numbers.

There are a number of techniques available to determine phytoplankton biomass in a given water sample. One option is to count all the phytoplankton cells in a sample and multiply the number counted by the average amount of carbon per individual cell. Another option relies on the fact that all phytoplankton possess chlorophyll *a*. Therefore, the concentration of chlorophyll *a* in a water sample is an estimate of the amount of photosynthetic biomass in the water sample. Chlorophyll *a* concentrations at different depths in the water column are commonly measured using two different approaches. The most direct, but also the most labor-intensive, way is to filter the phytoplankton cells from a volume of water, extract the chlorophyll *a* from all the cells trapped on the filter, and determine the concentration of chlorophyll *a* in the extract. An indirect, but more rapid method relies on the fact that, when relatively strong intensities of blue light are absorbed by chlorophyll *a*, some absorbed light energy is released as fluorescence (see fig. 12.24). Higher concentrations of chlorophyll *a* in a body of water result in higher levels of chlorophyll *a* fluorescence. In addition to detecting chlorophyll *a* fluorescence with a microscope, fluorescence can also be measured electronically using an instrument known as a fluorometer.

Some satellites can measure fluorescence directly, but it is much more common to detect **ocean color**. The satellite takes an image of (or "sees") the sea surface, and measures the wavelength, or color, of the reflected light. Since all photosynthesizing phytoplankton in the ocean contain chlorophyll, the more phytoplankton in the ocean, the more the color shifts because chlorophyll strongly absorbs blue and red light. The green light is reflected (because it is not absorbed), so the more chlorophyll, the greener the reflected light. This is the same reason that the leaves on trees typically look green. If there are little or no phytoplankton, then the water looks very blue, whereas if there is a lot of sediment or other materials, the water may look brown or reddish-brown. Ocean color satellites take advantage of these color shifts to estimate how much chlorophyll, or biomass, there is in the near-surface ocean (fig. 12.27).

(a)

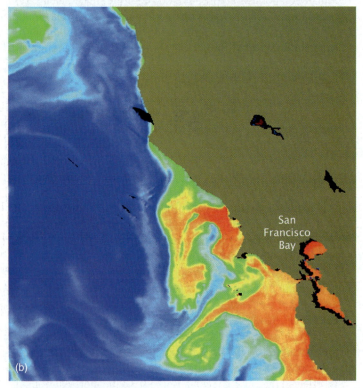

San Francisco Bay

(b)

Figure 12.27 Satellites can estimate the amount of chlorophyll in the surface ocean by measuring changes in ocean color. MODIS true-color (a) imagery off San Fransisco Bay, California shows turbid, muddy water in the bay with a plume of sediments extending into the coastal ocean. Away from shore, the water appears a deep blue. Using ocean color from the same image, chlorophyll (b) can be estimated; offshore the chlorophyll concentrations are lower (*blues*) while nearshore there is a phytoplankton bloom (*yellows* and *reds*) being mixed by the local currents.

Blooms of phytoplankton occur when phytoplankton biomass increases more rapidly than can be balanced by loss processes such as sinking, death and grazing. Blooms commonly occur when growth conditions for the phytoplankton are favorable (sufficient nutrients and sunlight, for example), but zooplankton numbers are still low. Blooms can be detected as a rapid increase in chlorophyll *a* concentrations at a given site or a given depth.

The relation between primary production rates and phytoplankton biomass can be examined by dividing the primary productivity values by the biomass values. Biomass normalized primary production rates are given in units of mg C/mg phytoplankton biomass/day.

Based on satellite images of chlorophyll *a* distributions across our planet, about 1000 times more carbon is stored in photosynthetic land-plant biomass at any given instant in time than is stored in phytoplankton biomass. In other words, the standing stock of plants is much greater than the standing stock of phytoplankton, yet the annual rate of net primary production in the ocean is comparable to the annual rate of net primary production on land. How can that be? Remember that primary productivity is the *rate* at which phytoplankton (or plants) are consuming carbon dioxide and producing oxygen, but biomass is the standing stock of carbon. If the productivity rate is high but loss processes (death, sinking, being eaten by other organisms) are also very high, the net result will be a low standing stock of carbon that turns over very quickly (the oceans). If, instead, the productivity rate is high but the standing stock is also very high, then the carbon turns over more slowly (the land; think of the amount of carbon associated with the nonleafy green parts of trees and shrubs). When we consider how fast the standing stock is replaced, we see that even the open ocean, where there is not much chlorophyll *a*, is incredibly vibrant. On average, the standing stock of phytoplankton biomass is completely replaced every week. In contrast, land plants are completely replaced approximately every ten years. So even in the low-biomass open ocean, the flow of energy between producers and consumers is very fast! A defining feature of life in the oceans is that carbon dioxide quickly becomes organic matter that serves as food.

QUICK REVIEW

1. Define biomass and explain how it relates to primary production.
2. Describe the processes that might allow a phytoplankton bloom to occur.
3. Describe the rates at which biomass is replaced on land versus in the ocean. Based on your answer, are the open ocean gyres devoid of life?

12.9 Controls on Productivity and Biomass

Phytoplankton primary productivity is controlled by temperature, light levels, and inorganic nutrient concentrations. The combination of these factors determines the maximum productivity possible in a given region of the ocean. The impact of these factors on phytoplankton biomass is referred to as **bottom-up control**. These factors determine how quickly phytoplankton grow and reproduce. Grazing by heterotrophic consumers, in addition to temperature, nutrients, and light, determines the maximum amount of biomass possible in a given region of the ocean. The influence of grazing on phytoplankton biomass is known as **top-down control**. The interactions between bottom-up and top-down controls determine phytoplankton standing stock. To illustrate how nutrients, light, and grazing impact the biomass and productivity of phytoplankton, three regions of the world's oceans will be considered: polar, temperate, and tropical regions.

Summer days are long in polar regions. Relative to lower latitudes, however, the intensity of light in polar regions is low, and the depth of light penetration is shallow. There is no sunlight for a significant portion of the year at the highest latitudes. Most of the Arctic is frozen throughout the winter and only begins to thaw in the spring. By summer, the days lengthen and light intensities reach a maximum. Much of the ice and snow melts, and nutrients are released into the waters. During this period of adequate sunlight and nutrients, the phytoplankton bloom and create high biomass. Growth of zooplankton lags behind growth of phytoplankton (fig. 12.28c). In late summer/early fall, temperatures and light intensities decrease rapidly, zooplankton abundance is at a maximum, and sea ice begins to fill the open water. Phytoplankton abundance during this period declines

dramatically. At these latitudes, the availability of light controls phytoplankton growth. Even though the burst of primary productivity associated with the phytoplankton bloom is short-lived, the high amount of organic matter generated during this period supports numerous larger organisms, including seals, whales, and polar bears.

The situation in the subtropics and tropics (fig. 12.28b) is very different from that at the poles. Abundant high-intensity sunlight is available year-round in the tropics, but nutrient concentrations are rarely high enough to support high phytoplankton biomass. The water column in the tropics is density stratified due to surface warming. The phytoplankton remain within the stratified surface waters because they are not easily displaced into the underlying denser water. The same density stratification that keeps the phytoplankton in the surface waters means that the upwelling and mixing processes for renewing nutrient supply to surface waters are weak and localized. This poor nutrient supply to the euphotic zone limits phytoplankton production in the tropics and subtropics despite high levels of sunlight. Any increases in phytoplankton biomass are quickly consumed by the zooplankton.

The situation in temperate (fig. 12.28a) regions is more complicated than in polar regions or the tropics. In temperate regions, the intensity and duration of sunlight, as well as nutrient concentrations, vary with season. During the winter, solar irradiance is low, winter storms are relatively frequent, and the

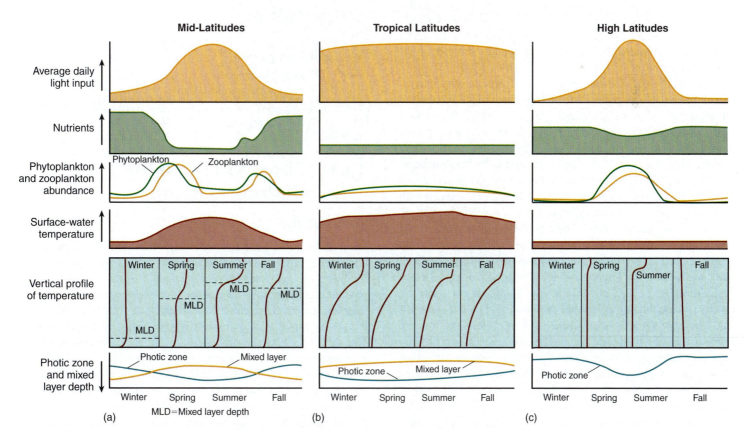

Figure 12.28 Vertical distribution of physical, chemical, and biological properties during the seasonal cycle in (a) temperate, (b) tropical, and (c) polar oceans.

water column is generally well mixed. Temperate surface waters in winter are characterized by high concentrations of inorganic nutrients and low levels of solar energy. Because phytoplankton require both light and nutrients for growth, phytoplankton biomass is low during the winter months of temperate regions (fig. 12.28). During spring, solar radiation increases and warms the surface waters, which increases the density stability of the water column. Just as described above for tropical waters, density stratification in temperate waters also helps to maintain the phytoplankton in the surface waters. During spring in these waters, there is sufficient nutrients and sufficient sunlight for photosynthesis in the surface waters. Under these conditions, the phytoplankton bloom, and biomass increases dramatically. Phytoplankton bloom during the spring in temperate waters because they are able to grow faster than the zooplankton grazers in the water column can consume them. The downturn in the biomass curve during the later months of spring happens for two reasons. First, the amount of nutrients in the stratified surface waters declines as the phytoplankton use the nutrients for their growth. The density stratification also minimizes renewal of nutrients to the surface waters. Nutrient limitation slows the growth of the phytoplankton. As phytoplankton growth slows, the zooplankton grazers begin to "catch up," and their feeding causes a decline in phytoplankton biomass. Towards the end of summer, solar radiation begins to decline, surface waters cool, density stratification begins to break down, and nutrients are renewed in surface waters. A second, smaller bloom often occurs in the fall. Ultimately the fall bloom declines as winter sets in and light levels decrease. In addition, enhanced winter mixing means that the phytoplankton are mixed deep in the water column where they are unable to obtain sufficient light for photosynthesis.

In addition to the seasonal cycle of productivity, phytoplankton are also strongly controlled by the vertical distribution of light within the water column. Vertical mixing—the upward and downward movement of seawater—carries the phytoplankton up and down into higher and lower light. If the mixed layer is shallow, the phytoplankton are always exposed to more sunlight. As the mixed layer deepens, the phytoplankton can be carried for some period of time to deeper depths where there is not enough sunlight to support photosynthesis. If we ignore mixing for the moment, then at some depth, the amount of oxygen production from photosynthesis will equal the oxygen consumption from respiration. This is called the **compensation depth**, or the depth below which phytoplankton cannot survive for long periods of time (because of lack of sunlight; fig. 12.29). If we allow the water to mix, the phytoplankton can be carried below that critical depth for some period of time, as long as they spend more time in the well-lit surface waters. On average, they are producing more oxygen than respiration is consuming. If the mixed layer keeps deepening, at some depth the phytoplankton spend more time in the dark than in the light; this is called the **critical depth**. If mixing is deeper than the critical depth, phytoplankton cannot **bloom**, or rapidly increase in biomass. This mixing helps to explain the difference in productivity patterns. At high latitiudes, mixing can be very deep and phytoplankton are typi-

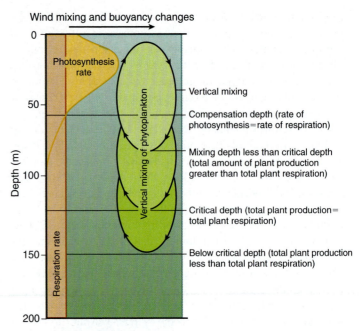

Figure 12.29 Vertical mixing, the compensation depth, and the critical depth.

cally light-limited. At low latitudes, mixing is reduced because of thermal stratification, and phytoplankton are primarily limited by the lack of nutrients.

QUICK REVIEW

1. How does light limit productivity as a function of depth, latitude, and season?
2. Explain what role nutrients play in controlling primary production of polar, mid-latitude, and tropical waters.
3. Describe the difference between the compensation depth and the critical depth.

12.10 Food Webs and the Biological Pump

Food Webs

Food webs describe the flow of nutrients and food between different groups of organisms. The majority of the organisms in any food web are less than a few centimeters in size. Photosynthetic phytoplankton form the base of the food web because they generate organic matter available for consumption by other organisms. A pyramid is the simplest way to visualize the flow of organic matter between organisms (fig. 12.30). Organic matter is generated at the base of the pyramid by the primary producers. A variety of zooplankton directly consume the phytoplankton, and they are referred to as herbivorous zooplankton by analogy to plant-eating animals on land. The zooplankton that consume the phytoplankton are known as primary consumers. The herbivorous zooplankton are in turn grazed upon by carnivorous zooplankton. The zooplankton that feed upon the herbivores are

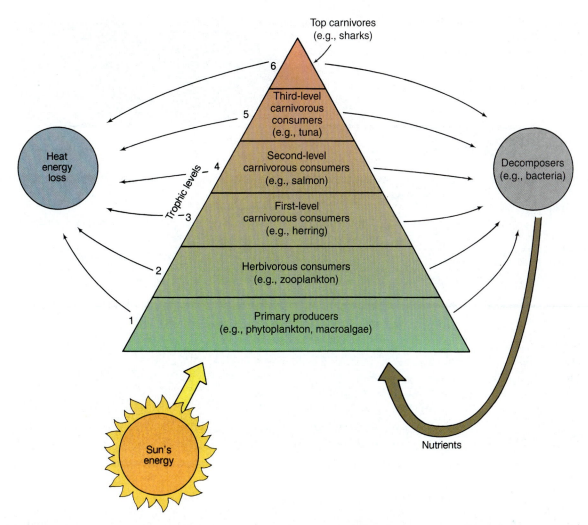

Figure 12.30 A trophic pyramid. Trophic levels are numbered from *base* to *top*. The first trophic level requires nutrients and energy. Nutrients are recycled at each level; energy is lost as heat at each level. Heat is lost from the ocean by radiation to a cooler atmosphere.

known as secondary consumers. The flow of energy and food between different groups of organisms in the pyramid construction is fairly direct: smaller organisms are consumed by bigger organisms, which in turn are consumed by even bigger organisms. The primary producers, primary consumers, secondary consumers, and so on represent different steps in the transfer of carbon and nutrients. Each step represents a different **trophic level.** Each transfer between trophic levels results in a loss of organic carbon and energy. Therefore, less total biomass is supported at each trophic level.

This pyramid-shaped transfer of energy and organic material from primary producers to apex predators can be described as a **food chain**, where energy is linearly transferred from one link in the chain (trophic level) to the next. Because some energy is lost at each step, the net result is that it takes a lot of primary producers to support a few apex organisms, or top predators (forming our pyramid). Typically, 10% is transferred from one trophic level to the next; this is called the **trophic efficiency**.

A great deal of research over the years indicates that the flow of nutrients and food energy between different groups

of organisms is actually more complicated than is depicted by a simple pyramid. All food webs reflect complex interactions between different groups of organisms. But some food webs are more complex than others. For example, the food web that supports herring is complicated because prey-predation levels change as the fish matures (fig. 12.31). As will be described in the following paragraphs, food webs in the open ocean are more like loops, since much of the material is "recycled" between the different groups, with little new input of nutrients into the ecosystem. These types of food webs are dominated by microbial processes and do not support many higher trophic levels. To illustrate different influences on food webs, the world's oceans will be divided into two general categories: high-nutrient input into surface waters as occurs in coastal waters and low-nutrient input into surface waters as occurs in the gyres of the open oceans.

The vast majority of the world's oceans, such as the open-ocean gyres, are low-nutrient regions with low rates of primary productivity (table 12.2). Phytoplankton in these low-nutrient regions tend to be very small. The size of a typical open-ocean phytoplankton is about 1 micron. The size of a coastal phytoplankton

Figure 12.31 Simplified schematic illustrating the complexity of marine food webs. This food web is for herring at various stages in the herring's life.

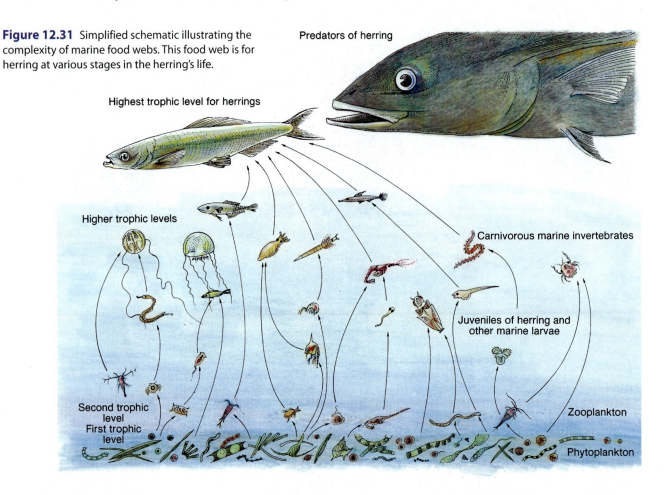

Table 12.2   Gross Primary Productivity of Land and Ocean

Ocean Area	Range (gC/m²/yr)	Average (gC/m²/yr)	Land Area	Amount (gC/m²/yr)
Open ocean	50–160	130 ± 35	Deserts, grasslands	50
Coastal ocean	100–500	300 ± 40	Forests, common crops, pastures	25–150
Estuaries	200–500	300 ± 100	Rain forests, moist crops, intensive agriculture	150–500
Upwelling zones	300–800	640 ± 150	Sugarcane and sorghum	500–1500
Salt marshes	1000–4000	2471		

can be almost 500 microns. Under the low-nutrient conditions of the open-ocean, phytoplankton benefit from being small because their greater surface area allows them to take up more nutrients per cell volume. Smaller phytoplankton tend to be consumed by smaller zooplankton, and these smaller zooplankton in turn tend to be consumed by only slightly larger zooplankton. Numerous trophic transfers, with a loss of carbon and nutrients at each step, are required in open oceans to generate large heterotrophic organisms (table 12.3). Much of the organic and inorganic matter is instead recycled between the different small organisms. The open-ocean environment is characterized by a tight loop of nutrient regeneration and nutrient uptake dominated by microbes. This type of food web is referred to as a **microbial loop.**

Biological Pump

The "biological pump" refers to the composite of food web processes that result in a draw-down, or transfer, of carbon dioxide from the atmosphere to the ocean due to phytoplankton photosynthesis and the generation of organic carbon that can be exported from surface waters to fuel life in the ocean. The way this works is that, as phytoplankton take up dissolved carbon dioxide for photosynthesis, more carbon dioxide from the atmosphere dissolves into the seawater. The organic matter generated from the phytoplankton is transferred through the food web, and a portion of the organic matter sinks through the water column to serve as food for other organisms. As the organic matter sinks to depth, the carbon dioxide that was used to originally generate the organic

Table 12.3 Oceanic Food Production

Area	Phytoplankton Production (metric tons of carbon/yr)	Efficiency of Mass and Energy Transfer per Trophic Level	Trophic Level Harvested	Estimated Fish Production (metric tons of carbon/yr)
Open ocean	40.3×10^9	10%	5	4.0×10^6
Coastal regions	10.8×10^9	15%	3	243×10^6
Upwelling areas	1.8×10^9	20%	2	360×10^6

matter is drawn-down to depth as well. A more efficient biological pump provides more organic compounds to deep waters than an inefficient pump. A large proportion of the organic matter generated by large phytoplankton can be exported to deeper waters. The nutrient-poor waters of the open ocean tend to support an inefficient biological pump that is dominated by smaller phytoplankton; most of the organic matter and nutrients are recycled in the surface waters. Less of the organic matter generated by phytoplankton in the open ocean is exported to deeper waters.

Marine Bacteria and Nutrients

In any food web, some fraction of the organic material and energy is not transferred to the next higher trophic level. Some of this organic material will be exported to the deep ocean via the biological pump to form sediments, but the vast majority of the material is broken down by a process called **remineralization**, mediated by heterotrophic bacteria (fig. 12.32). As the organic material sinks through the ocean, bacteria will colonize it to use the organics as a source of energy. Phytoplankton remove inorganic nutrients from seawater in a fairly constant ratio (the Redfield Ratio) if growth conditions are optimal. When the organisms decompose, the nutrients are released back to seawater in this same ratio, leading to **nutrient regeneration**.

It takes many different types of microorganisms to carry out this process, each adapted to consume a particular type of material. For example, inorganic nitrogen in seawater is typically found in several common forms: as nitrate (NO_3), nitrite (NO_2), ammonium (NH_4), and nitrogen gas (N_2). Nitrogen gas is generally not biologically available, except to specialized organisms called nitrogen fixers (see chapter 11). As organic matter decomposes, NH_4 is often produced. This is then oxidized to NO_2, which in turn is oxidized to NO_3. Each step (and there are many other possible reactions) requires a specialized type of microbe. Other bacteria can reverse the process, converting NO_3 back into N_2 gas through a series of intermediate compounds; this occurs most commonly in regions of the ocean called low-oxygen zones. The end result is the formation of large pools of inorganic nutrients (primarily NO_3 for nitrogen) in the deep ocean with proportional decreases in oxygen from respiration and remineralization, and increases in carbon dioxide. The microbial loop (or web) thus completes the circuit between photosynthesis in the surface ocean and the biogeochemical cycling of organic and inorganic materials throughout the ocean's depths.

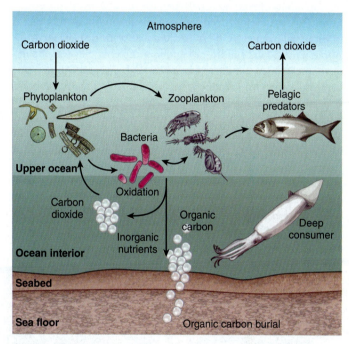

Figure 12.32 Organic matter produced in the surface ocean is exported to depth via the biological pump. Only a small fraction makes it to the deep-ocean sediment. The rest is consumed and respired by other organisms or is remineralized by bacteria, converting the organics back into carbon dioxide and inorganic nutrients.

QUICK REVIEW

1. Compare a food chain and a food web.
2. Explain how upwelling ecosystems are capable of supporting the largest animal in the world, the blue whale, in terms of trophic efficiency.
3. Sketch a trophic pyramid of four levels. Label primary producers, herbivores, and carnivores. How are energy and nutrients related to these trophic levels?
4. If we assume a trophic efficiency of 10% (in other words, 10% of the energy is passed from one trophic level to the next), how many kilograms of phytoplankton would it take to support 1 kilogram of herbivorous zooplankton? How many kilograms of phytoplankton would it take to support 1 kilogram of tuna, if the tuna were three trophic levels higher than the primary producers?
5. Assuming 10% trophic efficiency, what happens to the other 90% of the energy or biomass?

12.11 Global Patterns of Productivity

Given the links among photosynthesis, food webs, and the microbial loop, global patterns of productivity are more easily understood. The vast majority of the open ocean exhibits low productivity due to nutrient limitation, typically due to insufficient nitrogen or iron. The concept of **iron limitation** was championed by John Martin (1935–1993), who argued that **high nitrate, low chlorophyll *a* (HNLC) regions** of the equatorial Pacific, subarctic Pacific, and Southern Ocean exhibited low phytoplankton standing stock because of low iron concentrations. Phytoplankton need very small quantities of this metal to grow, and typically there is plenty of iron in seawater from Aeolian dust, land runoff, or benthic sediments. Several experiments have been conducted in these HNLC regions to demonstrate that when iron is added, phytoplankton (typically diatoms) will bloom, turning these low-productivity oceanic regions into biological "hotspots," at least until the iron is consumed (fig. 12.33). There has been a great deal of scientific and public interest in these experiments because iron fertilization could increase transfer of carbon dioxide to the deep ocean by increasing the efficiency of the biological pump. Oceanographers have raised concerns about the feasibility of iron fertilization on scales large enough to have a global impact, and have questioned the negative impacts to the marine food webs that might result

from such fertilization efforts. To date, there have been thirteen large-scale scientific iron fertilization experiments and two commercial trials. It is still unclear whether iron fertilization will result in long-term enhancement of organic carbon to the deep ocean. The utility and consequences of such global oceanic manipulations are still a hot topic of debate, and much remains to be answered about this form of geo-engineering.

When there is sufficient sunlight, regions that receive higher nutrient input display higher levels of net primary productivity (fig. 12.34). Coastal areas are generally much more productive than the open ocean. Rivers and land runoff supply nutrients to coastal waters and estuaries. Fresh water from rivers can also stabilize the water column by creating a low-density (less saline) surface layer that helps to maintain the phytoplankton in well-illuminated surface waters.

Narrow regions of particularly high productivity are found in the major upwelling zones along the west coasts of North and South America, the west coast of Africa, and the west side of the Indian Ocean (review chapter 8 on upwelling). The upwelled deep water brings nutrient-rich water to the surface, and when sufficient sunlight is available, phytoplankton blooms occur. The abundant populations of phytoplankton in these regions form a first step in the food webs that result in large numbers of commercially valuable fish.

The Equatorial Pacific demonstrates the influence of open-ocean upwelling on primary productivity. The equatorial divergence is where upwelling occurs, bringing nutrient-rich waters to the surface. Phytoplankton abundance is much higher at the equatorial divergence than at other open-ocean areas that do not experience sustained upwelling. The same holds true for the surface divergence that surrounds Antarctica. In contrast, surface convergence results in downwelling and can create vast areas of low-nutrient availability such as the open-ocean gyres.

On average, upwelling regions are almost twice as productive as coastal regions and about six times as productive as the open ocean (table 12.2). The total area covered by the open ocean is about ninety times greater than the area covered by upwelling regions and about nine times greater than the area covered by coastal regions (table 12.3). If we combine the two parameters (productivity and size), it becomes clear that most of the world's oceans are characterized by relatively low levels of productivity (table 12.2). Despite the low levels of productivity per unit area, the size of the ocean is so large—the ocean covers about 70% of Earth's surface—that about half of annual primary productivity on the planet is carried out by the microscopic phytoplankton.

High-productivity regions and low-productivity regions can be found both on land and in the oceans (table 12.2). For example, primary productivity per square meter in the open ocean is about the same as that of deserts on land. Upwelling regions are comparable to pastureland and lush forestland, and certain estuaries are comparable to the productivity of the most heavily cultivated land.

The highly productive coastal and upwelling regions result in efficient trophic transfer and enhanced fluxes of

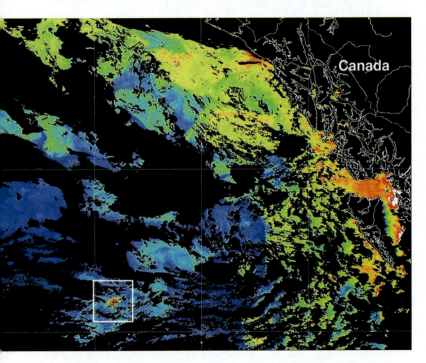

Figure 12.33 A SeaWiFS image of chlorophyll *a* distributions on day twenty of an iron fertilization experiment in the North Pacific. The warm colors (*reds* and *yellows*) indicate regions of high chlorophyll *a* concentrations, the cool colors (*blues*) indicate regions of low chlorophyll *a* concentrations. *Black areas* over the ocean indicate cloud cover, which prevents the satellite from detecting ocean color. The area highlighted with the *white box* is the site of the high chlorophyll *a* concentrations resulting from iron fertilization.

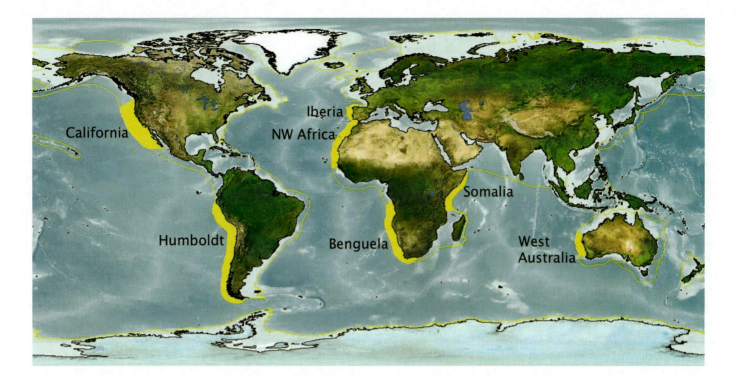

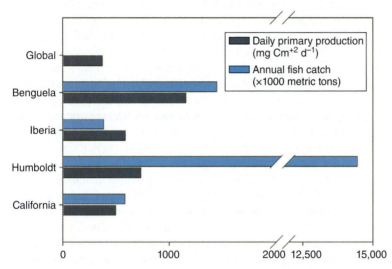

Figure 12.34 High-productivity regions of the world's oceans are associated with the major upwelling centers (*top panel*), shown in *yellow*. The coastal boundaries indicate large marine ecosystems (see chapter 11). These very productive coastal upwelling systems account for less than 5% of the surface area of the ocean, but more than 25% of the annual fish catch (*lower panel*).

organic material via the biological pump. In contrast, the vast open ocean regions, like the deserts and grasslands, are dominated by picoplankton that are efficiently grazed by microzooplankton such as ciliates and tintinnids (see fig. 12.18), leading to enhanced recycling within the microbial loop in the surface ocean. This, in turn, results in lower trophic transfer (fewer large predators) and decreased efficiency of the biological pump. Although these open ocean waters may seem less biologically rich than the coast, keep in mind that, on average, the entire standing stock of phytoplankton is replaced every week. The vast size of the open ocean means that even these low-productivity regions are critical components of the ocean's (and Earth's) biogeochemical cycles (fig. 12.35).

QUICK REVIEW

1. Describe the large-scale spatial patterns of biomass and productivity.

2. Compare the oceans to land in terms of productivity. Where is the highest productivity? The lowest?

3. How do primary productivity and fish catch relate to one another?

4. Explain why it was initially surprising that scientists found that large areas of the ocean are characterized by high macronutrient concentrations such as nitrogen and phosphorus, but low concentrations of chlorophyll.

5. Explain why the equatorial Pacific has high levels of chlorophyll and primary productivity, but the subtropical Pacific does not.

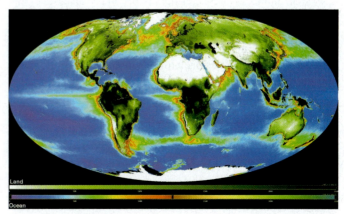

Figure 12.35 Global net primary production for the year 2006.

Summary

Plankton are the drifting organisms. The microscopic plankton are divided into groups by size. Phytoplankton are the dominant photosynthetic autotrophs in the sea. These organisms, together with marine plants and macroalgae, use sunlight and inorganic compounds to generate the organic matter that serves as food for life in the sea. *Sargassum* is the only commonly observed planktonic macroalgae.

The diatoms are found in cold, upwelled water; centric diatom forms are round, whereas pinnate diatoms are elongate. Diatoms reproduce rapidly by cell division and make up the first trophic level in marine ecosystems. Diatoms use silica to form a hard, transparent frustule. Dinoflagellates are single cells that can be both autotrophic and heterotrophic. Their cell walls are smooth or heavily armored with cellulose plates. Coccolitho-phorids and silicoflagellates are small autotrophic members of the phytoplankton. Cyanobacteria and green algae are important members of the picophytoplankton communities. Some cyano-bacteria can use nitrogen gas as a source of inorganic nitrogen and are important players in the global marine nitrogen cycle. Blooms of dinoflagellates and some other kinds of phytoplankton can produce harmful algal blooms (HABs) and ecologically disruptive algal blooms. Some, but not all, HABs are toxic, and red tides may or may not be HABs.

Carnivorous zooplankton are important in the recycling of nutrients to the phytoplankton. Many zooplankton migrate toward the sea surface at night and away from it during the day, forming the deep scattering layer. Although zooplankton can be strong swimmers, they are controlled by horizontal currents and still belong to the plankton. Zooplankton that spend their entire lives in the water column are called holoplankton. Copepods and euphausiids (krill) are the most common members of the holoplankton. Other holoplankton include chaetognaths (arrow-worms), foraminifera, radiolarians, tintinnids, and pteropods. Large zooplankton include the comb jellies, salps, and jellyfish. The meroplankton are the juvenile (or larval) stages of non-planktonic adults. This group includes fish eggs, very young fish, and the larvae of barnacles, snails, crabs, sea stars, and many other nonplanktonic animals.

Great numbers of bacterioplankton and viruses live free in seawater and coat every surface. They are important as food sources for other organisms and also mediate decay and break-down of organic matter. The microbial loop mediates the con-version of organic matter back to carbon dioxide and inorganic nutrients, remineralizing the majority of organic matter that leaves the surface ocean as part of the biological pump.

The process of generating organic carbon from carbon dioxide is referred to as carbon fixation, or primary production. Oxygen is formed as a by-product of photosynthesis. Respiration breaks down organic matter to extract chemical energy, and results in consumption of oxygen and production of carbon dioxide. Gross primary productivity is the total amount of organic produced by photosynthesis per volume of seawater per unit time. Net primary production is the gain of organic matter from photosynthesis minus the reduction of organic carbon from respiration from all the organisms in a volume of seawater. Common ways to measure primary productivity include measuring oxygen production and consumption, or by using the radio-tracer carbon-14.

Phytoplankton blooms occur when phytoplankton reproduce more rapidly than they are consumed by heterotrophs. Standing stock is the total phytoplankton biomas present at a given site at a given instant of time, and is often related to chlorophyll concentrations, since only photosynthetic autotrophs possess chlorophyll. Phytoplankton growth rate (primary productivity) is dependent on temperature, sunlight, and inorganic nutrients. At polar latitudes, the availability of light controls productivity; in the tropics, sunlight is available year-round, but thermal strati-fication of the water column limits the availability of nutrients from depth. At temperate latitudes, light, nutrients, and water column stability vary over the seasons, typically leading to phyto-plankton blooms in the spring and fall. Phytoplankton are also limited by vertical mixing on shorter time scales. Blooms can-not develop if the water column mixes below the critical depth.

Food webs describe the interconnections among different types of organisms within an ecosystem. The phytoplankton are primary producers. In the open ocean where nutrients are limit-ing, the dominant phytoplankton are small. In nutrient-rich wa-ters such as upwelling zones and coastal regions, the dominant phytoplankton are large. The organisms that consume primary producers are primary consumers; the organisms consuming the primary consumers are secondary consumers, and so on. Food chains describe linear relationships between producers and consumers. Trophic pyramids can be used to represent these relationships in terms of organic matter and nutrients. Food webs are similar to food chains, but take into account the complex interactions between trophic levels.

Global patterns of productivity largely reflect the avail-ability of light and nutrients. The open ocean is dominated by small phytoplankton, with many trophic steps leading to apex consumers, resulting in efficient recycling and a reduction in the efficiency of the biological pump. Coastal waters and upwell-ing systems in particular are dominated by large phytoplankton, leading to efficient trophic transfer and enhanced fisheries. Large parts of the ocean exhibit high-nutrient, low chlorophyll (HNLC) conditions, and are limited by the availability of iron. Several decades' worth of large-scale iron addition experiments have demonstrated that adding iron can stimulate phytoplankton blooms, but it is not clear whether this also leads to enhanced efficiency of the biological pump.

Key Terms

sea grasses, 308
cordgrasses, 308
halophytes, 308
mangroves, 308
macroalgae, 308
phytoplankton, 308
zooplankton, 308
bacterioplankton, 308
picoplankton, 309
nanoplankton, 309
microplankton/net
    plankton, 309
macroplankton, 309
diatom, 309
dinoflagellate, 309
coccolithophorid, 309
cyanobacteria, 309
green algae, 309
frustule, 309

radial symmetry, 310
bilateral symmetry, 310
centric, 310
pennate, 310
harmful algal bloom
    (HAB), 313
red tide, 313
ecologically disruptive algal
    blooms (EDABs), 314
holoplankton, 315
meroplankton, 315
crustacean, 315
copepod, 315
euphausiid, 315
krill, 315
baleen, 315
chaetognath, 316
foraminafera, 316
radiolarian, 316

tintinnid, 317
cilia, 317
pteropod, 317
ctenophore, 317
salp, 317
Coelenterata/Cnidaria, 318
colonial organism, 318
larva, 318
spore, 319
microbial loop, 319
virus, 320
pigments, 322
chlorophyll *a*, 322
extremophile, 323
carbon fixation, 324
fluorescence, 324
primary production, 324
gross primary production, 324
net primary production, 325

Redfield ratio, 326
standing stock, 327
ocean color, 327
bloom, 328, 330
bottom-up control, 329
top-down control, 329
compensation depth, 330
critical depth, 330
food webs, 330
trophic level, 331
food chain, 331
trophic efficiency, 331
microbial loop, 332
remineralization, 333
nutrient regeneration, 333
iron limitation, 334
high nitrate low chlorophyll *a*
    (HNLC) regions, 334

Study Problems

1. In the oceans, size and abundance are generally inversely related (in other words, the smaller the organism, the more abundant). Graph the size versus abundance of the following organisms:

   a. marine viruses (average size $<$ 1 micron, average abundance 1×10^7/ml)

   b. bacteria (average size ~0.5 micron, average abundance 1×10^6/ml)

   c. small phytoplankton (average size 10 microns, average abundance 1×10^4/ml)

   d. small zooplankton (average size 1–10 millimeters, average abundance 1000/liter)

   Is it easier to plot this on a linear or a logarithmic graph? Based on your results, how does this relate to the abundance of really large organisms like whales?

2. Refer to figure 12.26. Plot the net photosynthesis and respiration rates as a function of depth. At approximately what depth is the compensation depth, based on your data? Connect the points for net photosynthesis versus depth, and for respiration versus depth to create two lines. Are they similar or different? Why?

3. The Redfield ratio describes typical proportions of elements used by biological organisms in the ocean. If you measured a concentration of 1 μM phosphate in the surface ocean, how much nitrate would you expect to find? If nitrogen was limiting in the region where you sampled, would you expect the N:P ratio to be higher or lower than the Redfield ratio?

4. If there are 10,000 units of energy converted to carbon by primary producers, how much of that energy (as carbon) would you expect to find in the herbivores? How much would remain if the apex predator were at trophic level 5?

5. Using table 12.3, determine what percentage of total primary productivity is accounted for by upwelling systems. Now convert the productivity associated with upwelling from carbon to nitrogen and phosphorus, assuming Redfield ratio proportions. Note that the values in table 12.3 are in weight units, not molar units.

CHAPTER

13

The Nekton: Swimmers of the Sea

Learning Outcomes

After studying the information in this chapter students should be able to:

1. *explain* the concept of keystone predators and *describe* how they impact species diversity within an ecosystem,

2. *explain* the impacts of top-down and bottom-up controls on marine nekton,

3. *compare* and *contrast* how cartilaginous and bony fishes have adapted to the marine environment,

4. *describe* similarities and differences in the major groups of marine nekton, and

5. *describe* how threats to marine mammals have changed over the past century, and *explain* the roles that marine mammals play in marine ecosystems.

School of blue stripe sea perch or snapper.

The nekton are the free swimmers of the oceans. Thousands of species of nekton swim freely through the pelagic and neritic regions of the oceans. Nekton are fishes, marine mammals, ocean-living reptiles, and squid. Members of the nekton are able to move toward their food and away from their predators; many occupy the top trophic levels of marine food webs. Sizes range from the smallest fish of the tropical reefs to the largest animal ever to have existed on Earth, the blue whale. This chapter examines representative swimmers in coastal waters and in the open sea; marine birds are included in this chapter.

13.1 The Nekton Defined

The majority of animals living in the ocean spend their lives suspended in seawater; Chapter 12 discussed the plankton— those (mostly small) organisms capable of vertical migration but at the mercy of horizontal currents. The **nekton** include approximately 5000 species—**invertebrates** and **vertebrates**—capable of swimming freely through the neritic and pelagic environments. The invertebrate group is quite small, including some large shrimp, crabs, and the mollusks. Fish have been phenomenally successful and dominate the nekton. And, while very popular, marine reptiles, birds, and mammals are also a small but very visible percentage of the nekton.

All nekton must solve similar issues. They need to swim through the water, must avoid sinking, and must deal with the physical reality of their environment including temperature, salinity, light, and pressure (see chapter 11 for review). Additionally, they must be able to find food, avoid predators, and reproduce. Animals of the nekton have adapted in many different ways to solve these issues, but we will find some common solutions. Many unrelated organisms have become streamlined, for example, to reduce drag as they swim through the water. Through programs such as the *Census of Marine Life* and the **Tagging of Pacific Predators (TOPP)**, we are also learning that many of the nekton migrate over great distances to make a living (fig. 13.1). Some, particularly reptiles, birds, and mammals, spend part of their lives on land; to be considered "marine" they must spend at least 50% of their lives in the marine environment.

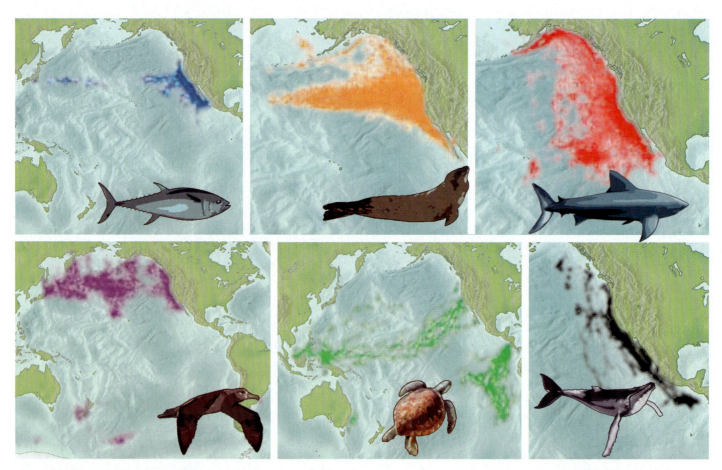

Figure 13.1 The Tagging of Pacific Predators project studies the movement of marine nekton. *Top*, left to right: (yellowfin, bluefin, albacore); pinnipeds (northern elephant seals, California sea lions, northern fur seals); sharks (salmon, white, blue, common thresher, mako). *Bottom*, left to right: seabirds (Laysan and black-footed albatrosses, sooty shearwaters); sea turtles (leatherback, loggerhead); cetaceans (blue, fin, sperm and humpback whales). Color shading indicates the daily mean position of electronically tagged animals from 2000–2009.

QUICK REVIEW

1. Explain the difference between nekton and plankton.
2. Why do we classify some animals as "marine" when they spend part of their lives on land?
3. Describe how technology has changed our understanding of marine nekton lifestyles.

Figure 13.2 The pelagic tuna crab is an important food source for fish (especially tuna), turtles, birds, and mammals. These crabs feed on small zooplankton and diatoms.

13.2 Swimming Marine Invertebrates

This group is best known for including squid, but it also includes other cephalopods, large shrimp, and some crabs such as the pelagic tuna crab (*Pleuroncodes planipes*; fig. 13.2). Squid are abundant in the world's oceans. Until recently, little was known about many of the 300 species of squid, particularly those inhabiting the ocean depths. They are elusive, swim rapidly, and are known to not only float with neutral buoyancy but also to rest on the bottom. Their wide range of bioluminescence (see chapter 11) and coloration allows them to change color and disappear rapidly.

Squid have eight short arms and two thin, long tentacles (fig. 13.3); the tentacles have a number of suckers at the ends, and in some species, bioluminescent lures to attract prey. They range from a few centimeters in length to the giant squid (*Architeuthis*) and colossal squid (*Mesonychoteuthis*), which can reach sizes of at least 15 m. Some, such as the Humboldt squid (*Dosidicus gigas*) are aggressive predators that feed on other members of the nekton. In contrast, the colossal squid, despite its large size, is thought to be quite sluggish and slow, conserving energy and feeding by ambushing prey. The large eyes of giant and colossal squid, which are typically 27 cm (11 in) across, help them to identify and avoid equally large predators such as the sperm whale (*Physeter macrocephalus*) within the nearly dark waters of the mesopelagic. It is thought that these very large eyes—the largest of any animal—allow the squid to detect at a great distance the bioluminescence produced by smaller organisms disturbed by the passage of the sperm whale.

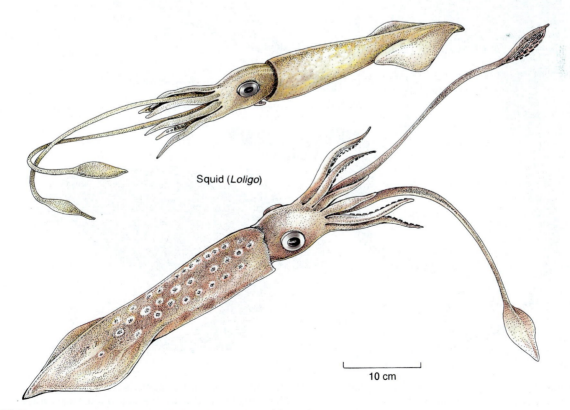

Squid (*Loligo*)

10 cm

Figure 13.3 The squid is a swimming mollusk and a member of the nekton.

QUICK REVIEW

1. Why are jellyfish considered plankton rather than nekton?
2. Describe some of the invertebrates that are part of the nekton.
3. What is the purpose of the large eyes found in giant and colossal squid?

13.3 Marine Reptiles

Although many land reptiles visit the shore to feed, mainly on crabs and shellfish, few reptiles are found in today's seas. Examples of marine reptiles are shown in figure 13.4. All reptiles are poikilotherms, which means their metabolism rates vary with temperature (see chapter 11 on temperature). The only modern marine lizard is the big, gregarious marine iguana of the Galápagos Islands. It lives along the shore and dives into the water at low tide to feed on the algae. This iguana has evolved a flattened tail for swimming and has strong legs with large claws for climbing back up on the cliffs. It regulates its buoyancy by expelling air, allowing it to remain underwater. The large monitor lizards of the Indian Ocean islands, known as the Komodo dragons, are capable of swimming but do so only under duress.

Alligators may enter shallow shore water, and crocodiles are known to go to sea. The estuarine crocodile of Asia is found in the coastal waters of India, Sri Lanka, Malaysia, and Australia. The Indian gavial and false gavial are slender-nosed, fish-eating crocodilians found in nearshore waters.

Marine iguana

Gavial

1 m

Sea snake

Green turtle

¼ m

Figure 13.4 Marine reptiles. The marine iguana is the only modern marine lizard; it inhabits the Galápagos Islands. The gavial is a fish-eating crocodilian from India. Sea snakes are venomous; they are found only in the Pacific and Indian Oceans. The green turtle feeds on sea grasses and seaweeds in tropical coastal waters.

Sea Snakes

About fifty kinds of sea snakes are found in the warm waters of the Pacific and Indian Oceans; there are no sea snakes in the Atlantic Ocean. Sea snakes are extremely poisonous; they have small mouths, flattened tails for swimming, and nostrils on the upper surface of the snout that can be closed when the snakes are submerged. The sea snakes' skin is nearly impervious to salt, but it is permeable to gases, passing nitrogen gas as well as carbon dioxide and oxygen. The snake is able to lose nitrogen gas to the water, allowing it to dive as deep as 100 m (328 ft), stay submerged as long as two hours, and surface rapidly with no decompression problems. Sea snakes eat fish, and most reproduce at sea by giving birth to live young.

Sea Turtles

Sea turtles can be distinguished from land turtles because sea turtles can not retract their heads into their shells and because their front limbs have been modified into flippers for swimming. Male and female turtles mate offshore and then the females return to shore to lay their eggs; females may return to shore several times during breeding season, each time laying as many as 100 eggs that are buried in the sand. Six of seven species of sea turtle are found in U.S. and Caribbean waters, including the green (*Chelonia mydas*), hawksbill (*Eretmochelys imbricate*), loggerhead (*Caretta caretta*), Kemp's ridley (*Lepidochelys kempii*), olive ridley (*L. olivacea*), and leatherback turtle (*Dermochelys coriacea*). The flatback sea turtle (*Natator depressus*) lives exclusively in Australian coastal waters.

Green turtles (fig. 13.4) are generally found in tropical and subtropical waters between 30°N and 30°S. Once they reach sexual maturity, females begin returning to the beaches where they were born. Green turtles occupy three different types of habitats. Nesting occurs on high-energy beaches, juveniles congregate in open-ocean convergence zones, and the adults occupy shallow coastal areas. They are primarily herbivorous, feeding mainly on sea grass and algae, although the juveniles can also feed on small animals. Major nesting colonies in the United States are on the east coast of Florida, in the U.S. Virgin Islands, and in Puerto Rico. Hawksbill turtles are found throughout the tropical oceans. Their name refers to their prominent hooked beak. Major nesting colonies in the United States are in Puerto Rico, the Virgin Islands, Florida, and Hawaii. As with green turtles, hawksbill juveniles are found in oceanic convergence zones, where they feed on algae. As they mature, they feed primarily on sponges found around coral reefs or rocky outcroppings. Loggerhead sea turtles are distributed throughout subtropical and temperate waters, primarily on continental shelves and in estuaries. Loggerhead hatchlings move offshore to convergence zones where *Sargassum* is located. After a few years at sea, they move into continental shelf regions to feed on crabs and shellfish. Nesting is concentrated in the temperate zone and subtropics. In U.S. waters, nesting sites are found along the Atlantic coast of Florida to North Carolina. The Kemp's ridley turtle is found primarily in the Gulf of Mexico. They nest in large aggregations in Rancho Nuevo, on the northeastern coast of Mexico. The hatchlings also

congregate on *Sargassum*. Adults appear to feed on crabs and mollusks. The olive ridley lives in tropical waters of the Pacific, Atlantic, and Indian Oceans as well as the Caribbean Sea. While olive ridley turtles are the most abundant, their population is still considered endangered because of few remaining nesting sites. They prefer relatively shallow coastal waters, where they feed on jellyfish, shrimp, crabs, shellfish, and other invertebrates.

The leatherback sea turtle is distinctive from other sea turtles. It is the largest of all sea turtles, with an average weight of about 500 kg (0.5 ton); the largest known leatherback was 3 m from beak tip to tail and weighed 1 ton. These turtles also do not have a hard shell as do other sea turtles. Instead they have a shell of distinctive, thick, leathery skin. The leatherback is almost entirely pelagic and can be found from the tropics to the poles. Adults can migrate across entire ocean basins. Nesting occurs only in the tropics, which is the main time that these animals enter coastal waters. They feed primarily on jellyfish, which they suck into their mouths with a large amount of water; specialized spines in the throat allow the leatherback to expel the water but retain the jellyfish.

Sea turtle population numbers are being dramatically reduced due to pollution, habitat degradation, and the demand for turtle products. Turtle eggs and turtle meat are prized throughout the Pacific, and turtle nests are raided by poachers. Sea turtle eggs incubate for up to a few months, during which time they are vulnerable to harvesting by humans, dogs, rats, or other carnivores. In addition, sea turtles reach sexual maturity at a relatively late age, between eight and fifty years. All of the turtles just discussed are on the U.S. endangered species list. The Kemp's ridley sea turtle is the most endangered of all, primarily because the females nest in large groups at the single known nesting site in Rancho Nuevo and are therefore particularly vulnerable to disruption. Several steps have been taken to eliminate or reduce threats to turtles. Turtle excluder devices (TEDs) are a grid of bars with an opening in shrimp trawl nets; small animals such as shrimp pass through the bars to be captured in the net. Large animals such as turtles and sharks hit the bars and are ejected. Fishery closures are also routinely employed to minimize accidental capture of turtles in specific regions, and there are several international agreements protecting the highly migratory turtle populations as part of global conservation efforts.

QUICK REVIEW

1. Describe some adaptations of marine snakes and iguanas that separate them from their relatives on land.
2. Why are marine turtles threatened by extinction?
3. How are marine snakes able to dive so deeply without getting "the bends"?

13.4 Marine Birds

About 3% of the estimated 9000–10,000 species of birds are marine species. Seabirds are birds that spend a significant portion of their lives at sea. Some are so well adapted to oceanic life that they come ashore only to reproduce; others move into coastal waters to feed; but all return to shore to nest. Most have definite breeding sites and seasons, and they may migrate thousands of kilometers as they travel from feeding ground to breeding area. Birds are homeotherms, which means that they can maintain a constant body temperature (see chapter 11 for a discussion of temperature). In general, seabirds have life histories characterized by low numbers of offspring, delayed maturity, and longer life spans than terrestrial birds. Many seabirds breed on islands, cliffs, or headlands in colonies as large as a million or more birds. This provides protection from land predators. In most seabird colonies there are multiple species nesting together, with some **niche separation** (selection of different nesting locations, such as ledges versus burrows) between species.

Seabirds swim at the sea surface and underwater; they use their webbed feet, their wings, or a combination of wings and feet. They float by using fat deposits in combination with light bones and air sacs developed for flight. Most seabirds have feathers waterproofed by an oily secretion called preen, and the air trapped under their feathers helps keep the birds afloat, insulates their bodies, and prevents heat loss. When diving, the birds reduce their buoyancy by exhaling the air from their lungs and air sacs and pulling in their feathers close to their bodies to squeeze out the trapped air. The underwater swimmers such as cormorants and penguins have thicker, solid bones, and penguins have no air sacs.

The eyesight of seabirds is highly developed. Their senses of hearing and smell are less developed, and their least developed sense is taste; they have few taste buds and swallow quickly. They drink seawater or obtain water from their food. Excess salt is concentrated by a gland over each eye, and a salty solution drips down or is blown out through their nasal passages. To conserve water, the birds reduce and concentrate their urine, forming uric acid, a nearly nontoxic white paste that is mixed with feces and excreted. Because of their great activity and high metabolic rate, birds are huge eaters. Seabirds feed on fish, squid, krill, egg masses, and bottom invertebrates, as well as on carrion and garbage. They are most plentiful where food is abundant, and their presence is a good indicator of high productivity in surface water.

There are four general related groups, or orders, of seabirds: (1) albatrosses, petrels, fulmars, and shearwaters; (2) penguins; (3) pelicans, cormorants, boobies, and frigate birds; and (4) gulls, terns, and alcids. The wandering albatross (*Diomedea exulans*) of the southern oceans is the most truly oceanic of marine birds with the largest wing spread of all birds, 3.5 m (11 ft). These great white birds with black-tipped wings spend four to five years at sea before returning to their nesting sites. The smaller, North Pacific, black-footed albatross (*Phoebastria nigripes*) is a pelagic scavenger, searching the sea surface for edible refuse, including scraps from ships and fishing boats. The smallest of the oceanic birds, Wilson's petrel (*Oceanites oceanicus*), is a swallowlike bird that breeds in Antarctica and flies 16,000 km (10,000 mi) along the Gulf Stream to Labrador during the Southern Hemisphere winter, returning to the Southern Hemisphere for the austral summer, another 16,000 km.

Penguins (fig. 13.5*a*) cannot fly. Their wings have been modified into flippers that allow them to swim underwater, using their feet for steering. All but one species of penguin live in the Southern Hemisphere, primarily in Antarctic and Subantarctic waters. They are adapted to life in the cold, with a thick layer of fat beneath their feathers. Penguins establish breeding colonies that for the Adélie penguins (*Pygoscelis adeliae*) may contain up to a million mating pairs. Emperor penguins (*Aptenodytes forsteri*) huddle together to keep warm; the male incubates the single laid egg on his feet against his body for over two months while the female returns to sea to hunt. Chicks are born in the spring when the plankton blooms. The Galápagos penguin (*Spheniscus mendiculus*) lives along the equator, far from Antarctica. This penguin is restricted to regions where cold water upwells.

Pelicans and cormorants are large fishing birds with big beaks. They are strong fliers found mostly in coastal areas, but some venture far out to sea. A pelican (fig. 13.5*b*) has a particularly large beak from which hangs a pouch used in catching fish. White pelicans of North America (*Pelecanus erythrorhynchos*) fish in groups, herding small schools of fish into shallow water and then scooping them up in their large pouches. The Pacific's brown pelican (*Pelecanus occidentalis*) does a spectacular dive from up to 10 m (30 ft) above the water to capture its prey. Cormorants are black, long-bodied birds with snakelike necks and moderately long bills that are hooked at the tip. They settle on the water and dive from the surface, swimming primarily with their feet but also using their wings. Because they do not have the water-repellent feathers of other seabirds, cormorants must return to land periodically to dry out.

Terns (fig. 13.5*c*) and gulls are members of the fourth group of seabirds and are found all over the world except in the South Pacific between South America and Australia. Gulls are strong flyers and will eat anything; they forage over beach and open water. The terns are small, graceful birds with slender bills and forked tails. The Arctic tern (*Sterna paradiseae*) breeds in the Arctic and each winter migrates south of the Antarctic Circle, a round trip of 35,000 km (20,000 mi).

Puffins, murres, and auks are heavy-bodied, short-winged, short-legged diving birds. They feed on fish, crustaceans, squid, and some krill. All are limited to the North Atlantic, North Pacific, and Arctic areas, where most nest on isolated cliffs and islands.

Shorebirds differ from seabirds in that they do not swim much and thus have a greater dependency on land. They include sandpipers, plovers, stilts, avocets, snipes, oystercatchers, turnstones, and phalaropes (fig. 13.5*d*). Many migrate long distances between winter feeding grounds and spring nesting grounds. For example, the semipalmated sandpiper (*Calidris pusilla*) flies about 2000 miles without stopping from the North Atlantic coast to Suriname in South America. The red knot (*Calidris canutus*) flies from the Arctic Circle to Tierra del Fuego at the tip of South America. Shorebirds commonly feed on small crustaceans, clams, and snails using a variety of styles: running and stabbing consists of capturing prey by stabbing it; chiseling and hammering describe how bivalves are opened; and pecking and probing are used to find prey beneath the mud or sand surface—the length of the bill determines what can be gathered as food. For example, the great blue herons (*Ardea herodias*) and snowy egrets (*Egretta thula*) are waders; their long necks enable them to strike at small fish in the water.

(a)

(b)

(d)

QUICK REVIEW

1. About what fraction (percentage) of birds are considered marine?
2. Describe some adaptations of the albatross and the penguin. How do they differ? How are they similar?
3. How are marine birds are able to drink seawater without becoming dehydrated?
4. What advantages do large breeding colonies on islands and headlands provide to birds?

13.5 Fish

Fish dominate the nekton. Fish are found at all depths and in all the oceans. Their distribution patterns are determined directly or indirectly by their dependency on the ocean's primary producers. Fish are concentrated in upwelling areas, shallow coastal areas, and estuaries. The surface waters support much greater populations per unit of water volume than the deeper zones, where food resources are sparser.

(c)

(e)

Figure 13.5 (a) These Magellanic penguins are part of a 500,000-bird colony at Punta Tombo, Argentina. (b) A brown pelican takes off from the water. (c) Crested terns face the wind on the beach of Heron Island, a part of the Great Barrier Reef, Australia. (d) A mixed flock of dunlins and sanderlings take flight along a sandy beach. (e) Kittiwakes (gull-like birds) and common murres nesting on an island in Kachemak Bay, Alaska.

There are three major groups of fish: jawless fishes, which include hagfish and lampreys; cartilaginous fishes, which include sharks, rays, skates, and ratfish; and bony fishes, which include the great majority of fishes. Fish come in a wide variety of shapes related to their environment and behavior. Some are streamlined, designed to move rapidly through the water (tuna and marlin); some are laterally compressed and swim more slowly around reefs and shorelines (snapper and tropical butterfly fish), others are bottom-dwelling fish that are flattened for life on the sea floor (sole and halibut); while still others are elongated for living in soft sediments and under rocks (some eels). Fins provide the push or thrust for locomotion and occur in a variety of shapes and sizes. Fins are used to change direction, turn, balance, and brake. Flying fish use their fins to glide above the sea surface; mudskippers and sculpins walk on their fins. A sample of the variety of marine fishes is shown in figure 13.6.

Schooling has evolved independently in the majority of fish species; almost 80% of the more than 30,000 freshwater and marine fish species exhibit this behavior at some point during their life cycle, including predatory fish such as tuna and sharks. Schools may consist of a few fish in a small area, or they may cover several square kilometers; for example, herring in the North Sea have been seen in schools 15 km long and 5 km wide (9 × 3 mi). Why has schooling evolved so many times? There are several benefits. First, it reduces individual encounters with predators and may also confuse predators, making it more difficult to target individuals. Schooling may also enhance foraging success by making it more difficult for prey to evade the school. Fish may also save energy when swimming together, similar to

(d)

(e)

Figure 13.6 The bony fishes dominate the world's aquatic environments. Their diversity is enormous, and they are found in almost every conceivable aquatic environment. Two of the thousands of species of reef fish: an emperor angel (a) and a butterflyfish (b). The wolf eel (c) is a carnivore with strong jaws and long teeth. The scales of the seahorse (d) are modified to form protective armor. Seahorse populations are declining, and their slow reproduction rate is not likely to support increasing harvests for aquariums, traditional medicine, and gourmet dishes. The red snapper (e) inhabits rocky coastal areas; it is used as a food fish.

(a)

(b)

(c)

drafting used by automobile drivers and bicyclists. Schooling also presumably increases reproduction success by maintaining a population in a constrained region. Usually fish in a school are of the same species and are similar in size. Schooling fish keep their relationships to one another constant as the school moves and changes direction. The patterns and movements of these schools result in **emergent behavior**, the development of higher levels of organization (the school) without intentional communication or leadership by the individual fish within the school.

Jawless Fish

The jawless fish include hagfish and lampreys. They have a sucking mouth with teeth or dental plates, and lack paired fins and stomachs. They are evolutionarily the most primitive of fishes. They are predators, scavengers, and parasites. Whereas hagfish are exclusively marine, lampreys spawn in fresh water before returning to the sea and can tolerate low salinity. Although there are many species of lamprey that are not parasitic, the accidental introduction of sea lampreys to the Great Lakes via shipping canals has resulted in significant damage to native fish populations, which the lampreys parasitize. The hagfish are perhaps best known for producing copious amounts of slime (they are sometimes called *slime eels*), and for being the only vertebrate that is iso-osmotic (same salinity as the surrounding seawater). Despite the poor reputation of lampreys and hagfish, they serve an important ecological role. Hagfish, for example, are consumed by pinnipeds, seabirds, and crustaceans.

Sharks and Rays

Sharks, skates, and rays belong to the class **Chondrichthyes**, or cartilagenous fish. The shark is an ancient fish; it predates the mammals, having first appeared in Earth's oceans 450 million years ago. Unlike most fishes, most sharks bear live young. They also differ from other fish by their skeletons of cartilage rather than bone and by their toothlike scales. Shark scales have a covering of dentine similar to that on the teeth of vertebrates, and they are extremely abrasive; sharkskin has been used as a sandpaper and polishing material. The shark's teeth are modified scales; they are replaced rapidly if lost, and they occur in as many as seven overlapping rows. Sharks are actively aware of their environment through good eyesight; excellent senses of smell, hearing, and mechanical reception; and electrical sense.

Sharks see well under dim-light conditions. They have the ability to sense chemicals in their environment through smell, taste, a general chemical sense, and unique pit-organs distributed over their bodies. These pit-organs contain clusters of sensory cells resembling taste buds. The shark's sense of smell is acute; it has a pair of nasal sacs located in front of its mouth, and when water flows into the sacs as the shark swims, it passes over a series of thin folds with many receptor cells. Sharks are most sensitive to chemicals associated with their feeding, and all are able to detect such chemicals in amounts as dilute as one part per billion.

Receptors along the shark's sides are sensitive to touch, vibration, currents, sound, and pressure. The movement of water from currents or from an injured or distressed fish are sensed by the shark's lines of receptors, which communicate with the watery environment by a series of tubes; water displacement stimulates nerve impulses along these systems. The shark uses its hearing in the location of its prey. Pores in the shark's skin, especially around the head and mouth, are sensitive to small electrical fields. Fish and other small marine organisms produce electrical fields around themselves, and the shark uses its electroreception sense to locate prey and recognize food. As a shark swims through Earth's magnetic field, an electrical field is produced that varies with direction, giving the shark its own compass.

More than 460 species of shark are known, and scientists are still discovering new species. Sharks are widely spread through the oceans and are found in rivers more than a hundred miles from the sea. Some of these sharks are shown in figure 13.7. The whale shark (*Rhicodon typus*) is the world's largest fish, reaching lengths of more than 15 m (50 ft). This shark filters plankton from the water and is harmless to other fish and mammals. The basking shark (*Cetorhinus maximus*), 5–12 m (15–40 ft) long, is another plankton feeder. It is found commonly off the California coast and in the North Atlantic, where it has been harvested for its oil-rich liver. A third species of plankton-feeding shark, 4 m (14 ft) long and weighing approximately 680 kg (1500 lbs), nicknamed megamouth (*Megachasma pelagios*), was discovered in 1976. This shark is hypothesized to migrate vertically through the water column over the day/night cycle, spending the day in deep waters (~150 m) and at shallower depths (~15 m) at night. This pattern of migrating to shallow depths at night is also seen with the deep scattering layer of zooplankton and small fish.

Many sharks are swift and active predators, attacking quickly and efficiently, using their rows of serrated teeth to remove massive amounts of tissue or whole limbs and body portions. They also play an important role as scavengers and, like wolves and the large cats on land, eliminate the diseased and aged animals. Sharks can and do attack humans, although the reasons for these attacks and the periodic frenzied feeding observed in groups of sharks are not understood. A human swimming inefficiently at the surface may look like a struggling, ailing animal and be attacked; a diver swimming completely submerged may appear as a more natural part of the environment and be ignored. Worldwide, there are only fifty to seventy confirmed shark attacks and five to fifteen fatalities annually.

Skates and rays (fig. 13.7) are flattened, sharklike fish that live near the sea floor; there are some 450–550 species of skates and rays. They move by undulating their large side fins, which gives them the appearance of flying through the water. Their five pairs of gill slits are on the underside of their bodies, not along their sides. The large manta rays are plankton feeders, but most rays and skates are carnivorous, eating fish but preferring crustaceans, mollusks, and other benthic organisms. Their tails are usually thin and whiplike, and in the case of stingrays, carry a poisonous barb at the base. Some skates and a few rays have shock-producing electric organs that can deliver shocks of up to 200 volts; these are located along the side of the tail in skates and on the wings of the rays. Their purpose appears to be mainly defensive. Like the sharks, most rays bear their young live. Skates

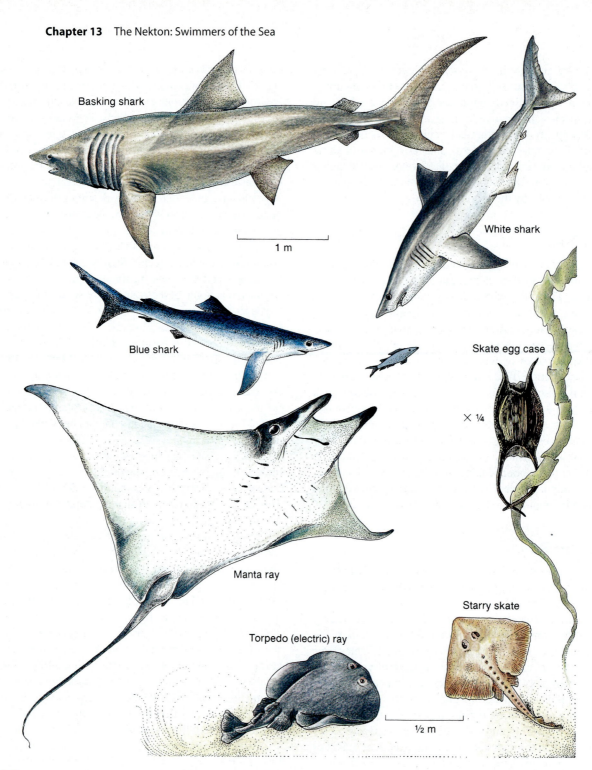

Basking shark

1 m

White shark

Blue shark

Skate egg case

× ¼

Manta ray

Starry skate

Torpedo (electric) ray

½ m

Figure 13.7 Sharks, skates, and rays are all cartilaginous fishes. The leathery egg case of a skate is known as a mermaid's purse.

enclose the fertilized eggs in a leathery capsule called a sea purse or mermaid's purse that is ejected into the sea and from which the young emerge in a few months.

Bony Fish

There are more than 16,000 species of marine fishes, with the vast majority belonging to the class **Osteichthyes** (bony fish). More than three-quarters of the fish in the ocean live in coastal waters. Given the incredible diversity of bony fishes, it is perhaps not surprising that this group displays an incredible array of adaptations to various environments. Most fish are ectotherms, but there are exceptions; both the tuna and the mackerel shark family (which includes great white, mako, and salmon sharks) are capable of maintaining warmer core body temperatures than the surrounding seawater, making them endothermic, or warm-blooded. They accomplish this by minimizing heat loss through counter-current heat exchange (see chapter 12) and heat generated from muscular activity.

All bony fish possess gills (fig. 13.8) that allow exchange of oxygen and carbon dioxide with seawater (jawless fish have gill-like structures called gill pouches; sharks, skates, and rays have gills, but not gill covers). Most bony fish also possess a swim bladder to maintain buoyancy. Cartilaginous fish lack a swim bladder and must use the pectoral fins to provide lift, somewhat like the wings of an airplane (sharks) or bird wings (skates and rays). The swim bladder also provides the bony fish with extraordinary maneuverability; they can hover in the water or even swim backward, something cartilaginous fishes cannot do.

Bony fish can be categorized based on body shape. The fast swimmers such as tuna and marlin have streamlined bodies that allow them to move quickly through water; fish such as angelfishes have laterally compressed bodies that allow them to maneuver around coral reefs or through kelp beds; fish such as flounders and halibuts are flat and adapted to living on the bottom—they lie on one side with both eyes on top. Fish that live on or near the bottom are known as **demersal fish.**

Most commercially valuable food fish are bony fish (fig. 13.9) and are found between the surface and depths of about 200 m (600 ft). Among the most important fisheries are those for the enormously abundant small, herring-type fish, such as sardine, anchovy, menhaden, and herring. These fish feed directly on plankton and are components of short, efficient food webs (see chapter 12). They are found in large schools and areas of high primary productivity such as upwelling zones.

Deep-Sea Species of Bony Fish

The fish of the deep sea are not well known because the depths below the epipelagic zone are difficult and expensive to sample. Representative fish are shown in figure 13.10.

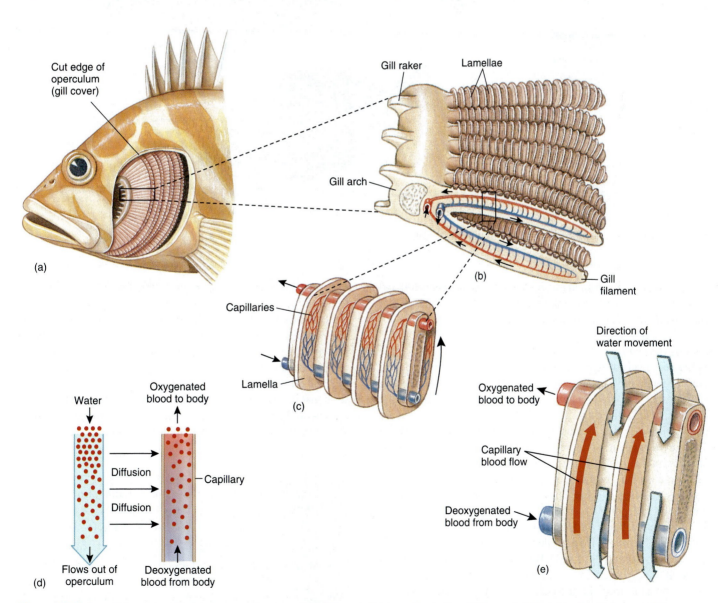

Figure 13.8 Bony fishes have gills (a) that contain rows of gill filaments (b). Lamellae in the gill filaments (c) increase the surface area for gas exchange. Oxygen from seawater diffuses in, while carbon dioxide from fish respiration diffuses out (d), and the concentration of oxygen (indicated by dots) is always higher in seawater (e) because of the reverse flow of blood and water.

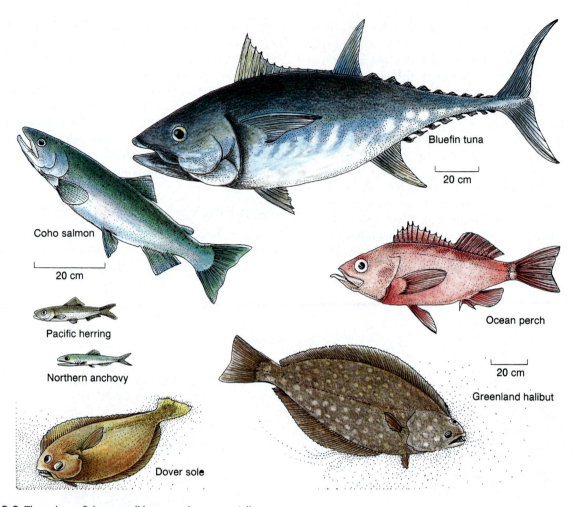

Figure 13.9 These bony fishes are all harvested commercially.

In the dim, transitional mesopelagic layer, between 200 and 1000 m (660 and 3300 ft), the waters support vast schools of small, luminous fish. Fishes of the mesopelagic zone are small, with large mouths, hinged jaws, and needlelike teeth. The bristle-mouths (*Cyclothone*) are believed to be the most common fish in the sea. Each of its species lives at a relatively fixed depth; the deeper-living species are black, and the shallower-living species are silvery to blend with the dim light. At this depth, the lanternfish has a worldwide distribution; some 200 species are distinguished by the pattern of light organs along their sides. They are a major item in the diet of tuna, squid, and porpoises. *Stomias,* a fish with a huge mouth, long, pointed teeth, and light organs along its sides, and the large-eyed hatchetfish prey on the great clouds of euphausiids and copepods found at the mesopelagic depth.

In the perpetually dark bathypelagic zone, there is little food; only about 5% of the food produced in the photic zone is avail-able at these depths (fig. 13.11). Here, the fish are small, between 2 and 10 cm (1 and 5 in), and have no spines or scales but are fierce and monstrous in appearance. They are mostly black in color with small eyes, huge mouths, and expandable stomachs. They breathe slowly, and the tissues of their small bodies have a high water and low protein content. Floating at constant depth without using energy for swimming, these fish go for long periods between feedings by using their food for energy rather than for increased tissue production. Most have bioluminescent organs or photophores on their undersides that allow them to blend with any light from above. These photophores may show different patterns among species and between sexes; they may also identify preda-tors and are used as lures to attract prey. Some species have large teeth that incline back toward the gullet so the prey cannot escape, and others have gaping mouths with jaws that unhinge to allow the catching and eating of fish larger than themselves.

QUICK REVIEW

1. Describe how sharks, skates, and rays differ from other fish.

2. At what trophic level do the largest sharks feed?

3. Explain why not all sharks are considered carnivores.

5. Why is schooling advantageous for fish? Describe the concept of "emergent behavior."

6. Describe the primary limiting factor(s) for deep-sea fishes and some adaptations that allow them to survive in the deep sea.

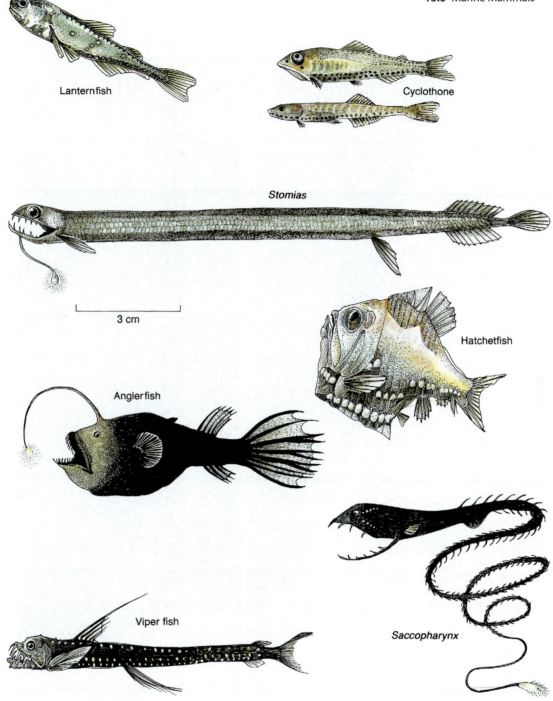

Lanternfish

Cyclothone

Stomias

3 cm

Hatchetfish

Anglerfish

Viper fish

Saccopharynx

Figure 13.10 Fishes from the deep sea.

13.6 Marine Mammals

Marine **mammals** are a small group of nekton compared to the large number of bony fish species, but this group includes the largest, most charismatic, and best known of marine animals. All mammals, including marine mammals, share basic characteristics: they are warm-blooded, they breathe air, they have hair (or fur) at some point during their development, and females produce milk and bear live young. Some marine mammals spend their entire lives at sea; others return to land to give birth.

Although all marine mammals spend at least half their lives at sea, the three major orders of marine mammals evolved from terrestrial ancestors. The fossil record indicates that land mammals re-entered the ocean at least seven separate times over a period from about 5–300 million years ago. These terrestrial ancestors gave rise to the order **Carnivora** (sea otters, walruses, polar bears, seals, sea lions), **Sirenia** (dugongs and manatees), and **Cetacea** (porpoises, dolphins, whales). While we don't know why mammals returned to the sea, it is likely that it was in response to plentiful food and perhaps to the opening of environmental

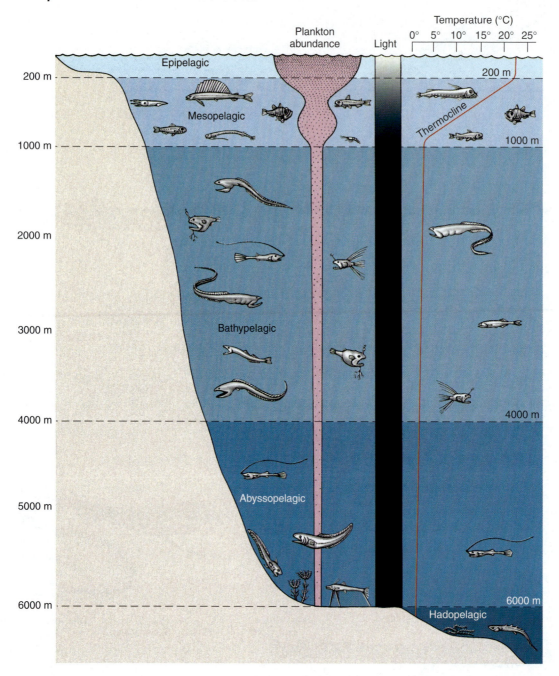

Figure 13.11 Life in the mesopelagic and deep sea is closely linked to the abundance of plankton and light intensity in the water column.

niches when the dinosaurs declined, although the specific reasons probably varied across the 300-million-year-period that led to modern marine mammals.

Marine mammals are homeotherms, which means they can maintain a near-constant body temperature, frequently well above ambient water temperature. Marine mammals can also make dives of spectacular depth and duration (see table 13.1). The northern elephant seal, for example, not only can dive to depths of more than 1500 m, but can stay submerged for up to two hours at a time! The best-diving mammals have streamlined shapes that reduce the drag on their bodies as they move through the water. Myoglobin helps regulate oxygen concentrations and is primarily

found in muscle tissue. Diving animals (including birds) appear to rely on oxygen bound to myoglobin while diving and are able to regulate its flow as needed during a dive. The concentration of myoglobin in marine mammals is three to ten times higher than that found in terrestrial mammals. Other diving adaptations include muscles that are tolerant of very low oxygen concentrations and, in cetaceans, weak, flexible rib cages that allow the lungs to collapse during dives. To conserve heat, marine mammals maintain a high metabolic rate combined with thick fur and/or layers of blubber. Marine mammals also minimize intake of seawater and excrete highly concentrated urine. To meet their water requirements, they extract water from the oxidation of their food.

Table 13.1 Record Diving Depth of Some Vertebrates

Species	Dive Depth (m)	(ft)
Yellow-bellied sea snake (*Pelamis platurus*)	50	164
Human (*Homo sapiens*), unassisted	101	331
Human (*Homo sapiens*), assisted	214	702
Common porpoise (*Delphinus delphis*)	226	741
Killer whale (*Orcinus orca*)	264	866
California sea lion (*Zalophus californianus*)	274	899
Harbor seal (*Phoca vitulina*)	308	1010
Bottlenose dolphin (*Tursiops truncatus*)	535	1755
Emperor penguin (*Apenodytes forsteri*)	564	1850
Beluga whale (*Delphinopterus leucas*)	647	2123
Weddell seal (*Leptonychotes weddellii*)	741	2431
Pilot whale (*Globicephala melanena*)	828	2716
Leatherback turtle (*Dermochelys coriacea*)	1280	4199
Elephant seal (*Mirounga angustiorostris*)	1653	5423
Cuvier's beaked whale (*Ziphius cavirostris*)	1888	6194
Sperm whale (*Physeter catodon*)	2035	6676

Sea Otters

Sea otters (figs. 13.12c and 13.13) are related to river otters but they live in salt water. Sea otters inhabit coastal areas, taking shellfish and other foods from the bottom in relatively shallow waters. Sea otters are the smallest marine mammals. An average male sea otter that weighs about 36 kg (80 lb) will consume almost 9 kg (20 lb) of food each day, about 20–25% of his body weight. This food intake allows sea otters to maintain a metabolic rate almost 2.5 times higher than terrestrial mammals of the same size. That amount of food consumption would be comparable to a 120 lb human consuming 30 lb of food each day without gaining weight! They depend on their thick fur, which traps air to keep them warm; unlike other marine mammals, they do not have an insulating layer of blubber or fat. Sea otters spend almost all of their time in the water, where they breed and give birth. Sea otters live in and around kelp beds, where they often use rocks to help them break open huge quantities of sea urchins, mussels, abalone, and crabs that they consume for food. They often float on their backs and use their bodies as a table to work on while they eat.

Sea otters are considered keystone predators in kelp forests. A keystone predator is one that maintains species diversity within an ecosystem. An important food source for sea otters is the sea urchins that graze on kelp. If the numbers of sea otters decline, the numbers of sea urchins increase. Increased numbers of sea urchins impact the ability of kelp forests to grow. Kelp forests serve as nurseries for small fish and other animals and a reduced density of kelp reduces the size of fish nurseries. This example of a keystone predator illustrates the complicated interactions that define food webs (review food webs in chapter 13).

Walrus

The walrus (see figs. 13.12d and 13.13) is placed in a separate taxonomic subgroup. It has no external ears and is able to rotate its hind flippers so that it can walk on a hard surface. Its heavy canine teeth (or tusks) are unique, and both males and females have them. These tusks help the walrus haul itself out of the water onto the ice and are probably used to glide over the bottom, like sled runners, while the animals forage for clams with their heavy, muscular whisker pads.

The Pacific walrus (*Odobenus rosmarus*) lives in the waters and on both coasts of the Bering and Chukchi Seas. It is thought that the population before the eighteenth-century exploitation was 200,000–250,000 animals. The 1950 population was 50,000–150,000. The last survey in 1990 estimated a population size of 201,000 individuals. The current population size of walruses is unknown.

Polar Bears

The polar bear (*Ursus maritimus*) is the top predator in the Arctic (fig. 13.14). They are long-lived (up to twenty-five years) carnivores with dense fur and blubber for insulation. They live only in the Northern Hemisphere and in a region where winter temperatures can be as low as −46°C (−50°F); they maintain a body temperature of 37°C (98.6°F). Polar bears actually have more problems with overheating than with getting too cold. Adult males weigh 250–800 kg (550–1700 lbs) and females weigh 100–300 kg (200–700 lbs). They feed primarily on ringed seals. They hunt seals by waiting for them to emerge from the openings in the ice that the seals make so they can breathe or climb out to rest. The polar bear may wait hours for the seal to emerge. Polar bears also feed on the carcasses of dead whales or walruses.

Polar bears travel long distances over the ice, 30 km (19 mi) or more each day. They are also strong swimmers, swimming continuously for 100 km (62 mi). They have partially webbed feet to aid swimming. Polar bears commonly breed in April. The males spend the winter on the pack ice; pregnant females dig large dens either on the mainland or on sea ice to spend the winter. They commonly give birth to two cubs in December or January. In spring, mother and cubs migrate to the coast near the open sea. Cubs remain with their mother for at least two years.

The world's total population of polar bears is estimated at 20,000 to 25,000. They are found in the United States, Canada, Greenland, Norway (Svalbard Islands), and Russia. The primary threat to polar bears is the decrease in sea ice coverage that is occurring. These bears are completely dependent on sea ice for all aspects of their lives.

(a)

(b)

(c)

(d)

(e)

Figure 13.12 Marine mammals. (a) The harbor seal *(Phoca vitulina)* is a friendly, curious animal that coexists well with humans. (b) This large male northern fur seal *(Callorhinus ursinus)* stands vigilant over his harem of females in the Pribilof Islands. (c) A sea otter *(Enhydra lutris)* and her pup dine on crab. (d) The walrus *(Odobenus rosmarus)* feeds from the bottom and relaxes on the Arctic ice floes. (e) Pacific white-sided dolphins *(Lagenorhynchus obliguidens)* keep pace with the bow wave of a research vessel.

Russia and Norway prohibit all hunting of polar bear. The U.S. Marine Mammal Protection Act of 1972 prohibits the killing of all marine mammals, including polar bear, except by native people for subsistence purposes.

Seals and Sea Lions

Seals and sea lions belong to the **pinnipeds,** or "feather-footed" animals, so named for their four characteristic flippers. Representative pinnipeds are shown in figures 13.12 and 13.13. Both seals and sea lions spend most of their time at sea, hunting for fish and

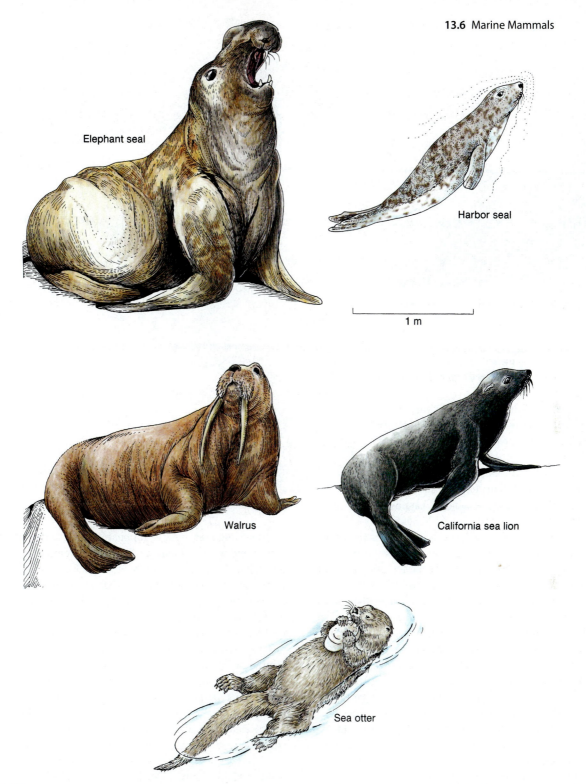

Figure 13.13 Marine mammals. The elephant seal, harbor seal, walrus, and California sea lion are pinnipeds. The sea otter, related to river otters, feeds on clams and sea urchins while floating on its back.

squid, but need to rest and breed on land. To stay warm, they have hair and blubber, although not nearly as much blubber as whales. They are also large animals with a relatively small surface area to volume. Seals are divided into true seals and eared seals. True seals have small ear holes and their rear flippers cannot be moved forward. Therefore, they do not get around well on land but move quickly through the water. Common true seals include the harbor

seal *(Phoca vitulina),* the harp seal *(Pagophilus groenlandicus),* the Weddell *(Leptonychotes weddellii)* and leopard *(Hydrurga leptonyx)* seals of the Antarctic, and the northern elephant seal *(Mirounga angustirostris),* whose males can reach 3 tons. Common eared seals include the California sea lion *(Zalophus californianus)* and the northern fur seal *(Callorhinus ursinus),* once almost hunted to extinction for their fur. Eared seals have external

Figure 13.14 A polar bear on the ice in the central Arctic Ocean.

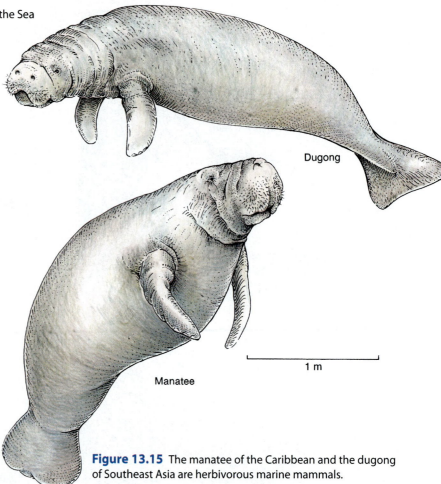

Figure 13.15 The manatee of the Caribbean and the dugong of Southeast Asia are herbivorous marine mammals.

ears and they can move their rear flippers forward so they can move quickly on land and can rear up into a partially erect position.

Pinnipeds display similar mating habits. Some breed on ice and some on land. They frequently migrate long distances to arrive at the breeding grounds (fig. 13.1). Commonly, the largest males arrive on shore first to establish territories, which they defend with fierce fighting. The females and subordinate males arrive later. Dominant males can have harems of as many as fifty females that they mate with. Females give birth to their pups on shore, as the pups cannot swim and must be nursed. The mothers leave them for times to hunt offshore.

Between 1870 and 1880, hunters reduced the population of northern elephant seals *(Mirounga angustirostris)* to as few as fifty animals. Because these animals spend most of their lives at sea, their entire populations were not wiped out. The most recent population estimates suggest population sizes of as many as 150,000 animals. Their rookeries in Mexico and U.S. waters are now fully protected. The Guadalupe fur seal *(Arctocephalus townsendi)* was also hunted to near extinction, until in 1892 only seven animals were thought to survive. Population size has recovered to about 10,000 animals, although almost all pups appear to be raised on a single island close to highly developed areas. Intensive hunting of seal populations can be considered as examples of the top-down control of organism abundance discussed in chapter 12. In these examples, humans are the predators.

Sea Cows

Manatees and **dugongs** are also known as **sea cows** (fig. 13.15); they are members of the order **Sirenia** and are thought by some to be the source of mermaid stories. Manatees and dugongs are the world's only herbivorous marine mammals. Manatees are found in the brackish coastal bays and waterways of the warm southern Atlantic coasts and in the Caribbean, and dugongs are found in the seas of Southeast Asia, Africa, and Australia. Like whales, they never leave the water.

The dugongs have disappeared from much of the Indian Ocean and South China Sea due to degradation of habitat. This 3 m (10 ft) long animal weighing about 360 kg (800 lbs) depends on sea grass beds for food, and these beds are being cleared for development and smothered by the silt and mud from eroding, overgrazed, and deforested lands. The dugongs are also hunted for their tusks, which are considered to have aphrodisiac properties and are used to make amulets. Many of the animals are also injured or killed by boat propellers.

Manatees in the coastal water of the Caribbean and South Atlantic are frequently injured and killed by collisions with the propellers of large and small vessels. The destruction of Florida's manatee habitat is also accelerating as salt marshes, sea grass beds, and mangrove areas are drained, reclaimed, or otherwise destroyed.

Long extinct, the Steller sea cow existed in the shallow waters off the Commander Islands in the Bering Sea. These sea cows were slow-moving, docile, totally unafraid of humans, and present in only limited numbers. These characteristics, coupled with a low reproduction rate, made them unable to withstand the human hunting pressure. The last Steller sea cow was killed for its meat in about 1768.

Whales

Whales belong to the mammal group called **cetaceans;** see table 13.2 for information about representative whales. Some cetaceans are toothed, pursuing and catching their prey with their teeth and jaws (for example, the killer whale, the sperm whale, and the small whales known as dolphins and porpoises). These whales belong to the suborder **Odontocetes**. Other whales have mouths fitted with strainers of **baleen**, or whalebone, through which they filter seawater to capture krill and other plankton. The blue, finback, right, sei, gray, and humpback whales are baleen whales, members of the suborder **Mysticetes**. The mouths of toothed and baleen whales are compared in figure 13.16. The blue, finback, and right whales possess baleen and swim open-mouthed to engulf water and plankton. The tongue pushes the water through the baleen, and the krill are trapped. The sei whale swims with its mouth partly open and uses its tongue to remove the organisms trapped in the baleen. The humpbacks circle an area rich in krill and expel air to form a circular screen, or a net of bubbles. The krill bunch together toward the center of this net, and the whales pass through the dense cloud of krill and scoop them up. The gray whale feeds mainly on small bottom crustaceans and worms.

Some whales migrate seasonally over thousands of miles; other whales stay in cold water and migrate over relatively short distances. The California gray whale and the humpback whale make long migratory journeys. Gray whales are found only in the North Pacific and adjacent seas. They are the most coastal of all great whales and are now composed of two populations—a large Eastern Pacific Stock and a small remnant Western Pacific Stock (table 13.3). In summer, the California gray whale is found in the shallow waters of the Bering Sea and the adjacent Arctic Ocean, where they feed all summer, building up layers of fat and blubber. In October, when the northern seas begin to freeze, the whales begin to move south, and in December, the first gray whales arrive off the west coast of Baja California. Here, they spend the winter in the warm, calm waters of sheltered lagoons, where the gray whales calve and mate but find little food; by the time they leave for their northward migration in February and March, they have lost 20–30% of their body weight. Moving at about 5 knots day and night, the whales make their annual 18,000 km (11,000 mi) migratory journey to link areas that provide abundant food with areas that ensure reproductive success (fig. 13.17a). Protection of the gray whales' calving and winter grounds in Baja California has allowed the population to build, and in 1994, the gray whale became the first marine mammal to be removed from the U.S. endangered species list.

The humpback whale also has well-defined migration patterns. Humpbacks are found in three geographically and reproductively isolated populations: in the North Pacific, North Atlantic, and Southern Ocean and a poorly studied population in the Arabian Sea. The North Pacific humpback spends the summer feeding in the Gulf of Alaska, along the northern islands of Japan, and in the Bering Sea; in winter, these North Pacific humpbacks migrate to the Mariana Islands in the west Pacific, the Hawaiian Islands in the central Pacific, and along the west coast of Baja California in the eastern Pacific. At these warmer latitudes, calves are born and mating takes place (fig. 13.17b). Globally, there may be as many as 40,000 humpback whales (table 13.3). Some populations are showing signs of recovery from exploitation.

Table 13.2 Principal Characteristics of Some Whales

Common Name	Scientific Name	Distribution	Weight (tons)	Length (m)	Food
Toothed					
Sperm	*Physeter catadon*	Worldwide	35	18	Squid, octopus, deep-sea fish
Narwhale	*Monodon monoceros*	Artic Circle	2	5	Fish, squid, shrimp
Killer	*Orcinus orca*	Worldwide	11	10	Mammals, fish, squid
Baleen					
Blue	*Balaenoptera musculus*	Worldwide; north-south migrations	80–150	33	Krill
Fin	*Balaenoptera physalus*	20°–75°N, 20°–75°S; north-south migrations	50–70	24–27	Krill, fish
Humpback	*Megaptera novaeangliae*	Worldwide; north-south migrations along coasts	35	11–17	Krill, copepods, fish, squid
Southern right	*Eubalaena australis*	Southern Hemisphere, 20°–55°S	80	17	Copepods, krill
Sei	*Balaenoptera borealis*	Mid-latitudes; seasonal migrations	30	15–18	Copepods, other plankton
Gray	*Eschrichtius robustus*	North Pacific; commonly within few kilometers of shore	35	11–15	Benthic invertebrates: amphipods, polychaete worms
Bowhead	*Balaena mysticetus*	Arctic and subarctic	100	18–20	Krill, copepods
Antarctic minke	*Balaenoptera bonaerensis*	Southern Hemisphere, 10°S to ice pack	14	9–11	Krill

(a)

(b)

(c)

Figure 13.16 (a) The killer whale *(Orcinus orca)* is a toothed whale. (b) The humpback whale *(Megaptera novaeangliae)* is a baleen whale. These two humpbacks with their mouths open illustrate the location of the baleen. (c) Bowhead whales *(Balaena mysticetus)* have the longest baleens of any whales and they are shown clearly here. This dead bowhead whale was hauled onto the ice and is shown lying on its back.

The bowhead, the beluga, and the single-tusked narwhal remain in cold water but still migrate over short distances each year. The beluga and narwhal are toothed whales; the bowhead is a baleen whale (fig. 13.16*c*). The largest population of bowhead whales is found in the Bering, Chukchi, and Beaufort Seas; this whale spends nearly all its life near the edge of the Arctic ice pack. Bowheads, singly or in pairs, often accompanied by belugas, migrate north from the Bering Sea to feed in the Beaufort and Chukchi Seas as the ice recedes in the spring, returning south to the Bering Sea in groups of up to fifty as the ice begins to extend in the winter (fig. 13.17*c*).

The narwhal is the most northerly whale and is found only in Arctic waters, most commonly on both sides of Greenland. This toothed whale possesses only two teeth. In males, one tooth grows out through the front of the head and becomes a tusk up to 3 m long. These tusks are thought to result from sexual selection, similar to the plume of male peacocks. In summer, these whales move north along the coasts of Ellesmere and Baffin Islands, and in the autumn, they return south to the waters along the Greenland coast (fig. 13.17*d*).

The whales of stories and songs are the "great whales": the blue, sperm, humpback, finback, sei, and right whales (fig. 13.18 and table 13.2); they are also the whales of the whaling industry.

Whaling

The earliest-known European whaling was done by the Norse between A.D. 800 and 1000. The Basque people of France and Spain hunted whales first in the Bay of Biscay, and then in the 1500s, the Basque whalers crossed the Atlantic to Labrador. They set up whaling stations along the Labrador coast to process the blubber of bowhead and right whales into oil for transport back across the Atlantic. In Red Bay, Labrador, the operation reached its peak in the 1560s and 1570s, when 1000 people gathered seasonally to hunt whales and produce 500,000 gallons of whale oil each year. By 1600, whaling had become a major commercial activity among the Dutch and the British, and at about the same time, the Japanese independently began harvesting whales. In the 1700s and 1800s, whalers from the United States, Great Britain, and the Scandinavian and other northern European countries pursued the whales far from shore, hunting them for their oil and baleen. Kills were made using handheld harpoons, and the whales were cut up and processed on land or onboard the ships at sea. Long voyages, intense effort, and dangerous combat between whalers and whales characterized these whale hunts.

In 1868, Svend Føyn, a Norwegian, invented the harpoon gun with its explosive harpoon and changed the character of

Table 13.3 Recent World Population Estimates of Representative Whale Populations Based on IWC Assessments

	Population	Year(s) to Which Estimate Applies	Estimated Population Size	Status
Minke				Most abundant whale
	Southern Hemisphere	1982/83–1988/89	761,000	
	North Atlantic	1996/2001	174,000	
	West Greenland	2005	10,800	
	North West Pacific and Okhotsk Sea	1989/90	25,000	
Fin				Endangered
	North Atlantic	1996/2001	30,000	
	West Greenland	2005	3200	
Gray				Endangered
	Western North Pacific	2007	121	
	Eastern North Pacific	1997/98	26,300	
Bowhead				Endangered
	Bering-Chukchi-Beaufort Seas	2001	10,500	
	West Greenland	2006	1230	
Humpback				Endangered
	North Pacific	2007	At least 10,000	
	Southern Hemisphere	1997/98	42,000	
	Western North Atlantic	1992/93	11,600	
Right				Endangered
	Western North Atlantic	2001	300	
	Southern Hemisphere	1997	7500	

whaling. Ships were motorized, and in 1925, harvesting was increased further by the addition of great factory ships, to which the small, high-speed whale-hunting vessels brought the dead whales for processing. This system freed the fleets, centered in the Antarctic, from dependence on shore stations. These methods continued into the twentieth century, greatly increasing the efficiency of the hunt and rapidly depleting the whale stocks of the world.

In the 1930s, the annual blue whale harvest reduced the population to less than 4% of its original numbers, threatening the species with extinction. In 1946, representatives from Australia, Argentina, Britain, Canada, Denmark, France, Iceland, Japan, Mexico, New Zealand, Norway, Panama, South Africa, the former Soviet Union, and the United States met in Washington, D.C., to establish the International Whaling Commission (IWC), which was set up under the International Convention for the Regulation of Whaling. The purpose of the convention was to ensure proper conservation of whale stocks and thus make possible the orderly development of the whaling industry. Regulations prohibited the killing of the blue, gray, bowhead, and right whales and of cows with calves. Opening and closing dates for whaling and minimum size data were set for each

species harvested. The most recent assessment of whale stocks provided by the IWC is given in table 13.3. Note that in 1989, the IWC decided to provide numbers only for populations that had been assessed in detail.

The IWC permits whaling by native peoples (in Alaska, Greenland, and the former Soviet Union) in order to balance the conservation of whales with the cultural and subsistence needs of these peoples.

While populations of California gray whales have recovered and some populations of other whale species are slowly increasing, the majority of whale species are still present in low numbers when compared to their estimated original populations. The eastern North Pacific right whale was nearly exterminated during the intensive whaling of the 1940s–60s, and it is considered the most endangered population of the large whales. Some researchers believe that the failure of whale populations to recover is due to the difficulty of finding mates in such small populations; the possibility also exists that the noise produced by increasing ship traffic interferes with whale communication. Other scientists are concerned that krill harvesting and the global depletion of fish species are affecting the whale populations; pollution may also play a role. Population estimates prior to large-scale whaling

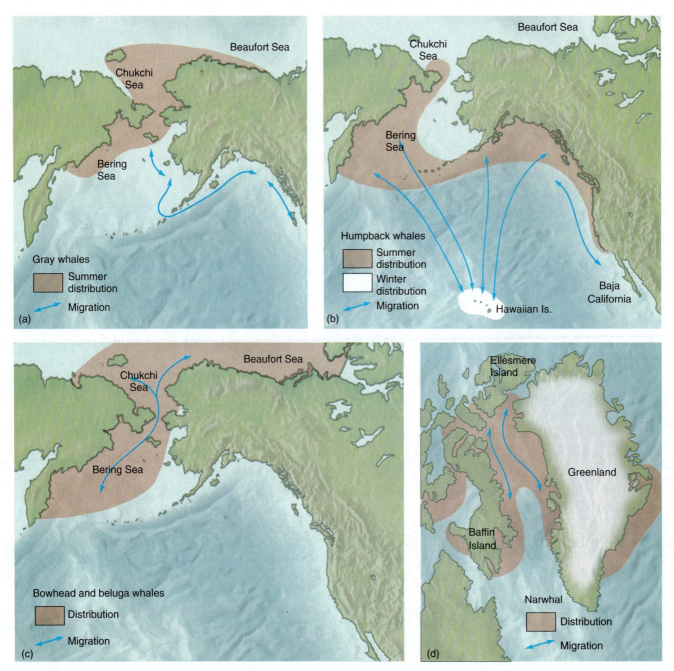

Figure 13.17 Migration paths and seasonal distribution of whales. The California gray whale (a) and the humpback whale (b) travel long distances between cold-water feeding and warm-water calving and mating areas. The bowhead and beluga whales (c) and the narwhal (d) remain in cold water and migrate over short distances.

vary considerably, but the biomass of whales is currently less than 25% of pre-whaling levels, perhaps much lower. Although whales are a small part of the nekton, they are long-lived animals and could have a large impact on the ocean's biogeochemical cycles. A recent estimate suggests that restoring all whale populations would result in an additional 160,000 tons of carbon export to the deep ocean per year as whale falls (see *Diving In* box), equivalent to preserving 843 hectares of forest per year, while the total biomass of the restored whale populations would be the equivalent of 110,000 hectares of forest. Whales have also been linked to natural iron fertilization of Southern Ocean waters. Whales consume iron-rich prey at depth, and defecate at the surface, making the iron available to the phytoplankton. This

iron fertilization could account for 400,000 metric tons of carbon export per year; after accounting for respiration by the whales, the net sink of carbon is still 200,000 tons/year. In other regions, such as the Gulf of Maine, whales may be contributing as much as 2300 tons of nitrogen per year to the phytoplankton in the euphotic zone, acting as a "whale pump" that counterbalances the biological pump, bringing nutrients back to the surface ocean.

Marine Mammal Protection Act

In 1972, the U.S. Congress established the Marine Mammal Protection Act, Public Law 92–522. This act includes a ban on the taking or importing of any marine mammals or marine mammal

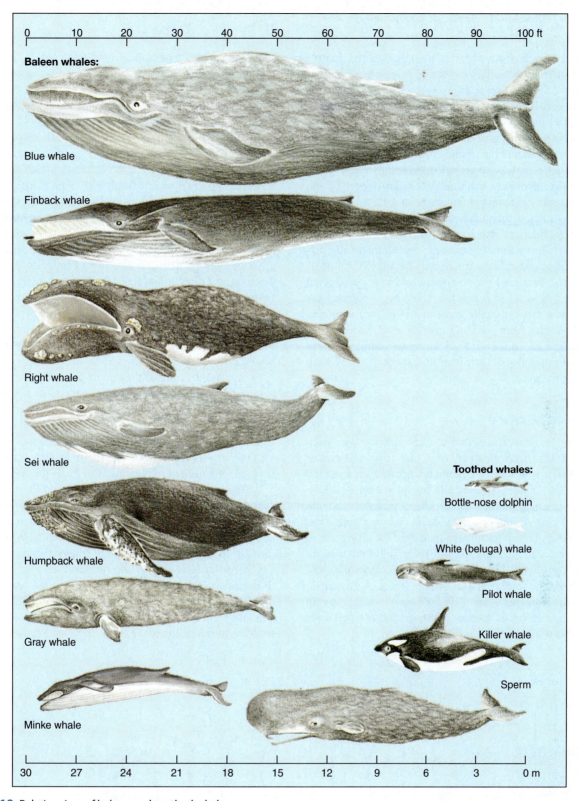

Figure 13.18 Relative sizes of baleen and toothed whales.

product. "Taking" is defined as the act of harvesting, hunting, capturing, or killing any marine mammal or attempting to do so. The act covers all U.S. territorial waters and fishery zones. It is also unlawful "for any person subject to the jurisdiction of the United States or any vessel or any convoy once subject to the jurisdiction of the United States to take any marine mammals on the high seas," except as provided under preexisting international treaty.

The act effectively removed the animals and their products from commercial trade in the United States. Only under strict permit procedures and with the approval of the Marine Mammal Commission can a few individual marine mammals be caught for scientific research and public display.

In 1994, the Marine Mammal Protection Act was amended to provide for exemption for Native subsistence, for permits for scientific research, for a program to authorize and control the taking of marine mammals incidental to commercial fishing operations, for the preparation of stock assessments for all marine mammals in U.S. waters, and for studies of fishery/pinniped interactions.

An important overall goal of the act is to maintain marine mammals as a "significant functioning element in the ecosystem of which they are a part." The act has dramatically reduced the death and injury of marine mammals. A recent report by the National Academy of Sciences indicates that today, the biggest threats to marine mammals are habitat degradation and the cumulative effects of harassment. Many effects of human activities on marine mammals can occur over dramatically different time scales. Individual animals may be affected immediately or over the course of years; populations may be affected over the course of years to generations; and ecosystem effects may not become obvious until generations or even centuries have passed. As has been seen in previous sections, trends in population sizes are difficult to determine—many of these animals spend a majority of their lives underwater or away from easy sightings by humans.

Communication

Many marine mammals use sound to communicate with each other and sound instead of sight to picture their underwater environment. The best-known communication between marine mammals are the "songs" of the male humpback whales. Different humpback populations have different songs, and the songs are transmitted from one individual to another within the population. Songs last up to thirty minutes and are changed and modified during each breeding season. These songs are thought to be announcements of presence and territory, although some scientists believe the singing is a secondary sexual characteristic of males in the breeding season. Female gray whales stay in contact with their calves by a series of grunts, and Weddell seals are known to communicate by audible squeaks.

Using sound to picture the environment is known as **echolocation.** The ability to make the sharp sounds required to produce the echoes that allow marine mammals to orient themselves and locate objects is suspected in all toothed whales, some pinnipeds (Weddell seal, California sea lion), and possibly the walrus. A few baleen whales—the gray, blue, and minke—also have this capacity. Although these animals can produce a range of sounds, the most useful sound for echolocation appears to be clicks of short duration released in single pulses or trains of pulses. The bottle-nose dolphin produces clicks in frequencies audible to the human ear and higher, each lasting less than a millisecond and repeated up to 800 times per second. When each click hits its target, part of the sound is reflected back; the animal continually evaluates the time and direction of return to learn the speed, distance, and direction of the reflecting target (fig. 13.19). Low-frequency clicks are used to scan the general surroundings, and higher frequencies are used for distinguishing specific objects.

Porpoises and dolphins move air in their nasal passages to vibrate the structures that produce the clicks; the whistles and squeaks are made by forcing air out of nasal sacs. The bulbous, fatty, rounded structure on the forehead of the porpoise, the **melon**, acts as a lens to concentrate the clicks into a beam and direct them forward. Sperm whales produce shorter, more powerful, long-range pulses at lower frequencies; these sounds may travel several kilometers. Each pulse from a sperm whale is compound, made up of as many as nine separate clicks. The sperm whale's massive forehead is filled with oil that may be used to focus the sound pulses.

These animals must be able to pick up the faint incoming echoes of their own clicks and screen out the louder outgoing clicks and other sea noises. Sounds enter through the lower jaw and travel through the skull by bone conduction. Within the lower jaw, fat and oil bodies channel the sound directly to the middle ear. Areas on each side of the forehead are also very sensitive to incoming sound. The hearing centers in the brains of marine mammals are extremely well developed, presumably to analyze and interpret returning sound messages. Their vision centers are less developed, and they are believed to have no sense of smell.

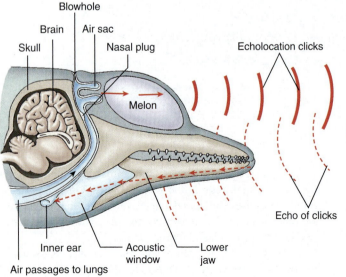

Figure 13.19 Many marine mammals, such as this dolphin, echolocate by emitting bursts of sound waves, or clicks, as air is pushed through internal passages. The clicks are focused into a beam by the melon. The reflected echoes are received by the lower jaw, and enable the animal to determine the speed, distance, and direction of the target.

Whale Falls

The death of a whale suddenly sends a huge, localized source of food to the sea floor. In 1987, Craig Smith and colleagues from the University of Hawaii accidentally discovered the carcass of a blue whale on the floor of the Santa Catalina basin off California (box fig. 1a). Their studies show that a whale carcass, or whale fall, supports a large community of organisms. This occurs in four stages. First, the mobile scavengers such as hagfish, crabs, and sharks reduce the body to bones in as little as four months. After the bones have been picked clean, the organic matter left behind enriches the surrounding sediment. Small worms, mollusks, and bacteria take over. This stage can last up to two years, and ends when the easily digestible nutrients are gone. Whale bones are rich in fats and oils that give the animal buoyancy in life; after death, they provide nutrition for the third and final stage, anaerobic bacteria. These bacteria decompose the fatty substances and generate hydrogen sulfide and other compounds that diffuse out through the bone. Sulfophilic ("sulfur-loving") chemosynthetic bacteria then metabolize the sulfides and form bacterial mats over the skeleton. The bacterial mats are grazed by worms, mollusks, crustaceans, and other organisms. Other animals may be attracted, and they get their nutrition from the sulfides, the bacterial mats, the fatty substances in the bones, or other animals at the site. This sulfophilic stage can last for perhaps 50–100 years for very large whale falls. Finally a reef stage may develop, during which the skeletal remains may be colonized by suspension feeders exploiting enhanced currents.

The number of species found on a single skeleton is surprising: 5098 animals from 178 species were isolated from five vertebral bones recovered from one whale, and it is estimated that whale falls average 185 species per skeleton during the third (sulfophilic) stage. Small mussels are shown on re-covered whalebone in box figure 1b. The bone surface area totaled 0.83 m² (9 ft²). Ten of these species, including worms and limpets, have been found only associated with whale skeletons. By comparison, the most fertile hydrothermal vent areas have yielded only 121 species and hydrocarbon seeps just thirty-six. Recent studies of whale falls have also identified two new types of worms in the genus Osedax ("bone-devourer") that burrows into the whale bones with green "roots" filled with symbiotic bacteria. The worm/bacterial assemblage breaks down fats and oils found in the bones, feeding both the worm and its symbiotic bacteria. Interestingly, these worms will also colonize fish bones, suggesting that these worms are important scavengers of the sea floor.

How abundant are these whale falls? At this time, that is difficult to say, for they can be anywhere on the ocean floor and are difficult to locate. In 1993, the U.S. Navy searched 20 km² (8 mi²) of the Pacific Missile Range off California with side-scan sonar while seeking a lost missile. Eight whale falls were videotaped, and one of them was subsequently located and examined by Smith and his team using a submersible. Scientists in Japan, New Zealand, and Iceland are also looking for whale-fall sites. With permission from the National Marine Fisheries Service, Smith has taken two dead stranded whales out to sea and sunk them to observe the colonizing of the carcasses and learn more about the diversity of the organisms associated with whale falls.

At least fifteen of the whale-fall species have also been found in other sulfide-rich habitats such as the deep-sea hydrothermal vents. Smith has suggested that the whale falls may serve as "stepping-stones" for the dispersal of organisms that depend on chemosynthesis, as from one

(a)

(b)

Box Figure 1   (a) The skull, jawbones, and vertebrae of a 21 m (70 ft) blue whale on the floor of the Santa Catalina basin off southern California, depth 1240 m (4000 ft). The skull is about 1.5 m (5 ft) long, and each vertebra is about 40 cm (16 in) long. (b) Sulfide-loving mussels (each about 1 cm [0.4 in] long) clustered on a recovered whalebone. Up to 178 species of animals have been found living on a single carcass.

hydrothermal vent to another. Objections to this theory include that the number of species that have been found to overlap whale falls and vents is small, that vents are found in more than 1500 m (5000 ft) of water, while most whales live and die in shallower water along the edge of the continental shelf, and that the increasing number of vents being found indicates that dispersal from vent to vent is not a problem. Whether or not the whale falls act as stepping-stones, it is of great interest that these whale-fall communities exist, that hydrothermal vents are not as isolated as was thought, and that another new kind of community has been discovered on the sea floor, a place once thought to be cold, dark, and unsuited to life.

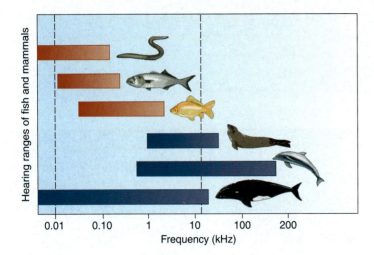

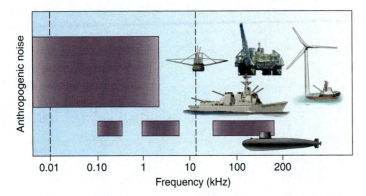

Figure 13.20 The hearing ranges of different kinds of fish (*red*) and mammals (*blue*) are shown, with some of the major sources of human-generated noise.

In recent years, there has been much debate on effects of human-generated noise on whale behavior. Sources of noise pollution in the ocean include military use of sonar, oil and gas exploration employing seismic air gun technology (which produce high intensity, low frequency bursts of sound), the gradual

increase of ambient noise from tankers and other ships, coastal jet ski traffic, and even offshore wind farms used to develop alternative energy. The navy uses sonar, which can travel for miles underwater, to detect the presence of submarine and other potential underwater hazards. A number of years ago, fourteen beaked whales were found beached on islands in the Bahamas about thirty-six hours after navy vessels used sonar for training exercises, and the strandings were linked to the sonar. While marine mammals are clearly susceptible to noise pollution, it is now clear that other organisms are also impacted (fig. 13.20). For example, squid and other cephalopods can "hear" low-frequency noise and have been affected by air guns, and many crustaceans apparently use environmental sound as cues for habitat selection. What can we do about noise pollution? Some of the noise sources are being addressed. The U.S. Navy has developed research programs with scientists to determine how best to avoid harming marine animals, and the United States and other countries have begun using "quiet research vessels" for oceanographic studies. Other long-term solutions are also being considered. In 2011, an open science meeting in support of the International Quiet Ocean Experiment was held to map out how best to monitor the effects of human noise on the ocean and minimize negative impacts. While there is no simple solution, efforts such as these are the first step in documenting the problems and addressing the impacts that human-generated noise is creating in the marine environment.

QUICK REVIEW

1. Describe the major groups (orders) of marine mammals.
2. Explain how sea otters and polar bears maintain warm body temperatures in their environments.
3. Describe the two kinds of whales; give examples of each.
4. What are the main characteristics of the two major groups of pinnipeds?
5. What has been the effect of the U.S. Marine Mammal Protection Act?
6. What role do whales play in ocean biogeochemical cycles?

Summary

The nekton swim freely and independently. Members of the nekton include some invertebrates, such as squid and other cephalopods, large shrimp, and some crabs. Sea snakes, the marine iguana, seagoing crocodiles, and sea turtles are the reptile members of the nekton. Many sea turtles are endangered. Female sea turtles return to specific beaches to lay their eggs. The buried eggs incubate for up to a few months. During this time they are susceptible to hunting and poaching.

Fish are by far the largest group comprising the nekton, and are found at all depths in the oceans. The jawless fish are the

most evolutionary primitive, and include hagfish and lampreys. Sharks, skates, and rays have cartilaginous skeletons. All other fish have bony skeletons, including the commercially fished species and the highly specialized types of the deeper ocean. The bony fish typically have swim bladders and exhibit a wide range of swimming behaviors. A majority of fish school at some point during their life cycle.

Marine birds are included with the nekton, even though not all of them swim. Only 3% of total bird species are marine; these birds are specialized for life at sea. Many marine birds are migratory and travel long distances each year. Seabirds spend a significant portion of their lives at sea, but must return to land to breed. Marine birds often form large breeding colonies where multiple species may be present. Shorebirds do not swim much and have a greater dependence on land.

The marine mammals include the orders Carnivora (sea otters, polar bears, walruses, seals, and sea lions), Sirenia (dugongs and manatees), and Cetacea (whales, dolphins, and porpoises). Marine mammals breathe air and maintain near-constant body temperatures. Many marine mammals have developed adaptations for deep diving. The U.S. Marine Mammal Protection Act was passed in 1972 to protect all marine mammals and to prohibit commercial trade in marine mammal products. Today, some of the biggest threats to marine mammals are habitat degradation and the cumulative effects of harassment.

Seals, walruses, sea lions, and sea otters belong to the pinnipeds. They spend most of their time at sea, hunting for food, but they still need to come ashore to rest and breed. Sea otters are considered keystone predators in kelp forests. The polar bear is the top predator of the Arctic's marine food web and is completely dependent on the presence of sea ice. The primary threat to polar bears is diminishing sea ice coverage. Manatees and dugongs feed on seagrass and are found in warm waters of the Indian and Atlantic Oceans where they are caught between increasing human contact and loss of habitat; the stellar sea cow is extinct.

Whales belong to the cetaceans, and are further subdivided into Mysticetes (baleen whales) and Odontocetes (toothed whales). Some whales migrate seasonally over thousands of miles. Many whales are endangered. Many whales communicate using sound. Dolphins, porpoises, and many whales also employ echolocation.

Key Terms

nekton, 340

invertebrates, 340

vertebrate, 340

Tagging of Pacific Predators (TOPP), 340

niche separation, 344

schooling, 346

emergent behavior, 347

Chondrichthyes, 347

Osteichthyes, 348

demersal fish, 349

mammal, 351

Carnivora, 351

Sirenia, 351, 356

Cetacea, 351

pinniped, 354

manatee, 356

dugong, 356

sea cow, 356

cetacean, 356

Odontocetes, 357

baleen, 357

Mysticetes, 357

echolocation, 362

melon, 362

Study Problems

1. Many of the larger nekton (turtles, sharks, whales) routinely travel great distances to forage and reproduce. As an example, a white shark tagged with a tracking device off South Africa swam to Australia and back in ninety-nine days, a distance of almost 7000 miles. If the shark swam continuously, what would its average speed be in km/h?

2. One of the fastest fish in the world is the Atlantic sailfish, which is capable of short bursts at up to 110 km/h. What is that speed in nautical miles per hour? How much faster is that than the average speed you calculated for problem 1?

3. Sea otters do not have a thick layer of blubber to maintain body temperature, unlike other marine mammals. To compensate, otters typically consume 20–25% of their body weight per day. If an otter weighs 25 kg, how much food must it consume per day? Assuming that the otter prefers to each shellfish (filter feeders such as clams and mussels), how much is the otter consuming in terms of phytoplankton productivity?

4. A typical blue whale is approximately 30 m in length and 180 metric tons in weight. Blue whales can get so large because of very efficient trophic transfer: They tend to eat small crustaceans called krill (about 5 cm in length),

which in turn eat phytoplankton such as diatoms (fractions of a millimeter in length). Calculate the ratio of predator to prey for the blue whale and krill. If humans were to eat food of a similar size, and we assume an average height of 1.7 m, what size prey (food) would we be eating?

5. A blue whale can consume about 6 metric tons of krill per day during the summer, and a typical blue whale weighs 115 metric tons. In contrast, a killer whale consumes about 3% of its body weight per day, and a typical killer whale weighs about 11 metric tons. If a killer whale feeds at about trophic level 4.5, what is the metric ton equivalent of phytoplankton productivity consumed by the blue whale and by the killer whale? How does this compare to the difference in sizes?

6. Whales in the Gulf of Maine have been estimated to add an additional 2300 metric tons per year of nitrogen to the surface ocean. If that nitrogen was used by phytoplankton, and all of the material was removed from the surface ocean via the biological pump, how much extra carbon would be exported to depth in the Gulf of Maine? How much phosphorus would the phytoplankton require to utilize this carbon and nitrogen?

CHAPTER 14

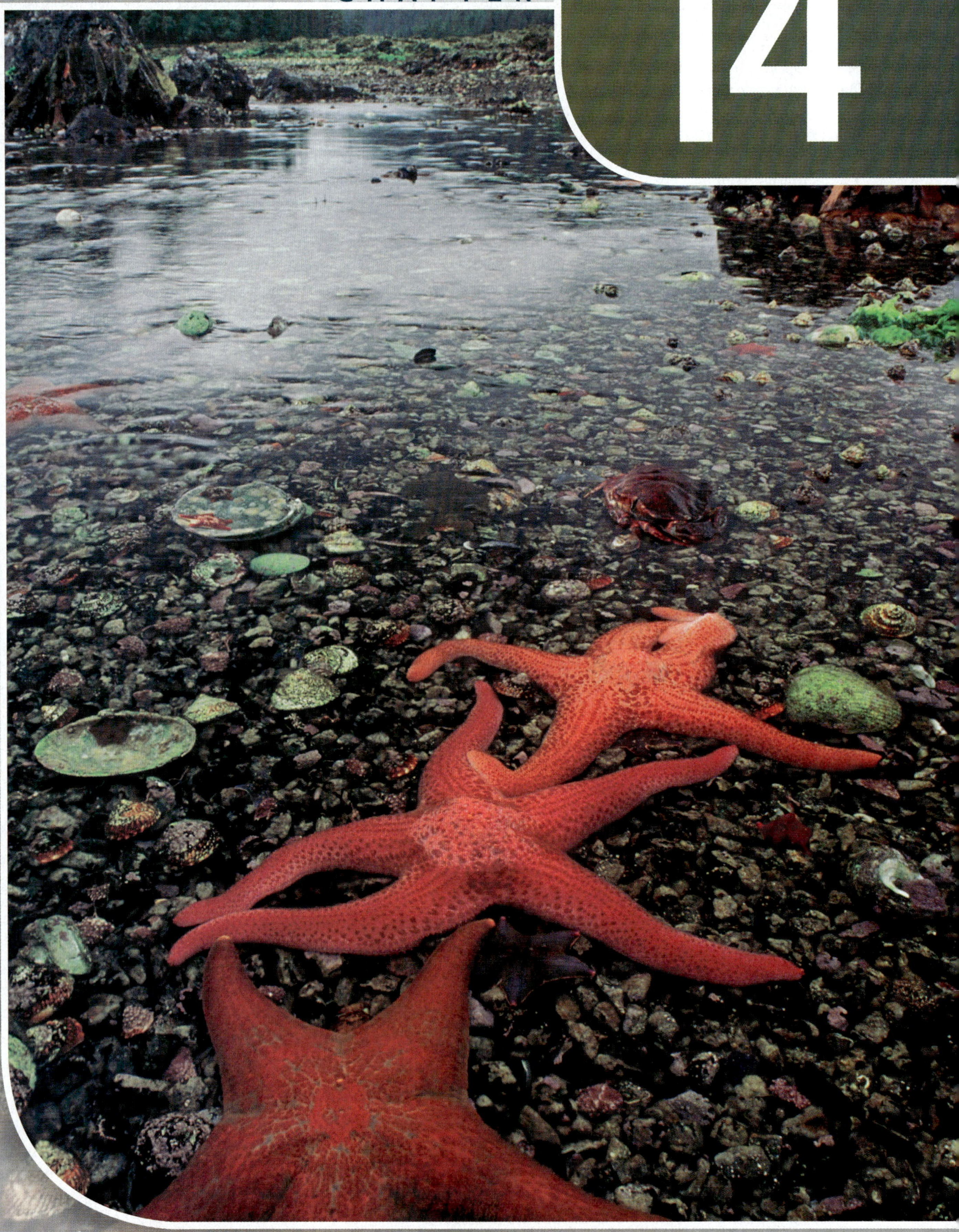

The Benthos:
Living on the Sea Floor

Learning Outcomes

After studying the information in this chapter students should be able to:

1. *describe* which factors limit where seaweeds can grow,

2. *understand* the role of keystone predators in intertidal regions,

3. *describe* the factors that dictate where benthic animals live, and

4. *understand* the role of symbioses in different ecosystems.

Sea stars and other organisms on a rocky shore at low tide.

The animals and algae that live on the sea floor or in the sediments are members of the benthos. The algae and plants are found in the sunlit shallow coastal areas and in the intertidal zones, while the benthic animals are found at all ocean depths. The benthos include remarkably rich and diverse groups of organisms, among them the luxuriant, colorful tropical coral reefs, the newly discovered organisms surrounding deep-sea hydrothermal vents, the great cold-water kelp forests, and the hidden life below the surface of a mud flat or a sandy beach. Benthic organisms are important food resources and provide valuable commercial harvests: for example, oysters, clams, crabs, and lobsters. In this chapter, we first present an overview of the benthos by group and habitat, with special focus on the world's more intriguing benthic communities, and then we consider the harvesting of the benthos, its problems, and its potential. This chapter is intended to serve only as an introduction to this topic, for no single chapter can do justice to the diversity of organisms present in this group.

14.1 The Benthic Environment

The benthos is a remarkably diverse grouping of algae and animals (fig. 14.1). It is easily accessible in the shallow littoral zone, and has been studied by scientists and students for hundreds of years. Today, submersibles, remotely operated vehicles (ROVs), and autonomous underwater vehicles (AUVs) survey previously unsampled marine habitats. As we continue to explore, we may expect to make discoveries as suprising and unexpected as the vent communities in the rift areas. More than 200,000 benthic species have been catalogued by scientists—many times greater than the number of pelagic species. About 80% of benthic animals belong to the **epifauna**, and live on or attached to rocky areas and firm sediments. Animals that live buried in the substrate belong to the **infauna** and are associated with soft sediments such as mud or sand.

In shallow waters where light penetrates to the bottom, photosynthetic autotrophs are prevalent and provide much of the primary production for secondary consumers. At least 60% of the Earth's surface lies deeper than 2 km (1.2 mi) of water in darkness where photosynthetic primary producers cannot survive. These deep environments rely instead on food (organic material) raining down from the surface, including **food falls** such as whale carcasses (see the *Diving In* box in chapter 13), or on chemosynthetic communities. As a result, benthic biomass tends to follow similar patterns to surface primary productivity, with high biomass in shallow water near land, and rapidly decreasing biomass with depth (compare fig. 14.2 with surface productivity shown in fig. 12.35), with an obvious exception being coral reef and deep-sea chemosynthetic habitats (see sections 14.5 and 14.7).

Some animals of the sea floor are **sessile**, or attached to the sea floor, as adults (for example, barnacles, sea anemones, and oysters), whereas others are motile all their lives (for example, crabs, sea stars, and snails). Most benthic forms produce motile larval stages that spend a few weeks of their lives as meroplankton (review chapter 12). These organisms have a **bipartite lifestyle** (fig. 14.3). By taking advantage of the ocean currents, offspring can disperse over large distances, increasing the potential number of sites where suitable habitat for adults can be found while reducing competition among closely related individuals.

The distribution of benthic animals is not governed by any single factor such as light or pressure. Animal distribution is controlled by a complex interaction of factors, creating living conditions that are extremely variable. Temperature is nearly constant in deep water but can change abruptly in shallow areas covered and uncovered by the daily tidal cycle. Salinity, pH, exposure to air, oxygen content of the water, and water turbulence also change abruptly in the intertidal zone. Benthic animals exist at all depths and are as diverse as the conditions under which they live. Their lifestyles are related to their varied habitats, as discussed in the following sections.

QUICK REVIEW

1. Explain the relationship between surface productivity and benthic biomass.
2. Describe some differences between adaptations of benthic versus pelagic organisms.
3. Why is a bipartite lifestyle common in the marine environment?

14.2 Seaweeds and Marine Plants

Seaweeds are members of a large group called **algae**. Photosynthetic autotrophs were introduced in chapter 11, and the unicellular planktonic algae were included in chapter 12. Here we consider the role of seaweeds and marine plants as part of the benthic community. Algae are distinguished from plants by body form, reproduction, accessory pigments, and storage products. Algae have simple tissues; they do not produce flowers or seeds (but some do produce planktonic spores), and they do not have roots (table 14.1).

General Characteristics of Benthic Algae

Seaweeds are benthic organisms; they grow attached to rocks, shells, or any solid object. Seaweeds are attached by a basal organ known as a **holdfast** that anchors the seaweed firmly to a solid base or substrate. The holdfast is not a root; it does not absorb water or nutrients. Above the holdfast is a stemlike portion known as the **stipe.** The stipe may be so short that it is barely identifiable, or it may be up to 35 m (115 ft) in length. It acts as a flexible connection between the holdfast and the **blades,** the alga's photosynthetic organs. Seaweed blades are thin and are bathed on all sides by water. They serve the same purpose as leaves, but they do not have the specialized tissues

Figure 14.1 The benthos is a large, varied group of animals living on or in the sea floor, especially the organisms of the tide pools and the intertidal areas of the rocky coasts. (a) The anemone *Tealia crassicornis* is a sessile carnivore. The graceful and beautiful nudibranchs, or sea slugs, include (b) the white *Dirona albolineata*, (c) the orange-flecked *Triopha carpenteri*, and (d) *Hermissenda crassicornis* with its orange-and-white-striped tips. Sea stars come in a remarkable diversity of shapes and sizes: (e) the bright orange blood star *(Henrica levinscula)*, the slender-armed *Evasterias troschelli,* the many-armed *Solaster dawsoni, Mediaster aequalis* with its wide disk and broad arms, the leather star *(Dermasterias imbricata),* and the purple, rough-skinned *Pisaster ochraceus.*

(f)

(g)

(h)

(i)

Figure 14.1 (continued) (f) All sea stars are carnivores and use their tube feet to hold and open the shellfish on which they feed. (g) The purple sea urchin *(Strongylocentrotus purpuratus)* and the green urchin *(S. droebachiensis)* are closely related to the sea stars but are herbivores, clipping off the algae with their specially constructed mouthparts (h). (i) The pink sea scallop *(Chlamys hastata hericia)* lies open as it filters organic particles from the seawater.

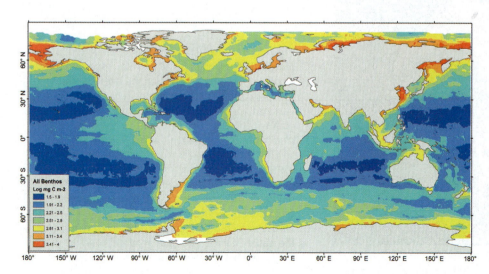

Figure 14.2 Using data from the Census of Marine Life, scientists have modeled the biomass (mg carbon per square meter) in the benthic environment. High biomass (*reds, oranges*) is associated with coastal waters; very low biomass (*dark blue*) is found in the deep ocean below the mid-ocean gyres.

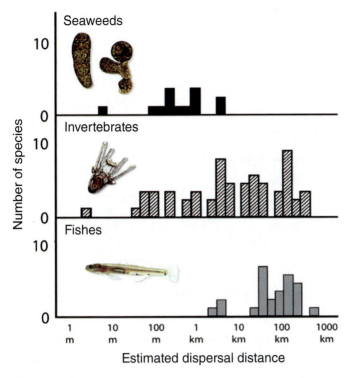

Figure 14.3 Many benthic marine organisms include a pelagic larval stage. This allows them to disperse over great distances. Here, estimates are derived for more than 100 species using genetic variation among populations.

and veins of leaves because algae do not require that water be conducted from the ground, up a stem, and through the veins to leaf cells. The blades may be flat, ruffled, feathery, or even

encrusted with calcium carbonate. The general characteristics of a benthic alga are shown in figure 14.4.

Seaweeds are not found in areas of mud or sand where their holdfasts have nothing to which they can attach. Sometimes during a storm seaweeds are dislodged, taking with them the rocks to which the holdfasts cling. Because the benthic algae are dependent on sunlight, they are confined to the shallow depths of the ocean, where they are surrounded by water, dissolved carbon dioxide, and nutrients. They are efficient primary producers, exposing a large blade area to both the water and the Sun.

In the sea, the quality and quantity of light change with depth. On land and at the sea surface, algae receive the full spectrum of visible light. The green algae have the same chlorophyll pigments as land plants; they are able to absorb the same wavelengths of light and they are found in shallow water. Algae found at moderate depths have a brown pigment that is more efficient at trapping the shorter wavelengths. At maximum growing depths, the algae are red, for the red pigment can best absorb the remaining blue-green light. The characteristic pattern for seaweed growing on a rocky shore is the green algae in shallow water, then the brown algae, and the red algae at greater depths.

Seaweeds provide food and shelter for many animals. They act in the sea much as forests and shrubs do on land. Some fish and other animals, such as sea urchins, limpets, and some snails, feed directly on the algae; other animals feed on shreds and pieces of algae as they settle to the bottom. Some organisms use large seaweeds as a place of attachment; some of the smaller algae grow on the larger forms.

Table 14.1 Important Characteristics of Seaweeds and Marine Plants

Group		Distinguishing Features	Photosynthetic Pigments	Major Food Reserves	Major Cell-Wall Components	Significance in the Marine Environment
Green algae		Eukaryotic, unicellular and multicellular; mostly bottom-dwelling	Chlorophyll *a*, *b*, carotenoids	Starch	Cellulose, calcium carbonate in calcareous algae	Primary producers; calcareous algae are important sources of calcareous deposits in coral reefs
Brown algae		Eukaryotic, multicellular; bottom-dwelling	Chlorophyll *a*, *c*, carotenoids (flucoxanthin and others)	Laminarin, oil	Cellulose, alginates	Primary producers; dominant components of kelp forests
Red algae		Eukaryotic, multicellular; bottom-dwelling	Chlorophyll *a*, phycobilins (phycocyanin, phycoerythrin), carotenoids	Starch	Agar, carageenan, cellulose, calcium carbonate in coralline algae	Primary producers; coralline algae are important sources of calcareous deposits in coral reefs
Flowering plants		Eukaryotic, multicellular; bottom-dwelling	Chlorophyll *a*, *b*, carotenoids	Starch	Cellulose	Dominant primary producers in seagrass beds, salt marshes, and mangrove forests; nursery grounds for many species; help stabilize soft bottoms, protect coast from turbulence

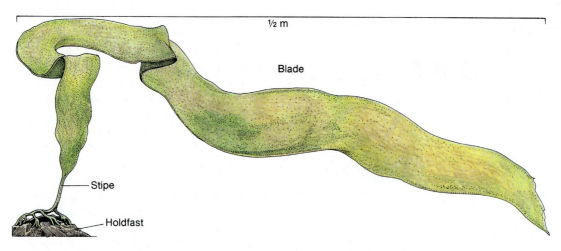

½ m

Blade

Stipe

Holdfast

Figure 14.4 Benthic algae are attached to the sea floor by a holdfast. A stipe connects the holdfast to the blade. *Laminaria* is a genus of kelp and a member of the brown algae.

Kinds of Seaweeds

The color a plant or an algae appears are the colors of light that are not well absorbed by their pigments. These colors are instead reflected back to our eyes. As described below, categorization of algae is based on additional characteristics besides their suite of pigments. Therefore, relying on visible color alone to categorize the algae can sometimes be misleading. Some red algae appear brown, green, or violet, and some brown algae appear black or greenish. Representative algae are shown in figure 14.5.

Green algae are mostly freshwater organisms, but a small number occur in the sea, including *Ulva,* the sea lettuce, and *Codium,* known as dead man's fingers. Green algae are related to the land plants; they have the same green chlorophyll pigments as land plants and they also store starch as a food reserve.

All brown algae are marine and range from simple microscopic chains of cells to the **kelps,** which are the largest of the algae. Kelps have more tissue structure than most algae but are much simpler than flowering land plants. The kelps have strong stipes and effective holdfasts that allow them to colonize rocky points in fast currents or heavy surf, a habitat favored by the sea palm, *Postelsia.* Other kelps grow with holdfasts well below the depth of wave action and float their blades at the surface supported by gas-filled floats; for example, the bull kelp, *Nereocystis,* is especially abundant along the Alaskan, British Columbian, and Washington coasts, and the great kelp, *Macrocystis,* is found along the California coast. Other species are found off Chile, New Zealand, northern Europe, and Japan. Storage products include the carbohydrates laminarin and mannitol but not starch.

The red algae are almost exclusively marine; they are the most abundant and widespread of the large algae. Their body forms are varied and often beautiful, flat, ruffled, lacy, or intricately branched. Their life histories are specialized and complex, and they are considered the most advanced of the algae. Their storage product is floridean starch.

Although the diatoms were discussed as part of the plankton (see chapter 12), there are also benthic diatoms. These diatoms are usually of the pennate type and grow on rocks, muds, docks, and blades of kelp, where they produce a slippery brown coating.

Marine Plant Communities

A few flowering plants with true roots, stems, and leaves have made a home in the sea. Seagrasses grow, often completely submerged, in patches along muddy beaches, where they help to stabilize the sediments. Eelgrass, with its strap-shaped leaves, is found on mud and sand in the quiet waters of bays and estuaries along the Pacific and Atlantic coasts, and turtle grass is common along the Gulf Coast. Surf grass flourishes in more turbulent areas exposed to waves and tidal action. Seagrasses are important primary producers, and their decomposing leaves add large quantities of vegetative material and nutrients to shores and estuaries. They provide a place of attachment for sponges, small worms, and tunicates and a feeding ground for many benthic organisms.

Temperate-area salt marshes are dominated by marsh grasses able to tolerate the brackish water. Marsh grasses are partly consumed by marsh herbivores, but much of the vegetation breaks down in the marsh and is washed into the estuaries by tidal creeks. There the vegetative remains are broken down by bacteria that release nutrients to the water to be reused by the plants and algae.

The mangroves grow in the intertidal zone along humid, tropical coasts. They are salt-tolerant, woody trees with special adaptations that allow them to thrive in oxygen-deficient muds. Their waxy leaves reduce water loss, and they excrete excess salt through glands located on their leaves. Some mangroves grow prop roots from overhanging branches to extend the root system above the water; others send up vertical roots from roots below the water's surface. Birds, insects, and other animals live in the mangrove's leafy canopy, and its intertwining prop roots provide shelter for small marine organisms. The roots also trap sediments and organic material, which help to build and extend the shore seaward.

QUICK REVIEW

1. Describe some adaptations of true marine plants.
2. Explain how seaweeds are distributed with depth. What limits the distribution of seaweeds?
3. Describe three general classes of seaweeds.

Figure 14.5 Representative benthic algae. *Ulva* and *Codium* are green algae. *Postelsia, Nereocystis,* and *Macrocystis* are kelps. The kelps and *Fucus* are brown algae. The red algae are *Corallina, Porphyra,* and *Polyneura. Corallina* has a hard, calcareous covering.

14.3 Animals of the Rocky Shore

The rocky coast is a region of rich and complex algal, plant, and animal communities living in an area of environmental extremes. It is a meeting place between the more variable land conditions and the more stable sea conditions. As the water moves in and out on its daily tidal cycle, rapidly changing combinations of temperature, salinity, moisture, pH, dissolved oxygen, and food supply are encountered. At the top of the littoral zone, organisms must cope with long periods of exposure, heat, cold, rain, snow, and predation by land animals and seabirds as well as turbulence of waves and seaward water flow. At the littoral zone's lowest reaches, organisms are rarely exposed but have their own problems of competition for space as well

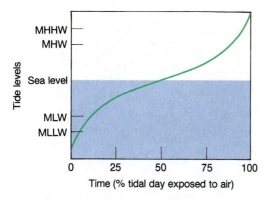

Figure 14.6 The time of exposure to air for intertidal benthic organisms is determined by their location above and below the sea level and by the tidal range. (MLLW = mean lower low water; MLW = mean low water; MHW = mean high water; MHHW = mean higher high water.)

as predation. The exposure endured by marine life at different levels in the littoral zone is shown in figure 14.6.

The distribution of the algae and animals is governed by their ability to cope with the stresses that accompany exposure, turbulence, and loss of water, as well as food web dynamics. Biologists have noted that patterns form as the algae and animals sort themselves out over the intertidal zone. This grouping is called zonation; **vertical zonation,** or **intertidal zonation,** is shown in figure 14.7. The distribution of the seaweeds, with the green algae in shallow water, the brown algae in the intertidal zone, and the red algae in the subtidal area, is an example of such vertical zonation.

In chapter 13, the concept of a keystone predator was introduced. The example discussed was the role of the sea otter as a keystone predator, whose presence maintains diversity within kelp forests. Another example of a keystone predator comes from intertidal community structure. The dominant mussels *(Mytilus californianus)* along the coast of Washington State are susceptible to desiccation and need to be covered by water for most of the day. Therefore, water height determines the upper limit of the mussels. The lower limit of the mussels is determined by their predator, the sea star *(Pisaster ochraceous)*. This sea star is larger than mussels and even more sensitive to desiccation so its upper limit is shallower than that of the mussel. Therefore, a narrow band of mussels grows in this intertidal region, with the upper limit determined by water height and the lower limit determined by predators. When scientists removed by hand all the sea stars from the zone below the mussels, the mussels invaded the lower zone and grew so rapidly that they eliminated other species within the area and the biodiversity of the region dropped dramatically. This drop in biodiversity with removal of a single predator is the foundation of the concept of keystone predator. Keep this concept in mind when considering distributions of organisms along the rocky shore.

In the supralittoral (or splash) zone, which is above the high-water level and is covered with water only during storms and the highest tides, the animals and algae occupy an area

that is as nearly as much land as it is ocean bed. The width of the zone varies with the slope of the rocky shore, variations in light and shade, exposure to waves and spray, tidal range, and the frequency of cool days and fogs. At the top of this area, patches of dark lichens and algae appear as crusts on the rocks; these crusts are often nearly indistinguishable from the rock itself. Scattered tufts of algae provide grazing for the small herbivorous snails and limpets of the supralittoral zone. The periwinkle snail *Littorina* is well adapted to an environment that is more dry than wet; it is an air-breather, and some species will drown if caught underwater. At low tide, snails withdraw into their shells to prevent moisture loss and seal themselves off from the air with a horny disk called an operculum. Limpets are able to use their single muscular foot to press themselves tightly to the rocks to prevent drying out.

Just below the zone of snails and limpets, the small acorn barnacles filter food from the seawater and are able to survive even though they are covered with water only briefly during the few days of the spring tides each month. Barnacles are crustaceans, related to crabs and lobsters but cemented firmly in place. They have been described as animals that lie on their backs and spend their lives kicking food into their mouths with their feet. In some areas, the rocky splash zone is the home of another crustacean, the large (3–4 cm, or 1–2 in) isopod *Ligia*. Organisms of the supralittoral zone are shown in figure 14.8.

Conspicuous members of the upper midlittoral zone are illustrated in figure 14.9. These organisms include several other species of barnacles, limpets, snails, and two other **mollusks:** the bivalved (or two-shelled) mussels and chitons, which may appear to resemble limpets but on closer inspection will be seen to have shells of eight separate plates. Chitons, like limpets, are grazers that scrape algae from hard surfaces. Mussels are filter feeders; food strained from the water is trapped in a heavy mucus and moved to the mouth by liplike palps. A muscular foot anchors chitons and limpets; strong cement secures barnacles; and special threads attach mussels to the rocks. Tightly closed shells protect many of the organisms from drying out during periods of low tide, and their rounded profiles present little resistance to the breaking waves. Species of brown algae in this zone, typically rockweed (*Fucus*), have strong holdfasts and flexible stipes.

Gooseneck barnacles are found attached to rocks where wave action is strong. They have evolved an interesting feeding style, facing shoreward and feeding by taking particulate matter from the runback of the surf rather than facing the sea, as might be expected. Mussel beds promote shelter for less conspicuous animals, such as the segmented sea worm *Nereis* and small crustaceans. Shore crabs of varied colors and patterns are found in the moist shelter of the rocks. The crabs are active predators as well as important scavengers. Small sea anemones, huddled together in large groups to conserve moisture, are also found in the higher regions of the midlittoral zone.

The area is crowded; the competition for space appears extreme. The free-swimming juveniles (or larval forms) of these animals settle and compete for space, each with its own special set of requirements. New space in an inhabited area becomes available as the whelks, which are carnivorous snails, prey on some species of thinner-shelled barnacles, and the sea stars move up with the tide to feed on the mussels. This predation restricts the thinner-shelled barnacles to the upper levels of the littoral zone, which are too dry for the predator snails. Other species of barnacles inhabit the lower midlittoral zone in association with the predator snails, which cannot pierce the heavier plates of the mature individuals. Seasonal die-offs of algae and

the battering action of strong seas and floating logs also act to clear space for newcomers.

A selection of organisms from the lower littoral zone is found in figure 14.10. The larger anemones are common inhabitants of the midlittoral and lower littoral zones. These delicate-looking, flowerlike animals attach firmly to the rocks and spread their tentacles, loaded with poisonous darts called nematocysts. The darts are fired when small fish, shrimp, or worms brush the tentacles. The prey is paralyzed, the tentacles grasp, and the prey is pushed down into the anemone's central mouth. Some snails and sea slugs are unaffected by the nematocysts and prey on the anemones. Certain sea slugs store the anemone's nematocysts in their tissues and use them for their own defense.

Sea stars of many colors and sizes make their home in the lower littoral zone; these slow-moving but voracious carnivores prey on shellfish, sea urchins, and limpets. Their mouths are on their undersides, at the center of their central disks, surrounded by strong arms that are equipped with hundreds of tiny suction cups, or tube feet. The tube feet are operated by a water-vascular system, a kind of hydraulic system that attaches the animal very firmly to a hard surface. In feeding, the tube feet attach to the shell of the intended prey and, by a combination of holding and pulling, open the shell sufficiently to insert the sea star's stomach, which can be extruded through its mouth; enzymes are released, and digestion begins. Shellfish sense the approach of a sea star by substances it liberates into the water, and some execute violent escape maneuvers. Scallops swim jerkily; clams and cockles jump away; even the slow-moving sea urchins and limpets move as rapidly as possible.

The filter-feeding sponges, some flat and some vase-shaped, encrust the rocks. On a minus tide, delicate, free-living flatworms and long **nemerteans,** or ribbon worms, armed with poison-injecting mouthparts, are found keeping moist under the mats of algae. Snails and crabs inhabit this zone; the scallop, another filter-feeding bivalve, and red algae are found as well, along the beds of kelp, eelgrass, and surf grass. Calcareous red algae encrust some rocks and are seen as tufts on others. Occasionally,

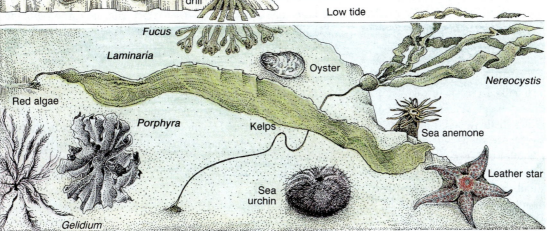

Figure 14.7 A typical distribution of benthic algae and animals on a rocky shore at temperate latitudes. Vertical zonation is the result of the relationships of the organisms to their intertidal environment.

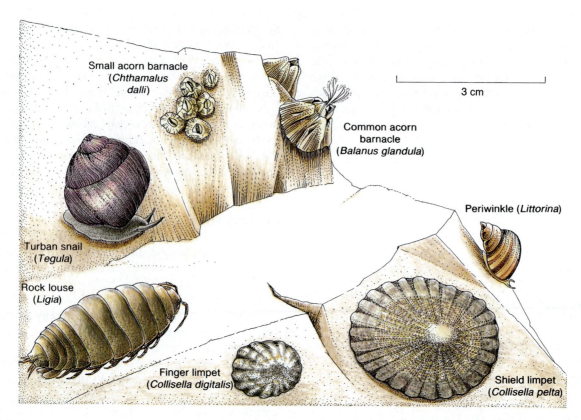

Figure 14.8 Organisms of the supralittoral zone. The limpet and the snail *Littorina* are herbivores. The barnacles feed on particulate matter in the water. *Ligia* is a scavenger.

brachiopods, or lampshells, are found in the lower littoral zone. They resemble clams but are completely unrelated to them. Their shells enclose a coiled ridge of tentacles used in feeding.

Beautiful, graceful, and colorful sea slugs, or **nudibranchs,** are active predators, feeding on sponges, anemones, and the spawn of other organisms. Although soft-bodied, they have few if any enemies because they produce poisonous acidic secretions. Herbivores are also present in the lower intertidal region; species of chitons and limpets as well as the sea urchin graze on the algae covering the rocks. Sea cucumbers are found wedged in cracks and crevices; some types are identifiable by their brightly colored tentacles, which act as mops to remove food particles from the water and thrust them into the animal's mouth. Tube worms secrete the leathery or calcareous tubes in which they live and extend only their graceful, feathered tentacles to strain their food from the water.

Octopuses are seen occasionally from shore on very low tides. These eight-armed carnivorous animals are soft-bodied mollusks. They feed on crabs and shellfish and live in caves or dens identifiable by the piles of waste shells outside. They are known for their ability to flash color changes and move gracefully and swiftly over the bottom and through the water. The world's largest octopus is found in the coastal waters of the eastern North Pacific. It commonly measures 2–3 m (10–16.5 ft) in diameter and weighs 20 kg (45 lb), but specimens in excess of 7 m (23 ft) and 45 kg (100 lb) have been observed. They are shy and nonaggressive, although they are curious and have been shown to have learning ability and memory.

Tide Pools

Tide pools are often our first introduction to the beauty and diversity of the rocky shore. The zonation of benthic forms varies with local conditions; zones are generally narrow where the beach is steep and the tidal range is small, while zones are wide in areas where the beach is flat and the range of tides is large.

The deeper the tide pool and the greater the volume of water, the more stable the environment during isolation by the receding tide. The larger the tide pool, the more slowly it will change temperature, salinity, pH, and the carbon dioxide–oxygen balance. Subtidal animals such as sea stars, sea urchins, and sea cucumbers can only survive in large, deep tide pools. These animals do not tolerate significant changes in their chemical and physical environment. A few fish species, such as the small sculpins, can be found in tide pools. These fish are patterned and colored to match the rocks and the algae within the pool. They spend much of their time resting on the bottom, swimming in short spurts from one resting place to another. Each tide pool is a specialized environment populated with organisms that are able to survive under the conditions established in that particular pool.

Some small, isolated tide pools provide a very specialized habitat of increased salinity and temperature due to evaporation and solar heating. On a summer day, the water in a tide pool may feel quite warm to the touch. Other tide pools act as catch basins for rainwater, lowering the salinity of the water and its

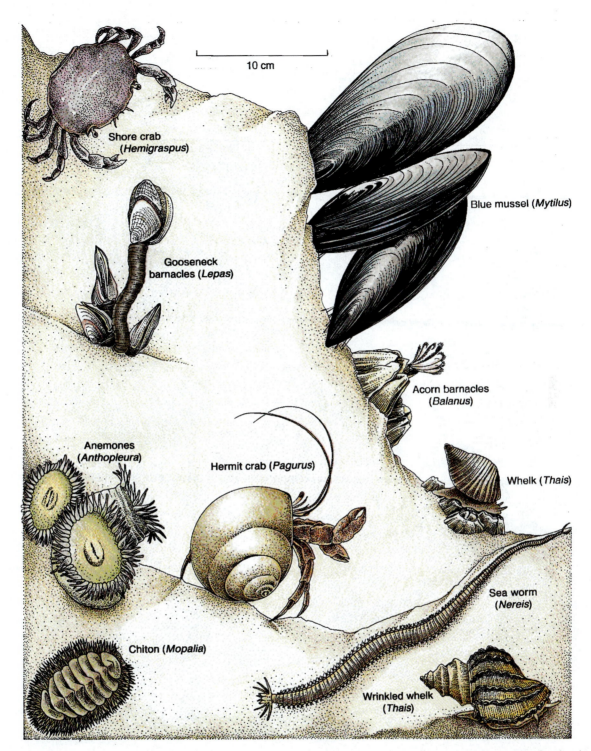

Figure 14.9 Representative organisms from the midlittoral zone. The mussels and barnacles filter their food from the water. The chitons graze on the algae covering the rocks. *Thais,* a snail, is a carnivore. Small shore crabs and hermit crabs are scavengers. *Balanus cariosus* is a larger and heavier barnacle than the barnacles of the supralittoral zone. *Nereis* is often found in the mussel beds. The anemones huddle together to keep moist when exposed.

temperature in the fall and winter. Isolated tide pools often support blooms of microscopic algae that give the water the appearance of pea soup, and in various parts of the world, a tiny, bright-red copepod, *Tigripus,* is also found.

Submerged Rocky Bottoms

Hard bottoms make up a small but important part of the continental shelf. Unlike the intertidal, these regions are always submerged, and dessication is never a problem. Seaweeds (see section 14.2) dominate, but the algae must compete with sessile

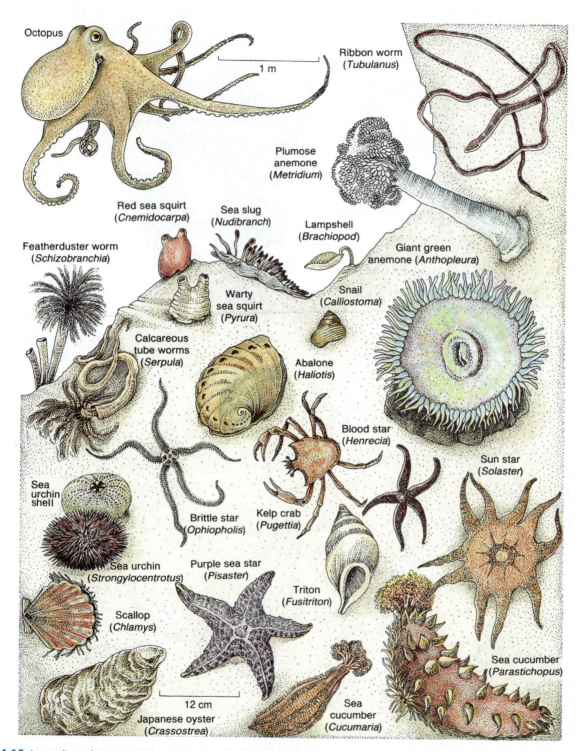

Figure 14.10 Lower littoral zone organisms. A variety of related organisms inhabit the area. The sea stars feed on the oysters; the related sea urchins are herbivores; and the sea cucumbers feed on detritus suspended in the water. Among the mollusks are the oysters, scallops, snails, abalone, nudibranchs, and octopuses. The oysters and scallops are filter feeders. *Calliostoma* and the abalone are grazers; the nudibranch, octopus, and triton snail are predators. Note the size of the anemones in this zone.

organisms for places to attach. In regions where kelp beds form, vertical and horizontal (depth) zonation is again apparent (fig. 14.11). Closest to the intertidal, feather-boa kelp (*Egregia*), which can tolerate strong wave action, tends to dominate. In deeper waters, bull kelp (*Nereocystis*) and giant kelp (*Macro-* *cystis*) take over, interspersed with other seaweeds. Offshore, elk kelp (*Pelagophycus*) extend to depths where light becomes limiting. This mixture of macroalgae forms a **kelp forest**, which has similarities to terrestrial forests. At the surface, the kelp form a **canopy**, with an understory below that, and low-light-adapted

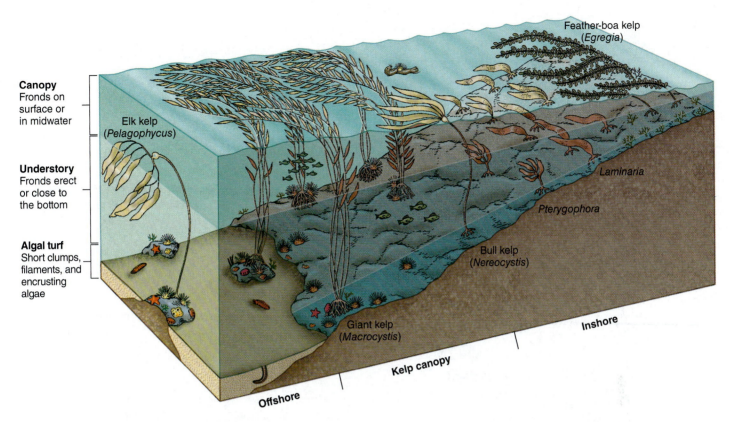

Figure 14.11 Kelp forests form a complex distribution of algae and animals resulting from the effects of factors such as light, substrate, wave action, depth, and the number and type of grazers. Depicted is a generalized schematic for a kelp forest along the west coast of North America.

(typically red) algae growing along the bottom. The kelp forest supports a tremendous variety of both pelagic and benthic organisms. Grazers are usually small, slow-moving invertebrates such as sea urchins, chitons, limpets, and abalone, as well as several types of fishes. Carnivores, including gastropods, sea stars, lobsters, and eels are also common, as well as many more fishes and mammals, such as the sea otter, that are more loosely associated with the kelp forest.

The bottom of the littoral (or intertidal) zone merges into the beginning of the sublittoral (or subtidal) zone extending across the continental shelf. If the shallow areas of the subtidal zone are rocky, many of the same lower littoral zone organisms will be found. When soft sediments begin to collect in protected areas or deeper water, the populations change, and animals of the rocky bottom are replaced by those of the mud and sand substrates.

QUICK REVIEW

1. Define *intertidal zonation* and explain what causes it.
2. Give examples of animals living in each region of the rocky intertidal.
3. Give examples of adaptations to the physical environment of the rocky intertidal.
4. How does zonation differ in the rocky intertidal and in a kelp forest?
5. Explain how predation might lead to increased biodiversity.

14.4 Animals of the Soft Substrates

The distribution of life in soft sediments is shown in figure 14.12. A selection of animals from this region is found in figure 14.13. Along exposed gravel and sand shores, waves produce an unstable benthic environment. Few algae can attach to the shifting substrate, and few grazing animals are found. Sands and muds deposited in coves and bays with reduced water motion provide a more stable habitat. Here, the size and shape of the sediment particles and the organic content of the sediment determine the quality of the environment. The size of the spaces between the particles regulates the flow of water and the availability of dissolved oxygen. Beach sand is fairly coarse and porous, gaining and losing water quickly, while fine particles of mud hold more water and replace the water more slowly. Sand beaches exchange water, dissolved wastes, and organic particles more quickly than mud flats. The finer the mud particles, the tighter they pack together and the slower the exchange of water; oxygen is not resupplied quickly, and wastes are removed slowly. Digging into the mud will generally show a black layer 1 or 2 cm (0.5 or 1 in) below the surface. Above this layer, the water between the sediment particles contains dissolved oxygen; below the black layer, organisms (mainly bacteria) function without oxygen, producing hydrogen sulfide (the rotten egg smell). Lack of oxygen restricts the depth to which infauna species can be found, but some animals,

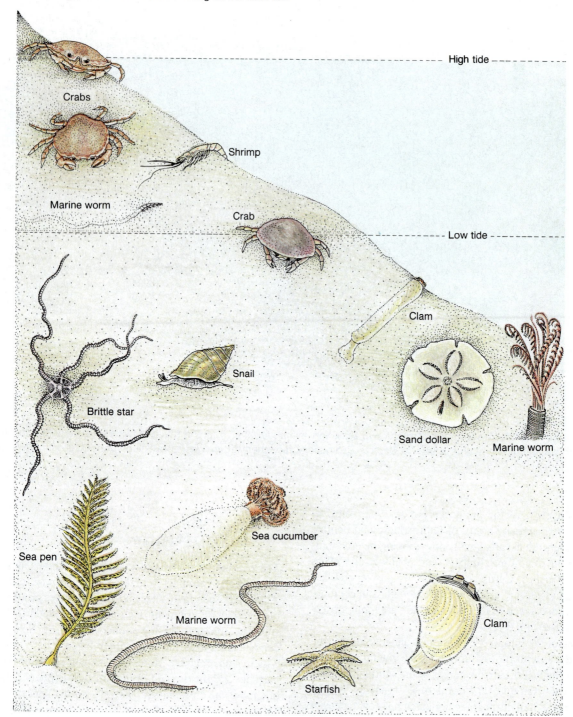

High tide

Crabs

Shrimp

Marine worm

Crab

Low tide

Clam

Snail

Brittle star

Sand dollar

Marine worm

Sea pen

Sea cucumber

Marine worm

Clam

Starfish

Figure 14.12 Zonation on a soft-sediment beach is less conspicuous than that found on a rocky beach. Animals living at the higher-tide levels burrow to stay moist.

such as clams, live below the oxygenated level. Clams use long extensions called siphons to obtain food and oxygen from the water above the sediments.

In locations protected from waves and currents, eelgrass and surf grass help stabilize the small-particle sediments and provide shelter, substrate, and food, creating a special community of plants and animals.

Most sand and mud animals are **detritus** feeders. Most detritus is formed from plant and algal material that is degraded

by bacteria and fungi. The sand dollar feeds on detritus particles found between sand grains. Clams, cockles, and some worms are filter feeders, feeding on the detritus and microscopic organisms suspended in the water. Other animals are deposit feeders that engulf the sediment and process it in their gut to extract organic matter in a manner similar to that of earthworms. These deposit feeders are usually found in muds or muddy sands that have a high organic content: for example, burrowing sea cucumbers and the lugworm *Arenicola,* which produces the coiled castings

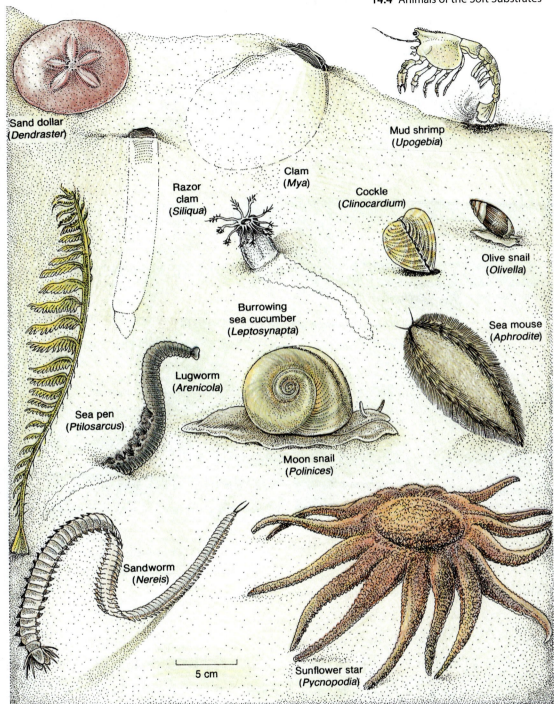

Figure 14.13 Organisms of the soft sediments. Infauna types include the shrimp (*Upogebia*), lugworm (*Arenicola*), clam, cockle, and burrowing sea cucumber. The sand dollar feeds on detritus; the Moon snail drills its way into shellfish; the sea pens feed from the water above the soft bottom. The sand worm, like *Nereis*, is a polychaete worm.

seen outside its burrow. The process of sediment disruption by feeding or burrowing organisms is known as **bioturbation.** Small crustaceans, crabs, and some worm species are scavengers, preying on any available algal or animal material, while still other worms and snails are carnivores. The Moon snail is a clam eater that drills a hole in the shell of its prey and then sucks out the flesh.

Bacteria not only play the major role in the decomposition of organic material; they also serve as a major protein source. It is estimated that 25–50% of the material the bacteria decompose is converted into bacterial cell material, which is consumed by other microorganisms, which in turn serve as food for tiny worms, clams, and crustaceans (fig. 14.14). Areas of mud that are high in organic detritus produce large quantities of bacteria.

The intertidal area of a soft-sediment beach shows some zonation of benthic organisms, but it is not nearly as clear-cut as the zonation along a rocky cliff. In temperate latitudes, small crustaceans called sandhoppers are found at the high intertidal

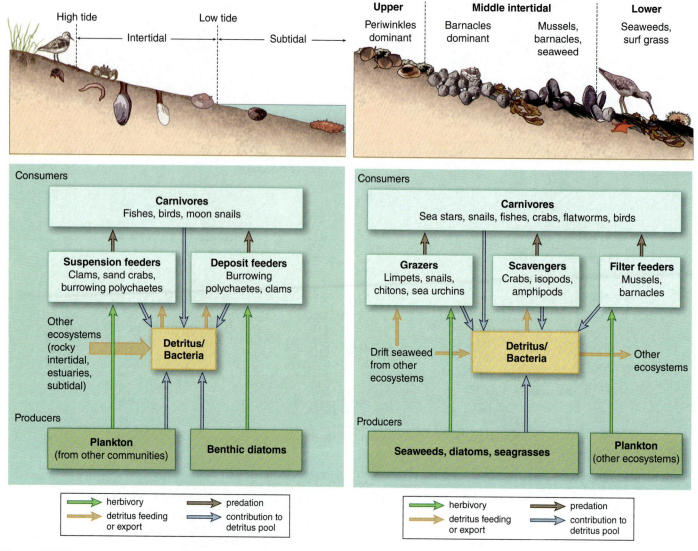

Figure 14.14 Generalized features of a rocky shore (right) and a sandy or muddy shore (left).

region; they are replaced by ghost crabs in the tropics. Lugworms, mole crabs, and ghost shrimp occupy the midbeach area, while clams, cockles, polychaete worms, and sand dollars are found in the lower intertidal region. The subtidal zone is home to sea cucumbers, sea pens, more crabs and clams, and some species of worms, snails, and sea slugs.

QUICK REVIEW

1. Compare the environmental conditions leading to a mudflat versus a sandy beach.

2. Why is detritus important in mudflat and sandy shore environments?

3. Compare the zonation of a sandy beach versus the rocky intertidal.

4. Explain why few benthic organisms live on a beach made of noncohesive sediments in a wave and surf area.

14.5 Animals of the Deep-Sea Floor

The deep-sea floor includes the flat abyssal plains, the trenches, and the rocky slopes of seamounts and mid-ocean ridges. This region is by far the largest marine benthic habitat. The seafloor sediments are more uniform and their particle size is smaller than those of the shallower regions close to land sources. The environment of the bathyal, abyssal, and hadal zones is uniformly cold and dark, but not devoid of life.

The stable conditions of the deep-sea floor appear to have favored deposit-feeding infaunal animals of many species. Many members of the deep-sea infauna are small. They are known as the **benthic meiofauna** and measure 2 mm or less. This group of organisms include nematode worms, burrowing crustaceans, and segmented worms. At 7000 m (23,000 ft), tusk shells lie buried in the ooze with tentacles at the sediment surface to feed on the foraminifera. Acorn worms are found frequently in samples taken at 4000 m (13,000 ft). Hagfish burrow into the sediment at the 2000 m (6600 ft) depth. Detritus-eating worms and bivalved mollusks have been found on the sea floor in all the oceans.

Among the epifauna, protozoans are abundant and are widely distributed. Glass sponges attach to the scattered rocks on oceanic ridges and seamounts (fig. 14.15), as do sea squirts and sea anemones. Their stalks lift them above the soft sediments into the water, where they feed by straining out organic matter. Stalked barnacles attach to the stalks of glass sponges and sea squirts as well as to shells and boulders. Tube worms are common, ranging in size from a few millimeters to 20 cm (8 in), and sea spiders with four pairs of very long legs that span up to 60 cm (27 in) are found at depths to 7000 m (23,000 ft). Snails

are found to the greatest of depths; those in the deepest trenches frequently have no eyes or eyestalks.

The beard worms, or **pogonophora,** are found in more productive areas at depths to 10,000 m (33,000 ft). They secrete a close-fitting tube and stand erect, with only their lower portion buried in the sediment. They have no mouth, no gut, and no anus and absorb their needed molecules through their skin.

Horny corals, or sea fans, which resemble algae more than animals, grow at depths of 5000–6000 m (16,000–20,000 ft); so do solitary stone corals, which grow larger at these depths than do the coral organisms in the surface waters. Sea lilies, or crinoids, which are related to sea stars, are also found at this depth,

Figure 14.15 (a) A vase-shaped glass sponge 640 m (2100 ft) under the surface on the Brown Bear Seamount in the northeastern Pacific Ocean. (b) A deep-sea crab photographed at a depth of 2000 m (6550 ft) on the Juan de Fuca Ridge. (c) A group of deep-sea sponges and an anemone are seen at 684 m (2244 ft) on the Brown Bear Seamount. (d) A deep-sea amphipod (*Alicella gigantea*) provides an example of deep-sea gigantism.

as are brittle stars and sea cucumbers. Sea cucumbers live in areas of sediments that are rich in organic substances; they are a dominant and widely spread organism of the deep-sea floor.

Perhaps one of the most unusual features of deep-sea animals is the development of **deep-sea gigantism**. Giant amphipod and isopod crustaceans are common members of the deep benthic community. Some, such as *Bathynomus giganteus*, a type of isopod, can reach 50 cm in length (compared to typical isopod lengths of 1–5 cm (~0.5–2 in)! Gigantism is not confined to benthic animals; the giant and colossal squid (*Architeuthis* and *Mesonychoteuthis*) are pelagic examples of the same phenomenon. Why do these organisms get so large? Several theories exist, including changes in metabolism linked to cold temperatures and high pressure, or continued growth in very long-lived species, but it is still not clear how or why this occurs.

QUICK REVIEW

1. Describe the environment of the deep-sea benthos. What kinds of organisms make a living there?
2. What are some sources of food for deep-sea organisms?
3. Describe the species diversity of the deep-sea benthos. Is it lower or higher than expected?
4. Why do some deep-sea animals become unusually large?

14.6 Coral Reefs

Tropical coral reefs are the most diverse and structurally complex of all marine communities. Coral reefs fringe one-sixth of the world's coastlines and provide habitat for tens of thousands of fish and other organisms. The largest coral reef in the world, the Great Barrier Reef, stretches more than 2000 km (1200 mi) from New Guinea southward along the eastern coast of Australia. Reef-building corals require warm, clear, shallow, clean water and a firm substrate to which they can attach. Because the water temperature must not go below 18°C and the optimal temperature is 23°–25°C, their growth is restricted to tropical waters between 30°N and 30°S and away from cold-water currents. Most Caribbean corals are found in the upper 50 m (164 ft) of lighted water; Indian and Pacific corals are found to depths of 150 m (500 ft) in the more transparent water of those oceans. Reefs usually are not found where sediments limit water transparency.

Tropical Corals

Corals are colonial animals, and individual coral animals are called **polyps** (fig. 14.16). A coral polyp is very similar to a tiny sea anemone with its tentacles and stinging cells, but, unlike the anemone, a coral polyp extracts calcium carbonate from the water and forms a calcareous skeletal cup. Large numbers of these polyps grow together in colonies of delicately branched forms or rounded masses.

Clear, shallow water is required by the reef-building coral, because within the tissues of the polyps are masses of single-celled dinoflagellate algae called **zooxanthellae** that require light for photosynthesis and therefore are limited to the photic

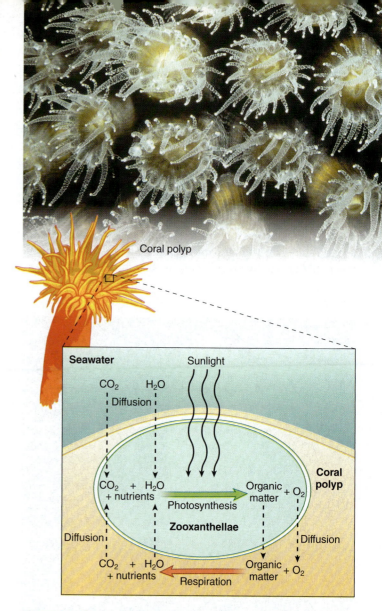

Figure 14.16 A colony of star coral polyps on a reef in the Caribbean Sea, Bonaire, Lesser Antilles. The inset shows the symbiosis between the coral polyp (animal) and the zooxanthellae. Most reef-building corals are colonies of interconnected polyps that use their tentacles to capture food and the zooxanthellae to photosynthesize.

zone. Polyps and zooxanthellae have a **symbiotic** relationship, in which the coral provides the algal cells with a protected environment, carbon dioxide, and nitrogen and phosphorus nutrients, and the algal cells photosynthesize, returning oxygen and removing waste. **Symbiosis** is a close ecological relationship between organisms of two different species. The zooxanthellae supply the corals with substantial amounts of their photosynthetic products; some coral species receive as much as 60% of their nutrition from the algae. Zooxanthellae also enhance the ability of the coral to extract the calcium carbonate from the seawater and increase the growth of their calcareous skeletons. The degree of interdependence between zooxanthellae and coral is thought to vary from species to species. The polyps feed actively at night, extending their tentacles to feed on zooplankton, but during the day, their tentacles are contracted, exposing the outer layer of cells containing zooxanthellae to the sunlight.

Tropical Coral Reefs

The corals require a firm base to which they can cement their skeletons. The classic reef types of the tropical sea—fringing reefs and barrier reefs—are attached to existing islands or landmasses. Atolls are attached to submerged seamounts. (For a review of reef formation around a seamount, see chapter 3.) Corals are slow-growing organisms; some species grow less than 1 cm (0.4 in) in a year, and others add up to 5 cm (2 in) a year. The same corals may be found in different shapes and sizes, depending on the depth and wave action. Environmental conditions vary over a reef, forming both horizontal and vertical zonation patterns that are the product of wave action and water depth, as shown in figure 14.17. On the sheltered (or lagoon) side of the reef, the shallow **reef flat** is covered with a large variety of branched corals and other organisms. Fine coral particles broken off from the reef top produce sand, which fills the sheltered lagoon floor. On the reef's windward side, the reef's highest point, or **reef crest,** may be exposed at low tide and is pounded by the breaking waves of the surf zone. Here, the more massive rounded corals grow. Below the low-tide line to a depth of 10–20 m (35–65 ft) on the seaward side is a zone of steep, rugged buttresses, which alternate with grooves on the reef face. Masses of large corals grow here, and many large fish frequent the area. The buttresses dissipate the wave energy, and the grooves drain off fine sands and debris, which would

smother the coral colonies. At depths of 20–30 m (65–100 ft), there is little wave energy, and the light intensity is only about 25% of its surface value; still, it is adequate to support reef algae and corals. The corals are less massive at this depth, and more delicately branched forms are found here. Between 30 and 40 m (100 and 130 ft), the slope is gentle and the level of light is very reduced; sediments accumulate at this depth, and the coral growth becomes patchy. Below 50 m (164 ft), the slope drops off sharply into the deep water.

Coral reefs are complex assemblages of many different types of algae and animals (see fig. 14.18). Tropical coral reefs account for somewhere between one-quarter and one-third of all marine species. A conservative estimate suggests that as many as 3000 species may occupy a single reef. A recent study of just 6.3 square meters (20.6 sq ft) from a variety of tropical reefs identified 525 crustacean species, suggesting that the biodiversity of tropical reefs is seriously underestimated. Competition for space and food is intense. Algae, sponges, and corals are constantly overgrowing and competing with each other. Some species are active only at night, such as some fishes, snails, shrimp, the octopus, fireworm, and moray eel. During the day, other species depend on color and vision to make their way. The giant reef clam *Tridacna* can measure up to a meter in length and weigh over 150 kg (330 lb). It also possesses zooxanthellae in large numbers in the colorful tissues that line the edges

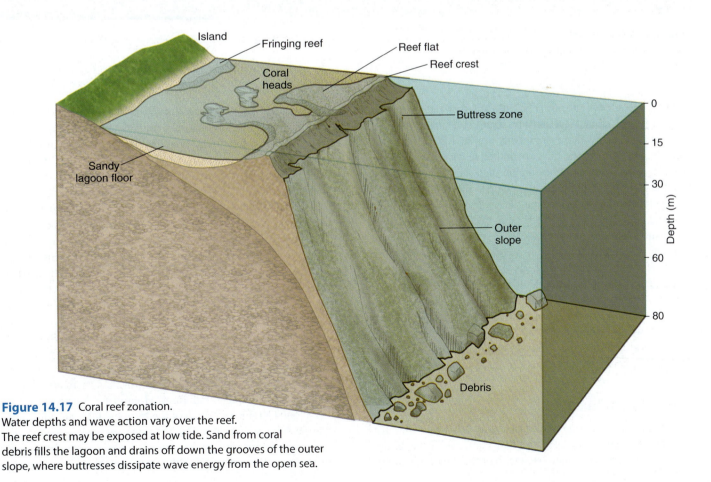

Figure 14.17 Coral reef zonation.
Water depths and wave action vary over the reef.
The reef crest may be exposed at low tide. Sand from coral debris fills the lagoon and drains off down the grooves of the outer slope, where buttresses dissipate wave energy from the open sea.

Figure 14.18 Coral reefs are rich communities of organisms. (a) Corals and a sea fan at Mana Island, Fiji. (b) A butterfly fish *(left)* and a moorish idol *(right)* at Soma Soma Straits, Taveuni Island, Fiji.

of the shell. Crabs, moray eels, colorful reef fish, poisonous stonefish, long-spined sea urchins, seahorses, shrimp, lobsters, sponges, and many more organisms are all found living here together. On some reefs, the zooxanthellae have been shown to produce several times more organic material per unit of space than the phytoplankton, probably owing to the rapid recycling of nutrients between the corals and the zooxanthellae.

The reefs are not formed exclusively from the calcium carbonate skeletons of the coral. Encrusting algae that produce an outer calcareous covering also contribute; so do the minute shells of foraminifera, the shells of bivalves, the calcareous tubes of polychaete worms, and the spines and plates of sea urchins. All are compressed and cemented together to form new places for more organisms to live. At the same time, some sponges, worms, and clams bore into the reef; some fish graze on the coral and the algae; and the sea cucumbers feed on the broken fragments, reducing them to sandy sediments.

Coral Bleaching

Oceanographers, marine biologists, and all those who have enjoyed the experience of a tropical coral reef have become increasingly concerned with the health of these biologically rich but delicately balanced regions of the oceans. Coral bleaching episodes, in which corals expel their zooxanthellae, exposing the coral's white calcium carbonate skeleton, have occurred from time to time, but unless the episodes were especially severe, the corals regained their algae and recovered. Between 1876 and 1979, only three bleaching events occurred, but over sixty bleach events were documented between 1979 and 1990. In the last two decades, this bleaching has become more frequent and more severe, and not all reefs have recovered. Large amounts of bleaching occurred at the time of the 1982–83 El Niño, during which shallow reefs of the Java Sea lost 80–90% of their living coral cover. In 1990–91, large-scale bleaching occurred in both the Caribbean and French Polynesia. During the 1997–98 El Niño event, which lasted for over a year, surface waters reached the highest temperatures on record throughout the Indian and western Pacific Oceans. 2005 saw the most extreme coral bleaching event to hit the Caribbean and Atlantic coral reefs. However, in 2005, an extensive monitoring system coordinated through the Global Coral Reef Monitoring Network was in place. The Coral Reef Watch of the National Oceanic and Atmospheric Administration (NOAA) utilizes remote sensing and in situ tools to monitor and report on the physical environmental conditions of coral reefs. Although warm temperatures are a common cause for bleaching events, there are other triggers. For example, in January 2010 cold water temperatures in the Florida Keys caused a coral bleaching event when water temperatures dropped 6.7°C (20°F). Bleaching can also be triggered by other environmental stresses, including bacterial infection, drops in salinity, extreme light, and various toxins (see *Diving In* box on Undersea Ultraviolet Radiation).

Predation and Disease

Periodically, the population of the sea star *Acanthaster*, known as the crown of thorns, increases dramatically; this sea star feeds on the coral polyps. Fossil evidence points to outbreaks having occurred along the Great Barrier Reef for at least the last 8000 years. Why the *Acanthaster* population increases so rapidly is unknown, but evidence points to a correlation between rainy weather (low salinity) and runoff (increased nutrients), allowing large numbers of *Acanthaster* larvae to survive. The clearing of land for agriculture and the development of coastal areas may also be related. Concern also exists that the harvesting of the large conchs that prey on the sea star may upset the population balance. One outbreak began in 1995 and by 1997 was affecting 40% of the reef. Outbreaks in individual reefs can last one to five years; outbreaks throughout reef systems can last up to a decade or longer. It takes ten to fifteen years for reef areas to recover from an *Acanthaster* outbreak, but the opening up of areas on the reef by the sea star may also allow slower-growing coral species to expand.

Reef-building algae as well as corals have recently been found under attack by several previously unknown diseases. One of these is coralline lethal orange disease, known as CLOD. CLOD is caused by a bright orange bacterial pathogen that is lethal to the encrusting red algae (corallines) that deposit calcium carbonate on the reefs. These algae cement together sand, dead algae, and other debris to form a hard, stable substrate. The disease was initially found in 1993 in the Cook Islands and Fiji; by 1994, it had spread to the Solomon Islands and New Guinea; and by 1995, it was found over a 6000 km (3600 mi) range of the South Pacific. It is unclear whether CLOD has been recently introduced from some obscure location or whether it has been present on the reefs but has now evolved into a more virulent form.

Human Activities

Humans and their activities are among the greatest threats to the reefs. Reefs are damaged by careless sport divers trampling delicate corals and by boats grounding or dragging their anchors. Deforestation and development increase land runoff, carrying sediments, nutrients, and human and animal waste into coastal waters. Additional nutrients favor the growth of algae; the algae out-compete the corals for space, smothering the existing corals and preventing new coral colonies from starting.

Coral reefs are despoiled by shell collectors and mined for building materials. In French Polynesia, Thailand, and Sri Lanka, tons of coral are used as construction material each year. Low-lying coastal areas that lose their coral reefs have no protection against storm surges that accompany extreme weather events. This is particularly important to island nations in the Pacific.

Where catching reef fish with a net is difficult, fishers may use sodium cyanide and explosives. Whether it is cyanide squirted into the water or explosives detonated in the water, the fish are temporarily stunned, making them easy to capture. These methods also kill many other reef organisms, including the reef-building organisms. Although banned in Indonesia and the Philippines, these methods are often the first choice of those in the very lucrative international aquarium and restaurant businesses.

At present, the most seriously threatened areas are considered to be the coasts of East Africa, Indonesia, Philippines, south China, Haiti, and the U.S. Virgin Islands. Ensuring the continued existence of these beautiful and productive areas requires an increased understanding of the complex nature of reef communities and the development of policies designed to protect them from human interference. The United Nations Environment Program, the Intergovernmental Oceanographic Commission, the World Meteorological Organization, and scientists from many different countries are involved in monitoring stations that cover the world's major reefs. Additional marine protected areas, increased scientific monitoring, and more global observer systems to regularly collect data are needed to ensure the survival of these complex and fragile areas.

Deep-Water Corals

Not all corals are found in shallow, warm tropical waters. Deep-water corals, or "cool corals," lack zooxanthellae; they form large reefs at temperatures down to 4°C and at depths of up to 2000 meters. Researchers believe these corals feed on periodic drifts of dead plankton or sift food from passing currents. Some reefs appear to be hundreds of years old. Many fish and invertebrate species aggregate at these deep reefs. Reefs have been located off Norway, Scotland, Ireland, Nova Scotia, Alaska's Aleutian Islands, the Southeast Atlantic coast of the United States, New Zealand, and Tasmania. There is concern that modern fishing with bottom trawling nets may crush these formations; in some places, newly discovered reefs have already been damaged by nets. Canada created a coral preserve in 2001, and Norway has barred trawling in some areas; several other countries have enacted similar measures.

QUICK REVIEW

1. Explain the symbiotic relationship between zooxanthellae and coral polyps.
2. How do coral reefs survive in low-productivity, tropical waters?
3. Describe the conditions necessary for reef-building and deep-water corals.
4. Describe some of the threats to tropical and deep-water corals.
5. What is coral bleaching? Why does it occur?

Dr. Chris Shank is an Assistant Professor at the University of Texas at Austin Marine Science Institute. His research interests are primarily coastal geochemistry and marine photochemistry.

The spectrum of solar irradiance reaching the Earth's surface consists of wavelengths in the ultraviolet (300–400) and visible region (400–700 nm), and both UV and visible light waves are capable of penetrating the ocean surface. While lower-energy visible light is beneficial to many underwater marine organisms (photosynthetic and nonphotosynthetic), high-energy UV is potentially very hazardous, just as it can cause acute (sunburns) and chronic (skin cancer) problems to humans. UV radiation is categorized into UVC (<280 nm), UVB (280–315 nm), and UVA (315–400 nm), but luckily for Earth's inhabitants, no wavelengths <300 nm reach the Earth's surface. UVB carries more energy and has greater destructive potential to organismal physiologic processes per photon than does UVA, but the ocean surface receives much higher exposure doses of UVA that also are potentially hazardous to marine organisms including corals.

Many important coral physiologic processes are influenced by solar radiation. For example, symbiotic zooxanthellae living within coral tissues require sufficient levels of visible light for photosynthesis. However, solar UV inhibits photosynthesis and can ultimately lead to coral bleaching. High ocean temperatures are often a primary cause of these bleaching episodes, but excessive UV irradiance has also been shown to induce coral bleaching. In many tropical areas, there is a synergistic effect of stressors, as highest ocean temperatures also commonly occur during periods of excessive underwater UV. Another reason UV is so dangerous is that it induces the production of reactive oxygen species

(ROS), radical molecules such as hydrogen peroxide that are extremely reactive and highly destructive to cell tissues. Because solar UVB and UVA may penetrate into ocean waters to depths of tens of meters, marine organisms such as corals that live in "clear" shallow waters are especially susceptible to UV damage.

In coastal ocean waters, including along many tropical regions with coral reefs, the primary attenuator of underwater UV is a pool of naturally occurring organic compounds known as chromophoric dissolved organic matter (CDOM). CDOM is very effective at limiting the penetration of UVA and UVB because it strongly absorbs solar radiation in the UV wavelengths. The presence of CDOM can be easily observed by noting the green to brown coloration of the water as opposed to a blue ocean color in waters with little CDOM. CDOM is not one particular organic molecule, but rather a heterogeneous pool of dissolved organic compounds. Further, it should not be confused with suspended phytoplankton or detrital particles that may also appear to have color (and can also absorb UV). There are numerous sources of CDOM in coastal waters, including decaying vegetation such as mangrove leaf litter, sea grasses, and phytoplankton blooms. Another important natural sunscreen for corals is a class of organic compounds called mycosporine-like amino acids (MAAs), which are produced by the zooxanthallae and are also capable of strongly absorbing UV wavelengths. Collectively, CDOM and MAAs (as well as light absorption and scattering by particles to some extent) control the magnitude of UV exposure by corals in a fashion similar to how sun lotions protect humans against sunburns.

The Florida Keys contain the only coral reefs in the contiguous United States. Interestingly, average coral cover along the patch reefs located nearer to the Keys island chain appears to be higher than coral cover along much of the semi-continuous offshore fringing reef habitats. Nearer to shore along the patch reef habitats, CDOM levels are much higher than in the offshore regions; thus, the water is considerably more opaque to visible light and UV closer to shore. One key facet of research into the health of coral reefs in the Florida Keys is determining the spatial and temporal (daily, seasonal, interannual) dynamics of underwater UV irradiances experienced by corals in the region and the influence of

14.7 Deep-Ocean Chemosynthetic Communities

Hot Vents

In March 1977, an expedition from Woods Hole Oceanographic Institution using the research submersible *Alvin* discovered densely populated communities of animals living around hydrothermal vents along the Galápagos Rift of the East Pacific Rise. Before then, all deep-sea benthic communities were assumed to consist of small numbers of deposit-feeding, slow-growing

animals living on and in the soft sediments of the sea floor and depending for nutrition on the slow descent of decayed organic material from the surface layers. Vent communities have now been found scattered throughout the world at seafloor spreading centers, and more than 700 new species have been identified, including in Antarctica (Fig. 14.19).

The animals living in these deep-sea vent areas include filter-feeding clams and mussels, in addition to anemones, worms, barnacles, limpets, crabs, and fish. The clams are very large and show one of the fastest growth rates known for deep-sea animals, up to 4 cm (2 in) per year. The tube worms

Box Figure 1 Deploying UV sensors at Sombrero Tower near Marathon Key, Florida.

CDOM in controlling corals' UV exposure (Box figure 1). A key question guiding this research is: Could corals' lower exposure to UV along the patch reef habitats be an important factor in lowering incidence of coral bleaching relative to the offshore reefs?

Throughout the middle and lower Keys region, several years of research have shown that CDOM accounts for approximately 90% of the UVB and 80% of the UVA attenuation, with absorption and scattering of UV by particles relatively minor. The overall importance of CDOM to the region's reefs is highlighted by optical models revealing that damaging UV levels with the potential to cause DNA damage in Keys' corals are more influenced by CDOM fluctuations than by atmospheric ozone fluctuations. The major source of CDOM to the middle Keys region is the discharge of water from Florida Bay, which flows through a couple of major channels between Keys, moving generally in a southwestward motion towards the lower Keys. Florida Bay is shallow (typical depth

~1–2 m), with a large seagrass community and numerous mangrove fringed islands. Florida Bay also receives pulses of freshwater and organic material from south Florida, including CDOM-rich waters flowing from the Everglades. Thus, the hydrology of the Everglades-Florida Bay system is intimately linked to the optical clarity of waters surrounding the coral reefs in the Florida Keys. This is of special concern because freshwater flows in the Everglades have been altered substantially over the past several decades as the population of south Florida has grown, concurrently with the decline of corals in the Keys. Yet, to this point, the direct links between declining coral health and Everglades hydrology have been difficult to ascertain in the middle and lower Keys.

Discharge of water through Florida Bay exhibits a distinct seasonal signal, with highest discharges typically during winter and summer, and lowest discharges during fall and spring. In turn, CDOM levels along the Keys also exhibit a similar seasonal pattern, although there can be large daily CDOM fluctuations due to tidal and wind-driven current patterns. During periods of low Florida Bay discharge, seagrasses and mangrove expanses along the Keys islands provide important quantities of CDOM to reef waters. As mentioned above, Florida Bay CDOM discharge is usually high during summer, thus it would seem that Keys' corals would be reasonably well protected against summertime's high UV levels. However, because CDOM strongly absorbs UV, it is also subject to a process called photobleaching (not to be confused with coral bleaching), where the pool of UV-absorbing organic material pool is photochemically altered, reducing its ability to further absorb UV. Thus, over time, in the absence of replacement CDOM, the water column will become optically clearer to UV (and visible light) as photobleaching lowers CDOM levels. Photobleaching is especially strong during summer months when the highest atmospheric UV levels occur, and water transparency can become especially clear in late summer after periods of extended photobleaching. Late summer is also the time of highest ocean temperatures in the Keys, so there is a potentially very hazardous synergy between temperature and UV. This setup is especially pronounced during El Niño years when low winds cause water stratification along the Keys, providing ideal conditions for both water heating and loss of CDOM sunscreen from photobleaching. Not coincidentally, coral bleaching episodes in this region have been shown to increase markedly during El Niño years, and evidence also suggests that incidence of bleaching appears much lower in the nearshore waters where underwater UV is mitigated by higher CDOM levels. Research clearly shows that the future of the Florida Keys corals are intimately linked to both human alterations of the Everglades-Florida Bay ecosystem and alterations in global climate that may raise ocean temperatures and influence the severity of El Niño episodes.

(fig. 14.20) are startling in size, up to 3 m (10 ft). They have been placed in the phylum Vestimentifera (a phylum is the most basic of taxonomic categories). These crowded communities have a biomass per unit area that is 500–1000 times greater than the biomass of the usual deep-sea floor.

The tube worms in the vent areas have no mouths and no digestive systems; the soft tissue mass of their internal body cavity is filled with bacteria. The tube worms have the ability to transport sulfide made nontoxic by a binding protein, carbon dioxide, and oxygen to the bacteria held in their body tissues; the synthesis of organic molecules is done within the bacteria.

These kinds of bacteria are the primary producers in the communities surrounding vents. They use chemosynthesis to fix carbon dioxide into organic molecules. Despite their size and abundance, these tube worms are completely dependent on an internal symbiosis with the bacteria for their nutrition. Both clams and mussels have large numbers of bacteria in their gills. The mussels have only a rudimentary gut, and the clams and mussels have red flesh and red blood; the color is due to the oxygen-binding molecule hemoglobin. The oxygen is needed for oxidation of the hydrogen sulfide and to maintain body tissues and high growth rates.

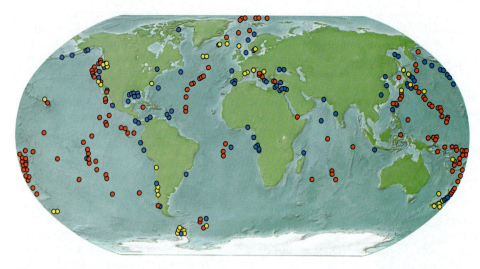

Figure 14.19 Global distribution of hydrothermal vent (*red*), cold seep (*blue*), and whale fall (*yellow*) sites that have been studied by the Census of Marine Life with respect to their fauna.

Figure 14.20 Tube worms crowd around deep-sea hydrothermal vents. The internal body cavity of each worm is filled with bacteria that synthesize organic molecules by chemosynthesis. *Photo © Richard A. Lutz.*

The vent plankton and the bacteria in the water provide nutrition for a variety of filter feeders. Other bacteria form mats that surround the vents, and these bacterial mats are the food for snails and other grazers. The vent microorganisms are the base of a broad-based, self-contained trophic system in which nutrition passes from animal to animal by symbiosis, grazing, filtering, and predation.

In 1991, a submersible monitoring vents on the East Pacific Rise, west of Acapulco and at a depth of 2500 m (1.5 mi), came upon a vent area immediately after a volcanic eruption that killed the larger organisms and left a white carpet of bacteria over the sea floor. Returning in 1992, researchers found the bacterial mats being grazed by large groups of fish, crabs, and other species. By 1993, giant tube worms were more than a meter (3.3 ft)

high. Growing at nearly 1 m per year, these tube worms are another extremely fast-growing marine invertebrate. At this time, researchers found that metal-rich sulfide deposits had grown into 10 m (30 ft) chimneys. Previously, geologists had believed that such formations took decades to form. Between 1995 and 1997, the number of species more than doubled, from twelve to twenty-nine. Mussels and small worms called serpulids moved in and so did clouds of tiny crustaceans, but no giant white clams had yet arrived.

How vents are first colonized by various animals is still not fully understood. Initially, the distances between vent areas appeared to be a barrier, but as more and more vents are discovered, distance seems less of an obstacle when combined with the fast maturation of the organisms and the large numbers of free-swimming larvae they produce. For another way in which organisms might move between vents, see the Diving In box titled "Whale Falls" in chapter 13.

Cold Seeps

Perhaps as surprising as the discovery of life at deep hydro-thermal vents, other slower-growing communities based on chemosynthesis have also been found associated with cold seepage areas. These include coastal seeps off Alaska, Oregon, California, Florida, Japan, and Africa. Bacteria are the primary producers of the communities, using methane and hydrogen sulfide as sources of energy. Similar to what has occured in hydrothermal vents, more than 600 new species have been identified at these cold seeps. Along the continental slope of Louisiana and Texas in the Gulf of Mexico, faulting has frac-tured the sea bottom and produced environments where oil and gas seep up and onto the surface of the sea floor. Clams, mussels, and large tube worms were first collected from these sites (fig. 14.21) in 1985. A variety of other animals, including fishes, crustaceans, and mollusks, commonly found along the continental shelf are abundant around these seeps, attracted by the food supply.

In 1990, salt seeps were reported on the floor of the Gulf of Mexico. The escape of gas through surface sediments has formed depressions that filled with salt brine more than 3.5 times the usual salinity of seawater. These extremely dense brine lakes also contain methane and are surrounded by large communities of mussels.

14.8 Symbiosis

Our early view of the deep ocean was that it is cold, dark, and inhospitable to life. The deep benthos is, in fact, host to an amazing species richness and diversity of lifestyles. Some communities rely on the rainfall of organic matter from above,

(a)

(b)

(c)

Figure 14.21 Photographs from Gulf of Mexico cold seeps.
(a) Deep-sea gorgonian corals, crabs, and tubeworms.
(b) *Bathymodiolus* mussels. (c) Tube worms.

and many more communities are driven by chemosynthesis. Wherever there is a source of chemical energy, ocean benthic communities have found a way to make a living. Throughout this chapter, there have been many examples of symbiosis, which allows benthic organisms to survive in regions where they might not otherwise exist. Symbioses are very common and integral to the normal functioning of ocean ecosystems.

There are three different categories of symbiosis. In **commensalism**, one partner benefits from the relationship and the other is unaffected. For example, barnacles that live on whales benefit from movement through the water, but the whale is thought to be unaffected. In **mutualism**, both partners benefit from the relationship (fig. 14.22). The hydrothermal vent worm *Riftia* and its symbiotic chemosynthetic bacteria that provide organic matter, the angler fish and its symbiotic bacteria that provide light, and the close association between coral polyps and zooxanthallae that provide the corals with the organic matter necessary for growth are all examples of mutualism. In each case, the host provides waste products to the symbiont (used as food by the symbiont), as well as a sheltered habitat. In exchange, the symbiotic partner provides a benefit to the host. In **parasitism**, one partner lives at the expense of the other. Parasites obtain food and shelter by damaging, but not killing, the hosts. For example, parasitic worms are found in most fish.

QUICK REVIEW

1. Explain how symbioses allow benthic animals to live in otherwise inhospitable habitats.
2. Give examples of commensalism, mutualism, and parasitism.

Figure 14.22 Clownfish have a mutualistic relationship with the sea anemone.

Summary

Benthic biomass largely follows the same pattern as primary productivity, with most benthic organisms found in shallow, near-shore environments. Benthic algae are anchored to firm substrates. These algae have a holdfast, a stipe, and photosynthetic blades but no roots, stems, or leaves. Algal growth along a rocky beach ranges from green algae at the surface to brown algae at moderate depths to red algae, which are found primarily below the low-tide level. Each group's pigments trap the available sunlight at these depths. Algae are generally classified by their principal pigment. The brown algae include the large kelps. Seaweeds provide food, shelter, and substrate for other organisms in the area. There are also benthic diatoms and a few seed plants, including seagrasses and mangroves.

Benthic animals are subdivided into the epifauna, which live on or attached to the bottom, and the infauna, which live buried in the substrate. Animals that inhabit the rocky littoral region are sorted by the stresses of the area into a series of zones. Organisms that live in the supralittoral (or splash) zone spend long periods of time out of water. The animals of the midlittoral zone experience nearly equal periods of exposure and submergence. These animals have tight shells or live close together to prevent drying out. The area is crowded and competition for space is great. The lower littoral zone is a less stressful environment. It is home to a wide variety of animals. The organisms of the littoral zone are herbivores and carnivores, and each has its specialized lifestyle and adaptations for survival.

The zonation of the organisms in the benthic region varies with local conditions. Tide pools provide homes for lower littoral zone organisms; they can also become extremely specialized habitats.

Mud, sand, and gravel areas are less stable than rocky areas. The size of the spaces between the substrate particles determines the porosity of the sediments. Some beaches have a higher organic content than others. Few algae can attach to soft sediments; thus, few grazers are found here. Eelgrass and surf grass provide food and shelter for specialized communities. Most organisms that live on soft sediments are detritus feeders or deposit feeders. Zonation patterns are not conspicuous along soft bottoms. Bacteria play an important role in the decomposition of plant material and its reduction to detritus. The bacteria themselves represent a large food resource.

The environment of the deep-sea floor is very uniform. The diversity of species increases with depth, but the population density decreases. The microscopic members of the deep-sea infauna are the benthic meiofauna. Larger burrowers such as sea cucumbers continually rework the sediments. The organisms of the epifauna are found at all depths.

Tropical coral reefs are specialized, self-contained systems. The coral animals require warm, clear, clean, shallow water and a firm substrate. Photosynthetic dinoflagellates, called zooxanthellae, live in the cells of the corals and giant clams. The reef exists in a complex but delicate biological balance that can easily be upset. Reefs have a typical zonation and structure associated with depth and wave exposure. Coral reefs are presently under great stress from coral bleaching, predation, and disease. Human activities are among the greatest threats; damage is due to overfishing, pollution, and the use of dynamite and poison. An international network of monitoring stations is being implemented.

A number of deep, cold-water coral reefs have been located. These corals lack zooxanthellae, and much has yet to be learned about them.

Self-contained, deep-ocean benthic communities of large, fast-growing animals depend on chemosynthetic bacteria as the first step in the food web. Communities are associated with hot-water vents and cold seeps.

Key Terms

epifauna, 368
infauna, 368
food falls, 368
sessile, 368
bipartite lifestyle, 368
algae, 368
holdfast, 368
stipe, 368

blade, 368
kelp, 372
vertical zonation, 374
intertidal zonation, 374
mollusk, 374
nemertean, 375
brachiopod, 376
nudibranch, 376

kelp forest, 378
canopy, 378
detritus, 380
bioturbation, 381
benthic meiofauna, 382
pogonophora, 383
deep-sea gigantism, 384
polyp, 384

zooxanthellae, 384
symbiosis/symbiotic, 384
reef flat, 385
reef crest, 385
commensalism, 391
mutualism, 391
parasitism, 391

Study Problems

1. Animals within the intertidal experience dramatic swings in temperature, salinity, and depth as function of the tides. In contrast, animals in the deep ocean experience nearly constant conditions. Calculate the difference in hydrostatic pressure that a sea star living in a tide pool would experience as the tides go from low (the sea star is just below the water) to 10 ft. Now calculate the pressure a sea star leaving in the deep ocean at 3800 m experiences. You can use the following:

   P (pressure) $= P_{atm}$ (atmospheric pressure) $+ \rho \times g \times h$

   ρ = the density of seawater, 1.03×10^3 kg/m^3
   g = gravity, 9.8 m/s^2
   h = height meters
   $P_{atm} = 1.01 \times 10^5$ N/m^2

2. If 1 atmosphere is 1.01×10^5 N/m^2, how many atmospheres of pressure would the seastar experience at 3800 m depth? How does this compare to the shortcut that is often used— that 10 m (33 ft) increases pressure by 1 atm? If the ocean was freshwater instead of saltwater, would the pressure at 3800 m increase or decrease? By what percentage?

3. In the oceans, most of the deep-sea benthos must rely on particulate organic material sinking from above to provide food. However, a very small fraction of it reaches the bottom. Measurements of how much material gets to the bottom have been used to develop an equation describing surface productivity (biomass), and the expected fraction at any given depth:

   $F_z = F_o(Z/Z_o)^b$

   Where F_z = the amount of organic material at depth
   F_o = the amount at the bottom of the surface mixed layer
   Z_o = the mixed layer depth
   Z = the depth you are interested in
   b = an exponential decay factor

   For the offshore Pacific, $b = -0.858$, and a typical mixed layer depth is 100 m.

   You start with 100% of the organic material (food) at the bottom of the mixed layer (100 m) and you want to estimate how much reaches a depth of 200, 500, 1000, 2000, and 4000 m. Calculate the percentage that gets to 4000 m and describe or draw how this amount changes as a function of depth.

4. Deep-sea gigantism is displayed by many marine organisms. The isopod *Bathynomus giganteus* can reach sizes up to 500 cm (0.5 m), compared to a typical size for closely related isopods of about 2.5 cm. *Bathynomus* is often found at depths of 2000 m. How many times larger is *Bathynomus* compared to other isopods? The typical common market squid (*Loligo opalescens*) is approximately 30 cm in length and lives in near-surface waters. If size is scaled as a function of depth, using *Bathynomus* as a guide, how large would we expect the giant squid (*Architeuthis*) to be if it were also found at 2000 m depth?

NORTH CORMORANT

Environmental Issues

Learning Outcomes

After studying the information in this chapter students should be able to:

1. *discuss* the problems posed by dumping solid wastes, sewage, and other toxicants in the oceans,

2. *describe* the origin of the Gulf of Mexico Dead Zone and the factors that control its size from year to year,

3. *evaluate* the hazards of plastic trash in the oceans,

4. *review* the trend in average number of marine oil spills per decade during the past four decades,

5. *explain* the importance of marine wetlands,

6. *review* the problem of marine invaders, and

7. *discuss* the problem of overfishing.

Oil platform in the North Sea.

The oceans have always had a profound influence on the people who have settled along their shores. The oceans provide food, clothing, and waste disposal. They have fascinated and inspired generations of artists, writers, poets, and musicians. They have provoked the imaginations of adventurers and explorers and stimulated the curiosity of scientists.

For centuries, the impact of humans on the ocean has been difficult to detect because human populations and technology continued at low levels in comparison to the vastness of the ocean. Today, human-induced changes are easier to identify: water pollution, plastics, oil spills, and extraction of energy, food, and raw materials. The results of our actions include decreasing wetlands, lower biodiversity, introduction of invasive species, declining fish populations, and expansion of dead zones. Science documents changes in the marine environment and identifies causes, but science alone is not able to make the required corrections. Problems, once identified, may become issues that require policy changes, and making those changes may require making choices that are expensive or unpopular. Whether you live by the ocean or not, the ocean is part of your world, and you need to consider the situations that are discussed in this chapter.

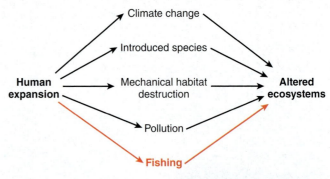

"Then" → "Now"

Figure 15.1 Coastal ecosystems have been impacted by a number of human-induced changes. Overfishing is almost always the first impact to be observed, followed by some or all of the other impacts, ultimately leading to alterations of marine ecosystems.

15.1 Human Impacts Through Time

Although we are perhaps more aware of our influence on the ocean today, this is not a new issue. Human impacts have, however, increased dramatically with rising human population over the last 150–300 years. As part of the Census of Marine Life and other scientific programs, scientists have evaluated the role humans have played in modifying the oceans over the course of documented history. There is strong evidence for overfishing starting with freshwater fish at least 1000 years ago. This increasing pressure on freshwater fishing led to the expansion of marine harvests, with subsequent declines of coastal habitats worldwide. It has also been suggested that overfishing is just the first and most obvious of a range of impacts humans have on ocean environments. These include increasing pollution, **eutrophication** (artificial enrichment of waters with a previously scarce nutrient), physical destruction of habitat, outbreak of disease, and introduction of invasive species. This sequence of events is linked to expansion of human populations. As a result, the pattern of ocean exploitation is similar regardless of when it started (fig. 15.1), and some regions of the ocean are more impacted (because of earlier use by humans) than others.

A recent compilation of human impacts was used to create a global map indicating which regions of the world's oceans are most affected (fig. 15.2). Today, no area is unaffected by human influence. The most heavily populated regions with the longest history of human civilization (such as along the Mediterranean Sea) are highly impacted by a number of factors, as predicted (fig. 15.1). What does this mean? Although humans started to change marine ecosystems thousands of years ago, our cumulative impacts have now reached every region of the world's oceans.

QUICK REVIEW

1. How have human impacts on the ocean changed through time?
2. Are human impacts on the ocean limited to coastal areas?

15.2 Marine Pollution

In the past, Earth's natural waters were assumed to be infinite in their ability to absorb and remove the by-products of human populations. However, too much use and too many wastes discharged into too small an area at too rapid a rate have produced problems that cannot be ignored. These discharges can and do exceed the ability of natural systems to flush themselves, and, consequently, disperse wastes into the open ocean. There are many possible definitions for pollution, but the most straightforward comes from the World Health Organization. **Marine pollution** is "the introduction by man, directly or indirectly, of substances or energy into the marine environment which results in deleterious effects." This broad definition includes everything from noise pollution (see chapter 13) and oil spills to floating objects accidentally lost at sea (**flotsam**) or deliberately discarded (**jetsam**), discharge of excess nutrients, and contamination of sediments or waters by heavy metals and man-made compounds.

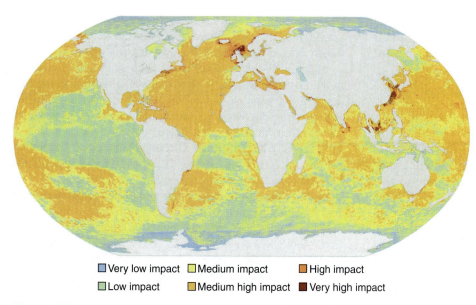

| ◻ Very low impact | ◻ Medium impact | ◻ High impact |
| ◻ Low impact | ◻ Medium high impact | ◼ Very high impact |

Figure 15.2 Global map of cumulative human impact across twenty ocean ecosystem types. Very highly impacted regions are *red*; very low impact regions are *blue*.

disposal is dumping at sea. After each modern war, obsolete military hardware and munitions have been disposed of at sea.

Marine disposal of waste material and subsequent pollution of the marine environment must be regulated internationally. Once pollutants enter the marine environment, they can be transported anywhere in the oceans. In 1972, the U.S. Congress passed the Marine Protection, Research, and Sanctuaries Act (MPRSA, also known as the Ocean Dumping Act) to regulate the intentional disposal of waste at sea and authorize research related to marine pollution. The MPRSA prohibits all ocean dumping, except as allowed by permits awarded by the Environmental Protection Agency (EPA), in any ocean waters under U.S. jurisdiction by any U.S. vessel or any vessel sailing from a U.S. port. It also bans any dumping of radiological, chemical, and biological warfare agents and any high-level radioactive waste, as well as medical wastes. Today, nearly all of the solid waste dumped at sea is dredge material.

The deleterious effects can be immediate, or may take hundreds of years (or longer) to have an impact. While progress has been made in eliminating some types of pollution, potential problems will only increase in the future as global population increases from just over 7 billion in 2012 to an estimated 8 to 10.5 billion people by 2050.

Solid Waste Dumping

Using the sea as a dump for trash and garbage was and is a common practice around the world. Probably more than 25% of the mass of all material dumped at sea is dredged material from ports and waterways, and one of the major methods of industrial waste

Other than dredge material, marine debris is often the result of accidental release or poor management of waste from human activities. In 2010, 250 million tons of municipal solid waste (MSW) was generated in the United States (fig. 15.3), or about 2.01 kg (4.43 lb) per person per day, of which 34% is recycled. Based on a five-year study of marine debris trends on U.S. beaches, three of the top five items discarded on beaches can be recycled: plastic bottles, plastic bags, and cans. Worldwide, the most common waste item found on beaches is cigarette butts.

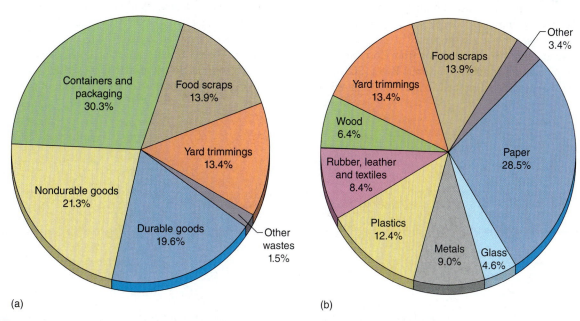

(a)

(b)

Figure 15.3 Total MSW generation by material (a) and category (b) for the United States in 2010.

The problems associated with the production and disposal of waste materials will only increase in the future as global population grows from about 7 billion as of 2012 to a projected 10 billion before 2200.

Sewage Effluent

Most urban sewer systems in the United States were built in the late 1800s and early 1900s; these systems carried raw sewage to the closest body of water: river, lake, or ocean. Since then, the systems have grown by adding to or combining old systems, building new systems, and adding treatment facilities. Until 1972, the law allowed the discharge of anything into a body of water until the water was polluted; "polluted" was defined by the individual states. In 1972, the Federal Clean Water Act, in an effort to make U.S. water "fishable and swimmable," mandated the upgrade of sewage treatment to the secondary treatment level by 1977.

For years, Boston Harbor has received sewage effluent in various stages of treatment from an enormous urban area, and the harbor was seriously polluted.

Construction of new sewage collecton systems and a sewage treatment plant began in 1989 at a cost of $3.8 billion. Important components of the new Deer Island Sewage Treatment Plant have been placed in operation in a series of steps since January 1995. The plant was constructed to remove human, household, business, and industrial pollutants from wastewater that originates in homes and businesses in forty-three different communities with a total population of roughly 2.5 million in the greater Boston area. It is the second-largest sewage treatment plant in the United States. The plant's peak capacity is 1270 million gallons of wastewater per day (mgd). Average daily flow is about 390 mgd. After the sewage is treated in the plant, it is discharged through an outfall tunnel 7 m (24 ft) in diameter beneath the sea floor that extends roughly 15.3 km (9.5 mi) offshore. The Boston Harbor cleanup has been a tremendous success story. Since completion, toxins and waste products have declined in the sediments, beaches have reopened, and marine life, from lobsters to whales, has returned to the area.

The southern California coastal region from Los Angeles to San Diego is known as the Southern California Bight (SCB). The SCB was home to almost 20 million people in 2006. Ocean-related tourism in this area generates an estimated $9 billion annually. As important as this coastal region is to tourism, it is also used for a variety of other, seemingly incompatible purposes, many of which result in the discharge of pollutants to coastal waters. Sources of pollution include discharges from municipal wastewater facilities and power-generating stations, oil platforms, industrial effluents, and dredging operations. Nineteen municipal wastewater treatment facilities discharge treated water directly to the SCB; four of these are very large facilities discharging more than 100 million gallons of treated water per day. Historically, these four facilities have been the leading source of point source contaminants to the SCB. Over the past thirty years, the cumulative flow from these four facilities has increased an overall 16%, but the discharge of suspended solids, oxygen-consuming constituents (often referred to as "biological oxygen demand" constituents, or BODs), and oil and grease have decreased significantly (fig. 15.4). These reductions are due largely to increased and improved treatment, greater control of these materials at their source, and land disposal of some solids.

Toxicants

The emission of **toxicants** (substances that degrade the environment and are harmful to organisms, including heavy metals, chemical compounds, and excessive concentrations of nutrients) into the SCB has also declined dramatically.

Surface runoff from coastal urban areas feeds directly into the marine environment via storm sewers. Storm sewers carry a wide mix of materials, including silt, hydrocarbons from oil, residues from industry, pesticides and fertilizers from residential areas, and coliform bacteria from animal wastes. Even the chlorine added to drinking water and used to treat sewage effluent as a bactericide may form a complex with organic compounds in the water to produce chlorinated hydrocarbons that are toxic in the marine environment.

From rural and agricultural lands, runoff finds its way through lakes, streams, and rivers to the coast. This runoff supplies pesticides and nutrients, which can poison or overfertilize the waters. Excess nutrients can be destructive because the nutrients stimulate plant growth that eventually dies and decays, removing large quantities of oxygen. Lack of oxygen then kills

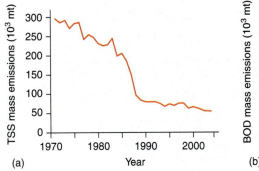
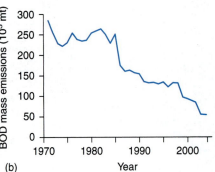
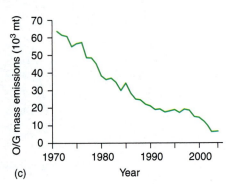

Figure 15.4 Estimated annual emissions from large municipal wastewater treatment facilities into the Southern California Bight between 1971 and 2004 of (a) total suspended solids (TSS), (b) biological oxygen demand substances (BOD), and (c) oil and grease (O/G) in thousands of metric tons.

other organisms, which in turn decay and continue to remove oxygen from the water. Animal wastes and failed septic tank systems also contribute contamination to rural runoff.

Many toxicants reaching the coast do not remain in the water but become adsorbed onto the small particles of matter suspended in the water column. These particles clump together and settle out because of their increased size, and toxicants become concentrated in the sediments.

The concentration of heavy metals in marine sediments in the SCB generally increased from 1845 through 1970, until concentrations were two to four times their natural levels. Improved wastewater treatment led to a dramatic decrease in the discharge of most heavy metals to the SCB from 1970 through 1990, when emission levels reached a steady-state, low level (fig. 15.5a–d).

Two additional toxicants that can have a major impact on the environment are the pesticide DDT (dichloro-diphenyltrichloroethane) and PCBs (polychlorinated biphenyls). DDT came into wide agricultural and commercial usage in this country in the late 1940s. Over the next twenty-five years, approximately 675,000 tons were applied domestically. The peak year for use in the United States was 1959, when nearly 40,000 tons were applied. From that high point, usage declined steadily until 1971,

after which DDT use in the United States was banned by the Environmental Protection Agency. PCBs are a group of over 200 synthetic compounds, some mutagenic (known to produce mutations in developing organisms) and carcinogenic (cancer-promoting). PCBs are among the most stable organic compounds known. Congress passed the Toxic Substances Control Act in 1976, which led to a ban in 1979 on the manufacture and use of PCBs at concentrations above 50 ppm. By 1984, additional restrictions were in place to extend the ban to most cases where the concentration was less than 50 ppm. Figure 15.6 shows concentrations of DDT and PCBs in sediments in Puget Sound from about 1890 to 1990. DDT and PCB contamination in the sediments peaked in 1960 and declined significantly thereafter through 1990, with the exception of a small peak in DDT measured in the 1986 sample.

In 1998, a global pact (Protocol on Persistent Organic Pollutants) was negotiated by the United States and twenty-eight other countries; the pact requires most countries in western and central Europe, North America, and the former Soviet Union to ban production of compounds such as DDT and PCBs.

Some of the particulate matter that adsorbs toxicants has a high organic content and forms a food source for marine creatures. In this way, heavy metals and organic toxicants associated with

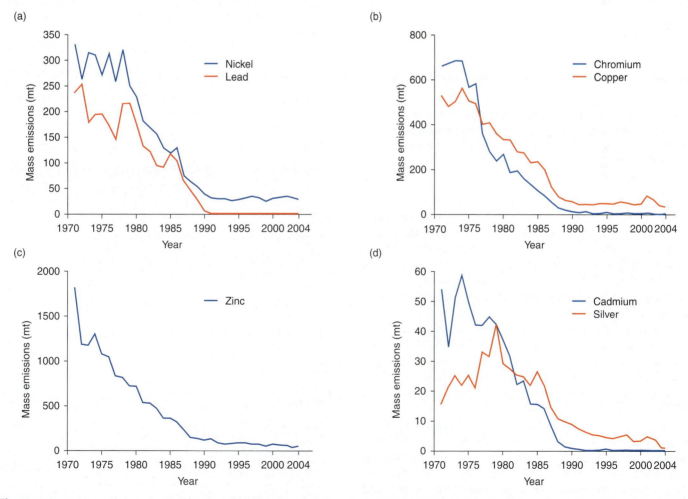

Figure 15.5 Estimated annual emissions from large municipal wastewater treatment facilities into the Southern California Bight between 1971 and 2004 of trace metals in metric tons: (a) nickel and lead, (b) chromium and copper, (c) zinc, and (d) cadmium and silver.

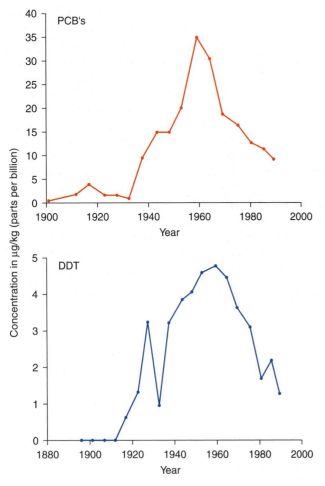

Figure 15.6 Concentration of PCBs and DDT in sediment cores from a deep-water Puget Sound station located due west of Seattle. Concentrations peaked in 1960 and then declined following the enactment of legislation controlling their use.

A tragic example of what can happen when humans ingest organisms that had accumulated a toxin occurred between 1952 and 1968 in Minamata, Japan. An estimated 200–600 tons of waste mercury from an industrial source were released into the coastal bay from which villagers gathered their shellfish harvest. The mercury formed an organic complex that was readily taken up by the marine life; the accumulation of mercury in the shellfish led to severe mercury poisoning and death among those eating the shellfish. The physical and mental degenerative effects were especially severe on children whose mothers had eaten large amounts of shellfish during pregnancy. The condition produced is named Minamata disease.

Today, mercury poisoning is still very much an issue. Mercury contamination of sediments and water is a long-lived problem, and methyl mercury can **bioaccumulate** (accumulation of toxic substances in organisms) and **biomagnify** (concentration of the toxic substance increases when one organism eats another; see table 15.1). Long-lived fish and apex predators such as sharks and swordfish are more likely to have bioaccumulated methyl mercury. The amount of mercury in the human diet depends on both the type of fish consumed and the amount consumed over a period of time. In the United States, the Food and Drug Administration and the EPA recommend eating fish as part of a healthy diet but limiting consumption of moderate and high-risk fish (fig. 15.7), particularly for pregnant or nursing mothers and young children.

Considering all the pathways, sources, and types of materials that can be classed as toxicants or pollutants, managing and eventually excluding them from the marine environment are extremely difficult and complicated. As seen in this discussion, the particles find their way into the body tissues of organisms, where they may accumulate and be passed on to predators. Due to their mass, these particulates often deposit on the sea floor. Thus, throughout the United States, toxic residues are found in estuarine bottom fish. Shellfish have been found to have concentrations of heavy metals at levels that are many thousands of times over the levels found in the surrounding waters. Scallops may have levels of cadmium about 2 million times over the water concentration, and oysters may have DDT levels 90,000 times over the water concentration. A classic 1967 study conducted after DDT had been sprayed on Long Island marshes to control mosquitoes documented the effect of this long-lived toxin as it was concentrated by the food chain (table 15.1).

The impact of DDT on the marine environment was most apparent in bird populations. DDT causes thinning of eggshells, leading to reproductive failure when the shell cannot support the weight of the incubating bird. DDT nearly wiped out the brown pelicans, ospreys, and bald eagles in large swaths of the coastal United States. Since DDT was banned in the 1970s, these bird populations have made a remarkable comeback from the brink of extinction.

Table 15.1   Food Chain Concentration of DDT

	DDT Residues in Parts per Million (ppm)
Water	0.00005
Plankton	0.04
Silverside minnow	0.23
Sheepshead minnow	0.94
Pickerel (predator)	1.33
Needlefish (predator)	2.07
Heron (small animal predator)	3.57
Tern (small animal predator)	3.91
Herring gull (scavenger)	6.00
Osprey egg	13.8
Merganser (fish-eater)	22.8
Cormorant (fish-eater)	26.4

Reprinted with permission from George M. Woodwell, et al., *DDT Residues in an East Coast Estuary, Science* 156 (1967): 821–24. Copyright 1967 American Association for the Advancement of Science.

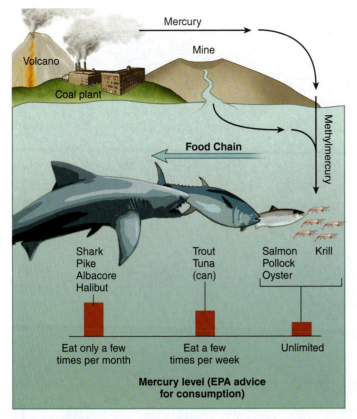

Figure 15.7 Mercury from coal-fired power plants and other sources travels though the atmosphere and water. Some is changed to methylmercury, which can enter the food chain to be concentrated at each step on that chain. Large, old predators like sharks and pike, or scavengers like halibut, hold the greatest concentrations of mercury.

efforts to reduce the discharge of toxic materials are showing signs of success, but it will take continued, long-term effort on the part of scientists, citizens, and governments to achieve a cleaner environment for the ocean world and the land world.

QUICK REVIEW

1. What is solid waste?
2. Why should pregnant and nursing women avoid certain types of fish?
3. Explain the difference between a pollutant and toxicant.
4. Why do some toxicants accumulate in the sediment, resulting in higher concentrations compared to the overlying water?
5. Describe how small levels of toxicants accumulate to selectively impact animals at higher trophic levels in food webs.

15.3 Plastic Trash

Walk any beach in any estuary or along any open coast and see the tide of plastics being washed ashore. It is estimated that every year, more than 135,000 tons of plastic trash have been routinely dumped by naval, merchant, and fishing vessels. The National Academy of Sciences estimates that the commercial

fishing industry yearly loses or discards about 149,000 tons of fishing gear (nets, ropes, traps, and buoys) made mainly of plastic and dumps another 26,000 tons of plastic packaging materials. Recreational and commercial vessels and oil and gas drilling platforms add their share. A total of 7 million tons of cargo and crew wastes were added during each year of the 1980s. Plastics are a worldwide problem; they come from many sources and are distributed by the currents to even the remotest ocean areas (fig. 15.8a).

In 1960, the U.S. production of plastics was 3 million tons; 31 million tons of plastic waste was generated in 2010, with more than 50 million tons of plastic produced annually. Plastic is inexpensive, strong, and durable; these characteristics make it the most widely used manufacturing material in the world today and a major environmental problem.

Thousands of marine animals are crippled and killed each year by these materials. Lost lobster and crab traps made entirely or partially of plastic continue to trap animals; one year 25% of the 96,000 traps that had been set off of Florida's western coast were lost. Seabirds die entangled in plastic fishing line; both seabirds and marine mammals swallow plastics. Porpoises and whales have been suffocated by plastic bags and sheeting, and the sheeting clings to coral and to rocky beaches, smothering plant and animal life. Fish are trapped in discarded netting; sea turtles eat plastic bags and die. Plastics are as great a source of mortality to marine organisms as oil spills, toxic wastes, and heavy metals (fig. 15.8b–d).

The Marine Plastic Pollution and Control Act of 1987 is the U.S. law that implements an international convention for the prevention of pollution from ships. This law prohibits the dumping of plastic debris everywhere in the oceans; other types of trash may be dumped at specific distances from shore, and ports are required to provide waste facilities for debris. The U.S. Coast Guard is the agency that must enforce this law in U.S. waters; no international enforcement exists.

Many plastics take hundreds of years to degrade in the environment. Making matters worse, plastics often break down into smaller and smaller pieces, but still remain in the ocean. This flotsam can be concentrated by the surface currents into large floating patches; recently the **Eastern Pacific garbage patch** has attracted a lot of attention to this problem. The Pacific gyre circulation tends to concentrate these floating bits of plastic and other debris, forming "patches" of elevated plastic concentration that expand and contract in response to the ocean currents (fig. 15.9). There is some debate about exactly how large the patch is, but there is no question that plastics are concentrated in these regions. Because most plastics break down into smaller pieces over time, it is not possible to simply remove the plastics from these regions. Making things worse, as plastics degrade (particularly Styrofoam), a toxic brew of chemicals is released that includes bisphenol A (BPA). The small bits of plastic act to concentrate other contaminants, including PCBs and DDT, making these small bits of plastic thousands to millions of times more toxic than the same chemicals in seawater. Eventually the plastic debris can sink. A 2010 study of the Southern California Bight documented about 3% of the ocean floor was covered in

(b)

Figure 15.8 The impact of plastics on the marine environment. (a) Persistent litter includes plastic nets, floats, and containers. (b) A common murre entangled in a six-pack yoke. (c) A young gray seal caught in a trawl net. (d) A sea turtle with a partially ingested plastic bag. Photos (a), (b), and (c) are from Sable Island in the North Atlantic, approximately 240 km (150 mi) east of Nova Scotia, Canada.

(a)

(c)

(d)

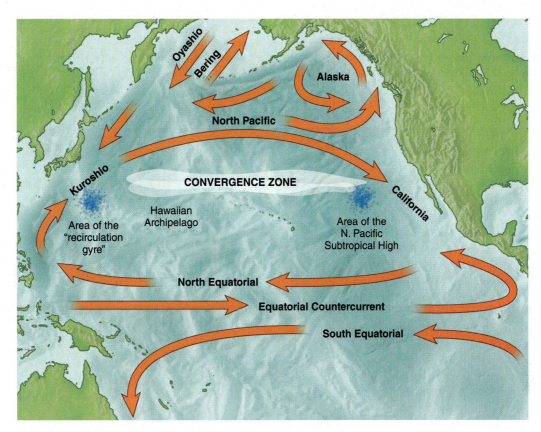

Figure 15.9 Plastics and other floating debris are concentrated into garbage patches at the eastern and western edges of oceanic gyres.

plastic. Today, there is no region of the world's oceans where plastics or chemicals associated with plastics cannot be found. It may be that only people educated to act responsibly will be able to reduce the tide of plastics pollution rising around the world.

QUICK REVIEW

1. Describe the impacts of plastics on marine birds and mammals.

2. What is the most common pollutant on U.S. beaches?

3. Why is the Pacific garbage patch located where it is? What keeps it in place?

4. Plastics eventually degrade in the ocean. What happens to the small plastic particles that allows them to continue to be hazardous to marine organisms?

15.4 Eutrophication and Hypoxia

Each year off the mouth of the Mississippi River, a low-oxygen, or **hypoxic** area forms in the northern Gulf of Mexico. It begins to appear in February, peaks in the summer, and dissipates in the fall, when storms stir up the gulf waters and increase mixing. It has been called the *dead zone* because the oxygen level in the water is too low to support most marine life. Fully oxygenated waters contain as much as 12 parts per million (ppm) of oxygen. Once oxygen falls to 5 ppm, fish and other aquatic animals

can have trouble breathing. As the dead zone grows in size and oxygen levels drop, highly mobile organisms such as fish and shrimp flee the area. More slowly moving bottom-dwellers such as crabs, snails, clams, and worms can be overtaken by the hypoxic waters and suffocate. Sediment dwellers that can't leave a hypoxic zone begin dying at around 2 ppm. In some dead zones, oxygen levels remain near 0.5 ppm or lower for months. The low-oxygen water does not just hug the bottom; it can affect 80% of the water column in shallow areas, extending upward to within a few meters of the surface.

The dead zone is caused by the introduction of large amounts of nitrogen as nitrate, much of it from fertilizers, into the gulf by the Mississippi and Atchafalaya Rivers. This input of excess nutrients leads to eutrophication. The Mississippi River alone delivers an average 580 km³ (140 mi³) of fresh water to the Gulf of Mexico each year. Thirty-one states (41% of the area of the continental United States) and half of the nation's farmlands have rivers that drain into the Mississippi. Records show that from the 1950s onward, the use of fertilizers in these areas increased dramatically. The amount of nitrate from fertilizer that flowed into the gulf tripled between 1960 and the late 1990s, and the amount of another fertilizer, phosphate, doubled. It is estimated that about 30% of the nitrogen entering the gulf comes from fertilizers used in agriculture, another 30% comes from natural soil decomposition, and the remainder comes from a variety of sources, including animal manure, sewage treatment plants, airborne nitrous oxides from the burning of fossil fuels, and industrial emissions.

When nitrates are added to the water, they trigger the rapid growth of tiny marine algae. This phenomenon is commonly referred to as a bloom. Massive blooms can produce a biological chain reaction in which there is a corresponding increase in small animals that feed on the algae, producing more plant and animal material than can be consumed by fish and other predators. Billions of these small organisms die and sink to the sea floor, where they decompose and are consumed by bacteria. These bacteria use up oxygen in the water in the process.

The seasonal appearance of the dead zone was first detected in 1972 but it was not until over twenty years later that scientists began to systematically map its location. The dead zone occupies different regions of the northern gulf from one year to the next, making it difficult to predict exactly what areas will be affected (fig. 15.10). The size of the dead zone seems to correlate with both the amount of agricultural fertilizer used in the Mississippi River drainage basin and with annual changes in rainfall, which cause variable water runoff into the gulf. The hypoxia formed in 1988 but did not persist because of a drought and very little freshwater runoff into the gulf. In 2000, when dry conditions prevailed and runoff from the Mississippi River was very low, the zone shrank to its smallest recorded size. Then it rebounded in 2001 and in 2002. In 2003, it shrank again to about 7000 km² (2700 mi²) before growing again (fig. 15.11).

The dead zone in the Gulf of Mexico is not an isolated situation, but it is the second-largest human-caused coastal dead zone in the world. Dead zones have been reported from more than 400 systems worldwide. These regions cover at least 245,000 square km (94,595 sq mi), and are related to coastal eutrophication driven by riverine runoff of fertilizers and the burning of fossil fuels (fig. 15.12). The largest dead zone ever documented occurred in the northwestern shelf of the Black Sea. It first formed in the 1960s with the introduction of large amounts of nutrients from fertilizers. When the former Soviet Union collapsed 1991, government-supported agriculture also decreased significantly, and by 1996, the hypoxic zone in the Black Sea disappeared for the first time in thirty years. Unfortunately, recovery from hypoxic conditions can take much longer than it takes to create a dead zone.

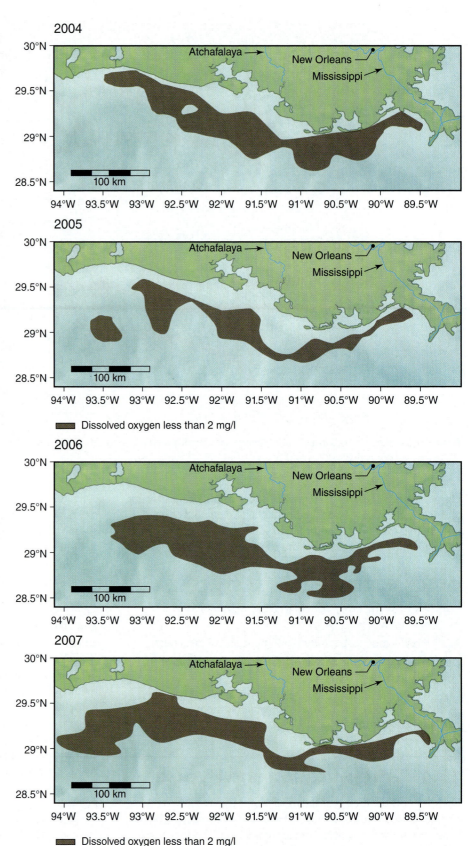

Figure 15.10 The location and size of the dead zone in the northern Gulf of Mexico vary from year to year. Examples of the size and shape of the dead zone (*brown*) as measured in July in the years 2004 through 2007. Research continues in an effort to better understand the factors controlling the pattern of hypoxic water.

Area of bottom-water less than 2 mg/l in mid-summer (km^2)

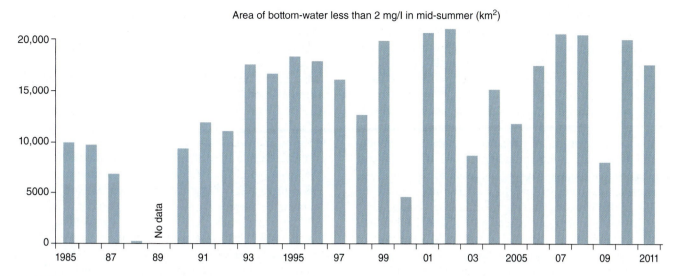

Figure 15.11 The size of the Gulf of Mexico dead zone. The dead zone grew during the particularly wet year of 1993, remained high through the rest of the decade, and then shrank significantly during the dry year of 2000. In 2001 and 2002, it grew once again before shrinking to smaller sizes from 2003 through 2005 and then growing in size again in 2006 through 2011.

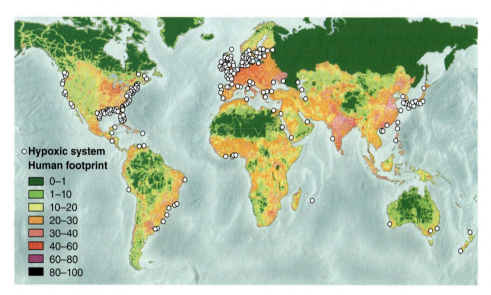

Figure 15.12 Distribution of more than 400 eutrophication-related dead zones. Their distribution is closely related to the global human footprint (expressed as a percentage) in the Northern Hemisphere. Dead zones from the Southern Hemisphere are likely underreported.

15.5 Oil Spills

Humans' activities in the twenty-first century depend heavily on oil, and this dependence requires the bulk transport of crude oil by sea to the land-based refineries and centers of use. This transport creates the potential for accidents that release large volumes of oil and expose the world's coasts and estuaries to spills associated with vessel casualties and transfer procedures. Drilling offshore wells exposes marine areas to the risks of blowouts, spills, and leaks. Because industry, agriculture, and private and commercial transportation require petroleum and petroleum products, oil is constantly being released into the environment, to find its way directly or indirectly to the sea.

Annual oil spills over 700 metric tons (205,800 gal) from tanker accidents from 1970 to 2011 are shown in figure 15.13a. The average number of large tanker spills per decade decreased significantly from over twenty-four per year in the 1970s to just over three per year in the 2000s. The quantity of oil spilled annually during this time period is shown in figure 15.13b. Individual major spills are typically responsible for a high percentage of the total oil spilled. For example, from 1990 to 1999, 346 spills over 7 metric tons each occurred, totaling nearly 1.1 million metric

In the case of the Black Sea, the benthic community has still not returned to its original state, more than fifteen years after hypoxic conditions disappeared.

QUICK REVIEW

1. Explain the relationship among hypoxia, anoxia, and organic matter.

2. How does a dead zone form?

3. Explain why reductions in use of fertilizers in the U.S. midwest would potentially improve the Gulf of Mexico Dead Zone.

4. How does the Gulf of Mexico dead zone expand and contract from year to year?

tons (322.2×10^6 gal), but 830,000 metric tons (244×10^6 gal) (75% of the total) were spilled in just ten incidents (just over 1% of the accidents). Consequently, the volume of oil spilled in any given year can be strongly influenced by a single large spill, as seen in figure 15.13*b*. Table 15.2 lists selected marine oil spills from tanker accidents.

The damage caused by spills far out at sea is difficult to assess because no direct visual or economic impact occurs on coastal areas and damage to marine life cannot be accurately evaluated. Spills due to the grounding of vessels or accidents during transfer or storage happen in coastal areas, where the environmental degradation and loss of marine life can be observed. Spills that occur in estuaries and along coasts affect regions that are oceanographically complex, biologically sensitive, and economically important. Four oil spills between 1978 and 2010 have become ecological landmarks; these are the sinking of the *Amoco Cadiz,* the grounding of the *Exxon Valdez,* the discharge of crude oil into the Persian Gulf during the Gulf War, and the *Deepwater Horizon* blowout in the Gulf of Mexico (see Diving In box).

In March 1978, the *Amoco Cadiz* lost its steering in the English Channel and broke up on the rocks of the Brittany coast of France (fig. 15.14*a* and *b*). Gale-force winds and high tides spread the oil over more than 300 km (180 mi) of the French coast; more than 3000 birds died; oyster farms and fishing suffered severely. Of the approximately 210,000 metric tons (61.7×10^6 gal) of oil spilled, roughly 30% evaporated and was carried over the French countryside; 20% was cleaned up by the army and volunteers; another 20% penetrated into the sand of the beaches to stay until winter storms washed it away; 10% (the lighter, more toxic fraction) dissolved in the seawater; and 20% sank to the sea floor in deeper water, where it continued to contaminate the area for an unknown period of time.

The United States experienced a large-volume oil spill in March 1989, when, 40 km (25 mi) out of Valdez, Alaska, the *Exxon Valdez,* loaded with 170,000 metric tons (50×10^6 gal) of crude oil, ran onto a reef, tearing huge gashes in its hull and spilling 35,000 metric tons (almost 10.3×10^6 gal) of oil into Prince William Sound. Local contingency plans were not adequate for cleaning up oil spills of this magnitude; delays, caused in part

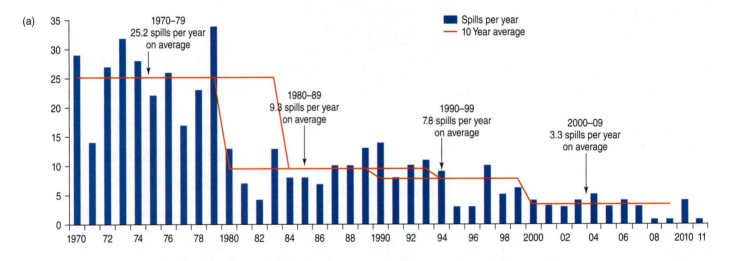

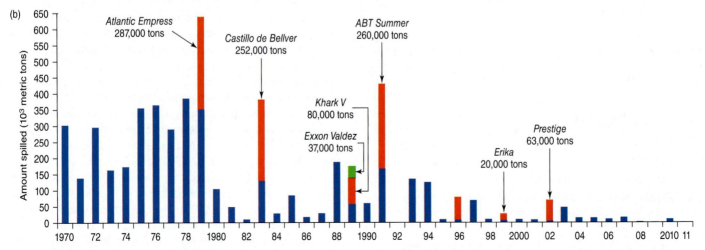

Figure 15.13 Annual (a) number of oils spills over 700 metric tons (205,800 gal) and (b) amount of oil spilled in thousands of metric tons in tanker accidents from 1970 to 2011.

Table 15.2 Selected Major Oil Spills from Tankers

Ship	Year	Location	Oil Lost (metric tons)	(10⁶ gal)
Atlantic Empress	1979	Off Tabago, West Indies	287,000	84.4
ABT Summer	1991	700 nautical miles off Angola	260,000	76.4
Castillo de Beliver	1983	Off Saldanha Bay, South Africa	252,000	74.1
Amoco Cadiz	1978	Off Brittany, France	223,000	65.6
Haven	1991	Genoa, Italy	144,000	42.3
Odyssey	1988	700 nautical miles off Nova Scotia, Canada	132,000	38.8
Torrey Canyon	1967	Scilly Isles, United Kingdom	119,000	35.0
Sea Star	1972	Gulf of Oman	115,000	33.8
Urquiola	1976	La Coruna, Spain	100,000	29.4
Irenes Serenade	1980	Navarino Bay, Greece	100,000	29.4
Hawaiian Patriot	1977	300 nautical miles off Honolulu, Hawaii	95,000	27.9
Independenta	1979	Bosphorus, Turkey	95,000	27.9
Jakob Maersk	1975	Oporto, Portugal	88,000	25.9
Braer	1993	Shetland Islands, United Kingdom	85,000	25.0
Khark 5	1989	120 nautical miles off Atlantic coast of Morocco	80,000	23.5
Aegean Sea	1992	La Coruna, Spain	74,000	21.8
Sea Empress	1996	Milford Haven, United Kingdom	72,000	21.2
Nova	1985	Off Kharg Island, Gulf of Iran	70,000	20.6
Katina P.	1992	Off Maputo, Mozambique	66,700	19.6
Prestige	2002	150 nautical miles off the Atlantic coast of Spain	63,000	18.5
Exxon Valdez	1989	Prince William Sound, Alaska	37,000	10.9

by out-of-service equipment, as well as a lack of equipment and personnel, the rugged coastline, the weather, and the tidal currents of the enclosed area in which the spill occurred, combined to intensify the problem. The oil spread quickly and was distributed unevenly over more than 2300 km² (900 mi²) of water, taking a severe toll on seabirds, marine mammals, fish, and other marine organisms. The oil moving out of Prince William Sound was not washed out to the open sea but was captured by the nearshore currents and moved westward parallel with the coast, repeatedly oiling the rocky wilderness beaches in the weeks that followed (fig. 15.14c and d). In the subarctic conditions of Prince William Sound, the photochemical and microbial degradation of the oil proceeded more slowly than it would have in warmer temperate regions. Plant and animal populations also recover more slowly in the cold water, where organisms tend to live longer and reproduce more slowly. Researchers inspecting the area in 1992, three years after the disaster, reported that the area was recovering and repopulating with organisms as the oil aged and degraded. They

also noted that the areas most intensively cleaned to remove oil are recovering more slowly and with less-balanced populations than the areas in which the oil has been allowed to degrade naturally. In the 1992 inspection, estimates were made of the dispersal of the spilled oil: 13% in the sediments, 2% in the beaches, 50% degraded naturally in the water, 20% degraded in the atmosphere, and 14% recovered and removed.

Much of the world's crude oil comes from wells of the Persian Gulf, and it must be transported the length of the gulf on its way to refineries around the world. Each year, a quarter of a million barrels of oil are routinely spilled into the gulf's shallow waters, making it one of the world's most polluted bodies of water. However, it is also an area of vigorous plant growth and supports fisheries of shrimp, mackerel, mullet, snapper, and grouper.

During the eight-year Iran-Iraq war, the bombing of oil facilities produced major spills, including one from an oil rig that poured out 172 metric tons (54,350 gal) per day for nearly three months, but the greatest catastrophe came in the 1991 Gulf War, when

Diving in

Impacts from the *Deepwater Horizon* Spill

BY WAWEISE SCHMIDT

Waweise Schmidt holds an M.S. degree in Marine Studies with an emphasis in Marine Biology. She is currently a Professor in the Science Department at Palm Beach State College in Palm Beach Gardens, Florida.

On April 20, 2010 the BP deep ocean well Macando experienced a blowout, and then on April 22, a fire continued to decimate the remains, which eventually led to the capsizing and sinking of the rig the *Deepwater Horizon*/BP Mississippi Canyon Block 252 (MS252) South Louisiana drilling platform located in the Gulf of Mexico (box fig. 1). The immediate outcome of this event was the death of eleven men and a leak from the well which lasted for approximately eighty-seven days. In the middle of July, the well was capped but not before releasing an estimated 4.9 million barrels of oil (approximately 206 million gallons) into the Gulf. Not only were the results of this event detrimental to the environment, but the oil spill also had a large socioeconomic impact on commercial fisheries and related employment areas, consumer expenditures, tourism and associated sales, property home sales, and vacation rentals within the areas affected. The size of this spill, which impacted the states of Mississippi, Texas, Alabama, Louisiana, and Florida, has placed it in the unique and unenviable category of being the largest oil spill in the U.S. history.

In response to oil and other chemical spills within coastal areas, the Office of Response and Restoration (OR&R), an interdisciplinary scientific team of the National Oceanic and Atmospheric Administration (NOAA), which is the lead federal trustee for coastal and marine resources, was responsible for evaluating what was spilled, where the material would travel, and what impacts there would be to the natural resources caused by the spill. To accomplish its goals, NOAA worked cooperatively with other trustee members, including federal, state, and local, to conduct a natural resource damage assessment (NRDA) to assess the injury, quantify the impacts, plan what could be done to restore the resources, and follow through with the restoration of resources to their prior conditions before the spill.

Immediate oil spill impacts included the known or visible signs such as dead finfish (red snapper, menhaden, mullet, amberjacks) and shrimp (pink, brown, and white); oil-saturated birds (brown pelicans, turtles (Kemp's Ridley; box fig. 2), and mammals (bottlenose dolphins); as well as oil-covered wetland plants, sea grass beds, coral reefs, and beaches. However, not all effects could immediately be seen due to the nature of the oil itself. The oil separated into three pools—the subsurface underwater plume, the surface slick, and the airborne plume. Some of the components of the plumes were known to be readily soluble in water; others were known to be volatile organic compounds (VOCs), which would evaporate quickly. Some heavier components were not expected to dissolve in water but to remain on the surface and weather or biodegrade.

Box figure 1 *Deepwater Horizon* rig on fire in the Gulf of Mexico.

Oil from this wellhead, like many other wells, was a mixture of thousands of other chemicals. The groups that were of particular interest in relation to the environment were those that were known to be toxic and persistent. The *Deepwater Horizon* wellhead produced a sweet crude that had low sulfur and low polyaromatic hydrocarbon (PAH) content and it was also high in alkanes. The low sulfur and PAH made the crude less toxic than other crude oils and the alkanes made the oil more desirable for some microorganisms to use as a food source, possibly aiding in biodegradation of the oil. While the VOCs (benzene, toluene, and xylene) were thought to have evaporated quickly and were of more concern when the oil was fresh, the lighter oil components were expected to form a thin layer of oil on the ocean surface. Some of the oil was also expected to emulsify to form a water-oil mixture and eventually coalesce to form tarballs, which can persist within the environment (box fig. 3). The immediate cleanup methods that were used to remove fresh oil from the land and ocean involved the use of absorbent pads, booms, and chemical shoreline cleaning agents; flushing with water; and the manual recovery of oil, as well as the use of chemical dispersants.

The response to this disaster did not involve only assessing damages to components of the environment but it also involved rescuing trapped or injured organisms and the continued efforts to determine possible long-term impacts on their future populations. Some impacts require time to study and assess, such as the possible loss of nesting sites and nurseries, the alteration of migration patterns, unusual mortality events associated with whale and dolphin strandings, illness from eating contaminated food, or effects of long-term exposure to PAH (some are identified as carcinogenic, mutagenic, or teratogenic) because the full effect may not become noticeable until years after the initial exposure.

Typically, seafood safety testing is conducted by each state; however, since the *Deepwater Horizon* oil spill in 2010, the Gulf Coast states have adopted the identical test for seafood safety that was utilized by the federal government. By adopting this seafood testing protocol, the states ensure consumers that the finfish and shellfish that were caught in their state waters were safe to eat. The majority of state waters are currently open; however, some areas, such as the Barataria Bay Basin area of Louisiana, have remained closed to commercial fishing industries.

Box figure 2   Kemp's Ridley turtle before (a) and after (b) cleaning.

Box figure 3   Florida beach with tarballs.

409

(a)

(b)

(c)

(d)

(f)

(e)

Figure 15.14 (a) The tanker *Amoco Cadiz* aground and broken in two off the coast of France in March 1978. (b) A bay along the French coast was fouled with oil. (c) Hot water under high pressure was used to clean Prince William Sound beaches after the 1989 *Exxon Valdez* oil spill. (d) Booms collect oil washed from a beach and hold it for pickup. (e) Oil remained along the Persian Gulf coast long after the 1991 Gulf War. (f) Cleaning up a 1993 oil spill along the beaches of Tampa Bay one more time in 1994.

an estimated 800,000 metric tons (252.8 × 10⁶ gal) of crude oil gushed into gulf waters, the world's greatest oil spill to date. Some of this oil was deliberately released, some came from a refinery at a gulf battle site, and bombing contributed additional quantities.

The average depth of the Persian Gulf is 40 m (131 ft), and the circulation of its high-salinity water is sluggish. Changes in the wind in the weeks after the spill stopped the oil from drifting the entire length of the gulf; still, some 570 km (350 mi) of

Saudi Arabia's shoreline were oiled. Oil spill experts from the United States, the United Kingdom, the Netherlands, Germany, Australia, and Japan rushed to help. They were able to protect the water-intake pipes of desalination plants and refineries, but other cleanup efforts were less successful (fig. 15.10e). About half of the oil evaporated, and about 300,000 metric tons (94.8 × 10⁶ gal) were recovered. Much sank to the bottom of the gulf, where oil may still be seeping from sunken tankers.

The tragedies of these spills continue to demonstrate that no adequate technology exists to cope with large oil spills, particularly under difficult weather and sea conditions, along irregular coastlines, or far from land-based supplies. On the average, only between 8% and 15% of oil spilled is recovered, and these may be inflated estimates because the recovered oil has a high water content.

The onshore cleanup is often as destructive as the spill, especially in wilderness situations, where the numbers of people, their equipment, and their wastes further burden the environment. The toxicity of weathered crude oil is low, and many oil-spill experts consider that cleanup efforts should be concerned not with removing oil from the beaches but with moving oil seaward to prevent it from reaching the beaches. Follow-up studies indicate that many of the cleanup methods used along shorelines cause more immediate and long-term ecological damage than leaving the oil to degrade naturally.

The immediate damage from a large spill is obvious and dramatic; by contrast, the effect of the small but continuous additions of oil that occur in every port and harbor are much more difficult to assess because they produce a chronic condition from which the environment has no chance to recover. Refined products such as gasoline and diesel fuel are more toxic to marine life than crude oil, but they evaporate rapidly and disperse quickly. Crude oil is slowly broken down by the action of water, sunlight, and bacteria, but the portion that settles on the sea floor moves down into the sediments, which it continues to contaminate for years.

Large volumes of crude oil also enter the marine environment from natural seeps. The current global rate of natural seepage of oil into the oceans is estimated to be 600,000 metric tons (176.4 × 10⁶ gal), with a range between 200,000 and 2 million metric tons (58.8–588 × 10⁶ gal) per year. Natural seeps are believed to be the single greatest source of oil in the marine environment.

QUICK REVIEW

1. What happens to petroleum products such as oil when they are released into the ocean?
2. Are oil spills from tankers the largest source of oil in the ocean?
3. How do the marine biota respond to oil spills?

15.6 Marine Wetlands

The value of shore and estuary areas as centers of productivity and nursery areas for the coastal marine environment is well known to oceanographers and biologists. Saltwater and brackish marshes and swamps, known as marine **wetlands**, border estuaries and provide nutrients, food, shelter, and spawning areas for marine species, including such commercially important organisms as crabs, shrimp, oysters, clams, and many species of fish. Nevertheless coastal countries have long histories of filling wetlands and modifying coasts to provide croplands, port facilities, and industrial space for their growing populations (fig. 15.15).

Another type of wetland is being destroyed along the muddy shores of tropical and subtropical lagoons and estuaries where several species of mangrove trees grow. These salt-tolerant trees protect the shore from wave erosion and storm damage and provide specialized habitats for fish, crustaceans, and shellfish on and among their tangled roots. In recent years, mangroves have been logged for timber, wood chips, and fuel; mangrove swamps have been cleared and filled to provide land for crops, shrimp ponds, and resorts. Eighty percent of the mangroves in the Philippines, 73% of Bangladesh's trees, and more than 50% of coastal stands in Africa have been removed. Over 50% of the world's mangroves are gone.

In the United States, more than 164 million people, about 52% of the national total, live along the coast. This population is expected to increase by an average of 4000 people per day, reaching 171 million by the year 2015. This rate of growth is faster than that for the nation as a whole.

Areas of recreation and retirement replace wetlands with waterfront homes, each with its own individual pleasure-craft moorage (fig. 15.16). Between the mid-1950s and the mid-1980s, approximately 20,000 acres of coastal wetlands were lost each year in the contiguous United States. Estuarine wetland losses were greatest in six states: Louisiana, Florida, Texas, New Jersey, New York, and California.

While there was an increase in wetlands in the United States from 1998–2004, this was due primarily to the establishment or recovery of freshwater areas. The loss of coastal wetlands accelerated during this period, with 59,000 acres lost each year in the coastal watersheds of the Great Lakes, Atlantic Ocean, and Gulf of Mexico. This is mirrored by global trends of 35% loss

Figure 15.15 The industrialized estuary of the Duwamish River at Seattle, Washington. The commercially developed flat areas adjacent to the river were once tidelands and salt marshes.

Figure 15.16 The wetlands of Barnegat Bay, New Jersey, were replaced by a housing and recreational complex.

of mangrove forests since the 1980s, and 29% loss of seagrass beds. These loss rates are comparable to those for coral reefs and tropical rainforests, making coastal wetlands some of the most threatened habitats on Earth.

QUICK REVIEW

1. Explain why wetlands are important parts of the coastal ocean.
2. Discuss why wetlands are being lost in the United States and globally.

15.7 Biological Invaders

A combination of physical and biological barriers sets the geographical limits of a species. These limits have changed with time as climate patterns have altered and as plate tectonics has shifted the configurations of the oceans and continents. For example, the opening of the Bering Strait between Asia and North America allowed interchange between the marine organisms of the North Pacific and the Arctic basins, and the separation of South America from Antarctica allowed the currents associated with the Southern Ocean's West Wind Drift to carry organisms from one cold-water ocean to another.

When humans began to cross oceans, they carried many organisms, plant and animal, intentionally and unintentionally. More than three and a half centuries ago, settlers and traders coming to the North American continent brought with them large communities of widely varied organisms that had attached to and bored into the wooden hulls of their ships in the harbors or bays where their journeys began. Many of the organisms transported across the oceans in this way were certainly swept from the ships' sides by the waves and currents, but a few invaders were present when the ships anchored at their journeys' ends. Hundreds of years later, there is no way of knowing when the first European barnacles or periwinkle snails began to colonize the eastern coasts of the United States and Canada; these two organisms are now considered typical of this region.

The steel hulls of today's ships do not carry such communities for they are protected by antifouling paints and the ships move through the water at speeds sufficient to sweep away many of the organisms carried by wooden hulls. However, the ships of the present do carry ballast water that is loaded and unloaded to preserve the stability of the ship as it unloads and loads cargo. Tens of thousands of vessels with ballast tanks ranging in capacity from hundreds to thousands of gallons move across our oceans. These vessels rapidly transport this ballast water and its populations of small floating organisms across natural oceanic barriers; ballast water and living organisms may be released days or weeks later, thousands of kilometers from their point of origin.

Some biological invaders remain unnoticed for many years; others begin almost immediately to seriously disrupt the ecology of their adopted areas. In 1985, an Asian clam (*Potamocorbula amurensis*) (fig. 15.17) was discovered in a northern arm of San Francisco Bay. This clam was previously unknown in the area and appears to have arrived in its juvenile or larval form in the

Figure 15.17 Asian clam (*Potamocorbula amurensis*).

ballast water of a cargo ship from China. During the next six years, the clam spread southward into the bay and formed dense colonies, as many as 10,000 clams per square meter. The Asian clam feeds on plantlike, single-celled diatoms and the larvae of crustaceans. The food requirements of these huge populations reduce amounts available for native species, stressing the system's species balance and food chains. In 1990, a second intruder, the European green crab, or green shore crab (*Carcinus maenas*) (fig. 15.18), was recognized in the southern part of San Francisco Bay. This crab moved from Europe to the East Coast of the United States in the 1820s and to Australia in the 1950s. The green shore crab is less than 6.5 cm (3 in) broad and is voracious and belligerent. It feeds on clams and mussels and may spawn several times from a single mating. It feeds enthusiastically on the Asian clam, but it can also feed on the native species and can outcompete them for food. *Carcinus maenas* is now moving northward along the West Coast of the United States. In 1997, it had reached Coos Bay, Oregon; in 1998 was discovered in Willapa Bay and Grays Harbor, Washington; and in 1998 was found on the west coast of Vancouver Island, British Columbia.

In 1982, the ballast water of a ship from the coast of America carried a jellyfish-like organism known as a comb jelly (*Mnemiopsis leidyi*) into the Black Sea. From the Black Sea, it spread to the Azov Sea and has recently moved into the Mediterranean. It has no predator in these areas and devours huge quantities of plankton, small crabs, shrimp, fish eggs, and fish larvae. Fish can also be transported; for example, Japanese sea bass were introduced into Australia's Sydney Harbor region in 1982–83. A small crab (*Hemigrapsus sanguineus*) common in Japanese waters was identified in New Jersey in 1988 and has since been found as far south as Chesapeake Bay and as far north as Cape Cod. The North American razor clam (*Ensis directus*) was detected in Germany in 1979, and it has since spread to France, Denmark, the Netherlands, and Belgium.

Not all invaders are animals. In the 1980s, an attractive, fast-growing, bright green tropical seaweed (*Caulerpa taxifolia*) was introduced into European aquariums. In 1985, some plants escaped into the Mediterranean Sea during a routine tank-cleaning at Monaco's aquarium. The seaweed rapidly spread and grew to cover more than 44.5 km² (11,000 acres) of the northern Mediterranean coastline by 1997. It has recently been reported off the coast of North Africa. In June 2000, *Caulerpa taxifolia* was discovered in a coastal lagoon just north of San Diego, California, and later in Huntington Harbor in Orange County, California, just 125 km (75 mi) further north. In areas where it becomes well established, it is capable of causing ecological and economic devastation by overgrowing and eliminating native species of seaweeds, seagrasses, reefs, and other organisms. In the Mediterranean, *Caulerpa taxifolia* has had a negative impact on tourism, recreational boating, and diving, and has harmed commercial fishing by causing changes in the distribution of fish and impeding net fishing.

As more and more alien species are recognized, biological oceanographers and marine biologists realize that an ecological revolution is taking place in estuaries and bays and along the rocky shores of all the continents. Many biologists refer to the introduction of these alien species as "biological pollution"; others are increasingly alarmed by the breakdown of natural barriers and the worldwide "biological homogenization" that may result.

In the case of ballast water, the introduction of zebra mussels into the Great Lakes and their subsequent invasion of the rivers and streams of the central United States sounded the alarm for the freshwater environment. In 1990, Congress enacted the Nonindigenous Aquatic Nuisance Prevention and Control Act (NANPCA). Under this law, the United States adopted voluntary regulations for exchanging ballast water on the high seas for vessels bound for the Great Lakes; this provision became law in 1993.

Controlling ballast water will not close all doors to invading marine species and will be costly. However, it will lead to fewer foreign invasions. Keep in mind that "No introduced marine organism, once established, has ever been successfully removed or contained, or the spread successfully slowed" (James T. Carlton, Maritime Studies Program, Williams College).

QUICK REVIEW

1. What is meant by the term *biological invaders*?
2. Explain the consequences of ballast water transport of marine organisms.
3. Explain why "biological homogenization" would be detrimental to marine ecosystems.

Figure 15.18 European green crab (*Carcinus maenas*). The width of the crab's back (carapace) is about 4 cm (1.5 in).

15.8 Overfishing and Incidental Catch

Around the world, too many fishing boats are taking too many fish, too fast. As discussed in section 15.1, this is not a new issue, but the diminishing fish populations are now victims of our ability to harvest fish from virtually anywhere in the oceans at an unprecedented rate. Declining fisheries are the result of **overfishing**—the removal of fish from a population faster than the population can reproduce—and **incidental bycatch**—the removal of fish or other marine organisms that were not the intended targets of fishing (turtles, for example). The Food and Agriculture Organization of the United Nations (FAO) tracks global fisheries practices, and reported that for 2008, every person on the planet consumed an average of 17.2 kg (about 38 lb) of fish per year, accounting for at least 20% of their animal protein intake.

Current estimates of the **maximum sustainable yield (MSY)** (the maximum fishery biomass that can be removed annually while maintaining a stable standing stock) are between 100–120 million metric tons per year. The FAO began keeping track of fisheries yields in 1950. At the time, global fisheries production was about 20 million metric tons, or 20% of the MSY. This value has increased steadily (fig. 15.19) until it plateaued in the early 1990s. Between 2000 and 2010, humans harvested between 89–94 million metric tons. In other words, we have nearly reached the MSY for global fisheries. This number does not include bycatch, which the FAO estimates at another 20 million metric tons annually. This suggests that humans are removing more than the MSY. The FAO estimates that 80% of the world's fish stocks for which data are available are fully exploited or overexploited (fig. 15.20). The world's oceans are being overfished.

What are the results of this fishing pressure? Several fisheries that were once plentiful have now collapsed, while others have been increasing rapidly. The Alaskan pollock fishery is

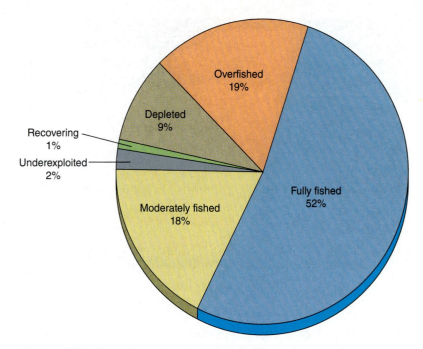

Figure 15.20 Status of global marine fisheries, based on FAO data. Fully 80% of the marine fisheries are fully fished or overexploited.

an example of a recent fishery that has responded to consumer demand. Pollock is a bottom fish; it is processed to remove the fats and oils that actually give the fish its flavor, and a highly refined fish protein called **surimi** is produced. Surimi is processed again and flavored to form artificial crab, shrimp, and scallops. The processing of pollock to form surimi is shown in figure 15.21.

Trends in Fishing Pressure

The Atlantic cod (*Gadus morhua*) is an example of a fishery that was over-exploited and subsequently collapsed. Cod are bottom-dwellers, or ground fish; they feed on small fish, crab, squid, and clams. They live twenty to twenty-five years, and mature in three to seven years. How could the fish that had fed the Pilgrims and produced catches of 50,000 metric tons by hand fishing a hundred years ago have failed so completely?

For nearly 400 years, cod were fished with hand lines; these gave way to long lines with hundreds of hooks and net traps in the nineteenth century. Following World War II, there was a virtual explosion of international fishing activities in the northwest Atlantic, with huge factory ships and trawlers from Great Britain, the former Soviet Union, Germany, and other nations joining the fishing. Between 1966 and 1976, the fishing capacity of the international fleet increased 500%. In contrast, the amount of fish caught for a given level of fishing effort (**catch per unit effort**) increased only 15%. In other words, despite increasingly efficient fishing, fewer fish were being caught. To manage this region, the International Commission for the Northwest Atlantic Fisheries was formed, but was largely ineffective at curbing fishing pressure. The United States extended its **Exclusive Economic Zone (EEZ)** from 12 to 200 miles in 1976. Canada followed in 1977. Now, the most productive regions of the northwest Atlantic, including the Grand Banks and Georges Bank, were under national control,

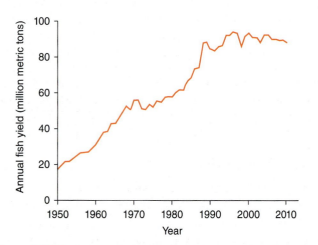

Figure 15.19 World fish landings from 1950–2010 in millions of metric tons per year. These values do not include incidental bycatch.

Figure 15.21 (a) Freshly caught pollock are poured from a trawl net onto the deck of a seafood-processing vessel. (b) Fishes are cleaned and boned below decks in the processing plant. After the fishes have been washed to remove blood and oils and then dried, the fish meat becomes a flavorless, odorless fish product, surimi. (c) The surimi base is made into analogs such as crab shapes.

forcing the international fleet out of those waters. Despite these legal changes, overfishing continued with virtually no limits. Finally, in response to continued decline in catch per unit effort and fish stocks, the 200-year-old Newfoundland cod fishery was closed in Canada in 1992, and in 1993, the National Marine Fisheries Service closed large parts of the U.S. cod fishery. Despite these closures, the cod fishery has still not recovered.

A third example highlights the impact on large, slow-growing fish such as the Atlantic bluefin tuna (*Thunnus thynnus*). Bluefin can exceed 450 kg (990 lb), averaging 2–2.5 m in length, and can live to at least thirty-five years. The population of bluefin dropped 80% between 1970 and 1993. The population is managed as two separate spawning stocks in the eastern and western Atlantic, despite evidence that the populations intermix. The western stock reproduces in the Gulf of Mexico, and has dropped 90% since 1975. Many observers believe that this species is doomed, for it is the world's most valuable fish (fig. 15.22). In 2007, a bluefin was sold at Tokyo's Tsukiji fish market for $35,000. In 2008, the record was $55,700. In 2009, it was $100,000. By January 2012, a record bluefin sold for $724,000, or $1238 per pound. At these prices, there is incentive to fish for the last remaining bluefin in the oceans.

Indirect Impacts

Fish populations are not the only casualties; other marine animals and marine birds are being affected as they compete for their share of the catch. There is increasing evidence that overfishing has other long-term consequences. For both sport and

Figure 15.22 Tuna being prepared for auction at the Tsukiji Fish Market, Tokyo, Japan.

commercial fishing, the largest and oldest fish are usually targeted first. Since 1950, 90% of the large fish have been removed from the oceans. This can also cause shifts in food webs as the apex predators are removed. Scientists have also shown a gradual decrease since the 1950s toward more fishing pressure at lower trophic levels (moving from sharks and tuna to squid and krill, for example). This phenomenon is called **fishing down the food web** (fig. 15.23). Recent scientific studies suggest that fishing pressure is more complicated than simply removing fish from successively lower trophic levels. For example, in some regions, fishing pressure has increased at multiple levels, resulting in fishing through the food web (simultaneous exploitation of many fish species). Nevertheless, the trend toward increasing fishing pressure and declining size and abundance of many fish is evident.

In the Shetland Islands, nesting seabirds failed to breed in the mid- to late 1980s, apparently in response to lack of food when the sand eels in the area were overfished. In Kenya, over-harvesting of the trigger fish on coral reefs allowed abnormal growth of sea urchins, thereby damaging the reef ecosystem. Alaska's Steller sea lion populations have plummeted. An estimated 170,000 Steller sea lions existed in 1970. By 1990, the count was 60,000, and by 2000, only about 45,000 remained. Studies in the Gulf of Alaska and the Bering Sea show that more than 50% of the Steller sea lion's diet is pollock, but the heavily fished pollock stock in the Bering Sea was down from an estimated 12.2 million tons in 1988 to 6.5 million tons in 1995.

The oceanic drift-net fishery became a serious problem in the 1980s, laying out nets up to 65 km (40 mi) long each night. These gill nets of almost invisible nylon hang like walls in the water, trapping and killing nearly everything that swims into them. In 1990, the U.S. Marine Mammal Commission estimated the aggregate length of these nets at 40,000 km (25,000 mi), enough to ring Earth. Three Asian nations—Japan, Taiwan, and South Korea—have used the nets to catch squid and fish in the Pacific and Indian Oceans; in the Atlantic, several European nations have drift-netted, mainly for albacore tuna. These nets do not rot; sections torn free in storms float in the ocean for months, even years, as ghost nets catching everything they encounter; no estimate exists on the animal life they destroy.

Large numbers of marine animals die each year only because they are caught incidentally by people fishing for other species. Incidental catch, or bycatch, or what are often called "trash fish," represents a tremendous waste of marine resources. The FAO estimates that each year about 25% of all reported commercial marine landings are caught as bycatch and discarded. Alaskan trawlers for pollock and cod throw back to the sea some 25 million pounds of halibut, worth about $30 million, as well as salmon and king crab because they are prohibited from keeping or selling these fish. Another 550 million pounds of bottom fish are discarded in Alaskan waters to save space for larger or more valuable fish.

The discard rate on bycatch varies from place to place; if incidental catch does not bring a high enough price and if processors or markets are not available, these "trash fish" will be returned to the sea, usually dead. Whereas in the Gulf of Mexico 1 pound of shrimp results in 10.3 pounds of bycatch that is nearly all discarded, in Southeast Asia and other areas with local fisheries and fresh fish markets much of the bycatch is used.

Fish Farming

The world's demand for seafood is increasing at the same time that the ocean harvest of wild fish is decreasing. An alternative way to increase the fish harvest is by fish farming, known as **aquaculture**, the growing or farming of animals and plants in a water environment. The term **mariculture** is used for the growing or farming of marine plants and animals. Currently, aquaculture provides 38% the world's fish market. Fish farming is the fastest-growing area of global food production, increasing at an annual rate of about 6.6%.

Farming the water began in China some 4000 years ago. The Chinese were culturing common carp in 1000 B.C., and in 500 B.C., a book was written giving directions on fish farming,

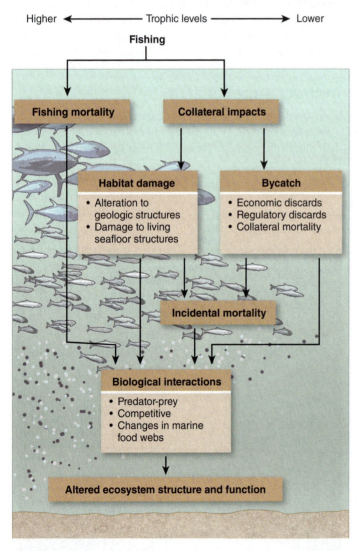

Figure 15.23 Fishing and its impacts have severely depleted higher trophic levels. Fishermen, to sustain a living, must fish on less-desirable species at lower trophic levels. The trend toward fishing on lower trophic levels has been termed "fishing down the food web."

including methods for building the pond, selecting the stock, and harvesting it. In China, Southeast Asia, and Japan, fish farming in fresh and salt water has continued to the present as a practical and productive method of raising large quantities of fish such as carp, milkfish, *Tilapia,* and catfish. Most of these farms are family-run, labor-intensive operations.

As commercial fishing has declined, mariculture has responded by more than tripling its worldwide output. In 1987, mariculture provided 4 million metric tons of fish and shellfish; in 2009, it supplied about 20.1 million metric tons. Marine fish are grown in floating cages or pens kept in calm, shallow, protected, coastal areas, which are becoming less and less available as shorelines are being increasingly built over and used for recreation. The lack of suitable sites, the impact of fish pens on water quality (increasing nutrients, decreasing oxygen), the release of antibiotics into the marine environment, and the possibility of the transmission of fish disease to wild populations are pushing the development of offshore, submersible systems that are more efficient and that distribute waste and excess food over larger areas. Additional research is going into high-energy feeds that promote faster growth and therefore decrease the time needed to bring the fish to market size.

Farming of Atlantic and Pacific salmon has surged past the catch of wild salmon. More than 60% of store-bought salmon is now farmed. A majority of the farmed salmon is raised in open net pens (fig. 15.24) in coastal areas near migration paths of wild salmon as they move to and from the ocean. A number of reports has suggested that farmed salmon can infect wild salmon with sea lice and in early 2008, a virus known as infectious salmon anemia was documented to be killing Chilean farmed salmon. Farmed salmon depend on a diet that is ~45% fish meal and 25% fish oil, commonly derived from menhaden and anchoveta. On average, every pound of farmed salmon requires 3 pounds of wild-caught fish as food, which acts to increase pressures on wild fisheries rather than decrease pressures.

Ecosystem-Based Fishery Management

The limitations of MSY-based fisheries management has led to adoption of **ecosystem-based fishery management**, which acknowledges "uncertainties in the biotic, abiotic, and human components of ecosystems." This approach views the entire ecosystem as relevant to managing a fishery, and allows for the diverse societal uses of fish stocks, including fishing, recreation, tourism, and overall ecosystem health. Major goals of this approach include selective fishing (targeting particular species) and reduction of incidental bycatch. The goal is to sustain healthy ecosystems while reducing habitat destruction, alteration of marine food webs, and impacts to nontarget organisms.

Ecosystem-based management has been widely accepted by the international community. However, a recent study suggests that while the methods are widely endorsed, there is insufficient enactment. It has been suggested that global marine fishing pressure needs to decrease by 20–50% to reach sustainable yields. It is projected that fishing pressure will continue to increase, however, by another 35 million metric tons by 2030. In 2006, 43.5 million people were directly engaged in fishing and aquaculture, and world exports of fish and fishery products reached $85.9 billion, increasing by more than 100% in two decades. Overall, fish production provided more than 2.9 billion people with at least 15% of their animal protein intake in 2010. Regardless of the fishery management strategy, solving the overfishing problem requires a sustained, international effort addressing not only direct fishing pressure but also the social, economic, and ecological implications of the growing human population's reliance on fish protein.

QUICK REVIEW

1. How has the use of fish catch changed since the 1950s? How has it resulted in increasing fishing pressure?

2. Explain what is meant by "bycatch."

3. Discuss how the maximum sustainable yield is related to global fish catch, and what alternatives there are for managing fishing pressure of wild stocks.

4. Why has fish farming increased production dramatically in the last few decades?

5. Describe what changes would need to be made socially and economically to effectively use the world's incidental bycatch.

15.9 Afterthoughts

Earth and its environment are works in progress. Since its beginning, Earth has constantly reworked and modified itself; living things have interacted with their environments, making changes and achieving new balances. Humans, however, have acquired

Figure 15.24 An aerial view of salmon pens, in which fish are raised to market size. Typical net pens are 100 m across and 30 m deep.

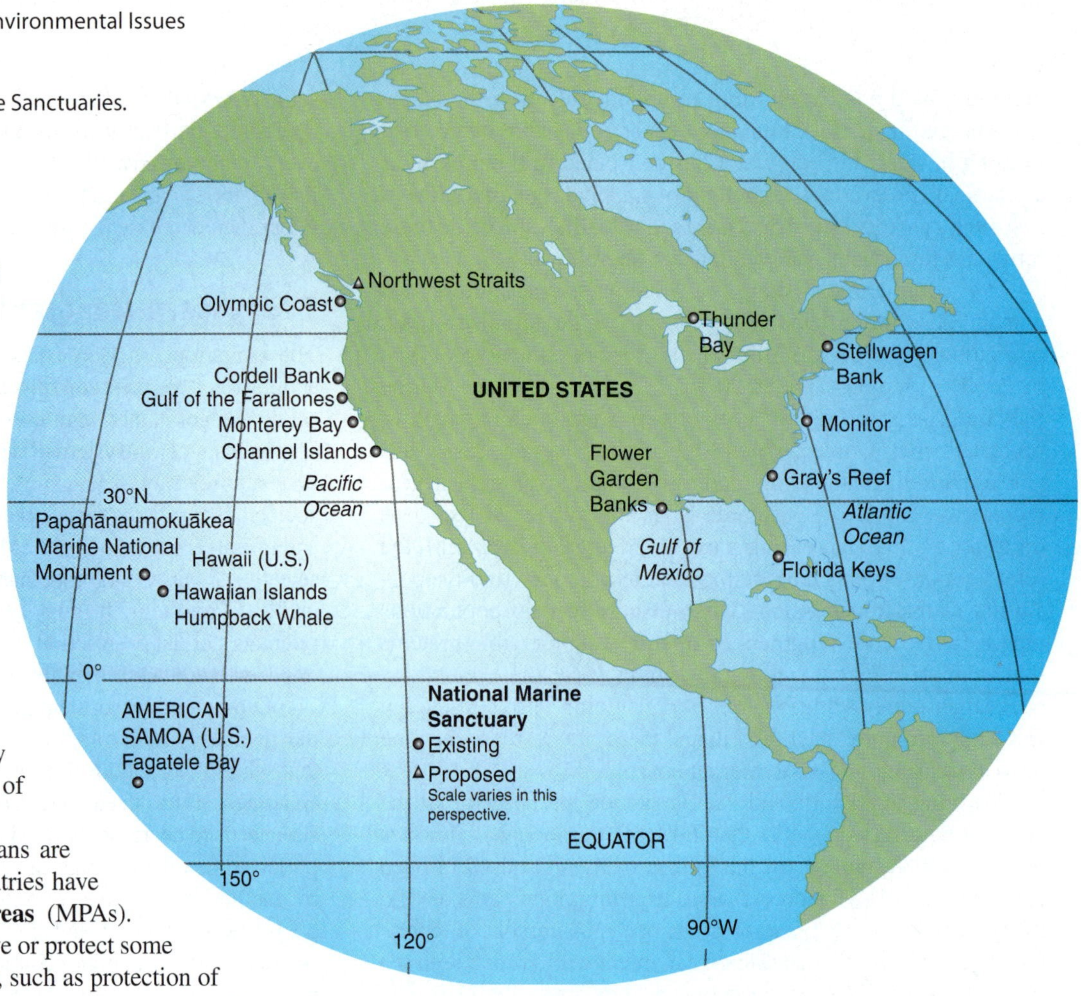

Figure 15.25 U.S. National Marine Sanctuaries.

the power to accelerate natural change and make fundamental environmental alterations for their own purposes. Few organisms compete with us successfully, and few environments are able to resist our presence. Humans will never be a zero impact factor, but we can challenge ourselves to understand and consider the implications of our choices. Science can help us understand the consequences resulting from our choices, but each of us, individually, must carefully define and protect the process of making the "best" choices.

To mitigate the impact humans are having on the oceans, many countries have developed **marine protected areas** (MPAs). These areas are created to conserve or protect some aspect of the marine environment, such as protection of historical artifacts, threatened and endangered species, vulnerable ecosystems, breeding grounds for marine organisms, and other conservation goals. MPAs include a wide range of legal restrictions on human activity. In some areas, fishing and other activities are completely prohibited. In the United States, these are referred to as **marine reserves**. In other areas such as the National Marine Sanctuaries (fig. 15.25), there are restrictions specific to each sanctuary, such as limits on vessel traffic or banning of oil and gas exploration. Globally there are almost 6000 designated MPAs ranging in size from a few square kilometers to the Phoenix Island Protected Area located in the Republic of Kiribati in the Pacific, which covers 408,250 square kilometers. All together, MPAs still account for less than 2% of the world's surface oceans, but they serve as an important tool for protecting ocean ecosystems.

The more we understand our Earth system, the better the choices that we can make. Each of us—by continuing our education (in school or out), by participating in the political process, by working with others—can help to make the intelligent, informed decisions that are required to maintain a healthy and productive planet.

QUICK REVIEW

1. Describe the difference between an MPA and a marine reserve.
2. How do the choices that humans make influence the state of ocean ecosystems?

Summary

Water quality is affected by dumping solid waste and liquid pollutants into coastal and offshore waters. Land runoff carries a mixture of toxicants, oil and gasoline residues, industrial wastes, pesticides, and fertilizers to estuaries and seacoasts. An expanding low-oxygen area in the Gulf of Mexico has been linked to increasing amounts of nitrates and phosphates flowing into the Gulf primarily from the Mississippi River. Toxicants are adsorbed onto silt particles and become concentrated in coastal sediments. Organisms further concentrate toxicants and pass them on to other members of marine food webs. Reducing the discharge of toxicants is showing success, with declines in lead, DDT, and PCBs.

Plastics are an increasing problem to marine life, killing thousands of fishes, birds, mammals, and turtles each year. Laws prohibit dumping of plastics at sea, but there is no international enforcement.

Oil spills are a special problem for inshore waters. Four significant oil spills are the sinking of the *Amoco Cadiz*, the grounding of the *Exxon Valdez*, the 1991 spills into the Persian Gulf, and the *Deepwater Horizon* blowout in the Gulf of Mexico. All have had devastating ecological effects that continue to demonstrate that no adequate cleanup technology exists at this time.

Wetlands border estuaries and coasts; they are important as areas of nutrients, food, and shelter for many marine species. Many wetlands have been filled, dredged, developed, and lost.

Organisms move across the oceans in the ballast water of thousands of cargo vessels, to be discharged into new environments far from their points of origin. Significant ecological disruptions have been found in San Francisco Bay, the Black Sea, the Azov Sea, coastal Australia, and coastal Europe. Many scientists refer to the introduction of alien species as biological pollution.

Overfishing in all the oceans is devastating the world's fisheries. People who make their living from fishing are losing their jobs, and the marine species that depend on the fish are declining as the fish populations decrease. The establishment of Exclusive Economic Zones has promoted overexpansion of fishing fleets. Incidental catch is the cause of a tremendous loss to fisheries and wildlife. Global efforts to minimize human impacts on the ocean have led to ecosystem-based management and the establishment of marine protected areas.

Key Terms

eutrophication, 396
marine pollution, 396
flotsam, 396
jetsam, 396
toxicants, 398
bioaccumulate, 400
biomagnify, 400
Eastern Pacific garbage patch, 401

hypoxic, 403
wetland, 411
overfishing, 414
incidental bycatch, 414
maximum sustainable yield (MSY), 414

surimi, 414
catch per unit effort, 414
Exclusive Economic Zone (EEZ), 414
fishing down the food web, 416

aquaculture, 416
mariculture, 416
ecosystem-based fishery management, 417
marine protected areas, 418
marine reserves, 418

Study Problems

1. Each year, about 150 square kilometers of wetland disappear under the Gulf of Mexico. Louisiana has about 4 million acres of wetland. What is the area of wetland lost each year in square kilometers? Square meters?

2. At the current rate of wetland loss, how long would it take to lose the equivalent of all the Louisiana wetlands?

3. Two of the most infamous oil spills in U.S. waters were caused by the *Exxon Valdez* oil tanker in Alaska, and the *Deepwater Horizon* oil platform in the Gulf of Mexico. They were estimated to release about 44 million liters and about 662 million liters of oil, respectively. How many times larger than the *Exxon Valdez* spill was the *Deepwater Horizon* spill? Despite the size of these spills, they are ranked fifty-fourth and fifty-fifth globally, in terms of size. The largest oil spills to date were caused by the Kuwaiti oil fires, releasing approximately 189 billion liters. What percentage of the Kuwaiti oil spills is represented by the *Deepwater Horizon* spill?

4. For many pollutants and contaminants, the level of concern is at very low concentrations—parts per billion (ppb) or less. If an automobile leaked a quart of oil into a storm drain and the entire amount flowed into the ocean, what volume would be required to dilute that oil to 1 ppb by volume? Express your answer in liters of ocean water. If a typical olympic-sized swimming pool has a volume of 2.5 million liters, how many swimming pool equivalents of ocean water would this be?

5. Fish farming is increasing at about 6.6% per year. Currently, aquaculture accounts for about 38% of the world's fish market, and the total world fisheries (wild-caught and farmed) is approximately 145.1 million metric tons per year. If the rate of increase were to continue at the same pace, how many years would it be before at least half of all the fish consumed on Earth came from fish farms?

The Oceans and Climate Disruption

Learning Outcomes

After studying the information in this chapter students should be able to:

1. *describe* what is meant by Earth's climate system,

2. *explain* how Earth's climate system is moderated by positive and negative feedback mechanisms,

3. *review* how Earth's climate has changed on long (thousands to millions of years) time scales,

4. *describe* how scientists measure climate change on short and long time scales and how climate predictions are made for the future,

5. *review* the trend in greenhouse gases and Earth's average air and surface ocean temperatures for the past century,

6. *identify* the major changes to the ocean environment in response to the abrupt climate change taking place,

7. *review* proposed solutions to mitigate the accumulation of greenhouse gases in the atmosphere, and

8. *discuss* the international response to climate change.

This composite image of the Earth taken from the VIIRS instrument aboard NASA's most recently launched Earth-observing satellite—Suomi NPP—uses a number of swaths of the Earth's surface taken on January 4, 2012. The NPP satellite was renamed "Suomi NPP" on January 24, 2012 to honor the late Verner E. Suomi of the University of Wisconsin.

Human activities have been linked to many changes in the ocean, both good and bad (review chapter 15). Perhaps the single most important change to the oceans and the planet is global warming and, more broadly, climate change. There are ongoing and vigorous debates about whether climate change is natural or amplified by human activities, whether the Earth is getting warmer, and whether humans should seek to mitigate these effects. We know from a historical perspective that the Earth's climate does change through time. If climate had no direct impact on Earth's ecosystems, these fluctuations would be of minor interest to most people. However, even small changes in climate can affect where organisms thrive in both the oceans and on land. The focus of this chapter is on **climate disruptions**—rapid (on geological time scales) changes in the Earth's atmosphere, temperature, precipitation, and wind. These changes lead to direct and indirect responses by ocean ecosystems. Regardless of your personal opinions about the role of human activities in modern climate change, it is important to understand the scientific evidence describing the causes and consequences of climate disruptions and the scientific evidence linking these changes to human activities.

Figure 16.1 The spheres of influence affecting Earth's climate system.

16.1 Earth as a Whole: The Oceans and Climate

We can define **climate** as the average conditions of the atmosphere, temperature, precipitation, and wind for a given ecosystem or for the planet as a whole. Climate is driven by interactions of six spheres: the atmosphere, the hydrosphere (primarily the oceans), the cryosphere (the ice and snow on the surface of the planet), the geosphere (the geological structure of the planet), and the biosphere, as well as the **anthrosphere**, which encompasses human modifications to the environment (fig. 16.1). Climate is often referred to as the average weather, but it encompasses more than that; climate also includes information about how likely we are to experience extremes (how much variability there is in the average). Weather systems come and go, and we generally expect winters to be cooler and summers to be warmer. We understand that the weather is changing all the time. If the climate is stable, we would expect variability (weather) around some long-term average. Climate disruptions occur when interactions among the various spheres result in an abrupt change in these long-term patterns. Climate is driven by changes to the **climate system**—the interactions of the six spheres.

Because of the complexity of the climate system, it is not always obvious how climate will respond to disturbances in any one component. There are also many feedback mechanisms that can modulate the response to perturbations. A feedback mechanism is a response to a disturbance that acts to enhance or minimize the disturbance. A **positive feedback mechanism** enhances the change, whereas a **negative feedback mechanism** works to keep the system (climate, in this case) in the status quo, or unchanging relative to the long-term average. As an example, we can examine the feedbacks on Earth's climate system from the cryosphere. Sea ice has enormous consequences for the exchange of heat, momentum, and gases across the air-sea interface. Sea ice also regulates Earth's radiation balance because sea ice and snow have a high **albedo**, or reflectance, while the ocean itself has a very low albedo (table 16.1). Albedo is the ratio of reflected radiation to incident radiation, and is usually expressed as a percentage (0% is nonreflective or black and 100% is perfectly reflective). On average, Earth's surface albedo is about 31%, but it changes with location and time of year. During periods of Earth's history when there has been a lot of sea ice, such as during ice ages, the increased albedo cools the Earth even more (positive feedback). When the snow and ice melts, albedo decreases and more solar radiation is absorbed by the surface and converted to heat, accelerating the warming (fig. 16.2).

A related example of a negative feedback mechanism is cloud formation. As the Earth warms, snow and ice albedo decreases, but the increased heating of the oceans results in enhanced evaporation. This enhances the water vapor content in the atmosphere, leading to more clouds and greater albedo, which cools the planet (negative feedback). These two examples focus on the cryosphere and atmosphere, but there can also be feedback mechanisms driven by biological interactions. Clouds form when water vapor coalesces on **cloud condensation nuclei (CCN)**, tiny particles that can come from dust, ocean aerosols,

Table 16.1 Average Albedo Values for Various Parts of Earth's Surface

Part of Earth's Surface	Albedo Range (%)
Fresh snow	60–90
Old snow	40–70
Bare sea ice	50–70
Clouds (average)	50–55
Beach sand	30–50
Grassland	24–27
Forest	5–20
Ocean	5–10
Asphalt	4–12

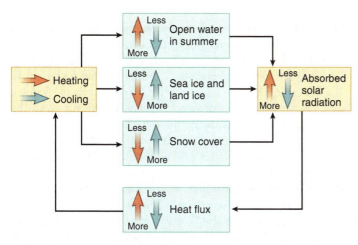

Figure 16.2 The ice-albedo positive feedback loop. As climate cools, reflective snow cover and sea ice grow in extent. In turn, Earth's albedo increases, leading to even more cooling. Melting of ice caps and sea ice can have the opposite effect. Decreases in ice extent accelerate warming by lowering Earth's albedo.

volcanic eruptions, or pollutants. **Dimethyl sulfide (DMS)** can also act as condensation nuclei. This compound is produced by many marine phytoplankton and bacteria. In the 1980s, scientists proposed that DMS from marine organisms was the dominant form of CCN. It was proposed that increasing ocean temperatures would lead to these organisms producing more DMS, thus producing more clouds, resulting in a negative feedback (fig. 16.3). This is sometimes called the **CLAW hypothesis** after the initials of the author's last names. Much work has gone into studying this hypothesis in the last few decades. Today the role of marine organisms is still considered important, but scientists have identified many other important sources of CCN such as sea salt, other organics, and anthropogenic pollutants. These are just a few examples of feedback mechanisms; others will be discussed. The combination of these forcing mechanisms and responses lead to the complex interactions that drive both short- and long-term changes in Earth's climate.

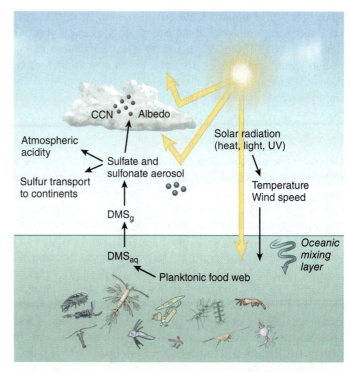

Figure 16.3 One potential climate feedback mechanism involves the production of DMS by marine organisms in response to warming of the surface ocean. The DMS compounds act as cloud condensation nuclei, producing more clouds, which cools the ocean. Lower levels of sunlight are proposed to lead to reductions in DMS production, setting up a negative feedback and stabilizing the climate.

QUICK REVIEW

1. Describe the six major components of the Earth's climate system. How are they linked?
2. Explain the difference between climate and weather.
3. Explain and give an example of positive and negative feedback mechanisms.
4. Describe the CLAW hypothesis; is this positive or negative feedback?

16.2 Earth's Climate: Always Changing

Earth's climate is not static and has been changing on short and long time scales since the formation of the planet. Human civilizations have only existed for a small fraction of this history, so clearly there are natural components to these changes. Climate change can occur gradually, but during some periods of Earth's history, climate disruptions have been quite dramatic, such as the various ice ages or more fundamental shifts such as the accumulation of oxygen in the atmosphere when photosynthesis evolved on the planet (review chapter 1). To put modern climate change into perspective, it is helpful to understand how climate has changed in the past. Scientists typically use three sources of information to document changes in Earth's climate: (1) direct measurements (of temperature, for example) for our modern climate, (2) models based on scientist's understanding of how Earth's six spheres interact for both the past and future climate scenarios, and (3) indirect measurements, particularly for understanding past climate.

Earth's Past Climate

One powerful tool for understanding current and future climate is to examine periods in Earth's past history when similar changes have occurred. To create these records scientists need to rely on indirect evidence, or **proxies**. Proxies record responses to changes in temperature, ocean salinity, precipitation, atmospheric composition, biological productivity, and ocean circulation. Some proxies rely on physical or chemical records, such as changes in the chemical composition of marine sediments, ice, or snowpack (fig. 16.4). Others identify changes recorded by biological organisms; for example, tree rings, coral banding, or changes in the chemical composition and abundance of fossils. Scientists that use these proxy records to understand Earth's historical climate work in the field of **paleoclimatology**. By combining different proxies it is possible to reconstruct records over millions of years of history. Because proxies are indirect records of the past, paleoclimatologists must often compare multiple data sets to properly interpret the results. Using these very long records, scientists have identified past climate disruptions and linked them to underlying causes or forcing mechanisms.

Scientists often use isotopes of oxygen preserved in marine fossils to reconstruct climate, extending hundreds of millions of years into the past (fig. 16.5). These long records show periodic oscillations between warmer and cooler periods, with the coldest periods associated with ice ages. By comparing

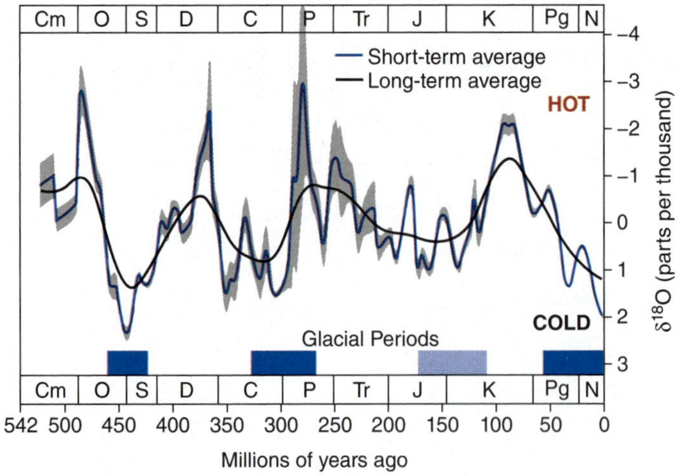

Figure 16.5 Earth's temperature over millions of years can be reconstructed by looking at the concentration of oxygen isotopes in marine fossils.

these climate records to observations of other natural factors scientists have determined that Earth's climate is influenced by, among other things, changes in solar energy and variations in Earth's orbit. This latter idea was first proposed by the Serbian astrophysicist Milutin Milankovich, and is referred to as the **Milankovich cycle**. As the Earth orbits the sun, it moves closer and farther away because the orbit is not perfectly circular (this is referred to as *eccentricity*). The tilt of the Earth also changes over long time periods (this is called *obliquity*; it is currently at 23.5°), and the Earth also wobbles around its axis (this is *precession*). These changes occur semi-periodically on time scales ranging from about 23,000 years to 400,000 years and result in varying levels of solar energy striking different parts of the planet, leading to periodic oscillations in warming and cooling (fig. 16.6).

Over these same long time periods, plate tectonics have shifted the location of continental landmasses and modified the depth and shape of the ocean basins (fig. 16.6). These changes also result in gradual changes in climate as ocean circulation is modified, changing the global heat balance of the world's oceans. It is important to note that changes in solar intensity, Milankovich cycle oscillations, and plate tectonic changes interact over many thousands to millions of years to influence the climate and are distinct from more abrupt climate change events that occur over much shorter time scales.

Earth's Present Climate

Although we must use proxies to examine much of Earth's past climate history, we have very good instrumental records, particularly for temperature, starting in about 1860. Based on these records as well as historical docu-

Figure 16.4 Researchers sample sea ice at the South Pole to investigate the history of fossil fuel emissions by sampling the air trapped in the polar ice sheets in Antarctica.

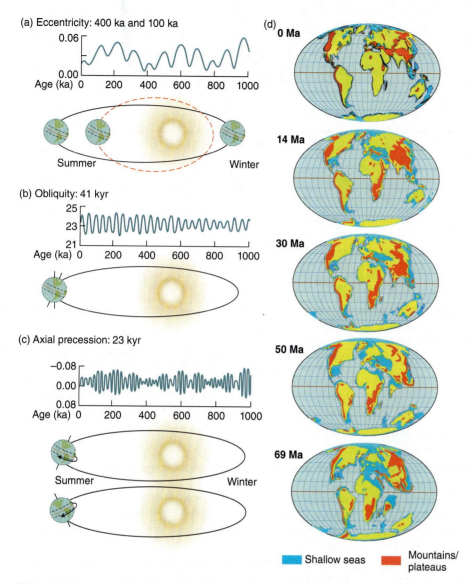

Figure 16.6 Earth's climate changes gradually on very long time scales due to changes in Earth's orbit caused by (a) eccentricity, (b) obliquity, (c) precession, and (d) modifications to ocean circulation caused by plate tectonics.

was maintained by sudden sea ice growth and subsequent sea ice/ocean feedbacks on heat transport in the North Atlantic, resulting in the prolonged cooling of much of the Northern Hemisphere for several centuries.

The Earth is currently warming at a pace faster than has occurred in the last 1000 years. While historical evidence for many rapid climate disruption events is readily available from both proxy records and from modern instrument data, there is no natural forcing function that would explain the rapidly rising global temperatures we are now experiencing. To identify the causes and consequences of this rapid change, the **Intergovernmental Panel on Climate Change (IPCC)** was established in 1988. This international body of experts is coordinated by the United Nations Environment Program and the World Meteorological Organization to provide a clear scientific consensus on climate change. The panel currently includes representatives from 195 countries. The IPCC has issued a series of reports on the status of Earth's climate. The most recent, the 4th Assessment Report, was released in 2007; the 5th Assessment Report will be released in 2014.

Based on evidence from a combination of paleoproxies, modern instrument records, and models, the most recent report states that "warming of the climate system is unequivocal, as is now evident from observations of increases in global average air and ocean temperatures, widespread melting of snow and ice and rising global average sea level." It is highly likely that this rapid temperature rise is directly related to increased **greenhouse gas** emissions, starting at about the time of the Industrial Revolution (ca. 1750–1850). When added to the atmosphere, these gases absorb solar energy and reemit it as heat. This leads to the **greenhouse effect**—the heating of Earth's atmosphere caused by the absorption and re-radiation of solar energy. The most important greenhouse gases (table 16.2) include water vapor, carbon dioxide (CO_2), methane (CH_4), nitrous oxide (N_2O), tropospheric ozone (O_3), and chlorofluorocarbons (CFCs). All of these greenhouse gases have natural sources (except for CFCs, which are manmade). Greenhouse gases are not inherently bad. If we did not have these gases in our atmosphere, the Earth's average temperature would be about −18°C (0°F) rather than about 15°C (59°F). The issue is that with increasing concentrations caused by human activity, the greenhouse effect also increases, warming the average temperature of the planet.

Using ice core data, scientists have been able to reconstruct Earth's atmospheric composition with very high temporal resolution for many thousands of years. Ice is formed by the gradual

ments prior to this period, scientists know there are other natural events that can result in much more rapid warming or cooling events. Perhaps one of the best known is the Little Ice Age (ca. 1350–1850 A.D.) when parts of North America and much of Europe experienced colder than average temperatures. Recent evidence suggests that this cool period was triggered by explosive volcanic activity. The massive quantities of aerosols injected into the atmosphere by volcanic eruptions can filter out a portion of the incoming solar energy, resulting in cooling of the planet. Four such events occurred during the past 150 years, coinciding with accurate temperature records, and clearly show a decrease in Earth's average air temperature (fig. 16.7). Normally these rapid cooling events last only a few years—the amount of time it takes the aerosols and particulates from the volcanic eruptions to leave the atmosphere. In the case of the Little Ice Age, the rapid cooling initiated by volcanic eruptions

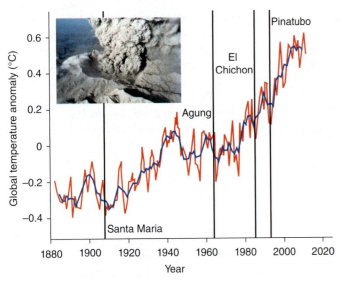

Figure 16.7 Volcanic activity can cause rapid cooling of Earth's temperature. Four volcanic eruptions are indicated on the temperature anomaly graph. After each eruption, the Earth cooled for several months to a few years. The inset shows the 1991 eruption of Mount Pinatubo in the Philippines.

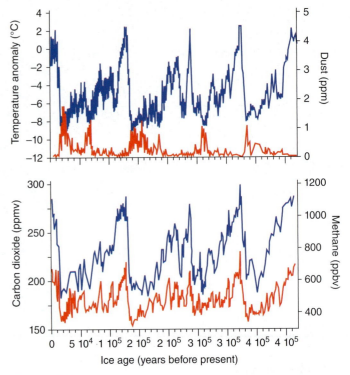

Figure 16.8 Vostok ice core data showing the trends in dust, carbon dioxide, methane, and changes in air temperature for the last 420,000 years. The amount of dust is recorded as parts per million (ppm); the carbon dioxide and methane are parts per million by volume (ppmv) and parts per billion by volume (ppbv).

compaction of snow. As the ice forms, tiny bubbles are trapped inside that preserve the atmospheric composition for that time period. Embedded in the ice are other potential paleoproxies such as atmospherically deposited dust and plant pollen. By carefully analyzing the chemistry of these trapped air bubbles and the amount of dust in the core at different depths (ages), it is possible to reconstruct the historical atmospheric composition. The most well-known climate record comes from the Vostok ice core in Antarctica (fig. 16.8). Data from that core show clear relationships between dust in the atmosphere (which reduces solar intensity; see the discussion of volcanoes), the concentration of carbon dioxide and methane, and Earth's temperature obtained from proxy records.

The evidence for a direct link between greenhouse gas emissions and the rapid rise in Earth's temperature is based on many sources of data. From the Vostock ice core it is apparent

that periods in Earth's history that had higher levels of greenhouse gases in the atmosphere correspond to periods of warmer temperature; this is an example of natural oscillations in the forcing of Earth's climate. Compared to these historical oscillations, the concentrations of greenhouse gases have risen rapidly in the last several decades, to levels that have not been seen in the past 10,000 years (fig. 16.9). The IPCC and other scientific organizations identify this link between greenhouse gas emissions and rising temperature that falls outside the normal range of variability as clear evidence for human-induced, abrupt climate change. Despite this evidence questions have remained about both the accuracy of the modern temperature record and

Table 16.2   Main Greenhouse Gases (excluding water vapor) and Their Contribution to the Greenhouse Effect (ppbv = parts per billion by volume)

Gas	Concentration (ppbv)	Relative Contribution to Increasing the Greenhouse Effect (%)	Greenhouse Potential (strength relative to CO_2)	Lifetime (years)
Carbon dioxide (CO_2)	391,570	60	1	~100
Methane (CH_4)	1811	16	25	12
Nitrous oxide (N_2O)	322	6	200	114
Tropospheric ozone (O_3)	10–50	11	2000	Hours-days
CFC-11	0.24	2	12,000	45
CFC-12	0.53	5	15,000	100

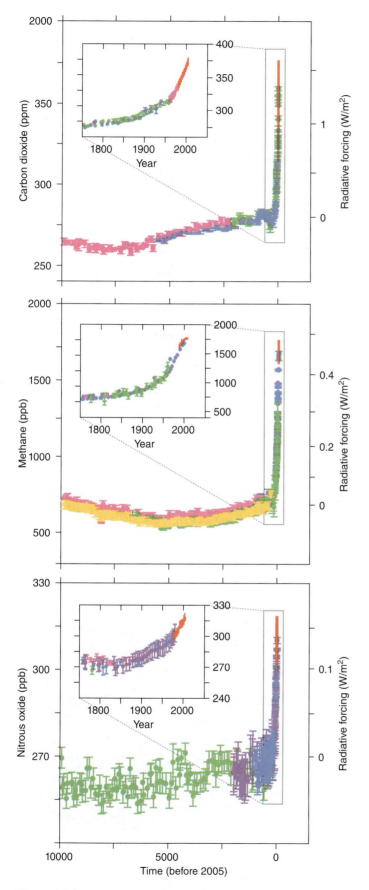

Figure 16.9 Atmospheric concentrations of greenhouse gases over the last 10,000 years and since 1750 (inset panels), and the resulting radiative forcing (heating), showing that these gases have risen rapidly following the industrial revolution, particularly after the 1950's.

the link to human-induced changes. To address these questions, an independent analysis was recently conducted by the Berkeley Earth Team. They compiled 1.6 billion land temperature records from 1850 to 2012 and independently confirmed that Earth's temperatures have risen by about 1°C since the 1950s. This rapid change in temperature is consistent with the view that human-induced climate change is the dominant factor and cannot be explained by either uncertainties and errors in the data or by known natural forcing of Earth's climate system.

Earth's Future Climate

While scientists use paleoclimatological proxies and the instrumental record to study Earth's past and present climate, future climate predictions must rely on projection of historical trends and model predictions. Prior to the establishment of the IPCC, global climate models were fairly primitive. They used simplified assumptions about how Earth's six spheres interact and had limited global data to test the model predictions. Models have improved dramatically in the past few decades (fig. 16.10). Scientists have worked to improve the predictions by including more complex feedbacks and higher spatial and temporal resolution. Future climate predictions are only as accurate as the data and models being used, so there is always uncertainty, particularly when looking at individual components of Earth's climate system or local and regional effects. To address this issue the IPCC runs multiple models with many scenarios and provides estimates of uncertainty for each prediction. Regardless of the scenario, the IPCC predicts continued warming of about 0.2°C per decade for the next two decades (fig. 16.11). The difference in predicted future temperatures is largely dependent on choices society makes about how we respond to the threat of climate change. Because of the high specific heat of water (review chapter 6), the oceans have accounted for about 84% of the increase in total heat content of the Earth during the period of rapid rise in CO_2 and temperature (fig. 16.12). As a result, the oceans have and will continue to experience dramatic changes in physical, chemical, and biological properties as the oceans and atmosphere continue to equilibrate.

QUICK REVIEW

1. Define *paleoclimatology*. Why must scientists use paleoproxies to study Earth's past climate?

2. Describe some of the changes in Earth's climate that can occur on decadal and millennial scales.

3. Why is Earth's climate is always changing, with or without human activity?

4. What is a greenhouse gas?

5. Explain how the IPCC and other scientific bodies have related the recent rapid warming of our climate to human activities.

6. Describe how climate models have improved over the past few decades.

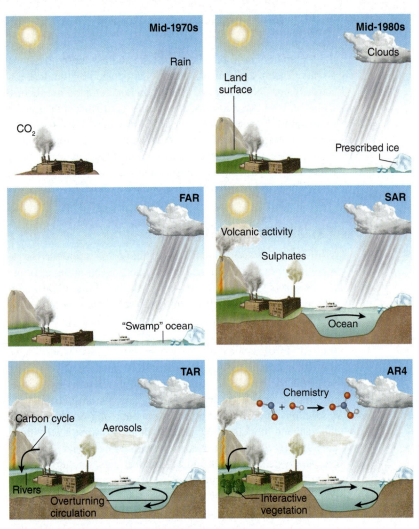

Figure 16.10 The complexity of climate models has increased over the last few decades. The additional physics incorporated into the models are shown pictorially by the different features of the modelled world. FAR, SAR, and TAR refer to the First, Second, and Third IPCC Assessment Reports.

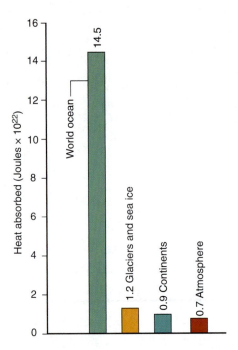

Figure 16.12 Estimates of the heat input to various components of Earth's heat budget as a result of global warming (1955–98). The 14.5×10^{22} J by the world oceans represents 84% of the total increase in global heat content.

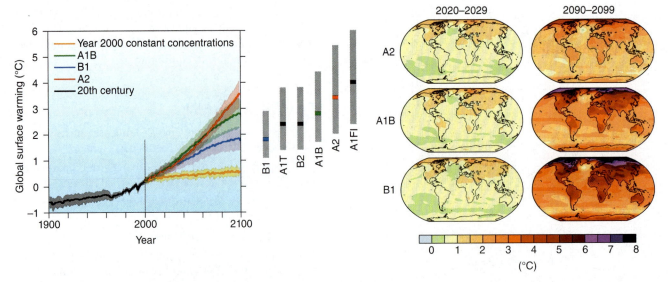

Figure 16.11 *Left panel*: Solid lines are multi-model global averages of surface warming (relative to 1980–99) for various Special Report on Emissions Scenarios used by the IPCC. The *orange line* is for the experiment where concentrations were held constant at year 2000 values. The *bars* indicate the best estimate (*solid black*) and uncertainty (*colored bar*) for each model scenario. *Right panels*: Projected surface temperature changes for the early and late twenty-first century relative to the period 1980–99.

16.3 The Oceans in a Warmer World

Rising atmospheric temperatures of about 0.2°C per decade have resulted in the average temperature of the upper ocean (700 m depth) increasing by about 0.6°C over the past century. By comparing data from modern instrumented floats with temperatures obtained from the *HMS Challenger* expedition 135 years ago (see Prologue), scientists have also reported increasing ocean temperatures at depths greater than 1600 m in the North Atlantic where thermohaline circulation rapidly transports surface water to depth. While rising temperatures are one of the most easily observed consequences of global warming, there are many other direct and indirect impacts.

Ocean Acidification

Ocean acidification, sometimes referred to as "the other CO_2 problem," results from increases in carbon dioxide in the ocean caused by equilibration between the ocean and atmosphere. As the CO_2 concentration increases, there is a shift in ocean pH resulting from the carbonate buffering system, which favors an equilibrium toward the right side of the equation (i.e., formation of carbonic acid and bicarbonate; review section 5.4). As a result, the absorption of anthropogenic CO_2 has acidified the surface layers of the ocean, causing an overall decrease since the pre-industrial period of 0.1 pH units. Although these increases appear small, they represent a precipitous decline compared to the conditions that have prevailed in the global ocean for hundreds of thousands to millions of years. Indeed, recent estimates suggest that the change in ocean pH is occurring more rapidly than has occurred for the last 300 million years.

The decrease in oceanic pH will most directly impact calcifying organisms such as corals and pteropods (see fig. 5.12), but there are other consequences as well. For example, a decline in pH of only 0.3 causes a 40% decrease in the sound absorption properties of surface seawater. Sound at frequencies important for marine mammals and for naval and industrial applications will travel 70% farther with the ocean pH change expected from a doubling of atmospheric CO_2, leading to more noise pollution (review chapter 13). Phytoplankton growth may also be influenced by changes in acid-base chemistry and trace metal availability; as pH drops many metals such as iron and copper become more soluble in the ocean. In some cases, this can enhance biological productivity, while in other cases this can lead to enhanced metal toxicity. The pH gradient across cell membranes is also coupled to numerous critical biochemical reactions within marine organisms, ranging from photosynthesis, to nutrient transport, to respiration. The impact of ocean acidification on these processes is barely understood.

Rising Sea Level

Sea level is on the rise around the coasts of the world. Since August 1992, satellite altimeters have been measuring sea level on a global basis with unprecedented accuracy. Data from satellite sensors such as *TOPEX/Poseidon* and *Jason* (see *Oceanography from Space*) provide observations of sea-level change globally. Over the past century, sea level has risen about 25 cm (10 in), but recently the rate of rise has increased. Sea level is presently rising at a rate of approximately 3.1 mm (0.12 in) per year (fig. 16.13). It is difficult to predict rates of sea-level rise for the future but current thought is that sea levels will rise at rates two to three times faster in the next 100 years than they did in the twentieth century due to **thermal expansion** of seawater, the increase in volume due to a decrease in density caused by heating. By the year 2100, sea level is expected to be 80–105 cm (2.6–3.6 ft) higher than at present, although considerably higher projections have also been made (fig. 16.14).

The global average change in sea level may not be reflected in the gradual changes observed at any particular location. In the tropics, sea temperatures are nearly constant, so changes attributable to thermal expansion are small, but the seasonal oscillation of the meteorological equator (the intertropical convergence zone) between approximately 0° and 10°N causes sea level to vary in response to atmospheric pressure changes. Changes in sea surface temperature associated with El Niño affect sea level, and in the North Atlantic and North Pacific, the annual seasonal shift from high- to low-pressure cells over the oceans causes a related change in sea level (rising with low atmospheric pressure and falling with high pressure). Onshore winds elevate and offshore winds depress the coastal water level. Ocean currents

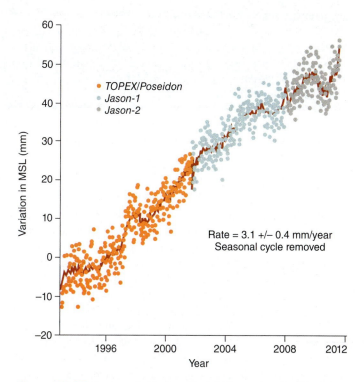

Figure 16.13 Temporal variations in global mean sea level (MSL) computed from *TOPEX/Poseidon*, *Jason-1*, and *Jason-2* satellite altimeters. Each *dot* is a single, ten-day estimate of the difference between MSL during those ten days and the mean MSL during the entire period from December 1992 through 2012. The *black line* is a 120-day smoothing function.

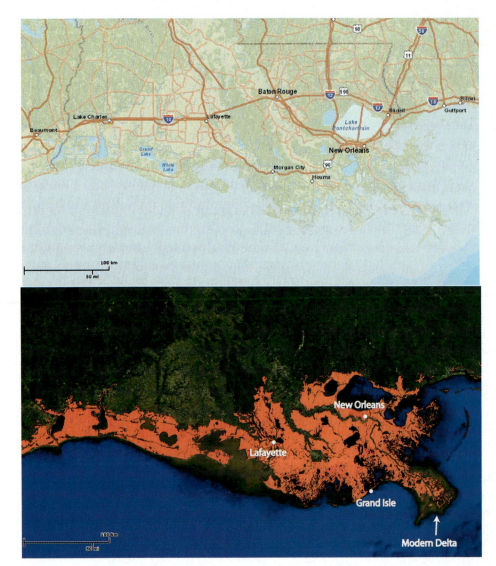

Figure 16.14 Sea level could rise as much as 12–21 m (40–70 ft) in response to global warming. This image shows the effect of that sea level rise on the coast of Louisiana.

consistent with observations of a warming Arctic and predictions from IPCC models that the Arctic is particularly susceptible to global warming. In contrast, the area covered by Antarctic sea ice has shown a small but statistically insignificant increasing trend. That is not the whole story, however. Gravity data collected from space show that Antarctica has been losing more than a 100 cubic kilometers (24 cubic miles) of ice each year since 2002, and that the rate of loss is increasing. This has been attributed to warming ocean water around West Antarctica. This warm pool of water has been thinning the underside of the West Antarctic Ice Sheet, leading to the potential collapse. If this were to occur, global sea level could rapidly rise 3.3–4.8 m (~11–16 ft). Thus, while the areal extent has slightly increased, the total mass of ice is decreasing, also consistent with climate change predictions.

As a result of these complex interactions among the atmosphere, hydrosphere, and cryosphere, rising sea level is a global phenomenon, but the problems it causes are not distributed evenly around the world. Areas in mid-ocean, such as small tropical atolls with elevations of only 1–2 m (3–7 ft) may be severely affected. Low coastal marshlands will undergo greater changes than coasts with steep, rocky coastlines. Each year, 100–130 km^2 (40–50 mi^2) of Louisiana's Mississippi River wetland marsh disappear under the Gulf of Mexico. Twelve thousand years ago, Louisiana's shoreline extended 200 km (120 mi) farther into the Gulf of Mexico. In the mid-1880s, Bailize was a busy river town at the tip of the delta; today, it is under 4.5 m (15 ft) of water. Current projections of sea level rise in response to continued warming suggest that New Orleans and other coastal communities could suffer the same fate (fig. 16.14).

Winds, Waves, and Storms

Ocean wind speeds and wave heights are directly related to the energy balance of Earth's climate system. Winds and waves help to balance the flux of energy between the atmosphere and the oceans and are therefore expected to be sensitive to increasing global air temperature. Although the recent landfall of several major hurricanes in the Gulf of Mexico and Eastern Atlantic (see *Oceanography from Space*) have been suggestive of climate change-induced increases, it is unclear whether this is natural variability in the cycle of tropical storms or a direct response to

under the influence of the Coriolis effect create a sea surface topography that varies by as much as a meter. Any climate-generated changes in speed or location of winds or currents alter sea level. Climate changes associated with atmospheric warming can be expected to alter the winds, the currents, and atmospheric pressure distributions as well as the thermal expansion of seawater. The atmospheric changes affect the sea surface, resulting in sea-level changes of as much as 1–2 m (3–7 ft).

Sea-level rise is affected by more than just thermal expansion. As the oceans heat up, sea ice extent would be expected to decrease, as would the volume of glaciers and ice caps. Both Arctic and Antarctic sea ice extent are characterized by fairly large variations from year to year. The monthly average extent can vary by as much as 1 million square kilometers (386,102 square miles) from the long-term average. Based on satellite estimates, both the thickness and extent of summer sea ice in the Arctic have shown a dramatic decline over the past thirty years (fig. 16.15),

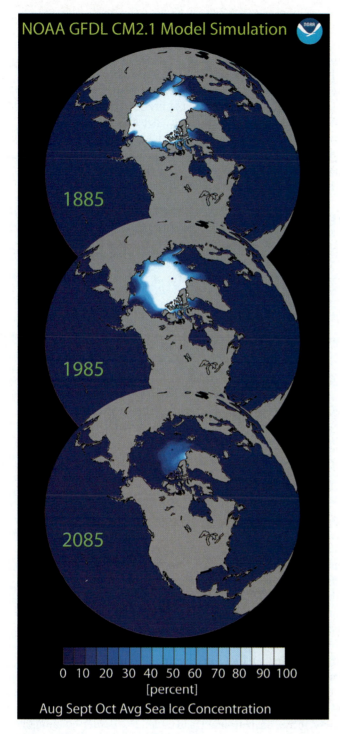

NOAA GFDL CM2.1 Model Simulation

1885

1985

2085

0 10 20 30 40 50 60 70 80 90 100
[percent]
Aug Sept Oct Avg Sea Ice Concentration

Figure 16.15 Sea ice concentrations simulated by the GFDL CM2.1 global coupled climate model averaged over August, September, and October (the months when Arctic sea ice concentrations generally are at a minimum). Three years (1885, 1985, and 2085) are shown to illustrate the model-simulated trend. A dramatic reduction of summertime sea ice is projected, with the rate of decrease being greatest during the twenty-first century portion. The colors range from *dark blue* (ice free) to *white* (100% sea ice covered).

climate change. There has been an increase in the total number and intensity of hurricanes in the Atlantic since at least the 1970s. Model scenarios suggest, however, that storm tracks should shift poleward resulting in fewer landfalls.

While hurricanes and cyclones capture the public's attention, climate change scenarios also suggest that oceanic winds and waves will also increase, particularly at higher latitudes. A recent twenty-three-year analysis indicates that this increase is already detectable and that extreme events are increasing faster than the mean conditions. In the northwest Pacific, the waves from "100-year event" storms are now predicted to exceed 14–17 m (46–55 ft), up from predictions of 10 m (33 ft) less than fifteen years ago. Similar increases have been reported for the Atlantic. The increased coastal erosion, flooding, and other damage caused by these extreme storms is expected to dwarf the impacts of more gradual sea level rise (fig. 16.16).

Thermohaline Circulation

IPCC models consistently predict increased warming and increased precipitation at high latitudes including the North Atlantic. This will result in decreased density and increased water column stability, potentially impacting formation of North Atlantic Deep Water (NADW; review section 7.3). Abrupt climate changes in Earth's past, based on paleoclimatology, are most often associated with changes in deep-water formation. Climate simulations suggest that deep-water formation may become more sluggish, but is unlikely to shut down within the next 100 years. There is some evidence that the most recent IPCC report does not adequately account for the effect of rapid melting of the Greenland ice shelf. It is therefore at least possible that NADW formation could be more sensitive to climate change than the models suggest. When combined with the general lack of knowledge about global warming impacts on deep-water formation in the Antarctic, there is considerable uncertainty about the potential consequences of a warmer and less-saline ocean on global thermohaline circulation. If modifications to thermohaline circulation were to act as a "tipping point," abruptly pushing Earth's climate into a new state, we could experience rapid and unexpected consequences that we cannot predict, similar to the positive feedback that pushed North Atlantic temperatures into a cool state for several centuries during the Little Ice Age.

Figure 16.16 Global warming is increasing ocean winds and waves, resulting in increasing levels of coastal erosion and flooding.

Biological Responses

Predicted responses to climate change include some direct impacts to biological organisms. Changes in sea ice timing and extent will obviously have impacts on high-latitude ecosystems, including apex predators such as polar bears and organisms that rely on the ice-edge community, such as migratory whales. Many arctic mammals face serious declines, with polar bears projected to lose 68% of their summer habitat by 2100. Antarctic organisms that rely on the presence of ice, such as penguins and seals, are declining and, in some cases, face extinction (fig. 16.17). Decreasing pH directly impacts calcifying organisms, including commercially important shellfish and crustaceans, and also has the potential to indirectly impact many other marine organisms because pH impacts many physiological processes. A growing number of studies suggest that a warmer ocean is also linked to increases in disease outbreaks as well as more rapid expansion of invasive species. At the same time, many species are migrating poleward and deeper to maintain an optimal temperature environment. Many models predict a global decline in species diversity and an increase in the rate of extinctions of marine organisms but the range and uncertainty of projected changes is very large due to both opportunities for intervention through conservation efforts and lack of robust predictions for population responses to climate change.

A warmer ocean may also modify phytoplankton primary production, the base of the food chain. Warmer temperatures should speed up the "microbial loop," leading to faster regeneration of nutrients as well as increased phytoplankton growth rates. Increased stratification of the surface ocean would ultimately lead to decreases in nutrients from deep-water sources,

resulting in long-term decreases in productivity. These changes to the base of the food chain would impact both higher trophic levels and the biogeochemical balance of the oceans. To detect these changes, scientists will require thirty to fifty years of global productivity data. To date, ocean color satellites can provide about fifteen years of data, making it difficult to prove or disprove long-term, climate-induced trends because of the inherent "noise" introduced by seasonal and interannual cycles such as El Niño and other multi-decadal oscillations.

QUICK REVIEW

1. How do increasing temperatures result in increased sea level?

2. Describe why ocean acidification is sometimes called "the other CO_2 problem."

3. Explain why coastal communities may experience more frequent and more intense storms.

4. Describe the global deep-water circulation pattern, and explain how it might change with increasing temperatures.

5. Why will migrating organisms such as grey whales be impacted by changes in climate?

6. Describe how climate change would impact the distribution and health of marine organism populations.

7. Explain why phytoplankton primary productivity might decrease in a warmer ocean. Why is it difficult to tell if this is happening?

16.4 Mitigation Strategies

Scientists overwhelmingly agree that Earth's climate is undergoing an abrupt increase in temperature and that this increase is linked to human production of greenhouse gases, primarily carbon dioxide. There are two general strategies to slow down or mitigate these changes: either reduce the amount of greenhouse gases entering the atmosphere or remove CO_2 from the atmosphere and sequester (store) that carbon someplace else. To address the first strategy, it is tempting to suggest that humanity should stop consuming fossil fuels since these lead to the majority of greenhouse gas emissions (fig. 16.18). In the short term, it is unrealistic to simply stop using fossil fuels because modern society relies on petroleum and other fossil fuel compounds for energy, transportation, and production of asphalt, fertilizer, plastics, and many other items that are necessary to maintain modern society. For the foreseeable future we will continue to extract fossil fuels including from the deep ocean (see *Diving In*: *Extracting Energy Resources from the Ocean*) while we develop alternative energy solutions.

Ocean Energy

The search for alternative sources of energy has raised interest in harvesting "clean" power from the oceans by utilizing wind, wave, current, and tidal power, salinity and thermal gradients, as well as novel sources such as biofuels produced from marine algae. Some of these technologies are already in place. For

Figure 16.17 Emperor penguins have a high probability of becoming "quasi-extinct" (reduced to less than 95% of their current population) by 2100 due to changes in Antarctic habitat in response to IPCC climate change predictions.

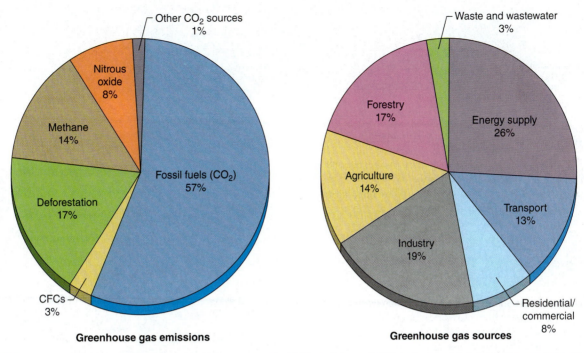

Greenhouse gas emissions

Greenhouse gas sources

Figure 16.18 Share of different anthropogenic greenhouse gases in 2004, and share of different sectors in total anthropogenic greenhouse gas emissions in ("Forestry" includes deforestation). Percentages are in CO_2 equivalents.

example, tidal power has been used since at least the eleventh century. Despite this long history, 90% of today's worldwide tidal energy production comes from a single site—the Rance Tidal Power Plant in France. **Ocean thermal energy conversion (OTEC)** was first proposed in 1881. It relies on a heat gradient between the surface and deep ocean of about 20°C (36°F) to drive heat exchangers. The first functional OTEC plant was built in Cuba in 1930. In the United States, the Natural Energy Laboratory of Hawaii Authority (NELHA) was established in 1974 to explore the use of this technology. While both tidal and OTEC power sources are well proven, the very high infrastructure costs have limited their use to a handful of sites, although NELHA and other groups are currently planning new facilities.

Most other forms of ocean energy production, such as turbines that harvest wave or current energy, are primarily in the pilot or demonstration stage, but are actively being pursued (fig. 16.19). Wave energy is being actively explored along the U.S. West Coast. California's Energy Commission estimates that 23% of current electricity consumption could be met by wave energy. The first West Coast wave energy buoy was deployed off Oregon in 2007. While promising, there are many issues to work through, including the technology to harness wave or current energy, permitting issues, and environmental concerns about the impacts to coastal fishermen, and the effects of large-scale "wave farms" on the environment. In contrast to these novel energy sources, wind power has been developed extensively in the marine environment by countries such as Denmark, where wind power is scheduled to provide 50% of the country's electricity by 2020 (fig. 16.20). Offshore wind farms are attractive for coastal countries because the heating and cooling difference

Figure 16.19 A tidal power turbine is being deployed as part of the Roosevelt Island Tidal Energy (RITE) project in New York City. The RITE project could support as many as 300 turbines and generate nearly 10 megawatts of power.

Diving in

Extracting Energy Resources from the Oceans

BY DR. ROBERT MCDOWELL

Dr. Robert McDowell teaches oceanography at the American Public University System. He is a former commercial diver in the offshore oil and gas industry, a Navy oceanographic and diving officer, and has worked for the Naval Research Laboratory.

Oil and gas production are the most used and economically essential resources that humans recover from the world's oceans, in terms of total monetary value worldwide. The total value of oil and gas extracted from the ocean floors and continental shelf areas far surpasses the value of the global seafood catch and mariculture production combined.

However, as the CEO of one of the major oil companies told a Congressional Committee during a hearing on oil prices and oil production: "*remember all of the oil that was easy to get to and recover has already been pumped out and used; what remains will be more difficult and more costly to produce.*" Thus, as the world's supply of oil diminishes, the petroleum industry will have to drill in deeper water and more hostile environments to identify and recover new oil and natural gas sources in order to keep pace with the expected increases in world demand.

Offshore oil and gas drilling rigs and production platforms are among the largest and most complex man-made structures ever built. In the past, huge stationary platforms that rival the size of the Empire State Building, and actually contain more steel, enabled companies to drill at water depths of up to 1650 ft (~500 m). Advances in technology in recent years have allowed floating drilling/production platforms to extract hydrocarbons at depths of up to 10,000 ft (~3000 m). These structures are able to remain over the drilling site by the use of cables and deep-sea anchors around the wellhead and GPS-guided dynamic positioning systems (box fig. 1).

Box Figure 1   A dynamic positioning system (DPS) stabilized rig in the North Sea off Aberdeen, Scotland, in over 1000 m (3280 ft) of water.

Box Figure 2   This ROV is working on a deep-water wellhead. It is grasping with one of its mechanical arms in order to stabilize itself against the current.

Designing and constructing equipment to work in the deep ocean requires state-of-the-art engineering. The deep sea floor is a hostile environment for multiple reasons—seawater is very corrosive, the deep ocean is very cold, and the pressure in deep water is very high. Electrical connections must be properly encased to avoid being intruded by water at these intense pressures.

In addition, it is difficult to communicate with instruments on the sea floor. NASA can communicate using radio or microwave transmissions with spacecraft millions of miles away. Neither of these forms of electromagnetic energy penetrates water very far; consequently, GPS signals don't work more than about 1 foot underwater without a surface antenna. So deep-ocean equipment must either have a connecting cable to the mother ship or platform (like ROVs: remotely operated vehicles) or some type of inertial navigation and control system like AUVs (autonomous underwater vehicles) (box fig. 2).

Other complicating factors for ocean technology are currents and tides. The underwater environment is constantly in motion, thus ROVs and AUVs must have strong enough motors and propeller thrusters to be able to maneuver underwater and/or stay in one place long enough to perform the assigned task.

These are exactly the types of robotic devices that the oil industry and its contractors had to use to assist in killing the *Deepwater Horizon* oil spill in the Gulf of Mexico in 2010. These machines are extremely expensive and they require lots of support equipment. They have mechanical arms, lights, sensors, and cameras. But controlling one is just as tricky as trying to control a rover robot on the surface of Mars, with one added problem: The control cable can become tangled or snagged on something underwater.

AUVs and some ROVs are primarily used for underwater surveying and location and to operate and repair subsea equipment. One of the primary sensors in these devices is side-scan sonar, which is a type of sonar that has a sonar beam that is directionalized and concentrated. Interestingly, it was invented by the same person who invented the medical sonogram. The technology applies the same principles of directing concentrated sound waves through water instead of the human body (which is mostly water). This side-scan sonar has revolutionized underwater location because it can actually "see" into different types of bottom substrates like sand and mud.

Box figure 3 is a graphic depiction of how these systems are used and how they interact, particularly around subsea oil and gas production equipment. Notice that the manned diver is shown very close to the top of the graphic. This is because ambient-pressure manned diving is limited to about 1000 ft (~300 m) maximum. The vast majority of

commercial divers work in the oil and gas industry. For depths below 1000 ft, robotic systems are required.

Modern drilling technology also allows for wells to be drilled directionally from a single platform. Today, multiple wells can be drilled and produced from a single stationary or floating drilling rig, thus minimizing the need to relocate the drilling rig or install a new platform (box fig. 4). This helps in reducing the impact of drilling operations on the environment.

The blowout and subsequent massive oil spill in the Gulf of Mexico, partly due to a technology failure, graphically illustrates the complexity and the limitations of modern undersea drilling technology. It took the full cooperation of the U.S. government, including NASA, NOAA, and the National Laboratories, to develop the capability to stop a leaking oil well 5000 ft (~1500 m) beneath the sea surface. The only organizations that normally have this type of technology (and it is as advanced as any technology developed by NASA or the military) are the multinational oil companies and their subcontractors.

Oil and gas (methane) deposits tend to occur together in the rock. These deposits are pressurized as a result of the weight of rock

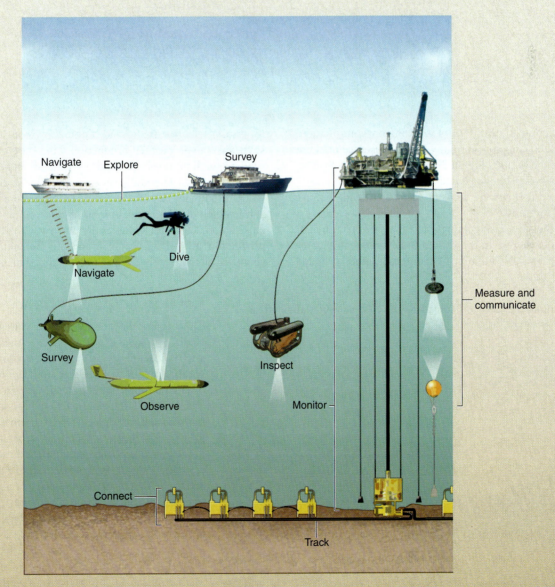

Box Figure 3   Deployment depths and interactions of subsea technology supporting oil and gas recovery.

Continued next page—

Diving in *Continued—*

layers above them, causing them to compress. While drilling a well, it is important that the weight of the column of drilling fluid in the drill pipe (called *drilling mud*) be kept at a higher pressure than that of the pressure found in the layers of rock being drilled through. Failure to keep these two opposing pressure forces in balance may result in a blowout. Blowouts are the result of uncontrolled releases of pressure from the oil and/or gas deposit that is being drilled into. Large and mechanically complex devices called blowout preventers (BOPs) are installed on top of the wellheads as an additional barrier to help to control the subsurface pressures. However, unexpected and/or unrecognized levels of gas pressure can overwhelm these devices. When they fail, the results can be catastrophic. On a positive note, there are over 40,000 oil and gas wells drilled in and around the Gulf of Mexico with relatively few blowouts. In most instances, these wells are brought under control quickly, resulting in minimal damage to the environment.

In an increasingly challenging quest to retrieve hydrocarbon energy resources from the world's oceans, the petroleum industry will have to employ ever-more advanced technology while minimizing environmental and financial risks in order to extract these increasingly scarce and in very high demand resources from the Earth.

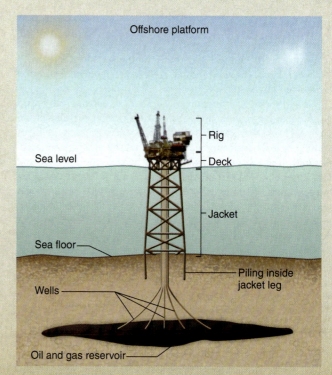

Box Figure 4 Diagram of the technique of multiple wells drilled directionally from a single stationary platform.

---

between the land and ocean results in strong onshore-offshore winds. The United Kingdom currently has 568 installed offshore wind turbines and is planning on 20% total electricity production by 2020. As of 2012, the United States has no offshore wind power, but several states are seeking permits to begin testing offshore wind turbines.

Algal biofuels are also gaining interest as an alternative to petroleum-based energy. In the United States today, it is common to see up to 10% addition of corn-based ethanol added to gasoline. Unlike traditional biofuel crops such as corn and soybeans, algal biofuel production does not compete with food production, does not require farmland, and can be grown on wastewater, potentially mitigating coastal eutrophication while producing alternative sources of petroleum products. While much of this production takes place on land in ponds or closed bioreactors, pilot studies are underway to utilize offshore production of both macroalgae and phytoplankton in several countries including the United States, China, Japan, and Ireland.

Figure 16.20 Countries such as Denmark rely on offshore wind turbines for a large fraction of their electricity.

Carbon Sequestration

The oceans play an important role in regulating Earth's climate. The oceans act as the largest reservoir of heat energy on the planet because of the high specific heat of water. On geological time scales, the ocean's biological pump also serves to remove vast quantities of atmospheric CO_2 by producing organic material and carbonates that are ultimately deposited in marine sediments. There have been several proposals to enhance this natural **carbon sequestration** (removal of carbon from the atmosphere to some reservoir) by enhancing the ocean's natural sequestration capabilities. Proposals to directly manipulate Earth's climate are called **geo-engineering**, and are highly controversial. In chapter 12 we introduced large-scale iron fertilization. There have also been smaller-scale phosphorus additions in the Eastern Mediterranean and off the coast of Northwest Africa, and artificial upwelling experiments conducted off

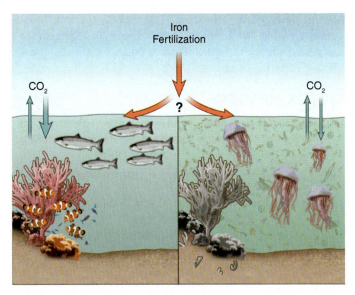

Figure 16.21 The consequences of nutrient fertilization experiments have been difficult to predict. While some experiments have resulted in beneficial blooms, others have stimulated more bacterial production, potentially leading to negative ecological consequences.

and at least two commercial fertilization experiments have taken place. In response to this commercial interest, a statement was issued by the International Maritime Organization, which administers the London Convention, the main legal instrument for controlling marine pollution. This group declared that ocean fertilization activities other than legitimate scientific research are prohibited. There is ongoing debate among both scientists and the public about whether and how these geo-engineering solutions should be regulated.

A second proposed mitigation strategy involves removal of CO_2 emissions from fossil fuel-burning power plants using chemical scrubbers. The accumulated CO_2 can be concentrated and pumped directly into the deep ocean (fig. 16.22). Although technically possible, some of the same concerns arise for this proposed mitigation strategy, including cost, the legal basis for injecting carbon into the deep ocean, the potential acidification of the deep ocean, uncertainty in how long the carbon would remain at depth, and the potential impacts to deep-sea ecosystems.

Hawaii. The experiments to date show that the biological and chemical responses to nutrient fertilization are variable and difficult to predict (fig. 16.21). Concerns about large-scale fertilization efforts include the potential disruption of the ecosystem, stimulation of toxic harmful algal blooms, formation of deep-water anoxic conditions, and the likelihood that any enhanced carbon drawdown would not mitigate ocean acidification.

While most of the mesoscale, open-ocean nutrient enrichment experiments were scientific, several commercial companies have expressed interest in large-scale fertilization with iron, urea, and phosphorus to stimulate the biological pump,

The International Response

Given the growing evidence that small increases in the concentration of greenhouse gases are triggering a wide array of unpredictable and possibly irreversible changes to marine ecosystems, avoiding further increases and mitigating existing impacts has become an international priority. The United Nations Framework Convention on Climate Change developed the **Kyoto Protocol** to voluntarily reduce greenhouse gas emissions by 5% relative to 1990 levels by 2012. This was signed and ratified by sixty countries in 1997. The United States never ratified the Protocol, citing potential damage to the global economy; Canada signed and ratified the Protocol but renounced it in 2011. Following on from this attempt at international cooperative regulation, the United Nations Climate Change Conference was held in 2009 in Copenhagen, Denmark. Led by the United States, this

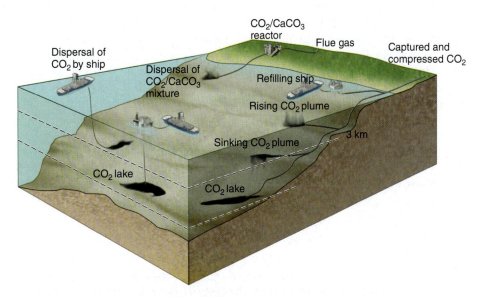

Figure 16.22 One proposal for sequestering carbon dioxide involves direct injection into the deep ocean.

meeting established the **Copenhagen Accord**. This nonbinding agreement endorsed the Kyoto Protocol, recognized the need to reduce anthropogenic greenhouse gas emissions, and recommended a target of no more than a 2°C rise in global temperature. The Copenhagen Accord did not establish specific targets for greenhouse gas emissions, nor did it provide a mechanism for enforcement.

Despite these international commitments, greenhouse gases continue to rise, leading to direct and indirect impacts on the ocean. There are no easy solutions to this problem and, despite the preponderance of scientific evidence, many people still disagree that our current period of disruptive climate change is either linked to human activities or of particular concern. According to a 2011 report,[1] 63% of the public surveyed in the United States agreed that global warming is occurring, and 50% of respondents believed that it is caused primarily by human activities. Everyone on the planet is intimately connected to the oceans; denying that this problem exists does not change the fact that the oceans are changing, and that many of these changes are directly linked to human activities.

QUICK REVIEW

1. Explain how fertilizing parts of the ocean with iron could reduce atmospheric carbon dioxide. Where in the oceans would iron fertilization be most effective?

2. Describe some sources of alternative energy that rely on the oceans.

3. Explain some of the pros and cons for direct injection of carbon dioxide in the deep ocean.

4. How would the Kyoto Protocol help to stabilize Earth's climate?

Summary

Earth's climate system is comprised of interactions among the atmosphere, hydrosphere, cryosphere, geosphere, biosphere, and anthrosphere. *Climate* is definied as the long-term average of conditions in a region or of the planet as a whole, including the likelihood of extreme events. Earth's climate is constantly changing on both short and long time scales. Climate disruptions refer to geologically rapid changes in climate.

Changes in Earth's climate can be amplified or dampened by positive and negative feedback mechanisms such as changes in Earth's albedo due to changes in cloud cover and presence and extent of snow and ice. Scientists study changes in climate by using a combination of proxy records to study the past, instrumental records to study the present, and models and predictions to study future climate scenarios. Paleoclimatology is the study of Earth's historical climate hundreds to millions of years in the past. Using proxies, paleoclimatologists have identified climate drivers that operate at cycles that span thousands to millions of years. These include variations in solar intensity, variations in Earth's orbit, and changes in ocean circulation driven by plate tectonics. Climate change driven by these events is generally gradual, taking thousands to millions of years to develop. On shorter time scales, the Earth can cool due to rapid changes in solar insolation caused by volcanic activity.

Paleoclimatological records indicate that we are currently experiencing a period of abrupt climate change. The Intergovernmental Panel on Climate Change was established to provide a scientific consensus on the causes and consequences of this disruptive climate change. Based on a preponderance of evidence, this rapid warming has been linked to human-induced release of greenhouse gases to the atmosphere. There have been multiple direct and indirect consequences for the oceans. Rising atmospheric CO_2 and temperature have resulted in warming of the surface ocean and a significant decrease in pH, leading to ocean acidification. This impacts marine organisms both directly (e.g., through dissolution of calcium carbonate) and indirectly (e.g., changes in the distance that sounds travel in the ocean, changes in metal solubility). Rising ocean temperatures also result in thermal expansion of seawater and melting of ice sheets and polar ice caps. This leads to steadily increasing sea level, which will disproportionately impact low-lying coastal regions. Rising temperatures also results in more wind and wave energy, leading to changes in the intensity of seasonal storms, maximum size of ocean waves, and the intensity and frequency of hurricanes and cyclones. Model predictions suggest that thermohaline circulation may slow but not stop in the immediate future, but there is some question about the accuracy of these models. These combined effects of global warming will lead to changes in many ecosystems, particularly at high latitudes. Outbreaks of disease and spread of invasive species are likely to increase, and global primary productivity may be impacted by changes in the stratification of the surface ocean, leading to changes in nutrient availability.

Several responses have been proposed to mitigate these impacts. Use of fossil fuels will continue for the foreseeable future, but alternative energy sources such as tidal, wind, wave, and current generators are being developed. Ocean thermal energy

---

[1]Leiserowitz, A., Maibach, E., Roser-Renouf, C., Smith, N. & Hmielowski, J. D. (2011). *Climate change in the American Mind: Americans' global warming beliefs and attitudes in November 2011.* Yale University and George Mason University. New Haven, CT: Yale Project on Climate Change Communication. http://environment .yale.edu/climate/files/ClimateBeliefsNovember2011.pdf

conversion (OTEC) is also being revitalized. As a way of reducing CO_2 in the atmosphere, enhancements to the biological pump have been proposed that generally involve fertilization of the open ocean with a limiting nutrient. Direct injection of CO_2 into the deep ocean has also been explored. Both mitigation strategies are controversial because of political, economic, and ecological uncertainties about the effectiveness of these strategies. In recent years, a series of international conventions have been proposed to limit further greenhouse gas emissions, but no single plan has been endorsed or enforced by the majority of countries.

Key Terms

climate disruption, 422
climate, 422
anthrosphere, 422
climate system, 422
positive/negative feedback
   mechanism, 422
albedo, 422

cloud condensation nuclei
   (CCN), 422
dimethyl sulfide (DMS), 423
CLAW hypothesis, 423
proxy, 424
paleoclimatology, 424
Milankovich cycle, 424

Intergovernmental Panel
   on Climate Change
   (IPCC), 425
greenhouse gas, 425
greenhouse effect, 425
thermal expansion, 429

Ocean thermal energy
   conversion (OTEC), 433
carbon sequestration, 436
geo-engineering, 436
Kyoto Protocol, 437
Copenhagen Accord, 438

Study Problems

1. The average concentration of dissolved inorganic carbon (DIC) in the ocean is 0.024 g/kg of seawater and the mass of water in the oceans is 1.36×10^{21} kg. What is the total amount of carbon stored in the ocean as DIC? Give your answer in units of grams.

2. The atmosphere contains 7×10^{17} g of carbon as carbon dioxide. Does the ocean or the atmosphere hold more carbon? Expressed as a percentage, what is the amount of carbon in the atmosphere?

3. The present atmosphere contains approximately 700 Gtons of carbon in the form of CO_2 [Gton = 10^{15} grams]. Earth's total recoverable fossil fuel reserves contain at least 4200 Gtons of carbon, mostly in the form of coal. At present, about half the CO_2 produced by fossil fuel burning stays in the atmosphere. The other half dissolves in the ocean or is taken up by the terrestrial biosphere. If this ratio remained constant and we burned up all of our fossil fuels instantaneously, by how much would atmospheric CO_2 concentrations rise? (Express your answer in terms of the new CO_2 level divided by the current one—in other words, by what factor CO_2 in the atmosphere changes.)

4. Climate models predict that doubling of the atmospheric CO_2 concentration from 700 to 1400 Gton carbon will cause the mean global temperature to increase by about 3°C. How much would temperature increase as a result of the scenario described in the previous question, assuming the warming was linear with increasing carbon dioxide?

5. As the ocean warms, it expands and sea level rises. Calculate the change in sea level due to a warming of the oceans of 0.5°C. For this problem, assume that the oceans of the world can be approximated by a basin with vertical sides—like a bathtub whose bottom has a constant area. As water warms, its volume increases. Since the bathtub does not increase in size, the water must expand upward (water level rise = eustatic adjustment).

You must carry forward at least five significant digits to do this problem because the magnitude of the sea level change is so small compared to the dimensions of the ocean.

Here are things that you will need to do this problem:

Radius of Earth $(r) = 6.37 \times 10^6$ m

Surface area of Earth $= 4\pi r^2$

Thermal heat expansion $= V_{final} = V_{initial}(1 + \beta\,\Delta T)$

Where V_{final} is the final volume, $V_{initial}$ is the initial volume, β is the volume expansion coefficient, and ΔT is the change in temperature in degrees C.

$\beta = 0.00021°C^{-1}$ (fractional volume change per degree C)

Appendix A

Scientific (or Exponential) Notation

Writing very large and very small numbers is simplified by using exponents, or powers of 10, to indicate the number of zeroes required to the left or to the right of the decimal point. The numbers that are equal to some of the powers of 10 are as follows:

$$1,000,000,000. = 10^9 = \text{one billion}$$
$$1,000,000. = 10^6 = \text{one million}$$
$$1000. = 10^3 = \text{one thousand}$$
$$100. = 10^2 = \text{one hundred}$$
$$10. = 10^1 = \text{ten}$$
$$1. = 10^0 = \text{one}$$
$$0.1 = 10^{-1} = \text{one tenth}$$
$$0.01 = 10^{-2} = \text{one hundredth}$$
$$0.001 = 10^{-3} = \text{one thousandth}$$
$$0.000001 = 10^{-6} = \text{one millionth}$$
$$0.000000001 = 10^{-9} = \text{one billionth}$$

149,000,000 is rewritten by moving the decimal point eight places to the left and multiplying by the exponential number 10^8, to form 1.49×10^8. In the same way, 605,000 becomes 6.05×10^5.

A very small number such as 0.000032 becomes 3.2×10^{-5} by moving the decimal point five places to the right and multiplying by the exponential number 10^{-5}. Similarly, 0.00000372 becomes 3.72×10^{-6}.

To add or subtract numbers written in exponential notation, convert the numbers to the same power of 10. For example,

$$
\begin{array}{r}
1.49 \times 10^3 \\
+6.05 \times 10^2 \\
\hline
\end{array}
=
\begin{array}{r}
14.90 \times 10^2 \\
6.05 \times 10^2 \\
\hline
20.95 \times 10^2
\end{array}
=
\begin{array}{r}
1.490 \times 10^3 \\
0.605 \times 10^3 \\
\hline
2.095 \times 10^3
\end{array}
$$

$$
\begin{array}{r}
2.36 \times 10^3 \\
-1.05 \times 10^2 \\
\hline
\end{array}
=
\begin{array}{r}
23.60 \times 10^2 \\
-1.05 \times 10^2 \\
\hline
22.55 \times 10^2
\end{array}
=
\begin{array}{r}
2.360 \times 10^3 \\
-1.05 \times 10^3 \\
\hline
2.255 \times 10^3
\end{array}
$$

To multiply, the exponents are added and the numbers are multiplied:

$$
\begin{array}{r}
4.6 \times 10^3 \\
\times 2.2 \times 10^2 \\
\hline
10.12 \times 10^5 = 1.012 \times 10^6
\end{array}
$$

To divide, subtract the exponents and divide the numbers:

$$\frac{6.0 \times 10^8}{2.5 \times 10^3} = 2.4 \times 10^5$$

The following prefixes correspond to the powers of 10 and are used in combination with metric units:

Exponential Notation

Exponential Value	Prefix	Symbol
10^{18}	exa	E
10^{15}	peta	P
10^{12}	tera	T
10^{9}	giga	G
10^{6}	mega	M
10^{3}	kilo	k
10^{2}	hecto	h
10^{1}	deka	da
10^{-1}	deci	d
10^{-2}	centi	c
10^{-3}	milli	m
10^{-6}	micro	μ
10^{-9}	nano	n
10^{-12}	pico	p
10^{-15}	femto	f
10^{-18}	atto	a

Appendix B

SI Units

The *Système international d' unités*, or International System of Units, is a simplified system of metric units (known as SI units) adopted by international convention for scientific use.

Basic SI Units

Quantity	Unit	Symbol
length	meter	m
mass	kilogram	kg
time	second	s
temperature	Kelvin	K

Derived SI Units

Quantity	Unit	Symbol	Expression
area	meter squared	m^2	m^2
volume	meter cubed	m^3	m^3
density	kilogram per cubic meter	kg/m^3	kg/m^3
speed	meter per second	m/s	m/s
acceleration	meter per second per second	m/s^2	m/s^2
force	newton	N	$(kg)(m)/s^2$
pressure	pascal	Pa	N/m^2
energy	joule	J	(N)(m)
power	watt	W	J/s; (N)(m)/s

Length: *The Basic SI Unit Is the Meter*

Quantity	Metric Equivalent	English Equivalent	Other
meter (m)	100 centimeters 1000 millimeters	39.37 inches 3.281 feet	0.546 fathom
kilometer (km)	1000 meters	0.621 land mile	0.540 nautical mile
centimeter (cm)	10 millimeters 0.01 meter	0.394 inch	
millimeter (mm)	0.1 centimeter 0.001 meter	0.0394 inch	
land mile (mi)	1609 meters	5280 feet	0.869 nautical mile
nautical mile (nm)	1852 meters	1.151 land miles 6076 feet	1 minute of latitude
fathom (fm)	1.8288 meters	6 feet	

Area: *Derived from Length*

Unit	Metric Equivalent	English Equivalent	Other
square meter (m^2)	10,000 square centimeters	10.76 square feet	
square kilometer (km^2)	1,000,000 square meters	0.386 square land mile	0.292 square nautical mile
square centimeter (cm^2)	100 square millimeters	0.151 square inch	

Volume: *Derived from Length*

Unit	Metric Equivalent	English Equivalent	Other
cubic meter (m³)	1,000,000 cubic centimeters	35.32 cubic feet	
	1000 liters	264 U.S. gallons	
cubic kilometer (km³)	1,000,000,000 cubic meters	0.2399 cubic land mile	0.157 cubic nautical mile
liter (L) or (l)	1000 cubic centimeters	1.06 quarts, 0.264 U.S. gallon	
milliliter (mL) or (ml)	1.0 cubic centimeter		

Mass: *The Basic SI Unit Is the Kilogram*

Unit	Metric Equivalent	English Equivalent
kilogram (kg)	1000 grams	2.205 pounds
gram (g)		0.035 ounce
metric ton, or	1000 kilograms	2205 pounds
tonne (t)	1,000,000 grams	
U.S. ton	907 kilograms	2000 pounds

Time: *The Basic SI Unit Is the Second*

Unit	Metric and EnglishEquivalent	
minute	60 seconds	
hour	60 minutes; 3600 seconds	
day	86,400 seconds	} mean solar day
	24 hours	
year	31,556,880 seconds	} mean solar year
	8756.8 hours	
	365.25 solar days	

Temperature: *The Basic SI Unit Is the Kelvin*

Reference Point	Kelvin (K)	Celsius (°C)	Fahrenheit (°F)
absolute zero	0	−273.2	−459.7
seawater freezes	271.2	−2.0	28.4
fresh water freezes	273.2	0.0	32.0
human body	310.2	37.0	98.6
fresh water boils	373.2	100.0	212.0
conversions	$K = °C + 273.2°$	$°C = \dfrac{(°F - 32)}{1.8}$	$°F = (1.8 \times °C) + 32$

Speed (Velocity): *The Derived SI Unit Is the Meter per Second*

Unit	Metric Equivalent	English Equivalent	Other
meter per second (m/s)	100 centimeters per second	3.281 feet per second	1.944 knots
	3.60 kilometers per hour	2.237 land miles per hour	
kilometer per hour (km/h)	0.277 meter per second	0.909 foot per second	0.55 knot
knot (kt)	0.51 meter per second	1.151 land miles per hour	1 nautical mile per hour

Acceleration: *The Derived SI Unit Is the Meter per Second per Second*

Unit	Metric Equivalent	English Equivalent
meter per second per second (m/s²)	100 centimeters per second per second	3.281 feet per second per second
	12,960 kilometers per hour per hour	8048 miles per hour per hour

Force: *The Derived SI Unit Is the Newton*

Unit	Metric Equivalent	English Equivalent
newton (N)	100,000 dynes	0.2248 pound force
dyne (dyn)	0.00001 newton	0.000002248 pound force

Pressure: *The Derived SI Unit Is the Pascal*

Unit	Metric Equivalent	English Equivalent	Other
pascal (Pa)	1 newton per square meter 10 dynes per square centimeter		
bar	100,000 pascals 1000 millibars	14.5 pounds per square inch	0.927 atmosphere 29.54 inches of mercury
standard atmosphere (atm)	1.013 bars 101,300 pascals	14.7 pounds per square inch	29.92 inches of mercury or 76 cm of mercury

Energy: *The Derived SI Unit Is the Joule*

Unit	Metric Equivalent	English Equivalent
joule (J)	1 newton-meter 0.2389 calorie	0.0009481 British thermal unit
calorie (cal)	4.186 joules	0.003968 British thermal unit

Power: *The Derived SI Unit Is the Watt*

Unit	Metric Equivalent	English Equivalent
watt (W)	1 joule per second 0.2389 calorie per second 0.001 kilowatt	0.0569 British thermal unit per minute 0.001341 horsepower

Density: *The Derived SI Unit Is the Kilogram per Cubic Meter*

Unit	Metric Equivalent	English Equivalent
kilogram per cubic meter	0.001 gram per cubic centimeter (g/cm³)	0.0624 pound per cubic foot

Equations and Quantitative Relationships

Distance, Rate, Time Problems

Distance = speed × time

Area = area spreading rate × time

Volume = volume flow rate × time

Mass transferred = mass transfer rate × time

Rotation = rotation rate × time

Volume flow rate = current speed × area

Equilibrium, Steady-State Problems

Volumes, masses, and energy are held constant in time

All inflows equal outflows. $(\text{rate} \times \text{time})_{in} = (\text{rate} \times \text{time})_{out}$

Residence time = mass or volume / supply or removal rate

Exponential Problems

$X = Y\, e^{\pm(rz)}$ Requires calculator

In time-dependent problems z = time, r = (A/time), a rate

X is the value of Y after a given time. If rz is +, then X > Y

If rz is −, then X < Y

When X = 0.5 (Y), (rz is negative) and (rz) = 0.693

Radioactive decay example. Half-life calculation:

$$0.5\,(Y) = 1\,(Y)\, e^{-(rt)}$$

Light attenuation example:

$$I_z = I_o\, e^{-kz}$$

where I_o = the light intensity at the sea surface, I_z = the light intensity at a depth z, and k is the attenuation coefficient (see discussion in chapter 5). The attenuation coefficient, k, is about $1.7/D$, where D is the Secchi disk depth. k can be determined for total available light or for individual wavelengths of light.

Population example:

$$P_{time=t} = P_{t=0}\, e^{+(rt)}$$

P is the number of individuals or mass. r in this case is a combination of reproduction rate minus the (death rate + grazing rate).

Three-Dimensional Problems

Volume = length × width × depth

Area = length × width

Slope Problems

Slope = rise/run

Horizontal distance × slope = elevation or height change

Small-Particle Settling Velocity: Stokes Law

$V = (2/9)\, [g\,(\rho_1 - \rho_2)/\mu]r^2$

V = settling speed, terminal velocity in cm/s

2/9 = shape factor constant for all small spheres

g = acceleration due to Earth's gravity, 981 cm/s²

ρ_1 = particle density (of quartz), g/cm³

ρ_2 = seawater density, g/cm³

μ = viscosity of seawater

r = radius of particle in cm

$$V_{cm/s} = (2.62 \times 10^4)r^2$$

Used to determine settling rate of small particles with diameters less than 0.125 mm. Can be used in reverse to determine size of small particles by measuring settling rates.

Heat Problems

Latent heat of fusion of water at 0°C = 80 cal/g

Latent heat of vaporization of water at 100°C = 540 cal/g

Heat capacity of water = 1 cal/g/°C

Calories = mass in g × heat capacity × temperature change, °C

Interpolation of Data Tables

Known values are axes values (a, c, d, and f) and corresponding table values (P, R, V, and X). Chosen axes values are b (between a and c) and e (between d and f). Find the table value T at chosen axes values b and e.

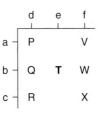

Step 1: Find the table value Q for axes values b and d.

$$Q = P + (b - a)(R - P)/(c - a)$$

Step 2: Find the table value W for axes values b and f.

$$W = V + (b - a)(X - V)/(c - a)$$

Step 3: Find the table value T for axes values b and e.

$$T = Q + (e - d)(W - Q)/(f - d)$$

Seawater, Constancy of Composition

$$Cl_1\text{‰}/Cl_2\text{‰} = S_1\text{‰}/S_2\text{‰} = ionA_1/ionA_2 = ionB_1/ionB_2$$

Subscripts 1 and 2 refer to two different concentrations.

Hydrostatic Pressure

$$P = \rho gz$$

P is pressure, ρ is fluid density, g is Earth's gravity, and z is the height of the fluid.

Progressive Surface Water Waves

Simple sinusoidal wave theory:

Wave speed, $C = L/T$, where L = wavelength and T, the wave period, is conserved.

$C^2 = (g/2\pi) L \tanh[(2\pi/L)D]$, where g = Earth's gravity, D = water depth, L = wavelength, and $\tanh[\varnothing]$ = the hyperbolic tangent of angle$[\varnothing]$. Here, $\varnothing$ equals $(2\pi/L)D$ in radians. Use calculator.

Deep-water approximation:

When $D > L/2$, $\tanh[(2\pi/L)D] \approx 1.0$, $C^2 = (g/2\pi) L$

Shallow-water approximation:

When $D < L/20$, $\tanh[(2\pi/L)D] \approx (2\pi/L)D$, $C = \sqrt{gD}$

Internal Waves

The celerity, C, or wave speed of internal waves along a boundary of a two-layer system is determined by a relationship between the density of the layers and their thicknesses. When the wavelength, L, is long compared with the water depths:

$$C^2 = (g/2\pi)L\left[\frac{\rho - \rho'}{\rho\coth(2\pi h/L) + \rho'\coth(2\pi h'/L)}\right]$$

h and ρ are the thickness and density of the denser lower layer, and h' and ρ' are the thickness and density of the less-dense upper layer. Cotanh is the hyperbolic cotangent. When L is small compared to the water depths, cotanh $(2\pi h/L)$ and cotanh $(2\pi h'/L)$ approach the value of 1.0 and

$$C^2 = (g/2\pi)L\left[\frac{\rho - \rho'}{\rho + \rho'}\right]$$

This equation relates to the deep-water wave equation at the sea surface, which is the boundary of two fluids—air and water. In this later case, ρ' of air is about 1/1000 the value of ρ of water and is neglected.

Standing Waves, Seiches

Based on $L/T = \sqrt{gD}$, shallow-water wave condition.
Closed basin:

The period of oscillation is $T = (1/n)\,(2l/\sqrt{gD})$ where n = the number of wave nodes; $L = 2 \times$ basin length, l; $L = 2l$; and D = basin depth.

Open-end basin:

The period of oscillation is

$$T = (1/n)\,(4l/\sqrt{gD}); L = 4 \times \text{basin length}; L = 4l$$

Refraction of Light, Sound, and Waves

Snell's Law:
The change of speed of propagation causes refraction, bending, of wave rays and wave fronts.

$$C_1 \sin(a_2) = C_2 \sin(a_1)$$

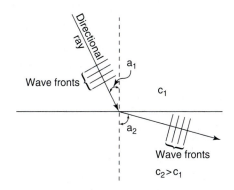

Tide-Raising Forces

The force between two masses due to their mutual gravitational attraction.

$$F = G\,(M_1 M_2)/R^2$$

where M_1 = mass 1, M_2 = mass 2, R = distance between centers of masses M_1 and M_2, and G = Newton's constant of gravitation = $6.67 \times 10^{-8}\,cm^3/g/s^2 = 6.67 \times 10^{-8} dyne \cdot cm^2/g^2$.

The gravitational force per unit mass on Earth, M_1, caused by the Sun or the Moon, mass M_2, is,

$$F/M_1 = G\,(M_2/R_2)$$

where R is the distance between the unit Earth mass M_1 and the Sun or the Moon center.

The tide-raising force, $\Delta F/M_1$, is the difference between the gravitational force acting on a unit mass at Earth's surface directly in line with the tide-raising body and a unit mass at Earth's center. (F/M_1)sur. $- (F/M_1)$cen. $= \Delta F/M_1$

$$\Delta F/M_1 = [GM_2/(R - r)^2] - [GM_2/(R)^2]$$

where R is the distance between the center of Earth and center of the Sun or Moon and r is Earth's radius.

$$\Delta F/M_1 = (-GM_2/R^2)\,[1 - (1 / (1 - r/R)^2)]$$
$$= (-GM_2/R^2)\,[-2r/R + (r/R)^2] / (1 - r/R)^2$$

If the magnitude of r/R and $(r/R)^2$ is calculated, then $(r/R)^2 << 2r/R$ and $(r/R)^2$ is neglected in the numerator; $(r/R) << 1.0$ and (r/R) is neglected in the denominator.

$$\Delta F/M_1 = (-GM_2/R^2)\,[(-2r/R) / 1] = 2GM_2\,r/R^3$$

Calculating Great Circle Distance

The great circle distance between two points on Earth can be calculated knowing the latitude and longitude of each point, and the average radius of Earth.

Let (ϕ_1, θ_1) and (ϕ_2, θ_2) be the geographical latitude and longitude expressed in radians of two locations. Let the difference in their longitude be $\Delta\theta$ expressed in radians.

Then the geocentric angular distance between the two points expressed in radians, $\Delta\sigma$, can be calculated from the following formula:

$$\Delta\sigma = \arctan\left(\frac{\sqrt{(\cos\phi_2\sin\Delta\theta)^2 + (\cos\phi_1\sin\phi_2 - \sin\phi_1\cos\phi_2\cos\Delta\theta)^2}}{\sin\phi_1\sin\phi_2 + \cos\phi_1\cos\phi_2\cos\Delta\theta}\right)$$

The great circle distance in kilometers between the two points is equal to the geocentric angular distance between the two points multiplied by Earth's average radius of curvature, which is about 6372.8 km.

For example, calculate the distance between Los Angeles International Airport (LAX, point 1) in Los Angeles, CA (N 33° 56.4′, W 118°24.0′) and O'Hare Airport (ORD, point 2) in Chicago, IL (N 41°58.7′, W 87°54.3′).

First, convert these coordinates to decimal degrees (remember that North latitude is positive and West longitude is negative); then convert them to radians by multiplying the decimal degrees by $(\pi/180)$.

LAX (point 1): $\phi_1 = +33.94°$ or 0.5924 rad
$\theta_1 = -118.40°$ or -2.0665 rad

ORD (point 2): $\phi_2 = +41.98°$ or 0.7327 rad
$\theta_2 = -87.91°$ or -1.5343 rad

then: $\Delta\theta = (-2.0665) - (-1.5343) = -0.5322$

Input these coordinates in radians into the geocentric angular distance formula to calculate $\Delta\sigma \approx 0.4406$. Multiply this angular distance in radians by the average radius of Earth to obtain the distance between the two points.

$$(0.4406) \times (6371 \text{ km}) = 2807 \text{ km (1744 mi)}$$

Water and Salt Budgets for Partially Mixed Estuaries

Water in = water out = water budget, volume constant

Salt in = salt out = salt budget, salt content constant

T_o = volume transport of mixed water out, vol/time
T_i = volume transport of seawater in, vol/time
R = river flow into estuary, vol/time
S_o = average salt content of T_o water, mass/vol
$\overline{S}_i$ = average salt content of T_i seawater, mass/vol

Water budget: $T_o = T_i + R$

Salt budget: $T_o\,\overline{S}_o = T_i\,\overline{S}_i$

Combined budgets: $T_o = [\overline{S}_i /(\overline{S}_i - \overline{S}_o)]\,R$

Estuary flushing time or estuary water residence time equals estuary volume/To.

$(\overline{S}_i - \overline{S}_o)/\overline{S}_i$ = fraction of fresh water in T_o type flow
$(\overline{S}_i - \overline{S}_a)/\overline{S}_i$ = fraction of fresh water in estuary volume
where $\overline{S}_a$ = average salt concentration of estuary water
$[(\overline{S}_i - \overline{S}_a)/\overline{S}_i] \times$ estuary volume = freshwater volume stored in the estuary
(Freshwater volume of estuary/R) = residence time of fresh water in the estuary

Plant Production of Carbon by Phytoplankton

Production of carbon (C), release of oxygen (O_2), and use of the nutrients nitrogen (N) and phosphorus (P) by phytoplankton in photosynthesis occurs with fixed ratios on a mass basis. If the use of or production rate of one of the terms is known, the others are calculated from mass-to-mass ratios.

$$O_2:C:N:P = 109:41:7.2:1$$

Glossary

A

absorption taking in of a substance by chemical or molecular means; change of sound or light energy into some other form, usually heat, in passing through a medium or striking a surface.

abyssal pertaining to the great depths of the ocean below approximately 4000 m.

abyssal clay lithogenous sediment on the deep-sea floor composed of at least 70% clay-sized particles by weight.

abyssal hill low, rounded submarine hill less than 1000 m high.

abyssal plain flat ocean basin floor extending seaward from the base of the continental slope and continental rise.

abyssopelagic oceanic zone from 4000 m to the deepest depths.

accretion natural or artificial deposition of sediment along a beach, resulting in the buildup of new land.

acoustic profiling the use of seismic energy to measure sediment thickness and layering on the sea floor.

active margin *see* leading margin.

adsorption attraction of ions to a solid surface.

advection horizontal or vertical transport of seawater, as by a current.

albedo the ratio of reflected radiation to incident radiation, typically expressed as percentage.

algae marine and freshwater organisms (including most seaweeds) that are single-celled, colonial, or multicelled, with chlorophyll but no true roots, stems, or leaves and with no flowers or seeds.

alluvial plain flat deposit of terrestrial sediment eroded by water from higher elevations.

amphidromic point point from which cotidal lines radiate on a chart; the nodal, or low-amplitude, point for a rotary tide.

amplitude for a wave, the vertical distance from sea level to crest or from undisturbed sea level to trough, or one-half the wave height.

anadromous migratory pattern in which juvenile fish migrate down rivers to mature as adults in the open oceans.

anaerobe an organism that lives in anaerobic environments.

anaerobic living or functioning in the absence of oxygen.

andesite a volcanic rock intermediate in composition between basalt and granite; associated with subduction zones.

anion negatively charged ion.

anoxic deficient in oxygen.

Antarctic Circle *see* Arctic and Antarctic Circles.

anthropogenic carbon dioxide carbon dioxide produced by human activities such as the burning of carbonaceous fuels, including wood, coal, oil, and natural gas.

anthrosphere the human component of Earth's climate system.

antinode portion of a standing wave with maximum vertical motion.

aphotic zone that part of the ocean in which light is insufficient to carry on photosynthesis.

aquaculture (mariculture) cultivation of aquatic organisms under controlled conditions.

Arctic and Antarctic Circles latitudes 66 1/2°N and 66 1/2°S, respectively, marking the boundaries of light and darkness during the summer and winter solstices.

armored beach a beach that is protected from wave and water erosion by coarse-size lag deposits.

aseismic ridge *see* transverse ridge.

asthenosphere upper, deformable portion of Earth's mantle, the layer below the lithosphere; probably partially molten; may be site of convection cells.

atmospheric pressure pressure, at any point on Earth, exerted by the atmosphere as a consequence of gravitational force exerted on the column of air lying directly above the point.

atoll ring-shaped coral reef that encloses a lagoon in which there is no exposed preexisting land and which is surrounded by the open sea.

attenuation decrease in the energy of a wave or beam of particles occurring as the distance from the source increases; caused by absorption, scattering, and divergence from a point source.

autotrophic pertaining to organisms able to manufacture their own food from inorganic substances. *See also* chemosynthesis and photosynthesis.

autumnal equinox *see* equinoxes.

B

backshore beach zone lying between the foreshore and the coast, acted on by waves only during severe storms and exceptionally high water.

bacterioplankton composed of members of the domains Bacteria and Archaea.

baleen whalebone; horny material growing down from the upper jaw of plankton-feeding whales; forms a strainer, or filtering organ, consisting of numerous plates with fringed edges.

bar offshore ridge or mound of sand, gravel, or other loose material that is submerged, at least at high tide; located especially at the mouth of a river or estuary or lying a short distance from and parallel to the beach.

barrier island deposit of sand, parallel to shore and raised above sea level; may support vegetation and animal life.

barrier reef coral reef that parallels land but is some distance offshore, with water between reef and land.

basalt fine-grained, dark igneous rock, rich in iron and magnesium, characteristic of oceanic crust.

basin large depression of the sea floor having about equal dimensions of length and width.

bathyal pertaining to ocean depths between approximately 1000 and 4000 m.

bathymetry study and mapping of seafloor elevations and the variations of water depth; the topography of the sea floor.

bathypelagic oceanic zone from 1000–4000 m.

beach zone of unconsolidated material between the mean low-water line and the line of permanent vegetation, which is also the effective limit of storm waves; sometimes includes the material moving in offshore, onshore, and longshore transport.

beach face section of the foreshore normally exposed to the action of waves.

Beaufort scale scale of wind forces by range of velocity; scale of sea state created by winds of these velocities.

benthic of the sea floor, or pertaining to organisms living on or in the sea floor.

benthic meiofauna small invertebrates that live in benthic environments such as between grains of sand.

benthos organisms living on or in the ocean bottom.

berm nearly horizontal portion of a beach (backshore) with an abrupt face; formed from the deposition of material by wave action at high tide.

berm crest ridge marking the seaward limit of a berm.

bilateral symmetry having right and left halves that are approximate mirror images of each other.

bioaccumulate/bioaccumulation the accumulation of any substance, such as toxicants, in the tissue of a living organism.

biodiversity the number of species in an area compared to the number of individuals.

biogenous sediment sediment derived from organisms.

biogeochemical provinces large regions of the ocean that exhibit similar biological and chemical properties.

biological pump photosynthetic transfer of carbon as CO_2 from the atmosphere to the ocean in the form of organic molecules; carbon is transferred to intermediate and deep-ocean water when organic material sinks and decays.

bioluminescence production of light by living organisms as a result of a chemical reaction either within certain cells or organs or outside the cells in some form of excretion.

biomagnify the concentration of impurities or contaminants in biological organisms through transfer from lower to higher trophic level organisms during consumption.

biomass the total mass of all or specific living organisms, usually expressed as dry weight in grams of carbon per unit area or unit volume.

biomechanics the field of research that looks for answers to how chemistry and physics affect basic biological characteristics such as size and shape.

bioturbation reworking of sediments by organisms that burrow into them and ingest them.

bipartite lifestyle describes the common practice of many marine organisms that spend part of their lives as pelagic and part of their lives as benthic organisms.

blade flat, photosynthetic, "leafy" portion of an alga or a seaweed.

bloom high concentration of phytoplankton in an area, caused by increased reproduction; often produces discoloration of the water. *See also* red tide.

body wave a seismic wave that travels beneath Earth's serface.

bottom-up control a flow of energy and materials in food webs where the primary producers control the growth and regulation of higher trophic levels.

brachiopod marine animals that have hinged shells (valves) similar to mollusks, typically found in cold water at depth or near the poles.

breaker sea surface water wave that has become too steep to be stable and collapses.

breakwater structure protecting a shore area, harbor, anchorage, or basin from waves; a type of jetty.

buffer substance able to neutralize acids and bases, therefore able to maintain a stable pH.

bulkhead structure separating land and water areas; primarily designed to resist earth sliding and slumping or to reduce wave erosion at the base of a cliff.

buoy floating object anchored to the bottom or attached to another object; used as a navigational aid or surface marker.

buoyancy ability of an object to float due to the support of the fluid the body is in or on.

by-catch *see* incidental catch.

C

caballing mixing of two water types with identical densities but different temperatures and salinities; the resulting mixture is denser than its components.

calcareous containing or composed of calcium carbonate.

calcareous ooze fine-grained deep-ocean biogenous sediment containing at least 30% calcareous tests, or the remains of small marine organisms.

calorie amount of heat required to raise the temperature of 1 g of water 1°C.

calving breaking away of a mass of ice from its parent glacier, iceberg, or sea-ice formation.

canopy the surface expression of a kelp forest, composed of kelp fronds.

capillary wave wave with wavelength less than 1.5 cm in which the primary restoring force is surface tension.

carbonate a sediment or rock formed from the accumulation of carbonate minerals ($CaCO_3$) precipitated organically or inorganically.

carbonate compensation depth (CCD), also known as the calcite compensation depth, is the depth at which the amount of calcium carbonate ($CaCO_3$) preserved falls below 20% of the total sediment. This is also commonly defined as the depth at which the amount of calcium carbonate produced by organisms as skeletal material in the overlying water column is equal to the rate at which it is dissolved in the water. No calcium carbonate will be deposited below this depth.

carbon fixation process of generating organic carbon from carbon dioxide.

carbon sequestration removal of atmospheric carbon dioxide to some reservoir, with storage typically occurring for long time scales.

Carnivora the order of marine mammals that includes otters, polar bears, and pinnipeds.

carnivore flesh-eating organism.

catadromous migratory pattern in which juvenile fish spawn in the open ocean but mature in fresh water.

catch per unit effort the commercial fishing catch per fishing boat per unit time.

cation positively charged ion.

cat's-paw patch of ripples on the water's surface, related to a discrete gust of wind.

centrifugal force outward-directed force acting on a body moving along a curved path or rotating about an axis; an inertial force.

centripetal force inward-directed force necessary to keep an object moving in a curved path or rotating about an axis.

Cetacea the order of marine mammals that includes whales, dolphins, and porpoises.

chaetognaths free-swimming, carnivorous, pelagic, wormlike, planktonic animals; arrowworms.

chemoautotrophs organisms that use energy derived from inorganic chemicals to drive production of organics.

chemosynthesis formation of organic compounds with energy derived from inorganic substances such as ammonia, methane, sulfur, and hydrogen.

chloride atom of chlorine in solution, forming an ion with a negative charge.

chlorinity (Cl‰) measure of the chloride content of seawater in grams per kilogram.

chlorophyll group of green pigments that are active in photosynthesis.

chloroplasts used to harvest sunlight by photosynthesis in cells.

Chondrichthyes fish that use cartilage rather than bone.

chromosome one of the bodies in a cell that carries the genes in a linear order.

chronometer portable clock of great accuracy used in determining longitude at sea.

ciguatera toxin found in fish of tropical regions; produced by dinoflagellates.

cilia microscopic, hairlike projections of living cells that beat in coordinated fashion and produce movement.

classification a method for grouping organisms into similar categories on the basis of size, lifestyle, chemical composition, anatomy, or other characteristics.

CLAW hypothesis proposal that DMS emitted by marine phytoplankton sets up a negative feedback mechanism to cool the climate by promoting formation of more clouds. Originally proposed by Robert Charlson, James Lovelock, Meinrat Andreae, and Stephan Warren, the acronym comes from the first initials of their surnames.

climate the average conditions of the atmosphere, temperature, precipitation, and wind for a given ecosystem or for the planet as a whole.

climate disruption geologically rapid changes in the Earth's atmosphere, temperature, precipitation, and wind.

climate system the combination of factors that drive climate, including the atmosphere, cryosphere, lithosphere, hydrosphere, biosphere, and anthrosphere.

cloud condensation nuclei (CCN) very small particles that serve as the initation (nucleus) point for condensation of water vapor in the atmosphere.

cluster a group of galaxies. A cluster may contain thousands of galaxies.

coast strip of land of indefinite width that extends from the shore inland to the first major change in terrain that is unaffected by marine processes.

coastal circulation cell (drift sector, littoral cell) longshore transport cell pattern of sediment moving from a source to a place of deposition.

coccolithophorid microscopic, planktonic alga surrounded by a cell wall with embedded calcareous plates (coccoliths).

cohesion molecular force between particles within a substance that acts to hold the particles together.

colonial organism organism consisting of semi-independent parts that do not exist as separate units; groups of organisms with specialized functions that form a coordinated unit.

commensalism an intimate association between different organisms in which one is benefited and the other is neither harmed nor benefited.

common ancestor the universal ancestor that all subsequent organisms are related to evolutionarily.

compensation depth depth at which there is a balance between the oxygen produced by algae through photosynthesis and that consumed through respiration; net oxygen production is zero.

condensation process by which a vapor becomes a liquid or a solid.

conduction transfer of heat energy through matter by internal molecular motion.

conservative constituent component or property of seawater whose value changes only as a result of mixing, diffusion, and advection and not as a result of biological or chemical processes; for example, salinity.

consumer animal that feeds on plants (primary consumer) or on other animals (secondary consumer).

continental crust crust forming the continental land blocks; mainly granite and its derivatives.

continental drift the movement of continents; the name of Alfred Wegener's theory, preceding plate tectonics.

continental margin zone separating the continents from the deep-sea bottom, usually subdivided into shelf, slope, and rise.

continental rise gentle slope formed by the deposition of sediments at the base of a continental slope.

continental shelf zone bordering a continent, extending from the line of permanent immersion to the depth at which there is a marked or rather steep descent to the great depths.

continental shelf break zone along which there is a marked increase of slope at the outer margin of a continental shelf.

continental slope relatively steep downward slope from the continental shelf break to depth.

contour line on a chart or graph connecting points of equal elevation, temperature, salinity, or other property.

convection transmission of heat by the movement of a heated gas or liquid; vertical circulation resulting from changes in density of a fluid.

convection cell circulation in a fluid, or fluidlike material, caused by heating from below. Heating the base of a fluid lowers its density, causing it to rise. The rising fluid cools, becomes denser, and sinks, creating circulation.

convergence situation in which substances come together, usually resulting in the sinking, or downwelling, of surface water and the rising of air.

convergent plate boundary a boundary between two plates that are converging or colliding with one another.

Copenhagen Accord a nonbinding document agreed to as part of the United Nations Framework Convention on Climate Change that recognizes human influences on climate through global warming, endorses the Kyoto Protocol, and recommends reducing greenhouse gas emissions to hold the global increase of temperature below 2°C.

copepod small, shrimplike member of the zooplankton; in the class Crustacea.

coral colonial animal that secretes a hard outer calcareous skeleton; the skeletons of coral animals form in part the framework for warm-water reefs.

corange lines in a rotary tide, lines of equal tidal range about the amphidromic point.

cordgrass marine grasses from the genus *Spartina*.

core vertical, cylindric sample of bottom sediments, from which the nature of the bottom can be determined; also the central zone of Earth, thought to be liquid or molten on the outside and solid on the inside.

corer device that plunges a hollow tube into bottom sediments to extract a vertical sample.

Coriolis effect apparent force acting on a body in motion, due to the rotation of Earth, causing deflection to the right in the Northern Hemisphere and to the left in the Southern Hemisphere; the force is proportional to the speed and varies with latitude of the moving body.

cosmogenous sediment sediment particles with an origin in outer space; for example, meteor fragments and cosmic dust.

cotidal lines lines on a chart marking the location of the tide crest at stated time intervals.

countercurrent heat exchange a mechanism in which there is a crossover of heat between two liquids flowing in opposite directions to each other.

covalent bond chemical bond formed by the sharing of one or more pairs of electrons.

cratered coast a primary coast formed when the seaward side of a volcanic crater is eroded away or is blown away by a volcanic eruption; opening the interior of the crater to the sea and creating a concave bay.

cratons large pieces of Earth's crust that form the centers of continents.

crest *see* berm crest, reef crest, wave crest.

critical depth the depth to which a photosynthetic organism can be mixed and still maintain zero or positive oxygen production when community respiration processes are taken into account.

crust outer shell of the solid Earth; the lower limit is usually considered to be the Mohorovičić discontinuity.

crustacean member of a class of primarily aquatic organisms with paired jointed appendages and a hard outer skeleton; includes lobsters, crabs, shrimps, and copepods.

cryoprotectants materials that lower the freezing points of organisms' internal fluids.

cryptic species morphologically indistinguishable species that are incapable of reproducing with one another.

ctenophore transparent, planktonic animal, spherical or cylindrical with rows of cilia; comb jelly.

Curie temperature temperature at which the magnetic signature is frozen into an igneous rock during cooling.

current horizontal movement of water.

current meter instrument for measuring the speed and direction of a current.

cusp one of a series of evenly spaced, crescent-shaped depressions along sand and gravel beaches.

cyanobacteria member of the phytoplankton that can dominate in open-ocean environments.

cyclone *see* typhoon, hurricane.

D

deadweight ton (DWT) capacity of a vessel in tons of cargo, fuel, stores, and so on; determined by the weight of the water displaced.

declinational tide *see* diurnal tide.

decomposer heterotrophic; microorganisms (usually bacteria and fungi) that break down nonliving organic matter and release nutrients, which are then available for reuse by autotrophs.

deep exceptionally deep area of the ocean floor, usually below 6000 m.

deep scattering layer (DSL) layer of organisms that move away from the surface during the day and toward the surface at night; the layer scatters or returns vertically directed sound pulses.

deep-sea gigantism the common occurrence of deep-sea species exhibiting unusually large body size compared to the same or closely related species found at shallower ocean depths.

deep-water wave wave in water, the depth of which is greater than one-half the wavelength.

degree an arbitrary measure of temperature. Temperature is measured in degrees using one of three different scales: Fahrenheit (°F), Celsius (°C), and Kelvin (K). See Appendix B for conversions among these scales.

delta area of unconsolidated sediment deposits, usually triangular in outline, formed at the mouth of a river.

demersal fish fish living near and on the bottom.

density property of a substance defined as mass per unit volume and usually expressed in grams per cubic centimeter or kilograms per cubic meter.

depth recorder *see* echo sounder.

desalination process of obtaining fresh water from seawater.

detritus any loose material, especially decomposed, broken, and dead organic materials.

dew point highest temperature at which water vapor condenses from an air mass. At the dew point temperature, the relative humidity is 100%.

diatom microscopic unicellular alga with an external skeleton of silica.

diatomaceous ooze sediment made up of more than 30% skeletal remains of diatoms.

diffraction process that transmits energy laterally along a wave crest.

diffusion movement of a substance from a region of higher concentration to a region of lower concentration (movement along a concentration gradient may be due to molecular motion or turbulence).

dimethyl sulfide (DMS) an organic compound produced by many marine phytoplankton and bacteria; the most abundant biologically derived sulfur compound emitted into the atmosphere.

dinoflagellate member of the plankton. Some species are photosynthetic and others are not.

dipole a magnetic field like Earth's, with two opposite poles.

dispersion (sorting) sorting of waves as they move out from a storm center; occurs because long-period waves travel faster in deep water than short-period waves.

disphotic zone see twilight zone.

diurnal inequality difference in height between the two high waters or two low waters of each tidal day; the difference in speed between the two flood currents or two ebb currents of each tidal day.

diurnal tide (declinational tide) tide with one high water and one low water each tidal day.

divergence horizontal flow of substances away from a common center, associated with upwelling in water and descending motions in air.

divergent plate boundary a boundary between two plates that are diverging or moving apart from one another.

Dobson unit ozone unit, defined as 0.01 mm thickness of ozone at Standard Temperature and Pressure (0°C and 1 atmosphere pressure). If all the ozone over a certain area is compressed to 1 atmosphere pressure at 0°C, it forms a slab of a thickness corresponding to a number of Dobson units.

doldrums nautical term for the belt of light, variable winds near the equator.

domain the highest level for grouping organisms. There are three domains of life—Bacteria, Archaea, and Eukarya.

downwelling zone sinking of water, usually the result of a surface convergence or an increase in density of water at the sea surface.

dredge cylindric or boxlike sampling device made of metal, net, or both that is dragged across the bottom to obtain biological or geologic samples.

drift bottle bottle released into the sea for use in studying currents; contains a card identifying date and place of release and requesting the finder to return it with date and place of recovery.

drift ice sea ice that floats freely without being attached to land or the seafloor.

drift sector *see* coastal circulation cell.

dugong *see* sea cow.

dune wind-formed hill or ridge of sand.

dune coast a primary coast formed by the deposition of sand in dunes by the wind.

dynamic equilibrium state in which the sums of all changes are balanced and there is no net change.

E

Earth sphere depth uniform depth of Earth below the present mean sea level, if the solid Earth surface were smoothed off evenly (2440 m).

Eastern Pacific garbage patch an area in the eastern part of the Pacific Ocean where floating plastic and other trash is concentrated by surface currents.

ebb current movement of a tidal current away from shore or seaward in a tidal stream as the tide level decreases.

ebb tide falling tide; the period of the tide between high water and the next low water.

echolocation use of sound waves by some marine animals to locate and identify underwater objects.

echo sounder (depth recorder) instrument used to measure the depth of water by measuring the time interval between the release of a sound pulse and the return of its echo from the bottom. *See also* precision depth recorder (PDR).

ecology the study of interactions between organisms and their environment and of interactions among organisms.

ecologically disruptive algal bloom (EDAB) a type of Harmful Algal Bloom that causes damage by disrupting the functioning of an ecosystem.

ecosystem the organisms in a community and the nonliving environment with which they interact.

eddy circular movement of water.

Ekman layer the surface layer of water approximately 100–150 m (330–500 ft) thick in which there is wind-driven motion. *See also* Ekman spiral and Ekman transport.

Ekman spiral in a theoretical ocean of infinite depth, unlimited extent, and uniform viscosity, with a steady wind blowing over the surface, the surface water moves 45° to the right of the wind in the Northern Hemisphere. At greater depths, the water moves farther to the right with decreased speed, until at some depth (approximately 100 m), the water moves opposite to the wind direction. Net water transport is 90° to the right of the wind in the Northern Hemisphere. Movement is to the left in the Southern Hemisphere.

Ekman transport the total volume of water in the Ekman layer transported at right angles to the wind direction.

electrodialysis separation process in which electrodes of opposite charge are placed on each side of a semipermeable membrane to accelerate the diffusion of ions across the membrane.

electromagnetic radiation waves of energy formed by simultaneous electrical and magnetic oscillations; the electromagnetic spectrum is the continuum of all electromagnetic radiation from low-energy radio waves to high-energy gamma rays, including visible light.

electromagnetic spectrum *see* electromagnetic radiation.

El Niño wind-driven reversal of the Pacific equatorial currents, resulting in the movement of warm water toward the coasts of the Americas, so called because it generally develops just after Christmas.

emergent behavior the development of higher levels of organisation (such as schooling in fish) without intentional communication of leadership by individuals.

endotherms maintain a higher temperature than the surrounding seawater but do not have the same level of temperature control as homeotherms.

entrainment mixing of salt water into fresh water overlying salt water, as in an estuary.

epicenter point on Earth's surface directly above an earthquake location, specified by identifying the latitude and longitude of the earthquake. *See also* focus, hypocenter.

epifauna animals living attached to the sea bottom or moving freely over it.

epipelagic upper portion of the oceanic pelagic zone, extending from the surface to about 200 m.

episodic wave abnormally high wave unrelated to local storm conditions.

equator 0° latitude, determined by a plane that is perpendicular to Earth's axis and is everywhere equidistant from the North and South Poles.

equilibrium tide theoretical tide formed by the tide-producing forces of the Moon and Sun on a nonrotating, water-covered Earth.

equinoxes days of the year when the Sun stands directly above the equator, so that day and night are of equal length around the world. The vernal equinox occurs about March 21, and the autumnal equinox occurs September 22–23.

escarpment nearly continuous line of cliffs or steep slopes caused by erosion or faulting.

estuary semi-isolated portion of the ocean that is diluted by freshwater drainage from land.

eukaryote grouping of organisms based on cell morphology. Eukaryotes are either unicellular or multicellular. All eukaryotic cells contain internal membrane-bound cell structures, including the nucleus that contains cellular DNA.

euphausiid planktonic, shrimplike crustacean. *See also* krill.

euphotic zone depth of the water column where there is sufficient sunlight for growth of photosynthetic organisms.

eustatic change global change in sea level that affects all of the world's coastlines.

eutrophication the overenrichment of waters with biologically important nutrients that leads to excessive algal growth.

evaporation process by which liquid becomes vapor. Returns moisture to the atmosphere in the hydrologic cycle; water warmed by the Sun's heat evaporates, becomes vapor or gas, and rises into the atmosphere.

evaporite deposit of minerals left behind by evaporating water, especially salt.

evapotranspiration the combined effect of evaporation and transpiration.

Exclusive Economic Zone (EEZ) the coastal zone that generally extends 200 nautical miles (370 miles) perpendicular from shore around the coastal boundaries of all countries; within the EEZ, each country regulates its own mineral resources, fishing stocks, and pollution controls.

extremophile microorganism that thrives under extreme conditions of temperature, lack of oxygen, or high acid or salt levels; these conditions kill most other organisms.

F

fast ice sea ice that is attached to land or shallow parts of the continental shelf.

fathom a unit of length equal to 1.8 m (6 ft); used to measure water depth.

fault break or fracture in Earth's crust in which one side has been displaced relative to the other.

fault bay a bay formed by faulting along a primary coast.

fault coast a primary coast formed by tectonic activity and faulting.

feedback mechanism a response to a disturbance that acts to enhance (positive feedback) or minimize (negative feedback) the disturbance.

fetch continuous area of water over which the wind blows in essentially a constant direction.

filament chain of living cells.

fishing down the food web the practice of targeting lower trophic levels when higher trophic levels have been depleted.

fjord narrow, deep, steep-walled inlet formed by the submergence of a mountainous coast or by the entrance of the ocean into a deeply excavated glacial trough after the melting of the glacier. A deep, small-surface-area estuary with moderately high river input and little tidal mixing.

floe discrete patch of sea ice moved by the currents or wind.

flotsam anything that is floating at sea; often refers to marine wreckage or cargo.

flood current movement of a tidal current toward the shore or up a tidal stream as the tide height increases.

flood tide rising tide; the period of the tide between low water and the next high water.

fluorescence the emission of light by a substance that has absorbed light.

flushing time length of time required for an estuary to exchange its water with the open ocean.

focus the location of an earthquake within Earth. Focus is specified by identifying latitude, longitude, and depth of the earthquake. *See also* epicenter.

fog visible assemblage of tiny droplets of water formed by condensation of water vapor in the air; a cloud with its base at the surface of Earth.

food chain sequence of organisms in which each is food for the next higher members in the sequence. *See also* food web.

food falls episodic settling of large food items to the sea floor.

food web complex of interacting food chains; all the feeding relations of a community taken together; includes production, consumption, decomposition, and the flow of energy.

foraminifera minute, one-celled animals that usually secrete calcareous shells.

foraminifera ooze sediment made up of 30% or more of skeletal remains of foraminifera.

forced wave wave generated by a continuously acting force and caused to move at a speed faster than it freely travels.

foreshore portion of the shore that includes the low-tide terrace and the beach face.

fouling attachment or growth of marine organisms on underwater objects, usually objects that are made or introduced by humans.

fracture zone long, linear zone of irregular bathymetry of the sea floor, characterized by asymmetric ridges and troughs; commonly associated with fault zones.

free wave wave that continues to move at its natural speed after its generation by a force.

friction resistance of a surface to the motion of a body moving (for example, sliding or rolling) along that surface.

fringing reef reef attached directly to the shore of an island or a continent and not separated from it by a lagoon.

frustule siliceous external shell of a diatom.

G

gabbro a coarse-grained, dark igneous rock, rich in iron and magnesium, the slow-cooling equivalent of basalt.

galaxy a huge aggregate of stars held together by mutual gravitation.

generating force disturbing force that creates a wave, such as wind or a landslide entering water.

geo-engineering deliberate changes to Earth's climate through human manipulation of the climate system.

geomorphology study of Earth's land forms and the processes that have formed them.

geostrophic flow horizontal flow of water occurring when there is a balance between gravitational forces and the Coriolis effect.

glacier mass of land ice, formed by the recrystallization of compacted old snow, flowing slowly from an accumulation area to an area of ice loss by melting, sublimation, or calving.

Global Positioning System (GPS) a worldwide radio-navigation system consisting of twenty-four navigational satellites and five ground-based monitoring stations. GPS uses this system of satellites as reference points for calculating accurate positions on the surface of Earth with readily available GPS receivers.

Gondwanaland an ancient landmass that fragmented to produce Africa, South America, Antarctica, Australia, and India.

graben a portion of Earth's crust that has moved downward and is bounded by steep faults; a rift.

grab sampler instrument used to remove a piece of the ocean floor for study.

granite crystalline, coarse-grained, igneous rock composed mainly of quartz and feldspar.

gravitational force mutual force of attraction between particles of matter (bodies).

gravity Earth's gravity; acceleration due to Earth's mass is 981 cm/s^2; in equations it is represented by g.

gravity wave water wave form in which gravity acts as the restoring force; a wave with wavelength greater than 2 cm.

great circle the intersection of a plane passing through the center of Earth with the surface of Earth. Great circles are formed by the equator and any two meridians of longitude 180° apart.

greenhouse effect the gradual increase in average global temperature caused by the absorption of infrared radiation from Earth's surface by "greenhouse gases" in the atmosphere such as water vapor and carbon dioxide.

Greenwich Mean Time (GMT) solar time along the prime meridian passing through Greenwich, England; also known as Universal Time or ZULU Time.

groin protective structure for the shore, usually built perpendicular to the shoreline; used to trap littoral drift or to retard erosion of the shore; a type of jetty.

gross primary production the rate at which primary producers capture and store chemical energy, such as through photosynthesis; does not account for respiration.

group speed speed at which a group of waves travels (in deep water, group speed equals one-half the speed of an individual wave); the speed at which the wave energy is propagated.

guyot submerged, flat-topped seamount. Also known as a tablemount.

gyre circular movement of water, larger than an eddy; usually applied to a bigger system.

H

habitat place where a plant or animal species naturally lives and grows.

hadal pertaining to the greatest depths of the ocean.

half-life time required for half of an initial quantity of a radioactive isotope to decay.

halocline water layer with a large change in salinity with depth.

halophyte salt-tolerant organism.

harmful algal bloom (HAB) term used to describe both toxic and nuisance phytoplankton blooms.

harmonic analysis process of separating astronomical tide-causing effects from the tide record, in order to predict the tides at any location.

heat a measure of total kinetic energy of atoms and molecules in a substance.

heat budget accounting for the total amount of the Sun's heat received on Earth during one year as being exactly equal to the total amount lost because of radiation and reflection.

heat capacity the quantity of heat required to produce a unit change of temperature in a unit mass of material. *See also* specific heat.

herbivore animal that feeds only on plants.

heterotrophic pertaining to organisms requiring organic compounds for food; unable to manufacture food from inorganic compounds.

higher high water higher of the two high waters of any tidal day in a region of mixed tides.

higher low water higher of the two low waters of any tidal day in a region of mixed tides.

high nitrate low chlorophyll *a* (HNLC) regions regions of the surface ocean where there are high nitrate concentrations, but iron limits phytoplankton productivity.

high water maximum height reached by a rising tide.

holdfast organ of a benthic alga that attaches the alga to the sea floor.

holoplankton organisms living their entire life cycle in the floating (planktonic) state.

homeotherm organism with a body temperature that varies only within narrow limits.

hook spit turned landward at its outer end.

horse latitudes regions of calms and variable winds that coincide with latitudes 30°–35°N and S.

hot spot surface expression of a persistent rising plume of hot mantle material.

hurricane severe, cyclonic, tropical storm at sea, with winds of 120 km (73 mi) per hour or more; generally applied to Atlantic Ocean storms. *See also* typhoon.

hydrogen bond in water, the weak attraction between the positively charged hydrogen end of one water molecule and the negatively charged oxygen end of another water molecule.

hydrogenous sediment sediment precipitated from substances dissolved in seawater.

hydrologic cycle movement of water among the land, oceans, and atmosphere due to vertical and horizontal transport, evaporation, and precipitation.

hydrothermal vent seafloor outlet for high-temperature groundwater and associated mineral deposits; a hot spring.

hypocenter *see* focus.

hypothesis initial explanation of data that is based on well-established physical or chemical laws.

hypoxic having low oxygen levels in the water; organisms may find survival in a hypoxic environment difficult or impossible.

hypsographic curve graph of land elevation and ocean depth versus area.

I

iceberg mass of land ice that has broken away from a glacier and floats in the sea.

ice floe a floating chunk of ice that is less than 10 km (6 mi) in its greatest dimension.

igneous rock rock formed by solidification of a molten magma.

incidental catch the portion of any catch or harvest taken in addition to the targeted species.

inertia the property of matter that causes it to resist any change in its motion.

infauna animals that live buried in the sediment.

inner core the innermost region of Earth. It is solid and consists primarily of iron with minor amounts of other elements that likely include nickel, sulfur, and oxygen.

Intergovernmental Panel on Climate Change (IPCC) the leading international body for the assessment of climate change, established by the United Nations Environment Programme (UNEP) and the World Meteorological Organization (WMO).

intermediate disturbance hypothesis the concept that maximum biodiversity is attained at intermediate levels of ecological disturbance.

internal wave wave created below the sea surface at the boundary between two density layers.

international date line an imaginary line through the Pacific Ocean roughly corresponding to 180° longitude, to the east of which, by international agreement, the calendar date is one day earlier than to the west.

intertidal *see* littoral.

intertidal volume in an embayment, the volume of water gained or lost owing to the rise and fall of the tide.

Intertropical Convergence Zone (ITCZ) a region of weak and variable winds where the trade winds of the two hemispheres converge. The ITCZ is generally associated with the zone of highest surface temperature and can be thought of as the climatic equator. The ITCZ is generally found between 3 and 10°N.

inverse estuary an embayment, often located in arid climates, with high evaporation and little freshwater input. Circulation is seaward at depth and inward at the surface, opposite that of a typical estuary. Salinity is generally higher than average seawater salinities.

invertebrate an animal without a backbone.

ion positively or negatively charged atom or group of atoms.

ionic bond the electrostatic force that holds together oppositely charged ions.

iron limitation control of primary producers by lack of the micronutrient iron.

island arc system chain of volcanic islands formed above the sinking plate at a subduction zone.

isobar line of constant pressure.

isobath contour of constant depth.

isohaline having a uniform salt content.

isopycnal having a uniform density.

isostasy mechanism by which areas of Earth's crust rise or subside until their masses are in balance, "floating" on the mantle.

isothermal having a uniform temperature.

isotope atoms of the same element having different numbers of neutrons.

J

jetsam anything intentionally discarded at sea, such as solid waste.

jet stream (polar) a stream of air, between 30° and 50°N and S and about 12 km above Earth, moving from west to east at an average speed of 100 km/h.

jetty structure located to influence currents or to protect the entrance to a harbor or river from waves (U.S. terminology). *See also* breakwater, groin.

K

Keeling curve a graph of carbon dioxide concentration in the atmosphere measured at Mauna Loa Observatory, Hawaii since the 1950s.

kelp any of several large, brown algae, including the largest known algae.

kelp forest an extensive bed of brown macroalgae and the biological community supported by the algal bed.

kinetic energy energy produced by the motion of an object.

knot a unit of speed equal to 0.51 m/s or 1 nautical mile per hour.

krill small, shrimplike crustaceans found in huge masses in polar waters and eaten by baleen whales.

Kyoto Protocol an international agreement linked to the United Nations Framework Convention on Climate Change; the major feature is that it sets binding targets for thirty-seven industrialized countries and the European community for reducing greenhouse gas emissions.

L

lag deposits large particles left on a beach after the smaller particles are washed away.

lagoon shallow body of water that usually has a shallow, restricted outlet to the sea.

Langmuir cells shallow wind-driven circulation; paired helixes of moving water form windrows of debris along convergence lines.

La Niña condition of colder-than-normal surface water in the eastern tropical Pacific.

large marine ecosystems (LMEs) areas of the ocean characterized by distinct bathymetry, hydrology, productivity, and trophic interations.

larva immature juvenile form of certain animals.

latent heat of fusion amount of heat required to change the state of 1 g of water from ice to liquid.

latent heat of vaporization amount of heat required to change the state of 1 g of water from liquid to gas.

latitude distance north or south of the equator. Latitude is the angle between the equatorial plane and a line drawn outward from the center of Earth to a point on the surface of Earth. Latitude varies from 0° to +90° north of the equator and 0° to −90° south of the equator. Together with longitude, it specifies the location of a point on the surface of Earth.

Laurasia an ancient landmass that fragmented to produce North America and Eurasia.

lava magma, or molten rock, that has reached Earth's surface; the same material solidified after cooling.

lava coast a primary coast formed by active volcanism producing lava flows that extend to the sea.

leading margin (active margin) the edge of the overriding plate at a trench or subduction zone.

lee shelter; the part or side sheltered from wind or waves.

light-year the distance light travels in one year. A light-year is equal to 9.46×10^{12} km, or 5.87×10^{12} mi.

Linnaen classification system the form of biological classification established by Carolus Linnaeus in the 1700s; the basis of modern taxonomic classification.

lithification the conversion of loose sediment to solid rock.

lithogenous sediment sediment composed of rock particles eroded mainly from the continents by water, wind, and waves.

lithosphere outer, rigid portion of Earth; includes the continental and oceanic crusts and the upper part of the mantle.

littoral area of the shore between mean high water and mean low water; the intertidal zone.

longitude distance east or west of the prime meridian. Longitude is the angle in the equatorial plane between the prime meridian and a second meridian that passes through a point on the surface of Earth whose location is being specified. Longitude may be specified in one of two ways; either from 0° to 360° east of the prime meridian, or 0° to 180° east and 0° to 180° west. Together with latitude, it specifies the location of a point on the surface of Earth.

longshore current current produced in the surf zone by the waves breaking at an angle with the shore; the current runs roughly parallel to the shoreline.

longshore transport (littoral drift) movement of sediment by the longshore current.

lower high water lower of the two high waters of any tidal day in a region of mixed tides.

lower low water lower of the two low waters of any tidal day in a region of mixed tides.

low-tide terrace flat section of the foreshore seaward of the sloping beach face.

low water lowest elevation reached by a falling tide.

lunar month time required for the Moon to pass from one new Moon to another new Moon (approximately twenty-nine days).

lysocline depth at which calcareous skeletal material first begins to dissolve.

M

macroplankton plankton from 0.2–2 mm.

magma molten rock material that forms igneous rocks upon cooling; magma that reaches Earth's surface is referred to as lava.

magnetic pole either of the two points on Earth's surface where the magnetic field force lines are vertical.

major constituent ions a total of eleven ions dissolved in seawater that account for 99.99% by weight of the ions in solution.

mammal air-breathing vertebrate animals with common characteristics including presence of hair, live birth, endothermy, and mammary glands.

manatee *see* sea cow.

manganese nodules rounded, layered lumps found on the deep-ocean floor that contain, on average, about 18% manganese, 17% iron, with smaller amounts of nickel, cobalt, and copper and traces of two dozen other metals; a hydrogenous sediment.

mangrove a tree that has developed multiple adaptations to living at the land-sea interface; dense groves form mangrove forests.

mantle main bulk of Earth between the crust and the core; increasing pressure and temperature with depth divide the mantle into concentric layers.

mariculture *see* aquaculture.

marine pollution the introduction of substances or energy that results in harm to the organisms living in the ocean or human use of the ocean.

maximum sustainable yield maximum number or amount of a species that can be harvested each year without steady depletion of the stock; the remaining stock is able to replace the harvested members by natural reproduction.

meander turn or winding curve of a current.

mean sea level average height of the sea surface, based on observations of all stages of the tide over a nineteen-year period in the United States.

melon a fatty organ used by some Odontocetes to aid echolocation.

Mercator projection a map projection in which the surface of Earth is projected onto a cylinder. Distortion is great at high latitudes and the poles cannot be shown. Mercator projections are frequently used for navigation because a straight line drawn on them is a line of true direction or constant compass heading.

meridian circle of longitude passing through the poles and any given point on Earth's surface.

meroplankton floating developmental stages (eggs and larvae) of organisms that as adults belong to the nekton and benthos.

mesopelagic oceanic zone from 200–1000 m.

mesosphere either the layer of the atmosphere above the stratosphere extending from about 50–90 km, or the region of the mantle beneath the asthenosphere.

metamorphic rock a rock created by the alteration of a preexisting rock by high temperature and/or pressure.

microbe a microscopic, single-celled living organism; includes bacteria and archaea.

microbial loop component of marine food webs in which dissolved organic material cycles through bacteria and nanoplankton then back to small members of the zooplankton.

microorganism synonym for microbe.

microplankton net plankton, composed of individuals from 0.07–1 mm in size but large enough to be retained by a small mesh net.

Milankovich cycle small, slow, periodic changes in Earth's climate caused by changes in Earth's orbit around the Sun.

minus tide low-tide level below the mean value of the low tides or the zero tidal depth reference.

mitigation coastal management concept requiring developers to replace developed areas with equivalent natural areas or to reengineer other areas to resemble areas prior to development.

mitochondria organelle (internal cell structure) used to derive energy from food in cells.

mixed layer (or mixed surface layer) a surface layer tens of meters thick with relatively constant temperature due to turbulent mixing by wind and waves.

mixed tide type of tide in which large inequalities between the two high waters and the two low waters occur in a tidal day.

mixotrophic plankton phytoplankton that can both photosynthesize and consume organic matter depending on environmental conditions.

Mohorovičić discontinuity (Moho) boundary between crust and mantle, marked by a rapid increase in seismic wave speed.

mollusks marine animals, usually with shells; includes mussels, oysters, clams, snails, and slugs.

monoculture cultivation of only one species of organism in an aquaculture system.

monsoon name for seasonal winds; first applied to the winds over the Arabian Sea, which blow for six months from the northeast and for six months from the southwest; now extended to similar winds in other parts of the world; in India, the term is popularly applied to the southwest monsoon and to the rains that it brings.

Moon tide portion of the tide generated solely by the Moon's tide-raising force, as distinguished from that of the Sun.

moraine glacial deposit of rock, gravel, and other sediment left at the margin of an ice sheet.

mutualism an intimate association between different organisms in which both organisms benefit.

Mysticetes the suborder of cetaceans that includes baleen whales.

N

nanoplankton plankton that passes through an ordinary plankton net but can be removed from the water by centrifuging water samples.

natural selection the differential survival and production of a population that results in evolution; the forces that cause evolution.

nautical mile unit of length equal to 1852 m, or 1.15 land miles or 1 minute of latitude.

neap tides tides occurring near the times of the first and last quarters of the Moon, when the range of the tide is least.

nebula a large dense cloud of gas and dust in space.

nekton pelagic animals that are active swimmers; for example, adult squid, fish, and marine mammals.

nemertean a phylum of marine worms known collectively as ribbon worms.

neritic zone shallow-water marine environment extending from low water to the edge of the continental shelf. *See also* pelagic.

net circulation the long-term transport of water out of an estuary at the surface and into the estuary at depth averaged over many tidal cycles.

net plankton *see* microplankton.

net primary production the rate at which primary producers capture and store energy after subtracting energy or biomass consumption due to respiration.

niche separation the geographical or temporal seperation of organisms that provide similar ecological roles within an ecosystem.

nitrogen fixation process of using nitrogen gas as a source of inorganic nitrogen.

node point of least or zero vertical motion in a standing wave.

nonconservative constituent component or property of seawater whose value changes as a result of biological or chemical processes as well as by mixing, advection, and diffusion; for example, nutrients and oxygen in seawater.

normalization standard approach dividing primary productivity values by biomass values.

nucleus part of a cell that houses the majority of the cell's DNA.

nudibranchs soft-bodied, gastropod mollusks; sea slugs.

nutrient in the ocean, any one of a number of inorganic or organic compounds or ions used primarily in the nutrition of primary producers; nitrogen and phosphorus compounds are examples.

nutrient regeneration the release of nutrients from organic matter through biological activity by consumers or decomposers.

O

ocean acidification a decrease in pH and consequent increase in acidity of seawater.

oceanic pertaining to the ocean water seaward of the continental shelf; the "open ocean." *See also* pelagic.

oceanic crust crust below the deep-ocean sediments; mainly basalt.

oceanic zone open ocean away from the direct influence of land.

Ocean Thermal Energy Consversion (OTEC) method of extracting energy from the ocean that relies on the temperature difference between surface water and water at depth.

Odontocetes the suborder of cetaceans that includes the toothed whales.

offshore direction seaward of the shore.

offshore current any current flowing away from the shore.

offshore transport movement of sediment or water away from the shore.

onshore direction toward the shore.

onshore current any current flowing toward the shore.

onshore transport movement of sediment or water toward the shore.

oolith small, rounded accretionary body of calcium carbonate. Created by precipitation of calcium carbonate in shallow, warm seawater due to a change in water temperature or acidity.

ooze fine-grained deep-ocean sediment composed of at least 30% calcareous or siliceous remains of small marine organisms, the remainder being clay-size particles.

orbit in water waves, the path followed by the water particles affected by the wave motion; also, the path of a body subjected to the gravitational force of another body, such as Earth's orbit around the Sun.

orographic effect precipitation patterns caused by the flow of air over mountains.

osmosis tendency of water to diffuse through a semipermeable membrane to make the concentration of water on one side of the membrane equal to that on the other side.

osmotic pressure pressure that builds up in a confined fluid because of osmosis.

Osteichtyes fish that have skeletons made of bone.

outer core a region surrounding the inner core. It is liquid and consists primarily of iron with minor amounts of other elements that likely include nickel, sulfur, and oxygen.

overfishing the act of removing adult fish faster than the population of that fish can accommodate, resulting in reduction of the total population.

overturn sinking of denser water and its replacement by less-dense water from below.

oxygen minimum zone in which respiration and decay reduce dissolved oxygen to a minimum, usually between 800 and 1000 m.

oxygen minimum zone a zone of low dissolved oxygen concentration in the ocean that typically occurs between about 700–1000 m, depending on location.

ozone a form of oxygen that absorbs ultraviolet radiation from the Sun; O_3.

P

pack ice large pieces of floating ice that have been driven together into a nearly continuous mass.

paleoceanography the study of ocean characteristics and processes in the past.

paleoclimatology the study of climate over the entire history of Earth.

paleomagnetism study of ancient magnetism recorded in rocks; includes study of changes in location of Earth's magnetic poles through time and reversals in Earth's magnetic field.

pancake a form of sea ice that consists of round pieces of ice with diameters ranging from a few inches to many feet, depending on the local conditions that affect ice formation. It may have a thickness of several inches. Pancake ice features elevated rims with a nearly uniform height of a few inches.

Pangaea an ancient landmass that consisted of all of the present-day continents; it fragmented into Laurasia and Gondwanaland.

Panthalassa the single great ocean that existed when all of the landmasses were combined in the ancient continent Pangaea.

parallel circle on the surface of Earth parallel to the plane of the equator and connecting all points of equal latitude; a line of latitude.

parasitism an intimate association between different organisms in which one is benefited and the other is harmed.

partially mixed estuary is one with a strong net seaward flow of fresh water at the surface and a strong inward flow of seawater at depth.

passive margin *see* trailing margin.

pelagic zone primary division of the sea, which includes the whole mass of water subdivided into neritic and oceanic zones; also pertaining to the open sea.

period *see* tidal period, wave period.

pH measure of the concentration of hydrogen ions in a water solution; the concentration of hydrogen ions determines the acidity of the solution; $pH = -\log_{10}(H^+)$, where H^+ is the concentration of hydrogen ions in gram atoms per liter.

phosphorite a sedimentary rock composed largely of calcium phosphate and largely in the form of concretions and nodules.

photic zone layer of a body of water that receives ample sunlight for photosynthesis; usually less than 100 m deep.

photoautotrophs organisms that harness light energy to generate organic matter.

photo cell a device that converts light energy to electrical energy; used to determine solar radiation below the sea surface.

photophore luminous organ found on fish.

photosynthesis manufacture by plants of organic substances and release of oxygen from carbon dioxide and water in the presence of sunlight and the green pigment chlorophyll.

phylogenies evolutionary connections between ancestor organisms and their descendents.

physiographic map portrayal of Earth's features by perspective drawing.

phytoplankton microscopic algal and photosynthetic forms of plankton.

picoplankton plankton less than 2–3 µm in size.

piezophile microorganisms that grow at high pressures such as those that live in the deep sea.

pinniped member of the marine mammal group that is characterized by four swimming flippers; for example, seals and sea lions.

plankton passively drifting or weakly swimming organisms.

plate one of roughly a dozen segments of lithosphere that move independently on the asthenosphere. Plates consist of oceanic and/or continental crust fused to the rigid upper mantle overlying the asthenosphere.

plate tectonics theory and study of Earth's lithospheric plates, their formation, movement, interaction, and destruction; the attempts to explain Earth's crustal changes in terms of plate movements.

pogonophora a phylum of marine worms ("beard worms") that live within tubes on the seafloor. The worms have no guts and rely on a commensal relationship with bacteria; typically found around hydrothermal vents.

poikilotherm organism with a body temperature that varies according to the temperature of its surroundings.

polar easterlies winds blowing from the poles toward approximately 60°N and 60°S; winds are northeasterly in the Northern Hemisphere and southeasterly in the Southern Hemisphere.

polar molecule a molecule with an unevenly distributed electrical charge. One end of the molecule has a slight positive charge and the other end has a slight negative charge.

polar reversal the periodic reversal of Earth's magnetic field where the north magnetic pole becomes the south magnetic pole and vice versa.

Polaris also known as the North Star, is located less than 1° from the celestial pole, a line corresponding to the extension of Earth's axis of rotation into the sky from the north geographic pole. The angular elevation of Polaris above the horizon corresponds to the latitude of an observer in the Northern Hemisphere.

polychaetes marine segmented worms, some in tubes, some free-swimming.

polyp sessile stage in the life history of certain members of the phylum Coelenterata (Cnidaria); sea anemones and corals.

potential energy energy that an object has because of its position or condition.

precipitation falling products of condensation in the atmosphere, such as rain, snow, or hail; also the falling out of a substance from solution.

precision depth recorder (PDR) instrument used to obtain a continuous pictorial record of the ocean bottom by timing the returning echoes of sound pulses. *See also* echo sounder.

primary coast coastline shaped primarily by terrestrial processes rather than marine processes.

primary production amount of living matter, or biomass, that is produced by photosynthetic or chemosynthetic organisms, usually expressed in grams of carbon per volume of seawater.

primary productivity rate at which biomass is produced by photosynthesis and chemosynthesis, usually expressed as grams of carbon produced per volume of seawater per day.

prime meridian meridian of 0° longitude, used as the origin for measurements of longitude; internationally accepted as the meridian of the Royal Naval Observatory, Greenwich, England.

principle of constant proportion the ratios between the abundance of major constituent ions in seawater remain constant regardless of seawater salinity.

producer an organism that manufactures food from inorganic substances.

progressive tide tide wave that moves, or progresses, in a nearly constant direction.

progressive wind wave wave that moves, or progresses, in a certain direction.

prokaryote grouping of organisms based on morphology. All prokaryotes are unicellular with no internal membrane structures. Prokaryotes are composed of members of the domains Archaea and Bacteria.

proxy an indirect measurement of another (unknown or difficult to measure) variable.

psychrophile microorganisms that grow in cold temperatures such as those that live in sea ice.

pteropod pelagic snail whose foot is modified for swimming.

pteropod ooze sediment made up of more than 30% shells of pteropods.

P-wave (or primary wave) a type of seismic wave in which material is alternately compressed and stretched in the direction of propagation of the wave.

pycnocline water layer with a large change in density with depth.

R

radial symmetry having similar parts regularly arranged around a central axis.

radiation energy transmitted as electromagnetic rays or waves without the need of a substance to conduct the energy.

radiolarian ooze sediment made up of more than 30% skeletal remains of radiolarians.

radiolarians single-celled protozoans with siliceous tests.

radiometric dating determining ages of geologic samples by measuring the relative abundance of radioactive isotopes and comparing isotopic systems.

rafting transport of sediment, rock, silt, and other land matter out to sea by ice, logs, and the like, with the deposition of the rafted material when the carrying agent melts or disintegrates.

rain shadow an area of low precipitation on the leeward (or sheltered) side of an island or mountain. Precipitation occurs on the windward side as the air is forced to ascend the side of the island or mountain so the descending air on the leeward side is dry.

range see tidal range.

red clay red to brown fine-grained lithogenous deposit, of predominantly clay size, which is derived from land, transported by winds and currents, and deposited far from land and at great depth; also known as brown mud and brown clay.

Redfield ratio the relative molar abundance of carbon, nitrogen, and phosphorus in marine organic matter, this ratio is C:N:P = 106:16:1. This has since been expanded to include silicon, to produce the ratio C:Si:N:P = 106:40:16:1.

red tide red coloration, usually of coastal waters, caused by large quantities of microscopic organisms (generally dinoflagellates); some red tides result in mass fish kills, others contaminate shellfish, and still others produce no toxic effects.

reef offshore hazard to navigation, made up of consolidated rock or coral, with a depth of 20 m or less.

reef coast a secondary coast formed by reef-building corals in tropical waters.

reef crest highest portion of a coral reef on the exposed seaward edge of the reef.

reef flat portion of a coral reef landward of the reef crest and seaward of the lagoon.

reflection rebounding of light, heat, sound, waves, and so on after striking a surface.

refraction change in direction, or bending, of a wave.

relict sediments sediments deposited by processes no longer active.

remineralization the transformation of organic matter to inorganic matter, typically mediated by biological oxidation.

reservoir a source or place of temporary residence for water, such as the oceans or atmosphere.

residence time mean time that a substance remains in a given area before being replaced, calculated by dividing the amount of a substance by its rate of addition or subtraction.

respiration metabolic process by which food or food-storage molecules yield the energy on which all living cells depend.

restoring force force that returns a disturbed water surface to the equilibrium level, such as surface tension and gravity.

reverse osmosis the process of producing fresh water from seawater by forcing the water molecules through a semipermeable membrane, leaving behind salt ions and other impurities.

ria coast a primary coast formed when rising sea level, caused by the melting of glaciers and ice sheets following the last ice age, flooded coastal river valleys.

ridge long, narrow elevation of the sea floor, with steep sides and irregular topography.

rift valley trough formed by faulting along a zone in which plates move apart and new crust is created, such as along the crest of a ridge system.

rift zone a region where the lithosphere splits and separates, allowing new crustal material to intrude into the crack or rift.

rip current strong surface current flowing seaward from shore; the return movement of water piled up on the shore by incoming waves and wind.

rise long, broad elevation that rises gently and generally smoothly from the sea floor.

Rodinia a supercontinent that formed approximately 1110 million years ago.

rotary current tidal current that continually changes its direction of flow through all points of the compass during a tidal period.

rotary standing tide tide that is the result of a standing wave moving around the central node of a basin.

S

salinity measure of the quantity of dissolved salts in seawater. It is formally defined as the total amount of dissolved solids in seawater in parts per thousand (‰) by weight when all the carbonate has been converted to oxide, all the bromide and iodide have been converted to chloride, and all organic matter is completely oxidized.

salinometer instrument for determining the salinity of water by measuring the electrical conductivity of a water sample of a known temperature.

salt budget balance between the rates of salt addition to and removal from a body of water.

salt marsh a relatively shallow coastal environment populated with salt-tolerant grasses. These are often found in temperate climates. They are extremely productive biologically and extend the shoreline seaward by trapping fine sediment.

salt wedge intrusion of salt water along the bottom; in an estuary, the wedge moves upstream on high tide and seaward on low tide.

sand spit see spit.

satellite body that revolves around a planet; a moon; a device launched from Earth into orbit around a planet or the Sun.

saturation concentration the maximum amount of any substance that can be held in solution without any of the substance coming out of solution. When a dissolved substance has reached its saturation concentration, the rate at which it enters the solution will equal the rate at which it leaves the solution.

scarp elongated and comparatively steep slope separating flat or gently sloping areas on the sea floor or on a beach.

scattering random redirection of light or sound energy by reflection from an uneven sea bottom or sea surface, from water molecules, or from particles suspended in the water.

schooling a behavioral adaptation common in marine fishes in which a group of individuals, typically of the same species, move as one.

sea same as the ocean; subdivision of the ocean; surface waves generated or sustained by the wind within their fetch, as opposed to swell.

sea cow (dugong, manatee) large herbivorous marine mammal of tropical and subtropical waters; includes the manatee and the dugong.

seafloor spreading movement of crustal plates away from the mid-ocean ridges; process that creates new crustal material at the mid-ocean ridges.

seagrass flowering plants living in the marine environment.

sea ice ice formed from freezing of seawater. The freezing process generally doesn't start until seawater temperature falls to about −1.8°C (28.8°F).

sea level height of the sea surface above or below some reference level. See also mean sea level.

seamount isolated volcanic peak that rises at least 1000 m from the sea floor.

sea stack isolated mass of rock rising from the sea near a headland from which it has been separated by erosion.

sea state numerical or written description of the roughness of the ocean surface relative to wave height.

Secchi disk white, or white and black, disk used to measure the transparency of the water by observing the depth at which the disk disappears from view.

secondary coast coastline shaped primarily by marine forces or marine organisms.

sediment particulate organic and inorganic matter that accumulates in loose, unconsolidated form.

sedimentary rock a rock formed by the cementation of mineral grains accumulated by wind, water, or ice transportation to the site of deposition or by chemical precipitation at the site.

seiche standing wave oscillation of an enclosed or semienclosed body of water that continues, pendulum fashion, after the generating force ceases.

seismic pertaining to or caused by earthquakes or Earth movements.

seismic sea wave see tsunami.

seismic tomography the use of seismic data to produce computerized, detailed, three-dimensional maps of the boundaries between Earth's layers.

seismic waves elastic disturbances, or vibrations, that are generated by earthquakes.

semidiurnal tide tide with two high waters and two low waters each tidal day.

semipermeable membrane membrane that allows some substances to pass through but restricts or prevents the passage of other substances.

sessile permanently fixed or sedentary; not free-moving.

set direction in which the current flows.

shallow-water wave wave in water whose depth is less than one-twentieth the average wavelength.

shingles flat, water-worn pebbles, or cobbles, found in beds along a beach.

shoal elevation of the sea bottom comprising any material except rock or coral (in which case it is a reef); may endanger surface navigation.

shore strip of ground bordering any body of water and alternately exposed and covered by tides and waves.

sidereal day time period determined by one rotation of Earth relative to a far-distant star, about four minutes shorter than the mean solar day.

sigma-t abbreviated value of the density of seawater; at a given temperature and salinity, neglecting pressure; $\sigma_t = (\text{density} - 1) \times 1000$.

siliceous containing or composed of silica.

siliceous ooze fine-grained deep-ocean biogenous sediment containing at least 30% siliceous tests, or the remains of small marine organisms.

sill shallow area that separates two basins from one another or a coastal bay from the adjacent ocean.

Sirenia the order of marine mammals that includes dugongs and manatees.

slack water state of a tidal current when its velocity is near zero; occurs when the tidal current changes direction.

slick area of smooth surface water.

SOFAR channel natural sound channel in the oceans in which sound can be transmitted for very long distances; the depth of minimum sound velocity; derived from the phrase "sound fixing and ranging."

solar constant rate at which solar radiation is received on a unit surface that is perpendicular to the direction of incident radiation just outside Earth's atmosphere at Earth's mean distance from the Sun; equal to 2 cal/cm²/min.

solar day time period determined by one rotation of Earth relative to the Sun; the mean solar day is twenty-four hours.

solstice times of the year when the Sun stands directly above 23 1/2°N or 23 1/2°S latitude. The winter solstice occurs about December 22, and the summer solstice occurs about June 22.

sonar method or equipment for determining, by underwater sound, the presence, location, or nature of objects in the sea; derived from the phrase "sound navigation and ranging."

sorting *see* dispersion.

sounding measurement of the depth of water beneath a vessel.

sound shadow zone area of the ocean into which sound does not penetrate because the density structure of the water refracts the sound waves.

Southern Oscillation a periodic reversal of the low- and high-pressure areas that typically dominate the eastern and western equatorial Pacific, respectively.

specific gravity ratio of the density of a substance to the density of 4°C water.

specific heat ratio of the heat capacity of a substance to the heat capacity of water.

spectrum the regular ordering of wave phenomena such as electromagnetic radiation by wavelength or frequency. A familiar spectrum is the regular ordering of visible light from the long-wavelength red to the short-wavelength violet when it passes through a prism.

sphere depth thickness of a material spread uniformly over a smooth sphere having the same area as Earth.

spit (sand spit) low tongue of land, or a relatively long, narrow shoal extending from the shore.

splash zone *see* supralittoral.

spoil dredged material.

spore minute, unicellular, asexual reproductive structure of an alga.

spreading center region along which new crustal material is produced.

spreading rate the rate at which two plates move apart. Spreading rates are generally between about 2 and 10 cm (0.8 and 4 in) per year.

spring tides tides occurring near the times of the new and full Moon, when the range of the tide is greatest.

standing stock biomass present at any given time.

standing wave type of wave in which the surface of the water oscillates vertically between fixed points called nodes, without progression; the points of maximum vertical rise and fall are called antinodes.

steepness of wave *see* wave steepness.

stipe portion of an alga between the holdfast and the blade.

storm berm *see* winter berm.

storm center area of origin for surface waves generated by the wind; an intense atmospheric low-pressure system.

storm surge elevation of the sea surface beneath the center of a hurricane or typhoon caused by the intense low pressure of the storm.

storm tide the combination of storm surge and high tide, creating the highest sea surface elevations beneath the center of a hurricane or typhoon.

stratosphere the layer of the atmosphere above the troposphere where temperature is constant or increases with altitude.

subduction zone plane descending away from a trench and defined by its seismic activity, interpreted as the convergence zone between a sinking plate and an overriding plate.

sublimation transition of a substance from its solid state to its gaseous state without becoming a liquid.

sublittoral benthic zone from the low-tide line to the seaward edge of the continental shelf; the subtidal zone.

submarine canyon relatively narrow, V-shaped, deep depression with steep slopes, the bottom of which grades continuously downward across the continental slope.

submersible a research submarine, designed for manned or remote operation at great depths.

subsidence sinking of a broad area of the crust without appreciable deformation.

substrate material making up the base on which an organism lives or to which it is attached.

subtidal *see* sublittoral.

summer berm a seasonal berm that is built by low-energy waves during the summer and removed by high-energy waves in the winter.

summer solstice *see* solstice.

Sun tide portion of the tide generated solely by the Sun's tide-raising force, as distinguished from that of the Moon.

supersaturation when the concentration of a dissolved substance is higher than its normal saturation value.

supralittoral benthic zone above the high-tide level that is moistened by waves, spray, and extremely high tides; also called splash zone.

surf wave activity in the area between the shoreline and the outermost limit of the breakers.

surface tension tendency of a liquid surface to contract owing to bonding forces between molecules.

surface wave a seismic wave that travels on Earth's surface.

surimi refined fish protein that is used to form artificial crab, shrimp, and scallop meat.

suture zone an area where two continental plates have joined together through continental collision. Suture zones are often marked by high mountain ranges, such as the Himalayas and the Alps.

swash the up-rush of water onto the beach from a breaking wave.

swash zone beach area where water from a breaking wave rushes.

S-wave (or secondary wave) a type of seismic wave in which material is sheared from side to side, perpendicular to the direction of propagation of the wave.

swell long and relatively uniform wind-generated ocean waves that have traveled out of their generating area.

symbiosis living together in intimate association of two dissimilar organisms.

T

Tagging of Pacific Predators (TOPP) an international program established by the Census of Marine Life designed to understand the ecology of top predators in the Pacific Ocean through the use of electronic tags.

taxonomy scientific classification of organisms.

tectonic pertaining to processes that cause large-scale deformation and movement of Earth's crust.

tektites particles with a characteristic round shape, formed from rock melting during meteorite impact.

temperature a measure of the rate of atomic or molecular motion in a substance.

terrigenous sediments of the land; sediments composed predominantly of material derived from the land.

test the shell of an organism.

Tethys Sea an ocean betwwen the continents of Gondwana and Laurasia before the opening of the Indian and Atlantic oceans.

theory a tested, reliable, and precise statement of the relationships among reproducible observations. A theory may be used to predict the existence of phenomena or relationships not previously recognized.

Theory of Evolution the theory describing how random changes in the genetic makeup of a species are acted upon by environmental forces over successive generations in a manner that forms new species.

thermal expansion changes in volume of a material (such as water) in response to changes in temperature; increasing temperature leads to increasing volume.

thermocline water layer with a large change in temperature with depth.

thermohaline circulation vertical circulation caused by changes in density; driven by variations in temperature and salinity.

thermosphere the layer of the atmosphere above the mesosphere; extends from 90 km to outer space.

tidal bore high-tide crest that advances rapidly up an estuary or river as a breaking wave.

tidal current alternating horizontal movement of water associated with the rise and fall of the tide.

tidal datum reference level from which ocean depths and tide heights are measured; the zero tide level.

tidal day time interval between two successive passes of the Moon over a meridian, approximately twenty-four hours and fifty minutes.

tidal period elapsed time between successive high waters or successive low waters.

tidal range difference in height between consecutive high and low waters.

tide periodic rising and falling of the sea surface that results from the gravitational attractions of the Moon and Sun acting on the rotating Earth.

tide wave long-period gravity wave that has its origin in the tide-producing force and is observed as the rise and fall of the tide.

tombolo deposit of unconsolidated material that connects an island to another island or to the mainland.

top-down control a food web in which the top predators control abundance and dynamics of lower trohpic levels.

topography general elevation pattern of the land surface (or the ocean bottom). *See also* bathymetry.

toxicant substance dissolved in water that produces a harmful effect on organisms, either by an immediate large dose or by small doses over a period of time.

trace element an element dissolved in seawater at a concentration less than one part per million.

trade winds wind systems that occupy most of the tropics and blow from approximately 30°N and 30°S toward the equator; winds are northeasterly in the Northern Hemisphere and southeasterly in the Southern Hemisphere.

trailing margin (passive margin) the continental margin closest to the mid-ocean ridge.

transform fault fault with horizontal displacement connecting the ends of an offset in a mid-ocean ridge. Some plates slide past each other along a transform fault.

transform plate boundary a boundary between two plates that are sliding past one another. This boundary is marked by a transform fault.

transverse ridge ridge running at nearly right angles to the main or principal ridge.

trench long, deep, and narrow depression of the sea floor with relatively steep sides, associated with a subduction zone.

trophic relating to nutrition; a trophic level is the position of an organism in a food chain or food (trophic) pyramid.

trophic efficiency the factional amount of energy or material transferred from one trophic level to the next.

Tropics of Cancer and Capricorn latitudes 23 1/2°N and 23 1/2°S, respectively, marking the maximum angular distance of the Sun from the equator during the summer and winter solstices.

troposphere the lowest layer of the atmosphere, where the temperature decreases with altitude.

trough long depression of the sea floor, having relatively gentle sides; normally wider and shallower than a trench. *See also* wave trough.

T-S diagram graph of temperature versus salinity, on which seawater samples taken at various depths are used to describe a water mass.

tsunami (seismic sea wave) long-period sea wave produced by a submarine earthquake, volcanic eruption, sediment slide, or seafloor faulting. It may travel across the ocean for thousands of miles unnoticed from its point of origin and build up to great heights over shallow water at the shore.

tube worm any worm or wormlike organism that builds a tube or sheath attached to a submerged substrate.

turbidite sediment deposited by a turbidity current, showing a pattern of coarse particles at the bottom, grading gradually upward to fine silt or mud.

turbidity loss of water clarity or transparency owing to the presence of suspended material.

turbidity current dense, sediment-laden current flowing downward along an underwater slope.

twilight zone region of the water column where there is some light but not enough for photosynthesis.

typhoon severe, cyclonic, tropical storm originating in the western Pacific Ocean, particularly in the vicinity of the South China Sea. *See also* hurricane.

U

Universal Time solar time along the prime meridian passing through Greenwich, England; also known as Greenwich Mean Time (GMT) or ZULU Time.

upwelling rising of water rich in nutrients toward the surface, usually the result of diverging surface currents.

V

vernal equinox *see* equinoxes.

vertebrates animals with backbones or a spinal column.

virus a noncellular infectious agent that reproduces only in living cells.

viscosity property of a fluid to resist flow; internal friction of a fluid.

W

Wadati-Benioff zone dipping patterns of earthquake activity that descend into the mantle along convergent plate boundaries.

water bottle device used to obtain a water sample at depth.

water budget balance between the rates of water added and lost in an area.

water mass body of water identified by similar patterns of temperature and salinity from surface to depth.

water type body of water identified by a specific range of temperature and salinity from a common source.

wave periodic disturbance that moves through or over the surface of a medium with a speed determined by the properties of the medium.

wave crest highest part of a wave.

wave height vertical distance between a wave crest and the adjacent trough.

wavelength horizontal distance between two successive wave crests or two successive wave troughs.

wave period time required for two successive wave crests or troughs to pass a fixed point.

wave ray line indicating the direction waves travel; drawn at right angles to the wave crests.

wave steepness ratio of wave height to wavelength.

wave train series of similar waves from the same direction.

wave trough lowest part of a wave.

well-mixed estuary is one in which there is strong wind-driven and tidal mixing. The salinity of the water in the estuary is relatively constant with depth and decreases from the ocean to the river.

westerlies wind systems blowing from the west between latitudes of approximately 30°N and 60°N and 30°S and 60°S; southwesterly in the Northern Hemisphere and northwesterly in the Southern Hemisphere.

western intensification the increase in speed of geostrophic currents as they move along the western boundary of an ocean basin.

wetland lands where saturation with water is a defining characteristic controlling the soil development and the plant and animal communities associated with the region.

Wilson Cycle a description of the life cycle of an ocean basin from opening to closing.

wind wave wave created by the action of the wind on the sea surface.

winter berm a relatively permanent berm that is formed by high-energy waves in the winter.

winter solstice *see* solstice.

Z

zenith point in the sky that is immediately overhead.

zonation parallel bands of distinctive plant and animal associations found within the littoral zones and distributed to take advantage of optimal conditions for survival.

zooplankton animal forms of plankton.

zooxanthellae symbiotic microscopic organisms (dinoflagellates) found in corals and other marine organisms.

ZULU Time solar time along the prime meridian passing through Greenwich, England; also known as Greenwich Mean Time (GMT) or Universal Time.

Credits

Photos

Prologue

Opener: © Brand X RF; **P.2**: © Anders Ryman/Corbis; **P.3**: Courtesy of Sea and Shore Museum, Port Gamble, WA; **P.9**: © National Maritime Museum, London; **P.11**: Franklin Folger, 1769; **P.12**: NASA; **Page 14**: Courtesy of Beth Simmons, Metropolitan State University Denver; **Box Figure 1a–c**: Challenger Expedition: Dec. 21, 1872–May 24, 1876. Engravings from Challenger Reports, vol. 1, 1885; **P.16a–b**: © The Norwegian Maritime Museum; **P.17a**: © Scripps Institution of Oceanography, University of California, San Diego; **P.17b**: © D. Weisman/Woods Hole Oceanographic Institution; **P.18a**: Courtesy of Joe Creager, University of Washington; **P.18b**: Integrated Ocean Drilling Program U.S. Implementing Organiztion (IODP-USIO), Texas A & M; **P.18c**: © Kyodo/Newscom; **P.19a**: Photo courtesy of Argo Program (http://www.argo.ucsd.edu, http://www.argo.net); **p. 21**: Courtesy of Dr. Gwynne Rife, University of Findlay; **Box Figure 1, 2a–b, 3**: © Scripps Institution of Oceanography, University of California, San Diego; **P.20**: CZCS Project/Goddard Space Flight Center.

Chapter 1

Opener: Keith A. Sverdrup; **1.1**: Image by Reto Stockli, NASA/Goddard Space Flight Center, Enhancements by Robert Simmon; **1.2**: © AP/Wide World Photos; **1.3**: NASA; **Box Figure 1**: © James A. Sugar/Corbis; **Box Figure 2**: © Louis A. Frank/University of Iowa; **1.5a–b**: NASA; **1.10**: © National Maritime Museum, London; **1.14**: U.S. Department of Defense; **1.17a–b, 1.18a–e**: Keith A. Sverdrup, Image generated from NOAA ETOPO2 data.

Chapter 2

Opener: NOAA; **2.10**: World Ocean Floor Panorama, Authors (Bruce C. Heezen and Marie Tharp), Date (1977): Copyright by Marie Tharp 1977/2003. Reproduced by permission of Marie Tharp Maps, LLC 8 Edward Street, Sparkill, New York 10976.; **2.13**: Dr. Robert M. Kieckhefer, Chevron Overseas Petroleum, Inc. San Ramon, CA; **2.17**: Integrated Ocean Drilling Program U.S. Implementing Organiztion (IODP-USIO), Texas A & M; **2.27**: © OAR/National Underseas Research; **Box Figure 1**: © Alyn & Alison Duxbury; **Box Figure 2, 3**: Courtesy Deborah Kelley, University of Washington; **Box Figure 4a–b**: © Alyn & Alison Duxbury; **Box Figure 5**: Courtesy Deborah Kelley, University of Washington; **2.32**: © P.W. Lipman, U.S. Geological Survey; **2.35**: National Geophysical Data Center/NOAA.

Chapter 3

Opener: © Keith A. Sverdrup; **3.5**: Courtesy Walter H. F. Smith/NOAA; **3.9**: © Joel Zatz/Alamy; **3.15**: National Geophysical Data Center/NOAA; **3.16a–b**: Courtesy Williamson & Associates, Seattle, WA; **3.18a–b**: Photo by Kenneth Adkins; **3.18c(both)**: Courtesy of Steve Nathan and R. Mark Leckie, University of Massachusetts; **3.21**: © Mila Zinkova; **3.22**: Courtesy of Joe Creager, University of Washington; **3.23**: NOAA; **3.25a**: Courtesy of Dean McManus, University of Washington; **3.25b**: Courtesy of Dawn Wright, Oregon State University; **3.26**: Courtesy of Kathy Newell, School of Oceanography, University of Washington; **3.27c–d**: Keith A. Sverdrup; **3.27e**: Courtesy Dr. Richard Sternberg, School of Oceanography, University of Washington; **3.28a**: © Science VU/NOAA/Visuals Unlimited; **3.28b**: © S. Scribner and J.C. Santamarina, Georgia Institute of Technology.

Chapter 4

Opener: © Keith A. Sverdrup; **4.6**: © Sandra Hines, News and Information, University of Washington; **4.16**: © Alyn & Alison Duxbury; **4.17**: Courtesy of Dennis K. Clark, MOBY Project Team Leader, NOAA; **4.19**: © Alyn & Alison Duxbury.

Chapter 5

Opener: © Jochem Wijnands/Picture Contact Bv/agefotostock.com; **5.11**: Plumbago via Wikipedia; **5.12a–e**: © David Liittschwager/National Geographic Stock; **5.13**: Courtesy of John Conomos, U.S. Geological Survey.

Chapter 6

Opener: NASA/GSFC/Laboratory for Atmospheres; **6.4**: NOAA/ESRL/PSD; **6.7a–b**: M. Chahine, Jet Propulsion Lab, Goddard Space Flight Center; **6.8**: © Mila Zinkova; **6.9**: Courtesy of Roger Andersen, Polar Science Center, Applied Physics Laboratory, University of Washington; **6.10a**: © Mila Zinkova; **6.10b**: NASA/JSC; **Box Figure 1**: © Alyn & Alison Duxbury; **6.15**: NASA; **6.25**: © SuperStock/agefotostock.com; **6.27**: NOAA **6.31**: NASA; **6.33**: U.S. Army Corps of Engineers photo by Alan Dooley; **6.34**: Photo taken on September 19, 2005 by Mark Wolfe/FEMA; **6.35**: Photo taken on October 4. 2005 by John Fleck/FEMA; **6.37a–c**: Scientific Visualization Studio, NASA, Goddard Space Flight Center.

Chapter 7

Opener: NOAA Geophysical Fluid Dynamics Laboratory; **7.24**: Courtesy NASA/JPL-Caltech; **7.26**: Courtesy Otis Brown, University of Miami/RSMAS; **7.27**: NASA; **7.28**: Mmelugin via Wikipedia; **7.32**: Courtesy Liesen Xie and William Hsieh, University of British Columbia; **7.34**: Courtesy of Mark Abbott/NASA; **7.38**: Courtesy of Will Patterson, School of Oceanography, University of Washington; **7.39a–b**: Courtesy of Kathy Newell, School of Oceanography, University of Washington; **Box Figure 1**: © Rowland Studio, Seattle, WA; **7.41**: Mark Holmes, School of Oceanography, University of Washington.

Oceanography from Space

OS.1: NASA; **OS.2**: NASA/Goddard Space Flight Center, The SeaWiFS Project and GeoEye, Scientific Visualization Studio; **OS.3**: NASA/GSFC/JPL-Caltech; **OS.4, OS.5**: Environmental Visualization laboratory/NOAA **OS.6**: NASA/JPL **OS.7**: NASA/Goddard Space Flight Center, The SeaWiFS Project and GeoEye; **OS.8**: Polovina et al; Forage and migration habitat of loggerhead (Caretta caretta) and olive ridley (Lepidochelys olivacea) sea turtles in the central North Pacific Ocean, Fisheries Oceanography Volume 13, Issue 1, pages 36–51, January 2004. DOI: 10.1046/j.1365-2419.2003.00270.x; **OS.9**: Ed Hanka/NASA.

Chapter 8

Opener: © Brian Sytnyk/Photographer's Choice/Getty Images; **8.1**: Photo by Fabrice Neyret;http://www-evasion.imag.fr/Membres/Fabrice.Neyret/; **8.2**: NOAA; **8.7**: © Eda Rogers/Sea Images; **8.8**: © PhotoLink/Getty Images RF; **8.13**: U.S. Army Corps; **8.21**: © Robert Harrison/Alamy; **8.22a–c**: Keith A. Sverdrup; **8.25a–b**: © Alyn & Alison Duxbury; **8.26**: © SuperStock; **8.27**: Courtesy of Lifeguard Captain Nick Steers, County of Los Angeles Fire Department & NOAA/USLA; **8.28**: National Weather Service (NWS) Collection/NOAA; **8.29a–c**: Photo by Guy Gelfenbaum/U.S. Geological Survey; **p. 232, Box Figure 1-5**: Courtesy of Dr. Eddie Bernard/NOAA/PMEL; **Box Figure 6**: Photo by Prof. Kenji Satake, University of Tokyo. Courtesy of Dr. Eddie Bernard/NOAA/PMEL; **Box Figure 7**:

Photo by L. Dengler. Courtesy of Dr. Eddie Bernard/NOAA/PMEL; **Box Figure 7**: The McGraw-Hill Companies, Inc.; **8.31a–b**: Courtesy U.S. National Tsunami Hazard Mitigation Program/NOAA; **8.32**: Courtesy of Dr. Eddie Bernard/NOAA/PMEL; **8.33b–c**: NASA/GSFC/METI/ERSDAC/JAROS, and U.S./Japan ASTER Science Team; **8.38**: European Marine Energy Test Centre.

Chapter 9

Opener: © Barrett & MacKay/All Canada Photos/Alamy; **9.17a–b**: Courtesy New Brunswick Department of Tourism and Parks; **9.18**: Courtesy Shubenacadie Tidal Bore Rafting Park; **9.20**: Courtesy Nova Scotia Power.

Chapter 10

Opener: © McGraw-Hill Companies/ Dr. Parvinder Sethi, Photographer; **10.1a**: Courtesy of Joe Creager, University of Washington; **10.1b**: NASA; **10.1c**: © Alyn & Alison Duxbury; **10.1d**: Courtesy Dr. Sherwood Maynard, University of Hawaii; **10.1e**: © TerraNOVA International; **10.1f**: U.S. Army Corps of Engineers/NASA/EOSAT Geological Survey, School of Oceanography, University of Washington; **10.2a**: Courtesy Dr. Richard Sternberg, School of Oceanography, University of Washington; **10.2b**: © Alyn & Alison Duxbury; **10.2c**: © 2012 Alex S. McLean/ Landslides; **10.2d–f**: © Alyn & Alison Duxbury; **10.2g**: P.R. Hoar/NOAA/NESDIS/ NCDDC; **10.3**: Courtesy of the Rosenberg Library, Galveston, TX; **10.5, 10.6, 10.7, 10.8**: Keith A. Sverdrup; **10.11**: © Carr Clifton; **10.15, 10.16**: Courtesy Dr. Richard Sternberg, School of Oceanography, University of Washington; **10.17**: S. Jeffress Williams/ USGS; **10.18b**: Photograph by Ken Winters, U.S. Army Corps of Engineers; **10.20**: © Bob Evans/La Mer Bleu Productions; **10.21**: U.S. Army Corps of Engineers.

Chapter 11

Opener: © Reinhard Dirscherl/agefotostock.com; **11.3, 11.4, 11.6a**: Photo courtesy of Karie Holtermann, University of Washington. Reprinted with permission; **11.6b**: © Comstock Images/PictureQuest RF; **11.7**: Courtesy of Kathy Newell, School of Oceanography, University of Washington; **11.8**: Courtesy Seattle Aquarium; **11.10**: © Royalty Free/ Corbis RF; **p. 296**: Courtesy of Dr. Anne M. Risser Lee, University of Findlay; **Box Figure 1**: © Doug Perrine/Alamy; **Box Figure 2**: Courtesy of Dr. Gwynne Rife, University of Findlay.

Chapter 12

Opener: © Mirko Zanni/Waterframe/Getty Images; **12.1**: © Melissa Booth; **12.2(both)**: Photo courtesy of Karie Holtermann, University of Washington. Reprinted with permission; **12.3**: © Science Photo Library RF/Getty

Images RF; **12.4a–e, 12.7a–d**: Photo courtesy of Karie Holtermann, University of Washington. Reprinted with permission; **12.9a**: Tiffany Vance/NOAA; **12.9b**: Image provided by the SeaWiFS Project, NASA, Goddard Space Center; **12.10**: Courtesy of Miriam Godfrey/ NIWA; **12.13**: Courtesy Kendra Daly, School of Oceanography, University of Washington; **12.18a–b**: Photo courtesy of Karie Holtermann, University of Washington. Reprinted with permission; **12.18c**: © Ingram Publishing/ agefotostock.com RF; **12.18d**: Courtesy Paulette Brunner, Seattle Aquarium; **Box Figure 1a–b**: © Dr. John Baross, School of Oceanography, University of Washington; **12.21**: Courtesy Dr. H. Paul Johnson, School of Oceanography, University of Washington; **12.22**: © Jed Furhman; **12.24a–b**: Photo courtesy of Karie Holtermann, University of Washington. Reprinted with permission; **12.27a–b**: Raphael Kudela, UCSC; **12.33**: Photo courtesy of Jim Gower, Institute of Ocean Sciences, Canada, SeaWiFS, GeoEye NASA; **12.35**: Raphael Kudela, UCSC.

Chapter 13

Opener: © Pete Atkinson/Stone/Getty Images; **13.2**: © Phillip Colla/oceanlight.com; **13.5a**: Courtesy Dee Boersma, Penguin Project; **13.5b–e**: © Alyn & Alison Duxbury; **13.6a–c**: Courtesy Leo J. Shaw, Seattle Aquarium; **13.6d**: Courtesy Paulette Brunner, Seattle Aquarium; **13.7e**: Courtesy Gail Scott, Seattle Aquarium; **13.12a**: Courtesy Leo J. Shaw, Seattle Aquarium; **13.12b**: Courtesy Kemper, Seattle Aquarium; **13.12c**: Courtesy Leo J. Shaw, Seattle Aquarium; **13.12d**: Courtesy Kemper, Seattle Aquarium; **13.12e**: © Alyn & Alison Duxbury; **13.14**: Courtesy Captain Lawson W. Brigham, Ret., USCG; **13.16a**: Courtesy Dr. Albert Erickson, Fisheries Research Institute, University of Washington; **13.16b**: © Getty Images RF; **13.16c**: Courtesy David Whitrow, National Marine Fisheries Service, NOAA; **Box Figure 1a**: © Dr. Craig R. Smith, Department of Oceanography, University of Hawaii; **Box Figure 1b**: © H. Kukert, Courtesy of Craig R. Smith.

Chapter 14

Opener: © David Nunuk/All Canada Photos/ Getty Images; **14.1a**: Courtesy Seattle Aquarium; **14.1b**: Courtesy Leo J. Shaw, Seattle Aquarium; **14.1c**: Courtesy Paulette Brunner, Seattle Aquarium; **14.1d–e**: Courtesy Seattle Aquarium; **14.1f–g**: Courtesy Leo J. Shaw, Seattle Aquarium; **14.1h**: Courtesy Tina Link, Seattle Aquarium; **14.1i**: Courtesy Leo J. Shaw, Seattle Aquarium; **14.2**: © 2010 Wei et al. Wei C-L, Rowe GT, Escobar-Briones E, Boetius A, Soltwedel T, et al. (2010) Global Patterns and Predictions of Seafloor Biomass Using Random Forests. PLoS ONE 5(12): e15323. doi:10.1371/journal.pone.0015323;

14.15a–c: Courtesy Verena Tunnicliffe, University of Victoria, B.C.; **14.15d**: © Robert Hessler, Scripps Institute of Oceanography; **14.16**: © David Hall/Photo Researchers, Inc.; **14.18a–b**: Courtesy Kathy Newell, School of Oceanography, University of Washington; **p. 388**: Courtesy of Dr. Chris Shank, University of Texas Austin Marine Science Institute; **Box Figure 1**: Photo courtesy of Mote Tropical Research Laboratory **14.20**: Photo © & Courtesy Richard A. Lutz; **14.21a–c**: NOAA; **14.22**: Courtesy Leo J. Shaw, Seattle Aquarium.

Chapter 15

Opener: © Digital Vision/Getty Images RF; **15.1(left)**: Library of Congress Prints and Photographs Division [LC-USZ62-13912]; **15.1(right)**: © Digital Vision/PunchStock RF; **15.8a–c**: Courtesy Joe Lucas/Marine Entanglement Research Program/NOAA; **15.8d**: Courtesy R. Herron/Marine Entanglement Research Program/NOAA; **p. 408**: © Sean Smith; **Box Figure 1**: NOAA; **Box Figure 2a**: NOAA and Georgia Department of Natural Resources; **Box Figure 2b, 3**: NOAA; **15.14a–f**: Courtesy Hazardous Materials Response Branch/Jerry Gault/NOAA; **15.15**: Courtesy Port of Seattle; **15.16**: © 2012 Alex S. McLean/Landslides; **15.17**: Dr. Fred Nichols/ U.S. Geological Survey, Menlo Park, CA; **15.18**: Photo courtesy of Thomas M. Niesen, San Francisco State University; **15.21a–b**: Courtesy Wayne A. Palsson; **15.21c**: Courtesy Uni Seafoods, Inc. Redmond, WA; **15.22**: Raphael Kudela, UCSC; **15.24**: Courtesy Northwest Sea Farm, Inc.

Chapter 16

Opener: NASA/NOAA/GSFC/Suomi NPP/ VIIRS/Norman Kuring; **16.1(Earth)**: NASA Goddard Space Flight Center. Image by Reto Stöckli. Enhancements by Robert Simmon. Data and technical support: MODIS Land Group; MODIS Science Data Support Team; MODIS Atmosphere Group; MODIS Ocean Group Additional data: USGS EROS Data Center; USGS Terrestrial Remote Sensing Flagstaff Field Center; Defense Meteorological Satellite Program; **16.1(Cryosphere)**: © Radius Images/Corbis RF; **16.1(Hydrosphere)**: © Purestock/SuperStock RF; **16.1(Atmosphere)**: NASA/Jeff Schmaltz, MODIS Land Rapid Response Team; **16.1(Lithosphere)**: © Janet S. Robbins; **16.1(Biosphere)**: © Lissa Harrison; **16.1(Anthrosphere)**: © Punchstock RF; **16.4**: © National Science Foundation; **16.7**: T. J. Casadevall/USGS; **16.14**: Data based on: Weiss, JL, JT Overpeck, and B Strauss (2011) *Implications of recent sea level rise science for low-elevation areas in coastal cities of the conterminous U.S.A.*, Climatic Change 105: 635–645. University of Arizona web map application, www.geo.arizona.edu/dgesl/; **16.15**: NOAA Geophysical Fluid Dynamics

Laboratory; **16.16**: © Dr. Parvinder Sethi; **16.17**: © Royalty-Free/Corbis RF; **16.19**: © Kris Unger/Verdant Power, Inc.; **p. 434**: Dr. Rob McDowell; **Box Figure 1**: U.S. Coast Guard photo by Petty Officer 1st Class Matthew Belson; **Box Figure 2**: © Leslie Garland Picture Library/Alamy; **16.20**: © Loren W. Linholm.

Text and Illustrations

Chapter 1

Figures 1.4, 1.6, 1.7: Chamberlin, Sean, Dickey, Tommy. From *Exploring the World Ocean*. Copyright © 2008 by The McGraw-Hill Companies.

Chapter 2

Figure 2.4: Abbott, Patrick, Leon. From *Natural Disasters*, 8th Edition. Copyright © 2012 by The McGraw-Hill Companies.; **Figures 2.6, 2.7, 2.9**: McConnell et al., David. From *Introduction to Earth Science*. Copyright © 2008 by The McGraw-Hill Companies.; 2.7: McConnell et al., David. From *Introduction to Earth Science*. Copyright © 2008 by The McGraw-Hill Companies; **Figure 2.11**: Plummer, Charles, Carlson, Diane, McGeary, David. From *Physical Geology*, 9th Edition. Copyright © 2004 by The McGraw-Hill Companies.; **Figure 2.15**: McConnell et al., David. From *Introduction to Earth Science*. Copyright © 2008 by The McGraw-Hill Companies.; **Figure 2.19**: D.H. Tarling and J.C. Mitchell. From *Geology*, vol. 4, No. 3, 1976. Copyright © 1976 by Geological Society of America. Reprinted by permission.; **Figure 2.20**: Larson, R.L. and Pitman, W.C. From *Geological Society of America Bulletin*. Copyright © by R.L. Larson and W.C. Pitman. Reprinted by permission of the authors.; **Figures 2.21, 2.22**: Plummer, Charles, Carlson, Diane, McGeary, David. From *Physical Geology*, 9th Edition. Copyright © 2004 by The McGraw-Hill Companies; **Figures 2.29, 2.30a,b,c**: Chamberlin, Sean, Dickey, Tommy. From *Exploring the World Ocean*. Copyright © 2008 by The McGraw-Hill Companies; **Figure 2.31**: Stern, Robert J.. From *Journal of Geoscience Education*, vol. 46, 1998, pp. 221–228. Copyright © 1998 by National Association of Geoscience Teachers. Reprinted by permission.; **Figures 2.33, 2.38, 2.39**: Chamberlin, Sean, Dickey, Tommy. From *Exploring the World Ocean*. Copyright © 2008 by The McGraw-Hill Companies; **Figure 2.40**: Chamberlin, Sean, Dickey, Tommy. From *Exploring the World Ocean*. Copyright © 2008 by The McGraw-Hill Companies.

Chapter 3

Figure 3.3: Chamberlin, Sean, Dickey, Tommy. From *Exploring the World Ocean*. Copyright © 2008 by The McGraw-Hill Companies.; **Figure 3.4**: McConnell et al., David. From *Introduction to Earth Science*. Copyright © 2008 by The McGraw-Hill Companies.; **Tables 3.4, 3.5**: Riley and Skirrow. From *Chemical Oceanography*, vol. 1, 1975. Copyright © 1975 by Elsevier Science Ltd. Reprinted by permission.; **Figure 3.6**: McConnell et al., David. From *Introduction to Earth Science*. Copyright © 2008 by The McGraw-Hill Companies.; **Figure 3.8a,b**: Shepard, Francis P. From *Submarine Geology* (Pearson, 1963). Copyright © 1963 by Pearson Education. Reprinted by permission; **Figures 3.17, 3.19, 3.20**: Chamberlin, Sean, Dickey, Tommy. From *Exploring the World Ocean*. Copyright © 2008 by The McGraw-Hill Companies; Figure **3.24a,b**: Hill, M.N. From *The Sea*, vol. 2 (Wiley-Blackwell, 1963). Copyright © 1963 by Wiley-Blackwell. Reprinted by permission.; **Figure 3.27a,b**: Chamberlin, Sean, Dickey, Tommy. From *Exploring the World Ocean*. Copyright © 2008 by The McGraw-Hill Companies.; **Figure 3.29a,b,c**: Reynolds, Stephen J. From *Exploring Geology*. Copyright © 2008 by The McGraw-Hill Companies. Reprinted by permission.

Chapter 4

Figures 4.2, 4.4, 4.11: Chamberlin, Sean, Dickey, Tommy. From *Exploring the World Ocean*. Copyright © 2008 by The McGraw-Hill Companies; **Table 4.5**: Riley and Chester. From *Chemical Oceanography*, vol. 8, 1983. Copyright © 1983 by Elsevier Science Ltd. Reprinted by permission.

Chapter 5

Table 5.3: Riley and Chester. From *Chemical Oceanography*, vol. 1, 1975. Copyright © 1975 by Elsevier Science Ltd. Reprinted by permission.; **Figures 5.7, 5.8**: Chamberlin, Sean, Dickey, Tommy. From *Exploring the World Ocean*. Copyright © 2008 by The McGraw-Hill Companies.; **Table 5.2**: Riley and Chester. From *Chemical Oceanography*, vol. 1, 1975. Copyright © 1975 by Elsevier Science Ltd. Reprinted by permission.

Chapter 6

Figure 6.6: Sverdrup, Keith, Duxbury, Alyn, Duxbury, Alison. From *Fundamentals of Oceanography*, 5th Edition. Copyright © 2008 by The McGraw-Hill Companies.; **Figures 6.11, 6.12, 6.17, 6.18**: McConnell et al., David. From *Introduction to Earth Science*. Copyright © 2008 by The McGraw-Hill Companies; **Figure 6.19**: Sverdrup, Keith, Duxbury, Alyn, Duxbury, Alison. From *Fundamentals of Oceanography*, 5th Edition. Copyright © 2008 by The McGraw-Hill Companies.

Chapter 7

Figure 7.1: Sverdrup, Johnson, Fleming. From *The Oceans* (Prentice Hall, 1970). Copyright © 1970 by Pearson Education. Reprinted by permission.; **Figures 7.12, 7.13, 7.14, 7.18, 7.19, 7.21, 7.22, 7.30, 7.33, 7.35**: Chamberlin, Sean, Dickey, Tommy. From *Exploring the World Ocean*. Copyright © 2008 by The McGraw-Hill Companies.; **Page 207, Box Figure 2**: Sverdrup, Keith, Duxbury, Alyn, Duxbury, Alison. From *Fundamentals of Oceanography*, 5th Edition. Copyright © 2008 by The McGraw-Hill Companies.; **Figure 7.9**: McConnell et al., David. From *Introduction to Earth Science*. Copyright © 2008 by The McGraw-Hill Companies.

Chapter 8

Figure 8.4: Chamberlin, Sean, Dickey, Tommy. From *Exploring the World Ocean*. Copyright © 2008 by The McGraw-Hill Companies.; **Figure 8.5**: McConnell et al., David. From *Introduction to Earth Science*. Copyright © 2008 by The McGraw-Hill Companies.; **Figures 8.8, 8.12, 8.21**: Chamberlin, Sean, Dickey, Tommy. From *Exploring the World Ocean*. Copyright © 2008 by The McGraw-Hill Companies.; **Figure 8.32c**: Brandt, W.P., Rubino, A. From *Remote Sensing of the European Seas* (Springer Science and Business Media, 2008). Copyright © 2008 by Springer Science and Business Media. Reprinted by permission.

Chapter 10

Figure 10.14: Inamn, D.L., Frautschy, J.D. From *Littoral Process and the Development of Shorelines* (American Society of Civil Engineers,1965). Copyright © 1965 by American Society of Civil Engineers. Reprinted by permission.; **Figures 10.17, 10.18**: Montgomery, Carla. From *Environmental Geology*, 9th Edition. Copyright © 2011 by The McGraw-Hill Companies. Reprinted by permission.

Chapter 11

Figure 11.1: Castro, Peter, Huber, Michael. From *Marine Biology*, 8th Edition. Copyright © 2011 by The McGraw-Hill Companies.; **Figure 11.14**: Castro, Peter, Huber, Michael. From *Marine Biology*, 8th Edition. Copyright © 2011 by The McGraw-Hill Companies.; **Figure 11.16**: Chamberlin, Sean, Dickey, Tommy. From *Exploring the World Ocean*. Copyright © 2008 by The McGraw-Hill Companies.

Chapter 12

Figures 12.11, 12.20, 12.25, 12.26, 12.28, 12.29: Chamberlin, Sean, Dickey, Tommy. From *Exploring the World Ocean*. Copyright © 2008 by The McGraw-Hill Companies; **Figure 12.34**: Kudela et al., Raphael. This article has been published in *Oceanography*, Volume 18, Number 2, a quarterly journal of The Oceanography Society. Copyright 2005 by The Oceanography Society. All rights reserved. Reproduction of any portion of this article by photocopy machine, reposting, or other means without prior authorization of

The Oceanography Society is strictly prohibited. Send all correspondence to: info@tos.org
or The Oceanography Society, PO Box 1931,
Rockville, MD 20849-1931, USA.

Chapter 13

Figure 13.8: Castro, Peter, Huber, Michael.
From *Marine Biology*, 8th Edition. Copyright
© 2011 by The McGraw-Hill Companies.;
Figure 13.11: Castro, Peter, Huber, Michael.
From *Marine Biology*, 8th Edition. Copyright
© 2011 by The McGraw-Hill Companies.;
Figure 13.19: Castro, Peter, Huber, Michael.
From *Marine Biology*, 8th Edition. Copyright
© 2011 by The McGraw-Hill Companies.

Chapter 14

Figure 14.3: Gaines et al., Steven. This
article has been published in *Oceanography*,
Volume 20, Number 3, a quarterly journal of
The Oceanography Society. Copyright 2007
by The Oceanography Society. All rights
reserved. Permission is granted to copy this
article for use in teaching and research.
Republication, systematic reproduction,
or collective redistribution of any portion of
this article by photocopy machine, reposting,
or other means is permitted only with the
approval of The Oceanography Society. Send
all correspondence to: info@tos.org or The
Oceanography Society, PO Box 1931, Rock-
ville, MD 20849-1931, USA; **Figure 14.11**:
Castro, Peter, Huber, Michael. From *Marine
Biology*, 8th Edition. Copyright © 2011 by
The McGraw-Hill Companies.; **Figure 14.14**:
Chamberlin, Sean, Dickey, Tommy. From
Exploring the World Ocean. Copyright © 2008
by The McGraw-Hill Companies.; **Table 14.1**:
Castro, Peter, Huber, Michael. From *Marine
Biology*, 8th Edition. Copyright © 2011 by
The McGraw-Hill Companies.

Chapter 15

Figure 15.7: Higman, Bretwood. From
Ground Truth Trekking . Copyright © 2012
by The McGraw-Hill Companies. Reprinted
by permission.; **Figure 15.23**: Chamberlin,
Sean, Dickey, Tommy. From *Exploring the
World Ocean*. Copyright © 2008 by The
McGraw-Hill Companies.

Chapter 16

Figure 16.2, **16.12**: Chamberlin, Sean,
Dickey, Tommy. From *Exploring the World
Ocean*. Copyright © 2008 by The McGraw-
Hill Companies.

Index